Springer Optimization and Its Applications

Volume 159

Aims and Scope

Optimization has continued to expand in all directions at an astonishing rate. New algorithmic and theoretical techniques are continually developing and the diffusion into other disciplines is proceeding at a rapid pace, with a spot light on machine learning, artificial intelligence, and quantum computing. Our knowledge of all aspects of the field has grown even more profound. At the same time, one of the most striking trends in optimization is the constantly increasing emphasis on the interdisciplinary nature of the field. Optimization has been a basic tool in areas not limited to applied mathematics, engineering, medicine, economics, computer science, operations research, and other sciences.

The series **Springer Optimization and Its Applications** (**SOIA**) aims to publish state-of-the-art expository works (monographs, contributed volumes, textbooks, handbooks) that focus on theory, methods, and applications of optimization. Topics covered include, but are not limited to, nonlinear optimization, combinatorial optimization, continuous optimization, stochastic optimization, Bayesian optimization, optimal control, discrete optimization, multi-objective optimization, and more. New to the series portfolio include Works at the intersection of optimization and machine learning, artificial intelligence, and quantum computing.

Volumes from this series are indexed by Web of Science, zbMATH, Mathematical Reviews, and SCOPUS.

More information about this series at http://www.springer.com/series/7393

Nicholas J. Daras • Themistocles M. Rassias
Editors

Computational Mathematics and Variational Analysis

Springer

Editors
Nicholas J. Daras
Department of Mathematics
Hellenic Military Academy
Vari Attikis, Greece

Themistocles M. Rassias
Department of Mathematics
National Technical University of Athens
Athens, Greece

ISSN 1931-6828 ISSN 1931-6836 (electronic)
Springer Optimization and Its Applications
ISBN 978-3-030-44624-6 ISBN 978-3-030-44625-3 (eBook)
https://doi.org/10.1007/978-3-030-44625-3

This Springer imprint is published by the registered company Springer Nature Switzerland AG.
The registered company address is: Gewerbestrasse 11, 6330 Cham, Switzerland

Preface

Computational Mathematics and Variational Analysis discusses a broad spectrum of computational methods and theories of an interdisciplinary nature. The book chapters are devoted to the study of research problems from pure and applied mathematical sciences. Among the topics treated are subjects of calculus of variations, optimization theory, operations research, game theory, differential equations, functional analysis, operator theory, approximation theory, numerical analysis, asymptotic analysis, and engineering.

The book presents 28 papers written by eminent scientists from the international mathematical community. These works are devoted to a number of significant advances in both classical and modern mathematical problems including algorithms for difference of monotone operators, variational inequalities in semi-inner product spaces, function variational principles and normed minimizers, equilibria of parametrized N-player nonlinear games, multi-symplectic numerical schemes for differential equations, time-delay multi-agent systems, computational methods in nonlinear design of experiments, unsupervised stochastic learning, asymptotic statistical results, global–local transformation, scattering relations of elastic waves, generalized Ostrowski and trapezoid type rules, numerical approximation, Szász–Durrmeyer operators and approximation, integral inequalities, behavior of the solutions of functional equations, functional inequalities in complex Banach spaces, and functional contractions in metric spaces.

The presentation of concepts and methods featured in this book make it an invaluable reference for both professors and researchers in mathematics, engineering, physics, and economics.

We would like to express our warm thanks to the contributors of chapters as well as to the staff of Springer for their excellent collaboration for this publication.

Vari Attikis, Greece — Nicholas J. Daras
Athens, Greece — Themistocles M. Rassias

Contents

Scattering Relations of Elastic Waves by a Multi-Layered Thermoelastic Body

Evagelia S. Athanasiadou, Vassilios Sevroglou, and Stefania Zoi

Abstract The scattering problem of a time-harmonic dependent plane elastic wave by a multi-layered thermoelastic body in an isotropic and homogeneous elastic medium is considered. The direct scattering problem is formulated. Integral representations for the total exterior elastic field and the total interior thermoelastic fields as well as expressions for the far-field patterns are obtained containing the physical parameters of the interior thermoelastic layers. A reciprocity type theorem, a general type scattering theorem and an optical type theorem for plane wave incidence are presented and proved.

Mathematical Subject Classification 35P25, 74J20, 35J57, 74B05

1 Introduction

In this paper, the scattering problem of time-harmonic dependent plane elastic waves by a multi-layered thermoelastic body is considered. A good introduction to the theory of thermoelasticity can be found in the books [15] and [16]. The expressions of the thermoelastic far-field patterns and cross sections have been presented in [11]. The low-frequency theory for thermoelastic wave scattering from an impenetrable scatterer has been studied in [12]. Uniqueness and existence results for the scattering of thermoelastic waves from a screen have been presented in [5] and for anisotropic bodies with cuts in [14]. The scattering problem of elastic waves for a penetrable thermoelastic body and for a two-layered thermoelastic body has been studied in

E. S. Athanasiadou · S. Zoi (✉)
Department of Mathematics, National and Kapodistrian University of Athens, Zographou, Athens, Greece
e-mail: eathan@math.uoa.gr; stefzoi@math.uoa.gr

V. Sevroglou
Department of Statistics & Insurance Science, University of Piraeus, Piraeus, Greece
e-mail: bsevro@unipi.gr

N. J. Daras, T. M. Rassias (eds.), *Computational Mathematics and Variational Analysis*, Springer Optimization and Its Applications 159,
https://doi.org/10.1007/978-3-030-44625-3_1

[13] and [6], respectively. Scattering relations for plane elastic waves for an elastic scatterer are presented in [8]. Corresponding results for a penetrable thermoelastic scatterer are presented in [4]. Scattering relations for spherical elastic waves for an elastic scatterer in $2D$ and $3D$ are treated in [2] and [1], respectively. Moreover, scattering theorems for acoustic and electromagnetic waves can be found in [10] and [3] as well as for complete dyadic fields in [9].

In Section 2, we formulate the scattering problem. In Section 3, we present the integral representations of the total exterior elastic and the total interior thermoelastic fields as well as integral representations for the three normalized spherical elastic far-field patterns. Moreover, we present and prove alternative integral representations for the total exterior elastic field as well as for the elastic far-field patterns, involving volume integrals over all thermoelastic layers and surface integrals over all surfaces, containing the displacement and temperature parts of the total interior thermoelastic fields as well as the elastic and thermal parameters of each thermoelastic layer. Also, we give the formulas of the scattering cross section, the absorption cross section and the extinction cross section. In Section 4, we present and prove scattering relations for our scattering problem, corresponding to the reciprocity type theorem, the general scattering theorem and the optical type theorem for plane wave incidence.

2 Formulation

Let D be a bounded, connected and smooth subset of $\mathbb{R}^3$ whose interior is divided by means of closed and non-intersecting surfaces S_j $(j = 1, 2, \ldots, N-1)$ of class C^2 into layers D_j $(j = 1, 2, \ldots, N)$ with $S_{j-1} = \partial D_{j-1} \cap \partial D_j$ while the surface S_{j-1} surrounds S_j. We refer to D as the multi-layered thermoelastic scatterer and we denote its boundary of class C^2 with S_0. Moreover, we denote with $D_0 = \mathbb{R}^3 \setminus \overline{D}$ the elastic medium of propagation and with $\hat{\mathbf{n}}$ the outward unit normal vector at each point of any surface S_j for $j = 0, 1, \ldots, N-1$ (Figure 1). Therefore:

$$D = \left(\cup_{j=1}^{N} D_j\right) \cup \left(\cup_{j=1}^{N-1} S_j\right), \quad D_0 = \mathbb{R}^3 \setminus \overline{D}, \quad \partial D = S_0, \quad \partial D_j = S_{j-1} \cup S_j,$$
$$\partial D_N = S_{N-1}. \tag{1}$$

for $j = 1, 2, \ldots, N-1$. The interior of the medium of propagation D_0 is filled up by an elastic material with elastic properties described by the positive Lamé constants $\lambda^{(0)}$, $\mu^{(0)}$ and the mass density $\rho^{(0)}$ and the interior of the layer D_j $(j = 1, 2, \ldots, N)$ is filled up by a thermoelastic material which is consisted by elastic properties described by the positive Lamé constants $\lambda^{(j)}$, $\mu^{(j)}$ and the mass density $\rho^{(j)}$ as well as by thermal properties described by the coefficient of thermal diffusivity $\kappa^{(j)}$. We consider that the above domains D_0 and D_j $(j = 1, 2, \ldots, N)$ are homogeneous and isotropic and that strong ellipticity conditions $\mu^{(0)} > 0,\ \lambda^{(0)} + 2\mu^{(0)} > 0$ and $\mu^{(j)} > 0,\ \lambda^{(j)} + 2\mu^{(j)} > 0$, for $j = 1, 2, \ldots, N$,

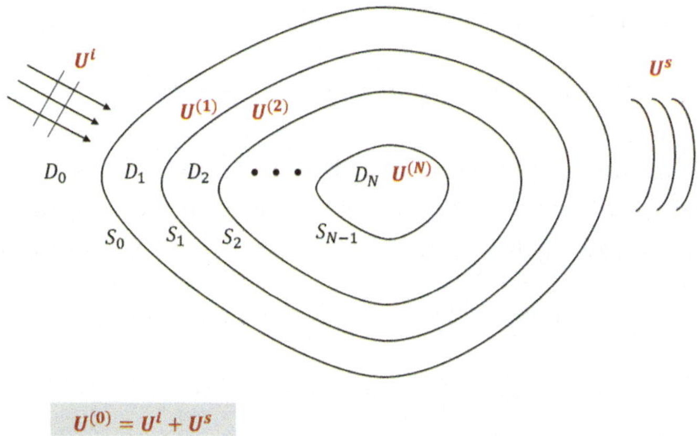

Fig. 1 Multi-layered thermoelastic scatterer

are also satisfied, in order for the mediums to sustain longitudinal as well as transverse waves.

In the sequel, assuming time-harmonic dependence in all wave fields, we denote an elastic field with $\mathbf{U} = (\mathbf{u}, 0)$ and a thermoelastic field with $\mathbf{U} = (\mathbf{u}, \theta)$, where $\mathbf{u} : \mathbb{R}^3 \to \mathbb{C}^3$ is the three-dimensional displacement field and $\theta : \mathbb{R}^3 \to \mathbb{C}$ is the scalar temperature field. Moreover, we denote the incident elastic field with $\mathbf{U}^i$, the scattered elastic field with $\mathbf{U}^s$, the total exterior elastic field in D_0 with $\mathbf{U}^{(0)}$ (which is a superposition of the incident and the scattered field) and the total interior thermoelastic field in D_j with $\mathbf{U}^{(j)}$, for $j = 1, 2, \ldots, N$. Therefore:

$$\mathbf{U}^i(\mathbf{r}) = \left(\mathbf{u}^i(\mathbf{r}), 0\right), \quad \mathbf{U}^s(\mathbf{r}) = \left(\mathbf{u}^s(\mathbf{r}), 0\right), \quad \mathbf{U}^{(0)}(\mathbf{r}) = \left(\mathbf{u}^{(0)}(\mathbf{r}), 0\right), \quad \mathbf{r} \in D_0 \tag{2}$$

where

$$\mathbf{U}^{(0)}(\mathbf{r}) = \mathbf{U}^i(\mathbf{r}) + \mathbf{U}^s(\mathbf{r}) \tag{3}$$

and

$$\mathbf{U}^{(j)}(\mathbf{r}) = \left(\mathbf{u}^{(j)}(\mathbf{r}), \theta^{(j)}(\mathbf{r})\right), \quad \mathbf{r} \in D_j, \quad j = 1, 2, \ldots, N. \tag{4}$$

We note that in all elastic fields the fourth coordinate is independent of any possible temperature of the medium of propagation and that it has been imposed equal to zero for uniformity reasons with the four-dimensional form of the thermoelastic fields. This four-dimensional form of the elastic fields will only be used in the unified relations presented in Sections 2 and 3.

We assume elastic plane wave incidence of the form:

$$\mathbf{u}^i(\mathbf{r};\hat{\mathbf{d}}) = a_p^{(0)}\hat{\mathbf{d}}e^{ik_p^{(0)}\hat{\mathbf{d}}\cdot\mathbf{r}} + a_s^{(0)}\hat{\mathbf{b}}e^{ik_s^{(0)}\hat{\mathbf{d}}\cdot\mathbf{r}}\,, \tag{5}$$

where $\hat{\mathbf{d}}$ is the direction of propagation, $\hat{\mathbf{b}}$ is the polarization (such that $\hat{\mathbf{d}}\cdot\hat{\mathbf{b}} = 0$), $k_p^{(0)}$ and $k_s^{(0)}$ are the wave numbers of the longitudinal and the transverse part of the incident elastic field, respectively, and $a_p^{(0)}$, $a_s^{(0)}$ are the corresponding real amplitudes.

The total exterior elastic field in D_0 satisfies the time-harmonic Navier equation:

$$\left(\left(\mu^{(0)}\Delta + \rho^{(0)}\omega^2\right)\widetilde{I}_3 + (\lambda^{(0)} + \mu^{(0)})\nabla\nabla\cdot\right)\mathbf{u}^{(0)}(\mathbf{r}) = \mathbf{0}\,, \quad \mathbf{r}\in D_0 \tag{6}$$

and the total interior thermoelastic field in D_j satisfies the time-harmonic Biot system of equations:

$$\begin{cases} \left(\left(\mu^{(j)}\Delta+\rho^{(j)}\omega^2\right)\widetilde{I}_3+(\lambda^{(j)}+\mu^{(j)})\nabla\nabla\cdot\right)\mathbf{u}^{(j)}(\mathbf{r})-\gamma^{(j)}\nabla\theta^{(j)}(\mathbf{r})=\mathbf{0} \\ i\omega\eta^{(j)}\nabla\cdot\mathbf{u}^{(j)}(\mathbf{r}) + \left(\Delta + q^{(j)}\right)\theta^{(j)}(\mathbf{r}) = 0 \end{cases}, \quad \mathbf{r}\in D_j\,, \tag{7}$$

where ω is the angular frequency and $\widetilde{I}_n$ is the identity dyadic in $\mathbb{R}^n$. The coupling constants $\gamma^{(j)}$, $\eta^{(j)}$ and the thermal constant $q^{(j)}$ are given by Cakoni and Dassios [6] and Dassios and Kostopoulos [11]:

$$\gamma^{(j)} = (\lambda^{(j)} + \frac{2}{3}\mu^{(j)})\alpha^{(j)}\,, \quad \eta^{(j)} = \frac{T_0\gamma^{(j)}}{\lambda_0^{(j)}}\,, \quad q^{(j)} = \frac{i\omega}{\kappa^{(j)}}\,, \tag{8}$$

for $j = 1, 2, \ldots, N$, where $\alpha^{(j)}$ is the volumetric thermal expansion coefficient, T_0 is the absolute temperature reference level and $\lambda_0^{(j)}$ is the coefficient of thermal conductivity of the medium. Equations (6)–(7) can be written in a four-dimensional unified form as follows:

$$\widetilde{\mathbf{L}}^{(j)}\mathbf{U}^{(j)}(\mathbf{r}) = \mathbf{0}\,, \quad \mathbf{r}\in D_j\,,\ j = 0, 1, 2, \ldots, N\,, \tag{9}$$

where $\widetilde{\mathbf{L}}^{(j)}$ are differential operators given in a 4×4 matrix form by

$$\widetilde{\mathbf{L}}^{(j)} = \begin{bmatrix} \left(\Delta^{(L,j)} + \rho^{(j)}\omega^2\right)\widetilde{I}_3 & -\gamma^{(j)}\nabla \\ i\omega\eta^{(j)}\nabla\cdot & \Delta + q^{(j)} \end{bmatrix}, \tag{10}$$

with

$$\Delta^{(L,j)} = \mu^{(j)}\Delta + \left(\lambda^{(j)} + \mu^{(j)}\right)\nabla\nabla\cdot\,, \tag{11}$$

for $j = 0, 1, \ldots, N$. We note that for $j = 0$ the volumetric thermal expansion coefficient $\alpha^{(0)}$ is assumed to be zero in order for the medium of propagation D_0 to sustain purely elastic fields uncoupled of any possible temperature of the medium, therefore $\gamma^{(0)} = 0$, $\eta^{(0)} = 0$ and the scalar block $\Delta + q^{(0)}$ is applied on the zeroth fourth coordinate of any elastic field in D_0.

On the surfaces S_j, for $j = 0, 1, 2, \ldots, N-1$, transmission conditions are imposed as follows [6]:

$$\left.\begin{aligned} \mathbf{u}^{(0)}(\mathbf{r}) &= \mathbf{u}^{(1)}(\mathbf{r}) \\ \widetilde{\mathbf{T}}^{(0)}\mathbf{u}^{(0)}(\mathbf{r}) &= \widetilde{\mathbf{T}}^{(1)}\mathbf{u}^{(1)}(\mathbf{r}) \\ \theta^{(1)}(\mathbf{r}) &= 0 \\ \frac{\partial}{\partial n}\theta^{(1)}(\mathbf{r}) &= 0 \end{aligned}\right\} , \quad \mathbf{r} \in S_0 , \tag{12}$$

and

$$\left.\begin{aligned} \mathbf{u}^{(j)}(\mathbf{r}) &= \mathbf{u}^{(j+1)}(\mathbf{r}) \\ \widetilde{\mathbf{T}}^{(j)}\mathbf{u}^{(j)}(\mathbf{r}) - \gamma^{(j)}\hat{\mathbf{n}}\theta^{(j)}(\mathbf{r}) &= \widetilde{\mathbf{T}}^{(j+1)}\mathbf{u}^{(j+1)}(\mathbf{r}) - \gamma^{(j+1)}\hat{\mathbf{n}}\theta^{(j+1)}(\mathbf{r}) \\ \theta^{(j)}(\mathbf{r}) &= \theta^{(j+1)}(\mathbf{r}) \\ \lambda_0^{(j)}\frac{\partial}{\partial n}\theta^{(j)}(\mathbf{r}) &= \lambda_0^{(j+1)}\frac{\partial}{\partial n}\theta^{(j+1)}(\mathbf{r}) \end{aligned}\right\} , \quad \mathbf{r} \in S_j , \tag{13}$$

for $j = 1, 2, \ldots, N-1$, with $\partial/\partial n = \hat{\mathbf{n}} \cdot \nabla$ be the exterior normal derivative on the corresponding surface and $\widetilde{\mathbf{T}}^{(j)}$ the surface traction operator defined by

$$\widetilde{\mathbf{T}}^{(j)} = 2\mu^{(j)}\hat{\mathbf{n}} \cdot (\nabla) + \lambda^{(j)}\hat{\mathbf{n}}(\nabla \cdot) + \mu^{(j)}\hat{\mathbf{n}} \times (\nabla \times) , \tag{14}$$

for $j = 0, 1, \ldots, N$. The transmission conditions (12) secure the continuity of the displacement field and the continuity of the traction field across the scatterer's boundary S_0 as well as the fact that there are no temperature fluctuations on S_0 and that it is thermally insulated. Also, the transmission conditions (13) secure the continuity of the displacement field, the traction field, the temperature field and the thermal flux field, respectively, on S_j for $j = 1, 2, \ldots, N-1$. Moreover, the transmission conditions (12) and (13) can be written equivalently in a four-dimensional unified form as:

$$\left.\begin{aligned} \mathbf{U}^{(0)}(\mathbf{r}) &= \mathbf{U}^{(1)}(\mathbf{r}) \\ \widetilde{\mathbf{B}}^{(0)}\mathbf{U}^{(0)}(\mathbf{r}) &= \widetilde{\mathbf{B}}^{(1)}\mathbf{U}^{(1)}(\mathbf{r}) \end{aligned}\right\} , \quad \mathbf{r} \in S_0 , \tag{15}$$

and

$$\begin{aligned}\mathbf{U}^{(j)}(\mathbf{r}) &= \mathbf{U}^{(j+1)}(\mathbf{r}) \\ \widetilde{\mathbf{B}}^{(j)}\mathbf{U}^{(j)}(\mathbf{r}) &= \widetilde{\mathbf{B}}^{(j+1)}\mathbf{U}^{(j+1)}(\mathbf{r}) \quad , \quad \mathbf{r} \in S_j \ ,\end{aligned} \tag{16}$$

for $j = 1, 2, \ldots, N-1$, respectively, where $\widetilde{\mathbf{B}}^{(0)}$ and $\widetilde{\mathbf{B}}^{(j)}$ are the boundary operators given in a 4×4 matrix form as:

$$\widetilde{\mathbf{B}}^{(0)} = \begin{bmatrix} \widetilde{\mathbf{T}}^{(0)} & \mathbf{0} \\ \mathbf{0} & 0 \end{bmatrix} , \tag{17}$$

$$\widetilde{\mathbf{B}}^{(j)} = \begin{bmatrix} \widetilde{\mathbf{T}}^{(j)} & -\gamma^{(j)}\hat{\mathbf{n}} \\ \mathbf{0} & \lambda_0^{(j)} \dfrac{\partial}{\partial n} \end{bmatrix} . \tag{18}$$

Moreover, the scattered elastic field:

$$\mathbf{u}^s(\mathbf{r}) = \mathbf{u}_p^s(\mathbf{r}) + \mathbf{u}_s^s(\mathbf{r}) , \tag{19}$$

where $\mathbf{u}_p^s$ is its longitudinal and $\mathbf{u}_s^s$ is its transverse part, satisfies the well-known Sommerfeld-Kupradze-type radiation conditions [15]:

$$\begin{aligned}\lim_{r\to\infty} r\left(\frac{\partial \mathbf{u}_p^s(\mathbf{r})}{\partial r} - ik_p^{(0)}\mathbf{u}_p^s(\mathbf{r})\right) &= \mathbf{0} \ , \\ \lim_{r\to\infty} r\left(\frac{\partial \mathbf{u}_s^s(\mathbf{r})}{\partial r} - ik_s^{(0)}\mathbf{u}_s^s(\mathbf{r})\right) &= \mathbf{0} \ ,\end{aligned} \tag{20}$$

uniformly for all directions $\hat{\mathbf{r}} = \mathbf{r}/r$ over the unit sphere S^2, where $r = |\mathbf{r}|$. The radiation conditions secure the radiative character of the scattered field and provide the appropriate decay for the problem to be well posed. Summarizing, the direct scattering problem consists of the field equations (3), (6), and (7), of the transmission conditions (12), (13) and of the radiation conditions (20).

3 Integral Representations and Far-Field Patterns

Starting this section, we present some formulas that we will use for the derivation of the integral representations of the total exterior and the total interior fields in the sequel as well as for the derivation of the scattering relations in the next section. The first formula is the known Betti's third identity and the second one is its thermoelastic analogue and they have been proved in [15].

Theorem 3.1 *Let* $\mathbf{u}\,,\ \mathbf{v} \in [C^2(D_j)]^3 \cap [C^1(\overline{D_j})]^3$. *Then the following formula holds true:*

$$\int_{D_j} \left(\mathbf{u} \cdot \Delta^{(L\,,\,j)} \mathbf{v} - \mathbf{v} \cdot \Delta^{(L\,,\,j)} \mathbf{u} \right) d\mathbf{r} = \int_{\partial D_j} \left(\mathbf{u} \cdot \widetilde{\mathbf{T}}^{(j)} \mathbf{v} - \mathbf{v} \cdot \widetilde{\mathbf{T}}^{(j)} \mathbf{u} \right) ds(\mathbf{r})\,. \tag{21}$$

Theorem 3.2 *Let* $\mathbf{U}\,,\ \mathbf{V} \in [C^2(D_j)]^4 \cap [C^1(\overline{D_j})]^4$. *Then the following formula holds true:*

$$\int_{D_j} \left(\mathbf{U} \cdot \widetilde{\mathbf{L}}^{(j*)} \mathbf{V} - \mathbf{V} \cdot \widetilde{\mathbf{L}}^{(j)} \mathbf{U} \right) d\mathbf{r} = \int_{\partial D_j} \left(\mathbf{U} \cdot \widetilde{\mathbf{R}}^{(j*)} \mathbf{V} - \mathbf{V} \cdot \widetilde{\mathbf{R}}^{(j)} \mathbf{U} \right) ds(\mathbf{r})\,, \tag{22}$$

with

$$\widetilde{\mathbf{L}}^{(j*)} = \begin{bmatrix} \left(\Delta^{(L,j)} + \rho^{(j)}\omega^2 \right) \widetilde{I}_3 & -\,i\omega\eta^{(j)}\nabla \\ \gamma^{(j)}\nabla\,\cdot & \Delta + q^{(j)} \end{bmatrix}, \tag{23}$$

$$\widetilde{\mathbf{R}}^{(j)} = \begin{bmatrix} \widetilde{\mathbf{T}}^{(j)} & -\,\gamma^{(j)}\hat{\mathbf{n}} \\ \mathbf{0} & \dfrac{\partial}{\partial n} \end{bmatrix}, \tag{24}$$

and

$$\widetilde{\mathbf{R}}^{(j*)} = \begin{bmatrix} \widetilde{\mathbf{T}}^{(j)} & -\,i\omega\eta^{(j)}\hat{\mathbf{n}} \\ \mathbf{0} & \dfrac{\partial}{\partial n} \end{bmatrix}, \tag{25}$$

for $j = 1, \ldots, N$.

Identity (22) is a consequence of identity (21), Green's second identity and Gauss theorem. Moreover, identity (21) and identity (22) can also be applied on a bounded region within D_0 where $\widetilde{\mathbf{L}}^{(0)} = \widetilde{\mathbf{L}}^{(0*)}$ and $\widetilde{\mathbf{R}}^{(0)} = \widetilde{\mathbf{R}}^{(0*)}$ are given by (10), (23) and (24), (25), respectively, by setting $\gamma^{(0)} = 0$ and $\eta^{(0)} = 0$.

The fundamental solution of the Navier equation, also known as the fundamental Kupradze dyadic, in D_0 denoted by $\widetilde{\mathbf{\Gamma}}(\mathbf{r}, \mathbf{r}')$ is given by

$$\widetilde{\mathbf{\Gamma}}(\mathbf{r}, \mathbf{r}') = -\frac{1}{\rho^{(0)}\omega^2} \nabla_{\mathbf{r}}\nabla_{\mathbf{r}} \frac{e^{ik_p^{(0)}|\mathbf{r}-\mathbf{r}'|}}{|\mathbf{r}-\mathbf{r}'|} + \frac{1}{\rho^{(0)}\omega^2} \left(\nabla_{\mathbf{r}}\nabla_{\mathbf{r}} + (k_s^{(0)})^2 \widetilde{I}_3 \right) \frac{e^{ik_s^{(0)}|\mathbf{r}-\mathbf{r}'|}}{|\mathbf{r}-\mathbf{r}'|} \tag{26}$$

and it solves the equation:

$$\left(\Delta_{\mathbf{r}}^{(L,0)} + \rho^{(0)}\omega^2 \right) \widetilde{\mathbf{\Gamma}}(\mathbf{r}, \mathbf{r}') = -4\pi\delta(\mathbf{r}-\mathbf{r}')\widetilde{I}_3\,, \tag{27}$$

where δ is the Dirac measure and the subscript $\mathbf{r}$ denotes the differentiation with respect to $\mathbf{r}$. Equivalently, in a unified form the fundamental solution of the Navier equation in D_0 can be written as:

$$\widetilde{\mathbf{G}}(\mathbf{r}, \mathbf{r}') = \widetilde{\mathbf{D}}_p^{(0)} \frac{e^{ik_p^{(0)}|\mathbf{r}-\mathbf{r}'|}}{|\mathbf{r}-\mathbf{r}'|} + \widetilde{\mathbf{D}}_s^{(0)} \frac{e^{ik_s^{(0)}|\mathbf{r}-\mathbf{r}'|}}{|\mathbf{r}-\mathbf{r}'|} \tag{28}$$

with

$$\widetilde{\mathbf{D}}_p^{(0)} = -\frac{1}{\rho^{(0)}\omega^2} \begin{bmatrix} \nabla_{\mathbf{r}}\nabla_{\mathbf{r}} & \mathbf{0} \\ \mathbf{0} & 0 \end{bmatrix}, \tag{29}$$

$$\widetilde{\mathbf{D}}_s^{(0)} = \frac{1}{\rho^{(0)}\omega^2} \begin{bmatrix} \nabla_{\mathbf{r}}\nabla_{\mathbf{r}} + \left(k_s^{(0)}\right)^2 \widetilde{I}_3 & \mathbf{0} \\ \mathbf{0} & 0 \end{bmatrix}. \tag{30}$$

Also, the fundamental exterior dyadic denoted by $\widetilde{\mathbf{F}}(\mathbf{r}, \mathbf{r}')$ is given by

$$\widetilde{\mathbf{F}}(\mathbf{r}, \mathbf{r}') = \widetilde{\mathbf{F}}^{\top}(\mathbf{r}, \mathbf{r}') = \begin{bmatrix} \widetilde{\mathbf{\Gamma}}(\mathbf{r}, \mathbf{r}') & \mathbf{0} \\ \mathbf{0} & \dfrac{e^{i\sqrt{q^{(0)}}|\mathbf{r}-\mathbf{r}'|}}{|\mathbf{r}-\mathbf{r}'|} \end{bmatrix}, \tag{31}$$

where $\top$ denotes transposition. Moreover, the fundamental solution of the Biot system of equations in D_j, for $j = 1, 2, \ldots, N$, denoted by $\widetilde{\mathbf{E}}^{(j)}(\mathbf{r}, \mathbf{r}')$ is given by

$$\widetilde{\mathbf{E}}^{(j)}(\mathbf{r}, \mathbf{r}') = \widetilde{\mathbf{D}}_1^{(j)} \frac{e^{ik_1^{(j)}|\mathbf{r}-\mathbf{r}'|}}{|\mathbf{r}-\mathbf{r}'|} + \widetilde{\mathbf{D}}_2^{(j)} \frac{e^{ik_2^{(j)}|\mathbf{r}-\mathbf{r}'|}}{|\mathbf{r}-\mathbf{r}'|} + \widetilde{\mathbf{D}}_s^{(j)} \frac{e^{ik_s^{(j)}|\mathbf{r}-\mathbf{r}'|}}{|\mathbf{r}-\mathbf{r}'|} \tag{32}$$

with

$$\widetilde{\mathbf{D}}_1^{(j)} = -\frac{1}{\rho^{(j)}\omega^2 \left(\left(k_1^{(j)}\right)^2 - \left(k_2^{(j)}\right)^2 \right)} \times \begin{bmatrix} \left(\left(k_p^{(j)}\right)^2 - \left(k_2^{(j)}\right)^2\right) \nabla_{\mathbf{r}}\nabla_{\mathbf{r}} & \gamma^{(j)} \left(k_p^{(j)}\right)^2 \nabla_{\mathbf{r}} \\ -i\omega\eta^{(j)} \left(k_p^j\right)^2 \nabla_{\mathbf{r}} & \left(\left(k_p^j\right)^2 - \left(k_1^j\right)^2\right)\rho^{(j)}\omega^2 \end{bmatrix} \tag{33}$$

$$\widetilde{\mathbf{D}}_2^{(j)} = \frac{1}{\rho^{(j)}\omega^2 \left(\left(k_1^{(j)}\right)^2 - \left(k_2^{(j)}\right)^2 \right)} \times \begin{bmatrix} \left(\left(k_p^{(j)}\right)^2 - \left(k_1^{(j)}\right)^2 \right) \nabla_{\mathbf{r}}\nabla_{\mathbf{r}} & \gamma^j \left(k_p^{(j)}\right)^2 \nabla_{\mathbf{r}} \\ -i\omega\eta^j \left(k_p^{(j)}\right)^2 \nabla_{\mathbf{r}} & \left(\left(k_p^{(j)}\right)^2 - \left(k_2^{(j)}\right)^2 \right) \rho^{(j)}\omega^2 \end{bmatrix} \tag{34}$$

$$\widetilde{\mathbf{D}}_s^{(j)} = \frac{1}{\rho^{(j)}\omega^2}\begin{bmatrix} \nabla_{\mathbf{r}}\nabla_{\mathbf{r}} + (k_s^{(j)})^2\,\widetilde{I}_3 & \mathbf{0} \\ \mathbf{0} & 0 \end{bmatrix}, \tag{35}$$

where $k_1^{(j)}$ and $k_2^{(j)}$ are the complex wave numbers, where $\mathrm{Im}\left(k_1^{(j)}\right) > 0$ and $\mathrm{Im}\left(k_2^{(j)}\right) > 0$ that satisfy the system:

$$\begin{cases} (k_1^{(j)})^2 + (k_2^{(j)})^2 = q^{(j)}(1+\varepsilon^{(j)}) + (k_p^{(j)})^2 \\ (k_1^{(j)})^2 (k_2^{(j)})^2 = q^{(j)}(k_p^{(j)})^2 \end{cases}, \tag{36}$$

with

$$\varepsilon^{(j)} = \frac{\gamma^{(j)}\eta^{(j)}\kappa^{(j)}}{\lambda^{(j)} + 2\mu^{(j)}}, \tag{37}$$

for $j = 1, 2, \ldots, N$ and it solves the equation:

$$\widetilde{\mathbf{L}}_{\mathbf{r}}^{(j)}\widetilde{\mathbf{E}}^{(j)}(\mathbf{r},\mathbf{r}') = \widetilde{\mathbf{L}}_{\mathbf{r}'}^{(j*)}\,\widetilde{\mathbf{E}}^{(j)\top}(\mathbf{r},\mathbf{r}') = -4\pi\delta(\mathbf{r}-\mathbf{r}')\widetilde{I}_4\,. \tag{38}$$

The integral representation of the elastic scattered field is given by Dassios and Kleinman [10]:

$$\mathbf{U}^s(\mathbf{r}) = \frac{1}{4\pi}\int_{S_0}\left(\mathbf{U}^{(0)}(\mathbf{r}')\cdot\widetilde{\mathbf{R}}_{\mathbf{r}'}^{(0)}\widetilde{\mathbf{G}}(\mathbf{r},\mathbf{r}') - \widetilde{\mathbf{G}}(\mathbf{r},\mathbf{r}')\cdot\widetilde{\mathbf{R}}_{\mathbf{r}'}^{(0)}\mathbf{U}^{(0)}(\mathbf{r}')\right)ds(\mathbf{r}')\,,\ \mathbf{r}\in D_0\,. \tag{39}$$

Therefore, the exterior integral representation of the total exterior elastic field in D_0, via (39), is given by Dassios and Kleinman [10]:

$$\begin{aligned}\mathbf{U}^{(0)}(\mathbf{r}) &= \mathbf{U}^i(\mathbf{r}) + \frac{1}{4\pi}\int_{S_0}\left(\mathbf{U}^{(0)}(\mathbf{r}')\cdot\widetilde{\mathbf{R}}_{\mathbf{r}'}^{(0)}\widetilde{\mathbf{G}}(\mathbf{r},\mathbf{r}') - \widetilde{\mathbf{G}}(\mathbf{r},\mathbf{r}')\cdot\widetilde{\mathbf{R}}_{\mathbf{r}'}^{(0)}\mathbf{U}^{(0)}(\mathbf{r}')\right)ds(\mathbf{r}') \\ &= \left(\mathbf{u}^i(\mathbf{r}) + \frac{1}{4\pi}\int_{S_0}\left(\mathbf{u}^{(0)}(\mathbf{r}')\cdot\widetilde{\mathbf{T}}_{\mathbf{r}'}^{(0)}\widetilde{\mathbf{\Gamma}}(\mathbf{r},\mathbf{r}')\right.\right. \\ &\qquad \left.\left. -\widetilde{\mathbf{\Gamma}}(\mathbf{r},\mathbf{r}')\cdot\widetilde{\mathbf{T}}_{\mathbf{r}'}^{(0)}\mathbf{u}^{(0)}(\mathbf{r}')\right)ds(\mathbf{r}'),\,0\right)\,,\ \mathbf{r}\in D_0\,. \end{aligned} \tag{40}$$

The interior integral representation of the total interior thermoelastic field in D_j is

$$\begin{aligned}\mathbf{U}^{(j)}(\mathbf{r}) &= \frac{1}{4\pi}\int_{S_j}\left(\mathbf{U}^{(j)}(\mathbf{r}')\cdot\widetilde{\mathbf{R}}_{\mathbf{r}'}^{(j*)}\widetilde{\mathbf{E}}^{(j)\top}(\mathbf{r},\mathbf{r}') - \widetilde{\mathbf{E}}^{(j)}(\mathbf{r},\mathbf{r}')\cdot\widetilde{\mathbf{R}}_{\mathbf{r}'}^{(j)}\mathbf{U}^{(j)}(\mathbf{r}')\right)ds(\mathbf{r}') \\ &\quad - \frac{1}{4\pi}\int_{S_{j-1}}\left(\mathbf{U}^{(j)}(\mathbf{r}')\cdot\widetilde{\mathbf{R}}_{\mathbf{r}'}^{(j*)}\widetilde{\mathbf{E}}^{(j)\top}(\mathbf{r},\mathbf{r}')\right. \\ &\qquad \left. -\widetilde{\mathbf{E}}^{(j)}(\mathbf{r},\mathbf{r}')\cdot\widetilde{\mathbf{R}}_{\mathbf{r}'}^{(j)}\mathbf{U}^{(j)}(\mathbf{r}')\right)ds(\mathbf{r}'),\quad \mathbf{r}\in D_j\,, \end{aligned} \tag{41}$$

for $j = 1, 2, \ldots, N-1$ and the interior integral representation of the total interior thermoelastic field in D_N is

$$\mathbf{U}^{(N)}(\mathbf{r}) = -\frac{1}{4\pi}\int_{S_{N-1}} \Big(\mathbf{U}^{(N)}(\mathbf{r}')\cdot \widetilde{\mathbf{R}}_{\mathbf{r}'}^{(N*)}\widetilde{\mathbf{E}}^{(N)\top}(\mathbf{r},\mathbf{r}') \\ -\widetilde{\mathbf{E}}^{(N)}(\mathbf{r},\mathbf{r}')\cdot \widetilde{\mathbf{R}}_{\mathbf{r}'}^{(N)}\mathbf{U}^{(N)}(\mathbf{r}')\Big)\, ds(\mathbf{r}'), \quad \mathbf{r}\in D_N . \tag{42}$$

We note that an alternative integral representation for the total exterior elastic field in D_0 can be derived making use of the following formula obtained via the transmission conditions on S_0 (12):

$$\int_{S_0} \left(\mathbf{U}^{(0)}(\mathbf{r}')\cdot \widetilde{\mathbf{R}}_{\mathbf{r}'}^{(0)}\widetilde{\mathbf{G}}\left(\mathbf{r},\mathbf{r}'\right) - \widetilde{\mathbf{G}}\left(\mathbf{r},\mathbf{r}'\right)\cdot \widetilde{\mathbf{R}}_{\mathbf{r}'}^{(0)}\mathbf{U}^{(0)}(\mathbf{r}')\right) ds(\mathbf{r}') \\ = \left(\widetilde{I}_3\cdot \int_{S_0} \left(\mathbf{U}^{(1)}(\mathbf{r}')\cdot \widetilde{\mathbf{R}}_{\mathbf{r}'}^{(0)}\widetilde{\mathbf{F}}\left(\mathbf{r},\mathbf{r}'\right) - \widetilde{\mathbf{F}}\left(\mathbf{r},\mathbf{r}'\right)\cdot \widetilde{\mathbf{R}}_{\mathbf{r}'}^{(1)}\mathbf{U}^{(1)}(\mathbf{r}')\right) ds(\mathbf{r}'),\, 0\right), \tag{43}$$

where the $(\widetilde{I}_3\cdot)$ product indicates that we keep the first three components of the integral on right-hand side of (43). Also, applying identity (22) for $\mathbf{U}^j$ and $\widetilde{\mathbf{F}}\left(\mathbf{r},\mathbf{r}'\right)$ in D_j for $j = 1, 2, \ldots, N$ and using the transmission conditions (16) we take

$$\int_{D_j} \left(\mathbf{U}^{(j)}(\mathbf{r}')\cdot \widetilde{\mathbf{L}}_{\mathbf{r}'}^{(j*)}\widetilde{\mathbf{F}}\left(\mathbf{r},\mathbf{r}'\right) - \widetilde{\mathbf{F}}\left(\mathbf{r},\mathbf{r}'\right)\cdot \widetilde{\mathbf{L}}_{\mathbf{r}'}^{(j)}\mathbf{U}^{(j)}(\mathbf{r}')\right) d\mathbf{r}' \\ = \int_{S_{j-1}} \left(\mathbf{U}^{(j)}(\mathbf{r}')\cdot \widetilde{\mathbf{R}}_{\mathbf{r}'}^{(j*)}\widetilde{\mathbf{F}}\left(\mathbf{r},\mathbf{r}'\right) - \widetilde{\mathbf{F}}\left(\mathbf{r},\mathbf{r}'\right)\cdot \widetilde{\mathbf{R}}_{\mathbf{r}'}^{(j)}\mathbf{U}^{(j)}(\mathbf{r}')\right) ds(\mathbf{r}') \\ - \int_{S_j} \left(\mathbf{U}^{(j)}(\mathbf{r}')\cdot \widetilde{\mathbf{R}}_{\mathbf{r}'}^{(j*)}\widetilde{\mathbf{F}}\left(\mathbf{r},\mathbf{r}'\right) - \widetilde{\mathbf{F}}\left(\mathbf{r},\mathbf{r}'\right)\cdot \widetilde{\mathbf{R}}_{\mathbf{r}'}^{(j)}\mathbf{U}^{(j)}(\mathbf{r}')\right) ds(\mathbf{r}') \\ = \int_{S_{j-1}} \left(\mathbf{U}^{(j)}(\mathbf{r}')\cdot \widetilde{\mathbf{R}}_{\mathbf{r}'}^{(j*)}\widetilde{\mathbf{F}}\left(\mathbf{r},\mathbf{r}'\right) - \widetilde{\mathbf{F}}\left(\mathbf{r},\mathbf{r}'\right)\cdot \widetilde{\mathbf{R}}_{\mathbf{r}'}^{(j)}\mathbf{U}^{(j)}(\mathbf{r}')\right) ds(\mathbf{r}') \\ - \int_{S_j} \Bigg(\mathbf{U}^{(j+1)}(\mathbf{r}')\cdot \widetilde{\mathbf{R}}_{\mathbf{r}'}^{(j*)}\widetilde{\mathbf{F}}\left(\mathbf{r},\mathbf{r}'\right) \\ -\widetilde{\mathbf{F}}\left(\mathbf{r},\mathbf{r}'\right)\cdot \left\{\widetilde{\mathbf{R}}_{\mathbf{r}'}^{(j+1)}\mathbf{U}^{(j+1)}(\mathbf{r}') - \begin{bmatrix} \mathbf{0} \\ \left(1-\dfrac{\lambda_0^{(j+1)}}{\lambda_0^{(j)}}\right)\dfrac{\partial\theta^{(j+1)}}{\partial n} \end{bmatrix}\right\}\Bigg) ds(\mathbf{r}'), \tag{44}$$

for $j = 1, 2, \ldots, N-1$, where $\mathbf{0} = (0, 0, 0)^\top$ and

$$\begin{aligned}&\int_{D_N}\left(\mathbf{U}^{(N)}(\mathbf{r}')\cdot\widetilde{\mathbf{L}}^{(N*)}_{\mathbf{r}'}\widetilde{\mathbf{F}}\left(\mathbf{r},\mathbf{r}'\right)-\widetilde{\mathbf{F}}\left(\mathbf{r},\mathbf{r}'\right)\cdot\widetilde{\mathbf{L}}^{(N)}_{\mathbf{r}'}\mathbf{U}^{(N)}(\mathbf{r}')\right)d\mathbf{r}'\\&=\int_{S_{N-1}}\left(\mathbf{U}^{(N)}(\mathbf{r}')\cdot\widetilde{\mathbf{R}}^{(N*)}_{\mathbf{r}'}\widetilde{\mathbf{F}}\left(\mathbf{r},\mathbf{r}'\right)-\widetilde{\mathbf{F}}\left(\mathbf{r},\mathbf{r}'\right)\cdot\widetilde{\mathbf{R}}^{(N)}_{\mathbf{r}'}\mathbf{U}^{(N)}(\mathbf{r}')\right)ds(\mathbf{r}').\end{aligned}\tag{45}$$

Adding equations (44) and (45) we obtain the following formula:

$$\begin{aligned}&\sum_{j=1}^{N}\int_{D_j}\left(\mathbf{U}^{(j)}(\mathbf{r}')\cdot\widetilde{\mathbf{L}}^{(j*)}_{\mathbf{r}'}\widetilde{\mathbf{F}}\left(\mathbf{r},\mathbf{r}'\right)\right)d\mathbf{r}'\\&=\int_{S_0}\left(\mathbf{U}^{(1)}(\mathbf{r}')\cdot\widetilde{\mathbf{R}}^{(1*)}_{\mathbf{r}'}\widetilde{\mathbf{F}}\left(\mathbf{r},\mathbf{r}'\right)-\widetilde{\mathbf{F}}\left(\mathbf{r},\mathbf{r}'\right)\cdot\widetilde{\mathbf{R}}^{(1)}_{\mathbf{r}'}\mathbf{U}^{(1)}(\mathbf{r}')\right)ds(\mathbf{r}')\\&+\sum_{j=1}^{N-1}\int_{S_j}\left(\mathbf{U}^{(j+1)}(\mathbf{r}')\cdot\left(\widetilde{\mathbf{R}}^{(j+1*)}_{\mathbf{r}'}-\widetilde{\mathbf{R}}^{(j*)}_{\mathbf{r}'}\right)\widetilde{\mathbf{F}}\left(\mathbf{r},\mathbf{r}'\right)-\widetilde{\mathbf{F}}\left(\mathbf{r},\mathbf{r}'\right)\right.\\&\left.\cdot\begin{bmatrix}\mathbf{0}\\\left(1-\dfrac{\lambda_0^{(j+1)}}{\lambda_0^{(j)}}\right)\dfrac{\partial\theta^{(j+1)}}{\partial n}\end{bmatrix}\right)ds(\mathbf{r}')\end{aligned}\tag{46}$$

Therefore, from relations (3), (39), (43) and (46) we obtain the following theorem.

Theorem 3.3 *The scattered elastic field admits the integral representation:*

$$\begin{aligned}\mathbf{U}^s(\mathbf{r})=\frac{1}{4\pi}&\left(\widetilde{I}_3\cdot\left\{\sum_{j=1}^{N}\int_{D_j}\left(\mathbf{U}^{(j)}(\mathbf{r}')\cdot\widetilde{\mathbf{L}}^{(j*)}_{\mathbf{r}'}\widetilde{\mathbf{F}}\left(\mathbf{r},\mathbf{r}'\right)\right)d\mathbf{r}'\right.\right.\\&-\sum_{j=1}^{N-1}\int_{S_j}\left(\mathbf{U}^{(j+1)}(\mathbf{r}')\cdot\left(\widetilde{\mathbf{R}}^{(j+1*)}_{\mathbf{r}'}-\widetilde{\mathbf{R}}^{(j*)}_{\mathbf{r}'}\right)\widetilde{\mathbf{F}}\left(\mathbf{r},\mathbf{r}'\right)-\widetilde{\mathbf{F}}\left(\mathbf{r},\mathbf{r}'\right)\right.\\&\left.\cdot\begin{bmatrix}\mathbf{0}\\\left(1-\dfrac{\lambda_0^{(j+1)}}{\lambda_0^{(j)}}\right)\dfrac{\partial\theta^{(j+1)}}{\partial n}\end{bmatrix}\right)ds(\mathbf{r}')\\&\left.\left.+\int_{S_0}\left(\mathbf{U}^{(1)}(\mathbf{r}')\cdot\left(\widetilde{\mathbf{R}}^{(0)}_{\mathbf{r}'}-\widetilde{\mathbf{R}}^{(1*)}_{\mathbf{r}'}\right)\widetilde{\mathbf{F}}\left(\mathbf{r},\mathbf{r}'\right)\right)ds(\mathbf{r}')\right\},0\right),\end{aligned}\tag{47}$$

or equivalently:

$$\mathbf{u}^s(\mathbf{r}) = \frac{1}{4\pi}\left\{\sum_{j=1}^{N}\int_{D_j}\left(\mathbf{u}^{(j)}(\mathbf{r}')\cdot\left(\Delta_{\mathbf{r}'}^{(L,j)}+\rho^{(j)}\omega^2\right)\widetilde{\mathbf{\Gamma}}(\mathbf{r},\mathbf{r}')+\gamma^{(j)}\theta^{(j)}(\mathbf{r}')\nabla_{\mathbf{r}'}\cdot\widetilde{\mathbf{\Gamma}}(\mathbf{r},\mathbf{r}')\right)d\mathbf{r}'\right.$$
$$\left.+\sum_{j=0}^{N-1}\int_{S_j}\left(\mathbf{u}^{(j+1)}(\mathbf{r}')\cdot\left(\widetilde{\mathbf{T}}_{\mathbf{r}'}^{(j)}-\widetilde{\mathbf{T}}_{\mathbf{r}'}^{(j+1)}\right)\widetilde{\mathbf{\Gamma}}(\mathbf{r},\mathbf{r}')\right)ds(r')\right\}. \tag{48}$$

The asymptotic form of the elastic scattered field for $r \to \infty$ is derived by using asymptotic analysis on the fundamental solution of the Navier equation in the integral representation (39) [7]:

$$\mathbf{u}^s(\mathbf{r};\hat{\mathbf{d}}) = u_r^\infty(\hat{\mathbf{r}};\hat{\mathbf{d}})\,\hat{\mathbf{r}}\,h(k_p^{(0)}r)+\left[u_\theta^\infty(\hat{\mathbf{r}};\hat{\mathbf{d}})\,\hat{\boldsymbol{\theta}}+u_\phi^\infty(\hat{\mathbf{r}};\hat{\mathbf{d}})\,\hat{\boldsymbol{\phi}}\right]h(k_s^{(0)}r)+\mathcal{O}\left(\frac{1}{r^2}\right), \tag{49}$$

where $h(x) = e^{ix}/ix$ is the Hankel function of the first kind and zeroth order and $\hat{\mathbf{r}}$, $\hat{\boldsymbol{\theta}}$, $\hat{\boldsymbol{\phi}}$ are the spherical unit vectors. Also, u_r^∞ is the normalized spherical radial far-field pattern and u_θ^∞, u_ϕ^∞ are the two normalized spherical tangential far-field patterns given in [6, 8]. The notation $\mathbf{u}^s(\mathbf{r};\hat{\mathbf{d}})$, $u_r^\infty(\hat{\mathbf{r}};\hat{\mathbf{d}})$, $u_\theta^\infty(\hat{\mathbf{r}};\hat{\mathbf{d}})$ and $u_\phi^\infty(\hat{\mathbf{r}};\hat{\mathbf{d}})$ indicate the dependence of the scattered field and of the far-field patterns on the direction of propagation $\hat{\mathbf{d}}$ of the incident field. Moreover, the on the surface of a sphere with radius that tends to infinity, asymptotic form of the corresponding surface traction field is given by Dassios et al. [8]:

$$\widetilde{\mathbf{T}}^{(0)}\mathbf{u}^s(\mathbf{r};\hat{\mathbf{d}}) = \frac{i\omega^2}{k_p^{(0)}}u_r^\infty(\hat{\mathbf{r}};\hat{\mathbf{d}})\,\hat{\mathbf{r}}\,h(k_p^{(0)}r)+\frac{i\omega^2}{k_s^{(0)}}\left[u_\theta^\infty(\hat{\mathbf{r}};\hat{\mathbf{d}})\,\hat{\boldsymbol{\theta}}+u_\phi^\infty(\hat{\mathbf{r}};\hat{\mathbf{d}})\,\hat{\boldsymbol{\phi}}\right]h(k_s^{(0)}r)$$
$$+\mathcal{O}\left(\frac{1}{r^2}\right). \tag{50}$$

We note that alternative expressions of the far-field patterns can be obtained, using asymptotic analysis on the fundamental solution of the Navier equation in (48). Specifically, using the following asymptotic forms for $r \to \infty$ [6]:

$$\widetilde{\mathbf{\Gamma}}(\mathbf{r},\mathbf{r}') = \hat{\mathbf{r}}\otimes\hat{\mathbf{r}}\,\frac{1}{\lambda^{(0)}+2\mu^{(0)}}\,e^{-ik_p^{(0)}\hat{\mathbf{r}}\cdot\mathbf{r}'}\,\frac{e^{ik_p^{(0)}r}}{r}+\left(\widetilde{I}_3-\hat{\mathbf{r}}\otimes\hat{\mathbf{r}}\right)\frac{1}{\mu^{(0)}}e^{-ik_s^{(0)}\hat{\mathbf{r}}\cdot\mathbf{r}'}\,\frac{e^{ik_s^{(0)}r}}{r}$$
$$+\mathcal{O}\left(\frac{1}{r^2}\right), \tag{51}$$

$$\nabla_{\mathbf{r}'}\cdot\widetilde{\mathbf{\Gamma}}(\mathbf{r},\mathbf{r}') = -\hat{\mathbf{r}}\,\frac{ik_p^{(0)}}{\lambda^{(0)}+2\mu^{(0)}}\,e^{-ik_p^{(0)}\hat{\mathbf{r}}\cdot\mathbf{r}'}\,\frac{e^{ik_p^{(0)}r}}{r}+\mathcal{O}\left(\frac{1}{r^2}\right), \tag{52}$$

$$\Delta_{\mathbf{r}'}^{(L,j)}\widetilde{\mathbf{\Gamma}}(\mathbf{r},\mathbf{r}') = -\hat{\mathbf{r}}\otimes\hat{\mathbf{r}}\,\frac{\lambda^{(j)}+2\mu^{(j)}}{\lambda^{(0)}+2\mu^{(0)}}(k_p^{(0)})^2 e^{-ik_p^{(0)}\,\hat{\mathbf{r}}\cdot\mathbf{r}'}\frac{e^{ik_p^{(0)}r}}{r}$$
$$-\left(\widetilde{I}_3-\hat{\mathbf{r}}\otimes\hat{\mathbf{r}}\right)\frac{\mu^{(j)}}{\mu^{(0)}}(k_s^{(0)})^2 e^{-ik_s^{(0)}\,\hat{\mathbf{r}}\cdot\mathbf{r}'}\frac{e^{ik_s^{(0)}r}}{r}+\mathcal{O}\left(\frac{1}{r^2}\right), \tag{53}$$

$$\left(\widetilde{\mathbf{T}}_{\mathbf{r}'}^{(j)}-\widetilde{\mathbf{T}}_{\mathbf{r}'}^{(j+1)}\right)\widetilde{\mathbf{\Gamma}}(\mathbf{r},\mathbf{r}')$$
$$= -\hat{\mathbf{n}}'\cdot\left[(\lambda^{(j)}-\lambda^{(j+1)})\widetilde{I}_3+2(\mu^{(j)}-\mu^{(j+1)})\hat{\mathbf{r}}\otimes\hat{\mathbf{r}}\right]$$
$$\otimes\,\hat{\mathbf{r}}\frac{ik_p^{(0)}}{\lambda^{(0)}+2\mu^{(0)}}e^{-ik_p^{(0)}\,\hat{\mathbf{r}}\cdot\mathbf{r}'}\frac{e^{ik_p^{(0)}r}}{r}$$
$$-2(\mu^{(j)}-\mu^{(j+1)})(\hat{\mathbf{n}}'\cdot\hat{\mathbf{r}})\left(\widetilde{I}_3-\hat{\mathbf{r}}\otimes\hat{\mathbf{r}}\right)\frac{ik_s^{(0)}}{\mu^{(0)}}e^{-ik_s^{(0)}\,\hat{\mathbf{r}}\cdot\mathbf{r}'}\frac{e^{ik_s^{(0)}r}}{r}$$
$$-(\mu^{(j)}-\mu^{(j+1)})\hat{\mathbf{n}}'\times\widetilde{I}_3\times\hat{\mathbf{r}}\frac{ik_s^{(0)}}{\mu^{(0)}}e^{-ik_s^{(0)}\,\hat{\mathbf{r}}\cdot\mathbf{r}'}\frac{e^{ik_s^{(0)}r}}{r}+\mathcal{O}\left(\frac{1}{r^2}\right), \tag{54}$$

for $j = 1, 2, \ldots, N$ in (53) and for $j = 1, 2, \ldots, N-1$ in (54), we obtain the following theorem.

Theorem 3.4 *The normalized spherical elastic far-field patterns admit the integral representations:*

$$u_r^{\infty}(\hat{\mathbf{r}};\hat{\mathbf{d}})=\sum_{j=1}^{N}\left\{i(k_p^{(0)})^3\left(\frac{\rho^{(j)}}{\rho^{(0)}}-\frac{\lambda^{(j)}+2\mu^{(j)}}{\lambda^{(0)}+2\mu^{(0)}}\right)\mathbf{N}_p^{(j)}\cdot\hat{\mathbf{r}}+\frac{(k_p^{(0)})^2}{\lambda^{(0)}+2\mu^{(0)}}\gamma^{(j)}\mathbf{J}_p^{(j)}\right.$$
$$\left.+\frac{(k_p^{(0)})^2}{\lambda^{(0)}+2\mu^{(0)}}\widetilde{\mathbf{H}}_p^{(j)}:\left[(\lambda^{(j-1)}-\lambda^{(j)})\widetilde{I}_3+2(\mu^{(j-1)}-\mu^{(j)})\hat{\mathbf{r}}\otimes\hat{\mathbf{r}}\right]\right\}, \tag{55}$$

$$u_\theta^{\infty}(\hat{\mathbf{r}};\hat{\mathbf{d}})=\sum_{j=1}^{N}\left\{i(k_s^{(0)})^3\left(\frac{\rho^{(j)}}{\rho^{(0)}}-\frac{\mu^{(j)}}{\mu^{(0)}}\right)\mathbf{N}_s^{(j)}\cdot\hat{\boldsymbol{\theta}}\right.$$
$$\left.+\frac{2(k_s^{(0)})^2}{\mu^{(0)}}(\mu^{(j-1)}-\mu^{(j)})\widetilde{\mathbf{H}}_s^{(j)}:\hat{\mathbf{r}}\otimes\hat{\boldsymbol{\theta}}+\frac{(k_s^{(0)})^2}{\mu^{(0)}}(\mu^{(j-1)}-\mu^{(j)})\mathbf{h}_s^{(j)}\cdot\hat{\boldsymbol{\phi}}\right\}, \tag{56}$$

$$u_\phi^\infty(\hat{\mathbf{r}};\hat{\mathbf{d}}) = \sum_{j=1}^{N}\Bigg\{ i(k_s^{(0)})^3\left(\frac{\rho^{(j)}}{\rho^{(0)}}-\frac{\mu^{(j)}}{\mu^{(0)}}\right)\mathbf{N}_s^{(j)}\cdot\hat{\boldsymbol{\phi}}$$
$$+\frac{2(k_s^{(0)})^2}{\mu^{(0)}}(\mu^{(j-1)}-\mu^{(j)})\widetilde{\mathbf{H}}_s^{(j)} : \hat{\mathbf{r}}\otimes\hat{\boldsymbol{\phi}}+\frac{(k_s^{(0)})^2}{\mu^{(0)}}(\mu^{(j-1)}-\mu^{(j)})\mathbf{h}_s^{(j)}\cdot\hat{\boldsymbol{\theta}}\Bigg\}, \tag{57}$$

where the double contraction operator ':' is given by

$$(\mathbf{a}\otimes\mathbf{b}) : (\mathbf{c}\otimes\mathbf{d}) = (\mathbf{a}\cdot\mathbf{d})(\mathbf{b}\cdot\mathbf{c}), \tag{58}$$

and

$$\mathbf{N}_p^{(j)} = \frac{1}{4\pi}\int_{D_j}\mathbf{u}^{(j)}(\mathbf{r}')e^{-ik_p^{(0)}\hat{\mathbf{r}}\cdot\mathbf{r}'}\,d\mathbf{r}', \tag{59}$$

$$\mathbf{N}_s^{(j)} = \frac{1}{4\pi}\int_{D_j}\mathbf{u}^{(j)}(\mathbf{r}')e^{-ik_s^{(0)}\hat{\mathbf{r}}\cdot\mathbf{r}'}\,d\mathbf{r}', \tag{60}$$

$$\mathbf{J}_p^{(j)} = \frac{1}{4\pi}\int_{D_j}\theta^{(j)}(\mathbf{r}')e^{-ik_p^{(0)}\hat{\mathbf{r}}\cdot\mathbf{r}'}\,d\mathbf{r}', \tag{61}$$

$$\widetilde{\mathbf{H}}_p^{(j)} = \frac{1}{4\pi}\int_{S_{j-1}}\mathbf{u}^{(j)}(\mathbf{r}')\otimes\hat{\mathbf{n}}'e^{-ik_p^{(0)}\hat{\mathbf{r}}\cdot\mathbf{r}'}\,ds(\mathbf{r}'), \tag{62}$$

$$\widetilde{\mathbf{H}}_s^{(j)} = \frac{1}{4\pi}\int_{S_{j-1}}\mathbf{u}^{(j)}(\mathbf{r}')\otimes\hat{\mathbf{n}}'e^{-ik_s^{(0)}\hat{\mathbf{r}}\cdot\mathbf{r}'}\,ds(\mathbf{r}'), \tag{63}$$

$$\mathbf{h}_s^{(j)} = \frac{1}{4\pi}\int_{S_{j-1}}\mathbf{u}^{(j)}(\mathbf{r}')\times\hat{\mathbf{n}}'e^{-ik_s^{(0)}\hat{\mathbf{r}}\cdot\mathbf{r}'}\,ds(\mathbf{r}'). \tag{64}$$

The scattering cross section is given by Cakoni and Dassios [6] and Dassios and Kleinman [10]:

$$\sigma_s = \frac{1}{\omega\rho^{(0)}\left(c_p^{(0)}(a_p^{(0)})^2+c_s^{(0)}(a_s^{(0)})^2\right)}\,Im\left[\int_{S_0}\overline{\mathbf{u}}^s(\mathbf{r})\cdot\widetilde{\mathbf{T}}^{(0)}\mathbf{u}^s(\mathbf{r})\,ds(\mathbf{r})\right]$$
$$= -\frac{1}{2i\omega\rho^{(0)}\left(c_p^{(0)}(a_p^{(0)})^2+c_s^{(0)}(a_s^{(0)})^2\right)}\int_{S_0}\left(\mathbf{u}^s(\mathbf{r})\cdot\widetilde{\mathbf{T}}^{(0)}\overline{\mathbf{u}}^s(\mathbf{r})-\overline{\mathbf{u}}^s\cdot\widetilde{\mathbf{T}}^{(0)}\mathbf{u}^s(\mathbf{r})\right)ds(\mathbf{r}), \tag{65}$$

where the overbar indicates complex conjugation, $a_p^{(0)}$ and $a_s^{(0)}$ are the real amplitudes of the longitudinal and of the transverse part of the plane incident elastic field, respectively, (see (5)) and $c_p^{(0)}$, $c_s^{(0)}$ are the corresponding phase velocities given by

$$c_p^{(0)} = \sqrt{\frac{\lambda^{(0)} + 2\mu^{(0)}}{\rho^{(0)}}} \quad , \quad c_s^{(0)} = \sqrt{\frac{\mu^{(0)}}{\rho^{(0)}}} \ . \tag{66}$$

Moreover, by using relations (49) and (50), the scattering cross section can be given in terms of the far-field patterns as follows [6, 10]:

$$\sigma_s = \frac{1}{\omega^2} \int_{S^2} \frac{(c_p^{(0)})^3 |u_r^\infty(\hat{\mathbf{r}};\hat{\mathbf{d}})|^2 + (c_s^{(0)})^3 \left(|u_\theta^\infty(\hat{\mathbf{r}};\hat{\mathbf{d}})|^2 + |u_\phi^\infty(\hat{\mathbf{r}};\hat{\mathbf{d}})|^2 \right)}{c_p^{(0)} (a_p^{(0)})^2 + c_s^{(0)} (a_s^{(0)})^2} ds(\hat{\mathbf{r}}) \ . \tag{67}$$

Also, the absorption cross section is given by Cakoni and Dassios [6] and Dassios and Kleinman [10]:

$$\begin{aligned} \sigma_\alpha &= - \frac{1}{\omega \rho^{(0)} \left(c_p^{(0)} (a_p^{(0)})^2 + c_s^{(0)} (a_s^{(0)})^2 \right)} Im \left[\int_{S_0} \overline{\mathbf{u}}^{(0)}(\mathbf{r}) \cdot \widetilde{\mathbf{T}}^{(0)} \mathbf{u}^{(0)}(\mathbf{r}) \, ds(\mathbf{r}) \right] \\ &= \frac{1}{2i\omega\rho^{(0)} \left(c_p^{(0)} (a_p^{(0)})^2 + c_s^{(0)} (a_s^{(0)})^2 \right)} \\ &\quad \int_{S_0} \left(\mathbf{u}^{(0)}(\mathbf{r}) \cdot \widetilde{\mathbf{T}}^{(0)} \overline{\mathbf{u}}^{(0)}(\mathbf{r}) - \overline{\mathbf{u}}^{(0)} \cdot \widetilde{\mathbf{T}}^{(0)} \mathbf{u}^{(0)}(\mathbf{r}) \right) ds(\mathbf{r}) \end{aligned} \tag{68}$$

Finally, the extinction cross section is given by Dassios and Kleinman [10]:

$$\sigma_e = \sigma_s + \sigma_\alpha \ . \tag{69}$$

Remark 3.5 The alternative integral representation (48) (or (47)) and the alternative expressions of the far-field patterns (55)–(57) are important since they involve volume integrals over the layers D_j for $j = 1, 2, \ldots, N$ as well as integrals over the surfaces S_j for $j = 0, \ldots, N-1$, which contain the physical parameters of the elastic material of the medium of propagation and of the thermoelastic materials of the layers as well as the interior total thermoelastic fields. Special cases can be derived via them. For example, by setting the parameters of the thermoelastic materials of all layers to be equal, the case of the (one-layer) penetrable thermoelastic scatterer is obtained [4, 13]. Also, for example, by setting $\alpha^{(j)} = 0$ (therefore $\gamma^{(j)} = 0$ and $\eta^{(j)} = 0$) for $j = 1, 2, \ldots, N$, the case of the multi-layered elastic scatterer can be obtained.

4 Scattering Relations

We start this section by presenting the following Twersky's notation: Let Ω be a bounded region with surface $S = \partial\Omega$. Then for every $\mathbf{u}, \mathbf{v} \in [C^2(\Omega)]^3 \cap [C^1(\overline{\Omega})]^3$ we define:

$$\{\mathbf{u}\,,\ \mathbf{v}\}_S = \int_S \left(\mathbf{u}\cdot\widetilde{\mathbf{T}}^{(0)}\mathbf{v} - \mathbf{v}\cdot\widetilde{\mathbf{T}}^{(0)}\mathbf{u}\right) ds\ . \tag{70}$$

The far-field patterns, via the Twersky's notation (70), satisfy the relations [8]:

$$u_r^\infty(\hat{\mathbf{r}};\hat{\mathbf{d}}) = \frac{i(k_p^{(0)})^3}{4\pi\omega^2}\left\{\mathbf{u}^s(\mathbf{r}';\hat{\mathbf{d}})\,,\ \hat{\mathbf{r}}e^{-ik_p^{(0)}\hat{\mathbf{r}}\cdot\mathbf{r}'}\right\}_{S_0}, \tag{71}$$

$$u_\theta^\infty(\hat{\mathbf{r}};\hat{\mathbf{d}}) = \frac{i(k_s^{(0)})^3}{4\pi\omega^2}\left\{\mathbf{u}^s(\mathbf{r}';\hat{\mathbf{d}})\,,\ \hat{\boldsymbol{\theta}}e^{-ik_s^{(0)}\hat{\mathbf{r}}\cdot\mathbf{r}'}\right\}_{S_0}, \tag{72}$$

$$u_\phi^\infty(\hat{\mathbf{r}};\hat{\mathbf{d}}) = \frac{i(k_s^{(0)})^3}{4\pi\omega^2}\left\{\mathbf{u}^s(\mathbf{r}';\hat{\mathbf{d}})\,,\ \hat{\boldsymbol{\phi}}e^{-ik_s^{(0)}\hat{\mathbf{r}}\cdot\mathbf{r}'}\right\}_{S_0}. \tag{73}$$

Moreover, the scattering cross section, the absorption cross section and the extinction cross section, via the Twersky's notation (70), satisfy the relations:

$$\sigma_s = -\frac{1}{2i\omega\rho^{(0)}\left(c_p^{(0)}(a_p^{(0)})^2 + c_s^{(0)}(a_s^{(0)})^2\right)}\left\{\mathbf{u}^s(\mathbf{r}';\hat{\mathbf{d}})\,,\ \overline{\mathbf{u}}^s(\mathbf{r}';\hat{\mathbf{d}})\right\}_{S_0}, \tag{74}$$

$$\sigma_a = \frac{1}{2i\omega\rho^{(0)}\left(c_p^{(0)}(a_p^{(0)})^2 + c_s^{(0)}(a_s^{(0)})^2\right)}\left\{\mathbf{u}^{(0)}(\mathbf{r}';\hat{\mathbf{d}})\,,\ \overline{\mathbf{u}}^{(0)}(\mathbf{r}';\hat{\mathbf{d}})\right\}_{S_0}, \tag{75}$$

$$\sigma_e = \frac{1}{2i\omega\rho^{(0)}\left(c_p^{(0)}(a_p^{(0)})^2 + c_s^{(0)}(a_s^{(0)})^2\right)}\left[\left\{\mathbf{u}^{(0)}(\mathbf{r}';\hat{\mathbf{d}})\,,\ \overline{\mathbf{u}}^{(0)}(\mathbf{r}';\hat{\mathbf{d}})\right\}_{S_0} - \left\{\mathbf{u}^s(\mathbf{r}';\hat{\mathbf{d}})\,,\ \overline{\mathbf{u}}^s(\mathbf{r}';\hat{\mathbf{d}})\right\}_{S_0}\right]. \tag{76}$$

We are now in position to prove the following scattering relations concerning the scattering of elastic waves by a multi-layered thermoelastic scatterer.

Lemma 4.1 *Let $\mathbf{u}_1^i$, $\mathbf{u}_2^i$ be two plane elastic waves incident upon the multi-layered thermoelastic scatterer D with $\mathbf{u}_a^s$, $\mathbf{u}_a^{(0)}$ and $\mathbf{U}_a^{(j)} = (\mathbf{u}_a^{(j)}, \theta_a^{(j)})$, for $a = 1, 2$ and $j = 1, 2, \ldots, N$, the corresponding scattered elastic, total exterior elastic and total interior in the layer D_j thermoelastic fields, respectively. Then the following relation holds true:*

$$\left\{\mathbf{u}_1^s\,,\,\mathbf{u}_2^i\right\}_{S_0} - \left\{\mathbf{u}_2^s\,,\,\mathbf{u}_1^i\right\}_{S_0} = I_{(0)}\,, \tag{77}$$

where

$$\begin{aligned}
I_{(0)} = & \int_{D_1}\left[\mathbf{u}_1^{(1)}\cdot\Delta^{(L,1)}\mathbf{u}_2^{(1)} - \mathbf{u}_2^{(1)}\cdot\Delta^{(L,1)}\mathbf{u}_1^{(1)}\right]d\mathbf{r} + I_{(B)} \\
& - \sum_{j=2}^{N}\int_{D_j}\left(\theta_1^{(j)}\Delta\theta_2^{(j)} - \theta_2^{(j)}\Delta\theta_1^{(j)}\right)d\mathbf{r} \\
& + \int_{S_1}\left[\left(\gamma^{(1)} - \gamma^{(2)} + i\omega\eta^{(2)}\right)\theta_2^{(2)}\mathbf{u}_1^{(2)}\cdot\hat{\mathbf{n}} - \gamma^{(1)}\theta_1^{(2)}\mathbf{u}_2^{(2)}\cdot\hat{\mathbf{n}}\right]ds \\
& + \sum_{j=3}^{N}\int_{S_{j-1}}\left[\left(\left(\gamma^{(j-1)} - i\omega\eta^{(j-1)}\right) - \left(\gamma^{(j)} - i\omega\eta^{(j)}\right)\right)\theta_2^{(j)}\mathbf{u}_1^{(j)}\cdot\hat{\mathbf{n}}\right]ds\,,
\end{aligned} \tag{78}$$

with

$$\begin{aligned}
I_{(B)} = & \sum_{j=2}^{N}\int_{D_j}\left(\mathbf{U}_1^{(j)}\cdot\widetilde{\mathbf{L}}^{(j*)}\mathbf{U}_2^{(j)}\right)d\mathbf{r} \\
= & \sum_{j=2}^{N}\int_{D_j}\left(\gamma^{(j)} - i\omega\eta^{(j)}\right)\left[\frac{1}{\gamma^{(j)}}\mathbf{u}_1^{(j)}\cdot\left(\Delta^{(L,j)} + \rho^{(j)}\omega^2\right)\mathbf{u}_2^{(j)}\right. \\
& \left. - \frac{1}{i\omega\eta^{(j)}}\theta_1^{(j)}\left(\Delta + q^{(j)}\right)\theta_2^{(j)}\right]d\mathbf{r}\,.
\end{aligned} \tag{79}$$

Proof In view of relation (3) and the bilinearity of Twersky's notation (70) we obtain

$$\left\{\mathbf{u}_1^{(0)}\,,\,\mathbf{u}_2^{(0)}\right\}_{S_0} = \left\{\mathbf{u}_1^i\,,\,\mathbf{u}_2^i\right\}_{S_0} + \left\{\mathbf{u}_1^i\,,\,\mathbf{u}_2^s\right\}_{S_0} + \left\{\mathbf{u}_1^s\,,\,\mathbf{u}_2^i\right\}_{S_0} + \left\{\mathbf{u}_1^s\,,\,\mathbf{u}_2^s\right\}_{S_0}\,. \tag{80}$$

For the calculation of the integral of the left-hand side of (80), using the transmission conditions (12), (13), the Biot system of equations (9) and applying successively identity (22) and the Green's second identity we obtain

$$\left\{\mathbf{u}_1^{(0)}\,,\,\mathbf{u}_2^{(0)}\right\}_{S_0} = I_{(0)}. \tag{81}$$

For the first integral of the right-hand side of (80), applying identity (21) in D and using the fact that the incident fields are regular solutions of the Navier equation (6) in D we obtain

$$\left\{\mathbf{u}_1^i\,,\,\mathbf{u}_2^i\right\}_{S_0} = 0\,. \tag{82}$$

For the calculation of the last integral of the right-hand side of (80), we assume a large sphere of surface S_R that contains the multi-layered thermoelastic scatterer D in its interior and we denote with $D_R = \left\{\mathbf{r}' \in D_0 : |\mathbf{r}'| < R\right\}$ the region within S_R but outside the scatterer (Figure 2). Then applying the identity (21) in D_R and using the fact that the scattered fields are solutions of the Navier equation (6) in D_R we take

$$\left\{\mathbf{u}_1^s\,,\,\mathbf{u}_2^s\right\}_{S_0} = \left\{\mathbf{u}_1^s\,,\,\mathbf{u}_2^s\right\}_{S_R}\,. \tag{83}$$

Next, letting $R \to \infty$ and using relations (49), (50) we obtain

$$\left\{\mathbf{u}_1^s\,,\,\mathbf{u}_2^s\right\}_{S_0} = \lim_{R\to\infty}\left\{\mathbf{u}_1^s\,,\,\mathbf{u}_2^s\right\}_{S_R} = 0\,. \tag{84}$$

Therefore, by substitution of (81), (82), and (84) in (80) and using the fact that $\left\{\mathbf{u}_1^i\,,\,\mathbf{u}_2^s\right\}_{S_0} = -\left\{\mathbf{u}_2^s\,,\,\mathbf{u}_1^i\right\}_{S_0}$ which is valid due to the definition of the Twersky's notation (70), Lemma 4.1 is proved. □

Theorem 4.2 *Let* $\mathbf{u}_1^i$ *and* $\mathbf{u}_2^i$ *be two plane elastic waves incident upon the multi-layered thermoelastic scatterer* D*, with directions of propagation* $\hat{\mathbf{d}}_1, \hat{\mathbf{d}}_2$ *and directions of polarization* $\hat{\mathbf{b}}_1, \hat{\mathbf{b}}_2$*, respectively. Then the following relations hold true.*

(i) If both incident fields are longitudinal plane waves, then:

Fig. 2 Sphere of surface S_R containing the multi-layered thermoelastic scatterer

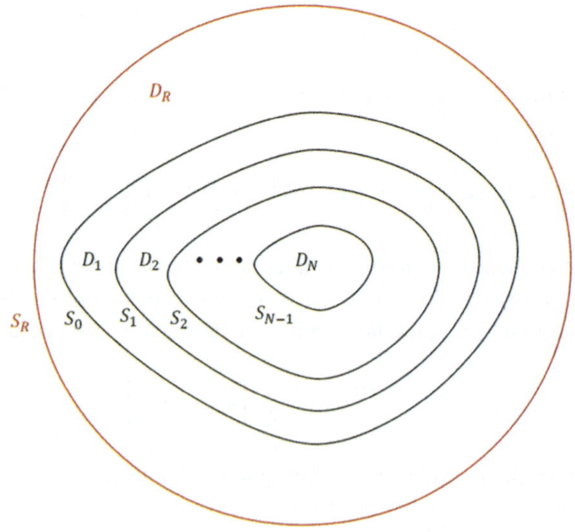

$$\frac{4\pi\omega^2}{i(k_p^{(0)})^3} u_r^\infty(-\hat{\mathbf{d}}_1; \hat{\mathbf{d}}_2) = \frac{4\pi\omega^2}{i(k_p^{(0)})^3} u_r^\infty(-\hat{\mathbf{d}}_2; \hat{\mathbf{d}}_1) + I_{(0)} . \tag{85}$$

(ii) If both incident fields are transverse plane waves, then:

$$- \frac{4\pi\omega^2}{i(k_s^{(0)})^3} \left[(\hat{\mathbf{b}}_1 \cdot \hat{\boldsymbol{\theta}})\, u_\theta^\infty(-\hat{\mathbf{d}}_1; \hat{\mathbf{d}}_2) + (\hat{\mathbf{b}}_1 \cdot \hat{\boldsymbol{\phi}})\, u_\phi^\infty(-\hat{\mathbf{d}}_1; \hat{\mathbf{d}}_2) \right]$$
$$= - \frac{4\pi\omega^2}{i(k_s^{(0)})^3} \left[(\hat{\mathbf{b}}_2 \cdot \hat{\boldsymbol{\theta}})\, u_\theta^\infty(-\hat{\mathbf{d}}_2; \hat{\mathbf{d}}_1) + (\hat{\mathbf{b}}_2 \cdot \hat{\boldsymbol{\phi}})\, u_\phi^\infty(-\hat{\mathbf{d}}_2; \hat{\mathbf{d}}_1) \right] + I_{(0)} . \tag{86}$$

(iii) If the incident field $\mathbf{u}_1^i$ is a longitudinal and the incident field $\mathbf{u}_2^i$ is a transverse plane wave, then:

$$\frac{4\pi\omega^2}{i(k_p^{(0)})^3} u_r^\infty(-\hat{\mathbf{d}}_1; \hat{\mathbf{d}}_2) = - \frac{4\pi\omega^2}{i(k_s^{(0)})^3} \left[(\hat{\mathbf{b}}_2 \cdot \hat{\boldsymbol{\theta}})\, u_\theta^\infty(-\hat{\mathbf{d}}_2; \hat{\mathbf{d}}_1) \right.$$
$$\left. + (\hat{\mathbf{b}}_2 \cdot \hat{\boldsymbol{\phi}})\, u_\phi^\infty(-\hat{\mathbf{d}}_2; \hat{\mathbf{d}}_1) \right) + I_{(0)} . \tag{87}$$

Where $I_{(0)}$ is given by (78).

Proof Theorem 4.2 is a consequence of Lemma 4.1 For a plane incident elastic wave with direction of propagation $\hat{\mathbf{r}}$ and polarization $\hat{\mathbf{b}}$ we have

$$\hat{\mathbf{b}} = \hat{\mathbf{b}} \cdot \widetilde{I}_3 = \hat{\mathbf{b}} \cdot (\hat{\mathbf{r}} \otimes \hat{\mathbf{r}} + \hat{\boldsymbol{\theta}} \otimes \hat{\boldsymbol{\theta}} + \hat{\boldsymbol{\phi}} \otimes \hat{\boldsymbol{\phi}}) = \left(\hat{\mathbf{b}} \cdot \hat{\boldsymbol{\theta}} \right) \hat{\boldsymbol{\theta}} + \left(\hat{\mathbf{b}} \cdot \hat{\boldsymbol{\phi}} \right) \hat{\boldsymbol{\phi}} . \tag{88}$$

Therefore, using relations (71)–(73) in the remaining Twersky-type integrals in (77) as well as relation (88) for the cases (ii) and (iii), the Theorem 4.2 is proved. □

Theorem 4.3 *Let $\mathbf{u}_1^i$ and $\mathbf{u}_2^i$ be two plane elastic waves incident upon the multi-layered thermoelastic scatterer D, with directions of propagation $\hat{\mathbf{d}}_1$,$\hat{\mathbf{d}}_2$ and directions of polarization $\hat{\mathbf{b}}_1$,$\hat{\mathbf{b}}_2$, respectively. Then the following relations hold true.*

(i) If both incident fields are longitudinal plane waves, then:

$$\frac{4\pi\omega^2}{i(k_p^{(0)})^3} \left[\overline{u_r^\infty(\hat{\mathbf{d}}_1; \hat{\mathbf{d}}_2)} + u_r^\infty(\hat{\mathbf{d}}_2; \hat{\mathbf{d}}_1) \right] = I_{(\overline{0})} - I_{(\overline{s})} . \tag{89}$$

(ii) If both incident fields are transverse plane waves, then:

$$\frac{4\pi\omega^2}{i\,(k_s^{(0)})^3}\Big[(\hat{\mathbf{b}}_1\cdot\hat{\boldsymbol{\theta}})\,\overline{u_\theta^\infty}(\hat{\mathbf{d}}_1;\hat{\mathbf{d}}_2)+(\hat{\mathbf{b}}_1\cdot\hat{\boldsymbol{\phi}})\,\overline{u_\phi^\infty}(\hat{\mathbf{d}}_1;\hat{\mathbf{d}}_2)$$
$$+(\hat{\mathbf{b}}_2\cdot\hat{\boldsymbol{\theta}})\,u_\theta^\infty(\hat{\mathbf{d}}_2;\hat{\mathbf{d}}_1)+(\hat{\mathbf{b}}_2\cdot\hat{\boldsymbol{\phi}})\,u_\phi^\infty(\hat{\mathbf{d}}_2;\hat{\mathbf{d}}_1)\Big]=I_{(\bar{0})}-I_{(\bar{s})}\,. \tag{90}$$

(iii) If the first incident field is a longitudinal and the second incident field is a transverse plane wave, then:

$$\frac{4\pi\omega^2}{i\,(k_p^{(0)})^3}\,\overline{u_r^\infty}(\hat{\mathbf{d}}_1;\hat{\mathbf{d}}_2)+\frac{4\pi\omega^2}{i\,(k_s^{(0)})^3}\Big[(\hat{\mathbf{b}}_2\cdot\hat{\boldsymbol{\theta}})\,u_\theta^\infty(\hat{\mathbf{d}}_2;\hat{\mathbf{d}}_1)+(\hat{\mathbf{b}}_2\cdot\hat{\boldsymbol{\phi}})\,u_\phi^\infty(\hat{\mathbf{d}}_2;\hat{\mathbf{d}}_1)\Big]$$
$$=I_{(\bar{0})}-I_{(\bar{s})}. \tag{91}$$

With

$$I_{(\bar{s})}=-2i\omega^2\int_{S^2}\Bigg[\frac{1}{(k_p^{(0)})^3}u_r^\infty(\hat{\mathbf{r}};\,\hat{\mathbf{d}}_1)\,\overline{u_r^\infty}(\hat{\mathbf{r}};\,\hat{\mathbf{d}}_2)$$
$$+\frac{1}{(k_s^{(0)})^3}\,u_\theta^\infty(\hat{\mathbf{r}};\,\hat{\mathbf{d}}_1)\,\overline{u_\theta^\infty}(\hat{\mathbf{r}};\,\hat{\mathbf{d}}_2)+\frac{1}{(k_s^{(0)})^3}\,u_\phi^\infty(\hat{\mathbf{r}};\,\hat{\mathbf{d}}_1)\,\overline{u_\phi^\infty}(\hat{\mathbf{r}};\,\hat{\mathbf{d}}_2)\Bigg]\,ds(\hat{\mathbf{r}})) \tag{92}$$

and

$$I_{(\bar{0})}=\int_{D_1}\Big[\mathbf{u}_1^{(1)}\cdot\Delta^{(L,1)}\overline{\mathbf{u}_2^{(1)}}-\overline{\mathbf{u}_2^{(1)}}\cdot\Delta^{(L,1)}\mathbf{u}_1^{(1)}\Big]\,d\mathbf{r}+I_{(\overline{B})}$$
$$-\sum_{j=2}^{N}\int_{D_j}\Big(\theta_1^{(j)}\Delta\overline{\theta_2^{(j)}}-\overline{\theta_2^{(j)}}\Delta\theta_1^{(j)}\Big)\,d\mathbf{r}$$
$$+\int_{S_1}\Big[\Big(\gamma^{(1)}-\gamma^{(2)}+i\omega\eta^{(2)}\Big)\,\overline{\theta_2^{(2)}}\mathbf{u}_1^{(2)}\cdot\hat{\mathbf{n}}-\gamma^{(1)}\theta_1^{(2)}\overline{\mathbf{u}_2^{(2)}}\cdot\hat{\mathbf{n}}\Big]\,ds(\mathbf{r})$$
$$+\sum_{j=3}^{N}\int_{S_{j-1}}\Big[\Big(\Big(\gamma^{(j-1)}-i\omega\eta^{(j-1)}\Big)-\Big(\gamma^{(j)}-i\omega\eta^{(j)}\Big)\Big)\,\overline{\theta_2^{(j)}}\mathbf{u}_1^{(j)}\cdot\hat{\mathbf{n}}\Big]\,ds(\mathbf{r})\,, \tag{93}$$

where

$$I_{(\overline{B})} = \sum_{j=2}^{N} \int_{D_j} \left(\mathbf{U}_1^{(j)} \cdot \widetilde{\mathbf{L}}^{(j*)} \overline{\mathbf{U}_2^{(j)}} \right) d\mathbf{r}$$

$$= \sum_{j=2}^{N} \int_{D_j} \left[\left(\gamma^{(j)} - i\omega\eta^{(j)} \right) \frac{1}{\gamma^{(j)}} \mathbf{u}_1^{(j)} \cdot \left(\Delta^{(L,j)} + \rho^{(j)}\omega^2 \right) \overline{\mathbf{u}_2^{(j)}} \right.$$

$$\left. + \left(\gamma^{(j)} + i\omega\eta^{(j)} \right) \frac{1}{i\omega\eta^{(j)}} \theta_1^{(j)} \Delta \overline{\theta_2^{(j)}} - \left(\gamma^{(j)} - i\omega\eta^{(j)} \right) \frac{1}{i\omega\eta^{(j)}} \theta_1^{(j)} q^{(j)} \overline{\theta_2^{(j)}} \right] d\mathbf{r}, \tag{94}$$

where the overbar indicates complex conjugation and S^2 is the unit sphere in $\mathbb{R}^3$.

Proof In view of relation (3) and the bilinearity of Twersky's notation (70) we obtain

$$\left\{ \mathbf{u}_1^{(0)}, \overline{\mathbf{u}}_2^{(0)} \right\}_{S_0} = \left\{ \mathbf{u}_1^{i}, \overline{\mathbf{u}}_2^{i} \right\}_{S_0} + \left\{ \mathbf{u}_1^{i}, \overline{\mathbf{u}}_2^{s} \right\}_{S_0} + \left\{ \mathbf{u}_1^{s}, \overline{\mathbf{u}}_2^{i} \right\}_{S_0} + \left\{ \mathbf{u}_1^{s}, \overline{\mathbf{u}}_2^{s} \right\}_{S_0}. \tag{95}$$

For the calculation of the integral of the left-hand side of (95), using the transmission conditions (12), (13), the Biot system of equations (9) and applying successively formula (22) and the Green's second formula we obtain

$$\left\{ \mathbf{u}_1^{(0)}, \overline{\mathbf{u}}_2^{(0)} \right\}_{S_0} = I_{(\overline{0})}. \tag{96}$$

For the first integral of the right-hand side of (95), applying the Betti's third formula (21) in D and using the fact that the incident field $\mathbf{u}_1^{i}$ as well as the field $\overline{\mathbf{u}}_2^{i}$ are regular solutions of the Navier equation (6) in D we obtain

$$\left\{ \mathbf{u}_1^{i}, \overline{\mathbf{u}}_2^{i} \right\}_{S_0} = 0. \tag{97}$$

For the calculation of the last integral of the right-hand side of (95), we assume a large sphere of surface S_R that contains the multi-layered scatterer D in its interior and we denote with $D_R = \left\{ \mathbf{r}' \in D_0 : |\mathbf{r}'| < R \right\}$ the region within S_R but outside the scatterer (Figure 2). Then applying the Betti's third formula (21) in D_R and using the fact that the scattered field $\mathbf{u}_1^{s}$ as well as the field $\overline{\mathbf{u}}_2^{s}$ are solutions of the Navier equation (6) in D_R we take

$$\left\{ \mathbf{u}_1^{s}, \overline{\mathbf{u}}_2^{s} \right\}_{S_0} = \left\{ \mathbf{u}_1^{s}, \overline{\mathbf{u}}_2^{s} \right\}_{S_R}. \tag{98}$$

Next, letting $R \to \infty$ and using relations (49), (50) we obtain

$$
\begin{aligned}
\{\mathbf{u}_1^s, \overline{\mathbf{u}_2^s}\}_{S_0} &= \lim_{R\to\infty} \{\mathbf{u}_1^s, \overline{\mathbf{u}_2^s}\}_{S_R} \\
&= -2i\omega^2 \int_{S_\infty} \Bigg[\frac{1}{k_p^{(0)}} |h(k_p^{(0)} r)|^2 u_r^\infty(\hat{\mathbf{r}};\, \hat{\mathbf{d}}_1)\, \overline{u_r^\infty}(\hat{\mathbf{r}};\, \hat{\mathbf{d}}_2) \\
&\quad + \frac{1}{k_s^{(0)}} |h(k_s^{(0)} r)|^2 u_\theta^\infty(\hat{\mathbf{r}};\, \hat{\mathbf{d}}_1)\, \overline{u_\theta^\infty}(\hat{\mathbf{r}};\, \hat{\mathbf{d}}_2) \\
&\quad + \frac{1}{k_s^{(0)}} |h(k_s^{(0)} r)|^2 u_\phi^\infty(\hat{\mathbf{r}};\, \hat{\mathbf{d}}_1)\, \overline{u_\phi^\infty}(\hat{\mathbf{r}};\, \hat{\mathbf{d}}_2) \Bigg] ds(\mathbf{r}) \\
&= -2i\omega^2 \int_{S^2} \Bigg[\frac{1}{(k_p^{(0)})^3} u_r^\infty(\hat{\mathbf{r}};\, \hat{\mathbf{d}}_1)\, \overline{u_r^\infty}(\hat{\mathbf{r}};\, \hat{\mathbf{d}}_2) \\
&\quad + \frac{1}{(k_s^{(0)})^3} u_\theta^\infty(\hat{\mathbf{r}};\, \hat{\mathbf{d}}_1)\, \overline{u_\theta^\infty}(\hat{\mathbf{r}};\, \hat{\mathbf{d}}_2) + \frac{1}{(k_s^{(0)})^3} u_\phi^\infty(\hat{\mathbf{r}};\, \hat{\mathbf{d}}_1)\, \overline{u_\phi^\infty}(\hat{\mathbf{r}};\, \hat{\mathbf{d}}_2) \Bigg] ds(\hat{\mathbf{r}})) = I_{(\overline{s})} .
\end{aligned}
\tag{99}
$$

Finally, via (71)–(73) we obtain

$$
\overline{u_r^\infty}(\hat{\mathbf{d}}_1;\, \hat{\mathbf{d}}_2) = -\frac{i\,(k_p^{(0)})^3}{4\pi\omega^2} \left\{ \overline{\mathbf{u}}^s(\mathbf{r}';\, \hat{\mathbf{d}}_2),\, \hat{\mathbf{d}}_1 e^{ik_p^{(0)}\, \hat{\mathbf{d}}_1\cdot\mathbf{r}'} \right\}_{S_0} , \tag{100}
$$

$$
\overline{u_\theta^\infty}(\hat{\mathbf{d}}_1;\, \hat{\mathbf{d}}_2) = -\frac{i\,(k_s^{(0)})^3}{4\pi\omega^2} \left\{ \overline{\mathbf{u}}^s(\mathbf{r}';\, \hat{\mathbf{d}}_2),\, \hat{\boldsymbol{\theta}} e^{ik_s^{(0)}\, \hat{\mathbf{d}}_1\cdot\mathbf{r}'} \right\}_{S_0} , \tag{101}
$$

$$
\overline{u_\phi^\infty}(\hat{\mathbf{d}}_1;\, \hat{\mathbf{d}}_2) = -\frac{i\,(k_s^{(0)})^3}{4\pi\omega^2} \left\{ \overline{\mathbf{u}}^s(\mathbf{r}';\, \hat{\mathbf{d}}_2),\, \hat{\boldsymbol{\phi}} e^{ik_s^{(0)}\, \hat{\mathbf{d}}_1\cdot\mathbf{r}'} \right\}_{S_0} . \tag{102}
$$

Therefore, by substitution of (96), (97), (99) in (95) and using (100) for (i) or (101)–(102) for (ii), Theorem 4.3 is proved. □

Theorem 4.4

(i) Let a longitudinal plane elastic wave with direction of propagation $\hat{\mathbf{d}}$ *incident upon the multi-layered thermoelastic scatterer D. Then the following relation holds true:*

$$
\sigma_e = -\frac{4\pi}{\rho^{(0)}\,(k_p^{(0)})^2}\, Re\left[u_r^\infty(\hat{\mathbf{d}};\, \hat{\mathbf{d}}) \right] . \tag{103}
$$

(ii) Let a transverse plane elastic wave with direction of propagation $\hat{\mathbf{d}}$ *and polarization* $\hat{\mathbf{b}}$ *incident upon the multi-layered thermoelastic scatterer D. Then the following relation holds true:*

$$\sigma_e = -\frac{4\pi}{\rho^{(0)}(k_s^{(0)})^2} Re\left[\left(\hat{\mathbf{b}}\cdot\hat{\boldsymbol{\theta}}\right)u_\theta^\infty(\hat{\mathbf{d}};\hat{\mathbf{d}}) + \left(\hat{\mathbf{b}}\cdot\hat{\boldsymbol{\phi}}\right)u_\phi^\infty(\hat{\mathbf{d}};\hat{\mathbf{d}})\right]. \quad (104)$$

Theorem 4.4 is proved via relation (76) and is a consequence of Theorem 4.3 when the two incident fields are assumed to be identical. Moreover, Theorem 4.2 is a reciprocity type theorem and Theorem 4.4 is an optical type theorem, for plane wave elastic incidence. Also, Theorem 4.3 is a general type scattering theorem where the case (i) corresponds to a radial scattering theorem, the case (ii) corresponds to an angular scattering theorem and the case (iii) corresponds to a radial-angular type theorem, for plane wave elastic incidence. Note that in Theorems 4.2 and 4.4, without loss of generality, the existing amplitudes of the incident fields have been taken equal to one. Analogue results for the case of a penetrable (one-layered) thermoelastic scatterer are given in [4] and analogue results for the case of an elastic scatterer (penetrable or impenetrable) are given in [8].

References

1. C. Athanasiadis, V. Sevroglou, I.G. Stratis, 3-D elastic scattering theorems for point sources: acoustic and electromagnetic waves. J. Math. Phys. **43**, 5683–5697 (2002)
2. C. Athanasiadis, V. Sevroglou, I.G. Stratis, Scattering relations for point-generated dyadic fields in two-dimensional linear elasticity. Q. Appl. Math. **LXIV**, 695–710 (2006)
3. C. Athanasiadis, P.A. Martin, A. Spyropoulos, I.G. Stratis, Scattering relations for point generated dyadic fields. Math. Methods Appl. Sci. **31**, 987–1003 (2008).
4. E. Athanasiadou, V. Sevroglou, S. Zoi, Scattering theorems of elastic waves for a thermoelastic body. Math. Methods Appl. Sci. **41**(3), 998–1004 (2016). https://doi.org/10.1002/mma.4051
5. F. Cakoni, Boundary integral methods for thermoelastic screen scattering problem in $\mathbb{R}^3$. Math. Methods Appl. Sci. **23**, 441–466 (2000)
6. F. Cakoni, G. Dassios, The coated thermoelastic body within a low frequency elastodynamic field. Int. J. Eng. Sci. **36**, 1815–1838 (1998)
7. G. Dassios, K. Kiriaki, The low-frequency theory of elastic wave scattering. Q. Appl. Math. **42**, 225–248 (1984)
8. G. Dassios, K. Kiriaki, D. Polyzos, On the scattering amplitudes of elastic waves. ZAMP **38**, 856–873 (1987)
9. G. Dassios, K. Kiriaki, D. Polyzos, Scattering theorems for complete dyadic fields. Eng. Sci. **33**, 269–277 (1995)
10. G, Dassios, R. Kleinman, *Low Frequency Scattering* (Clarendon Press, Oxford, 2000)
11. G. Dassios, V. Kostopoulos, The scattering amplitudes and cross sections in the theory of thermoelasticity. SIAM J. Appl. Math. **48**, 79–98 (1988)
12. G. Dassios, V. Kostopoulos, On Rayleigh expansions in thermoelastic scattering. SIAM J. Appl. Math. **50**, 1300–1324 (1990)
13. G. Dassios, V. Kostopoulos, Scattering of elastic waves by a small thermoelastic body. Eng. Sci. **32**, 1593–1603 (1994)
14. R. Duduchava, D. Natroshvili, E. Shargorodsky, Basic boundary value problems of thermoelasticity for anisotropic bodies with cuts I and II. Georgian Math. J. **2**, 123–140 and **3**, 259–276 (1995)
15. V.D. Kupradze, T.G. Gegelia, M.O. Basheleishvili, T.V. Burchuladze, *Three Dimensional Problems of the Mathematical Theory of Elasticity and Thermoelasticity*. North Holland Series in Applied Mathematics and Mechanics (North-Holland, Amsterdam, 1979)
16. W. Nowacki, *Thermoelasticity*, 2nd edn. (Revised and Enlarged PWN-Polish Scientific Publishers, Warsaw and Pergamon Press, Oxford, New York, 1986)

Blind Transfer of Personal Data Achieving Privacy

Alexis Bonnecaze and Robert Rolland

Abstract Exploitation of data for statistical or economic analyses is an important and rapidly growing area. In this article, we address the problem of privacy when data containing sensitive information are processed by a third party. In order to solve this problem, we propose a cryptographic protocol and we prove its security. The security analysis leads to introduce the new notion of generalized discrete logarithm problem. Our protocol has effectively been deployed within a network of more than 5000 pharmacies.

1 Introduction

With the development of the digital world, a growing number of data are created every day. Statistical information derived from these data can be very useful in many fields such as commerce, marketing, or medicine. These data have a market value and are likely to be sold or to be made available to organizations or companies specialized in data analysis. However, they often contain sensitive information that should not be leaked. Thus, data should be pre-treated in order to eliminate records which are to remain secret while preserving the consistency of the data. Moreover, statistical analysis of the data should not lead to the knowledge of any individual information. As an example, let us consider a data basis containing customer names with their purchases. It could be possible to deduce clients' profiles from statistical analysis. In order to achieve privacy, clients' name should be erased from the records but at the same time, it is required to be able to detect that two distinct records belonging to the same individual entity have been purchased by the same client.

The paper is organized as follows: The next section describes the problematic and the system infrastructure. Section 3 introduces our protocol to anonymize the identity of individual entities and fulfill the requirements given in the preceding

A. Bonnecaze · R. Rolland (✉)
Aix Marseille Univ, CNRS, Marseille, France
e-mail: alexis.bonnecaze@univ-amu.fr; robert.rolland@acrypta.fr

N. J. Daras, T. M. Rassias (eds.), *Computational Mathematics and Variational Analysis*, Springer Optimization and Its Applications 159,
https://doi.org/10.1007/978-3-030-44625-3_2

section. Section 4 is devoted to security considerations. Our analysis leads us to introduce the new notion of generalized discrete logarithm problem. Then in Section 5, we focus on a case study which addresses this general problematic in the particular case of medical field.

2 System Infrastructure

A processing center (PC) collects Data Records about individual entities from a set of Data ACquisition Centers (DACCs) in order to make a comprehensive study on this set of data. Each record has two parts: a header containing the personal data of an entity and a body which contains data related to this entity. This last part can be made public if it cannot be related to the identity defined in the header, whereas the header should be blinded. Remark that collecting data without headers does not constitute a solution since it fails to recognize whether two records involve the same entity or not. We therefore need to blind the headers while enabling to detect when two headers are identical. In order to solve this problem we will use a trusted third party TS. Note that in many situations, administrative rules impose the use of a trusted third party, to avoid a direct contact between entities and the PC, and to ensure that the protocol is fairly applied and that the cryptographic material is well-managed.

As illustrated in Figure 1, the architecture of the system is a network with four components: the network of individual entities, the Data ACquisition Centers (DACCs), the set of the trusted server (TS), and the processing center (PC) which collects records.

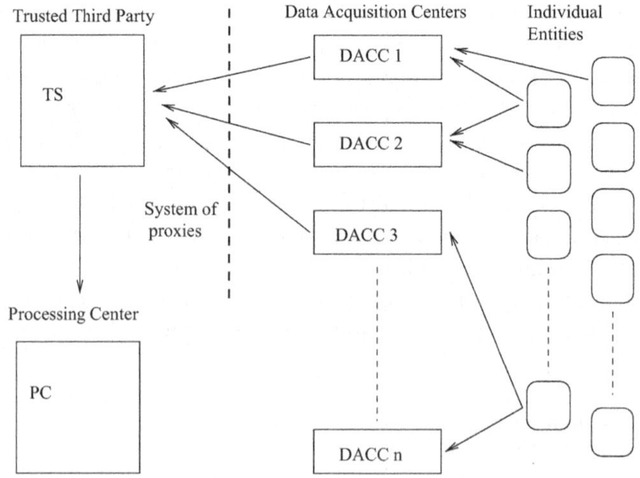

Fig. 1 Network infrastructure

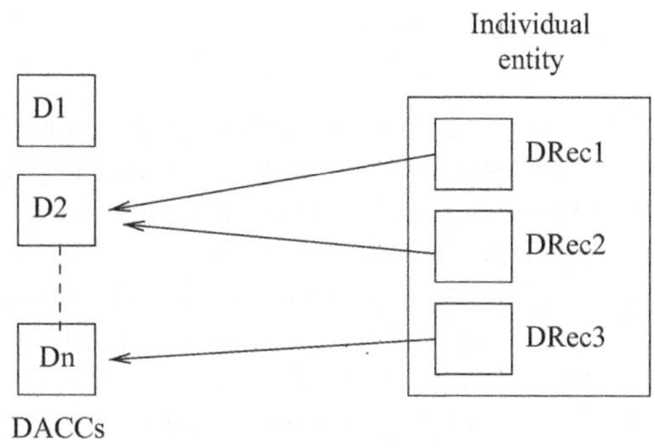

Fig. 2 Relationship between an individual entity and the acquisition centers

A Data Record (DRec) is given to only one Data ACquisition Center. Furthermore, two Data Records can be collected by the same Data ACquisition Center or by two different Data ACquisition Centers, as illustrated in Figure 2.

Data ACquisition Centers transmit the data records to a trusted server which relays this transmission and sends the Data Records to the processing center. The following constraints are imposed:

- the trusted server ensures that the protocol is fairly applied;
- the trusted server should not be able to link any data to an unblinded header;
- the processing center should not be able to link any data to an unblinded header;
- the processing center should be able to detect that two masked headers come from the same unblinded header.

Moreover, but this is not mandatory, in order to enhance the privacy, the trusted server should not be able to detect which Data ACquisition Center is transmitting the data.

3 The Proposed Protocol

Every DACC is able to encrypt using elliptic curve ElGamal algorithm. Key management is achieved by a certification authority. The trusted server needs to know the signature public key of the group of DACCs. It is trusted in regard to transmission and non-disclosure of data transmission. However, it is not entitle to manage sensitive information. Its main work is to forward data to the PC after having blinding it using a random number. In order to enhance privacy, TS is not authorized to know which entity sends the data. Thus, TS should not be able to link any data with an individual entity.

In terms of network transmission, it is assumed that a DACC can reach TS and any other entity, and TS can reach the PC.

3.1 Cryptographic Concerns

Our protocol makes use of cryptographic primitives. Every encryption uses elliptic curve ElGamal encryption. We introduce here the mathematical objects that we will use thereafter. Let us consider a cryptographic elliptic curve Γ over a prime field $\mathbb{F}_p$. Let Γ_p be the set of $\mathbb{F}_p$-rational points of Γ and $n = \#\Gamma_p$ the number of $\mathbb{F}_p$-rational points of Γ. We suppose that Γ is such that n is a prime number of size 256 bits. Let us denote by G the cyclic group of order n of rational points on Γ and let P be a public generator of G. The curve Γ is chosen in order that the discrete logarithm problem be hard. Examples of such curves can be found in [3, 4] or [1]. Let H be a public map-to-point function as defined, for example, in [5] or [2]. Namely H transforms a message m to a point $H(M)$ of the curve and acts as a hash function.

3.2 Anonymization of the Header

In this section, we describe the cryptographic protocol which allows the header m to be anonymized. First, let us consider the following simple solution: the header m of a record is hashed using a hash function. This solution is far from being secure, as it is vulnerable to dictionary attack. It is therefore necessary to provide a more comprehensive mechanism.

An overview of our protocol is illustrated in Figure 3, where the encryption function is denoted $E()$. The important property of this function is that $k.E(M) = E(k.M)$ for any message M. This property ensures the fourth constraint, the encryption ensures privacy with regard to TS and the masking by k ensures privacy with regard to the processing center.

We describe here in more detail the process to anonymize the header m of the record to transmit.

Fig. 3 Anonymization of the record

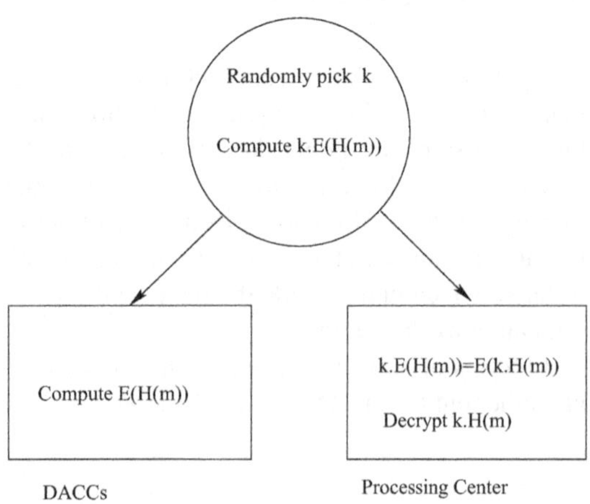

The cryptographic setup phase is as follows:

1. The trusted party TS picks at random a key k such that $0 \le k \le n-1$ and keeps it secret.
2. PC picks at random a key a such that $0 \le a \le n-1$ and keeps it secret.
3. PC computes the point $Q = aP$ and transmits it to the network of entities (this is the public key of PC).

When the setup is done, any entity can forward a data. Suppose a Data ACquisition Center wants to transmit a Data Record having a header M and a body C. The body part C of the document is transmitted without any modification, while the header part M is transmitted as follows:

1. The Data ACquisition Center draws at random an integer k_1 between 0 and $n-1$ and computes
$$P_1 = k_1 P \quad P_2 = H(m) + k_1 Q.$$
The points P_1 and P_2 are sent (with the body C) to the trusted third party TS.
2. The trusted third party TS computes, using its secret key k, the two following points:
$$R_1 = kP_1 \quad R_2 = kP_2$$
and sends R_1 and R_2 to PC.
3. Now PC computes the anonymous number AN associated with the header
$$AN = (R_2 - aR_1)_x,$$
where $(R_2 - aR_1)_x$ denotes the x-coordinate of the point $R_2 - aR_1$.

Remark 1 The random number k_1 must be recalculated for each record. However the secret key k of the trusted third party must remain the same throughout the study.

Proposition 1 *The anonymous number AN is*
$$AN = \left(kH(m)\right)_x.$$

Proof We compute $R_2 - aR_1$ and obtain successively:
$$R_2 - aR_1 = kH(m) + kk_1Q - akP_1$$
$$= kH(m) + kk_1aP - akk_1P = kH(m).$$

4 Security Considerations

In this section, we show that our protocol fulfills the security requirements.

4.1 Privacy in Regard to TS

It is required that TS be able to authenticate a message received from a DACC. Every message is dated and signed using DACCs' private key (all the DACCs use the same key). At the transport level, confidentiality is not mandatory since the header of the data is encrypted by an elliptic curve ElGamal ciphering. More precisely, the header $H(m)$ is masked by $k_1 Q$, where k_1 is random. It is well known that this type of ciphering is indistinguishable under Chosen Plaintext Attack (IND-CPA) in the random oracle model, as far as we work on a group where the decisional Diffie–Hellman problem is hard (see [6]). In particular, it means that TS is not able to distinguish whether two encrypted headers represent the same plaintext header or not.

4.2 Privacy in Regard to PC

PC needs to be sure that the sender is TS. Authentication is done by adding a timestamp and signing the message. Even though confidentiality is not required, a protocol like TLS may be used. When PC receives the plaintext message, it remains to treat the header. Using homomorphic properties of the ciphering, PC can eliminate the mask k_1. The header is now protected by the blinding factor k. The underlying security problem is the generalized discrete logarithm problem on the chosen elliptic curve. This problem is analyzed in the next subsection.

4.3 Generalized Discrete Logarithm of Order s

Suppose that an attacker knows some identities of entities and the set of corresponding blinded headers. Since the blinding value k is fixed, is he able to calculate k? Let s be an integer such that $1 \leq s \leq n-1$, let a set of rational points $A = \{A_1, \dots, A_s\}$ and let k be an integer such that $1 < k < n-1$. We denote kA the set $\{kA_1, kA_2, \cdots, kA_s\}$. The problem P_s of the generalized discrete logarithm of order s on the group Γ_p is the following: Given A and $A' = kA$, calculate k.

Remark 2 The knowledge of A and $A' = kA$ is equivalent to the knowledge of $B = \complement A$ and $B' = kB = \complement A'$. In particular, P_{n-1} is equivalent to P_1, the discrete logarithm problem (DLP).

In our case study, the value s is much smaller than n and in practice, we may assume that $500 \leq s \leq 10^6$. We will show that P_s is at least as much hard as DLP.

Theorem 1 *Suppose we know an algorithm $\mathcal{A}(\Gamma_p, s)$ which solves P_s in a time bounded by $T(s)$, then it is possible to construct an algorithm which solves DLP on Γ_p in a time bounded by $T(s)+st_0$, where t_0 is the time needed to choose an integer m and to calculate two scalar multiplications on Γ_p.*

Proof Let A_1, $A'_1 = kA_1$ be an instance of the DLP. Let us choose distinct integers $m_2, \cdots, m_s$ such that $1 < m_i < n$ in order to construct the points $A_i = m_i A_1$ and $A'_i = m_i A'_1$. We have $A'_i = m_i k A_1 = k m_i A_1 = k A_i$. Thus, if $A' := \{A'_1, A'_2, \cdots, A'_s\}$, we obtain $A' = kA$. By this way, we just constructed an instance of P_s. The time needed for this construction is bounded by st_0. Applying the algorithm $\mathcal{A}(\Gamma_p, s)$ to this instance of P_s, we can obtain k. We have therefore solved DLP in a time bounded by $T(s) + st_0$.

Consequently, if we had a practical algorithm to solve P_s, s being sufficiently small (in order that st_0 can be reached in practice), then we could solve DLP over Γ_p.

As an example, if we choose a curve over $\mathbb{Z}/p\mathbb{Z}$ where the size of p is around 256 bits, then from Weil's bound, the size of n is of the same order. This means that n is of order 2^{256} and the best known algorithms to solve DLP need about 2^{128} operations. If s is bounded by 10^6 (our case study), then s is negligible compared with 2^{128}. Thus, unless breaking the DLP for this size, we cannot obtain an algorithm to solve P_s with a number of operations significantly less than 2^{128}.

5 A Case Study

We consider here a company which collects data from a number of pharmacies for statistical or economic analyses. In particular, these data contain patients' names and information regarding the patients like the name of drugs they have bought. Each medical record has two parts: a header containing the patient's name and other personal data and a body which contains various medical data. This last part can be made public if it cannot be related to the identity defined in the header, whereas the header is blinded. Moreover, we need to detect when two headers are identical. The architecture of the system includes the network of pharmacies, the trusted server (TS), and the company which collects medical records. The objective of the system is to enable a pharmacy to forward a data to the company with the following requirements. Individual privacy must be preserved and two records involving the same patient must have the same header. Our protocol is perfectly adapted to this kind of problematic. Therefore, it has been deployed within a network of more than 5000 pharmacies and it enabled the company to exploit data while respecting administrative procedures in terms of privacy issues. Indeed, our solution could have similarly been applied to many other fields.

6 Conclusion

This article solves a problem which has effectively been encountered in an industrial framework. Our protocol has a wide range of applications since statistical analyses of data are used extensively and privacy is becoming a major concern. We showed that the solution fulfills all the requirements regarding privacy concerns. Moreover, the processing center is able to distinguish whether two records involve the same individual entity while this property is not allowed to the third party designed to forward the data and blind the header. Our analysis led us to introduce the new concept of generalized discrete logarithm problem of order s and we proved that this problem is at least as hard as the discrete logarithm problem.

References

1. ECC Brainpool, ECC Brainpool Standard Curves and Curve Generation (2005). http://www.ecc-brainpool.org/download/Domain-parameters.pdf.
2. T. Icart. How to hash into elliptic curves, in *Annual International Cryptology Conference* (2009), pp. 303–316
3. H. Ivey-Law, R. Rolland, Constructing a database of cryptographically strong elliptic curves, in *Proceeding of SAR-SSI* (2010). http://www.acrypta.com/index.php/telechargements#ARCANA
4. NIST, NIST-FIPS 186-3 (website) (2009). http://csrc.nist.gov/publications/fips/fips186-3/fips_186-3.pdf
5. D. Poulakis, R. Rolland, A signature scheme based on elliptic curve discrete logarithm and factoring. Cryptology ePrint Archive, 2012/134 (2012)
6. Y. Tsiounis, M. Yung, On the security of ElGamal based encryption, in *Proceedings of the First International Workshop on Practice and Theory in Public Key Cryptography: Public Key Cryptography, PKC '98*, London (Springer, Berlin, 1998)

Equilibria of Parametrized N-Player Nonlinear Games Using Inequalities and Nonsmooth Dynamics

Monica G. Cojocaru and Fatima Etbaigha

Abstract In this paper we present a combination of theoretical and computational results meant to give insights into the question of existence of non-unique Nash equilibria for N-players nonlinear games. Our inquiries make use of the theory of variational inequalities and projected systems to highlight cases where multiplayer Nash games with parametrized payoffs exhibit changes in the number of Nash equilibria, depending on given parameter values.

1 Introduction

The question of identifying existence and uniqueness results for equilibrium strategies of Nash games dates back to the last century, with the works [1] and [2]. Game theory is a vast area of research to date, where, depending on the type of game, number of players, payoff characteristics, and strategy sets characteristics, a large body of existence and uniqueness results are available (see, for instance, [3–7] and the references therein).

Theoretical results asserting uniqueness of a Nash equilibrium exist and some come from the direct relation between some classes of Nash games and variational inequality problems (see, for instance, [8]), i.e., the Nash equilibria coincide with the solution set of a variational inequality problem, thus uniqueness of solutions to the latter leads directly to a singleton set of Nash equilibria. Variational inequalities have been introduced in the last century as well, in relation to studying boundary value problems in partial differential equations. As with game theory, the variational inequality literature is vast (see [9–11] and the references therein). Some of the application areas of variational inequalities are equilibrium problems, which, apart from games, consist of market (Wardrop, Walras) equilibrium problems, network

M. G. Cojocaru (✉) · F. Etbaigha
Department of Mathematics and Statistics, University of Guelph, Guelph, ON, Canada
e-mail: mcojocar@uoguelph.ca; fetbaigh@uoguelph.ca

N. J. Daras, T. M. Rassias (eds.), *Computational Mathematics and Variational Analysis*, Springer Optimization and Its Applications 159,
https://doi.org/10.1007/978-3-030-44625-3_3

equilibrium problems, and generalized Nash games (see [10, 12, 13] and [11, 14, 15] and references therein).

In this work we introduce some theoretical results, paired with a computational method, to identify types of Nash games where the players' payoffs may be dependent on a parameter and where, due to the presence of this parameter, the game's set of Nash equilibria changes as the parameter takes on values across a given interval of interest. Payoff parametrization is interpreted as an adjustment of player's strategies to changing conditions in the game (as in [14]). In the current literature, there are results (and we reference a few of them below in detail) which are concerned with uniqueness of Nash equilibria, even in the presence of a parameter variation. In fact, we show in this paper several results leading to uniqueness of a Nash equilibrium by way of variational inequalities. One of the results has been previously arrived at in [33], though the proof is different than the one here.

There are no particular strategies, however, that practitioners may employ in situations where the payoff parametrization does not fall into any of the known theoretical results which may guarantee uniqueness. In this case we propose a straightforward computational method which can be employed to identify ranges of parameter values that lead to the presence of multiple Nash equilibria for the parametrized game.

Our method does not rely on variational inequalities anymore, but rather on projected dynamics [10, 13, 16], where in order to identify possible multiple Nash equilibria, we sweep the set of initial conditions of a projected equation for each sample value of the given parameter. We choose this approach as conditions for existence of critical points of projected equations do not depend on monotonicity properties of the equation's vector field [16], whereas existence results for solutions of variational inequality problems do [9].

Generally speaking, from a dynamical systems' perspective, to each Nash game in our context we can associate a dynamical system. If the game is parametrized, then we obtain a family of dynamical systems whose members may display one or multiple equilibria, depending on the parameter values under consideration. Thus in essence we look at a bifurcation-type problem for the projected dynamical system associated to the parametrized Nash game. The issue of bifurcations in constraint dynamics, along the lines of bifurcation theory in classical dynamical systems, is not well studied. The issue of bifurcations in variational inequalities has been tackled in [17], however, our results are new, as the authors in [17] consider variational problems with completely continuous fields only.

Last but not least, we would like to highlight that knowing whether a Nash game has nounique equilibria is of interest in applied problems. Once multiple equilibria are present, the question of selection comes forth: can the equilibria be compared in some meaningful fashion, and if they can be compared, is one "better" than others (see for instance [15] for examples of games with equilibrium sets).

The paper is structured as follows: in Section 2 we provide some mathematical background from Nash games, variational inequalities, and projected dynamics. In Section 3 we present some new theoretical results leading to conditions for uniqueness of Nash equilibria for parametrized Nash games. In the absence of any type of known conditions that guarantee uniqueness we propose here a computational method to gain insights into parameter specific values giving rise to multiple Nash equilibria. We illustrate all our results in Section 4 on three examples, two theoretical, and one applied. We finish with some conclusions and future ideas.

2 N-Player Games, Inequalities and Nonsmooth Dynamics

In general, a multiplayer game involves a finite number of players, denoted here by $N > 0$. A generic player $i \in \{1, \ldots, N\}$ is thought to have a strategy set $S_i \subset \mathbb{R}^{n_i}$, whose strategies are vectors $x_i \in S_i$, and a payoff function $\theta_i : K \to \mathbb{R}$, where we denote by $K := S_1 \times \ldots \times S_N \subseteq \mathbb{R}^{n_1+\ldots+n_N}$, assumed to be closed and convex. A Nash equilibrium of a multiplayer game is then defined as follows:

Definition 1 Assume each player is rational and wants to minimize their payoff function $\theta_i : K \to \mathbb{R}$. Then a Nash equilibrium is a vector $x^* \in K := S_1 \times \ldots \times S_N$ which satisfies the inequalities:

$$\forall i \in \{1, 2, \ldots, N\},\ \theta_i(x_i^*, x_{-i}^*) \le \theta_i(x_i, x_{-i}^*),\ \forall x_i \in S_i, \tag{1}$$

where in general we denote by $x_{-i} := (x_1, \ldots, x_{i-1}, x_{i+1}, \ldots, x_N)$.

It is known that Nash equilibrium points of a nonlinear N-player game can be obtained by the equivalent reformulation of such a game into a variational inequality (VI) problem defined below (see [8, 12] for a proof):

Theorem 1 *Assume a game as in Definition 1 above. Then if for each $i \in \{1, \ldots, N\}$ we have that θ_i is of class $\mathscr{C}^1$ and that $-\theta_i$ is concave with respect to the strategy x_i, then a Nash equilibrium of the game (1) is a solution of the variational inequality problem:*

$$\textit{find } x^* \in K \ \textit{s.t.}\ \langle F(x^*), y - x^* \rangle \ge 0,\ \forall y \in K \tag{2}$$

where the mapping $F := \left(\nabla_{x_1}\theta_1, \ldots, \nabla_{x_N}\theta_N\right)$. The converse also holds.

Furthermore, the Nash equilibria of the game in Definition 1 are also the same as the critical points of the nonsmooth dynamical system

$$\frac{dx(t)}{dt} = P_{T_K(x(t))}(-F(x(t))),\ x(0) \in K, \tag{3}$$

where $P_K : \mathbb{R}^n \to K$ is the closest element mapping of $v \in \mathbb{R}^n$ to K, for any K closed and convex subset in $\mathbb{R}^n$, and where $T_K(x)$ is the tangent[1] cone to K and a point $x \in K$. This interpretation of the Nash equilibria as critical points of a differential equation is made possible due to known results (see [10, 12, 16]) showing that critical points of the projected equation (3) are solutions of VI problem (2) and vice versa.

In this work, we consider the case of a parameter being introduced in both players' payoffs. We denote this parameter by $a \in [\alpha, \beta] \subseteq \mathbb{R}$, where $[\alpha, \beta]$ is a given interval in $\mathbb{R}$. In this case, the equivalent variational inequality problem (2) becomes dependent on the payoffs' parameters. Furthermore, system (3) associated to such a VI problem becomes a nonsmooth dynamical system whose right-hand side will depend on the payoffs' parameters.

Whenever a parameter $a \in [\alpha, \beta] \subseteq \mathbb{R}$ is introduced in the players' payoffs, the Nash equilibrium points will depend on the parameter a, thus to determine them we search for the critical points of the perturbed system:

$$\frac{dx(t)}{dt} = P_{T_K(x(t))}(-F(a, x(t))),\ a \in [\alpha, \beta],\ x(0) \in K \tag{4}$$

i.e., we search for critical points $x^* \in K$ so that $P_{T_K(x^*)}(-F(a, x^*)) = 0,\ \forall t \geq t_0$, or equivalently

$$\text{find } x^* \in K,\ a \in [\alpha, \beta] \text{ s.t. } \langle F(a, x^*), y - x^* \rangle \geq 0,\ \forall y \in K. \tag{5}$$

We note at this point that the values of the possible critical points x^* will depend on the parameter a; this further implies that the formulation of problem (5) is not complete. Since $x^* := x^*(a)$, then we should formulate the search of $x^*(\cdot)$ on a space of functions:

$$\text{find } x^* \in D \text{ s.t. } \langle\langle F(a, x^*(\cdot)), y(\cdot) - x^*(\cdot) \rangle\rangle \geq 0,\ \forall y \in D,$$

where $D := \Big\{ u \in L^2([\alpha, \beta], \mathbb{R}^n) \mid u(a) \in K,\ \text{a.e. } a \in [\alpha, \beta] \Big\}$, $\langle\langle \cdot, \cdot \rangle\rangle$ is the inner product on $L^2([\alpha, \beta], \mathbb{R}^n)$, and where K is given above in Definition 1. Problem (6) is known in the literature as an evolutionary variational inequality problem (EVI) (see [11]). An equivalent formulation is the so-called pointwise EVI problem (see [18]):

$$\text{find } x^*(a) \in D(a) \text{ s.t. } \langle F(a, x^*(a)), y(a) - x^*(a) \rangle \geq 0,\ \forall y(a) \in D(a), \tag{6}$$

[1] We consider the reader is familiar with the definitions of tangent and normal cones to a closed, convex subset of the Euclidean space at a given point in the set (see [9])

where $D(a) := K, \forall a$ in our case. Results on existence, uniqueness, and regularity of solutions $x^*(\cdot)$ for an EVI problem as above exist in the literature (see [11, 19] and the references therein).

In our context here is of little interest whether or not the solutions $x^*(a)$ are unique for a given parameter value a. We are interested in the dynamic behaviour of system (4), and thus we want to derive conditions under which the solutions of problem (6) are not unique, at least for some parameter values $a \in [\alpha, \beta]$.

Based on current literature on VI problems and dynamical systems in general, we know that there are clear instances where the answer to our question can be obtained by means of known results. First, if $x^*(a)$ belongs to the interior of K for some $a \in [\alpha, \beta]$, then the discussion on the effects of the payoff parameter takes place exactly as in the classical theory of dynamical systems whenever $-F$ is a function of class $\mathscr{C}^1$ (see [20]). Second, it is known from previous work [17] that certain conditions applied to the variational inequality problem of Definition 1 leads to bifurcation points for the VI solution set. Last but not least, it is also known from [14, 21, 22] that whenever F is monotone (as recalled below in Definition 2) but not strictly so (i.e., $\langle F(x^*) - F(x), x^* - x\rangle = 0$), the dynamics structure and behaviour of (3) changes, for instance, periodic cycles may appear around a stable unique critical point.

3 Results—Changes in Game Equilibria Using Variational Inequalities

In this section we propose new theoretical and computational results to test and visualize the behaviour of Nash equilibria of a parametrized Nash game of the type introduced above. Whenever available, we will define and prove the concepts and results in the larger context of a Hilbert space H.

3.1 Theoretical Approach

We recall first (see, for instance, [10, 12]) that uniqueness and stability of critical points of equations (3) are based on monotonicity (respectively, pseudo-monotonicity) type properties of the function F on neighbourhoods of the constraint set K. For the ease of reading, we recall first monotonicity-type conditions for mappings (see also [23] and [10]).

Definition 2 Let K be a closed, convex non-empty set of a Hilbert space H. Then $F : K \to H$ is called locally monotone at $x^* \in K$ if for every $x \in \mathscr{N}(x^*)$ a neighbourhood of x^* in K, we have that $\langle F(x^*) - F(x), x^* - x\rangle \geq 0$.

It is locally strictly monotone at x^* if for every $x \neq x^* \in \mathscr{N}(x^*)$, we have $\langle F(x^*) - F(x), x^* - x\rangle > 0$.

It is locally η-strongly monotone if for every $x \neq x^* \in \mathcal{N}(x^*)$, we have $\exists \eta > 0$ so that $\langle F(x^*) - F(x), x^* - x\rangle \geq \eta||x^* - x||^2$.

Finally, it is called locally pseudo-monotone at $x^* \in K$ if for every $x \in \mathcal{N}(x^*)$ a neighbourhood of x^* in K, we have that $\langle F(x), x^* - x\rangle \geq 0 \implies \langle F(x^*), x^* - x\rangle \geq 0$.

It is known that if F is strictly pseudo-monotone (or strongly pseudo-monotone, strongly pseudo-monotone of a given degree [10]) on a certain neighbourhood of F around a critical point $x^* \in K$, then x^* is unique and is a locally monotone attractor (respectively, locally exponentially stable, local finite-time attractor) on that neighbourhood.

Next, we use relaxed cocoercive mappings to prove two new results regarding the uniqueness of solutions to a VI problem of type (2). We mention that an equivalent result to our Theorem 2 below exists in the literature in [33], though we were not aware of its existence while working on this paper. We would like to thank the referee who brought this result to our knowledge. There are important differences in the way we arrive to our proof here versus the results in [33] though, as well as in the ways our result is used and exemplified here.

To proceed, we need to recall the definition of a relaxed cocoercive mapping, and its relation with monotone mappings (see [24–26, 33]).

Definition 3 Let H be a Hilbert space. A mapping $F : H \to H$ is said to be:
(s)-cocoercive if there exist a constant $s > 0$ such that

$$\langle F(x) - F(y), x - y\rangle \geq s\|F(x) - F(y)\|^2, \quad \forall x, y \in H.$$

It is called (m, γ)-relaxed cocoercive if there exist constants $m,\ \gamma > 0$ such that

$$\langle F(x) - F(y), x - y\rangle \geq (-m)\|F(x) - F(y)\|^2 + \gamma\|x - y\|^2, \quad \forall x, y \in H.$$

The following links between cocoercive and monotone-type mappings are known (see [25]): Each η-strongly monotone and b-Lipschitz continuous mapping is $\frac{\eta}{b^2}$-cocoercive for $\eta > 0$ and $b > 0$; Each η-strongly monotone mapping is $(1, \eta + \eta^2)$-relaxed cocoercive for $\eta > 0$.

We are now ready to state the main theoretical results of our paper.

Lemma 1 *Let K be a closed, convex subset of a Hilbert space H and let $F : K \to H$ be a continuous mapping. If the VI(F,K) problem:*

$$\langle F(x), y - x\rangle \geq 0,\ \forall y \in K, \tag{7}$$

has two distinct solutions $x_1 \neq x_2 \in K$, then F satisfies: $\langle F(x_1) - F(x_2), x_1 - x_2\rangle \leq 0$.

Proof In the context of a known existence result in [9], we have that our VI has at least one solution, say x. Thus we have that: $\langle F(x), y - x\rangle \geq 0,\ \forall y \in K$. If we assume that this VI has two solutions $x_1 \neq x_2 \in K$, then the following statements hold:

$$\langle F(x_1), y - x_1 \rangle \geq 0, \ \forall y \in K, \ \langle F(x_2), y - x_2 \rangle \geq 0, \ \forall y \in K. \tag{8}$$

Let $y := x_2$ in the first inequality in (8) and $y := x_1$ in the second inequality in (8). This implies

$$\langle F(x_1), x_2 - x_1 \rangle \geq 0 \text{ and } \langle F(x_2), x_1 - x_2 \rangle \geq 0 \Leftrightarrow \langle F(x_2), x_2 - x_1 \rangle \leq 0. \tag{9}$$

Now by (9) we obtain $\langle F(x_2), x_2 - x_1 \rangle \leq 0 \leq \langle F(x_1), x_2 - x_1 \rangle$. Hence this implies

$$\langle F(x_2), x_2 - x_1 \rangle \leq \langle F(x_1), x_2 - x_1 \rangle \implies \langle F(x_1) - F(x_2), x_1 - x_2 \rangle \leq 0. \tag{10}$$

Theorem 2 *Let K be a closed, convex subset of a Hilbert space H and $F : K \to H$ a Lipschitz continuous mapping with Lipschitz constant $b > 0$. If F is (m, γ)-relaxed cocoercive, then*

$$\langle F(x) - F(y), x - y \rangle > 0 \ \forall x \neq y \in K \ \textit{whenever} \ \gamma > mb^2.$$

Moreover, the VI(F,K) problem (7) has exactly one solution whenever $\gamma > mb^2$ which is an exponentially stable Nash equilibrium in the projected dynamics (3).

Proof If F is (m, γ)-relaxed cocoercive, then:

$$\langle F(x) + F(y), x - y \rangle \geq (-m)\|F(x) - F(y)\|^2 + \gamma \|x - y\|^2, \forall x, y \in K.$$

This is equivalent to

$$\langle F(x) - F(y), x - y \rangle \geq (-mb^2)\|x - y\|^2 + \gamma \|x - y\|^2, \forall x, y \in K, \tag{11}$$

since F is (b)-Lipschitz continuous. Then we have that

$$\langle F(x) - F(y), x - y \rangle \geq (\gamma - mb^2)\|x - y\|^2, \ \forall x, y \in K. \tag{12}$$

Assuming $x \neq y$ in (15) we get

$$\langle F(x) - F(y), x - y \rangle \geq (\gamma - mb^2)\|x - y\|^2. \tag{13}$$

Our result follows if $\gamma - mb^2 > 0 \Leftrightarrow \gamma > mb^2$. Assume now that there are at least two distinct solutions of $VI(F, K)$, $x_1 \neq x_2 \in K$. Then we have relation (10) for $x := x_1$ and $y := x_2$. But we also have that (13) holds at $x_1 \neq x_2$, which is a contradiction. Thus the VI(F,K) can have at most one solution. From [16], since F is L-continuous, critical points of PDS (3) exist, thus the VI(F,K) has exactly one solution. Last but not least, we note that whenever $\gamma > mb^2$, F becomes strongly monotone of degree 2, thus the VI(F,K) admits a unique solution which is proven to be an exponentially stable critical point of the system (3) (see, for instance, [10]).

Remark 1 Theorem 2 appears in [33] where it is proven using a constructive algorithm to reach a solution of the VI. We were not aware of this result at the time of our submission, thus we arrived at our result independently. Theorem 2 holds without the requirement to have F Lipschitz continuous, but under the assumptions that K is a convex polyhedral set (in $\mathbb{R}^n$) and F is a vector field with linear growth. In this case we have the following result below.

Theorem 3 *Let K be a convex bounded polyhedral subset of $\mathbb{R}^n$ and $F : K \to \mathbb{R}^n$ a mapping with linear growth (i.e. there exists $M > 0$ so that $||F(x)|| \leq M(1 + ||x||)$). If F is (m, γ)-relaxed cocoercive, then*

$$\langle F(x) - F(y), x - y\rangle > 0 \ \forall x \neq y \in K \ \textit{whenever} \ \gamma > mM^2(2 + 2B)^2.$$

Moreover, the VI(F,K) problem (7) has exactly one solution in this case which is an exponentially stable Nash equilibrium in the projected dynamics (3).

Proof If F is (m, γ)-relaxed cocoercive, this implies $\exists\, \gamma, m > 0$ s.t.

$$\langle F(x) + F(y), x - y\rangle \geq (-m)\|F(x) - F(y)\|^2 + \gamma\|x - y\|^2, \forall x, y \in K.$$

From the linear growth condition we have that

$$\|F(x) - F(y)\| \leq \|F(x)\| + \|F(y)\| \leq M(2 + ||x|| + ||y||) \leq M(2 + 2B) \implies$$
$$-mM^2(2 + 2B)^2 \leq (-m)\|F(x) - F(y)\|^2,$$

where B is the bounding constant for K. This is equivalent to

$$\langle F(x) - F(y), x - y\rangle \geq (-mM^2(2 + 2B)^2)\|x - y\|^2 + \gamma\|x - y\|^2, \forall x, y \in K, \tag{14}$$

since F is (b)-Lipschitz continuous. Then we have that

$$\langle F(x) - F(y), x - y\rangle \geq (\gamma - mM^2(2 + 2B)^2)\|x - y\|^2, \ \forall x, y \in K. \tag{15}$$

Assuming $x \neq y$, then our result follows if $\gamma - mM^2(2 + 2B)^2 > 0$, i.e.

$$\langle F(x) - F(y), x - y\rangle \geq (\gamma - mM^2(2 + 2B)^2)\|x - y\|^2 > 0, \ \forall x \neq y \in K.$$

Assume now that there are at least 2 distinct solutions of $VI(F, K)$, $x_1 \neq x_2 \in K$. Then we have relation (10) for $x := x_1$ and $y := x_2$. But we also have that (13) holds at $x_1 \neq x_2$, which is a contradiction. Thus the VI(F,K) can have at most one solution. From [12] since F is with linear growth and K is polyhedral, then critical points of

PDS (3) exist, thus the VI(F,K) has exactly one solution. Last but not least, we note that whenever $\gamma > mM^2(2+2B)^2$, F becomes strongly monotone of degree 2, thus the VI(F,K) admits a unique solution which is proven to be an exponentially stable critical point of the system (3) (see, for instance, [10, 12]).

The above result leaves open the possibility that for mappings F which are b-Lipschitz continuous and (m, γ)-relaxed cocoercive with $\gamma \leq mb^2$, the problem $VI(F, K)$ may admit more than one solution, which means that the game (1) may have non-unique equilibria. Let us look closely at these potential cases. In (13) let us assume that $\gamma \leq mb^2$ (or respectively, in case of Theorem 3: $\gamma \leq mM^2(2+2B)^2$). However we are also particularly interested in mappings F with

$$\langle F(x) - F(y), x - y \rangle \leq 0, \tag{16}$$

otherwise F becomes strictly monotone and our investigation is moot.

Putting (13) and the above together we get

$$0 \geq \langle F(x) - F(y), x - y \rangle \geq (\gamma - mb^2)\|x - y\|^2$$

Equivalently, we also see that

$$\gamma \|x - y\|^2 \geq 0 \geq \langle F(x) - F(y) + mb^2 x - mb^2 y, x - y \rangle \geq \gamma \|x - y\|^2 \Leftrightarrow$$

$$\langle F(x) - F(y) + mb^2 x - mb^2 y, x - y \rangle = \gamma \|x - y\|^2 \Leftrightarrow$$

$$\langle F(x) - F(y), x - y \rangle = (\gamma - mb^2)\|x - y\|^2. \tag{17}$$

We will introduce now the notion of monotone attractor.

Definition 4 An equilibrium x^* of (3) is a monotone attractor if $\exists r > 0$ such that, $\forall y \in B(x^*, r)$ and $x(t)$ the unique solution of (3) starting at the point y, the function $d(x, t) := ||x(t) - x^*||$ is non-increasing as a function of t. The point x^* is a **global** monotone attractor if all the above is true for any $y \in K$.

Definition 4 is used in [10, 27–29]. We note that an equilibrium x^* may be a monotone attractor, but not necessarily an attractor in the classical sense of dynamical systems, as shown in Figure 1.

There exists in the literature another formulation for a variational inequality problem, introduced by G. Minty [34].

Definition 5 Given a closed, convex subset K in a Hilbert space X, and a mapping $F : K \to X$, a variational inequality problem in the sense of Minty (MVI) is to:

$$\text{find } x \in K \text{ such that } \langle F(y), x - y \rangle \leq 0, \ \forall y \in K.$$

Fig. 1 A monotone attractor (left) versus an attractor (right)

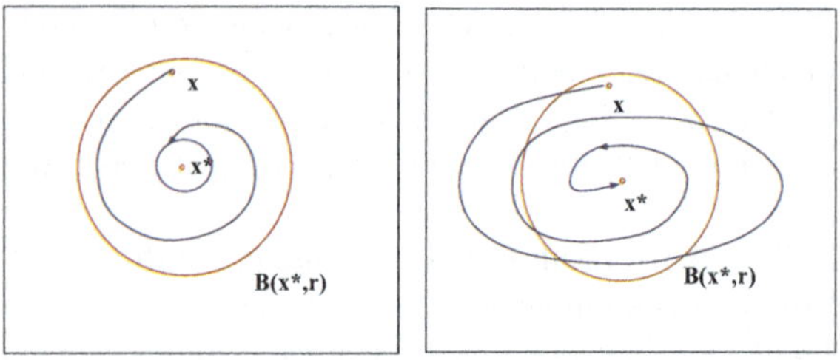

There exists a relation between solutions of VI(F,K) and those of MVI(F,K)—see, for instance, [30] (Chapter III, Lemma 1.5) for a proof:

Theorem 4

(1) If F is continuous on finite dimensional subspaces[2] *and x is a solution to the MVI, then it is a solution to the VI.*

(2) If F is pseudo-monotone and x is a solution to the VI, then it is a solution to the MVI.

Further, it is known that we can characterize the set of solutions to the MVI(F,K) using the projected dynamical system (3) as follows (see [30]):

Theorem 5 *A local monotone attractor for the projected system* (3) *is a solution of the MVI(F,K) associated to the same system. The converse is also true.*

We are now ready to prove the following:

Theorem 6 *Let K be a closed, convex subset of a Hilbert space H and $F : K \to H$ a Lipschitz continuous mapping with Lipschitz constant $b > 0$. Further, let F be (m, γ)-relaxed cocoercive with $\gamma \leq mb^2$ and satisfying assumption (16). Then the following hold:*

(1) Any solution of the MVI(F,K) is a solution of VI(F,K).

(2) The converse of statement 1) above is not true: any x^* solution of VI(F,K) with $x^* \in int(K)$ and $\gamma < mb^2$ is not a solution of MVI(F,K).

(3) If PDS (3) has more than one local monotone attractor, then $VI(F, K)$ has a non-unique solution.

Proof

(1) From the hypothesis and from (16) we have as in (17) that

$$\langle F(y) - F(x^*), y - x^* \rangle = (\gamma - mb^2)\|y - x^*\|^2, \ \ y \neq x^*.$$

[2]Let H be a Hilbert space and K a closed and convex subset in H. A mapping $F : K \to X$ is called continuous on finite dimensional subspaces if for any finite dimensional subspace $S \subset H$, the restriction of F to $K \cap S$ is weakly continuous.

If x^* solves MVI(F,K), then $\langle F(y), y - x^*\rangle \geq 0$ and

$$\langle F(y), y-x^*\rangle-(\gamma - mb^2)\|y-x^*\|^2 =\langle F(x^*), y-x^*\rangle \Longrightarrow$$
$$\langle F(x^*), y - x^*\rangle \geq 0, \ \forall y \neq x^* \in K.$$

(2) If x^* is a solution of VI, then it is a critical point of (3) and thus $P_{T_K(x^*)}(-F(x^*)) = 0$. Now if x^* is also in the interior of K, then $T_K(x^*) := H$ and so $P_{T_K(x^*)}(-F(x^*)) = -F(x^*) = 0$ which implies from (17) that:

$$\langle F(y), y-x^*\rangle=(\gamma - mb^2)\|y-x^*\|^2 < 0 \text{ for } \gamma < mb^2 \Longrightarrow \langle F(y), x^* - y\rangle$$
$$> 0, \ \forall y \neq x^*$$

which is a contradiction with the definition of a solution of MVI.

(3) Here we use Theorem 5 above to deduce that the set of solutions to MVI(F,K) is not a singleton. Then from the first part of this theorem we obtain the conclusion in this case.

3.2 Computational Approach for Non-unique Nash Equilibria

Returning now to our problem formulation, we note that if there are qualitative changes in the game dynamics due to the presence of the parameter a (as in 6), we propose to solve for the critical points of (4) using an approach tailored to solve the pointwise VI problems (6) at some fixed a values. We present below a computational approach for determining parameter values where the game dynamics may give rise to more than one Nash equilibrium by tracing local monotone attractors of the projected system, as shown in Theorem 5 above. Note that this computational method applies to the class of vector fields $F : K \to H$ which are merely L-continuous. We do not need to have any monotonicity-type conditions as in Theorems 6.

Let $a \in [\alpha, \beta] \subseteq \mathbb{R}$ as before.

1. Let Δ_k be a division of the interval $[\alpha, \beta]$ given by

$$a_0 := \alpha < a_1 < a_2 < \ldots < a_k := \beta$$

2. Solve the pointwise VI problem (6) at each a_i, $i \in \{0, \ldots, k\}$ as follows:

   (a) For a given a_i, we generate a uniform distribution of initial conditions $x(0, a_i)$ for the system (4).
   (b) We numerically integrate to obtain trajectories of the perturbed system (4) using a projection type method, based on the constructive proof in [16]; we

integrate all trajectories starting from all the initial points generated in **Step 2** (2a).[3]

(c) We stop integration along a trajectory when a critical point is reached (with a tolerance of 10^{-6})

3. Collect the solutions $x^*(a_i)$, $i \in \{0, \ldots, k\}$ found in Step 2.
4. Identify the values $\{a_i, a_{i+1}\}$ for which a change in dynamics behaviour (change in the number of critical points) takes place as the system (4) evolves between two consecutive sample parameter values.

4 Discussion of Results on Examples

In this section we show how all our results can be used to study changes in Nash equilibria for three games proposed below. The first two games are theoretical; the third one comes from previous applied work.

Example 1 Let us consider a 2-player game where each player has a fixed 1-dimensional strategy vector, denoted by x, respectively y, so that $x \in [0, 10]$ and $y \in [0, 10]$. Let the payoff functions be denoted by $\theta_1(x, y) = ax^2/4 + 3x$ and $\theta_2(x, y) = ay^2/4 - 7y$, $a \in \mathbb{R}$. We assume that players want to minimize their payoffs, subject to the other player's choices. Thus a vector of strategies $(x^*, y^*) \in [0, 10]^2$ is a Nash equilibrium of the game if the following are satisfied:

$$\theta_1(x^*, y^*) \leq \theta_1(x, y^*), \ \forall x \in [0, 10] \ \text{and} \ \theta_2(x^*, y^*) \leq \theta_2(x^*, y), \ \forall y \in [0, 10].$$

This game gives rise to the projected system:

$$\frac{d(x, y)(t)}{dt} = P_{T_K(x(t), y(t))}(-F(x(t), y(t)))$$

with initial conditions $(x(0), y(0)) \in [0, 10]^2$.

The vector field associated to this game is as follows: $F(x, y) = (ax/2 + 3, ay/2 - 7)$. We now check whether this example satisfies the assumptions of Theorem (2).

For any elements $(x_1, y_1), (x_2, y_2) \in [0, 10]^2$, we have

$$F(x_1, y_1) = (ax_1/2 + 3, ay_1/2 - 7), \ F(x_2, y_2) = (ax_2/2 + 3, ay_2/2 - 7).$$

It follows that

[3]This stage is very important to our approach and is different than other algorithmic solutions for VI problems, as we implement a projection type method to trace trajectories of the PDS which end up in local monotone attractors.

$$\begin{aligned}\|F(x_1, y_1) - F(x_2, y_2)\| =&\|(ax_1/2 + 3, ay_1/2 - 7) - (ax_2/2 + 3, ay_2/2 - 7)\|\\ =&\sqrt{(ax_1 - ax_2/2)^2 + (ay_1 - ay_2/2)^2}\\ =&(|a|/2)\|(x_1, y_1) - (x_2, y_2)\|.\end{aligned}$$

Then F is (b)-Lipschitz continuous with $b = (|a|/2)$ where $a \neq 0$. Further,

$$\begin{aligned}&\langle F(x_1, y_1) - F(x_2, y_2), (x_1, y_1) - (x_2, y_2)\rangle\\ =&\langle(ax_1/2 + 3, ay_1/2 - 7) - (ax_2/2 + 3, ay_2/2 - 7), (x_1, y_1) - (x_2, y_2)\rangle\\ =&\langle(a/2(x_1 - x_2), a/2(y_1 - y_2)), (x_1 - x_2, y_1 - y_2)\rangle\\ =&a/2(x_1 - x_2)^2 + a/2(y_1 - y_2)^2\\ =&(a/2)\|(x_1, y_1) - (x_2, y_2)\|^2. \end{aligned} \tag{18}$$

Case 1: Let $a > 0$ By (18), F is $(a/2)$-strongly monotone, which implies that F is $(1, a/2 + a^2/4)$-relaxed cocoercive.

Since $\gamma = (a/2 + a^2/4)$ and $m = 1 \Rightarrow (a/2 + a^2/4) > 1 \cdot (a^2/4)$, therefore we expect a unique solution of the $VI(F, K)$ for each value of a when $a > 0$, thus a unique Nash equilibrium for the game for each a. Recalling [19], the Nash equilibria will depend continuously on the values of a.

Case 2: Let $a < 0$ From (18), we have

$$\begin{aligned}&\langle F(x_1, y_1) - F(x_2, y_2), (x_1, y_1) - (x_2, y_2)\rangle\\ =&(a/2)\|(x_1, y_1)-(x_2, y_2)\|^2+\|F(x_1, y_1)-F(x_2, y_2)\|^2-\|F(x_1, y_1)-F(x_2, y_2)\|^2\\ =&(a/2)\|(x_1, y_1)-(x_2, y_2)\|^2+(a^2/4)\|(x_1, y_1)-(x_2, y_2)\|^2-\|F(x_1, y_1)-F(x_2, y_2)\|^2\\ =&(a/2 + a^2/4)\|(x_1, y_1) - (x_2, y_2)\|^2 - \|F(x_1, y_1) - F(x_2, y_2)\|^2.\end{aligned}$$

Since $(a/2+a^2/4) > 0 \,\forall\, a \in (-\infty, -2)$, we can say that F is $(1, a/2+a^2/4)$-relaxed cocoercive, where $a \in (-\infty, -2)$. In this case, we know that $a/2 < 0 \Rightarrow a/2 + a^2/4 < a^2/4$. Thus $\gamma < mb^2$ for all $a \in (-\infty, -2)$.

Case 3 The remaining interval is $a \in [-2, 0]$. On this interval we can say that F is L-continuous for $a \neq 0$, thus the projected system (3) has solutions.

For the last 2 cases and with $a \neq 0$, we use now our computational approach. Let Δ be a division of $[-2, 0)$ with 20 equally spaced points. Implementing our method gives rise to the following cases (in Figure 2).

The Nash equilibria values for parameter values of $a \in [-20, 0)$ for an equally spaced 20 points division has shown that the projected dynamics has two distinct local monotone attractors located at $(0, 10)$ and at $(10, 10)$. Figure 3 shows these attractors for 100 initial points and $a = -20$:

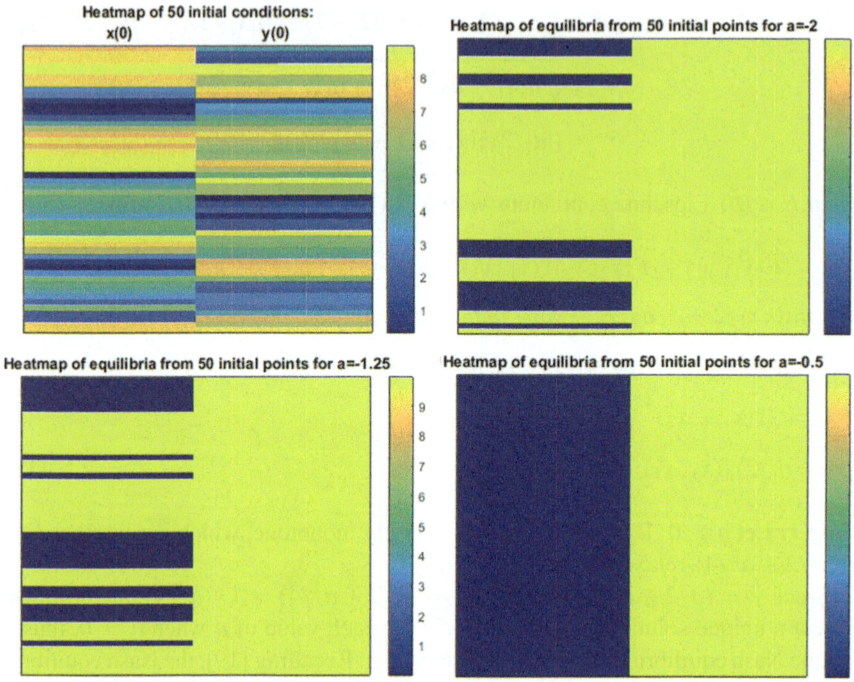

Fig. 2 Nash equilibria values for the game with $a = -2$ (upper right panel), $a = -1.25$ (lower left panel), and $a = -0.5$ (lower right panel) presented in heatmap format and computed using the projected dynamics starting at each value of the 50 initial points, illustrated in upper left panel

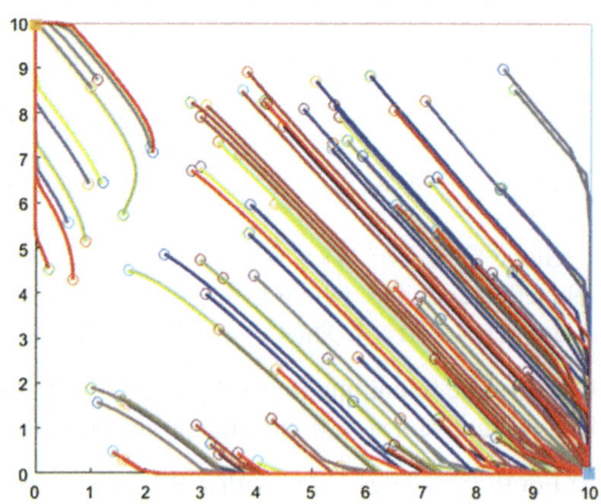

Fig. 3 Nash equilibria values for the game with $a = -20$ obtained as local monotone attractors of the associated projected dynamics

Example 2 Let us consider a 2-player game, where each player has a fixed 1-dimensional strategy vector, denoted by x, respectively, y, so that $x \in [0, 10]$ and $y \in [0, 10]$. Let the payoff functions be denoted by $\theta_1(x, y) = x^2 + (a - y)x$ for player 1, and $\theta_2(x, y) = -axy + \frac{y^2}{2} + 3$ for player 2, where $a \in [1, 5]$.

We assume that players want to maximize their payoffs, thus a vector of strategies $(x^*, y^*) \in [0, 10]^2$ is a Nash equilibrium of the game if the following are satisfied:

$$\theta_1(x^*, y^*) \geq \theta_1(x, y^*), \ \ \forall x \in [0, 10] \ \text{ and } \ \theta_2(x^*, y^*) \geq \theta_2(x^*, y), \ \ \forall y \in [0, 10].$$

The VI form of this game gives rise to the VI problem

$$\text{find } z \in [0, 10]^2 \text{ s. t. } \langle F(z), w - z\rangle \geq 0, \ \forall w \in [0, 10]^2$$

and the associated projected system: $\dfrac{d(x, y)}{dt} = P_{T_K(x(t),y(t))}(-F(x(t), y(t)))$, with initial conditions $(x(0), y(0)) \in [0, 10]^2$, where $F := (-\nabla_x\theta_1, -\nabla_y\theta_2)$.

We study the parametrized game according to conditions in Theorem 2.

First we can show that $F(x, y) := (-(2x + a - y), -(-ax + y))$ is Lipschitz continuous for any $a \in [1, 5]$. This is clearly the case as F is linear and is defined on a compact set $[0, 10]^2$. For a fixed a we can find a Lipschitz constant as follows:

$$||F(x_1, y_1) - F(x_2, y_2)||^2 = || - 2(x_1 - x_2) + (y_1 - y_2), a(x_1 - x_2) - (y_1 - y_2))||^2 =$$

$$(4 + a^2)(x_1 - x_2)^2 - 2(a + 2)(x_1 - x_2)(y_1 - y_2) + 2(y_1 - y_2)^2 \overset{4+a^2\leq 4+a^2+4a}{\leq}$$

$$\Big((4+a^2+4a)(x_1-x_2)^2-2(a+2)(x_1-x_2)(y_1-y_2)+(y_1-y_2)^2\Big)+(y_1-y_2)^2 \overset{-2uv\leq u^2+v^2}{\leq}$$

$$\Big(2(4 + a^2 + 4a)(x_1 - x_2)^2 + 2(y_1 - y_2)^2\Big) + (y_1 - y_2)^2 =$$

$$2(2 + a)^2(x_1 - x_2)^2 + 3(y_1 - y_2)^2 \overset{3\leq 2+a,\, a\in[1,5]\Rightarrow 3\leq(2+a)^2}{\leq}$$

$$2(2+a)^2\Big((x_1-x_2)^2+(y_1-y_2)^2\Big) = 2(2+a)^2||(x_1, y_1)-(x_2, y_2)||^2 \implies \tag{19}$$

$$b := (2 + a)\sqrt{2} \ \text{ for any } \ a \in [1, 5].$$

Further, we want to show that F is $(m, \gamma)-$ relaxed cocoercive for appropriate $m, \gamma > 0$ for any $a \in [1, 5]$. We first compute the left-hand side:

$$\langle F(x_1, y_1) - F(x_2, y_2), (x_1, y_1) - (x_2, y_2)\rangle = -2Z_1^2 + (a + 1)Z_1Z_2 - Z_2^2, \tag{20}$$

where $Z_1 = (x_1 - x_2)$ and $Z_2 = (y_1 - y_2)$. Further, using (19), we have that

$$(-m)\|F(x_1, y_1) - F(x_2, y_2)\|^2 \leq -2m(a + 2)Z_1Z_2 \implies$$

$$(-m)\|F(x_1, y_1)-F(x_2, y_2)\|^2+\gamma\|(x_1, y_1)-(x_2, y_2)\|^2 \leq (\gamma-2m(a+2))(Z_1^2+Z_2^2). \tag{21}$$

Now if we show that for any $a \in [1, 5]$, there exist $m, \gamma > 0$ so that (20) $\geq$ (21), then F is $(m, \gamma)-$relaxed cocoercive. Thus we want to find $m, \gamma > 0$ so that

$$-2Z_1^2 + (a + 1)Z_1Z_2 - Z_2^2 \geq (\gamma - 2m(a + 2))(Z_1^2 + Z_2^2) \Leftrightarrow$$

$$(2m(a + 2) - 2 - \gamma)Z_1^2 + (a + 1)Z_1Z_2 + (2m(a + 2) - 1 - \gamma)Z_2^2 \geq 0. \tag{22}$$

Now let us rewrite the middle term above as: $(a + 1)Z_1Z_2 = 2\frac{a+1}{2}Z_1Z_2$ and let us show that, for all $a \in [1, 5]$, there exist $m, \gamma > 0$ so that (22) is a sum of squares. For this purpose, take: $m := \frac{C}{(2+a)}$, $C > 0$ and find C so that

$$(2m(a + 2) - 2 - \gamma) = \frac{(a + 1)^2}{4} \text{ and } (2m(a + 2) - 1 - \gamma) > 1.$$

Using our value of m from above, we search to find $C > 0$ so that:

$$\gamma > 0 \text{ and } \gamma = 2C - 2 - \frac{(a + 1)^2}{4},\ \gamma < 2C - 2.$$

Since the maximal value of $\frac{(a+1)^2}{4} = 9$ for $a \in [1, 5]$, then let us choose $C := 6$. Then we have that there exist: $m := \frac{6}{(2+a)} > 0$ and $\gamma := 10 - \frac{(a+1)^2}{4} > 0$ which imply the following in (22):

$$(2m(a + 2) - 2 - \gamma)Z_1^2 + (a + 1)Z_1Z_2 + (2m(a + 2) - 1 - \gamma)Z_2^2 =$$

$$(\frac{a + 1}{2}Z_1 + Z_2)^2 + \frac{(a + 1)^2}{4}Z_2^2 \geq 0.$$

The last condition of Theorem 2 states that the $VI(F, K)$ (and thus the game) will have a unique solution if $\gamma > mb^2$, which in our case amounts to checking that

$$10 - \frac{(a + 1)^2}{4} > \frac{6}{2 + a}((2 + a)\sqrt{2})^2,\ a \in [1, 5]$$

which is false. To conclude our analysis, we showed that our example falls outside of the scope of Theorem 2, as F is $(m, \gamma)-$ relaxed cocoercive but without $\gamma > mb^2$. To identify values of $a \in [1, 5]$ which give rise to non-unique Nash pairs, we now employ our numerical method on the interval $[1, 5]$. We choose a division $\Delta_{k=17}$ values given by $a_0 = 1 < a_1 = 1.25 < a_2 = 1.5 < .... < a_{17} = 5$. As in Example 1 above, we include in Figure 4 a few instance of our results for parameter values $a = 1$, $a = 3$, $a = 5$, however, in all cases we found either 2 or 3 distinct Nash

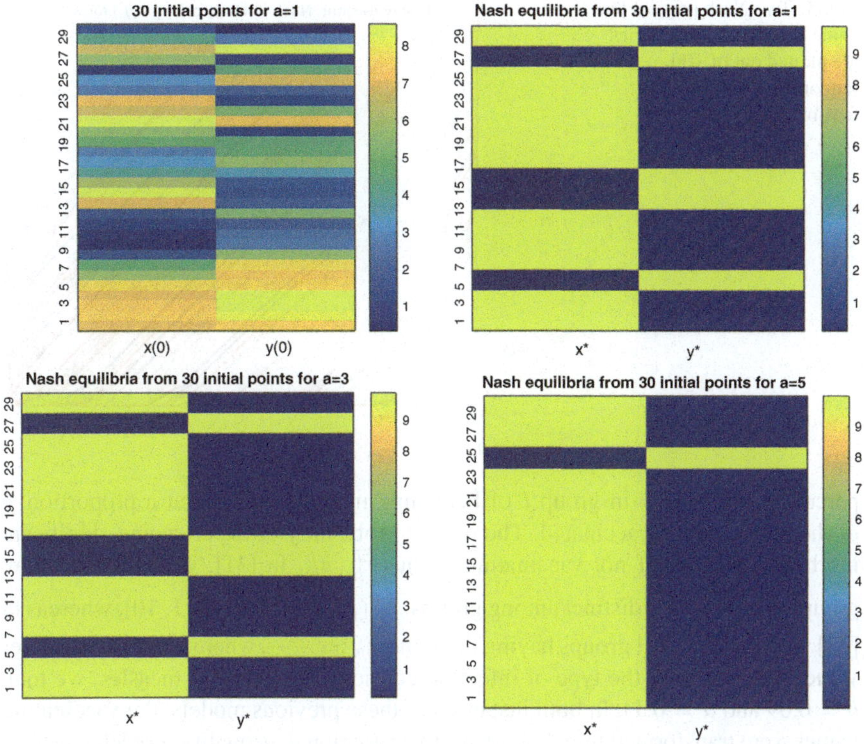

Fig. 4 Two Nash equilibrium pairs of the parametrized game. Values were obtained based on 30 randomly selected initial points of the projected dynamics (3) depicted in the upper left panel

strategy pairs (x^*, y^*). To clearly see the projected dynamics trajectories leading to some of the above cases we present our results in Figure 5.

Example 3 In [31, 32], we introduced a simple Nash vaccinating game played by cohorts of parents of babies who consider whether or not to vaccinate their offspring against paediatric diseases such as measles, mumps, rubella, polyo, etc. The game we proposed is played among a finite number of groups of parents ($k > 1$) where parents in a group are considered to hold the same risk perceptions regarding both vaccinating their offspring, as well as not vaccinating. We consider each group to represent a fixed proportion $\epsilon_i \in (0, 1)$ in the population, and we consider the population to be fixed, i.e. $\sum \epsilon_i = 1$. In general, each group has a mixed strategy given by their probability of vaccinating $P_i \in [0, 1]$, where the coverage in the population is considered to be (excluding time lags between vaccination and uptake of the vaccine) $p = \sum_{i=1}^{k} \epsilon_i P_i$. Each group is given a utility of vaccinating, defined as: $u_i(P) = -r_i P_i - \pi_p(1 - P_i)$, where by $r^i := \dfrac{r_v^i}{r_{inf}^i}$ we denote the relative perceived risk of vaccination versus infection in group i, and by π_p^i we denote the

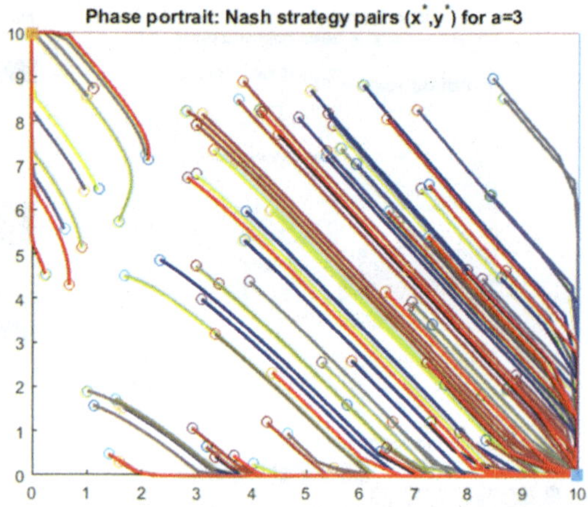

Fig. 5 Two Nash pairs of the game for $a = 3$. Trajectories are plotted out of 100 randomly selected initial feasible points in $[0, 10]^2$

perceived probability in group i of becoming infected, given that a proportion p of the population is vaccinated. The overall probability of experiencing significant morbidity because of not vaccinating is thus $r^i_{inf}\pi^i_p$. In [31], we assume π^i_p are group dependent and distinct among groups: $\pi^i_p(p) = e^{a^i p}, a_i \in [1, 10]$, whereas in [32] we considered all groups having the same $\pi_p := \frac{b}{\alpha+p}$, where α, b are parameter values dependent on the type of infection considered (in case of measles, we took $b = 0.09$ and $\alpha = 0.1$). In both instances of these previous models, the vaccinating games were transformed into their respective variational inequality problems, whose F fields were strongly monotone. The strong monotonicity guaranteed in turn the uniqueness of the Nash equilibrium strategies.

Here we consider the case where the groups have distinct and functionally different π^i_p expressions. Specifically, we consider

$$\pi^2_p := \frac{0.09}{0.1+p},\ \pi^1_p(p) = e^{-ap},\ \pi^3_p(p) = e^{-\frac{a}{2}p},\ a \in [0, 1] \tag{23}$$

with further values of $r_1 = 1$, $r_2 = 2$, $r_3 = 0.5$, $\epsilon_1 = 0.5$, $\epsilon_2 = 0.1$, $\epsilon_3 = 0.4$. The risk values highlight that group 1 are indifferent between risks ($r_1 = 1$), group 2 is a vaccine adverse group ($r2 > 1$), and group 3 is a vaccine inclined group ($r_3 < 1$).

The VI problem associated to this game (where players wish to maximize their utility) is driven by the vector field $F(P_1, P_2, P_3) := \left(-\frac{\partial u_1}{\partial P_1}, -\frac{\partial u_2}{\partial P_2}, -\frac{\partial u_3}{\partial P_3}\right)$, where

$$F(P_1, P_2, P_3) = \begin{cases} -\frac{\partial u_1}{\partial P_1} = e^{-ap} + (1-P_1)\epsilon_1 a e^{-ap} - 1 \\ -\frac{\partial u_2}{\partial P_2} = \frac{0.09}{0.1+p} + \frac{0.09\epsilon_2(1-P_2)}{(0.1+p)^2} - 2 \\ -\frac{\partial u_3}{\partial P_3} = e^{\frac{-ap}{2}} + (1-P_3)\frac{\epsilon_3 a}{2}e^{\frac{-ap}{2}} - 0.5. \end{cases}$$

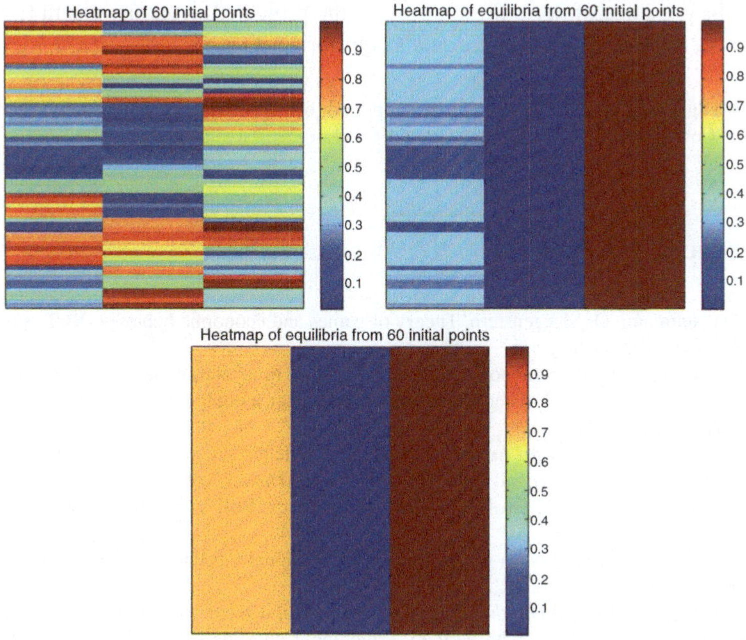

Fig. 6 Nash equilibrium triplets of the game for $a = 0.01$ (upper right panel) and for $a = 1$ (lower panel). The 60 randomly selected initial feasible points for the dynamics (3) are shown in the upper left panel

It is clear that the functional form of F in this example is a lot more complicated than pur previous Examples 1 and 2, which leads to higher complexities when testing for conditions of Theorem 2. However, we can apply our computational method and explore the interval $a \in [0, 1]$. We report our findings in Figure 6.

5 Conclusions

In this paper we discussed a larger class of monotone-like conditions giving rise to unique solutions of variational inequality problems, which in turn give rise to unique Nash equilibria of a differential game. We also introduced a computational method to identify parameter values giving rise to multiple Nash equilibria in a parameterized Nash game, based on the relation between local monotone attracting critical points of a projected system associated to the game. We showed that our theoretical results can be tested for and used in concrete examples. However, sometimes due to the complexity of the game's payoffs, the easier way to tackle the issue is by using our computational approach. As future ideas, our work could be expanded to bridge our results with known results in [17], and to see to what extent the presence of periodic solutions of the projected dynamics (as in [21]) can be related, meaningfully, with a game's equilibria set.

Acknowledgments This work has been supported by a National Science and Engineering Research Council (NSERC) Discovery Grant 400684 of the first author.

References

1. J. Von Neumann, O. Morgenstern, Theory of games and economic behavior. Bull. Am. Math. Soc. 51, 498–504 (1945)
2. J.F. Nash, et al., Equilibrium points in n-person games. Proc. Nat. Acad. Sci. **36**, 48–49 (1950)
3. K.J. Arrow, G. Debreu, Existence of an equilibrium for a competitive economy. J. Econom. Soc. **22**(3), 265–290 (1954)
4. T. Ichiishi, *Game Theory for Economic Analysis* (Academic Press, New York, 1983)
5. T. Basar, G.J. Olsder, *Dynamic Noncooperative Game Theory*, vol. 23 (SIAM, 1999)
6. J. Hofbauer, K. Sigmund, Evolutionary game dynamics. Bull. Am. Math. Soc. **40**, 479–519 (2003)
7. P.T. Harker, Generalized Nash games and quasi-variational inequalities. Eur. J. Oper. Res. **54**, 81–94 (1991)
8. D. Gabay, H. Moulin, *On the Uniqueness and Stability of Nash's Equilibrium in Non Cooperative Games* (Université Paris IX-Dauphine, Centre de Recherche de Mathématiques de la Décision, 1978)
9. D. Kinderlehrer, G. Stampacchia, *An Introduction to Variational Inequalities and Their Applications*, vol. 31 (SIAM, 1980)
10. G. Isac, M.G. Cojocaru, The projection operator in a Hilbert space and its directional derivative. Consequences for the theory of projected dynamical systems. J. Funct. Spaces Appl. **2**, 71–95 (2004)
11. M.G. Cojocaru, P. Daniele, A. Nagurney, Double-layered dynamics: a unified theory of projected dynamical systems and evolutionary variational inequalities. Eur. J. Oper. Res. **175**, 494–507 (2006)
12. A. Nagurney, D. Zhang, *Projected Dynamical Systems and Variational Inequalities with Applications*, vol. 2 (Springer, Berlin, 2012)
13. P. Dupuis, A. Nagurney, Dynamical systems and variational inequalities. Ann. Oper. Res. **44** 7–42 (1993)
14. M.G. Cojocaru, S. Greenhalgh, Evolution solutions of equilibrium problems: a computational approach, in *Mathematical Analysis, Approximation Theory and Their Applications* (Springer, Berlin, 2016), pp. 121–138
15. M.G. Cojocaru, E. Wild, A. Small, On describing the solution sets of generalized Nash games with shared constraints. Optim. Eng. **19**(4), 845–870 (2018)
16. M.G. Cojocaru, L. Jonker, Existence of solutions to projected differential equations in Hilbert spaces. Proc. Am. Math. Soc. **132**, 183–193 (2004)
17. V.K. Le, K. Schmitt, *Global Bifurcation in Variational Inequalities: Applications to Obstacle and Unilateral Problems*, vol. 123 (Springer, Berlin, 1997)
18. M.G. Cojocaru, S. Greenhalgh, Dynamic vaccination games and hybrid dynamical systems. Optim. Eng. **13**, 505–517 (2012)
19. A. Barbagallo, M.G. Cojocaru, Continuity of solutions for parametric variational inequalities in Banach space. J. Math. Anal. Appl. **351**, 707–720 (2009)
20. L. Perko, *Differential Equations and Dynamical Systems*, vol. 7 (Springer, Berlin, 2013)
21. M.G. Cojocaru, Monotonicity and existence of periodic orbits for projected dynamical systems on Hilbert spaces. Proc. Am. Math. Soc. **134**, 793–804 (2006)

22. M. Johnston, M. Cojocaru, Equilibria and periodic solutions of projected dynamical systems on sets with corners (2007). Preprint, University of Guelph
23. S. Karamardian, S. Schaible, Seven kinds of monotone maps. J. Optim. Theory Appl. **66**, 37–46 (1990)
24. D. O'Regan, R.P. Agarwal, *Set Valued Mappings with Applications in Nonlinear Analysis* (CRC Press, Boca Raton, 2003)
25. R.U. Verma, Sensitivity analysis for relaxed cocoercive nonlinear quasivariational inclusions. Int. J. Stochastic Anal. (2006)
26. I.K. Argyros, S. Hilout, *Aspects of the Computational Theory for Certain Iterative Methods* (Polimetrica sas, Milano, 2009)
27. A. Nagurney, D. Zhang, On the stability of an adjustment process for spatial price equilibrium modeled as a projected dynamical system. J. Econ. Dyn. Control **20**, 43–62 (1996)
28. M. Pappalardo, M. Passacantando, Stability for equilibrium problems: from variational inequalities to dynamical systems. J. Optim. Theory Appl. **113**, 567–582 (2002)
29. G. Isac, M.G. Cojocaru, Variational inequalities, complementarity problems and pseudo-monotonicity, in *Proceedings of the International Conference on Nonlinear Operators, Differential Equations and Applications* (Babes-Bolyai University of Cluj-Napoca III, 2002)
30. M.G. Cojocaru, *Projected Dynamical Systems on Hilbert spaces* (Queen's University, Canada, 2002)
31. M.G. Cojocaru, C.T. Bauch, Vaccination strategies of population groups with distinct perceived probabilities of infection. J. Inequal. Pure Appl. Math. **10**(1), 16 (2009)
32. M.G. Cojocaru, C.T. Bauch, M.D. Johnston, Dynamics of vaccination strategies via projected dynamical systems. Bull. Math. Biol. **69**, 1453–1476 (2007)
33. S.S. Chang, H.W. Lee Joseph and C.K. Chan, Generalized system for relaxed cocoercive variational inequalities in Hilbert spaces. Appl. Math. Lett. **20**(3), 329–334 (2007)
34. George J. Minty, On variational inequalities for monotone operators, I. Adv. Math. **30**(1) 1–7 1978

Numerical Approximation of a Class of Time-Fractional Differential Equations

Aleksandra Delić, Boško S. Jovanović, and Sandra Živanović

Abstract We consider a class of linear fractional partial differential equations containing two time-fractional derivatives of orders α, $\beta \in (0, 2)$ and elliptic operator on space variable. Three main types of such equations with α and β in the corresponding subintervals were determined. The existence of weak solutions of the corresponding initial-boundary value problems has been proved. Some finite difference schemes approximating these problems are proposed and their stability is proved. Estimates of their convergence rates, in special discrete energetic Sobolev's norms, are obtained. The theoretical results are confirmed by numerical examples.

1 Introduction

The theory of fractional derivatives became very useful in description of some natural phenomena in various fields of science. Specially, they are necessary in modeling diverse processes with memory effects such as anomalous diffusive and subdiffusive systems, turbulent flow, fractional kinetics, signal processing, viscoelasticity, and so on. The fundamental results about fractional differential equations one can find in [8, 11, 13]. It is well-known that many of them do not have the exact solution, so we are forced to use numerical methods for their solving. Also, fractional partial differential equations, such as one-dimensional time-fractional diffusion-wave equation describe relevant physical process. In recent years the fractional telegraph equation is the topic of studying for many authors [12, 16, 17], because they are applicable in several fields such as wave and electrical signal propagation, random walk theory, etc.

The layout of the paper is as follows. In Section 2 we expose some properties of fractional derivative and define some Sobolev-like functional spaces endowed with

A. Delić · B. S. Jovanović (✉) · S. Živanović
Faculty of Mathematics, University of Belgrade, Belgrade, Serbia
e-mail: adelic@matf.bg.ac.rs; bosko@matf.bg.ac.rs; sandra@matf.bg.ac.rs

N. J. Daras, T. M. Rassias (eds.), *Computational Mathematics and Variational Analysis*, Springer Optimization and Its Applications 159,
https://doi.org/10.1007/978-3-030-44625-3_4

norms and inner products that are used hereafter. In Section 2 we also formulate our problem, while in Section 3 we prove several a priori estimates. In Section 4 we propose numerical approximation for this problem based on the theory of finite difference schemes. We proved stability of discrete problem in discrete Sobolev-like spaces. Section 5 is about error analysis of proposed finite difference schemes, while Section 6 is equipped with numerical experiment in aim to demonstrate the convergence behavior.

2 Preliminaries

2.1 Fractional Derivatives

Here, we shall present some fundamental definitions of the fractional calculus theory. There are several ways to define fractional derivatives in sense of a generalization of ordinary derivatives. The most appropriate for the applications are the Riemann–Liouville and the Caputo ones. The left Riemann–Liouville fractional time-derivative is given by Kilbas et al. [8] and Podlubny [13]

$$\partial^{\sigma}_{t,0+}u(x,t)=\frac{1}{\Gamma(n-\alpha)}\frac{\partial^n}{\partial t^n}\int_0^t u(x,s)(t-s)^{n-\sigma-1}ds, \quad t>0,$$

where $n-1\leq\sigma<n$, $n\in\mathbb{N}$ and $\Gamma(\cdot)$ denotes the Gamma function. Analogously one define the left Caputo fractional derivative:

$${}^{C}\partial^{\sigma}_{t,0+}u(x,t)=\frac{1}{\Gamma(n-\alpha)}\int_0^t \frac{\partial^n u(x,s)}{\partial t^n}(t-s)^{n-\sigma-1}ds, \quad t>0.$$

For the sufficient number of zero initial conditions this two definitions of fractional derivatives are equivalent, otherwise the following relation is valid:

$${}^{C}\partial^{\sigma}_{t,0+}u(x,t)=\partial^{\sigma}_{t,0+}u(x,t)-\sum_{k=0}^{n-1}\frac{t^{k-\sigma}}{\Gamma(k-\sigma+1)}\frac{\partial^k u(x,0)}{\partial t^k}.$$

Particularly if $u(x,t)=t^k$, where k is positive integer, $k>\lceil\sigma\rceil$, one can obtain that:

$$\partial^{\sigma}_{t,0+}t^k=\frac{\Gamma(k+1)}{\Gamma(k+1-\sigma)}t^{k-\sigma}.$$

Also, if u has sufficient number of continuous partial derivatives of integer order and

$$\frac{\partial^j u}{\partial t^j}(x,0)=0, \quad j=0,1,\ldots,n-1,$$

where $n=\max\{n_1,n_2\}$, $n_i-1\le\sigma_i<n_i$, $n_i\in\mathbb{N}$, $i=1,2$, than the Riemann–Liouville derivatives $\partial^{\sigma}_{t,0+}$ satisfies the semigroup property (see [13]):

$$\partial^{\sigma_1}_{t,0+}\partial^{\sigma_2}_{t,0+}u=\partial^{\sigma_2}_{t,0+}\partial^{\sigma_1}_{t,0+}u=\partial^{\sigma_1+\sigma_2}_{t,0+}u.$$

2.2 Some Function Spaces

Let Ω be bounded domain in $\mathbb{R}^n$. As usual, we denote by $C^k(\Omega)$ and $C^k(\bar{\Omega})$, $k\in\mathbb{N}$, the space of k-fold differentiable functions. By $C_0^{\infty}(\Omega)$ we denote the space of infinitely differentiable functions with compact support in Ω. By $L^p(\Omega)$, $p\ge 1$, we denote Lebesgue spaces. The inner product and norm in the space of measurable functions whose square is integrable in Ω, denoted by $L^2(\Omega)$, are defined by

$$(u,v)_{L^2(\Omega)}=\int_{\Omega}uv\,dx, \quad \|u\|_{L^2(\Omega)}=(u,u)^{1/2}_{L^2(\Omega)}.$$

We also use $W_p^{\sigma}(\Omega)$ and $\mathring{W}_p^{\sigma}(\Omega)=W_{p,0}^{\sigma}(\Omega)$ to denote Sobolev spaces [1]. In particular, for $p=2$ we set $H^{\sigma}(\Omega)=W_2^{\sigma}(\Omega)$ and $\mathring{H}^{\sigma}(\Omega)=\mathring{W}_2^{\sigma}(\Omega)$. We denote with $\mathring{H}_l^{\sigma}$ the set of functions u defined in Ω whose extensions by zero $\tilde{u}$ are in $H^{\sigma}(-\infty,1)$. For $u\in H_l^{\sigma}(\Omega)$, we set $\|u\|_{\mathring{H}_l^{\sigma}(\Omega)}:=\|\tilde{u}\|_{H^{\sigma}(-\infty,1)}$. Let $C_l^{\sigma}(\Omega)$ denote the closure of the set of functions $u\in C^{\infty}(\bar{\Omega})$ satisfying $u^{(k)}(0)=0$ for $k\in\mathbb{N}_0$, $k\le\sigma-1/2$ [6]. In all of the above formulas, the domain Ω can be replaced by I.

It is worthy to mention here that the operator $\partial^{\sigma}_{t,0+}$ satisfies the following fundamental properties.

Lemma 1 ([5, 9]) *Let* $\sigma>0$, $u\in C^{\infty}(\mathbb{R})$ *and* $\operatorname{supp}u\subset(0,T]$. *Then*

$$\left(\partial^{\sigma}_{t,0+}u,\partial^{\sigma}_{t,T-}u\right)_{L^2(0,T)}=\cos\pi\sigma\,\|\partial^{\sigma}_{t,0+}u\|^2_{L^2(0,+\infty)}. \tag{1}$$

Lemma 2 ([9, 10]) *Let* $0<\sigma<1$, $u\in\mathring{H}_l^{\sigma}(0,T)$, $v\in C^{\infty}(0,T)$. *Then*

$$\left(\partial^{\sigma}_{t,0+}u,v\right)_{L^2(0,T)}=\left(u,\partial^{\sigma}_{t,T-}v\right)_{L^2(0,T)}. \tag{2}$$

Finally, for $n\in\mathbb{R}$, $\sigma>0$, and $p\ge 1$, we introduce the Banach spaces $C(\bar{I},\mathscr{U})$, $C^n(\bar{I},\mathscr{U})$, $C_+^{\sigma}(\bar{I},\mathscr{U})$, $H^n(I,\mathscr{U})$, $H_+^{\sigma}(I,\mathscr{U})$, and $L^p(I,\mathscr{U})$ of vector valued functions $u:\bar{I}\to\mathscr{U}$ (or $u:I\to\mathscr{U}$) equipped, respectively, with the norms [7]:

$$\|u\|_{C(\bar{I},\mathscr{U})} = \max_{t\in\bar{I}} \|u(t)\|_{\mathscr{U}}\,, \quad \|u\|_{C^n(\bar{I},\mathscr{U})} = \max_{t\in\bar{I}} \left(\|u(t)\|_{\mathscr{U}} + \|u^{(n)}(t)\|_{\mathscr{U}}\right),$$

$$\|u\|_{C_+^\sigma(\bar{I},\mathscr{U})} = \max_{t\in\bar{I}} \|\partial_{t,0+}^\sigma u(t)\|_{\mathscr{U}}\,, \quad \|u\|_{H^n(I,\mathscr{U})} = \left\{\int_I \left(\|u(t)\|_{\mathscr{U}}^2 + \|u^{(n)}(t)\|_{\mathscr{U}}^2\right) dt\right\}^{1/2},$$

$$\|u\|_{H_+^\sigma(I,\mathscr{U})} = \left\{\int_I \left(\|u(t)\|_{\mathscr{U}}^2 + \|\partial_{t,0+}^\sigma u(t)\|_{\mathscr{U}}^2\right) dt\right\}^{1/2}, \quad \|u\|_{L^p(I,\mathscr{U})} = \|\, \|u(\cdot)\|_{\mathscr{U}}\, \|_{L^p(I)}.$$

In particular, we set

$$C^{m,n}(\bar{Q}) = C(\bar{I}, C^m(\bar{\Omega})) \cap C^n(\bar{I}, C(\bar{\Omega})) \qquad L^{2,1}(Q) = L^1(I, L^2(\Omega))\,,$$

and for $\theta > 0$, we define anisotropic Sobolev space

$$H_\pm^{\sigma,\theta} = L^2(I, H^\sigma(\Omega)) \cap H_\pm^\theta(I, L^2(\Omega)).$$

Finally, we can formulate our problem.

2.3 *Problem Formulation*

Let $\Omega = (0, 1)$ and $I = (0, T)$ be the space and time domain, respectively, and $Q = \Omega \times I$. We consider the following partial differential equation:

$$\partial_{t,0+}^\alpha u + a(x)\, \partial_{t,0+}^\beta u - \frac{\partial^2 u}{\partial x^2} = f(x, t), \quad (x, t) \in Q, \tag{3}$$

subject to boundary

$$u(0, t) = 0, \quad u(1, t) = 0, \quad \forall t \in I, \tag{4}$$

and initial conditions (the first one or both)

$$u(x, 0) = 0, \quad \forall x \in \Omega, \tag{5}$$

$$\frac{\partial u}{\partial t}(x, 0) = 0, \quad \forall x \in \Omega. \tag{6}$$

Throughout the paper we assume that the coefficient $a(x)$ satisfies the following conditions:

$$a(x) \in L^\infty(0, 1), \quad 0 < a_0 \le a(x) \le a_1 \text{ a.e.} \tag{7}$$

Also, depending on the order of fractional derivative, there are three cases that we shall investigate:

- $0 < \beta < \alpha < 1$,
- $1 < \beta < \alpha < 2$,
- $0 < \beta < 1 < \alpha < 2$.

3 A Priori Estimates

3.1 Case $0 < \beta < \alpha < 1$

Theorem 1 *Let $f \in L^2(Q)$. Then the problem (3)–(5) is well posed in the space $H_+^{1,\alpha/2}(\Omega)$ and its solution satisfies the following a priori estimate:*

$$\|u\|^2_{H_+^{1,\alpha/2}(Q)} \equiv \left\|\partial^{\alpha/2}_{t,0+}u\right\|^2_{L^2(Q)} + \left\|\frac{\partial u}{\partial x}\right\|^2_{L^2(Q)} + \|u\|^2_{L^2(Q)} \leq C \, \|f\|^2_{L^2(Q)} \, . \tag{8}$$

Proof By multiplying equation (3) with u and after integration over the domain Q, it follows:

$$\int\int_Q \left(\partial^{\alpha}_{t,0+}u\right) u\,dxdt + \int\int_Q a \left(\partial^{\beta}_{t,0+}u\right) u\,dxdt - \int\int_Q \frac{\partial^2 u}{\partial x^2} u\,dxdt = \int\int_Q f\,u\,dxdt.$$

Further,

$$\int\int_Q \left(\partial^{\alpha}_{t,0+}u\right) u\,dxdt \geq \cos\frac{\pi\alpha}{2} \left\|\partial^{\alpha/2}_{t,0+}u\right\|^2_{L^2(Q)},$$

$$\int\int_Q a \left(\partial^{\beta}_{t,0+}u\right) u\,dxdt \geq a_0 \cos\frac{\pi\beta}{2} \left\|\partial^{\beta/2}_{t,0+}u\right\|^2_{L^2(Q)},$$

$$\int\int_Q \frac{\partial^2 u}{\partial x^2} u\,dxdt = -\left\|\frac{\partial u}{\partial x}\right\|^2_{L^2(Q)},$$

and

$$\int\int_Q f\,u\,dxdt \leq \|f\|_{L^2(Q)} \, \|u\|_{L^2(Q)} \, .$$

Thanks to zero boundary conditions, the following inequality is valid:

$$\left\|\frac{\partial u}{\partial x}\right\|_{L^2(Q)} \geq C \, \|u\|_{L^2(Q)}$$

and thus the a priori estimate (8) is obtained.

Remark On the left side of the inequality (3) we omit the norm $\left\| \partial_{t,0+}^{\beta/2} u \right\|_{L^2(Q)}^2$. Nevertheless, by using the property (see [5]):

$$\left\| \partial_{t,0+}^{\alpha/2} u \right\|_{L^2(Q)} \geq C \left\| \partial_{t,0+}^{\beta/2} u \right\|_{L^2(Q)} \geq C \, \|u\|_{L^2(Q)}$$

it follows that we have the equivalent norms:

$$\|u\|_{H_+^{1,\alpha/2}(Q)}^2 \asymp \left\| \partial_{t,0+}^{\alpha/2} u \right\|_{L^2(Q)}^2 + \left\| \partial_{t,0+}^{\beta/2} u \right\|_{L^2(Q)}^2 + \left\| \frac{\partial u}{\partial x} \right\|_{L^2(Q)}^2 + \|u\|_{L^2(Q)}^2 .$$

3.2 Case $1 < \beta < \alpha < 2$

Theorem 2 *Let $f \in L^2(Q)$. Then the problem* (3)–(6) *is well posed in the space $C^1(\bar{I}, L^2(\Omega)) \cap H^{(\alpha+1)/2}(I, L^2(Q))$ and its solution satisfies the following a priori estimate:*

$$\max_{t\in[0,T]} \left\| \frac{\partial u}{\partial x}(\cdot, t) \right\|_{L^2(\Omega)} + \left\| \partial_{t,0+}^{(\alpha+1)/2} u \right\|_{L^2(Q)} \leq C \, \|f\|_{L^2(Q)} . \tag{9}$$

Proof Knowing that initial and boundary conditions are equal to zero, note that we have $\partial_{t,0+}^{\alpha} u = \partial_{t,0+}^{\alpha-1} \frac{\partial u}{\partial t}$ and $\partial_{t,0+}^{\beta} u = \partial_{t,0+}^{\beta-1} \frac{\partial u}{\partial t}$. Multiply (3) with $\frac{\partial u}{\partial t}$ and take integral over the area $Q_t = (0, 1) \times (0, t)$, for fixed $t \leq T$:

$$\int_0^t \int_0^1 \left(\partial_{t,0+}^{\alpha-1} \frac{\partial u}{\partial t} \right) \frac{\partial u}{\partial t} \, dxdt' + \int_0^t \int_0^1 a \left(\partial_{t,0+}^{\beta-1} \frac{\partial u}{\partial t} \right) \frac{\partial u}{\partial t} \, dxdt' - \int_0^t \int_0^1 \frac{\partial^2 u}{\partial x^2} \frac{\partial u}{\partial t} \, dxdt'$$

$$= \int_0^t \int_0^1 f \, \frac{\partial u}{\partial t} \, dxdt'. \tag{10}$$

Now, for $0 < \beta - 1 < \alpha - 1 < 1$, the first two integrals of the left side, we estimate on similar way like in the case above:

$$\int_0^t \int_0^1 \left(\partial_{t,0+}^{\alpha-1} \frac{\partial u}{\partial t} \right) \frac{\partial u}{\partial t} \, dxdt' \geq \cos \frac{\pi(\alpha-1)}{2} \left\| \partial_{t,0+}^{(\alpha-1)/2} \frac{\partial u}{\partial t} \right\|_{L^2(Q_t)}^2 \geq 0,$$

$$\int_0^t \int_0^1 a \left(\partial_{t,0+}^{\beta-1} \frac{\partial u}{\partial t} \right) \frac{\partial u}{\partial t} \, dxdt' \geq a_0 \cos \frac{\pi(\beta-1)}{2} \left\| \partial_{t,0+}^{(\beta-1)/2} \frac{\partial u}{\partial t} \right\|_{L^2(Q_t)}^2 \geq 0.$$

Also,

$$-\int_0^t\int_0^1 \frac{\partial^2 u}{\partial x^2}\frac{\partial u}{\partial t}\,dxdt' = \int_0^t\int_0^1 \frac{\partial u}{\partial x}\frac{\partial^2 u}{\partial t\partial t}\,dxdt' = \int_0^t \frac{1}{2}\frac{\partial}{\partial t}\int_0^1 \left(\frac{\partial u}{\partial x}\right)^2 dxdt' =$$

$$= \frac{1}{2}\left\|\frac{\partial u}{\partial x}(\cdot,t)\right\|^2_{L^2(\Omega)} - \frac{1}{2}\left\|\frac{\partial u}{\partial x}(\cdot,0)\right\|^2_{L^2(\Omega)} = \frac{1}{2}\left\|\frac{\partial u}{\partial x}(\cdot,t)\right\|^2_{L^2(\Omega)} \geq 0$$

and

$$\left|\int_0^t\int_0^1 f\frac{\partial u}{\partial t}\,dxdt'\right| \leq \int_0^T\int_0^1 \left|f\frac{\partial u}{\partial t}\right| dxdt' \leq \|f\|_{L^2(Q)}\left\|\frac{\partial u}{\partial t}\right\|_{L^2(Q)}.$$

Let $t = T$. All the terms on the left side of (10) are nonnegative and we keep only the first of them. Since $Q_T = Q$ and

$$\left\|\frac{\partial u}{\partial t}\right\|_{L^2(Q)} \leq C\left\|\partial_{t,0+}^{(\alpha-1)/2}\frac{\partial u}{\partial t}\right\|_{L^2(Q)}$$

(see [5]), it follows:

$$\left\|\partial_{t,0+}^{(\alpha-1)/2}\frac{\partial u}{\partial t}\right\|_{L^2(Q)} \leq C\,\|f\|_{L^2(Q)}.$$

Now, let $t \leq T$. Using the previous inequalities we have

$$\left\|\frac{\partial u}{\partial x}(\cdot,t)\right\|^2_{L^2(\Omega)} \leq 2\,\|f\|_{L^2(Q)}\left\|\frac{\partial u}{\partial t}\right\|_{L^2(Q)} \leq C\,\|f\|^2_{L^2(Q)}.$$

The a priori estimate (9) is obtained from the last two inequalities.

We also can obtain another a priori estimate.

Theorem 3 *Let $f \in L^{2,1}(Q)$. Then the problem* (3)–(6) *is well posed in the space $C^{\alpha-1}(\bar{I}, L^2(\Omega)) \cap H^{(\alpha-1)/2}(I, L^2(Q))$ and its solution satisfies the following a priori estimate:*

$$\max_{t\in[0,T]}\left\|\partial_{t,0+}^{\alpha-1}u(\cdot,t)\right\|_{L^2(\Omega)} + \left\|\partial_{t,0+}^{(\alpha-1)/2}\left(\frac{\partial u}{\partial x}\right)\right\|_{L^2(Q)} \leq C\,\|f\|_{L^{2,1}(Q)}. \quad (11)$$

Multiplying equation (3) with $\partial_{t,0+}^{\alpha-1}u$ and after integration over the domain Q_t, where $t \leq T$, it follows:

$$\int_0^t\int_0^1 \left(\partial_{t,0+}^{\alpha}u\right)\left(\partial_{t,0+}^{\alpha-1}u\right) dxdt' + \int_0^t\int_0^1 a\left(\partial_{t,0+}^{\beta}u\right)\left(\partial_{t,0+}^{\alpha-1}u\right) dxdt'$$

$$-\int_0^t\int_0^1 \frac{\partial^2 u}{\partial x^2}\left(\partial_{t,0+}^{\alpha-1}u\right)dxdt' = \int_0^t\int_0^1 f\left(\partial_{t,0+}^{\alpha-1}u\right)dxdt'. \tag{9}$$

Further,

$$\int_0^t\int_0^1 \left(\partial_{t,0+}^{\alpha}u\right)\left(\partial_{t,0+}^{\alpha-1}u\right)dxdt' = \int_0^t\int_0^1 \left(\frac{\partial}{\partial t}\left(\partial_{t,0+}^{\alpha-1}u\right)\right)\left(\partial_{t,0+}^{\alpha-1}u\right)dxdt'$$

$$= \int_0^t\int_0^1 \frac{1}{2}\frac{\partial}{\partial t}\left(\left(\partial_{t,0+}^{\alpha-1}u\right)^2\right)dxdt'$$

$$= \frac{1}{2}\left\|\partial_{t,0+}^{\alpha-1}u(\cdot,t)\right\|^2_{L^2(\Omega)} - \frac{1}{2}\left\|\partial_{t,0+}^{\alpha-1}u(\cdot,0)\right\|^2_{L^2(\Omega)} = \frac{1}{2}\left\|\partial_{t,0+}^{\alpha-1}u(\cdot,t)\right\|^2_{L^2(\Omega)} \geq 0,$$

$$\int_0^t\int_0^1 a\left(\partial_{t,0+}^{\beta}u\right)\left(\partial_{t,0+}^{\alpha-1}u\right)dxdt' = \int_0^t\int_0^1 a\left(\partial_{t,0+}^{\beta+1-\alpha}\left(\partial_{t,0+}^{\alpha-1}u\right)\right)\left(\partial_{t,0+}^{\alpha-1}u\right)dxdt'$$

$$\geq a_0\cos\frac{\pi(\beta+1-\alpha)}{2}\left\|\partial_{t,0+}^{(\beta+1-\alpha)/2}\left(\partial_{t,0+}^{\alpha-1}u\right)\right\|^2_{L^2(Q_t)} \geq 0,$$

$$-\int_0^t\int_0^1 \frac{\partial^2 u}{\partial x^2}\left(\partial_{t,0+}^{\alpha-1}u\right)dxdt' = \int_0^t\int_0^1 \frac{\partial u}{\partial x}\left(\partial_{t,0+}^{\alpha-1}\frac{\partial u}{\partial x}\right)dxdt'$$

$$\geq \cos\frac{\pi(\alpha-1)}{2}\left\|\partial_{t,0+}^{(\alpha-1)/2}\left(\frac{\partial u}{\partial x}\right)\right\|^2_{L^2(Q_t)} \geq 0$$

and

$$\left|\int_0^t\int_0^1 f\left(\partial_{t,0+}^{\alpha-1}u\right)dxdt'\right| \leq \int_0^t \|f(\cdot,t')\|_{L^2(\Omega)}\left\|\partial_{t,0+}^{\alpha-1}u(\cdot,t')\right\|_{L^2(\Omega)}dt'$$

$$\leq \left(\int_0^T \|f(\cdot,t')\|_{L^2(\Omega)}\,dt'\right)\max_{t\in[0,T]}\left\|\partial_{t,0+}^{\alpha-1}u(\cdot,t')\right\|_{L^2(\Omega)}.$$

We can conclude that three integrals on the left side in (10) are nonnegative, and if we omit two of them, it follows:

$$\left\|\partial_{t,0+}^{\alpha-1}u(\cdot,t)\right\|^2_{L^2(\Omega)} \leq 2\left(\int_0^T \|f(\cdot,t')\|_{L^2(\Omega)}\,dt'\right)\max_{t\in[0,T]}\left\|\partial_{t,0+}^{\alpha-1}u(\cdot,t')\right\|_{L^2(\Omega)}.$$

Since this inequality is valid for every $t\in[0,T]$, we have

$$\max_{t\in[0,T]}\left\|\partial_{t,0+}^{\alpha-1}u(\cdot,t)\right\|_{L^2(\Omega)} \leq 2\left(\int_0^T \|f(\cdot,t')\|_{L^2(\Omega)}\,dt'\right) \equiv 2\,\|f\|_{L^{2,1}(Q)}. \tag{12}$$

On the other side, for $t = T$, if we keep only the third term on the left side in (10), it follows:

$$\left\| \partial_{t,0+}^{(\alpha-1)/2} \left(\frac{\partial u}{\partial x} \right) \right\|_{L^2(Q)}^2 \leq$$

$$\leq \frac{1}{\cos \frac{\pi(\alpha-1)}{2}} \left(\int_0^T \left\| f(\cdot, t') \right\|_{L^2(\Omega)} dt' \right) \max_{t \in [0,T]} \left\| \partial_{t,0+}^{\alpha-1} u(\cdot, t') \right\|_{L^2(\Omega)} .$$

So, by using (12), we have

$$\left\| \partial_{t,0+}^{(\alpha-1)/2} \left(\frac{\partial u}{\partial x} \right) \right\|_{L^2(Q)}^2 \leq C \, \| f \|_{L^{2,1}(Q)}^2 ,$$

and the result follows immediately from (12) and the last inequality.

3.3 Case $0 < \beta < 1 < \alpha < 2$

This case is considered in [4], where we shown that the a priori estimate (11) is valid.

In this case, we shall also obtain the a priori estimate (9). Since $0 < \beta < 1$, multiplying equation (3) with $\frac{\partial u}{\partial t}$ and take integral over the domain $Q_t = (0, 1) \times (0, t)$ we have

$$\int_0^t \int_0^1 \left(\partial_{t,0+}^{\alpha-1} \frac{\partial u}{\partial t} \right) \frac{\partial u}{\partial t} \, dx dt' + \int_0^t \int_0^1 a \left(\partial_{t,0+}^{\beta} u \right) \frac{\partial u}{\partial t} \, dx dt' - \int_0^t \int_0^1 \frac{\partial^2 u}{\partial x^2} \frac{\partial u}{\partial t} \, dx dt'$$

$$= \int_0^t \int_0^1 f \, \frac{\partial u}{\partial t} \, dx dt' .$$

The first, the third, and the forth integral in the last equality are almost estimated in the previous case. We shall proof that the second integral also nonnegative. It is obviously $0 < 1 - \beta < 1$, and thanks to zero initial condition, we have $\frac{\partial u}{\partial t} = D_{t,0+}^{1-\beta} \left(D_{t,0+}^{\beta} u \right)$. Using this facts, one can get:

$$\int_0^t \int_0^1 a \left(\partial_{t,0+}^{\beta} u \right) \frac{\partial u}{\partial t} \, dx dt' = \int_0^t \int_0^1 a \left(\partial_{t,0+}^{\beta} u \right) \partial_{t,0+}^{1-\beta} \left(\partial_{t,0+}^{\beta} u \right) dx dt'$$

$$\geq a_0 \cos \frac{\pi(1-\beta)}{2} \left\| \partial_{t,0+}^{(1-\beta)/2} \left(\partial_{t,0+}^{\beta} u \right) \right\|_{L^2(Q_t)}^2$$

$$= a_0 \cos\frac{\pi(1-\beta)}{2}\left\|\partial_{t,0+}^{(\beta+1)/2}u\right\|^2_{L^2(Q_t)} \ge 0\,.$$

The a priori estimate (9) follows from the last one and the previous proved inequalities.

4 Numerical Approximation

Let N and M be positive integers. We define uniform meshes

$$\bar{\omega}_h = \{x_i = ih|i = 0, 1\ldots, N;\ h = 1/N\}$$

and

$$\bar{\omega}_\tau = \{t_j = j\tau|j = 0, 1, \ldots, M;\ \tau = T/M\}$$

in intervals $[0, 1]$ and $[0, T]$, respectively. Also, we set $\omega_h := \bar{\omega}_h \cap (0, 1)$, $\omega_h^- := \omega_h \cup \{0\}$, $\omega_\tau := \bar{\omega}_\tau \cap (0, T)$, and $\omega_\tau^+ = \omega_\tau \cup \{T\}$.

For the grid function v defined on $\bar{\omega}_h \times \bar{\omega}_\tau$ we will use the standard notations from the theory of finite difference schemes (see [14])

$$v = v(x,t), \quad v_x = v_x(x,t) = \frac{v(x+h,t) - v(x,t)}{h} = v_{\bar{x}}(x+h,t),$$

$$v_t = v_t(x,t) = \frac{v(x,t+\tau) - v(x,t)}{\tau} = v_{\bar{t}}(x,t+\tau),$$

$$v_{\bar{x}x} = \frac{v_x - v_{\bar{x}}}{h}, \quad v_{\bar{t}t} = \frac{v_t - v_{\bar{t}}}{\tau},$$

$$v^j = v(x,t_j), \qquad \bar{v}^j = \frac{v(x,t_j) + v(x,t_{j-1})}{2}.$$

We use the standard approximation of the Caputo fractional derivative of order $\sigma \in (0, 1)$ for $t = t_j$, $j = 1, 2, \ldots, M$, given by Oldham and Spanier [11]

$${}^C D_{t,0+}^{\sigma} u(x,t_j) \approx \frac{\tau^{1-\sigma}}{\Gamma(2-\sigma)}\sum_{k=1}^{j} a_{j-k}u_{\bar{t}}^k =: \left(\Delta_{t,0_+}^{\sigma-1}u_{\bar{t}}\right)^j =: \left(\Delta_{t,0_+}^{\sigma}u\right)^j, \qquad (13)$$

where

$$a_{j-k} = a_{j-k}^{[1-\sigma]} = (j-k+1)^{1-\sigma} - (j-k)^{1-\sigma} > 0.$$

Let u be the function which satisfies zero initial conditions and $\sigma \in (1, 2)$. Then the Caputo fractional derivative we discretize by Sun and Wu [15]

$$\frac{1}{2}\left({}^{C}D^{\sigma}_{t,0+}u^{j} + {}^{C}D^{\sigma}_{t,0+}u^{j-1}\right) \approx \left(\Delta^{\sigma}_{t,0_+} u\right)^{j}, \tag{14}$$

where

$$\left(\Delta^{\sigma}_{t,0_+} u\right)^{j} := \begin{cases} \dfrac{\tau^{-\sigma}}{\Gamma(3-\sigma)} u^{1}, & j = 1, \\ \dfrac{\tau^{2-\sigma}}{\Gamma(3-\sigma)} \left[\displaystyle\sum_{k=2}^{j} a_{j-k} u^{k}_{\bar{t}t} + \frac{a_{j-1}}{\tau^{2}} u^{1}\right], & j = 2, 3, \ldots, M. \end{cases} \tag{15}$$

By introducing a fictitious time level $t_{-1} = -\tau$, the approximation (14)–(15) can be written in the following form:

$$\left(\Delta^{\sigma}_{t,0_+} u\right)^{j} = \left(\Delta^{\sigma-2}_{t,0_+} u_{\bar{t}}\right)^{j}_{\bar{t}} = \left(\Delta^{\sigma-2}_{t,0_+} u_{\bar{t}t}\right)^{j} = \frac{\tau^{2-\sigma}}{\Gamma(3-\sigma)} \sum_{k=1}^{j} a_{j-k} u^{k}_{\bar{t}t}, \quad j \geq 1, \tag{16}$$

where

$$u^{-1} = u^{0}_{\bar{t}} = \Delta^{\sigma-2}_{t,0+} u^{0}_{\bar{t}} = 0$$

and

$$a_{j-k} = a^{[2-\sigma]}_{j-k} = (j-k+1)^{2-\sigma} - (j-k)^{2-\sigma} > 0.$$

Further, we introduce the following Steklov averaging operators in the usual manner [7]:

$$T^{+}_{x} v(x,t) = \frac{1}{h}\int_{x}^{x+h} v(s,t)ds = T^{-}_{x} v(x+h,t),$$

$$T^{2}_{x} v(x,t) = T^{+}_{x} T^{-}_{x} v(x,t) = \frac{1}{h}\int_{x-h}^{x+h} \left(1 - \left|\frac{x-s}{h}\right|\right) v(s,t)ds.$$

Lemma 3 ([2]) *Let v be a function defined on the mesh $\bar{\omega}_{\tau}$ and $\sigma \in (0, 1)$. Then the following inequalities are valid:*

$$v^{j}\left(\Delta^{\sigma-1}_{t,0_+} v_{\bar{t}}\right)^{j} \geq \frac{1}{2}\left(\Delta^{\sigma-1}_{t,0_+}\left(v^{2}\right)_{\bar{t}}\right)^{j} + \frac{\tau^{\sigma}\Gamma(2-\sigma)}{2}((\Delta^{\sigma-1}_{t,0_+} v_{\bar{t}})^{j})^{2},$$

$$v^{j-1}\left(\Delta^{\sigma-1}_{t,0_+} v_{\bar{t}}\right)^{j} \geq \frac{1}{2}\left(\Delta^{\sigma-1}_{t,0_+}\left(v^{2}\right)_{\bar{t}}\right)^{j} - \frac{\tau^{\sigma}\Gamma(2-\sigma)}{2(2-2^{1-\sigma})}((\Delta^{\sigma-1}_{t,0_+} v_{\bar{t}})^{j})^{2}.$$

Lemma 4 ([3]) *For every function $v(t)$ defined on the mesh $\bar{\omega}_\tau$ which satisfies $v(0) = 0$ and $\sigma \in (0, 1)$ the following equality is valid:*

$$\tau \sum_{j=1}^{M} \left(\Delta_{t,0_+}^{\sigma-1} (v^2)_{\bar{t}} \right)^j = \frac{\tau^{1-\sigma}}{\Gamma(2-\sigma)} \sum_{j=1}^{M} a_{M-j} (v^j)^2, \quad a_{M-j} = a_{M-j}^{[1-\sigma]}.$$

Let us define discrete inner products and norms:

$$(u, v)_h = (u, v)_{L^2(\omega_h)} = h \sum_{x \in \omega_h} u(x) v(x), \quad \|v\|_h = \|v\|_{L^2(\omega_h)} = (v, v)_h^{1/2},$$

$$(u, v]_h = (u, v]_{L^2(\omega_h^+)} = h \sum_{x \in \omega_h^+} u(x) v(x), \quad ||v]|_h = \|v]|_{L^2(\omega_h^+)} = (v, v]_h^{1/2},$$

$$|v|_{H^1(\omega_h)} = \|v_{\bar{x}}]|_h, \quad \|v\|_{H^1(\omega_h)} = \left(|v|^2_{H^1(\omega_h)} + \|v\|^2_{L^2(\omega_h)} \right)^{1/2},$$

$$\|v\|_{L^{2,1}(Q_{h\tau})} = \|v\|_{L^1(\omega_\tau^+, L^2(\omega_h))} = \tau \sum_{t \in \omega_\tau^+} \|v(\cdot, t)\|_h,$$

$$\|v\|_{L^2(Q_{h\tau})} = \left(\tau \sum_{t \in \omega_\tau^+} \|v(\cdot, t)\|_h^2 \right)^{1/2}, \quad \|v]|_{L^2(Q_{h\tau})} = \left(\tau \sum_{t \in \omega_\tau^+} \|v(\cdot, t)]|_h^2 \right)^{1/2},$$

$$\|v\|^2_{B^{\sigma/2}(\omega_\tau^+, L^2(\omega_h))} = \tau^{1-\sigma} \sum_{j=1}^{M} a_{M-j} \|v(\cdot, t_j)\|_h^2, \quad \sigma \in (0, 1), \quad a_{M-j} = a_{M-j}^{[1-\sigma]}.$$

In the sequel we assume that $a \in C(\bar{\Omega})$.

4.1 Case $0 < \beta < \alpha < 1$

The finite difference scheme, approximating problem (3)–(5) for $0 < \beta < \alpha < 1$, has the form

$$(\Delta_{t,0_+}^{\alpha-1} v_{\bar{t}})^j + a(x)(\Delta_{t,0_+}^{\beta-1} v_{\bar{t}})^j - v_{\bar{x}x}^j = \varphi^j, \quad j = 1, 2, \ldots, M, \quad x \in \omega_h, \tag{17}$$

$$v(0, t) = v(1, t) = 0, \quad t \in \omega_\tau, \tag{18}$$

$$v(x, 0) = 0, \quad x \in \omega_h, \tag{19}$$

where

$$\varphi^j = T_x^2 f(x, t_j).$$

Theorem 4 *Let* $0 < \beta < \alpha < 1$. *Then the finite difference scheme* (17)–(19) *is absolutely stable and its solution satisfies the a priori estimate:*

$$\|v\|_{B^{\alpha/2}(\omega_\tau^+, L^2(\omega_h))} + \|v_{\bar{x}}]|_{L^2(Q_{h\tau})} + \|v\|_{L^2(Q_{h\tau})} \leq C\|\varphi\|_{L^2(Q_{h\tau})}. \tag{20}$$

Proof We multiply (17) by hv^j and sum over the mesh ω_h:

$$h \sum_{x \in \omega_h} v^j (\Delta_{t,0_+}^{\alpha-1} v_{\bar{t}})^j + h \sum_{x \in \omega_h} a(x) v^j (\Delta_{t,0_+}^{\beta-1} v_{\bar{t}})^j - h \sum_{x \in \omega_h} v^j v_{\bar{x}x}^j = (\varphi^j, v^j)_h. \tag{21}$$

The first and second terms can be evaluate using Lemma 3

$$h \sum_{x \in \omega_h} v^j (\Delta_{t,0_+}^{\alpha-1} v_{\bar{t}})^j \geq \frac{1}{2} h \sum_{x \in \omega_h} (\Delta_{t,0_+}^{\alpha-1} v_{\bar{t}}^2)^j,$$

$$h \sum_{x \in \omega_h} a(x) v^j (\Delta_{t,0_+}^{\beta-1} v_{\bar{t}})^j \geq \frac{1}{2} h \sum_{x \in \omega_h} a(x) (\Delta_{t,0_+}^{\beta-1} v_{\bar{t}}^2)^j \geq \frac{a_0}{2} h \sum_{x \in \omega_h} (\Delta_{t,0_+}^{\beta-1} v_{\bar{t}}^2)^j,$$

while for the third term, using the partial summation, one can obtain

$$h \sum_{x \in \omega_h} v^j v_{\bar{x}x}^j = -||v_{\bar{x}}^j]|^2.$$

For the estimate of the right-hand side, one can use ε-inequality

$$(\varphi^j, v^j)_h \leq \varepsilon ||v^j||_h^2 + \frac{1}{4\varepsilon} \|\varphi^j\|_h^2.$$

Because the discrete Poincaré–Friedrichs inequality

$$||v||_h \leq \frac{1}{\sqrt{8}} ||v_{\bar{x}}]|_h$$

is valid, by omitting the second positive term in (21), we obtain

$$\left(\Delta_{t,0_+}^{\alpha-1} \left(\|v\|_h^2 \right)_{\bar{t}} \right)^j + ||v_{\bar{x}}^j]|_h^2 + ||v^j||_h^2 \leq C \|\varphi^j\|_h^2.$$

Multiplying the last inequality by τ and summing over the mesh ω_{τ^+}, we obtain the a priori estimate

$$\tau\sum_{j=1}^{M}\left(\Delta_{t,0+}^{\alpha-1}\left(\|v\|_h^2\right)\right)_{\bar{t}}^{j}+\|v_{\bar{x}}]|_{L^2(Q_{h\tau})}^2+\|v\|_{L^2(Q_{h\tau})}^2\leq C\|\varphi\|_h^2,$$

i.e.

$$\tau^{1-\alpha}\sum_{j=1}^{M}a_{M-j}\|v(\cdot,t_j)\|_h^2+\|v_{\bar{x}}]|_{L^2(Q_{h\tau})}^2+\|v\|_{L^2(Q_{h\tau})}^2\leq C\|\varphi\|_{L^2(Q_{h\tau})}^2.$$

Remark Analogous to the continuous case, following discrete inequalities are valid:

$$\tau\sum_{j=1}^{M}\left(\Delta_{t,0+}^{\alpha-1}\left(\|v\|_h^2\right)\right)_{\bar{t}}^{j}\geq C\tau\sum_{j=1}^{M}\left(\Delta_{t,0+}^{\beta-1}\left(\|v\|_h^2\right)\right)_{\bar{t}}^{j}\geq\bar{C}\tau\sum_{j=1}^{M}\|v^j\|_h^2, \tag{22}$$

which follows directly from (4).

4.2 Case $1<\beta<\alpha<2$

Let $1<\beta<\alpha<2$. The problem (3)–(6) can be approximated by the following difference equation:

$$(\Delta_{t,0+}^{\alpha-2}v_{\bar{t}})_{\bar{t}}^{j}+a(x)(\Delta_{t,0+}^{\beta-2}v_{\bar{t}})_{\bar{t}}^{j}-\bar{v}_{\bar{x}x}^{j}=\varphi^j,\quad j=1,2,\ldots,M,\quad x\in\omega_h, \tag{23}$$

$$v(0,t)=v(1,t)=0,\quad t\in\omega_\tau, \tag{24}$$

$$v(x,0)=0,\quad v_{\bar{t}}^0=0,\quad x\in\omega_h, \tag{25}$$

where

$$\varphi^j=\frac{T_x^2f(x,t_j)+T_x^2f(x,t_{j-1})}{2}.$$

Theorem 5 *Let* $1<\beta<\alpha<2$. *Then the finite difference scheme* (23)–(25) *is absolutely stable and its solution satisfies the a priori estimate:*

$$\|v\|_{B^{(\alpha-1)/2}(\omega_\tau^+,L^2(\omega_h))}+\max_{1\leq k\leq M}||v_{\bar{x}}^k]|_h\leq C\|\varphi\|_{L^2(Q_{h\tau})}. \tag{26}$$

Proof We multiply (23) by $2\tau v_{\bar{t}}^j$ and sum over the mesh ω_h:

$$2\tau\left((\Delta_{t,0+}^{\alpha-2}v_{\bar{t}})_{\bar{t}}^{j},v_{\bar{t}}^{j}\right)_h+2\tau\left(a(\Delta_{t,0+}^{\beta-2}v_{\bar{t}})_{\bar{t}}^{j},v_{\bar{t}}^{j}\right)_h+||v_{\bar{x}}^{j}]|_h^2-||v_{\bar{x}}^{j-1}]|_h^2=2\tau(\varphi^j,v_{\bar{t}}^{j})_h.$$

Summing for $j = 1, 2, \ldots, k,\ k \le M$, we obtain

$$2\tau \sum_{j=1}^{k} \left((\Delta_{t,0_+}^{\alpha-2} v_{\bar{t}})_{\bar{t}}^{j},\, v_{\bar{t}}^{j} \right)_h + 2\tau \sum_{j=1}^{k} \left(a(\Delta_{t,0_+}^{\beta-2} v_{\bar{t}})_{\bar{t}}^{j},\, v_{\bar{t}}^{j} \right)_h + ||v_{\bar{x}}^{k}]|_h^2 \le 2\tau \sum_{j=1}^{k} (\varphi^j, v_{\bar{t}}^{j})_h. \tag{27}$$

Because $0 < \beta - 1 < \alpha - 1 < 1$, the first and second term can be evaluate using Lemmas 3 and 4

$$2\tau \sum_{j=1}^{k} \left((\Delta_{t,0_+}^{\alpha-2} v_{\bar{t}})_{\bar{t}}^{j},\, v_{\bar{t}}^{j} \right)_h \ge \tau \sum_{j=1}^{k} \left(\Delta_{t,0_+}^{\alpha-2} \|v_{\bar{t}}\|_h^2 \right)^j = \frac{\tau^{2-\alpha}}{\Gamma(3-\alpha)} \sum_{j=1}^{k} a_{k-j}^{[\alpha-2]} \|v_{\bar{t}}^{j}\|_h^2 \ge 0,$$

$$2\tau \sum_{j=1}^{k} \left(a(\Delta_{t,0_+}^{\beta-2} v_{\bar{t}})_{\bar{t}}^{j},\, v_{\bar{t}}^{j} \right)_h \ge a_0 \tau \sum_{j=1}^{k} \left(\Delta_{t,0_+}^{\beta-2} \|v_{\bar{t}}\|_h^2 \right)^j = \frac{a_0 \tau^{2-\beta}}{\Gamma(3-\beta)} \sum_{j=1}^{k} a_{k-j}^{[\beta-2]} \|v_{\bar{t}}^{j}\|_h^2 \ge 0.$$

The right-hand side one can estimate using Cauchy–Schwarz inequality

$$\left| 2\tau \sum_{j=1}^{k} (\varphi^j, v_{\bar{t}}^{j})_h \right| \le 2\tau \sum_{j=1}^{M} |(\varphi^j, v_{\bar{t}}^{j})_h| \le 2\|\varphi\|_{L^2(Q_{h\tau})} \|v_{\bar{t}}\|_{L^2(Q_{h\tau})}.$$

Omitting the last two positive terms on the left-hand side in (27) and taking $k = M$ we obtain

$$\|v\|_{B^{(\alpha-1)/2}(\omega_\tau^+, L^2(\omega_h))} \le C \|\varphi\|_{L^2(Q_{h\tau})}. \tag{28}$$

Omitting the first two positive terms on the left-hand side in (27) and using (28) and Lemma 4 we obtain

$$||v_{\bar{x}}^{k}]|_h^2 \le 2\|\varphi\|_{L^2(Q_{h\tau})} \|v_{\bar{t}}\|_{L^2(Q_{h\tau})} \le C \|\varphi\|_{L^2(Q_{h\tau})}^2. \tag{29}$$

Using inequalities (28) and (29) we obtain the a priori estimate (26).

4.3 *Case* $0 < \beta < 1 < \alpha < 2$

In this case the problem (3)–(6) can be approximated by the following difference equation:

$$\left(\Delta_{t,0_+}^{\alpha-2} v_{\bar t}\right)_{\bar t}^{j} + a\Delta_{t,0_+}^{\beta-\alpha}\left(\Delta_{t,0_+}^{\alpha-2} \bar v_{\bar t}\right)_{\bar t}^{j} - \bar v_{\bar x x}^{j} = \varphi^j, \quad j = 1,\ldots,M, \quad x \in \omega_h, \tag{30}$$

$$v(0,t) = 0, \quad v(1,t) = 0, \quad t \in \bar\omega_\tau, \tag{31}$$

$$v(x,0) = 0, \quad v_{\bar t}^{0} = 0, \quad x \in \omega_h. \tag{32}$$

Theorem 6 ([4]) *Let $\alpha \in (1,2)$, $\beta \in (0,1)$, and $\beta > \alpha - 1$. The finite difference scheme* (30)–(32) *is absolutely stable and its solution satisfies the a priori estimate:*

$$\max_{1\le j\le M}\left\|\left(\Delta_{t,0_+}^{\alpha-2} v_{\bar t}\right)^{j}\right\|_h + \|\bar v\|_{B^{(\alpha-1)/2}(\omega_\tau^+, H^1(\omega_h))} \le C\tau\sum_{j=1}^{M}\left\|\varphi^j\right\|_h. \tag{33}$$

5 Convergence of Finite Difference Schemes

In the sequel we will use the following lemmmas.

Lemma 5 ([15]) *Suppose that $\sigma \in (0,1)$, $w \in C^2[0,t]$, and $t \in \omega_\tau^+$. Then*

$$\left|\Delta_{t,0_+}^{\sigma-1} w_{\bar t} - {}^{C}\partial_{t,0+}^{\sigma} w\right| \le C\tau^{2-\sigma}\max_{0\le s\le t}|w''(s)|. \tag{34}$$

Lemma 6 ([15]) *Suppose that $\sigma \in (1,2)$, $w \in C^3[0,t_j]$, $w(0) = w'(0) = 0$, and $t_j \in \omega_\tau^+$. Then*

$$\left|(\Delta_{t,0_+}^{\sigma} w)^j - \frac{1}{2}\left({}^{C}\partial_{t,0+}^{\sigma} w^j + {}^{C}\partial_{t,0+}^{\sigma} w^{j-1}\right)\right| \le C\tau^{3-\sigma}\max_{0\le s\le t}\left|w'''(s)\right|. \tag{35}$$

5.1 *Case* $0 < \beta < \alpha < 1$

Let u be the solution of the initial-boundary value problem (3)–(5) and v the solution of the difference problem (17)–(19). Then, the error $z = u - v$ satisfies

$$\begin{gathered}(\Delta_{t,0_+}^{\alpha-1} z_{\bar t})^j + a(x)(\Delta_{t,0_+}^{\beta-1} z_{\bar t})^j - z_{\bar x x}^{j} = \psi^j,\\ j = 1,2,\ldots,M, \quad x \in \omega_h,\\ z(0,t) = z(1,t) = 0, \quad t \in \omega_\tau,\\ z(x,0) = 0, \quad x \in \omega_h,\end{gathered} \tag{36}$$

where

$$\psi = \xi + \eta = (\xi_1 + \xi_2) + (\eta_1 + \eta_2)$$

and

$$\xi_1 = \Delta^{\alpha-1}_{t,0_+} u_{\bar{t}} - \partial^{\alpha}_{t,0+} u,$$
$$\xi_2 = \partial^{\alpha}_{t,0+}(u - T^2_x u),$$
$$\eta_1 = a\left(\Delta^{\beta-1}_{t,0_+} u_{\bar{t}} - \partial^{\beta}_{t,0+} u\right),$$
$$\eta_2 = a\partial^{\beta}_{t,0+} u - T^2_x\left(a\partial^{\beta}_{t,0+} u\right).$$

Theorem 7 *Let $u \in C^{0,2}(\bar{Q}) \cap C^{\alpha}_{+}\left(\bar{I}, H^2(\Omega)\right)$, $a \in H^2(\Omega)$, and $0 < \beta < \alpha < 1$. Then the solution v of the finite difference scheme* (17)–(19) *converges to the solution u of the initial-boundary-value problem* (3)–(5) *and the following convergence rate estimate holds:*

$$\|z\|_{B^{\alpha/2}(\omega^+_\tau, L^2(\omega_h))} + \|z_{\bar{x}}]|_{L^2(Q_{h\tau})} + \|z\|_{L^2(Q_{h\tau})} \le C(\tau^{2-\alpha} + h^2). \tag{37}$$

Proof From the a priori estimate (20) directly follows inequality

$$\|z\|_{B^{\alpha/2}(\omega^+_\tau, L^2(\omega_h))} + \|z_{\bar{x}}]|_{L^2(Q_{h\tau})} + \|z\|_{L^2(Q_{h\tau})} \le C\left(\|\xi_1\|_{L^2(Q_{h\tau})} + \|\xi_2\|_{L^2(Q_{h\tau})}\right.$$
$$\left. + \|\eta_1\|_{L^2(Q_{h\tau})} + \|\eta_2\|_{L^2(Q_{h\tau})}\right). \tag{38}$$

Terms ξ_1 and η_1 can be estimated using inequality (34):

$$\|\xi_1\|_{L^2(Q_{h\tau})} \le C\tau^{2-\alpha} \max_{t\in[0,T]} \max_{x\in[0,1]} \left|\frac{\partial^2 u}{\partial t^2}\right| \le C\tau^{2-\alpha}\|u\|_{C^{0,2}(\bar{Q})}, \tag{39}$$

$$\|\eta_1\|_{L^2(Q_{h\tau})} \le C\|a\|_{C(\bar{\Omega})}\tau^{2-\beta} \max_{t\in[0,T]} \max_{x\in[0,1]} \left|\frac{\partial^2 u}{\partial t^2}\right| \le C\tau^{2-\beta}\|a\|_{C(\bar{\Omega})}\|u\|_{C^{0,2}(\bar{Q})}. \tag{40}$$

Using the integral representation of $u - T^2_x u$ like in [7], one easily obtains

$$\|\xi_2\|_{L^2(Q_{h\tau})} \le Ch^2\|u\|_{C^{\alpha}_{+}(I, H^2(\Omega))} \tag{41}$$

and

$$\|\eta_2\|_{L^2(Q_{h\tau})} \le Ch^2\|a\|_{H^2(\Omega)}\|u\|_{C^{\beta}_{+}(I, H^2(\Omega))}. \tag{42}$$

The results follows from (38)–(42).

5.2 Case $1 < \beta < \alpha < 2$

Let u be the solution of the initial-boundary value problem (3)–(6) and v the solution of the difference problem (23)–(25). Then, the error $z = u - v$ satisfies

$$\begin{aligned} &(\Delta_{t,0_+}^{\alpha-2} z_{\bar{t}})_{\bar{t}}^j + a(x)(\Delta_{t,0_+}^{\beta-2} z_{\bar{t}})_{\bar{t}}^j - \bar{z}_{\bar{x}x}^j = \phi^j, \\ &j = 1, 2, \ldots, M, \quad x \in \omega_h, \\ &z(0,t) = z(1,t) = 0, \quad t \in \omega_\tau, \\ &z(x,0) = 0, \quad z_{\bar{t}}^0 = 0, \quad x \in \omega_h, \end{aligned} \tag{43}$$

where

$$\phi = \chi_1 + \bar{\chi}_2 + \zeta_1 + \bar{\zeta}_2$$

and

$$\begin{aligned} \chi_1 &= \Delta_{t,0_+}^{\alpha-2} u_{\bar{t}\bar{t}} - \overline{\partial_{t,0+}^{\alpha} u}, \\ \chi_2 &= \partial_{t,0+}^{\alpha}(u - T_x^2 u), \\ \zeta_1 &= a\left(\Delta_{t,0_+}^{\beta-2} u_{\bar{t}\bar{t}} - \overline{\partial_{t,0+}^{\beta} u}\right), \\ \zeta_2 &= a\partial_{t,0+}^{\beta} u - T_x^2\left(a\partial_{t,0+}^{\beta} u\right). \end{aligned}$$

Theorem 8 *Let $u \in C^{0,3}(\bar{Q}) \cap C_+^{\alpha}\left(\bar{I}, H^2(\Omega)\right)$, $a \in H^2(\Omega)$, and $1 < \beta < \alpha < 2$. Then the solution v of the finite difference scheme (23)–(25) converges to the solution u of the initial-boundary-value problem (3)–(6) and the following convergence rate estimate holds:*

$$\|z\|_{B^{(\alpha-1)/2}(\omega_\tau^+, L^2(\omega_h))} + \max_{1\le j\le M} \|z_{\bar{x}}^j]|_{L^2(Q_{h\tau})} \le C(\tau^{3-\alpha} + h^2). \tag{44}$$

Proof From the a priori estimate (26) directly follows inequality

$$\begin{aligned} \|z\|_{B^{(\alpha-1)/2}(\omega_\tau^+, L^2(\omega_h))} + \max_{1\le j\le M} \|z_{\bar{x}}^j]|_{L^2(Q_{h\tau})} \le C\big(&\|\chi_1\|_{L^2(Q_{h\tau})} + \|\chi_2\|_{L^2(Q_{h\tau})} \\ &+ \|\zeta_1\|_{L^2(Q_{h\tau})} + \|\zeta_2\|_{L^2(Q_{h\tau})}\big). \end{aligned} \tag{45}$$

Terms χ_1 and ζ_1 can be estimate using inequality (35):

$$\|\chi_1\|_{L^2(Q_{h\tau})} \le C\tau^{3-\alpha} \max_{t\in[0,T]} \max_{x\in[0,1]} \left|\frac{\partial^3 u}{\partial t^3}\right| \le C\tau^{3-\alpha} \|u\|_{C^{0,3}(\bar{Q})}, \tag{46}$$

$$\|\zeta_1\|_{L^2(Q_{h\tau})} \le C\tau^{3-\beta} \|a\|_{C(\bar{\Omega})} \|u\|_{C^{0,3}(\bar{Q})}. \tag{47}$$

5.3 *Case* $0 < \beta < 1 < \alpha < 2$

Theorem 9 ([4]) *Let* $u \in C^{0,3}(\bar{Q}) \cap C_+^{\alpha}\left(\bar{I}, H^2(\Omega)\right)$, $a \in H^2(\Omega)$ *and let the conditions* $\alpha \in (1,2)$, $\beta \in (0,1)$, *and* $\beta > \alpha - 1$ *be satisfied. Then the solution* v *of the finite difference scheme* (30)–(32) *converges to the solution* u *of the initial-boundary-value problem* (3)–(6) *and the following convergence rate estimate holds:*

$$\max_{1\le j\le M} \left\| \left(\Delta_{t,0_+}^{\alpha-2} z_{\bar{t}}\right)^j \right\|_h + \|\bar{z}\|_{B^{(\alpha-1)/2}(\omega_\tau^+, H^1(\omega_h))} \le C(\tau^{\min\{3-\alpha, 2-\beta, 1+\alpha-\beta\}} + h^2). \tag{48}$$

6 Numerical Examples

6.1 *Case* $0 < \beta < \alpha < 1$

The finite difference scheme approximating problem (3)–(5) for $T = 1$ is given by (17)–(19), as we mentioned before. If we take

$$f(x,t) = t^2 x \sin(\pi x)\left[\frac{2t^{-\alpha}}{\Gamma(3-\alpha)} + \frac{2t^{-\beta}}{\Gamma(3-\beta)} + \pi^2\right] - 2\pi t^2 \cos(\pi x),$$

then the exact solution is $u(x,t) = t^2 x \sin(\pi x)$.

Tables 1 and 2 demonstrate the computational results for this scheme. We have computed the errors in the norm (37) (in the tables these are denoted by $\|\cdot\|_{N_1}$) and in the discrete maximum norm

$$\|v\|_{C(Q_{h\tau})} = \max_{x\in\bar{\omega}_h} \max_{t\in\bar{\omega}_\tau} |v(x,t)|. \tag{49}$$

The temporal convergence rate is $2 - \max\{\alpha, \beta\} = 2 - \alpha$, while the spatial convergence rate is 2.

Table 1 The error and order of convergence in the time direction with $h = 2^{-12}$

α	β	τ	$\|z\|_{N_1(Q_{h\tau})}$	$\log_2 \frac{\|z\|_{N_1(Q_{h\tau})}}{\|z\|_{N_1(Q_{h\tau/2})}}$	$\|z\|_{L_2(Q_{h\tau})}$	$\log_2 \frac{\|z\|_{L_2(Q_{h\tau})}}{\|z\|_{L_2(Q_{h\tau/2})}}$	$\|z\|_{C(Q_{h\tau})}$	$\log_2 \frac{\|z\|_{C(Q_{h\tau})}}{\|z\|_{C(Q_{h\tau/2})}}$
0.8	0.5	2^{-4}	3.4912e−03	1.24	9.9183e−04	1.24	1.6014e−03	1.25
		2^{-5}	1.4755e−03	1.24	4.1931e−04	1.24	6.7470e−04	1.24
		2^{-6}	6.2518e−04	1.23	1.7770e−04	1.23	2.8515e−04	1.24
		2^{-7}	2.6565e−04	1.23	7.5516e−05	1.23	1.2096e−04	1.23
		2^{-8}	1.1322e−04		3.2189e−05		5.1502e−05	
0.5	0.2	2^{-4}	8.0938e−04	1.50	2.3101e−04	1.50	3.6503e−04	1.51
		2^{-5}	2.8647e−04	1.50	8.1787e−05	1.50	1.2829e−04	1.51
		2^{-6}	1.0109e−04	1.51	2.8869e−05	1.51	4.4978e−05	1.51
		2^{-7}	3.5604e−05	1.51	1.0169e−05	1.51	1.5761.e−05	1.51
		2^{-8}	1.2532e−05		3.5792e−06		5.5323e−06	
0.9	0.4	2^{-4}	4.7225e−03	1.13	1.3406e−03	1.13	2.1502e−03	1.13
		2^{-5}	2.1581e−03	1.12	6.1280e−04	1.12	9.8001e−04	1.13
		2^{-6}	9.9061e−04	1.12	2.8135e−04	1.12	4.4900e−04	1.12
		2^{-7}	4.5642e−04	1.11	1.2964e−04	1.11	2.0664e−04	1.11
		2^{-8}	2.1094e−04		5.9920e−05		9.5440e−05	

Table 2 The error and order of convergence in the space direction with $\tau = 2^{-11}$

α	β	h	$\|z\|_{N_1(Q_{h\tau})}$	$\log_2 \frac{\|z\|_{N_1(Q_{h\tau})}}{\|z\|_{N_1(Q_{h/2\tau})}}$	$\|z\|_{L_2(Q_{h\tau})}$	$\log_2 \frac{\|z\|_{L_2(Q_{h\tau})}}{\|z\|_{L_2(Q_{h/2\tau})}}$	$\|z\|_{C(Q_{h\tau})}$	$\log_2 \frac{\|z\|_{C(Q_{h\tau})}}{\|z\|_{C(Q_{h/2\tau})}}$
0.8	0.5	2^{-3}	$9.4623e{-03}$	1.98	$1.8683e{-03}$	2.00	$7.0421e{-03}$	1.98
		2^{-4}	$2.3990e{-03}$	1.99	$4.6693e{-04}$	1.99	$1.7892e{-03}$	1.99
		2^{-5}	$6.0538e{-04}$	1.96	$1.1790e{-04}$	1.94	$4.4975e{-04}$	1.96
		2^{-6}	$1.5538e{-04}$	1.85	$3.0772e{-05}$	1.75	$1.1533e{-04}$	1.87
		2^{-7}	$4.3199e{-05}$		$9.1292e{-06}$		$3.1529e{-05}$	
0.5	0.2	2^{-3}	$9.5837e{-03}$	1.98	$1.9821e{-03}$	2.01	$7.3112e{-03}$	1.97
		2^{-4}	$2.4252e{-03}$	2.00	$4.9375e{-04}$	2.00	$1.8603e{-03}$	2.00
		2^{-5}	$6.0829e{-04}$	2.00	$1.2338e{-04}$	2.00	$4.6470e{-04}$	2.00
		2^{-6}	$1.5241e{-04}$	1.99	$3.0915e{-05}$	1.99	$1.1652e{-04}$	1.99
		2^{-7}	$3.8339e{-05}$		$7.8054e{-06}$		$2.9285e{-05}$	
0.9	0.4	2^{-3}	$9.5949e{-03}$	1.98	$1.8658e{-03}$	2.00	$7.0437e{-03}$	1.97
		2^{-4}	$2.4368e{-03}$	1.97	$4.6804e{-04}$	1.96	$1.7934e{-03}$	1.98
		2^{-5}	$6.2014e{-04}$	1.91	$1.1993e{-04}$	1.85	$4.5435e{-04}$	1.92
		2^{-6}	$1.6492e{-04}$	1.64	$3.3205e{-05}$	1.46	$1.2011e{-04}$	1.72
		2^{-7}	$5.2874e{-05}$		$1.2046e{-05}$		$3.6479e{-05}$	

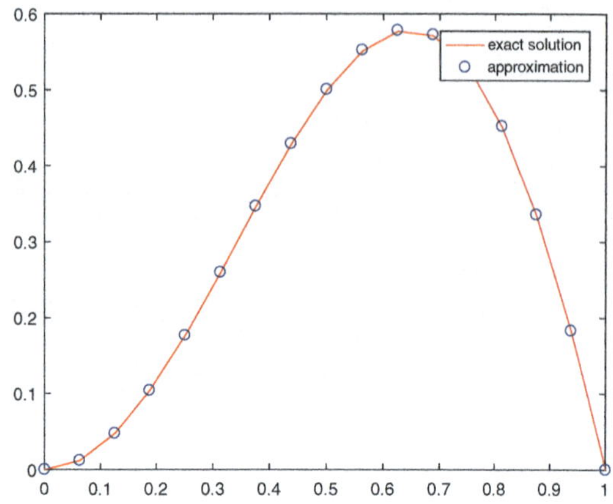

Fig. 1 The numerical and exact solution for $\alpha = 1.8$, $\beta = 1.5$ when $h = 2^{-4}$ and $\tau = 2^{-9}$

6.2 *Case* $1 < \beta < \alpha < 2$

The finite difference scheme approximating problem (3)–(6) for $T = 1$ is given by (23)–(25). For the right-hand side

$$f(x,t) = t^3 x \sin(\pi x)\left[\frac{6t^{-\alpha}}{\Gamma(4-\alpha)} + \frac{6t^{-\beta}}{\Gamma(4-\beta)} + \pi^2\right] - 2\pi t^3 \cos(\pi x),$$

then the exact solution is $u(x,t) = t^3 x \sin(\pi x)$.

Tables 3 and 4 demonstrate the computational results for the scheme (23). We have computed the errors in the norm (44) (in the tables these are denoted by $\|\cdot\|_{N_2}$) and (49).

We can deduce that the temporal convergence rate is $3 - \max\{\alpha, \beta\} = 3 - \alpha$ while the spatial convergence rate is 2. Figure 1 presents the exact and numerical solution which is obtained by (23)–(25).

6.3 *Case* $0 < \beta < 1 < \alpha < 2$

Numerical results for this case are completely presented in [4].

Table 3 The error and order of convergence in the time direction with $h = 2^{-12}$

α	β	τ	$\|z\|_{N_2(Q_{h\tau})}$	$\log_2 \frac{\|z\|_{N_2(Q_{h\tau})}}{\|z\|_{N_2(Q_{h\tau/2})}}$	$\|z\|_{L_2(Q_{h\tau})}$	$\log_2 \frac{\|z\|_{L_2(Q_{h\tau})}}{\|z\|_{L_2(Q_{h\tau/2})}}$	$\|z\|_{C(Q_{h\tau})}$	$\log_2 \frac{\|z\|_{C(Q_{h\tau})}}{\|z\|_{C(Q_{h\tau/2})}}$
1.8	1.5	2^{-4}	2.1708e−02	1.24	3.4509e−03	1.26	7.5908e−03	1.24
		2^{-5}	9.1743e−03	1.24	1.4383e−03	1.25	3.2109e−03	1.24
		2^{-6}	3.8780e−03	1.24	6.0395e−04	1.24	1.3579e−03	1.24
		2^{-7}	1.6438e−03	1.23	2.5519e−04	1.24	5.7574e−04	1.23
		2^{-8}	6.9930e−04		1.0839e−04		2.4496e−04	
1.6	1.4	2^{-4}	9.0807e−03	1.45	1.5951e−03	1.46	3.2117e−03	1.45
		2^{-5}	3.3223e−03	1.45	5.7911e−04	1.45	1.1760e−03	1.45
		2^{-6}	1.2177e−03	1.44	2.1145e−04	1.45	4.3121e−04	1.44
		2^{-7}	4.4761e−04	1.44	7.7571e−05	1.44	1.5854e−04	1.44
		2^{-8}	1.6504e−04		2.8571e−05		5.8460e−05	
1.3	1.1	2^{-4}	1.8667e−03	1.75	3.7749e−04	1.76	6.5976e−04	1.75
		2^{-5}	5.5319e−04	1.75	1.1108e−04	1.75	1.9576e−04	1.75
		2^{-6}	1.6472e−04	1.74	3.2952e−05	1.74	5.8328e−05	1.74
		2^{-7}	4.9246e−05	1.74	9.8308e−06	1.74	1.7445e−05	1.74
		2^{-8}	1.4787e−05		2.9465e−06		5.2397e−06	

Table 4 The error and order of convergence in the space direction with $\tau = 2^{-11}$

α	β	h	$\|z\|_{N_2(Q_{h\tau})}$	$\log_2 \frac{\|z\|_{N_2(Q_{h\tau})}}{\|z\|_{N_2(Q_{h/2\tau})}}$	$\|z\|_{L_2(Q_{h\tau})}$	$\log_2 \frac{\|z\|_{L_2(Q_{h\tau})}}{\|z\|_{L_2(Q_{h/2\tau})}}$	$\|z\|_{C(Q_{h\tau})}$	$\log_2 \frac{\|z\|_{C(Q_{h\tau})}}{\|z\|_{C(Q_{h/2\tau})}}$
1.8	1.5	2^{-3}	1.6322e−02	1.96	8.9855e−04	1.97	4.5010e−03	1.98
		2^{-4}	4.2093e−03	1.96	2.2961e−04	1.91	1.1425e−03	1.94
		2^{-5}	1.0790e−03	1.88	6.1139e−05	1.62	2.9729e−04	1.79
		2^{-6}	2.9288e−04	1.50	1.9822e−05	0.86	8.6143e−05	1.34
		2^{-7}	1.0326e−04		1.0731e−06		3.4065e−05	
1.6	1.4	2^{-3}	1.6936e−02	1.97	1.0017e−03	1.99	4.9269e−03	2.00
		2^{-4}	4.3374e−03	1.99	2.5207e−04	1.99	1.2352e−03	1.99
		2^{-5}	1.0938e−03	1.98	6.3657e−05	1.94	3.1158e−04	1.97
		2^{-6}	2.7713e−04	1.93	1.6538e−05	1.77	7.9794e−05	1.87
		2^{-7}	7.2878e−05		4.8603e−06		2.1796e−05	
1.3	1.1	2^{-3}	1.8206e−02	1.97	1.1968e−03	2.00	5.6657e−03	2.00
		2^{-4}	4.6410e−03	1.99	2.9948e−04	2.00	1.4151e−03	1.99
		2^{-5}	1.1659e−03	2.00	7.4904e−05	2.00	3.5725e−04	2.00
		2^{-6}	2.9200e−04	2.00	1.8760e−05	1.99	8.9398e−05	1.99
		2^{-7}	7.3197e−05		4.7239e−06		2.2440e−05	

References

1. R. Adams, *Sobolev Spaces* (Academic Press, New York, 1975)
2. A.A. Alikhanov, Boundary value problems for the diffusion equation of the variable order in differential and difference settings. Appl. Math. Comput. **219**, 3938–3946 (2012)
3. A. Delić, B. S. Jovanović, Numerical approximation of an interface problem for fractional in time diffusion equation. Appl. Math. Comput. **229**, 467–479 (2014)
4. A. Delić, B.S. Jovanović, S. Živanović, Finite difference approximation of a generalized time-fractional telegraph equation. Comput. Mathods Appl. Math. (2019)
5. V.J. Ervin, J.P. Roop, Variational formulation for the stationary fractional advection dispersion equation. Numer. Methods Partial Differ. Equ. **22**(3), 558–576 (2006)
6. B. Jin, R. Lazarov, J. Pasciak, W. Rundell, Variational formulation of problems involving fractional order differential operators. Math. Comp. **84**(296), 2665–2700 (2015)
7. B.S. Jovanović and E. Süli, in *Analysis of Finite Difference Schemes: For Linear Partial Differential Equations with Generalized Solutions*. Springer Series in Computational Mathematics, vol. 46 (Springer, London, 2014)
8. A. Kilbas, H. Srivastava, J. Trujillo, *Theory and Applications of Fractional Differential Equations* (Elsevier Science and Technology, Boston, 2006)
9. X. Li, C. Xu, A space-time spectral method for the time fractional diffusion equation. SIAM J. Numer. Anal. **47**(3), 2108–2131 (2009)
10. X. Li, C. Xu, Existence and uniqueness of the weak solution of the space-time fractional diffusion equation and a spectral method approximation. Commun. Comput. Phys. **8**(8), 1016–1051 (2010)
11. K. Oldham, J. Spanier, *The Fractional Calculus: Theory and Applications of Differentiation and Integration to Arbitrary Order* (Academic Press, New York, 1974)
12. E. Orsingher, L. Beghin, Time-fractional telegraph equations and telegraph processes with Brownian time. Probab. Theory Relat. Fields **128**, 141–160 (2004)
13. I. Podlubny, *Fractional Differential Equations* (Academic Press, London, 1998)
14. A.A. Samarskii, in *The Theory of Difference Schemes*. Monographs and Textbooks in Pure and Applied Mathematics, vol. 240 (Marcel Dekker, New York, 2001)
15. Z.-Z. Sun, X. Wu, A fully discrete difference scheme for a diffusion-wave system. Appl. Numer. Math. **56**(2), 193–209 (2006)
16. S. Yakubovich, M.M. Rodrigues, Fundamental solutions of the fractional two-parameter telegraph equation. Integral Transforms Spec. Funct. **23**(7), 509–519 (2012)
17. Y.H. Youssri, W.M. Abd-Elhameed, Numerical spectral Legendre-Galerkin algorithm for solving time fractional telegraph equation. Rom. J. Phys. **63**, 107 (2018)

Approximating the Integral of Analytic Complex Functions on Paths from Convex Domains in Terms of Generalized Ostrowski and Trapezoid Type Rules

Silvestru Sever Dragomir

Abstract In this paper we establish some results in approximating the integral of analytic complex functions on paths from convex domains in terms of generalized Ostrowski and Trapezoid type rules. Error bounds for these expansions in terms of p-norms are also provided. Examples for the complex logarithm and the complex exponential are also given.

1991 Mathematics Subject Classification 30A10, 26D15, 26D10

1 Introduction

Let $f : D \subseteq \mathbb{C} \to \mathbb{C}$ be an analytic function on the convex domain D and $z, x \in D$, then we have the following Taylor's expansion with integral remainder:

$$f(z) = \sum_{k=0}^{n} \frac{1}{k!} f^{(k)}(x)(z-x)^k + \frac{1}{n!}(z-x)^{n+1} \int_0^1 f^{(n+1)}[(1-s)x + sz](1-s)^n ds \tag{1.1}$$

for $n \geq 0$, see for instance [15].

Consider the function $f(z) = \operatorname{Log}(z)$ where $\operatorname{Log}(z) = \ln|z| + i \operatorname{Arg}(z)$ and $\operatorname{Arg}(z)$ is such that $-\pi < \operatorname{Arg}(z) \leq \pi$. Log is called the *"principal branch"* of the complex logarithmic function. The function f is analytic on all of $\mathbb{C}_\ell :=$

S. S. Dragomir (✉)
Mathematics, College of Engineering & Science, Victoria University, Melbourne, VIC, Australia

DST-NRF Centre of Excellence in the Mathematical and Statistical Sciences, School of Computer Science & Applied Mathematics, University of the Witwatersrand, Johannesburg, South Africa
e-mail: sever.dragomir@vu.edu.au; http://rgmia.org/dragomir

N. J. Daras, T. M. Rassias (eds.), *Computational Mathematics and Variational Analysis*, Springer Optimization and Its Applications 159,
https://doi.org/10.1007/978-3-030-44625-3_5

$\mathbb{C}\backslash\{x+iy : x \le 0,\ y=0\}$ and

$$f^{(k)}(z) = \frac{(-1)^{k-1}(k-1)!}{z^k},\ k \ge 1,\ z \in \mathbb{C}_\ell.$$

Using the representation (1.1) we then have

$$\operatorname{Log}(z) = \operatorname{Log}(x) + \sum_{k=1}^{n} \frac{(-1)^{k-1}}{k}\left(\frac{z-x}{x}\right)^k + (-1)^n (z-x)^{n+1} \int_0^1 \frac{(1-s)^n\, ds}{[(1-s)x+sz]^{n+1}} \tag{1.2}$$

for all $z, x \in \mathbb{C}_\ell$ with $(1-s)x+sz \in \mathbb{C}_\ell$ for $s \in [0,1]$.

Consider the complex exponential function $f(z) = \exp(z)$, then by (1.1) we get

$$\exp(z) = \sum_{k=0}^{n} \frac{1}{k!}(z-x)^k \exp(x) + \frac{1}{n!}(z-x)^{n+1} \int_0^1 (1-s)^n \exp[(1-s)x+sz]\, ds \tag{1.3}$$

for all $z, x \in \mathbb{C}$.

For various inequalities related to Taylor's expansions for real functions see [1–14].

Suppose γ is a *smooth path* parametrized by $z(t)$, $t \in [a,b]$ and f is a complex function which is continuous on γ. Put $z(a) = u$ and $z(b) = w$ with $u, w \in \mathbb{C}$. We define the integral of f on $\gamma_{u,w} = \gamma$ as

$$\int_\gamma f(z)\, dz = \int_{\gamma_{u,w}} f(z)\, dz := \int_a^b f(z(t))\, z'(t)\, dt.$$

We observe that the actual choice of parametrization of γ does not matter.

This definition immediately extends to paths that are *piecewise smooth*. Suppose γ is parametrized by $z(t)$, $t \in [a,b]$, which is differentiable on the intervals $[a,c]$ and $[c,b]$, then assuming that f is continuous on γ we define

$$\int_{\gamma_{u,w}} f(z)\, dz := \int_{\gamma_{u,v}} f(z)\, dz + \int_{\gamma_{v,w}} f(z)\, dz,$$

where $v := z(c)$. This can be extended for a finite number of intervals.

We also define the integral with respect to arc-length

$$\int_{\gamma_{u,w}} f(z)\,|dz| := \int_a^b f(z(t))\,|z'(t)|\,dt$$

and the length of the curve γ is then

$$\ell(\gamma) = \int_{\gamma_{u,w}} |dz| = \int_a^b |z'(t)|\,dt.$$

Let f and g be holomorphic in G, an open domain and suppose $\gamma \subset G$ is a piecewise smooth path from $z(a) = u$ to $z(b) = w$. Then we have the *integration by parts formula*

$$\int_{\gamma_{u,w}} f(z)\,g'(z)\,dz = f(w)\,g(w) - f(u)\,g(u) - \int_{\gamma_{u,w}} f'(z)\,g(z)\,dz. \tag{1.4}$$

We recall also the *triangle inequality* for the complex integral, namely

$$\left|\int_\gamma f(z)\,dz\right| \le \int_\gamma |f(z)|\,|dz| \le \|f\|_{\gamma,\infty}\,\ell(\gamma), \tag{1.5}$$

where $\|f\|_{\gamma,\infty} := \sup_{z\in\gamma} |f(z)|$.

We also define the p-norm with $p \ge 1$ by

$$\|f\|_{\gamma,p} := \left(\int_\gamma |f(z)|^p\,|dz|\right)^{1/p}.$$

For $p = 1$ we have

$$\|f\|_{\gamma,1} := \int_\gamma |f(z)|\,|dz|.$$

If $p, q > 1$ with $\frac{1}{p} + \frac{1}{q} = 1$, then by Hölder's inequality we have

$$\|f\|_{\gamma,1} \le [\ell(\gamma)]^{1/q}\,\|f\|_{\gamma,p}.$$

In this paper we establish some results in approximating the integral of analytic complex functions on paths from convex domains in terms of generalized Ostrowski and Trapezoid type rules. Error bounds for these expansions in terms of p-norms are also provided. Examples for the complex logarithm and the complex exponential are also given.

2 Ostrowski and Trapezoid Type Equalities

We have:

Theorem 1 *Let $f : D \subseteq \mathbb{C} \to \mathbb{C}$ be an analytic function on the convex domain D and $x \in D$. Suppose $\gamma \subset D$ is a smooth path parametrized by $z(t)$, $t \in [a, b]$ with $z(a) = u$ and $z(b) = w$ where $u, w \in D$. Then we have the Ostrowski type equality*

$$\int_\gamma f(z)\,dz = \sum_{k=0}^{n} \frac{1}{(k+1)!} f^{(k)}(x) \left[(w-x)^{k+1} + (-1)^k (x-u)^{k+1}\right] + R_n(x,\gamma), \tag{2.1}$$

where the remainder $R_n(x,\gamma)$ is given by

$$\begin{aligned} R_n(x,\gamma) &:= \frac{1}{n!}\int_\gamma (z-x)^{n+1}\left(\int_0^1 f^{(n+1)}[(1-s)x+sz](1-s)^n\,ds\right)dz \\ &= \frac{1}{n!}\int_0^1 \left(\int_\gamma (z-x)^{n+1} f^{(n+1)}[(1-s)x+sz]\,dz\right)(1-s)^n\,ds. \end{aligned} \tag{2.2}$$

Proof If we take the integral on the path $\gamma = \gamma_{u,w}$ in the equality (1.1), then we get

$$\begin{aligned} \int_\gamma f(z)\,dz &= \sum_{k=0}^{n} \frac{1}{k!} f^{(k)}(x) \int_{\gamma_{u,w}} (z-x)^k\,dz \\ &\quad + \frac{1}{n!}\int_{\gamma_{u,w}} (z-x)^{n+1}\left(\int_0^1 f^{(n+1)}[(1-s)x+sz](1-s)^n\,ds\right)dz \\ &= \sum_{k=0}^{n} \frac{1}{k!} f^{(k)}(x) \frac{(w-x)^{k+1}-(u-x)^{k+1}}{k+1} \\ &\quad + \frac{1}{n!}\int_{\gamma_{u,w}} (z-x)^{n+1}\left(\int_0^1 f^{(n+1)}[(1-s)x+sz](1-s)^n\,ds\right)dz \\ &= \sum_{k=0}^{n} \frac{1}{(k+1)!} f^{(k)}(x) \left[(w-x)^{k+1} + (-1)^{k+2}(x-u)^{k+1}\right] \\ &\quad + \frac{1}{n!}\int_{\gamma_{u,w}} (z-x)^{n+1}\left(\int_0^1 f^{(n+1)}[(1-s)x+sz](1-s)^n\,ds\right)dz, \end{aligned} \tag{2.3}$$

which proves the equality (2.1) with the first representation of the remainder from (2.2).

The second representation in (2.2) follows by Fubini's theorem. □

Corollary 1 *With the assumptions of Theorem 1 we have the midpoint equality*

$$\int_\gamma f(z)\,dz = \sum_{k=0}^{n} \frac{1}{2^k (k+1)!} f^{(k)}\left(\frac{u+w}{2}\right)\left[\frac{1+(-1)^k}{2}\right](w-u)^{k+1} + M_n(\gamma), \tag{2.4}$$

where the remainder $R_n(x,\gamma)$ *is given by*

$$\begin{aligned} M_n(\gamma) &:= \frac{1}{n!}\int_\gamma \left(z - \frac{u+w}{2}\right)^{n+1}\left(\int_0^1 f^{(n+1)}\left[(1-s)\frac{u+w}{2} + sz\right](1-s)^n\,ds\right)dz \\ &= \frac{1}{n!}\int_0^1 \left(\int_\gamma \left(z - \frac{u+w}{2}\right)^{n+1} f^{(n+1)}\left[(1-s)\frac{u+w}{2} + sz\right]dz\right)(1-s)^n\,ds. \end{aligned} \tag{2.5}$$

The proof follows from Theorem 1 by taking $x = \frac{u+w}{2} \in D$.

Corollary 2 *With the assumptions of Theorem 1 and if* $\lambda \in \mathbb{C}$, *then we have the weighted trapezoid equality*

$$\int_\gamma f(z)\,dz = \sum_{k=0}^{n} \frac{1}{(k+1)!}\left[\lambda f^{(k)}(u) + (1-\lambda)(-1)^k f^{(k)}(w)\right](w-u)^{k+1} + T_n(\lambda,\gamma), \tag{2.6}$$

where the remainder $T_n(\lambda,\gamma)$ *is given by*

$$\begin{aligned} T_n(\lambda,\gamma) &:= \frac{\lambda}{n!}\int_\gamma (z-u)^{n+1}\left(\int_0^1 f^{(n+1)}[(1-s)u + sz](1-s)^n\,ds\right)dz \\ &\quad + \frac{(1-\lambda)}{n!}\int_\gamma (z-w)^{n+1}\left(\int_0^1 f^{(n+1)}[(1-s)w + sz](1-s)^n\,ds\right)dz \\ &= \frac{\lambda}{n!}\int_0^1\left(\int_\gamma (z-u)^{n+1} f^{(n+1)}[(1-s)u + sz]\,dz\right)(1-s)^n\,ds \\ &\quad + \frac{(1-\lambda)}{n!}\int_0^1\left(\int_\gamma (z-w)^{n+1} f^{(n+1)}[(1-s)w + sz]\,dz\right)(1-s)^n\,ds. \end{aligned} \tag{2.7}$$

In particular, for $\lambda = \frac{1}{2}$ we have the trapezoid equality

$$\int_\gamma f(z)\,dz = \sum_{k=0}^{n} \frac{1}{(k+1)!} \left[\frac{f^{(k)}(u) + (-1)^k f^{(k)}(w)}{2} \right] (w-u)^{k+1} + T_n(\gamma), \tag{2.8}$$

where the remainder $T_n(\gamma)$ is given by

$$\begin{aligned} T_n(\gamma) &:= \frac{1}{2n!} \int_\gamma (z-u)^{n+1} \left(\int_0^1 f^{(n+1)}[(1-s)u+sz](1-s)^n\,ds \right) dz \\ &+ \frac{1}{2n!} \int_\gamma (z-w)^{n+1} \left(\int_0^1 f^{(n+1)}[(1-s)w+sz](1-s)^n\,ds \right) dz \\ &= \frac{1}{2n!} \int_0^1 \left(\int_\gamma (z-u)^{n+1} f^{(n+1)}[(1-s)u+sz]\,dz \right) (1-s)^n\,ds \\ &+ \frac{1}{2n!} \int_0^1 \left(\int_\gamma (z-w)^{n+1} f^{(n+1)}[(1-s)w+sz]\,dz \right) (1-s)^n\,ds. \end{aligned} \tag{2.9}$$

Proof We write the equality (2.1) for $x = u$ to get

$$\int_\gamma f(z)\,dz = \sum_{k=0}^{n} \frac{1}{(k+1)!} f^{(k)}(u)(w-u)^{k+1} + R_n(u,\gamma), \tag{2.10}$$

where the remainder $R_n(u,\gamma)$ is given by

$$\begin{aligned} R_n(u,\gamma) &:= \frac{1}{n!} \int_\gamma (z-u)^{n+1} \left(\int_0^1 f^{(n+1)}[(1-s)u+sz](1-s)^n\,ds \right) dz \\ &= \frac{1}{n!} \int_0^1 \left(\int_\gamma (z-u)^{n+1} f^{(n+1)}[(1-s)u+sz]\,dz \right) (1-s)^n\,ds, \end{aligned} \tag{2.11}$$

and for $x = w$ to get

$$\int_\gamma f(z)\,dz = \sum_{k=0}^{n} \frac{(-1)^k}{(k+1)!} f^{(k)}(w)(w-u)^{k+1} + R_n(w,\gamma), \tag{2.12}$$

where the remainder $R_n(w,\gamma)$ is given by

$$R_n(w,\gamma) := \frac{1}{n!}\int_\gamma (z-w)^{n+1}\left(\int_0^1 f^{(n+1)}[(1-s)w+sz](1-s)^n\, ds\right)dz$$

$$= \frac{1}{n!}\int_0^1 \left(\int_\gamma (z-w)^{n+1} f^{(n+1)}[(1-s)w+sz]\, dz\right)(1-s)^n\, ds. \tag{2.13}$$

Now, if we multiply the equality (2.10) by λ and the equality (2.13) by $1-\lambda$ and sum, then we obtain the desired result. □

Remark 1 Let $f : D \subseteq \mathbb{C} \to \mathbb{C}$ be an analytic function on the convex domain D and $x \in D$. Suppose $\gamma \subset D$ is a *smooth path* parametrized by $z(t)$, $t \in [a,b]$ with $z(a) = u$ and $z(b) = w$ where $u, w \in D$.

If we take $n = 0$ in Theorem 1, then we obtain the Ostrowski type equality

$$\int_\gamma f(z)\, dz = f(x)(w-u) + R(x,\gamma), \tag{2.14}$$

where the remainder $R(x,\gamma)$ is given by

$$R(x,\gamma) := \int_\gamma (z-x)\left(\int_0^1 f'[(1-s)x+sz]\, ds\right)dz$$

$$= \int_0^1 \left(\int_\gamma (z-x) f'[(1-s)x+sz]\, dz\right)ds. \tag{2.15}$$

In particular, for $x = \frac{u+w}{2}$ we have the midpoint type equality

$$\int_\gamma f(z)\, dz = f\left(\frac{u+w}{2}\right)(w-u) + R(\gamma), \tag{2.16}$$

where the remainder $R(\gamma)$ is given by

$$R(\gamma) := \int_\gamma \left(z - \frac{u+w}{2}\right)\left(\int_0^1 f'\left[(1-s)\frac{u+w}{2} + sz\right] ds\right)dz$$

$$= \int_0^1 \left(\int_\gamma \left(z - \frac{u+w}{2}\right) f'\left[(1-s)\frac{u+w}{2} + sz\right] dz\right)ds. \tag{2.17}$$

If we take $n = 0$ in Corollary 2, then we obtain the weighted trapezoid equality for $\lambda \in \mathbb{C}$

$$\int_\gamma f(z)\, dz = [\lambda f(u) + (1-\lambda) f(w)](w-u) + T(\lambda,\gamma), \tag{2.18}$$

where the remainder $T(\lambda,\gamma)$ is given by

$$\begin{aligned}T(\lambda,\gamma) &:= \lambda\int_\gamma (z-u)\left(\int_0^1 f'[(1-s)u+sz]\,ds\right)dz\\&\quad+(1-\lambda)\int_\gamma (z-w)\left(\int_0^1 f'[(1-s)w+sz]\,ds\right)dz\\&=\lambda\int_0^1\left(\int_\gamma (z-u)f'[(1-s)u+sz]\,dz\right)ds\\&\quad+(1-\lambda)\int_0^1\left(\int_\gamma (z-w)f'[(1-s)w+sz]\,dz\right)ds.\end{aligned}\tag{2.19}$$

In particular, for $\lambda=\frac{1}{2}$ we have the trapezoid type equality

$$\int_\gamma f(z)\,dz=\frac{f(u)+f(w)}{2}(w-u)+T(\gamma),\tag{2.20}$$

where the remainder $T(\gamma)$ is given by

$$\begin{aligned}T(\gamma) &:= \frac{1}{2}\int_\gamma (z-u)\left(\int_0^1 f'[(1-s)u+sz]\,ds\right)dz\\&\quad+\frac{1}{2}\int_\gamma (z-w)\left(\int_0^1 f'[(1-s)w+sz]\,ds\right)dz\\&=\frac{1}{2}\int_0^1\left(\int_\gamma (z-u)f'[(1-s)u+sz]\,dz\right)ds\\&\quad+\frac{1}{2}\int_0^1\left(\int_\gamma (z-w)f'[(1-s)w+sz]\,dz\right)ds.\end{aligned}\tag{2.21}$$

Let $f: D\subseteq\mathbb{C}\to\mathbb{C}$ be an analytic function on the convex domain D and $x\in D$. Suppose $\gamma\subset D$ is a *smooth path* parametrized by $z(t)$, $t\in[a,b]$ with $z(a)=u$ and $z(b)=w$ where $u,w\in D$.

For $n=1$ in (2.1) we get the perturbed Ostrowski's equality

$$\int_\gamma f(z)\,dz=f(x)(w-u)+f'(x)\left(\frac{w+u}{2}-x\right)(w-u)+R_1(x,\gamma),\tag{2.22}$$

where the remainder $R_1(x,\gamma)$ is given by

$$R_1(x,\gamma) := \int_\gamma (z-x)^2 \left(\int_0^1 f''[(1-s)x+sz](1-s)\,ds\right) dz$$
$$= \int_0^1 \left(\int_\gamma (z-x)^2 f''[(1-s)x+sz]\,dz\right)(1-s)\,ds. \tag{2.23}$$

In particular, for $x = \frac{w+u}{2}$ we get the midpoint equality

$$\int_\gamma f(z)\,dz = f\left(\frac{w+u}{2}\right)(w-u) + M_1(\gamma), \tag{2.24}$$

where the remainder $M_1(\gamma)$ is given by

$$M_1(\gamma) := \int_\gamma \left(z-\frac{w+u}{2}\right)^2 \left(\int_0^1 f''\left[(1-s)\frac{w+u}{2}+sz\right](1-s)\,ds\right) dz$$
$$= \int_0^1 \left(\int_\gamma \left(z-\frac{w+u}{2}\right)^2 f''\left[(1-s)\frac{w+u}{2}+sz\right] dz\right)(1-s)\,ds. \tag{2.25}$$

If we take $n = 1$ in Corollary 2, then we get the perturbed weighted trapezoid equality for $\lambda \in \mathbb{C}$

$$\int_\gamma f(z)\,dz = [\lambda f(u) + (1-\lambda) f(w)](w-u)$$
$$+ \frac{1}{2}\left[\lambda f'(u) - (1-\lambda) f'(w)\right](w-u)^2 + T_1(\lambda,\gamma), \tag{2.26}$$

where

$$T_1(\lambda,\gamma) := \lambda \int_\gamma (z-u)^2 \left(\int_0^1 f''[(1-s)u+sz](1-s)\,ds\right) dz$$
$$+ (1-\lambda)\int_\gamma (z-w)^2 \left(\int_0^1 f''[(1-s)w+sz](1-s)\,ds\right) dz$$
$$= \lambda \int_0^1 \left(\int_\gamma (z-u)^2 f''[(1-s)u+sz]\,dz\right)(1-s)\,ds$$
$$+ (1-\lambda)\int_0^1 \left(\int_\gamma (z-w)^2 f''[(1-s)w+sz]\,dz\right)(1-s)\,ds. \tag{2.27}$$

In particular, for $\lambda = \frac{1}{2}$ we have the perturbed trapezoid type equality

$$\int_{\gamma} f(z)\,dz = \frac{f(u) + f(w)}{2}(w-u) + \frac{1}{4}\left[f'(u) - f'(w)\right](w-u)^2 \tag{2.28}$$
$$+ T_1(\gamma),$$

where

$$T_1(\gamma) := \frac{1}{2}\int_{\gamma} (z-u)^2 \left(\int_0^1 f''[(1-s)u + sz](1-s)\,ds\right) dz$$
$$+ \frac{1}{2}\int_{\gamma} (z-w)^2 \left(\int_0^1 f''[(1-s)w + sz](1-s)\,ds\right) dz$$
$$= \frac{1}{2}\int_0^1 \left(\int_{\gamma} (z-u)^2 f''[(1-s)u + sz]\,dz\right)(1-s)\,ds$$
$$+ \frac{1}{2}\int_0^1 \left(\int_{\gamma} (z-w)^2 f''[(1-s)w + sz]\,dz\right)(1-s)\,ds. \tag{2.29}$$

Consider the function $f(z) = \frac{1}{z}$, $z \in \mathbb{C}\setminus\{0\}$. Then

$$f^{(k)}(z) = \frac{(-1)^k k!}{z^{k+1}} \text{ for } k \geq 0,\ z \in \mathbb{C}\setminus\{0\}$$

and suppose $\gamma \subset \mathbb{C}_{\ell}$ is a *smooth path* parametrized by $z(t)$, $t \in [a,b]$ with $z(a) = u$ and $z(b) = w$ where $u, w \in \mathbb{C}_{\ell}$. Then

$$\int_{\gamma} f(z)\,dz = \int_{\gamma_{u,w}} f(z)\,dz = \int_{\gamma_{u,w}} \frac{dz}{z} = \mathrm{Log}(w) - \mathrm{Log}(u)$$

for $u, w \in \mathbb{C}_{\ell}$.

Let D be a convex domain included in $\mathbb{C}_{\ell}$. Assume that $\gamma = \gamma_{u,w} \subset D$ and $x \in D$. Then by Theorem 1 we have

$$\mathrm{Log}(w) - \mathrm{Log}(u) = \sum_{k=0}^{n} \frac{(-1)^k}{(k+1)}\left[\left(\frac{w-x}{x}\right)^{k+1} + (-1)^k\left(\frac{x-u}{x}\right)^{k+1}\right] \tag{2.30}$$
$$+ R_n(x,\gamma),$$

where the remainder $R_n(x,\gamma)$ is given by

$$R_n(x,\gamma) := (n+1)(-1)^{n+1}\int_{\gamma_{u,w}} (z-x)^{n+1}\left(\int_0^1 \frac{(1-s)^n\,ds}{[(1-s)x + sz]^{n+2}}\right) dz$$

$$= (n+1)(-1)^{n+1} \int_0^1 \left(\int_{\gamma_{u,w}} \frac{(z-x)^{n+1}}{[(1-s)x+sz]^{n+2}} dz \right) (1-s)^n \, ds, \tag{2.31}$$

for $n \geq 0$.

Consider the function $f(z) = \mathrm{Log}(z)$, the *"principal branch"* of the complex logarithmic function. The function f is analytic on all of $\mathbb{C}_\ell := \mathbb{C} \backslash \{x+iy : x \leq 0,\ y = 0\}$ and

$$f^{(k)}(z) = \frac{(-1)^{k-1}(k-1)!}{z^k},\ k \geq 1,\ z \in \mathbb{C}_\ell.$$

Suppose $\gamma \subset \mathbb{C}_\ell$ is a *smooth path* parametrized by $z(t)$, $t \in [a,b]$ with $z(a) = u$ and $z(b) = w$ where $u, w \in \mathbb{C}_\ell$. Then

$$\begin{aligned}\int_\gamma f(z)\,dz &= \int_{\gamma_{u,w}} f(z)\,dz = \int_{\gamma_{u,w}} \mathrm{Log}(z)\,dz = \\ &= z\,\mathrm{Log}(z)|_u^w - \int_{\gamma_{u,w}} (\mathrm{Log}(z))' z\,dz \\ &= w\,\mathrm{Log}(w) - u\,\mathrm{Log}(u) - \int_{\gamma_{u,w}} dz \\ &= w\,\mathrm{Log}(w) - u\,\mathrm{Log}(u) - (w-u),\end{aligned}$$

where $u, w \in \mathbb{C}_\ell$.

Let D be a convex domain included in $\mathbb{C}_\ell$. Assume that $\gamma = \gamma_{u,w} \subset D$ and $x \in D$. Then by Theorem 1 we have

$$\begin{aligned}\int_\gamma f(z)\,dz &= f(x)(w-u) \\ &+ \sum_{k=1}^{n} \frac{1}{(k+1)!} f^{(k)}(x) \left[(w-x)^{k+1} + (-1)^k (x-u)^{k+1} \right] + R_n(x,\gamma),\end{aligned} \tag{2.32}$$

which gives

$$\begin{aligned}&w\,\mathrm{Log}(w) - u\,\mathrm{Log}(u) - (w-u) \\ &= (w-u)\,\mathrm{Log}(x) + x \sum_{k=1}^{n} \frac{(-1)^{k-1}}{(k+1)k} \left[\left(\frac{w-x}{x} \right)^{k+1} + (-1)^k \left(\frac{x-u}{x} \right)^{k+1} \right] \\ &\qquad + R_n(x,\gamma),\end{aligned} \tag{2.33}$$

where

$$R_n(x,\gamma) := (-1)^n \int_\gamma (z-x)^{n+1} \left(\int_0^1 \frac{(1-s)^n\, ds}{[(1-s)x+sz]^{n+1}} \right) dz$$
$$= (-1)^n \int_0^1 \left(\int_\gamma \frac{(z-x)^{n+1}}{[(1-s)x+sz]^{n+1}} dz \right) (1-s)^n\, ds, \tag{2.34}$$

for $n \geq 1$.

Consider the function $f(z) = \exp(z)$, $z \in \mathbb{C}$. Then

$$f^{(k)}(z) = \exp(z) \text{ for } k \geq 0,\ z \in \mathbb{C}$$

and suppose $\gamma \subset \mathbb{C}$ is a *smooth path* parametrized by $z(t)$, $t \in [a,b]$ with $z(a) = u$ and $z(b) = w$ where $u, w \in \mathbb{C}$. Then

$$\int_\gamma f(z)\, dz = \int_{\gamma_{u,w}} f(z)\, dz = \int_{\gamma_{u,w}} \exp(z)\, dz = \exp(w) - \exp(u).$$

By Theorem 1 we get

$$\exp(w) - \exp(u) = \exp(x) \sum_{k=0}^{n} \frac{1}{(k+1)!} \left[(w-x)^{k+1} + (-1)^k (x-u)^{k+1} \right] \tag{2.35}$$
$$+ R_n(x,\gamma),$$

where the remainder $R_n(x,\gamma)$ is given by

$$R_n(x,\gamma) := \frac{1}{n!} \int_\gamma (z-x)^{n+1} \left(\int_0^1 \exp[(1-s)x+sz](1-s)^n\, ds \right) dz$$
$$= \frac{1}{n!} \int_0^1 \left(\int_\gamma (z-x)^{n+1} \exp[(1-s)x+sz]\, dz \right) (1-s)^n\, ds \tag{2.36}$$

for $n \geq 0$.

3 Error Bounds for Ostrowski's Rule

We have the following error bounds:

Theorem 2 *Let $f : D \subseteq \mathbb{C} \to \mathbb{C}$ be an analytic function on the convex domain D and $x \in D$. Suppose $\gamma \subset D$ is a smooth path parametrized by $z(t)$, $t \in [a, b]$ with $z(a) = u$ and $z(b) = w$ where $u, w \in D$. Then we have the representation (2.1) where the remainder $R_n(x, \gamma)$, $n \geq 0$ satisfies the bounds*

$$|R_n(x,\gamma)| \leq \frac{1}{n!}\int_\gamma |z-x|^{n+1}\left(\int_0^1 \left|f^{(n+1)}[(1-s)x+sz]\right|(1-s)^n ds\right)|dz| \leq B_n(x,\gamma), \quad (3.1)$$

where

$$B_n(x,\gamma) := \frac{1}{n!}\begin{cases} \frac{1}{n+1}\int_\gamma |z-x|^{n+1}\left(\max_{s\in[0,1]}\left|f^{(n+1)}[(1-s)x+sz]\right|\right)|dz|; \\ \frac{1}{(qn+1)^{1/q}}\int_\gamma |z-x|^{n+1}\left(\int_0^1 \left|f^{(n+1)}[(1-s)x+sz]\right|^p ds\right)^{1/p}|dz| \\ p,\ q>1 \text{ with } \frac{1}{p}+\frac{1}{q}=1; \\ \int_\gamma |z-x|^{n+1}\left(\int_0^1 \left|f^{(n+1)}[(1-s)x+sz]\right| ds\right)|dz|. \end{cases} \quad (3.2)$$

Moreover, we have

$$B_n(x,\gamma) \leq \frac{1}{(n+1)!} \times \begin{cases} \max_{s\in[0,1],z\in\gamma}\left|f^{(n+1)}[(1-s)x+sz]\right|\int_\gamma |z-x|^{n+1}|dz| \\ \left(\int_\gamma |z-x|^{\alpha(n+1)}|dz|\right)^{1/\alpha}\left[\int_\gamma \left(\max_{s\in[0,1]}\left|f^{(n+1)}[(1-s)x+sz]\right|\right)^\beta |dz|\right]^{1/\beta} \\ \alpha,\ \beta>1 \text{ with } \frac{1}{\alpha}+\frac{1}{\beta}=1; \\ \max_{z\in\gamma}|z-x|^{n+1}\int_\gamma \max_{s\in[0,1]}\left|f^{(n+1)}[(1-s)x+sz]\right||dz|, \end{cases} \quad (3.3)$$

$$B_n(x,\gamma) \leq \frac{1}{n!\,(qn+1)^{1/q}}$$

$$\times \begin{cases} \max_{z\in\gamma} \left(\int_0^1 \left|f^{(n+1)}[(1-s)x+sz]\right|^p ds\right)^{1/p} \int_\gamma |z-x|^{n+1}|dz| \\ \left(\int_\gamma |z-x|^{\alpha(n+1)}|dz|\right)^{1/\alpha} \left[\left(\int_0^1 \left|f^{(n+1)}[(1-s)x+sz]\right|^p ds\right)^{\beta/p}|dz|\right]^{1/\beta} \\ \alpha,\ \beta>1 \text{ with } \frac{1}{\alpha}+\frac{1}{\beta}=1; \\ \max_{z\in\gamma}|z-x|^{n+1}\int_\gamma \left(\int_0^1 \left|f^{(n+1)}[(1-s)x+sz]\right|^p ds\right)^{1/p}|dz| \end{cases} \tag{3.4}$$

and

$$B_n(x,\gamma) \le \frac{1}{n!}\begin{cases} \max_{z\in\gamma}\int_0^1 \left|f^{(n+1)}[(1-s)x+sz]\right| ds \int_\gamma |z-x|^{n+1}|dz| \\ \left(\int_\gamma |z-x|^{\alpha(n+1)}|dz|\right)^{1/\alpha}\left[\left(\int_0^1 \left|f^{(n+1)}[(1-s)x+sz]\right| ds\right)^{\beta}|dz|\right]^{1/\beta} \\ \alpha,\ \beta>1 \text{ with } \frac{1}{\alpha}+\frac{1}{\beta}=1; \\ \max_{z\in\gamma}|z-x|^{n+1}\int_\gamma\left(\int_0^1 \left|f^{(n+1)}[(1-s)x+sz]\right| ds\right)|dz|. \end{cases} \tag{3.5}$$

Proof Taking the modulus in the first representation in (2.2) we get

$$\begin{aligned} |R_n(x,\gamma)| &= \frac{1}{n!}\left|\int_\gamma (z-x)^{n+1}\left(\int_0^1 f^{(n+1)}[(1-s)x+sz](1-s)^n ds\right)dz\right| \\ &\le \frac{1}{n!}\int_\gamma \left|(z-x)^{n+1}\left(\int_0^1 f^{(n+1)}[(1-s)x+sz](1-s)^n ds\right)\right||dz| \\ &= \frac{1}{n!}\int_\gamma |z-x|^{n+1}\left|\int_0^1 f^{(n+1)}[(1-s)x+sz](1-s)^n ds\right||dz| \\ &\le \frac{1}{n!}\int_\gamma |z-x|^{n+1}\left(\int_0^1 \left|f^{(n+1)}[(1-s)x+sz]\right|(1-s)^n ds\right)|dz| \\ &=: A_n(x,\gamma) \end{aligned}$$

for $x \in D$.

Using Hölder's integral inequality we get

$$\int_0^1 \left| f^{(n+1)}\left[(1-s)x+sz\right]\right| (1-s)^n\, ds$$

$$\leq \begin{cases} \max_{s\in[0,1]} \left|f^{(n+1)}\left[(1-s)x+sz\right]\right| \int_0^1 (1-s)^n\, ds; \\ \\ \left(\int_0^1 \left|f^{(n+1)}\left[(1-s)x+sz\right]\right|^p ds\right)^{1/p} \left(\int_0^1 (1-s)^{qn}\, ds\right)^{1/q} \\ p,\ q>1 \text{ with } \frac{1}{p}+\frac{1}{q}=1; \\ \\ \int_0^1 \left|f^{(n+1)}\left[(1-s)x+sz\right]\right| ds \end{cases}$$

$$= \begin{cases} \frac{1}{n+1} \max_{s\in[0,1]} \left|f^{(n+1)}\left[(1-s)x+sz\right]\right|; \\ \\ \frac{1}{(qn+1)^{1/q}} \left(\int_0^1 \left|f^{(n+1)}\left[(1-s)x+sz\right]\right|^p ds\right)^{1/p} \\ p,\ q>1 \text{ with } \frac{1}{p}+\frac{1}{q}=1; \\ \\ \int_0^1 \left|f^{(n+1)}\left[(1-s)x+sz\right]\right| ds. \end{cases}$$

Therefore

$$A_n(x,\gamma) \leq \frac{1}{n!} \begin{cases} \frac{1}{n+1} \int_\gamma |z-x|^{n+1} \left(\max_{s\in[0,1]} \left|f^{(n+1)}\left[(1-s)x+sz\right]\right|\right) |dz|; \\ \\ \frac{1}{(qn+1)^{1/q}} \int_\gamma |z-x|^{n+1} \left(\int_0^1 \left|f^{(n+1)}\left[(1-s)x+sz\right]\right|^p ds\right)^{1/p} |dz| \\ p,\ q>1 \text{ with } \frac{1}{p}+\frac{1}{q}=1; \\ \\ \int_\gamma |z-x|^{n+1} \left(\int_0^1 \left|f^{(n+1)}\left[(1-s)x+sz\right]\right| ds\right) |dz| \end{cases}$$

for $x \in D$, which proves the second bound in (3.1).

The bounds (3.3)–(3.5) follows by Hölder's integral inequality. □

For a recent survey on Ostrowski type inequalities for functions of a real variable, see [6].

In a similar way we can prove:

Theorem 3 *Let $f : D \subseteq \mathbb{C} \to \mathbb{C}$ be an analytic function on the convex domain D and $x \in D$. Suppose $\gamma \subset D$ is a smooth path parametrized by $z(t)$, $t \in [a,b]$ with $z(a) = u$ and $z(b) = w$ where $u,\ w \in D$. Then we have the representation (2.1) where the remainder $R_n(x,\gamma)$, $n \geq 0$ satisfies the bounds*

$$|R_n(x,\gamma)| \leq \frac{1}{n!}\int_0^1 \left(\int_\gamma |z-x|^{n+1} \left|f^{(n+1)}\left[(1-s)x+sz\right]\right| |dz|\right)(1-s)^n\, ds$$
$$\leq C_n(x,\gamma), \qquad (3.6)$$

where

$$C_n(x,\gamma) := \begin{cases} \frac{1}{(n+1)!}\max_{s\in[0,1]}\int_\gamma |z-x|^{n+1}\left|f^{(n+1)}[(1-s)x+sz]\right||dz| \\ \frac{1}{n!(nq+1)^{1/q}}\int_0^1\left(\int_\gamma |z-x|^{n+1}\left|f^{(n+1)}[(1-s)x+sz]\right||dz|\right)^{1/p}ds \\ p,\ q>1 \text{ with } \frac{1}{p}+\frac{1}{q}=1; \\ \frac{1}{n!}\int_0^1\left(\int_\gamma |z-x|^{n+1}\left|f^{(n+1)}[(1-s)x+sz]\right||dz|\right)ds. \end{cases} \quad (3.7)$$

Moreover, we have

$$C_n(x,\gamma) \le \frac{1}{(n+1)!} \times \begin{cases} \max_{s\in[0,1],z\in\gamma}\left|f^{(n+1)}[(1-s)x+sz]\right|\int_\gamma |z-x|^{n+1}|dz| \\ \left(\int_\gamma |z-x|^{\alpha(n+1)}|dz|\right)^{1/\alpha}\max_{s\in[0,1]}\left(\int_\gamma\left|f^{(n+1)}[(1-s)x+sz]\right|^\beta|dz|\right)^{1/\beta} \\ \alpha,\ \beta>1 \text{ with } \frac{1}{\alpha}+\frac{1}{\beta}=1; \\ \max_{z\in\gamma}|z-x|^{n+1}\max_{s\in[0,1]}\int_\gamma\left|f^{(n+1)}[(1-s)x+sz]\right||dz|, \end{cases} \quad (3.8)$$

$$C_n(x,\gamma) \le \frac{1}{n!(nq+1)^{1/q}} \times \begin{cases} \max_{z\in\gamma}|z-x|^{(n+1)/p}\int_0^1\left(\int_\gamma\left|f^{(n+1)}[(1-s)x+sz]\right||dz|\right)^{1/p}ds \\ \left(\int_\gamma |z-x|^{\alpha(n+1)}|dz|\right)^{1/(p\alpha)}\int_0^1\left(\int_\gamma\left|f^{(n+1)}[(1-s)x+sz]\right|^\beta|dz|\right)^{\beta/p}ds \\ \alpha,\ \beta>1 \text{ with } \frac{1}{\alpha}+\frac{1}{\beta}=1; \\ \left(\int_\gamma |z-x|^{n+1}|dz|\right)^{1/p}\int_0^1\max_{z\in\gamma}\left|f^{(n+1)}[(1-s)x+sz]\right|^{1/p}ds \end{cases} \quad (3.9)$$

and

$$C_n(x,\gamma) \le \frac{1}{n!}\begin{cases} \max_{z\in\gamma}|z-x|^{n+1}\int_0^1\left(\int_\gamma\left|f^{(n+1)}[(1-s)x+sz]\right||dz|\right)ds \\ \\ \left(\int_\gamma|z-x|^{\alpha(n+1)}|dz|\right)^{1/\alpha}\int_0^1\left(\int_\gamma\left|f^{(n+1)}[(1-s)x+sz]\right|^\beta|dz|\right)^{1/\beta}ds \\ \alpha,\ \beta>1 \text{ with } \frac{1}{\alpha}+\frac{1}{\beta}=1; \\ \\ \int_\gamma|z-x|^{n+1}|dz|\int_0^1\max_{z\in\gamma}\left|f^{(n+1)}[(1-s)x+sz]\right|ds. \end{cases} \tag{3.10}$$

The following particular case may be useful for applications:

Corollary 3 *Let $f : D \subseteq \mathbb{C} \to \mathbb{C}$ be an analytic function on the convex domain D and $x \in D$. Suppose $\gamma \subset D$ is a smooth path parametrized by $z(t)$, $t \in [a,b]$ with $z(a) = u$ and $z(b) = w$ where u, $w \in D$. If*

$$\left\|f^{(n+1)}\right\|_{D,\infty} := \sup_{z\in D}\left|f^{(n+1)}(z)\right| < \infty \text{ for some } n \ge 0, \tag{3.11}$$

then we have the representation (2.1) where the remainder $R_n(x,\gamma)$ satisfies the bound

$$|R_n(x,\gamma)| \le \frac{1}{(n+1)!}\left\|f^{(n+1)}\right\|_{D,\infty}\int_\gamma|z-x|^{n+1}|dz|. \tag{3.12}$$

Remark 2 Let $f : D \subseteq \mathbb{C} \to \mathbb{C}$ be an analytic function on the convex domain D and $x \in D$. Suppose $\gamma \subset D$ is a *smooth path* parametrized by $z(t)$, $t \in [a,b]$ with $z(a) = u$ and $z(b) = w$ where u, $w \in D$.

For $n = 0$ we get the Ostrowski type inequality

$$\left|\int_\gamma f(z)\,dz - f(x)(w-u)\right| \le \left\|f'\right\|_{D,\infty}\int_\gamma|z-x|\,|dz| \tag{3.13}$$

for $x \in D$, provided $\left\|f'\right\|_{D,\infty} < \infty$.

In particular, we have the midpoint type inequality

$$\left|\int_\gamma f(z)\,dz - f\left(\frac{u+w}{2}\right)(w-u)\right| \le \left\|f'\right\|_{D,\infty}\int_\gamma\left|z-\frac{u+w}{2}\right||dz| \tag{3.14}$$

provided $\left\|f'\right\|_{D,\infty} < \infty$.

For $n = 1$ we get the perturbed Ostrowski type inequality

$$\left|\int_\gamma f(z)\,dz - f(x)(w-u) - f'(x)\left(\frac{w+u}{2} - x\right)(w-u)\right| \le \frac{1}{2}\|f''\|_{D,\infty}\int_\gamma |z-x|^2\,|dz| \tag{3.15}$$

for $x \in D$, provided $\|f''\|_{D,\infty} < \infty$.

In particular, we have the midpoint type inequality

$$\left|\int_\gamma f(z)\,dz - f\left(\frac{u+w}{2}\right)(w-u)\right| \le \frac{1}{2}\|f''\|_{D,\infty}\int_\gamma \left|z - \frac{u+w}{2}\right|^2 |dz| \tag{3.16}$$

provided $\|f''\|_{D,\infty} < \infty$.

We also have:

Theorem 4 *Let $f : D \subseteq \mathbb{C} \to \mathbb{C}$ be an analytic function on the convex domain D and $x \in D$. Suppose $\gamma \subset D$ is a smooth path parametrized by $z(t)$, $t \in [a,b]$ with $z(a) = u$ and $z(b) = w$ where $u, w \in D$. If $\left|f^{(n+1)}\right|$ is convex on D, for some $n \ge 0$, then we have the representation (2.1) where the remainder $R_n(x,\gamma)$ satisfies the bounds*

$$|R_n(x,\gamma)| \le \frac{1}{n!(n+2)} \times \left[\left|f^{(n+1)}(x)\right|\int_\gamma |z-x|^{n+1}|dz| + \frac{1}{(n+1)}\int_\gamma |z-x|^{n+1}\left|f^{(n+1)}(z)\right||dz|\right]. \tag{3.17}$$

Proof We have by (3.6) and by the convexity of $\left|f^{(n+1)}\right|$ that

$$|R_n(x,\gamma)| \le \frac{1}{n!}\int_\gamma |z-x|^{n+1}\left(\int_0^1 \left|f^{(n+1)}[(1-s)x + sz]\right|(1-s)^n\,ds\right)|dz|$$

$$\le \frac{1}{n!}\int_\gamma |z-x|^{n+1}\left(\int_0^1 \left[(1-s)\left|f^{(n+1)}(x)\right| + s\left|f^{(n+1)}(z)\right|\right](1-s)^n\,ds\right)|dz|$$

$$= \frac{1}{n!}\int_\gamma |z-x|^{n+1}$$

$$\times \left[\left| f^{(n+1)}(x) \right| \int_0^1 (1-s)^{n+1} ds + \left| f^{(n+1)}(z) \right| \int_0^1 s(1-s)^n ds \right] |dz| =: C_n. \tag{3.18}$$

Since

$$\int_0^1 (1-s)^{n+1} ds = \int_0^1 s^{n+1} ds = \frac{1}{n+2}$$

and

$$\int_0^1 s(1-s)^n ds = \int_0^1 (1-s) s^n ds = \int_0^1 \left(s^n - s^{n+1}\right) ds = \frac{1}{n+1} - \frac{1}{n+2}$$
$$= \frac{1}{(n+1)(n+2)},$$

hence

$$C_n = \frac{1}{n!} \int_\gamma |z-x|^{n+1} \left[\left| f^{(n+1)}(x) \right| \frac{1}{n+2} + \left| f^{(n+1)}(z) \right| \frac{1}{(n+1)(n+2)} \right] |dz|$$
$$= \frac{1}{n!(n+2)} \int_\gamma |z-x|^{n+1} \left[\left| f^{(n+1)}(x) \right| + \left| f^{(n+1)}(z) \right| \frac{1}{(n+1)} \right] |dz|$$
$$= \frac{1}{n!(n+2)}$$
$$\times \left[\left| f^{(n+1)}(x) \right| \int_\gamma |z-x|^{n+1} |dz| + \frac{1}{(n+1)} \int_\gamma |z-x|^{n+1} \left| f^{(n+1)}(z) \right| |dz| \right]$$

and by (3.18) we get the desired result (3.18). □

Remark 3 Let $f : D \subseteq \mathbb{C} \to \mathbb{C}$ be an analytic function on the convex domain D and $x \in D$. Suppose $\gamma \subset D$ is a *smooth path* parametrized by $z(t)$, $t \in [a, b]$ with $z(a) = u$ and $z(b) = w$ where $u, w \in D$. If $|f'|$ is convex on D, then we have the Ostrowski type inequality

$$\left| \int_\gamma f(z) dz - f(x)(w-u) \right|$$
$$\leq \frac{1}{2} \left[\left| f'(x) \right| \int_\gamma |z-x| |dz| + \int_\gamma |z-x| \left| f'(z) \right| |dz| \right] \tag{3.19}$$

for any $x \in D$.

In particular, we have the midpoint inequality

$$\begin{aligned}&\left|\int_{\gamma} f(z)\,dz - f\left(\frac{u+w}{2}\right)(w-u)\right|\\&\quad\le \frac{1}{2}\left[\left|f'\left(\frac{u+w}{2}\right)\right|\int_{\gamma}\left|z-\frac{u+w}{2}\right||dz| + \int_{\gamma}\left|z-\frac{u+w}{2}\right|\left|f'(z)\right||dz|\right].\end{aligned} \tag{3.20}$$

If $|f''|$ is convex on D, then we have the perturbed Ostrowski type inequality

$$\begin{aligned}&\left|\int_{\gamma} f(z)\,dz - f(x)(w-u) - f'(x)\left(\frac{w+u}{2}-x\right)(w-u)\right|\\&\qquad\le \frac{1}{6}\left[\left|f''(x)\right|\int_{\gamma}|z-x|^{2}|dz| + \frac{1}{2}\int_{\gamma}|z-x|^{2}\left|f''(z)\right||dz|\right]\end{aligned} \tag{3.21}$$

for any $x \in D$.

In particular, we have the midpoint inequality

$$\begin{aligned}&\left|\int_{\gamma} f(z)\,dz - f\left(\frac{u+w}{2}\right)(w-u)\right|\\&\le \frac{1}{6}\left[\left|f''\left(\frac{w+u}{2}\right)\right|\int_{\gamma}\left|z-\frac{w+u}{2}\right|^{2}|dz| + \frac{1}{2}\int_{\gamma}\left|z-\frac{w+u}{2}\right|^{2}\left|f''(z)\right||dz|\right].\end{aligned} \tag{3.22}$$

4 Error Bounds for Trapezoid Rule

Similar inequalities may be stated for the trapezoid rule; however, here we present only the simplest case of bounded derivatives:

Theorem 5 *Let $f : D \subseteq \mathbb{C} \to \mathbb{C}$ be an analytic function on the convex domain D and $x \in D$. Suppose $\gamma \subset D$ is a smooth path parametrized by $z(t)$, $t \in [a, b]$ with $z(a) = u$ and $z(b) = w$ where $u, w \in D$. If $f^{(n+1)}$ satisfies the condition (3.11) for some $n \ge 0$ and $\lambda \in \mathbb{C}$, then we have the representation (2.6) and the remainder $T_n(\lambda, \gamma)$ satisfies the bound*

$$\begin{aligned}&|T_n(\lambda,\gamma)|\\&\quad\le \frac{1}{(n+1)!}\left\|f^{(n+1)}\right\|_{D,\infty}\left[|\lambda|\int_{\gamma}|z-u|^{n+1}|dz| + |1-\lambda|\int_{\gamma}|z-w|^{n+1}|dz|\right]\end{aligned}$$

$$\leq \max\{|\lambda|, |1-\lambda|\} \frac{1}{(n+1)!} \left\| f^{(n+1)} \right\|_{D,\infty} \left[\int_\gamma |z-u|^{n+1} |dz| + \int_\gamma |z-w|^{n+1} |dz| \right]. \tag{4.1}$$

In particular, if $\lambda = \frac{1}{2}$, *then we have the bound*

$$|T_n(\gamma)| \leq \frac{1}{2(n+1)!} \left\| f^{(n+1)} \right\|_{D,\infty} \left[\int_\gamma |z-u|^{n+1} |dz| + \int_\gamma |z-w|^{n+1} |dz| \right]. \tag{4.2}$$

Proof Using the representation (2.7) we have

$$\begin{aligned} |T_n(\lambda,\gamma)| &\leq \left| \frac{\lambda}{n!} \int_\gamma (z-u)^{n+1} \left(\int_0^1 f^{(n+1)} [(1-s)u + sz] (1-s)^n ds \right) dz \right| \\ &+ \left| \frac{(1-\lambda)}{n!} \int_\gamma (z-w)^{n+1} \left(\int_0^1 f^{(n+1)} [(1-s)w + sz] (1-s)^n ds \right) dz \right| \\ &\leq |\lambda| \frac{1}{n!} \int_\gamma |z-u|^{n+1} \left| \int_0^1 f^{(n+1)} [(1-s)u + sz] (1-s)^n ds \right| |dz| \\ &+ \frac{1}{n!} |1-\lambda| \int_\gamma |z-w|^{n+1} \left| \int_0^1 f^{(n+1)} [(1-s)w + sz] (1-s)^n ds \right| |dz| \end{aligned} \tag{4.3}$$

$$\begin{aligned} &\leq |\lambda| \frac{1}{n!} \int_\gamma |z-u|^{n+1} \left(\int_0^1 f^{(n+1)} |[(1-s)u + sz]| (1-s)^n ds \right) |dz| \\ &+ \frac{1}{n!} |1-\lambda| \int_\gamma |z-w|^{n+1} \left(\int_0^1 \left| f^{(n+1)} [(1-s)w + sz] \right| (1-s)^n ds \right) |dz| \\ &\leq |\lambda| \frac{1}{n!} \left\| f^{(n+1)} \right\|_{D,\infty} \int_\gamma |z-u|^{n+1} \left(\int_0^1 (1-s)^n ds \right) |dz| \\ &+ \frac{1}{n!} |1-\lambda| \left\| f^{(n+1)} \right\|_{D,\infty} \int_\gamma |z-w|^{n+1} \left(\int_0^1 (1-s)^n ds \right) |dz| \\ &= |\lambda| \frac{1}{(n+1)!} \left\| f^{(n+1)} \right\|_{D,\infty} \int_\gamma |z-u|^{n+1} |dz| \\ &+ \frac{1}{(n+1)!} |1-\lambda| \left\| f^{(n+1)} \right\|_{D,\infty} \int_\gamma |z-w|^{n+1} |dz|, \end{aligned}$$

which proves the desired result (4.1). □

For some inequalities of trapezoid type for functions of a real variable, see [4].

Remark 4 Let $f : D \subseteq \mathbb{C} \to \mathbb{C}$ be an analytic function on the convex domain D and $x \in D$. Suppose $\gamma \subset D$ is a *smooth path* parametrized by $z(t)$, $t \in [a, b]$ with $z(a) = u$ and $z(b) = w$ where $u, w \in D$.

For $n = 0$ we get the generalized trapezoid type inequality

$$\left|\int_\gamma f(z)\,dz - [\lambda f(u) + (1-\lambda) f(w)](w-u)\right| \leq \|f'\|_{D,\infty} \left[|\lambda| \int_\gamma |z-u|\,|dz| + |1-\lambda| \int_\gamma |z-w|\,|dz|\right] \tag{4.4}$$

for $\lambda \in \mathbb{C}$, provided $\|f'\|_{D,\infty} < \infty$.

In particular, we have the trapezoid inequality

$$\left|\int_\gamma f(z)\,dz - \frac{f(u)+f(w)}{2}(w-u)\right| \leq \frac{1}{2}\|f'\|_{D,\infty} \left[\int_\gamma |z-u|\,|dz| + \int_\gamma |z-w|\,|dz|\right] \tag{4.5}$$

provided $\|f'\|_{D,\infty} < \infty$.

We also have the perturbed generalized trapezoid inequality

$$\begin{aligned} &\left|\int_\gamma f(z)\,dz - [\lambda f(u) + (1-\lambda) f(w)](w-u) - \frac{1}{2}\left[\lambda f'(u) - (1-\lambda) f'(w)\right](w-u)^2\right| \\ &\leq \frac{1}{2}\|f''\|_{D,\infty} \left[|\lambda| \int_\gamma |z-u|^2\,|dz| + |1-\lambda| \int_\gamma |z-w|^2\,|dz|\right] \\ &\leq \frac{1}{2}\max\{|\lambda|, |1-\lambda|\}\|f''\|_{D,\infty} \left[\int_\gamma |z-u|^2\,|dz| + \int_\gamma |z-w|^2\,|dz|\right] \end{aligned} \tag{4.6}$$

for $\lambda \in \mathbb{C}$, provided $\|f''\|_{D,\infty} < \infty$.

In particular, we have the perturbed trapezoid inequality

$$\left|\int_\gamma f(z)\,dz - \frac{f(u)+f(w)}{2}(w-u) - \frac{1}{4}\left[f'(u) - f'(w)\right](w-u)^2\right|$$

$$\leq \frac{1}{4} \left\| f'' \right\|_{D,\infty} \left[\int_\gamma |z-u|^2 |dz| + \int_\gamma |z-w|^2 |dz| \right] \tag{4.7}$$

provided $\left\| f'' \right\|_{D,\infty} < \infty$.

We also have:

Theorem 6 *Let $f : D \subseteq \mathbb{C} \to \mathbb{C}$ be an analytic function on the convex domain D and $x \in D$. Suppose $\gamma \subset D$ is a smooth path parametrized by $z(t)$, $t \in [a,b]$ with $z(a) = u$ and $z(b) = w$ where $u, w \in D$. If $\left| f^{(n+1)} \right|$ is convex on D, for some $n \geq 0$, then we have the representation (2.6) and the remainder $T_n(\lambda, \gamma)$ satisfies the bound*

$$|T_n(\lambda,\gamma)| \leq \frac{1}{(n+2)\,n!}$$

$$\times \left\{ |\lambda| \left[f^{(n+1)} |(u)| \int_\gamma |z-u|^{n+1} |dz| + \frac{1}{(n+1)} \int_\gamma |z-u|^{n+1} f^{(n+1)} |(z)| \, |dz| \right] \right.$$

$$\left. + |1-\lambda| \left[f^{(n+1)} |(w)| \int_\gamma |z-w|^{n+1} |dz| + \frac{1}{(n+1)} \int_\gamma |z-w|^{n+1} f^{(n+1)} |(z)| \, |dz| \right] \right\}$$

$$\leq \frac{1}{(n+2)\,n!} \max\{|\lambda|, |1-\lambda|\}$$

$$\times \left[f^{(n+1)} |(u)| \int_\gamma |z-u|^{n+1} |dz| + f^{(n+1)} |(w)| \int_\gamma |z-w|^{n+1} |dz| \right.$$

$$\left. + \frac{1}{(n+1)} \int_\gamma \left[|z-u|^{n+1} + |z-w|^{n+1} \right] f^{(n+1)} |(z)| \, |dz| \right] \tag{4.8}$$

for $\lambda \in \mathbb{C}$.

Proof Using the representation (2.7) and the convexity of $\left| f^{(n+1)} \right|$ is on D we have

$$|T_n(\lambda,\gamma)|$$

$$\leq |\lambda| \frac{1}{n!} \int_\gamma |z-u|^{n+1} \left(\int_0^1 f^{(n+1)} |[(1-s)u + sz]| (1-s)^n \, ds \right) |dz|$$

$$+ \frac{1}{n!} |1-\lambda| \int_\gamma |z-w|^{n+1} \left(\int_0^1 \left| f^{(n+1)} [(1-s)w + sz] \right| (1-s)^n \, ds \right) |dz|$$

$$\leq \frac{1}{n!} \left\{ |\lambda| \int_\gamma |z-u|^{n+1} \left(\int_0^1 \left[(1-s) f^{(n+1)} |(u)| + s f^{(n+1)} |(z)| \right] (1-s)^n \, ds \right) |dz| \right.$$

$$+ |1-\lambda|$$

$$\times \int_\gamma |z-w|^{n+1} \left(\int_0^1 \left[(1-s) \left| f^{(n+1)}(w) \right| + s \left| f^{(n+1)}(z) \right| \right] (1-s)^n \, ds \right) |dz| \Bigg\} \tag{4.9}$$

for $\lambda \in \mathbb{C}$.

Since

$$\begin{aligned} &\int_\gamma |z-u|^{n+1} \left(\int_0^1 \left[(1-s) f^{(n+1)} |(u)| + s f^{(n+1)} |(z)| \right] (1-s)^n \, ds \right) |dz| \\ &\qquad = f^{(n+1)} |(u)| \int_\gamma |z-u|^{n+1} |dz| \int_0^1 (1-s)^{n+1} \, ds \\ &\qquad + \int_\gamma |z-u|^{n+1} f^{(n+1)} |(z)| \, |dz| \int_0^1 s(1-s)^n \, ds \\ &\qquad = \frac{1}{n+2} f^{(n+1)} |(u)| \int_\gamma |z-u|^{n+1} |dz| \\ &\qquad + \frac{1}{(n+1)(n+2)} \int_\gamma |z-u|^{n+1} f^{(n+1)} |(z)| \, |dz| \\ &= \frac{1}{n+2} \left[f^{(n+1)} |(u)| \int_\gamma |z-u|^{n+1} |dz| + \frac{1}{(n+1)} \int_\gamma |z-u|^{n+1} f^{(n+1)} |(z)| \, |dz| \right] \end{aligned}$$

and, similarly

$$\begin{aligned} &\int_\gamma |z-w|^{n+1} \left(\int_0^1 \left[(1-s) \left| f^{(n+1)}(w) \right| + s \left| f^{(n+1)}(z) \right| \right] (1-s)^n \, ds \right) |dz| \\ &= \frac{1}{n+2} \left[f^{(n+1)} |(w)| \int_\gamma |z-w|^{n+1} |dz| + \frac{1}{(n+1)} \int_\gamma |z-w|^{n+1} f^{(n+1)} |(z)| \, |dz| \right], \end{aligned}$$

hence by (4.9) we get the desired result (4.8). □

Remark 5 For $\lambda = \frac{1}{2}$ we have the representation (2.8) and the remainder satisfies the bound

$$\begin{aligned} |T_n(\gamma)| &\leq \frac{1}{2(n+2)\,n!} \\ &\times \left\{ \left[f^{(n+1)} |(u)| \int_\gamma |z-u|^{n+1} |dz| + f^{(n+1)} |(w)| \int_\gamma |z-w|^{n+1} |dz| \right] \right. \\ &\left. + \frac{1}{(n+1)} \left[\int_\gamma \left[|z-u|^{n+1} |dz| + |z-w|^{n+1} \right] f^{(n+1)} |(z)| \, |dz| \right] \right\}, \end{aligned} \tag{4.10}$$

provided $\left|f^{(n+1)}\right|$ is convex on D, for some $n \geq 0$.

Remark 6 If $\left|f'\right|$ is convex on D, then we have the inequality

$$\begin{aligned}
&\left|\int_\gamma f(z)\,dz - [\lambda f(u) + (1-\lambda) f(w)](w-u)\right| \\
&\quad\leq \frac{1}{2}\left\{|\lambda|\left[\left|f'(u)\right|\int_\gamma |z-u|\,|dz| + \int_\gamma |z-u|\left|f'(z)\right||dz|\right]\right. \\
&\quad\left. + |1-\lambda|\left[\left|f'(w)\right|\int_\gamma |z-w|\,|dz| + \int_\gamma |z-w|\left|f'(z)\right||dz|\right]\right\} \\
&\leq \frac{1}{2}\max\{|\lambda|, |1-\lambda|\}\left[\left|f'(u)\right|\int_\gamma |z-u|\,|dz| + \left|f'(w)\right|\int_\gamma |z-w|\,|dz|\right. \\
&\qquad\left. + \int_\gamma [|z-u| + |z-w|]\left|f'(z)\right||dz|\right] \qquad (4.11)
\end{aligned}$$

and for $\lambda = \frac{1}{2}$,

$$\begin{aligned}
&\left|\int_\gamma f(z)\,dz - \frac{f(u)+f(w)}{2}(w-u)\right| \\
&\quad\leq \frac{1}{4}\left[\left|f'(u)\right|\int_\gamma |z-u|\,|dz| + \left|f'(w)\right|\int_\gamma |z-w|\,|dz|\right. \\
&\qquad\left. + \int_\gamma [|z-u| + |z-w|]\left|f'(z)\right||dz|\right]. \qquad (4.12)
\end{aligned}$$

References

1. M. Akkouchi, Improvements of some integral inequalities of H. Gauchman involving Taylor's remainder. Divulg. Mat. **11**(2), 115–120 (2003)
2. G.A. Anastassiou, Taylor-Widder representation formulae and Ostrowski, Grüss, integral means and Csiszar type inequalities. Comput. Math. Appl. **54**(1), 9–23 (2007)
3. G.A. Anastassiou, Ostrowski type inequalities over balls and shells via a Taylor-Widder formula. J. Inequal. Pure Appl. Math. **8**(4), 106 (2007)
4. P. Cerone, S.S. Dragomir, Trapezoidal-type rules from an inequalities point of view, in *Handbook of Analytic-Computational Methods in Applied Mathematics* (Chapman & Hall/CRC, Boca Raton, 2000), pp. 65–134
5. S.S. Dragomir, New estimation of the remainder in Taylor's formula using Grüss' type inequalities and applications. Math. Inequal. Appl. **2**(2), 183–193 (1999)
6. S.S. Dragomir, Ostrowski type inequalities for Lebesgue integral: a survey of recent results. Aust. J. Math. Anal. Appl. **14**(1), 1–283 (2017)

7. S.S. Dragomir, H.B. Thompson, A two points Taylor's formula for the generalised Riemann integral. Demonstratio Math. **43**(4), 827–840 (2010)
8. H. Gauchman, Some integral inequalities involving Taylor's remainder. I. J. Inequal. Pure Appl. Math. **3**(2), 26 (2002)
9. H. Gauchman, Some integral inequalities involving Taylor's remainder. II. J. Inequal. Pure Appl. Math. **4**(1), 1 (2003)
10. D.-Y. Hwang, Improvements of some integral inequalities involving Taylor's remainder. J. Appl. Math. Comput. **16**(1–2), 151–163 (2004)
11. A.I. Kechriniotis, N.D. Assimakis, Generalizations of the trapezoid inequalities based on a new mean value theorem for the remainder in Taylor's formula. J. Inequal. Pure Appl. Math. **7**(3), 90 (2006)
12. Z. Liu, Note on inequalities involving integral Taylor's remainder. J. Inequal. Pure Appl. Math. **6**(3), 72 (2005)
13. W. Liu, Q. Zhang, Some new error inequalities for a Taylor-like formula. J. Comput. Anal. Appl. **15**(6), 1158–1164 (2013)
14. N. Ujević, Error inequalities for a Taylor-like formula. Cubo **10**(1), 11–18 (2008)
15. Z.X. Wang, D.R. Guo, *Special Functions* (World Scientific, Teaneck, 1989)

Szász–Durrmeyer Operators and Approximation

Vijay Gupta

Abstract The Szász–Durrmeyer operators were introduced three and half decades ago in order to approximate integrable functions on the positive real axis. Several approximation properties of these operators have been discussed by researchers. In the present paper, we discuss some of the approximation properties of these operators in terms of weighted modulus of continuity and also in terms of first-order modulus of continuity having exponential growth. In the end, we find the difference estimate of Szász–Durrmeyer operators from the Baskakov–Szász–Mirakyan operators in weighted approximation.

1 Some Operators

The well-known examples of discretely defined exponential type operators are Szász–Mirakyan, Baskakov, Bernstein, Ismail–May type operators, etc. The Szász–Mirakyan and Baskakov operators are, respectively, defined by

$$S_n(f,x) = \sum_{k=0}^{\infty} s_{n,k}(x) f\left(\frac{k}{n}\right), \tag{1}$$

where $s_{n,k}(x) = e^{-nx}\frac{(nx)^k}{k!}$ and

$$B_n(f,x) = \sum_{k=0}^{\infty} v_{n,k}(x) f\left(\frac{k}{n}\right), \tag{2}$$

V. Gupta (✉)
Department of Mathematics, Netaji Subhas University of Technology, New Delhi, India

N. J. Daras, T. M. Rassias (eds.), *Computational Mathematics and Variational Analysis*, Springer Optimization and Its Applications 159,
https://doi.org/10.1007/978-3-030-44625-3_6

where $v_{n,k}(x) = \binom{n+k-1}{k} x^k (1+x)^{-n-k}$. These operators preserve linear functions. Some general integral versions of these operators have been discussed in [1] and [3].

In the above two operators (1) and (2) of exponential type, if we replace $f\left(\frac{k}{n}\right)$ by the weights of Szász–Mirakyan basis $\frac{\langle s_{n,k}, f\rangle}{\langle s_{n,k}, 1\rangle}$, where $\langle g, h\rangle = \int_0^\infty g(t)h(t)dt$, we get, respectively, the Szász–Mirakyan–Durrmeyer operators M_n introduced by Mazhar and Totik [8] and Baskakov–Szász–Mirakyan operators proposed by Gupta and Srivastava [7] as follows:

$$M_n(f, x) = \sum_{k=0}^{\infty} s_{n,k}(x) \frac{\langle s_{n,k}, f\rangle}{\langle s_{n,k}, 1\rangle}. \tag{3}$$

$$L_n(f, x) = \sum_{k=0}^{\infty} v_{n,k}(x) \frac{\langle s_{n,k}, f\rangle}{\langle s_{n,k}, 1\rangle}. \tag{4}$$

The operators (3) and (4) preserve only constant functions and are not exponential type operators.

The present article discusses the approximation properties of the Szász–Mirakyan–Durrmeyer operators in terms of weighted modulus of continuity due to Păltănea [9] and also in terms of first-order modulus of continuity having exponential growth. Finally in the end, a difference estimate of Szász–Durrmeyer operators with the Baskakov–Szász–Mirakyan operators in weighted approximation is established.

2 Lemmas

Lemma 1 ([4]) *The following recurrence relation holds for moments of Szász–Mirakyan operators:*

$$S_n(e_{m+1}, x) = \frac{x}{n} S_n'(e_m, x) + x S_n(e_m, x).$$

The following recurrence relation holds for moments of Baskakov operators:

$$B_n(e_{m+1}, x) = \frac{x(1+x)}{n} B_n'(e_m, x) + x B_n(e_m, x).$$

The moments of the operators can be represented in terms of confluent hypergeometric function ${}_1F_1$ as follows:

$$M_n(e_s, x) = \frac{s!}{n^s}\; {}_1F_1(-s; 1; -nx),$$

where $e_s(t) = t^s$, $s = 0, 1, 2, \cdots$ (see [4, pp. 32] and the references therein).

Lemma 2 ([4, pp. 34]) *If we denote the central moment by* $\mu_{n,s}(x) = M_n((t-x)^s, x)$*, then we have*

$$n\mu_{n,s+1}(x) = x[(\mu_{n,s}(x))' + 2s\mu_{n,s-1}(x)] + (s+1)\mu_{n,s}(x).$$

In particular by Gupta et al. [5, pp. 21], we have

$$\mu_{n,0}(x) = 1,\ \mu_{n,1}(x) = \frac{1}{n},\ \mu_{n,2}(x) = \frac{2(nx+1)}{n^2}$$

$$\mu_{n,4}(x) = \frac{24+72nx+12n^2x^2}{n^4},\ \mu_{n,6}(x) = \frac{720+3600nx+2160n^2x^2+120n^3x^3}{n^6}.$$

Lemma 3 *For* $A > 0$ *we have*

$$M_n(e^{At}, x) = \frac{n}{n-A}\exp\left(\frac{nAx}{n-A}\right).$$

Further,

$$M_n(te^{At}, x) = \left[\frac{n}{(n-A)^2} + \frac{n^3x}{(n-A)^3}\right]\exp\left(\frac{nAx}{n-A}\right)$$

and

$$M_n(t^2e^{At}, x) = \left[\frac{2n}{(n-A)^3} + \frac{4n^3x}{(n-A)^4} + \frac{n^5x^2}{(n-A)^5}\right]\exp\left(\frac{nAx}{n-A}\right).$$

Proof We have

$$\begin{aligned} M_n(e^{At}, x) &= n\sum_{k=0}^{\infty} s_{n,k}(x)\int_0^\infty \exp(-t(n-A))\frac{(nt)^k}{k!}dt \\ &= \sum_{k=0}^{\infty}\exp(-nx)\frac{(nx)^k}{k!}\left(\frac{n}{n-A}\right)^{k+1} \\ &= \frac{n}{n-A}\exp(-nx)\exp\left(\frac{n^2x}{n-A}\right) \\ &= \frac{n}{n-A}\exp\left(\frac{nAx}{n-A}\right). \end{aligned}$$

Now differentiating partially the above with respect to A, we immediately get the other consequences.

Lemma 4 *In (3), if we denote by* $F_k^n(f) = \frac{\langle s_{n,k}, f\rangle}{\langle s_{n,k}, 1\rangle}$*, then we have* $F_k^n(e_r) = \frac{(k+r)!}{k!.n^r}$*. Also, for any* $m \in N$ *we have*

$$\begin{aligned} T_2^{F_k^n} &:= F_k^n[e_1 - F_k^n(e_1)]^2 \\ &= F_k^n(e_2) + \left(\frac{k+1}{n}\right)^2 - 2F_k^n(e_1, x)\left(\frac{k+1}{n}\right) \\ &= \frac{(k+2)(k+1)}{n^2} - \left(\frac{k+1}{n}\right)^2 \\ &= \frac{k+1}{n^2} \end{aligned}$$

and

$$\begin{aligned} T_6^{F_k^n} &= F_k^n\left(e_1 - b^{F_k^n} e_0\right)^6 \\ &= \frac{5(k+1)(k(3k+32)+53)}{n^6}. \end{aligned}$$

3 Estimates for Polynomially Bounded Functions

Păltănea in [9] proposed the following weighted modulus, defined as

$$\omega_\varphi(f; h) = \sup\left\{|f(x) - f(y)| : x \geq 0, y \geq 0, |x - y| \leq h\varphi\left(\frac{x+y}{2}\right)\right\}, h \geq 0,$$

where $\varphi(x) = \frac{\sqrt{x}}{1+x^m}$, $x \in [0, \infty)$, $m \in N$, $m \geq 2$.

By $W_\varphi[0, \infty)$ we mean the subspace of all real functions defined on $[0, \infty)$, for which the following two conditions hold true:

1. The function $f \circ e_2$ is uniformly continuous on $[0, 1]$.
2. The function $f \circ e_v$, $v = \frac{2}{2m+1}$ is uniformly continuous on $[1, \infty)$.

In [6] we studied positive linear operators $L_n : E \to C[0, \infty)$, where E is a subspace of $C[0, \infty)$, such that $C_k[0, \infty) \subset E$ with $k = \max\{m+3, 6, 2m\}$ and

$$C_k[0, \infty) = \{f \in C[0, \infty), \exists M > 0 : |f(x)| \leq M(1 + x^k), \forall x \geq 0\}, k \in N.$$

Theorem A ([6, Th. 2.2]) *Let* $L_n : E \to C[0, \infty)$, $C_k[0, \infty) \subset E$, $k = \max\{m + 3, 6, 2m\}$ *be sequence of linear positive operators, preserving the linear functions. If* $f \in C^2[0, \infty) \cap E$ *and* $f'' \in W_\varphi[0, \infty)$*, then we have for* $x \in (0, \infty)$ *that*

$$\left|L_n(f,x)-f(x)-\frac{1}{2}f''(x)\mu_{n,2}^L(x)\right|$$
$$\leq \frac{1}{2}\left[\mu_{n,2}^L(x)+\sqrt{2}\sqrt{L_n\left(\left[1+\left(x+\frac{|t-x|}{2}\right)^m\right]^2;x\right)}\right]\omega_\varphi\left(f'';\left(\frac{\mu_{n,6}^L}{x}\right)^{1/2}\right).$$

Here $\mu_{n,k}^L(x)=L_n((t-x)^k,x)$ is the k-th order central moment of L_n.

As an application of Theorem A, we have the following result for Szász–Durrmeyer operators:

Theorem 1 *For Szász–Durrmeyer operators $M_n : E \to C[0,\infty)$, $C_k[0,\infty) \subset E$, $k=\max\{m+3,6,2m\}$, if $f \in C^2[0,\infty)\cap E$ and $f'' \in W_\varphi[0,\infty)$, then we have for $x\in(0,\infty)$ that*

$$\left|M_n(f,x)-f(x)-\frac{1}{n}f'(x)-\frac{nx+1}{n^2}f''(x)\right|$$
$$\leq \frac{1}{2}\left[\frac{2(nx+1)}{n^2}+\sqrt{2}\sqrt{M_n\left(\left[1+\left(x+\frac{|t-x|}{2}\right)^m\right]^2;x\right)}\right]$$
$$\omega_\varphi\left(f'';\frac{\sqrt{720+3600nx+2160n^2x^2+120n^3x^3}}{\sqrt{x}n^3}\right).$$

We consider $M_{n,k}^L=L_n(|t-x|^k,x)$ as the k-th order absolute moments of operators L_n.
The next main result of [6], which we are going to apply for the operators M_n is the following quantitative Voronovskaja theorem:

Theorem B ([6, Th.]) *Let $L_n : E\to C[0,\infty)$, $C_k[0,\infty)\subset E$, $k=\max\{m+3,4\}$ be sequence of linear positive operators, preserving the linear functions. If $f\in C^2[0,\infty)\cap E$ and $f''\in W_\varphi[0,\infty)$, then we have for $x\in(0,\infty)$ that*

$$\left|L_n(f,x)-f(x)-\frac{1}{2}f''(x)\mu_{n,2}^L(x)\right|$$
$$\leq \frac{1}{2}\left[\mu_{n,2}^L(x)+\frac{\sqrt{2}}{\sqrt{x}}\mu_{n,2}^L(x).C_{n,2,m}(x)\right]\omega_\varphi\left(f'';\sqrt{\frac{\mu_{n,4}^L(x)}{\mu_{n,2}^L(x)}}\right),$$

where

$$C_{n,2,m}(x)=1+\frac{1}{M_{n,3}^L(x)}.\sum_{k=0}^{m}\binom{m}{k}x^{m-k}\frac{M_{n,k+3}^L(x)}{2^k}.$$

As an application of Theorem B we obtain the following quantitative asymptotic Voronovskaja theorem for M_n:

Theorem 2 *For the Szász–Durrmeyer operators $M_n : E \to C[0,\infty)$, $C_k[0,\infty) \subset E$, $k = \max\{m+3, 4\}$, if $f \in C^2[0,\infty) \cap E$ and $f'' \in W_\varphi[0,\infty)$, then we have for $x \in (0,\infty)$ that*

$$\left| M_n(f,x) - f(x) - \frac{1}{n} f'(x) - \frac{nx+1}{n^2} f''(x) \right|$$

$$\leq \frac{nx+1}{n^2} \left[1 + \frac{\sqrt{2}}{\sqrt{x}} . C_{n,2,m}(x) \right] \omega_\varphi \left(f''; \sqrt{\frac{6(2+6nx+n^2x^2)}{n^2(nx+1)}} \right),$$

where

$$C_{n,2,m}(x) = 1 + \frac{1}{M_{n,3}^{M_n}(x)} . \sum_{k=0}^{m} \binom{m}{k} x^{m-k} \frac{M_{n,k+3}^{M_n}(x)}{2^k}.$$

Corollary 1 *If f, f'' satisfy the same conditions as in the assumption of Theorem 2, then we have for $x \in (0,\infty)$ that*

$$\lim_{n\to\infty} n[M_n(f,x) - f(x)] = f'(x) + x f''(x).$$

4 Exponential Growth

For continuous functions on $[0,\infty)$ with exponential growth $\|f\|_A := \sup_{x\in[0,\infty)} |f(x)e^{-Ax}| < \infty$, $A > 0$, the first-order modulus of continuity (see [6, 10]) is defined as

$$\omega_1(f,\delta,A) = \sup_{|h|\leq\delta, 0\leq x<\infty} |f(x) - f(x+h)| e^{-Ax}.$$

Theorem 3 *Let $f \in C^2[0,\infty) \cap E$ and $f'' \in Lip(\beta, A)$, $0 < \beta \leq 1$, then for $n > 2Ax$, we have for $x \in [0,\infty)$*

$$\left| M_n(f,x) - f(x) - \frac{1}{n} f'(x) - \frac{nx+1}{n^2} f''(x) \right|$$

$$\leq \left[e^{2Ax} + \frac{C(A,x)}{2} + \frac{\sqrt{C(2A,x)}}{2} \right] \cdot \frac{2(nx+1)}{n^2} \cdot \omega_1 \left(f'', \sqrt{\frac{6(2+6nx+n^2x^2)}{n^2(nx+1)}}, A \right),$$

where $C(A,x) = 2(16 + 8xA^4 + xA^2)e^{2Ax}$ *and the spaces* $Lip(\beta, A), 0 < \beta \leq 1$ *consist of all functions such that* $\omega_1(f, \delta, A) \leq M\delta^\beta$ *for all* $\delta < 1$.

Proof For the function $f \in C^2[0,\infty)$, Taylor's expansion at the point $x \in [0,\infty)$ is given by

$$f(t) = f(x) + (t-x)f'(x) + \frac{(t-x)^2}{2!}f''(x) + \varepsilon_2(t,x), \tag{5}$$

where

$$\varepsilon_2(t,x) = \frac{f''(\xi) - f''(x)}{2}(t-x)^2$$

and ξ lies between x and t, also applying the operator M_n on (5) and using Lemma 2, we have

$$\left| M_n(f,x) - f(x) - \frac{1}{n}f'(x) - \frac{nx+1}{n^2}f''(x) \right| \leq M_n\left(|\varepsilon_2(t,x)|, x\right). \tag{6}$$

In order to estimate the proof of theorem we estimate $M_n\left(|\varepsilon_2(t,x)|, x\right)$. Following [6, pp.101], we have

$$|\varepsilon_2(t,x)| \leq \frac{1}{2}\left(e^{2Ax} + e^{At}\right)\left(1 + \frac{|t-x|}{h}\right)\omega_1\left(f'', h, A\right)|t-x|^2.$$

Consequently

$$M_n\left(|\varepsilon_2(t,x)|, x\right) \leq \frac{1}{2}\left[M_n\left(\left(e^{2Ax} + e^{At}\right).\left(|t-x|^2 + \frac{|t-x|^3}{h}\right); x\right)\right]\omega_1\left(f'', h, A\right).$$

By using linearity and Lemma 3, we get

$$\begin{aligned} M_n\left((t-x)^2 e^{At}, x\right) = \frac{n}{n-A}\Bigg[&\frac{4n^2x}{(n-A)^3} + \frac{n^4x^2}{(n-A)^4} + \frac{2}{(n-A)^2} \\ &- \frac{2x}{(n-A)} - \frac{2n^2x^2}{(n-A)^2} + x^2\Bigg]\exp\left(\frac{nAx}{n-A}\right). \end{aligned}$$

Let $n > 2A$, implying $n - A > \frac{n}{2}$, i.e. $n/(n-A) < 2$, thus the above equality becomes

$$\begin{aligned}M_n\left((t-x)^2e^{At},x\right) &\le 2\left[\frac{32x}{n}+\frac{8}{n^2}+\frac{(A^4+4n^2A^2-4nA^3)x^2}{(n-A)^4}\right]e^{2Ax}\\ &\le 2\left[\frac{8}{n^2}+\frac{2x}{n}(16+8xA^4+xA^2)\right]e^{2Ax}\\ &\le 2(16+8xA^4+xA^2)\left[\frac{2}{n^2}+\frac{2x}{n}\right]e^{2Ax}\\ &= 2(16+8xA^4+xA^2)e^{2Ax}\mu_{n,2}(x).\end{aligned}$$

Also, by Cauchy–Schwarz inequality we get

$$\begin{aligned}M_n\left(|t-x|^3e^{At},x\right) &\le \sqrt{M_n\left((t-x)^2e^{2At},x\right)}.\sqrt{\mu_{n,4}(x)}\\ &\le \sqrt{C(2A,x)\mu_{n,2}(x)}.\sqrt{\mu_{n,4}(x)}.\end{aligned}$$

Substituting $h:=\sqrt{\dfrac{\mu_{n,4}(x)}{\mu_{n,2}(x)}}$ and combining the above estimates, we get the desired result.

5 Difference Between Operators

For any two linear positive operators $\widetilde{M}_n$ and $\widetilde{L}_n$ having different basis and same weights, i.e.

$$\widetilde{M}_n(f,x)=\sum_{k\in K}p_{n,k}(x)F_k^n(f) \tag{7}$$

and

$$\widetilde{L}_n(f,x)=\sum_{k\in K}v_{n,k}(x)F_k^n(f), \tag{8}$$

where K be a set of non-negative integers. Also, we denote by $C_2[0,\infty)$ the class of all continuous functions on $[0,\infty)$ such that $f(x)=O(1+x^2)$. Further $\widetilde{C}_2[0,\infty)$ denotes the closed subspace of $C_2[0,\infty)$ for which $\lim_{x\to\infty}|f(x)|(1+x^2)^{-1}$ is bounded with norm $||.||_2=\sup_{x\in[0,\infty)}|f(x)|(1+x^2)^{-1}$.

The difference estimates having different basis the result was given in our recent paper [2]. If we take for different basis function and same weights then following [2], the general result takes the following form:

Theorem 4 *Let* $f\in C_2[0,\infty)$ *with* $f''\in\widetilde{C}_2[0,\infty)$. *Then*

$$\left|(\widetilde{M}_n - \widetilde{L}_n)(f,x)\right| \leq \frac{1}{2}||f''||_2(\alpha_1(x)+\alpha_2(x)) + 8\Omega\left(f'',\delta_1\right)(1+\alpha_1(x))$$
$$+8\Omega\left(f'',\delta_2\right)(1+\alpha_2(x)) + 16\Omega(f,\delta_3)(\gamma_1(x)+1)$$
$$+16\Omega(f,\delta_4)(\gamma_2(x)+1),$$

where

$$\alpha_1(x)=\sum_{k\in K} p_{n,k}(x)\left(1+\left(F_k^n(e_1)\right)^2\right)T_2^{F_k^n},\ \alpha_2(x)=\sum_{k\in K} v_{n,k}(x)\left(1+\left(F_k^n(e_1)\right)^2\right)T_2^{F_k^n},$$
$$\delta_1^4(x)=\sum_{k\in K} p_{n,k}(x)\left(1+\left(F_k^n(e_1)\right)^2\right)T_6^{F_k^n},\ \delta_2^4(x)=\sum_{k\in K} v_{n,k}(x)\left(1+\left(F_k^n(e_1)\right)^2\right)T_6^{F_k^n},$$
$$\delta_3^4(x) = \sum_{k\in K} p_{n,k}(x)\left(1+\left(F_k^n(e_1)\right)^2\right)(F_k^n(e_1)-x)^4,$$
$$\delta_4^4(x) = \sum_{k\in K} v_{n,k}(x)\left(1+\left(F_k^n(e_1)\right)^2\right)(F_k^n(e_1)-x)^4.$$
$$\gamma_1(x) = \sum_{k\in K} p_{n,k}(x)(1+(F_k^n(e_1))^2),\ \gamma_2(x) = \sum_{k\in K} v_{n,k}(x)(1+(F_k^n(e_1))^2)$$

and $e_r(t) = t^r$, $r = 0,1,2,\cdots$, $T_r^{F_k^n} = F_k^n\left[(e_1 - F_k^n(e_1)\right]^r$, $r \in N$. *We consider here that* $\delta_1(x) \leq 1, \delta_2(x) \leq 1, \delta_3(x) \leq 1, \delta_4(x) \leq 1$ *with* $\Omega(f,\delta) = \sup_{x\geq 0, |a|<\delta} |f(x+a) - f(x)|[(1+x^2)(1+a^2)]^{-1}$.

As application of Theorem 4, for Szász–Mirakyan–Durrmeyer operators and Baskakov-Szász–Mirakyan operators, we have the following quantitative estimate:

Theorem 5 *Let* $f \in C_2[0,\infty)$ *with* $f'' \in \widetilde{C}_2[0,\infty)$. *Then*

$$|(M_n - L_n)(f,x)| \leq \frac{1}{2}||f''||_2(\alpha_1(x)+\alpha_2(x)) + 8\Omega\left(f'',\delta_1\right)(1+\alpha_1(x))$$
$$+8\Omega\left(f'',\delta_2\right)(1+\alpha_2(x)) + 16\Omega(f,\delta_3)(\gamma_1(x)+1)$$
$$+16\Omega(f,\delta_4)(\gamma_2(x)+1),$$

where

$$\alpha_1(x) = \frac{1}{n^2}\left(1+\frac{1}{n^2}\right) + \frac{x}{n}\left(1+\frac{7}{n^2}\right) + \frac{6x^2}{n^2} + \frac{x^3}{n},$$
$$\alpha_2(x) = \frac{1}{n^2}\left(1+\frac{1}{n^2}\right) + \frac{x}{n}\left(1+\frac{7}{n^2}\right) + \frac{6x^2}{n^2}\left(1+\frac{1}{n}\right) + \frac{x^3}{n}\left(1+\frac{3}{n}+\frac{2}{n^2}\right),$$

$$\delta_1^4(x) = \frac{5}{n^8}\Big(53+651nx+n^4x^2(44+71x^2)+n^3x(123+479x^2)+n^2(53+1064x^2)$$
$$+3n^5(x^3+x^5)\Big),$$

$$\delta_2^4(x) = \frac{1}{8n^8}\Big[n^5x^3(x^2+120)+10n^4x^2(x^3+165x^2+36x+176)+5n^3x(7x^4$$
$$+1980x^3+3285x^2+352x+984)+5n^2(10x^5+3630x^4+9711x^3$$
$$+8155x^2+424)+nx(24x^4+9900x^3+32370x^2+40775x+25921)+2120\Big],$$

$$\delta_3^4(x) = \frac{1}{n^6}\Big(1+59nx+n^3x(11+76x^2)+n^2(1+183x^2)+3n^4(x^2+x^4)\Big),$$

$$\delta_4^4(x) = \frac{1}{n^6}\Big[1+3n^4x^2(1+x)^2(1+x^2)+n^3x(11+25x+96x^2+216x^3+195x^4+61x^5)$$
$$+nx(59+301x+700x^2+840x^3+504x^4+120x^5)$$
$$+n^2(1+183x^2+690x^3+1020x^4+690x^5+178x^6)\Big]$$

$$\gamma_1(x) = 1+x^2+\frac{3x}{n}+\frac{1}{n^2},$$

$$\gamma_2(x) = 1+x^2+\frac{x^2}{n}+\frac{3x}{n}+\frac{1}{n^2}.$$

Proof Applying Lemmas 1 and 4, we have

$$\begin{aligned}\alpha_1(x) &= \sum_{k=0}^{\infty} s_{n,k}(x)\left(1+\left(F_k^n(e_1)\right)^2\right)T_2^{F_k^n}\\ &= \sum_{k=0}^{\infty} s_{n,k}(x)\left(1+\frac{(k+1)^2}{n^2}\right)\frac{k+1}{n^2}\\ &= \frac{1}{n^2}\left(1+\frac{1}{n^2}\right)+\frac{x}{n}\left(1+\frac{7}{n^2}\right)+\frac{6x^2}{n^2}+\frac{x^3}{n}.\end{aligned}$$

$$\begin{aligned}\alpha_2(x) &= \sum_{k=0}^{\infty} v_{n,k}(x)\left(1+\left(F_k^n(e_1)\right)^2\right)T_2^{F_k^n}\\ &= \sum_{k=0}^{\infty} v_{n,k}(x)\left(1+\frac{(k+1)^2}{n^2}\right)\frac{k+1}{n^2}\end{aligned}$$

$$= \frac{1}{n^2}\left(1+\frac{1}{n^2}\right)+\frac{x}{n}\left(1+\frac{7}{n^2}\right)+\frac{6x^2}{n^2}\left(1+\frac{1}{n}\right)+\frac{x^3}{n}\left(1+\frac{3}{n}+\frac{2}{n^2}\right).$$

$$\begin{aligned}
\delta_1^4(x) &= \sum_{k=0}^{\infty} s_{n,k}(x)\left(1+\left(F_k^n(e_1)\right)^2\right)T_6^{F_k^n},\\
&= \sum_{k=0}^{\infty} s_{n,k}(x)\left(1+\left(\frac{k+1}{n}\right)^2\right)\left(\frac{5(k+1)(k(3k+32)+53)}{n^6}\right)\\
&= \frac{5}{n^8}\sum_{k=0}^{\infty} s_{n,k}(x)\Big(3k^5+41k^4+(158+3n^2)k^3+(258+35n^2)k^2\\
&\qquad +(191+85n^2)k+(53+53n^2)\Big)\\
&= \frac{5}{n^8}\Big(3n^5S_n(e_5,x)+41n^4S_n(e_4,x)+(158+3n^2)n^3S_n(e_3,x)\\
&\qquad +(258+35n^2)n^2S_n(e_2,x)\\
&\qquad +(191+85n^2)nS_n(e_1,x)+(53+53n^2)S_n(e_0,x)\Big)\\
&= \frac{5}{n^8}\Big(53+651nx+n^4x^2(44+71x^2)+n^3x(123+479x^2)\\
&\qquad +n^2(53+1064x^2)+3n^5(x^3+x^5)\Big).
\end{aligned}$$

$$\begin{aligned}
\delta_2^4(x) &= \sum_{k=0}^{\infty} v_{n,k}(x)\left(1+\left(F_k^n(e_1)\right)^2\right)T_6^{F_k^n}\\
&= \sum_{k=0}^{\infty} b_{n,k}(x)\left(1+\left(\frac{k+1}{n}\right)^2\right)\left(\frac{5(k+1)(k(3k+32)+53)}{n^6}\right)\\
&= \frac{1}{8n^8}\Big[n^5x^3(x^2+120)+10n^4x^2(x^3+165x^2+36x+176)+5n^3x(7x^4\\
&\qquad +1980x^3+3285x^2+352x+984)+5n^2(10x^5+3630x^4+9711x^3\\
&\qquad +8155x^2+424)+nx(24x^4+9900x^3+32370x^2+40775x+25921)+2120\Big].
\end{aligned}$$

$$\delta_3^4(x) = \sum_{k=0}^{\infty} s_{n,k}(x)\left(1+\left(F_k^n(e_1)\right)^2\right)\left(F_k^n(e_1)-x\right)^4$$

$$= \sum_{k=0}^{\infty} s_{n,k}(x) \left(1 + \left(\frac{k+1}{n}\right)^2\right) \left(\frac{k+1}{n} - x\right)^4$$

$$= \sum_{k=0}^{\infty} s_{n,k}(x) \Bigg[\left(\frac{k+1}{n}\right)^6 - 4x \left(\frac{k+1}{n}\right)^5 + (6x^2+1) \left(\frac{k+1}{n}\right)^4$$

$$-(4x^3+4x) \left(\frac{k+1}{n}\right)^3 + (6x^2+x^4) \left(\frac{k+1}{n}\right)^2 - 4x^3 \left(\frac{k+1}{n}\right) + x^4 \Bigg]$$

$$= \frac{1}{n^6} \Big[n^6 S_n(e_6, x) - 2(2nx-3) n^5 S_n(e_5, x)$$

$$+ (15 + n^2 - 20nx + 6n^2x^2) n^4 S_n(e_4, x)$$

$$-4(nx-1)(5+n^2-5nx+n^2x^2) n^3 S_n(e_3, x) + (nx-1)^2$$
$$(15 + 6n^2 - 10nx + n^2x^2) n^2 S_n(e_2, x)$$

$$-2(nx-1)^3(3+2n^2-nx) n S_n(e_1, x) + (1+n^2)(nx-1)^4 S_n(e_0, x) \Big]$$

$$= \frac{1}{n^6} \left(1 + 59nx + n^3x(11+76x^2) + n^2(1+183x^2) + 3n^4(x^2+x^4)\right).$$

$$\delta_4^4(x) = \sum_{k=0}^{\infty} v_{n,k}(x) \left(1 + \left(F_k^n(e_1)\right)^2\right) (F_k^n(e_1) - x)^4$$

$$= \sum_{k=0}^{\infty} v_{n,k}(x) \left(1 + \left(\frac{k+1}{n}\right)^2\right) \left(\frac{k+1}{n} - x\right)^4$$

$$= \frac{1}{n^6} \sum_{k=0}^{\infty} v_{n,k}(x) \Big[k^6 - 2(2nx-3)k^5 + (15 + n^2 - 20nx + 6n^2x^2)k^4$$

$$-4(nx-1)(5+n^2-5nx+n^2x^2)k^3 + (nx-1)^2(15+6n^2-10nx+n^2x^2)k^2$$

$$-2(nx-1)^3(3+2n^2-nx)k + (1+n^2)(nx-1)^4 \Big]$$

$$= \frac{1}{n^6} \Big[n^6 B_n(e_6, x) - 2(2nx-3) n^5 B_n(e_5, x)$$

$$+ (15 + n^2 - 20nx + 6n^2x^2) n^4 B_n(e_4, x)$$

$$-4(nx-1)(5+n^2-5nx+n^2x^2) n^3 B_n(e_3, x) + (nx-1)^2$$
$$(15 + 6n^2 - 10nx + n^2x^2) n^2 B_n(e_2, x)$$

$$-2(nx-1)^3(3+2n^2-nx)nB_n(e_1,x)+(1+n^2)(nx-1)^4B_n(e_0,x)\Bigg]$$

$$= \frac{1}{n^6}\Bigg[1+3n^4x^2(1+x)^2(1+x^2)+n^3x(11+25x+96x^2+216x^3+195x^4+61x^5)$$

$$+nx(59+301x+700x^2+840x^3+504x^4+120x^5)$$

$$+n^2(1+183x^2+690x^3+1020x^4+690x^5+178x^6)\Bigg].$$

$$\begin{aligned}\gamma_1(x) &= \sum_{k=0}^{\infty} s_{n,k}(x)(1+(F_k^n(e_1))^2)\\ &= \sum_{k=0}^{\infty} s_{n,k}(x)\left(1+\frac{(k+1)^2}{n^2}\right)\\ &= 1+x^2+\frac{3x}{n}+\frac{1}{n^2},\end{aligned}$$

$$\begin{aligned}\gamma_2(x) &= \sum_{k=0}^{\infty} v_{n,k}(x)(1+(F_k^n(e_1))^2)\\ &= \sum_{k=0}^{\infty} v_{n,k}(x)\left(1+\frac{(k+1)^2}{n^2}\right)\\ &= 1+x^2+\frac{x^2}{n}+\frac{3x}{n}+\frac{1}{n^2}.\end{aligned}$$

References

1. V. Gupta, Rate of approximation by new sequence of linear positive operators. Comput. Math. Appl. **45**(12), 1895–1904 (2003)
2. V. Gupta, Estimate for the difference of operators having different basis functions. Rend. Circ. Mat. Palermo, II. Ser (2019). https://doi.org/10.1007/s12215-019-00451-y
3. V. Gupta, G.C. Greubel, Moment estimations of a new Szasz–Mirakyan–Durrmeyer operators. Appl. Math. Comput. **271**, 540–547 (2015)
4. V. Gupta, M.T. Rassias, in *Moments of Linear Positive Operators and Approximation*. Series: SpringerBriefs in Mathematics (Springer, Basel, 2019)
5. V. Gupta, T.M. Rassias, P.N. Agrawal, A.M. Acu, *Recent Advances in Constructive Approximation Theory* (Springer, Cham, 2018)
6. V. Gupta, G. Tachev, *Approximation with Positive Linear Operators and Linear Combinations*, Series: Developments in Mathematics, vol. 50 (Springer, Berlin, 2017)

7. V. Gupta, G.S. Srivastava, Simultaneous approximation by Baskakov–Szász type operators. Bull. Math. Soc. Sci. Roumanie (N. S.) **37**(85), nr. (3–4), 73–85 (1993)
8. S.M. Mazhar, V. Totik, Approximation by modified Szász operators. Acta Sci. Math. (Szeged) **49**, 257–269 (1985)
9. R. Păltănea, Estimates of approximation in terms of a weighted modulus of continuity. Bull. Transilv. Univ. Brasov **4**(53), 67–74 (2011)
10. G. Tachev, V. Gupta, A. Aral, Voronovskaja's theorem for functions with exponential growth. Georgian Math. J. https://doi.org/10.1515/gmj-2018-0041

Leibniz's Rule and Fubini's Theorem Associated with a General Quantum Difference Operator

Alaa E. Hamza, Enas M. Shehata, and Praveen Agarwal

Abstract In this paper, we derive Leibniz's rule and Fubini's theorem associated with a general quantum difference operator D_β which is defined by $D_\beta f(t) = \frac{f(\beta(t))-f(t)}{\beta(t)-t}$, $\beta(t) \neq t$. Here β is a strictly increasing continuous function defined on a set $I \subseteq \mathbb{R}$ that has only one fixed point $s_0 \in I$ and satisfies the inequality $(t - s_0)(\beta(t) - t) \leq 0$ for all $t \in I$.

1 Introduction

Quantum difference operators have an interest role due to their applications in many fields in mathematics and physics. For the applications in mathematics, see [1, 17, 18, 20, 22] and for the applications in physics, see [19, 23]. It substitutes the classical derivative by a difference operator which allows one to deal with sets of non-differentiable functions. Sometimes, to solve differential equations, we use the finite-difference method, in which we replace the derivative $Dx(t) = \lim_{\omega\to 0} \dfrac{x(t+\omega) - x(t)}{\omega}$ by its approximation $D_\omega x(t) = \dfrac{x(t+\omega) - x(t)}{\omega}$, where ω is very close to zero. See [16]. The operator D_ω is called a forward (resp. backward)

A. E. Hamza
Department of Mathematics, Faculty of Science, Jeddah, Saudi Arabia

Department of Mathematics, Faculty of Science, Cairo University, Giza, Egypt

E. M. Shehata
Department of Mathematics, Faculty of Science, Menoufia University, Shibin El-Koom, Egypt

P. Agarwal (✉)
Department of Mathematics, Anand International College of Engineering, Jaipur, India

International Center for Basic and Applied Sciences, Jaipur, India

Harish-Chandra Research Institute (HRI), Jhusi, Uttar Pradesh, India

Netaji Subhas University of Technology, Dwarka, New Delhi, India

N. J. Daras, T. M. Rassias (eds.), *Computational Mathematics and Variational Analysis*, Springer Optimization and Its Applications 159,
https://doi.org/10.1007/978-3-030-44625-3_7

difference operator, when $\omega > 0$ (resp. $\omega < 0$). There are many other types of quantum difference operators such as the q-Jackson difference operator and the Hahn difference operator, which are defined by

$$D_q f(t) = \begin{cases} \dfrac{f(qt) - f(t)}{t(q-1)}, & t \neq 0, \\ \\ f'(0), & t = 0, \end{cases}$$

and

$$D_{q,\omega} f(t) = \begin{cases} \dfrac{f(qt+\omega) - f(t)}{t(q-1)+\omega}, & t \neq \omega_0, \\ \\ f'(\omega_0), & t = \omega_0, \end{cases}$$

respectively, where $q \in (0, 1)$, $\omega > 0$ are fixed and $\omega_0 = \dfrac{\omega}{1-q}$. See [3, 4, 15].

Hahn quantum difference operator unifies the Jackson q-difference operator, D_q, and the forward difference operator D_ω. For other types of quantum difference operators see also [2, 5, 21].

In this paper we consider a general quantum difference operator, D_β, which is defined by

$$D_\beta f(t) = \frac{f(\beta(t)) - f(t)}{\beta(t) - t}, \quad \beta(t) \neq t \tag{1}$$

and $D_\beta f(t) = f'(t)$ when $\beta(t) = t$ provided that $f'(t)$ exists in the usual sense. f is an arbitrary function defined, in general, on an interval $I \subseteq \mathbb{R}$. The function $\beta : I \longrightarrow I$ is a strictly increasing continuous function defined on I such that $\beta(t) \in I$ for any $t \in I$ and satisfies the inequality

$$(t - s_0)(\beta(t) - t) \leq 0 \quad \text{for all } t \in I,$$

where s_0 is the unique fixed point of the function β that belongs to I. For more details about the calculus based on D_β see [11]. The general quantum difference operator D_β yields the Hahn difference operator and the Jackson q-difference operator when $\beta(t) = qt + \omega$ and $\beta(t) = qt$, respectively, $q \in (0, 1)$ and $\omega \geq 0$, see [3, 4, 6, 9, 15]. Also, it yields the power quantum difference operator $D_{n,q}$, when $\beta(t) = qt^n, q \in (0, 1), n > 1$ is a fixed odd positive integer and $I = (-q^{\frac{1}{1-n}}, q^{\frac{1}{1-n}})$, see [2]. In [11], the β-difference operator D_β and its inverse integral operator d_β were defined. Also, some results of the calculus associated with D_β and d_β were proved such as Leibniz formula, chain rule, the fundamental theorem of β-calculus, and the mean value theorem. In [12], some inequalities based on D_β were presented. In [8] the theory of nth-order linear quantum difference equations associated with D_β was established. See also [7, 13, 14]. The general quantum difference operator and its calculus allow us to avoid repetition in proving results in each calculus, individually, associated with every quantum difference operator in the form of D_β. In [10], Leibniz's rule and Fubini's theorem associated with the power quantum operator $D_{n,q}$ was proved. In this paper, we generalize the results of [10] and prove some results concerning with the β-differentiation under the integral sign. This is known by Leibniz's rule. Also, we establish Fubini's theorem associated with the operator D_β. We organize this paper as follows. In Section 2, we present some needed preliminaries about the β-calculus, from [11]. In Section 3, we establish Leibniz's rule. We derive a formula for the β-derivative of $\int_{s_0}^{t} f(t, y) d_\beta t$. Also, we prove Fubini's theorem. That is, under appropriate conditions, the iterated integrals $\int_{s_0}^{a}\int_{s_0}^{b} f(t, y) d_\beta t d_\beta y$ and $\int_{s_0}^{b}\int_{s_0}^{a} f(t, y) d_\beta y d_\beta t$ are equal.

In the following section we present some results from [11] which are foundational for the current paper.

2 Preliminaries

In the following $\mathbb{X}$ is a Banach space, $\|.\|$ is the norm defined on $\mathbb{X}$, $\beta^k(t) := \underbrace{\beta \circ \beta \circ \ldots \circ \beta}_{k-times}(t)$, s_0 is the unique fixed point of β that belongs to I and $\mathbb{N}_0 := \{0, 1, 2, \ldots\}$.

Definition 1 For a function $f : I \longrightarrow \mathbb{X}$, we define the β-difference operator of f as

$$D_\beta f(t) = \begin{cases} \dfrac{f(\beta(t)) - f(t)}{\beta(t) - t}, & t \neq s_0, \\ f'(s_0), & t = s_0, \end{cases}$$

provided that the ordinary derivative f' exists at $t = s_0$. In this case, we say that $D_\beta f(t)$ is the β-derivative of f at t. We say that f is β-differentiable on I if $f'(s_0)$ exists.

Lemma 1 *The following statements are true.*

(i) The sequence of functions $\{\beta^k(t)\}_{k\in\mathbb{N}_0}$ converges uniformly to the constant function $\hat{\beta}(t) := s_0$ on every compact interval $J \subseteq I$ containing s_0.
(ii) The series $\sum_{k=0}^{\infty}|\beta^k(t) - \beta^{k+1}(t)|$ is uniformly convergent to $|t - s_0|$ on every compact interval $J \subseteq I$ containing s_0.

Lemma 2 *If $f : I \longrightarrow \mathbb{X}$ is continuous at s_0, then the sequence $\{f(\beta^k(t))\}_{k\in\mathbb{N}_0}$ converges uniformly to $f(s_0)$ on every compact interval $J \subseteq I$ containing s_0.*

Theorem 1 *If* $f : I \longrightarrow \mathbb{X}$ *is continuous at* s_0*, then the series* $\sum_{k=0}^{\infty}\|(\beta^k(t) - \beta^{k+1}(t))f(\beta^k(t))\|$ *is uniformly convergent on every compact interval* $J \subseteq I$ *containing* s_0*.*

Theorem 2 *Assume that* $f : I \longrightarrow \mathbb{X}$ *and* $g : I \longrightarrow \mathbb{R}$ *are* β*-differentiable functions at* $t \in I$*. Then:*

(i) *The product* $fg : I \longrightarrow \mathbb{X}$ *is* β*-differentiable at* t *and*

$$\begin{aligned} D_\beta(fg)(t) &= (D_\beta f(t))g(t) + f(\beta(t))D_\beta g(t) \\ &= (D_\beta f(t))g(\beta(t)) + f(t)D_\beta g(t). \end{aligned}$$

(ii) f/g *is* β*-differentiable at* t *and*

$$D_\beta\big(f/g\big)(t) = \frac{(D_\beta f(t))g(t) - f(t)D_\beta g(t)}{g(t)g(\beta(t))}, \qquad g(t)g(\beta(t)) \neq 0.$$

Definition 2 Let $s_0 \in [a, b] \subseteq I$. We define the β-interval by

$$[a, b]_\beta = \{\beta^k(a); k \in \mathbb{N}_0\} \cup \{\beta^k(b); k \in \mathbb{N}_0\} \cup \{s_0\},$$

and the class $[c]_\beta$ for any point $c \in I$ by

$$[c]_\beta = \{\beta^k(c); k \in \mathbb{N}_0\} \cup \{s_0\}.$$

Theorem 3 *Assume* $f : I \to \mathbb{X}$ *is continuous at* s_0*. Then, the function* F *defined by*

$$F(t) = \sum_{k=0}^{\infty}\big(\beta^k(t) - \beta^{k+1}(t)\big)f(\beta^k(t)),\ t \in I \tag{2}$$

is a β*-antiderivative of* f *with* $F(s_0) = 0$*. Conversely, a* β*-antiderivative* F *of* f *vanishing at* s_0 *is given by the formula* (2).

Definition 3 Let $f : I \longrightarrow \mathbb{X}$ and $a, b \in I$. We define the β-integral of f from a to b by

$$\int_a^b f(t)d_\beta t = \int_{s_0}^b f(t)d_\beta t - \int_{s_0}^a f(t)d_\beta t, \tag{3}$$

where

$$\int_{s_0}^x f(t)d_\beta t = \sum_{k=0}^{\infty}\big(\beta^k(x) - \beta^{k+1}(x)\big)f(\beta^k(x)), \quad x \in I, \tag{4}$$

provided that the series converges at $x = a$ and $x = b$. f is called β-integrable on I if the series converges at a, b for all $a, b \in I$. Clearly, if f is continuous at $s_0 \in I$, then f is β-integrable on I.

Lemma 3 *Let $f, g : I \longrightarrow \mathbb{X}$ be β-integrable on I and $a, b, c \in I$, then the following statements are true:*

(i) The β-integral is a linear operator.
(ii) $\int_a^a f(t)d_\beta t = 0$,
(iii) $\int_a^b f(t)d_\beta t = -\int_b^a f(t)d_\beta t$,
(iv) $\int_a^b f(t)d_\beta t = \int_a^c f(t)d_\beta t + \int_c^b f(t)d_\beta t$.

Theorem 4 *Let $f : I \longrightarrow \mathbb{X}$ be continuous at s_0. Define the function*

$$F(x) = \int_{s_0}^{x} f(t)d_\beta t, \quad x \in I. \tag{5}$$

Then F is continuous at s_0, $D_\beta F(x)$ exists for all $x \in I$ and $D_\beta F(x) = f(x)$.

Corollary 1 *If $f : I \longrightarrow \mathbb{X}$ is continuous at s_0. Then*

$$\int_{t}^{\beta(t)} f(\tau)d_\beta \tau = (\beta(t) - t)f(t), \qquad t \in I. \tag{6}$$

Theorem 5 *If $f : I \longrightarrow \mathbb{X}$ is β-differentiable on I, then*

$$\int_a^b D_\beta f(t)d_\beta(t) = f(b) - f(a), \quad for\ all\ a, b \in I. \tag{7}$$

Lemma 4 *Let $f : I \longrightarrow \mathbb{X}$, $g : I \longrightarrow \mathbb{R}$ be β-integrable functions on I. If*

$$\|f(t)\| \leq g(t) \quad for\ all\ \ t \in [a, b]_\beta,\ a, b \in I\ and\ a \leq b,$$

then for $\ x, y \in [a, b]_\beta,\ \ x < s_0 < y$, we have

$$\left\| \int_{s_0}^{y} f(t)d_\beta t \right\| \leq \int_{s_0}^{y} g(t)d_\beta t, \tag{8}$$

$$\left\| \int_{s_0}^{x} f(t)d_\beta t \right\| \leq -\int_{s_0}^{x} g(t)d_\beta t, \tag{9}$$

and

$$\left\| \int_{x}^{y} f(t)d_\beta t \right\| \leq \int_{x}^{y} g(t)d_\beta t. \tag{10}$$

Consequently, if $g(t) \geq 0$ *for all* $t \in [a, b]_\beta$*, then the inequalities* $\int_{s_0}^{y} g(t)d_\beta t \geq 0$ *and* $\int_x^y g(t)d_\beta t \geq 0$ *hold for all* $x, y \in [a, b]_\beta, x < s_0 < y$.

Lemma 5 *Let* $f : I \longrightarrow \mathbb{X}$ *and* $g : I \longrightarrow \mathbb{R}$ *be* β*-differentiable on* I*. If*

$$\|D_\beta f(t)\| \leq D_\beta g(t), \quad t \in [a, b]_\beta, \quad a, b \in I \text{ and } a \leq b,$$

then

$$\|f(y) - f(x)\| \leq g(y) - g(x), \tag{11}$$

for every $x, y \in [a, b]_\beta$, $x < s_0 < y$.

3 Quantum Leibniz's Rule and Fubini's Theorem

In this section, we prove the β-Leibniz's rule and β-Fubini's theorem. Throughout the section f is a real valued function on $I \times I$.

Definition 4 We say that $f(t, y)$ is continuous at $t = t_0$ uniformly with respect to $y \in A \subseteq I$ if

$$\lim_{t \to t_0} f(t, y) = f(t_0, y)$$

uniformly with respect to $y \in A$.

Definition 5 We say that $f(t, y)$ is uniformly partially differentiable at $t = s_0$ with respect to $y \in A \subseteq I$ if

$$\lim_{t \to s_0} \frac{f(t, y) - f(s_0, y)}{t - s_0}$$

exists uniformly with respect to $y \in A$. In this case this limit is denoted by $\frac{\partial_\beta f}{\partial_\beta t}(s_0, y)$. The partial derivative of f(t,y) with respect to t at $t \neq s_0$ is defined by

$$\frac{\partial_\beta f}{\partial_\beta t}(t, y) = \frac{f(\beta(t), y) - f(t, y)}{\beta(t) - t}.$$

Similarly, we can define the partial derivative of $f(t, y)$ with respect to y.

Theorem 6 *Let* $b \in I$*. Define the function* F *by*

$$F(t) := \int_{s_0}^{b} f(t, y)d_\beta y, \quad t \in I.$$

Then the following hold:

(i) *If the function f is continuous at $t = s_0$ uniformly with respect to $y \in [b]_\beta$, then $F(t)$ is continuous at $t = s_0$.*
(ii) *$D_\beta F(t)$ at $t \neq s_0$ exists and is given by*

$$D_\beta F(t) = \int_{s_0}^{b} \frac{\partial_\beta f}{\partial_\beta t}(t, y) d_\beta y.$$

(iii) *If $\frac{\partial_\beta f}{\partial_\beta t}(t, y)$ exists uniformly at $t = s_0$ with respect to $y \in I$, then F is β-differentiable and*

$$D_\beta F(s_0) = \int_{s_0}^{b} \frac{\partial_\beta f}{\partial_\beta t}(s_0, y) d_\beta y.$$

Proof

(i) Let $\epsilon > 0$. There exists $\delta > 0$ such that $|t - s_0| < \delta$ implies $\Big|f(t, \beta^k(b)) - f(s_0, \beta^k(b))\Big| < \frac{\epsilon}{|b - s_0|}$, $k \in \mathbb{N}_0$. For $t \in (s_0 - \delta, s_0 + \delta)$, we have

$$\begin{aligned}\Big|F(t) - F(s_0)\Big| &= \Big|\sum_{k=0}^{\infty} (\beta^k(b) - \beta^{k+1}(b))(f(t, \beta^k(b)) - f(s_0, \beta^k(b)))\Big| \\ &< |b - s_0| \frac{\epsilon}{|b - s_0|} \\ &= \epsilon.\end{aligned}$$

Then $F(t)$ is continuous at $t = s_0$.
(ii) At $t \neq s_0$, we have

$$\begin{aligned}D_\beta F(t) &= \frac{\sum_{k=0}^{\infty} (\beta^k(b) - \beta^{k+1}(b))\Big(f(\beta(t), \beta^k(b)) - f(t, \beta^k(b))\Big)}{\beta(t) - t} \\ &= \int_{s_0}^{b} \frac{\partial_\beta f}{\partial_\beta t}(t, y) d_\beta y.\end{aligned}$$

(iii) Let $\epsilon > 0$. There exists $\delta > 0$ such that $0 < |t - s_0| < \delta$ implies

$$\left|\frac{f(t, y) - f(s_0, y)}{t - s_0} - \frac{\partial_\beta f}{\partial_\beta t}(s_0, y)\right| < \frac{\epsilon}{|b - s_0|}, \quad \text{for all } y \in I.$$

For $0 < |t - s_0| < \delta$, we have

$$\begin{aligned}\Big|\frac{F(t)-F(s_0)}{t-s_0} - \int_{s_0}^{b}\frac{\partial_\beta f}{\partial_\beta t}(s_0,y)d_\beta y\Big| &\\ \leq \int_{s_0}^{b}\Big|\frac{f(t,y)-f(s_0,y)}{t-s_0} - \frac{\partial_\beta f}{\partial_\beta t}(s_0,y)\Big|d_\beta y &\\ \leq \frac{\epsilon}{|b-s_0|}|b-s_0| = \epsilon. &\end{aligned}$$

Therefore,

$$D_\beta F(s_0) = \int_{s_0}^{b}\frac{\partial_\beta f}{\partial_\beta t}(s_0,y)d_\beta y.$$

Theorem 7 (Leibniz's Rule) *Define F by*

$$F(t) := \int_{s_0}^{t} f(t,y)d_\beta y.$$

Then the following statements are true:

(i) If the sequence $\{f(t,\beta^k(t))\}_k$, $k \in \mathbb{N}_0$, is bounded uniformly in a neighborhood $(s_0-\delta_1, s_0+\delta_1)$ of s_0, then the function $F(t)$ is continuous at $t = s_0$.

(ii) $D_\beta F(t)$ at $t \neq s_0$ exists and is given by

$$D_\beta F(t) = f(\beta(t),t) + \int_{s_0}^{t}\frac{\partial_\beta f}{\partial_\beta t}(t,y)d_\beta y.$$

(iii) If $f(t,y)$ is continuous at (s_0,s_0), then F is β-differentiable at $t = s_0$ and

$$D_\beta F(s_0) = f(s_0,s_0).$$

Proof

(i) Assume there is $M > 0$ such that $|f(t,\beta^k(t))| \leq M$ for every $k \in \mathbb{N}_0$, $t \in (s_0-\delta_1, s_0+\delta_1)$. Let $\epsilon > 0$. Choose $0 < \delta \leq min(\delta_1, \frac{\epsilon}{M})$. For $t \in (s_0-\delta, s_0+\delta)$, we have

$$\begin{aligned}|F(t)-F(s_0)| &\leq \sum_{k=0}^{\infty}\Big|\beta^k(t)-\beta^{k+1}(t)\Big|\Big|f(t,\beta^k(t))\Big| \\ &\leq M\sum_{k=0}^{\infty}\Big|\beta^k(t)-\beta^{k+1}(t)\Big| \\ &= M|t-s_0| < M\frac{\epsilon}{M} = \epsilon.\end{aligned}$$

Then $F(t)$ is continuous at $t = s_0$.

(ii) For $t \neq s_0$, we have

$$D_\beta F(t) = \frac{1}{\beta(t) - t}\Big(\sum_{k=0}^{\infty}(\beta^{k+1}(t) - \beta^{k+2}(t))f(\beta(t), \beta^{k+1}(t))$$

$$- \sum_{k=0}^{\infty}(\beta^k(t) - \beta^{k+1}(t))f(t, \beta^k(t))\Big)$$

$$= \frac{\sum_{k=1}^{\infty}(\beta^k(t) - \beta^{k+1}(t))\Big(f(\beta(t), \beta^k(t)) - f(t, \beta^k(t))\Big)}{\beta(t) - t}$$

$$+ f(t, t)$$

$$= \frac{\sum_{k=0}^{\infty}(\beta^k(t) - \beta^{k+1}(t))\Big(f(\beta(t), \beta^k(t)) - f(t, \beta^k(t))\Big)}{\beta(t) - t}$$

$$- \frac{(t - \beta(t))(f(\beta(t), t) - f(t, t))}{\beta(t) - t} + f(t, t)$$

$$= f(\beta(t), t) + \int_{s_0}^{t} \frac{\partial_\beta f}{\partial_\beta t}(t, y) d_\beta y.$$

(iii) Assume that $f(t, y)$ is continuous at (s_0, s_0). For $\epsilon > 0$, there exists $\delta > 0$ such that $|f(t, \beta^k(t)) - f(s_0, s_0)| < \epsilon$, $k \in \mathbb{N}_0$ whenever $|t - s_0| < \delta$. In view of Lemma 1, we have

$$\left|\frac{F(t) - F(s_0)}{t - s_0} - f(s_0, s_0)\right|$$

$$= \left|\sum_{k=0}^{\infty} \frac{(\beta^k(t) - \beta^{k+1}(t))}{t - s_0}\Big(f(t, \beta^k(t)) - f(s_0, s_0)\Big)\right|$$

$$\leq \sum_{k=0}^{\infty}\left|\frac{(\beta^k(t) - \beta^{k+1}(t))}{t - s_0}\right|\left|f(t, \beta^k(t)) - f(s_0, s_0)\right|$$

$$< \epsilon,$$

whenever $|t - s_0| < \delta$. Then, we get the desired result.

Theorem 8 *Let $\phi : I \to I$ be bounded. Define the function F by*

$$F(t) := \int_{s_0}^{\phi(t)} f(t, y) d_\beta y.$$

Then the following statements are true:

(i) *$F(t)$ is β-differentiable at $t \neq s_0$ and*

$$D_\beta F(t) = \frac{1}{\beta(t)-t}\int_{\phi(t)}^{\phi(\beta(t))} f(\beta(t), y)d_\beta y + \int_{s_0}^{\phi(t)} \frac{\partial_\beta f}{\partial_\beta t}(t, y)d_\beta y, \quad t \neq s_0.$$

(ii) *If the function ϕ is β-differentiable, $f(t, y)$ is uniformly partially differentiable at $t = s_0$ with respect to $y \in I$ and the function $H(t) = \int_{s_0}^{t} f(s_0, y)d_\beta y$ is differentiable at $\phi(s_0)$, then F is differentiable at s_0 and*

$$D_\beta F(s_0) = f(s_0, \phi(s_0))\phi'(s_0) + \int_{s_0}^{\phi(s_0)} \frac{\partial_\beta f}{\partial_\beta t}(s_0, y)d_\beta y, \quad t = s_0.$$

Proof

(i) For $t \neq s_0$, we conclude that

$$\begin{aligned}
D_\beta F(t) = &\frac{1}{\beta(t)-t}\Big(\sum_{k=0}^{\infty}(\beta^k(\phi(\beta(t))) - \beta^{k+1}(\phi(\beta(t))))f(\beta(t), \beta^k(\phi(\beta(t))))\\
&- \sum_{k=0}^{\infty}(\beta^k(\phi(t)) - \beta^{k+1}(\phi(t)))f(t, \beta^k(\phi(t)))\Big)\\
= &\frac{1}{\beta(t)-t}\Big(\sum_{k=0}^{\infty}(\beta^k(\phi(\beta(t))) - \beta^{k+1}(\phi(\beta(t))))f(\beta(t), \beta^k(\phi(\beta(t))))\\
&- \sum_{k=0}^{\infty}(\beta^k(\phi(t)) - \beta^{k+1}(\phi(t)))\Big[f(t, \beta^k(\phi(t))) - f(\beta(t), \beta^k(\phi(t)))\Big]\Big)\\
&- \frac{1}{\beta(t)-t}\sum_{k=0}^{\infty}(\beta^k(\phi(t)) - \beta^{k+1}(\phi(t)))f(\beta(t), \beta^k(\phi(t)))\\
= &\frac{1}{\beta(t)-t}\Big(\int_{s_0}^{\phi(\beta(t))} f(\beta(t), y)d_\beta y - \int_{s_0}^{\phi(t)} f(\beta(t), y)d_\beta y\Big)\\
&+ \int_{s_0}^{\phi(t)} \frac{\partial_\beta f}{\partial_\beta t}(t, y)d_\beta y\\
= &\frac{1}{\beta(t)-t}\int_{\phi(t)}^{\phi(\beta(t))} f(\beta(t), y)d_\beta y + \int_{s_0}^{\phi(t)} \frac{\partial_\beta f}{\partial_\beta t}(t, y)d_\beta y \quad t \neq s_0.
\end{aligned}$$

(ii) Consider the case $t = s_0$. We write F as follows:

$$F(t) = G(t) + (H \circ \phi)(t),$$

where

$$G(t) = \int_{s_0}^{\phi(t)} \{f(t, y) - f(s_0, y)\}\, d_\beta y.$$

We have

$$\left| \frac{G(t) - G(s_0)}{t - s_0} - \int_{s_0}^{\phi(s_0)} \frac{\partial_\beta f}{\partial_\beta t}(s_0, y)\, d_\beta y \right|$$

$$\leq \int_{s_0}^{\phi(t)} \left| \frac{f(t, y) - f(s_0, y)}{t - s_0} - \frac{\partial_\beta f}{\partial_\beta t}(s_0, y) \right| d_\beta y + \left| \int_{s_0}^{\phi(t)} \frac{\partial_\beta f}{\partial_\beta t}(s_0, y)\, d_\beta y \right.$$

$$\left. - \int_{s_0}^{\phi(s_0)} \frac{\partial_\beta f}{\partial_\beta t}(s_0, y)\, d_\beta y \right|.$$

Since $f(t, y)$ is uniformly partially differentiable at $t = s_0$ with respect to I, and ϕ is bounded, then

$$\int_{s_0}^{\phi(t)} \left| \frac{f(t, y) - f(s_0, y)}{t - s_0} - \frac{\partial_\beta f}{\partial_\beta t}(s_0, y) \right| d_\beta y \to 0 \text{ as } t \to s_0.$$

The continuity of $\frac{\partial_\beta f}{\partial_\beta t}(s_0, y)$ at $y = s_0$ implies that $\psi(t) = \int_{s_0}^{t} \frac{\partial_\beta f}{\partial_\beta t}(s_0, y)\, d_\beta y$ is continuous at $t = s_0$ which in turn implies that $\psi(\phi(t))$ is continuous at $t = s_0$, that is

$$\left| \int_{s_0}^{\phi(t)} \frac{\partial_\beta f}{\partial_\beta t}(s_0, y)\, d_\beta y - \int_{s_0}^{\phi(s_0)} \frac{\partial_\beta f}{\partial_\beta t}(s_0, y)\, d_\beta y \right| \to 0 \text{ as } t \to s_0.$$

Consequently, we deduce that

$$\left| \frac{G(t) - G(s_0)}{t - s_0} - \int_{s_0}^{\phi(s_0)} \frac{\partial_\beta f}{\partial_\beta t}(s_0, y)\, d_\beta y \right| \to 0 \text{ as } t \to s_0. \tag{12}$$

Since H is differentiable at $\phi(s_0)$ and ϕ is differentiable at s_0, then $H \circ \phi$ is differentiable at $t = s_0$ and $D_\beta(H \circ \phi)(s_0) = (H \circ \phi)'(s_0) = H'(\phi(s_0))\phi'(s_0) = f(s_0, \phi(s_0))\phi'(s_0)$. Therefore, the result holds.

Corollary 2 *Define the function F by*

$$F(t) := \int_{\phi(t)}^{\psi(t)} f(t, y)\, d_\beta y.$$

Then, the following statements are true

(i) *Assume that $f(t, y)$ is a continuous function at $y = s_0, t \in I$. Then*

$$D_\beta F(t) = \frac{1}{\beta(t) - t}\left(\int_{\psi(t)}^{\psi(\beta(t))} f(\beta(t), y)\, d_\beta y - \int_{\phi(t)}^{\phi(\beta(t))} f(\beta(t), y)\, d_\beta y\right)$$
$$+ \int_{\phi(t)}^{\psi(t)} \frac{\partial_\beta f}{\partial_\beta t}(t, y)\, d_\beta y, \quad t \neq s_0.$$

(ii) *Assume that $\phi, \psi : I \to \mathbb{R}$ are continuous and β-differentiable, and $f(t, y)$ is uniformly differentiable at $t = s_0$ with respect to $[\phi(s_0)]_\beta$ and $[\psi(s_0)]_\beta$, then $D_\beta F(t)$ exists at $t = s_0$ and*

$$D_\beta F(s_0) = f(s_0, \psi(s_0))\, \psi'(s_0) - f(s_0, \phi(s_0))\, \phi'(s_0) + \int_{\phi(s_0)}^{\psi(s_0)} \frac{\partial_\beta f}{\partial_\beta t}(s_0, y)\, d_\beta y.$$

Proof According to

$$D_\beta\left(\int_{\phi(t)}^{\psi(t)} f(t, y)\, d_\beta y\right) = D_\beta\left(\int_{s_0}^{\psi(t)} f(t, y)\, d_\beta y - \int_{s_0}^{\phi(t)} f(t, y)\, d_\beta y\right),$$

and applying Theorem 8, we get the desired result.

Theorem 9 (Fubini's Theorem) *Let $[a, b]$ be a subset of I, $s_0 \in [a, b]$. Assume that $f : [a, b] \times [a, b] \to \mathbb{R}$ is continuous function at (s_0, s_0). Then,*

$$\int_{s_0}^{b} \int_{s_0}^{a} f(x, y) d_\beta x d_\beta y = \int_{s_0}^{a} \int_{s_0}^{b} f(x, y) d_\beta y d_\beta x. \tag{13}$$

Proof Let $F(t, y) = \int_{s_0}^{t} f(x, y) d_\beta x$, $G(x) = \int_{s_0}^{b} f(x, y) d_\beta y$, $\phi_1(t) = \int_{s_0}^{b} F(t, y) d_\beta y$, and $\phi_2(t) = \int_{s_0}^{t} G(x) d_\beta x$. One can see that ϕ_1 and ϕ_2 are β-differentiable functions. By Theorem 6,

$$D_\beta \phi_1(t) = \int_{s_0}^{b} \frac{\partial_\beta F}{\partial_\beta t}(t, y) d_\beta y.$$

For $t \neq s_0$, we have

$$\begin{aligned}\frac{\partial_\beta}{\partial_\beta t}F(t,y) =& \frac{1}{\beta(t)-t}\Big(\sum_{k=0}^{\infty}(\beta^{k+1}(t)-\beta^{k+2}(t))f(\beta^{k+1}(t),y)\\ &-\sum_{k=0}^{\infty}(\beta^{k}(t)-\beta^{k+1}(t))f(\beta^{k}(t),y)\Big)\\ =& \frac{1}{\beta(t)-t}\Big(\sum_{k=1}^{\infty}(\beta^{k}(t)-\beta^{k+1}(t))f(\beta^{k}(t),y)\\ &-(t-\beta(t))f(t,y)-\sum_{k=1}^{\infty}(\beta^{k}(t)-\beta^{k+1}(t))f(\beta^{k}(t),y)\Big)\\ =& f(t,y).\end{aligned}$$

Then, $D_\beta\phi_1(t) = \int_{s_0}^{b} f(t,y)d_\beta y$. Hence,

$$D_\beta\phi_2(t) = D_\beta\int_{s_0}^{t} G(x)d_\beta x = G(t) = \int_{s_0}^{b} f(t,y)d_\beta y = D_\beta\phi_1(t).$$

Therefore, $\phi_1(t) = \phi_2(t), t \in [a,b]$.

4 Conclusion

In this paper, we investigated a general form of Leibniz's rule and Fubini's theorem based on the general quantum difference operator D_β which is defined by $D_\beta f(t) = \dfrac{f(\beta(t)) - f(t)}{\beta(t) - t}$, $t \neq s_0$, where β is a strictly increasing continuous function defined on an interval $I \subseteq \mathbb{R}$ and has only one fixed point $s_0 \in I$.

Acknowledgments The 3rd author is thankful to his research grant supported by SERB, Project Number: TAR/2018/000001.

References

1. P. Agarwal, J. Choi, Fractional calculus operators and their image formulas. J. Korean Math. Soc. **53**(5), 1183–1210 (2016)
2. K.A. Aldowah, A.B. Malinowska, D.F.M. Torres, The power quantum calculus and variational problems. Dyn. Cont. Disc. Impul. Syst. **19**, 93–116 (2012)
3. M.H. Annaby, A.E. Hamza, K.A. Aldowah, Hahn difference operator and associated Jackson-Norlund integrals. J. Optim Theory Appl. **154**, 133–153 (2012)
4. M.H. Annaby, Z.S. Mansour, *q-Fractional Calculus and Equations* (Springer, Berlin, 2012)

5. T.J. Auch, *Development and Application of Difference and Fractional Calculus on Discrete Time Scales*, Ph.D. Thesis, University of Nebraska-Lincoln (2013)
6. M. Bohner, Calculus of variations on time scales. Dyn. Syst. Appl. **13**, 339–349 (2004)
7. N. Faried, E.M. Shehata, R.M. El Zafarani, On homogenous second order linear general quantum difference equations. J. Inequalities Appl. **2017**, 198 (2017). https://doi.org/10.1186/s13660-017-1471-3
8. N. Faried, E.M. Shehata, R.M. El Zafarani, Theory of nth-order linear general quantum difference equations. Adv. Differ. Equ. **2018**, 264 (2018). https://doi.org/10.1186/s13662-018-1715-7
9. W. Hahn, Über Orthogonalpolynome, die q-differenzenlgleichungen genügen. Math. Nachr. **2**, 4–34 (1949)
10. A.E. Hamza, M.H. Al-Ashwal, Leibniz's rule and Fubini's theorem associated with power quantum difference operators. Int. J. Math. Anal. **9**(55), 2733–2747 (2015)
11. A.E. Hamza, A.M. Sarhan, E.M. Shehata, K.A. Aldowah, A general quantum difference calculus. Adv. Differ. Equ. **2015**, 182 (2015). https://doi.org/10.1186/s13660-015-0518-3
12. A.E. Hamza, E.M. Shehata, Some inequalities based on a general quantum difference operator. J. Inequal. Appl. **2015**, 38 (2015). https://doi.org/10.1186/s13660-015-0566-y
13. A.E. Hamza, E.M. Shehata, Existence and uniqueness of solutions of a general quantum difference equations. Adv. Dyn. Syst. Appl. **11**(1), 45–58 (2016)
14. A.E. Hamza, A.M. Sarhan, E.M. Shehata, Exponential trigonometric and hyperbolic functions associated with a general quantum difference operator. Adv. Dyn. Syst. Appl. **12**(1), 25–38 (2017)
15. V. Kac, P. Cheung, *Quantum Calculus* (Springer, New York, 2002)
16. R.J. Leveque, *Finite Difference Methods for Ordinary and Partial Differential Equations* (SIAM, Philadelphia, 2007)
17. A.B. Malinowska, D.F.M. Torres, The Hahn quantum variational calculus. J. Optim Theory Appl. **147**, 419–442 (2010)
18. L. Nottale, *Fractal Space-Time and Microphysics: Towards a Theory of Scale Relativity* (World Scientific, Singapore, 1993)
19. D.N. Page, Information in black hole radiation. Phys. Rev. Lett. **71**(23), 3743–3746 (1993)
20. S. Salahshour, A. Ahmadian, N. Senu, D. Baleanu, P. Agarwal, On analytical solutions of the fractional differential equation with uncertainty: application to the Basset problem. Entropy **2**(17) , 885–902 (2015)
21. A.M. Sarhan, E.M. Shehata, On the fixed points of certain types of functions for constructing associated calculi. J. Fixed Point Theory Appl. **20**, 124 (2018). https://doi.org/10.1007/s11784-018-0602-x
22. J. Tariboon, S.K. Ntouyas, P. Agarwal, New concepts of fractional quantum calculus and applications to impulsive fractional q-difference equations. Adv. Differ. Equ. **2015**, 18 (2015). https://doi.org/10.1186/s13662-014-0348-8
23. D. Youm, q-deformed conformal quantum mechanics. Phys. Rev. D **62**(5), 095009 (2000)

Some New Ostrowski Type Integral Inequalities via General Fractional Integrals

Artion Kashuri and Themistocles M. Rassias

Abstract In this paper, authors discover an interesting identity regarding Ostrowski type integral inequalities. By using the lemma as an auxiliary result, some new estimates with respect to Ostrowski type integral inequalities via general fractional integrals are obtained. It is pointed out that some new special cases can be deduced from main results. Some applications to special means for different real numbers and new error estimates for the midpoint formula are provided as well. The ideas and techniques of this paper may stimulate further research.

1 Introduction

The following notations are used throughout this paper. We use I to denote an interval on the real line $\mathbb{R} = (-\infty, +\infty)$. For any subset $K \subseteq \mathbb{R}^n$, K° is the interior of K. The set of integrable functions on the interval $[a_1, a_2]$ is denoted by $L[a_1, a_2]$.

The following result is known in the literature as the Ostrowski inequality, see [28] and the references cited therein, which gives an upper bound for the approximation of the integral average $\frac{1}{b-a}\int_a^b f(t)dt$ by the value $f(x)$ at point $x \in [a, b]$.

Theorem 1 *Let $f : I \longrightarrow \mathbb{R}$ be a mapping differentiable on I° and let $a_1, a_2 \in I^\circ$ with $a_1 < a_2$. If $|f'(x)| \leq M$ for all $x \in [a_1, a_2]$, then*

A. Kashuri (✉)
Department of Mathematics, Faculty of Technical Science, University Ismail Qemali of Vlora, Vlorë, Albania

T. M. Rassias
Department of Mathematics, National Technical University of Athens, Athens, Greece
e-mail: trassias@math.ntua.gr

N. J. Daras, T. M. Rassias (eds.), *Computational Mathematics and Variational Analysis*, Springer Optimization and Its Applications 159,
https://doi.org/10.1007/978-3-030-44625-3_8

$$\left| f(x) - \frac{1}{a_2 - a_1} \int_{a_1}^{a_2} f(t)dt \right| \leq M(a_2 - a_1) \left[\frac{1}{4} + \frac{\left(x - \frac{a_1 + a_2}{2}\right)^2}{(a_2 - a_1)^2} \right], \quad \forall x \in [a_1, a_2]. \tag{1}$$

For other recent results concerning Ostrowski type inequalities, see [2–4, 17, 20–22, 26–29], [31–33, 36, 38, 42, 43, 46, 47]. Ostrowski inequality is playing a very important role in all the fields of mathematics, especially in the theory of approximations. Thus such inequalities were studied extensively by many researches and numerous generalizations, extensions and variants of them for various kind of functions like bounded variation, synchronous, Lipschitzian, monotonic, absolutely continuous, and n-times differentiable mappings, etc., appeared in a number of papers, see [11–13, 18]. In recent years, one more dimension has been added to these studies, by introducing a number of integral inequalities involving various fractional operators like Riemann–Liouville, Erdelyi–Kober, Katugampola, conformable fractional integral operators, etc., by many authors, see [1, 6–10, 23–25, 35, 37, 41]. Riemann–Liouville fractional integral operators are the most central between these fractional operators.

In numerical analysis many quadrature rules have been established to approximate the definite integrals. Ostrowski inequality provides the bounds for many numerical quadrature rules, see [14, 15]. In recent decades Ostrowski, Hermite–Hadamard, and Simpson type inequalities are studied in fractional calculus and generalized invexity analysis point of view by many mathematicians, see [5, 16, 19, 30, 34, 45].

Let us recall some special functions and evoke some basic definitions as follows.

Definition 1 For $k \in \mathbb{R}^+$ and $x \in \mathbb{C}$, the k-gamma function is defined by

$$\Gamma_k(x) = \lim_{n \longrightarrow \infty} \frac{n! k^n (nk)^{\frac{x}{k} - 1}}{(x)_{n,k}}. \tag{2}$$

Its integral representation is given by

$$\Gamma_k(\alpha) = \int_0^{\infty} t^{\alpha - 1} e^{-\frac{t^k}{k}} dt. \tag{3}$$

One can note that

$$\Gamma_k(\alpha + k) = \alpha \Gamma_k(\alpha). \tag{4}$$

For $k = 1$, (3) gives integral representation of gamma function.

Definition 2 ([44]) A set $S \subseteq \mathbb{R}^n$ is said to be an invex set with respect to the mapping $\eta : S \times S \longrightarrow \mathbb{R}^n$, if $x + t\eta(y, x) \in S$ for every $x, y \in S$ and $t \in [0, 1]$.

The invex set S is also termed an η-connected set.

Definition 3 Let $S \subseteq \mathbb{R}^n$ be an invex set with respect to $\eta : S \times S \longrightarrow \mathbb{R}^n$. A function $f : S \longrightarrow [0, +\infty)$ is said to be preinvex with respect to η, if for every $x, y \in S$ and $t \in [0, 1]$,

$$f\big(x + t\eta(y, x)\big) \leq (1 - t)f(x) + tf(y). \tag{5}$$

Also, let us define a function $\varphi : [0, \infty) \longrightarrow [0, \infty)$ satisfying the following conditions:

$$\int_0^1 \frac{\varphi(t)}{t} dt < \infty, \tag{6}$$

$$\frac{1}{A} \leq \frac{\varphi(s)}{\varphi(r)} \leq A \text{ for } \frac{1}{2} \leq \frac{s}{r} \leq 2 \tag{7}$$

$$\frac{\varphi(r)}{r^2} \leq B\frac{\varphi(s)}{s^2} \text{ for } s \leq r \tag{8}$$

$$\left|\frac{\varphi(r)}{r^2} - \frac{\varphi(s)}{s^2}\right| \leq C|r - s|\frac{\varphi(r)}{r^2} \text{ for } \frac{1}{2} \leq \frac{s}{r} \leq 2, \tag{9}$$

where $A, B, C > 0$ are independent of $r, s > 0$. If $\varphi(r)r^\alpha$ is increasing for some $\alpha \geq 0$ and $\frac{\varphi(r)}{r^\beta}$ is decreasing for some $\beta \geq 0$, then φ satisfies (6)–(9), see [40]. Therefore, we define the following left-sided and right-sided generalized fractional integral operators, respectively, as follows:

$$_{a_1^+}I_\varphi f(x) = \int_{a_1}^x \frac{\varphi(x-t)}{x-t} f(t)dt, \quad x > a_1, \tag{10}$$

$$_{a_2^-}I_\varphi f(x) = \int_x^{a_2} \frac{\varphi(t-x)}{t-x} f(t)dt, \quad x < a_2. \tag{11}$$

The most important feature of generalized fractional integrals is that they generalize some types of fractional integrals such as Riemann–Liouville fractional integral, k-Riemann–Liouville fractional integral, Katugampola fractional integrals, conformable fractional integral, Hadamard fractional integrals, etc., see [39].

Motivated by the above literatures, the main objective of this paper is to discover in Section 2, an interesting identity in order to establish some new bounds regarding Ostrowski type integral inequalities. By using the lemma as an auxiliary result, some new estimates with respect to Ostrowski type integral inequalities via general fractional integrals will be obtained. It is pointed out that some new special cases will be deduced from main results. In Section 3, some applications to special means for different real numbers and new error estimates for the midpoint formula will be given. The ideas and techniques of this paper may stimulate further research in the field of integral inequalities.

2 Main Result

Let $a_1 < a_2$ and $m \in (0, 1]$ be a fixed number. Throughout this study, for brevity, we define

$$\Lambda_m(t) := \int_0^t \frac{\varphi(\eta(a_2, ma_1)u)}{u} du < \infty, \quad \forall t \in [0, 1], \ \eta(a_2, ma_1) > 0 \tag{12}$$

and $\alpha(x) = \dfrac{x - ma_1}{\eta(a_2, ma_1)}$, $\forall x \in P = [ma_1, ma_1 + \eta(a_2, ma_1)]$.

For establishing some new results regarding general fractional integrals we need to prove the following lemma.

Lemma 1 *Let $f : P \longrightarrow \mathbb{R}$ be a differentiable mapping on $(ma_1, ma_1 + \eta(a_2, ma_1))$. If $f' \in L(P)$, then the following identity for generalized fractional integrals holds:*

$$f(x) - \frac{1}{\eta(a_2, ma_1)\left[\Lambda_m(\alpha(x)) + \Lambda_m(1 - \alpha(x))\right]}$$
$$\times \left[{}_{x^+}I_\varphi f(ma_1 + \eta(a_2, ma_1)) + {}_{x^-}I_\varphi f(ma_1)\right]$$
$$= \frac{1}{\Lambda_m(\alpha(x))} \int_0^{\alpha(x)} \Lambda_m(t) f'(ma_1 + t\eta(a_2, ma_1))\, dt \tag{13}$$
$$- \frac{1}{\Lambda_m(1 - \alpha(x))} \int_{\alpha(x)}^1 \Lambda_m(1 - t) f'(ma_1 + t\eta(a_2, ma_1))\, dt.$$

We denote

$$T_{f,\Lambda_m}(x; a_1, a_2) := \frac{1}{\Lambda_m(\alpha(x))} \int_0^{\alpha(x)} \Lambda_m(t) f'(ma_1 + t\eta(a_2, ma_1))\, dt \tag{14}$$
$$- \frac{1}{\Lambda_m(1 - \alpha(x))} \int_{\alpha(x)}^1 \Lambda_m(1 - t) f'(ma_1 + t\eta(a_2, ma_1))\, dt.$$

Proof Integrating by parts (14) and changing the variable of integration, we have

$$T_{f,\Lambda_m}(x; a_1, a_2) = \frac{1}{\Lambda_m(\alpha(x))} \times \left\{ \frac{\Lambda_m(t) f(ma_1 + t\eta(a_2, ma_1))}{\eta(a_2, ma_1)} \Bigg|_0^{\alpha(x)} \right.$$
$$\left. - \frac{1}{\eta(a_2, ma_1)} \times \int_0^{\alpha(x)} \frac{\varphi(\eta(a_2, ma_1)t)}{t} f(ma_1 + t\eta(a_2, ma_1))\, dt \right\}$$

$$-\frac{1}{\Lambda_m\left(1-\alpha(x)\right)} \times \left\{ \frac{\Lambda_m(1-t) f\left(ma_1+t\eta(a_2,ma_1)\right)}{\eta(a_2,ma_1)} \bigg|_{\alpha(x)}^{1} \right.$$

$$\left. +\frac{1}{\eta(a_2,ma_1)} \times \int_{\alpha(x)}^{1} \frac{\varphi\left(\eta(a_2,ma_1)(1-t)\right)}{(1-t)} f\left(ma_1+t\eta(a_2,ma_1)\right) dt \right\}$$

$$=\frac{1}{\Lambda_m\left(\alpha(x)\right)} \times \left\{ \Lambda_m\left(\alpha(x)\right) f(x) - \frac{1}{\eta(a_2,ma_1)} \times {}_{x^-}I_{\varphi} f\left(ma_1\right) \right\}$$

$$+\frac{1}{\Lambda_m\left(1-\alpha(x)\right)} \times \left\{ \Lambda_m\left(1-\alpha(x)\right) f(x) - \frac{1}{\eta(a_2,ma_1)} \times {}_{x^+}I_{\varphi} f\left(ma_1+\eta(a_2,ma_1)\right) \right\}$$

$$= f(x) - \frac{1}{\eta(a_2,ma_1)\left[\Lambda_m\left(\alpha(x)\right)+\Lambda_m\left(1-\alpha(x)\right)\right]}$$

$$\times \left[{}_{x^+}I_{\varphi} f\left(ma_1+\eta(a_2,ma_1)\right) + {}_{x^-}I_{\varphi} f\left(ma_1\right) \right].$$

This completes the proof of the lemma.

Remark 1 Taking $m=1$ and $\varphi(t)=t$ in Lemma 1, we get [[20], Lemma 2.1].

Theorem 2 *Let $f : P \longrightarrow \mathbb{R}$ be a differentiable mapping on $(ma_1, ma_1+\eta(a_2,ma_1))$. If $|f'|^q$ is preinvex on P for $q>1$ and $p^{-1}+q^{-1}=1$, then the following inequality for generalized fractional integrals holds:*

$$\left|T_{f,\Lambda_m}(x;a_1,a_2)\right| \le \frac{1}{\Lambda_m\left(\alpha(x)\right)} \sqrt[p]{B_{\Lambda_m}(x;p)} \times \sqrt[q]{A_2(x)|f'(ma_1)|^q + A_1(x)|f'(a_2)|^q} \tag{15}$$

$$+\frac{1}{\Lambda_m\left(1-\alpha(x)\right)} \sqrt[p]{C_{\Lambda_m}(x;p)} \times \sqrt[q]{\left(\frac{1}{2}-A_2(x)\right)|f'(ma_1)|^q + \left(\frac{1}{2}-A_1(x)\right)|f'(a_2)|^q},$$

where

$$B_{\Lambda_m}(x;p) := \int_0^{\alpha(x)} \left[\Lambda_m(t)\right]^p dt, \quad C_{\Lambda_m}(x;p) := \int_{\alpha(x)}^{1} \left[\Lambda_m(1-t)\right]^p dt \tag{16}$$

and

$$A_1(x) := \frac{\alpha^2(x)}{2}, \quad A_2(x) := \alpha(x) - A_1(x). \tag{17}$$

Proof From Lemma 1, preinvexity of $|f'|^q$, Hölder inequality, and properties of the modulus, we have

$$\left|T_{f,\Lambda_m}(x;a_1,a_2)\right| \leq \frac{1}{\Lambda_m(\alpha(x))}\int_0^{\alpha(x)} \Lambda_m(t)\left|f'(ma_1+t\eta(a_2,ma_1))\right|dt$$

$$+\frac{1}{\Lambda_m(1-\alpha(x))}\int_{\alpha(x)}^1 \Lambda_m(1-t)\left|f'(ma_1+t\eta(a_2,ma_1))\right|dt$$

$$\leq \frac{1}{\Lambda_m(\alpha(x))}\left(\int_0^{\alpha(x)}\left[\Lambda_m(t)\right]^p dt\right)^{\frac{1}{p}}\left(\int_0^{\alpha(x)}\left|f'(ma_1+t\eta(a_2,ma_1))\right|^q dt\right)^{\frac{1}{q}}$$

$$+\frac{1}{\Lambda_m(1-\alpha(x))}\left(\int_{\alpha(x)}^1\left[\Lambda_m(1-t)\right]^p dt\right)^{\frac{1}{p}}\left(\int_{\alpha(x)}^1\left|f'(ma_1+t\eta(a_2,ma_1))\right|^q dt\right)^{\frac{1}{q}}$$

$$\leq \frac{1}{\Lambda_m(\alpha(x))}\sqrt[p]{B_{\Lambda_m}(x;p)}\left(\int_0^{\alpha(x)}\left[(1-t)\left|f'(ma_1)\right|^q+t\left|f'(a_2)\right|^q\right]dt\right)^{\frac{1}{q}}$$

$$+\frac{1}{\Lambda_m(1-\alpha(x))}\sqrt[p]{C_{\Lambda_m}(x;p)}\left(\int_{\alpha(x)}^1\left[(1-t)\left|f'(ma_1)\right|^q+t\left|f'(a_2)\right|^q\right]dt\right)^{\frac{1}{q}}$$

$$=\frac{1}{\Lambda_m(\alpha(x))}\sqrt[p]{B_{\Lambda_m}(x;p)}\times\sqrt[q]{A_2(x)|f'(ma_1)|^q+A_1(x)|f'(a_2)|^q}$$

$$+\frac{1}{\Lambda_m(1-\alpha(x))}\sqrt[p]{C_{\Lambda_m}(x;p)}\times\sqrt[q]{\left(\frac{1}{2}-A_2(x)\right)|f'(ma_1)|^q+\left(\frac{1}{2}-A_1(x)\right)|f'(a_2)|^q}.$$

The proof of this theorem is complete.

We point out some special cases of Theorem 2.

Corollary 1 *Taking* $m=1$, $\eta(a_2,ma_1)=a_2-ma_1$, *and* $x=\frac{a_1+a_2}{2}$, *in Theorem 2, we get*

$$\left|T_{f,\Lambda_1}\left(\frac{a_1+a_2}{2};a_1,a_2\right)\right| \leq \frac{1}{\sqrt[q]{8}\Lambda_1\left(\frac{1}{2}\right)}\times\sqrt[p]{B^*_{\Lambda_1}\left(\frac{a_1+a_2}{2};p\right)} \tag{18}$$

$$\times\left\{\sqrt[q]{3|f'(a_1)|^q+|f'(a_2)|^q}+\sqrt[q]{|f'(a_1)|^q+3|f'(a_2)|^q}\right\},$$

where

$$B^{*}_{\Lambda_1}\left(\frac{a_1+a_2}{2};p\right):=\int_0^{\frac{1}{2}}\left[\Lambda_1(t)\right]^p dt. \tag{19}$$

Corollary 2 *Taking $p=q=2$ in Theorem 2, we get*

$$\left|T_{f,\Lambda_m}(x;a_1,a_2)\right|\le\frac{1}{\Lambda_m\left(\alpha(x)\right)}\sqrt{B_{\Lambda_m}(x;2)}\times\sqrt{A_2(x)|f'(ma_1)|^2+A_1(x)|f'(a_2)|^2} \tag{20}$$
$$+\frac{1}{\Lambda_m\left(1-\alpha(x)\right)}\sqrt{C_{\Lambda_m}(x;2)}\times\sqrt{\left(\frac{1}{2}-A_2(x)\right)|f'(ma_1)|^2+\left(\frac{1}{2}-A_1(x)\right)|f'(a_2)|^2}.$$

Corollary 3 *Taking $\varphi(t)=t$ in Theorem 2, we get*

$$\left|T_{f,\Lambda_m}(x;a_1,a_2)\right|\le\sqrt[p]{\frac{\alpha(x)}{p+1}}\times\sqrt[q]{A_2(x)|f'(ma_1)|^q+A_1(x)|f'(a_2)|^q} \tag{21}$$
$$+\sqrt[p]{\frac{1-\alpha(x)}{p+1}}\times\sqrt[q]{\left(\frac{1}{2}-A_2(x)\right)|f'(ma_1)|^q+\left(\frac{1}{2}-A_1(x)\right)|f'(a_2)|^q}.$$

Corollary 4 *Taking $\varphi(t)=\frac{t^\alpha}{\Gamma(\alpha)}$ in Theorem 2, we get*

$$\left|T_{f,\Lambda_m}(x;a_1,a_2)\right|\le\sqrt[p]{\frac{\alpha(x)}{p\alpha+1}}\times\sqrt[q]{A_2(x)|f'(ma_1)|^q+A_1(x)|f'(a_2)|^q} \tag{22}$$
$$+\sqrt[p]{\frac{1-\alpha(x)}{p\alpha+1}}\times\sqrt[q]{\left(\frac{1}{2}-A_2(x)\right)|f'(ma_1)|^q+\left(\frac{1}{2}-A_1(x)\right)|f'(a_2)|^q}.$$

Corollary 5 *Taking $\varphi(t)=\frac{t^{\frac{\alpha}{k}}}{k\Gamma_k(\alpha)}$ in Theorem 2, we get*

$$\left|T_{f,\Lambda_m}(x;a_1,a_2)\right|\le\sqrt[p]{\frac{\alpha(x)}{\frac{p\alpha}{k}+1}}\times\sqrt[q]{A_2(x)|f'(ma_1)|^q+A_1(x)|f'(a_2)|^q} \tag{23}$$
$$+\sqrt[p]{\frac{1-\alpha(x)}{\frac{p\alpha}{k}+1}}\times\sqrt[q]{\left(\frac{1}{2}-A_2(x)\right)|f'(ma_1)|^q+\left(\frac{1}{2}-A_1(x)\right)|f'(a_2)|^q}.$$

Corollary 6 *Taking $m = 1$, $\eta(a_2, ma_1) = a_2 - ma_1$, and $x = \frac{a_1+a_2}{2}$, in Corollary 5, we get*

$$\left|T_{f,\Lambda_1}\left(\frac{a_1+a_2}{2}; a_1, a_2\right)\right| \leq \frac{1}{2\sqrt[q]{4}}\sqrt[p]{\frac{k}{p\alpha+k}} \tag{24}$$

$$\times\left\{\sqrt[q]{3|f'(a_1)|^q + |f'(a_2)|^q} + \sqrt[q]{|f'(a_1)|^q + 3|f'(a_2)|^q}\right\}.$$

Corollary 7 *Taking $m = 1$, $\eta(a_2, ma_1) = a_2 - ma_1$, $\varphi(t) = t(a_2 - t)^{\alpha-1}$ for $\alpha \in (0, 1)$ and $f(x)$ is symmetric to $x = \frac{a_1+a_2}{2}$, in Theorem 2, we get*

$$\left|T_{f,\Lambda_1}\left(\frac{a_1+a_2}{2}; a_1, a_2\right)\right| \leq \frac{1}{2\sqrt[q]{4}}\left(\frac{a_2^\alpha - a_1^\alpha}{a_2^\alpha - \left(\frac{a_1+a_2}{2}\right)^\alpha}\right) \tag{25}$$

$$\times\left\{\sqrt[q]{3|f'(a_1)|^q + |f'(a_2)|^q} + \sqrt[q]{|f'(a_1)|^q + 3|f'(a_2)|^q}\right\}.$$

Corollary 8 *Taking $m = 1$, $\eta(a_2, ma_1) = a_2 - ma_1$, $\varphi(t) = \frac{t}{\alpha}\exp\left[\left(-\frac{1-\alpha}{\alpha}\right)t\right]$, for $\alpha \in (0, 1)$ and $x = \frac{a_1+a_2}{2}$, in Theorem 2, we get*

$$\left|T_{f,\Lambda_1}\left(\frac{a_1+a_2}{2}; a_1, a_2\right)\right| \leq \frac{1}{2\sqrt[q]{4}}\left(\frac{\exp\left[\left(-\frac{1-\alpha}{\alpha}\right)(a_2-a_1)\right]-1}{\exp\left[\left(-\frac{1-\alpha}{\alpha}\right)\frac{(a_2-a_1)}{2}\right]-1}\right) \tag{26}$$

$$\times\left\{\sqrt[q]{3|f'(a_1)|^q + |f'(a_2)|^q} + \sqrt[q]{|f'(a_1)|^q + 3|f'(a_2)|^q}\right\}.$$

Theorem 3 *Let $f : P \longrightarrow \mathbb{R}$ be a differentiable mapping on $(ma_1, ma_1 + \eta(a_2, ma_1))$. If $|f'|^q$ is preinvex on P for $q \geq 1$, then the following inequality for generalized fractional integrals holds:*

$$\left|T_{f,\Lambda_m}(x; a_1, a_2)\right| \leq \frac{1}{\Lambda_m(\alpha(x))}\left(B_{\Lambda_m}(x; 1)\right)^{1-\frac{1}{q}} \tag{27}$$

$$\times\sqrt[q]{D_{\Lambda_m}(x)|f'(ma_1)|^q + E_{\Lambda_m}(x)|f'(a_2)|^q}$$

$$+\frac{1}{\Lambda_m(1-\alpha(x))}\left(C_{\Lambda_m}(x; 1)\right)^{1-\frac{1}{q}} \times \sqrt[q]{F_{\Lambda_m}(x)|f'(ma_1)|^q + G_{\Lambda_m}(x)|f'(a_2)|^q},$$

where

$$D_{\Lambda_m}(x) := \int_0^{\alpha(x)} (1-t)\Lambda_m(t)dt, \quad E_{\Lambda_m}(x) := \int_0^{\alpha(x)} t\Lambda_m(t)dt \tag{28}$$

and

$$F_{\Lambda_m}(x) := \int_0^{1-\alpha(x)} t\Lambda_m(t)dt, \quad G_{\Lambda_m}(x) := \int_0^{1-\alpha(x)} (1-t)\Lambda_m(t)dt \tag{29}$$

and $B_{\Lambda_m}(x;1)$, $C_{\Lambda_m}(x;1)$ *are defined as in Theorem 2.*

Proof From Lemma 1, preinvexity of $|f'|^q$, the well-known power mean inequality and properties of the modulus, we have

$$\left|T_{f,\Lambda_m}(x;a_1,a_2)\right| \le \frac{1}{\Lambda_m(\alpha(x))}\int_0^{\alpha(x)} \Lambda_m(t)\left|f'(ma_1+t\eta(a_2,ma_1))\right|dt$$

$$+\frac{1}{\Lambda_m(1-\alpha(x))}\int_{\alpha(x)}^1 \Lambda_m(1-t)\left|f'(ma_1+t\eta(a_2,ma_1))\right|dt$$

$$\le \frac{1}{\Lambda_m(\alpha(x))}\left(\int_0^{\alpha(x)} \Lambda_m(t)dt\right)^{1-\frac{1}{q}}\left(\int_0^{\alpha(x)} \Lambda_m(t)\left|f'(ma_1+t\eta(a_2,ma_1))\right|^q dt\right)^{\frac{1}{q}}$$

$$+\frac{1}{\Lambda_m(1-\alpha(x))}\left(\int_{\alpha(x)}^1 \Lambda_m(1-t)dt\right)^{1-\frac{1}{q}}$$

$$\left(\int_{\alpha(x)}^1 \Lambda_m(1-t)\left|f'(ma_1+t\eta(a_2,ma_1))\right|^q dt\right)^{\frac{1}{q}}$$

$$\le \frac{1}{\Lambda_m(\alpha(x))}\left(B_{\Lambda_m}(x;1)\right)^{1-\frac{1}{q}}\left(\int_0^{\alpha(x)} \Lambda_m(t)\left[(1-t)\left|f'(ma_1)\right|^q+t\left|f'(a_2)\right|^q\right]dt\right)^{\frac{1}{q}}$$

$$+\frac{1}{\Lambda_m(1-\alpha(x))}\left(C_{\Lambda_m}(x;1)\right)^{1-\frac{1}{q}}$$

$$\left(\int_{\alpha(x)}^1 \Lambda_m(1-t)\left[(1-t)\left|f'(ma_1)\right|^q+t\left|f'(a_2)\right|^q\right]dt\right)^{\frac{1}{q}}$$

$$= \frac{1}{\Lambda_m(\alpha(x))}\left(B_{\Lambda_m}(x;1)\right)^{1-\frac{1}{q}} \times \sqrt[q]{D_{\Lambda_m}(x)|f'(ma_1)|^q + E_{\Lambda_m}(x)|f'(a_2)|^q}$$

$$+\frac{1}{\Lambda_m(1-\alpha(x))}\left(C_{\Lambda_m}(x;1)\right)^{1-\frac{1}{q}} \times \sqrt[q]{F_{\Lambda_m}(x)|f'(ma_1)|^q + G_{\Lambda_m}(x)|f'(a_2)|^q}.$$

The proof of this theorem is complete.

We point out some special cases of Theorem 3.

Corollary 9 *Taking* $m = 1$, $\eta(a_2, ma_1) = a_2 - ma_1$, *and* $x = \frac{a_1+a_2}{2}$, *in Theorem 3, we get*

$$\left|T_{f,\Lambda_1}\left(\frac{a_1+a_2}{2};a_1,a_2\right)\right| \le \frac{1}{\Lambda_1\left(\frac{1}{2}\right)}\left(B^*_{\Lambda_1}\left(\frac{a_1+a_2}{2};1\right)\right)^{1-\frac{1}{q}} \tag{30}$$

$$\times\left\{\sqrt[q]{D_{\Lambda_1}\left(\frac{a_1+a_2}{2}\right)|f'(a_1)|^q + E_{\Lambda_1}\left(\frac{a_1+a_2}{2}\right)|f'(a_2)|^q}\right.$$

$$\left.+\sqrt[q]{E_{\Lambda_1}\left(\frac{a_1+a_2}{2}\right)|f'(a_1)|^q + D_{\Lambda_1}\left(\frac{a_1+a_2}{2}\right)|f'(a_2)|^q}\right\},$$

where $B^*_{\Lambda_1}\left(\frac{a_1+a_2}{2};1\right)$ *is defined by Equation (19).*

Corollary 10 *Taking* $q = 1$ *in Theorem 3, we get*

$$\left|T_{f,\Lambda_m}(x;a_1,a_2)\right| \le \frac{1}{\Lambda_m(\alpha(x))}\times\left[D_{\Lambda_m}(x)|f'(ma_1)|+E_{\Lambda_m}(x)|f'(a_2)|\right] \tag{31}$$

$$+\frac{1}{\Lambda_m(1-\alpha(x))}\times\left[F_{\Lambda_m}(x)|f'(ma_1)|+G_{\Lambda_m}(x)|f'(a_2)|\right].$$

Corollary 11 *Taking* $\varphi(t) = t$ *in Theorem 3, we get*

$$\left|T_{f,\Lambda_m}(x;a_1,a_2)\right| \le \frac{\left(A_1(x)\right)^{1-\frac{1}{q}}}{\alpha(x)}\times\sqrt[q]{\beta_{\alpha(x)}(2,2)|f'(ma_1)|^q+\frac{2\alpha(x)A_1(x)}{3}|f'(a_2)|^q} \tag{32}$$

$$+\frac{\left(\frac{1}{2}-A_2(x)\right)^{1-\frac{1}{q}}}{1-\alpha(x)}\times\sqrt[q]{\frac{(1-\alpha(x))^3}{3}|f'(ma_1)|^q+\beta_{(1-\alpha(x))}(2,2)|f'(a_2)|^q},$$

where $\beta_a(\cdot,\cdot)$ *is the incomplete beta function and* $A_1(x)$, $A_2(x)$ *are defined by Equation (17).*

Corollary 12 *Taking* $\varphi(t) = \frac{t^{\alpha}}{\Gamma(\alpha)}$ *in Theorem 3, we get*

$$\left|T_{f,\Lambda_m}(x;a_1,a_2)\right| \le \left(\frac{\Gamma(\alpha+1)}{\Gamma(\alpha+2)}\right)^{1-\frac{1}{q}} \tag{33}$$

$$\times\left\{(\alpha(x))^{1-\frac{\alpha+1}{q}}\sqrt[q]{\beta_{\alpha(x)}(\alpha+1,2)|f'(ma_1)|^q+\frac{\alpha^{\alpha+2}(x)}{\alpha+2}|f'(a_2)|^q}\right.$$

$$\left.+(1-\alpha(x))^{1-\frac{\alpha+1}{q}}\times\sqrt[q]{\frac{(1-\alpha(x))^{\alpha+2}}{\alpha+2}|f'(ma_1)|^q+\beta_{(1-\alpha(x))}(\alpha+1,2)|f'(a_2)|^q}\right\}.$$

Corollary 13 *Taking* $\varphi(t) = \frac{t^{\frac{\alpha}{k}}}{k\Gamma_k(\alpha)}$ *in Theorem 3, we get*

$$\left|T_{f,\Lambda_m}(x;a_1,a_2)\right| \le \left(\frac{\Gamma_k(\alpha+k)}{\Gamma_k(\alpha+k+1)}\right)^{1-\frac{1}{q}} \tag{34}$$

$$\times\left\{(\alpha(x))^{1-\frac{\frac{\alpha}{k}+1}{q}}\sqrt[q]{\beta_{\alpha(x)}\left(\frac{\alpha}{k}+1,2\right)|f'(ma_1)|^q+\frac{\alpha^{\frac{\alpha}{k}+2}(x)}{\frac{\alpha}{k}+2}|f'(a_2)|^q}\right.$$

$$\left.+(1-\alpha(x))^{1-\frac{\frac{\alpha}{k}+1}{q}}\times\sqrt[q]{\frac{(1-\alpha(x))^{\frac{\alpha}{k}+2}}{\frac{\alpha}{k}+2}|f'(ma_1)|^q+\beta_{(1-\alpha(x))}\left(\frac{\alpha}{k}+1,2\right)|f'(a_2)|^q}\right\}.$$

Corollary 14 *Taking* $m = 1$, $\eta(a_2, ma_1) = a_2 - ma_1$, *and* $x = \dfrac{a_1+a_2}{2}$, *in Corollary 13, we get*

$$\left|T_{f,\Lambda_1}\left(\frac{a_1+a_2}{2};a_1,a_2\right)\right| \le \left(\frac{\Gamma_k(\alpha+k)}{\Gamma_k(\alpha+k+1)}\right)^{1-\frac{1}{q}}\left(\frac{1}{2}\right)^{1-\frac{\alpha+k}{kq}} \tag{35}$$

$$\times\left\{\sqrt[q]{\beta_{\frac{1}{2}}\left(\frac{\alpha}{k}+1,2\right)|f'(a_1)|^q+\frac{k}{2^{\frac{\alpha+2k}{k}}(\alpha+2k)}|f'(a_2)|^q}\right.$$

$$\left.+\sqrt[q]{\frac{k}{2^{\frac{\alpha+2k}{k}}(\alpha+2k)}|f'(a_1)|^q+\beta_{\frac{1}{2}}\left(\frac{\alpha}{k}+1,2\right)|f'(a_2)|^q}\right\}.$$

Corollary 15 *Taking* $m = 1$, $\eta(a_2, ma_1) = a_2 - ma_1$, $\varphi(t) = t(a_2-t)^{\alpha-1}$ *for* $\alpha \in (0,1)$ *and* $f(x)$ *is symmetric to* $x = \dfrac{a_1+a_2}{2}$, *in Theorem 3, we get the same inequality (25) as in Corollary 7.*

3 Applications to Special Means and Some New Error Estimates for Midpoint Formula

Consider the following special means for different real numbers α, β and $\alpha\beta \neq 0$, as follows:

1. The arithmetic mean:
$$A := A(\alpha, \beta) = \frac{\alpha + \beta}{2},$$
2. The harmonic mean:
$$H := H(\alpha, \beta) = \frac{2}{\frac{1}{\alpha} + \frac{1}{\beta}},$$
3. The logarithmic mean:
$$L := L(\alpha, \beta) = \frac{\beta - \alpha}{\ln|\beta| - \ln|\alpha|},$$
4. The generalized log-mean:
$$L_n := L_n(\alpha, \beta) = \left[\frac{\beta^{n+1} - \alpha^{n+1}}{(n+1)(\beta - \alpha)}\right]^{\frac{1}{n}}; \quad n \in \mathbb{Z} \setminus \{-1, 0\}.$$

It is well-known that L_n is monotonic nondecreasing over $n \in \mathbb{Z}$ with $L_{-1} := L$. In particular, we have the following inequality $H \leq L \leq A$. Now, using the theory results in Section 2, we give some applications to special means for different real numbers.

Proposition 1 *Let $a_1, a_2 \in \mathbb{R} \setminus \{0\}$, where $a_1 < a_2$. Then for $r \geq 2$, where $q > 1$ and $p^{-1} + q^{-1} = 1$, the following inequality holds:*

$$\left|A^r(a_1, a_2) - \frac{1}{a_2 - a_1} L_r^r(a_1, a_2)\right| \leq \frac{r}{2\sqrt[q]{2}\sqrt[p]{p+1}} \tag{36}$$
$$\times \left\{\sqrt[q]{A\left(3|a_1|^{q(r-1)}, |a_2|^{q(r-1)}\right)} + \sqrt[q]{A\left(|a_1|^{q(r-1)}, 3|a_2|^{q(r-1)}\right)}\right\}.$$

Proof Taking $m = 1$, $\eta(a_2, ma_1) = a_2 - ma_1$, $x = \dfrac{a_1 + a_2}{2}$, $f(t) = t^r$, and $\varphi(t) = t$, in Theorem 2, one can obtain the result immediately.

Proposition 2 *Let $a_1, a_2 \in \mathbb{R} \setminus \{0\}$, where $a_1 < a_2$. Then for $q > 1$ and $p^{-1} + q^{-1} = 1$, the following inequality holds:*

$$\left|\frac{1}{A(a_1,a_2)}-\frac{1}{(a_2-a_1)L(a_1,a_2)}\right|\le\sqrt[q]{\frac{3}{2}}\frac{1}{2\sqrt[p]{p+1}} \tag{37}$$

$$\times\left\{\sqrt[q]{\frac{1}{H\left(|a_1|^{2q},3|a_2|^{2q}\right)}}+\sqrt[q]{\frac{1}{H\left(3|a_1|^{2q},|a_2|^{2q}\right)}}\right\}.$$

Proof Taking $m=1$, $\eta(a_2,ma_1)=a_2-ma_1$, $x=\dfrac{a_1+a_2}{2}$, $f(t)=\dfrac{1}{t}$, and $\varphi(t)=t$, in Theorem 2, one can obtain the result immediately.

Proposition 3 *Let $a_1,a_2\in\mathbb{R}\setminus\{0\}$, where $a_1<a_2$. Then for $r\ge 2$, where $q\ge 1$, the following inequality holds:*

$$\left|A^r(a_1,a_2)-\frac{1}{a_2-a_1}L_r^r(a_1,a_2)\right|\le\left(\frac{1}{4}\right)^{1-\frac{3}{q}}\frac{1}{\sqrt[q]{3}} \tag{38}$$

$$\times\left\{\sqrt[q]{A\left(24\beta_{\frac{1}{2}}(2,2)|a_1|^{q(r-1)},|a_2|^{q(r-1)}\right)}+\sqrt[q]{A\left(|a_1|^{q(r-1)},24\beta_{\frac{1}{2}}(2,2)|a_2|^{q(r-1)}\right)}\right\}.$$

Proof Taking $m=1$, $\eta(a_2,ma_1)=a_2-ma_1$, $x=\dfrac{a_1+a_2}{2}$, $f(t)=t^r$ and $\varphi(t)=t$, in Theorem 3, one can obtain the result immediately.

Proposition 4 *Let $a_1,a_2\in\mathbb{R}\setminus\{0\}$, where $a_1<a_2$. Then for $q\ge 1$, the following inequality holds:*

$$\left|\frac{1}{A(a_1,a_2)}-\frac{1}{(a_2-a_1)L(a_1,a_2)}\right|\le\left(\frac{1}{2}\right)^{1+\frac{1}{q}}\sqrt[q]{3\beta_{\frac{1}{2}}(2,2)} \tag{39}$$

$$\times\left\{\sqrt[q]{\frac{1}{H\left(|a_1|^{2q},24\beta_{\frac{1}{2}}(2,2)|a_2|^{2q}\right)}}+\sqrt[q]{\frac{1}{H\left(24\beta_{\frac{1}{2}}(2,2)|a_1|^{2q},|a_2|^{2q}\right)}}\right\}.$$

Proof Taking $m=1$, $\eta(a_2,ma_1)=a_2-ma_1$, $x=\dfrac{a_1+a_2}{2}$, $f(t)=\dfrac{1}{t}$, and $\varphi(t)=t$, in Theorem 3, one can obtain the result immediately.

Remark 2 Applying our Theorems 2 and 3 for appropriate choices of function $\varphi(t)=\frac{t^{\alpha}}{\Gamma(\alpha)},\ \frac{t^{\frac{\alpha}{k}}}{k\Gamma_k(\alpha)};\ \varphi(t)=\frac{t}{\alpha}\exp\left[\left(-\frac{1-\alpha}{\alpha}\right)t\right]$ for $\alpha\in(0,1)$, such that $|f'|^q$ to be preinvex, we can deduce some new general fractional integral inequalities using above special means. The details are left to the interested reader.

Next, we provide some new error estimates for the midpoint formula.

Let Q be the partition of the points $a_1 = x_0 < x_1 < \ldots < x_k = a_2$ of the interval $[a_1, a_2]$. Let us consider the following quadrature formula:

$$\int_{a_1}^{a_2} f(x)dx = M(f, Q) + E(f, Q),$$

where

$$M(f, Q) = \sum_{i=0}^{k-1} f\left(\frac{x_i + x_{i+1}}{2}\right)(x_{i+1} - x_i)^2$$

is the midpoint version and $E(f, Q)$ denote their associated approximation error.

Proposition 5 *Let $f : [a_1, a_2] \longrightarrow \mathbb{R}$ be a differentiable function on (a_1, a_2), where $a_1 < a_2$. If $|f'|^q$ is convex on $[a_1, a_2]$ for $q > 1$ and $p^{-1} + q^{-1} = 1$, then the following inequality holds:*

$$\left|E(f, Q)\right| \leq \frac{1}{2\sqrt[q]{4}\sqrt[p]{p+1}} \times \sum_{i=0}^{k-1}(x_{i+1} - x_i)^3 \tag{40}$$

$$\times\left\{\sqrt[q]{3|f'(x_i)|^q + |f'(x_{i+1})|^q} + \sqrt[q]{|f'(x_i)|^q + 3|f'(x_{i+1})|^q}\right\}.$$

Proof Applying Theorem 2 for $m = 1$, $\eta(a_2, ma_1) = a_2 - ma_1$, $\varphi(t) = t$, and $x = \dfrac{a_1 + a_2}{2}$, on the subintervals $[x_i, x_{i+1}]$ $(i = 0, \ldots, k-1)$ of the partition Q, we have

$$\left|f\left(\frac{x_i + x_{i+1}}{2}\right) - \frac{1}{(x_{i+1} - x_i)^2}\int_{x_i}^{x_{i+1}} f(x)dx\right| \leq \frac{(x_{i+1} - x_i)^2}{2\sqrt[q]{4}\sqrt[p]{p+1}} \tag{41}$$

$$\times\left\{\sqrt[q]{3|f'(x_i)|^q + |f'(x_{i+1})|^q} + \sqrt[q]{|f'(x_i)|^q + 3|f'(x_{i+1})|^q}\right\}.$$

Hence from (41), we get

$$\left|E(f, Q)\right| = \left|\int_{a_1}^{a_2} f(x)dx - M(f, Q)\right|$$

$$\leq \left|\sum_{i=0}^{k-1}\left\{\int_{x_i}^{x_{i+1}} f(x)dx - f\left(\frac{x_i + x_{i+1}}{2}\right)(x_{i+1} - x_i)^2\right\}\right|$$

$$\leq \sum_{i=0}^{k-1}\left|\left\{\int_{x_i}^{x_{i+1}} f(x)dx - f\left(\frac{x_i + x_{i+1}}{2}\right)(x_{i+1} - x_i)^2\right\}\right|$$

$$\leq \frac{1}{2\sqrt[q]{4}\sqrt[p]{p+1}} \sum_{i=0}^{k-1}(x_{i+1}-x_i)^3$$

$$\times\left\{\sqrt[q]{3|f'(x_i)|^q+|f'(x_{i+1})|^q}+\sqrt[q]{|f'(x_i)|^q+3|f'(x_{i+1})|^q}\right\}.$$

The proof of this proposition is complete.

Proposition 6 *Let $f:[a_1,a_2]\longrightarrow\mathbb{R}$ be a differentiable function on (a_1,a_2), where $a_1<a_2$. If $|f'|^q$ is convex on $[a_1,a_2]$ for $q\geq 1$, then the following inequality holds:*

$$\left|E(f,Q)\right| \leq \left(\frac{1}{4}\right)^{1-\frac{3}{q}} \frac{1}{\sqrt[q]{3}} \times \sum_{i=0}^{k-1}(x_{i+1}-x_i)^3 \tag{42}$$

$$\times\left\{\sqrt[q]{24\beta_{\frac{1}{2}}(2,2)|f'(x_i)|^q+|f'(x_{i+1})|^q}+\sqrt[q]{|f'(x_i)|^q+24\beta_{\frac{1}{2}}(2,2)|f'(x_{i+1})|^q}\right\}.$$

Proof The proof is analogous as to that of Proposition 5 but use Theorem 3.

Remark 3 Applying our Theorems 2 and 3, where $m=1$ and $\eta(a_2,ma_1)=a_2-ma_1$, for appropriate choices of function $\varphi(t)=\frac{t^\alpha}{\Gamma(\alpha)}$, $\frac{t^{\frac{\alpha}{k}}}{k\Gamma_k(\alpha)}$; $\varphi(t)=t(a_2-t)^{\alpha-1}$, where $f(x)$ is symmetric to $x=\dfrac{a_1+a_2}{2}$, and $\varphi(t)=\frac{t}{\alpha}\exp\left[\left(-\frac{1-\alpha}{\alpha}\right)t\right]$ for $\alpha\in(0,1)$, such that $|f'|^q$ to be convex, we can deduce some new bounds for midpoint formula using above ideas and techniques. The details are left to the interested reader.

References

1. T. Abdeljawad, On conformable fractional calculus. J. Comput. Appl. Math. **279**, 57–66 (2015)
2. R.P. Agarwal, M.J. Luo, R.K. Raina, On Ostrowski type inequalities. Fasc. Math. **204**, 5–27 (2016)
3. M. Ahmadmir, R. Ullah, Some inequalities of Ostrowski and Grüss type for triple integrals on time scales. Tamkang J. Math. **42**(4), 415–426 (2011)
4. M. Alomari, M. Darus, S.S. Dragomir, P. Cerone, Ostrowski type inequalities for functions whose derivatives are s-convex in the second sense. Appl. Math. Lett. **23**, 1071–1076 (2010)
5. T. Antczak, Mean value in invexity analysis. Nonlinear Anal. **60**, 1473–1484 (2005)
6. Y.-M. Chu, M. Adil Khan, T. Ali, S.S. Dragomir, Inequalities for α-fractional differentiable functions. J. Inequal. Appl. **2017**, 12 (2017). Article no. 93
7. Z. Dahmani, New inequalities in fractional integrals. Int. J. Nonlinear Sci. **9**(4), 493–497 (2010)
8. Z. Dahmani, On Minkowski and Hermite-Hadamard integral inequalities via fractional integration. Ann. Funct. Anal. **1**(1), 51–58 (2010)
9. Z. Dahmani, L. Tabharit, S. Taf, Some fractional integral inequalities. Nonlinear. Sci. Lett. A **1**(2), 155–160 (2010)

10. Z. Dahmani, L. Tabharit, S. Taf, New generalizations of Grüss inequality using Riemann-Liouville fractional integrals. Bull. Math. Anal. Appl. **2**(3), 93–99 (2010)
11. S.S. Dragomir, On the Ostrowski's integral inequality for mappings with bounded variation and applications. Math. Inequal. Appl. **1**(2) (1998)
12. S.S. Dragomir, The Ostrowski integral inequality for Lipschitzian mappings and applications. Comput. Math. Appl. **38**, 33–37 (1999)
13. S.S. Dragomir, Ostrowski-type inequalities for Lebesgue integral: a survey of recent results. Aust. J. Math. Anal. Appl. **14**(1), 1–287 (2017)
14. S.S. Dragomir, S. Wang, An inequality of Ostrowski-Grüss type and its applications to the estimation of error bounds for some special means and for some numerical quadrature rules. Comput. Math. Appl. **13**(11), 15–20 (1997)
15. S.S. Dragomir, S. Wang, A new inequality of Ostrowski's type in L_1-norm and applications to some special means and to some numerical quadrature rules. Tamkang J. Math. **28**, 239–244 (1997)
16. T.S. Du, J.G. Liao, Y.J. Li, Properties and integral inequalities of Hadamard-Simpson type for the generalized (s, m)-preinvex functions. J. Nonlinear Sci. Appl. **9**, 3112–3126 (2016)
17. G. Farid, Some new Ostrowski type inequalities via fractional integrals. Int. J. Anal. Appl. **14**(1), 64–68 (2017)
18. G. Farid, A. Javed, A.U. Rehman, On Hadamard inequalities for n-times differentiable functions which are relative convex via Caputo k-fractional derivatives. Nonlinear Anal. Forum. **22**(2), 17–28 (2017)
19. H. Hudzik, L. Maligranda, Some remarks on s-convex functions. Aequationes Math. **48**, 100–111 (1994)
20. I. Işcan, Ostrowski type inequalities for functions whose derivatives are preinvex. Bull. Iranian Math. Soc. **40**, 373–386 (2014)
21. A. Kashuri, R. Liko, Ostrowski type fractional integral inequalities for generalized (s, m, φ)-preinvex functions. Aust. J. Math. Anal. Appl. **13**(1), 1–11 (2016). Article 16
22. A. Kashuri, R. Liko, Generalizations of Hermite-Hadamard and Ostrowski type inequalities forMT_m-preinvex functions. Proyecciones **36**(1), 45–80 (2017)
23. U.N. Katugampola, A new approach to generalized fractional derivatives. Bulletin Math. Anal. Appl. **6**(4), 1–15 (2014)
24. R. Khalil, M. Al Horani, A. Yousef, M. Sababheh, A new definition of fractional derivative. J. Comput. Appl. Math. **264**, 65–70 (2014)
25. A.A. Kilbas, H.M. Srivastava, J.J. Trujillo, *Theory and Applications of Fractional Differential Equations*. North-Holland Mathematics Studies, vol. 204 (Elsevier, New York, 2006)
26. Z. Liu, Some Ostrowski-Grüss type inequalities and applications. Comput. Math. Appl. **53**, 73–79 (2007)
27. Z. Liu, Some companions of an Ostrowski type inequality and applications. J. Inequal. Pure Appl. Math **10**(2), 12 (2009). Art. 52
28. W. Liu, W. Wen, J. Park, Ostrowski type fractional integral inequalities for MT-convex functions. Miskolc Math. Notes **16**(1), 249–256 (2015)
29. M. Matloka, Ostrowski type inequalities for functions whose derivatives are h-convex via fractional integrals. J. Sci. Res. Rep. **3**(12), 1633–1641 (2014)
30. D.S. Mitrinovic, J.E. Pečarić, A.M. Fink, *Classical and New Inequalities in Analysis* (Kluwer Academic Publishers, Dordrecht, 1993)
31. M.E. Özdemir, H. Kavurmac, E. Set, Ostrowski's type inequalities for (α, m)-convex functions. Kyungpook Math. J. **50**, 371–378 (2010)
32. B.G. Pachpatte, On an inequality of Ostrowski type in three independent variables. J. Math. Anal. Appl. **249**, 583–591 (2000)
33. B.G. Pachpatte, On a new Ostrowski type inequality in two independent variables. Tamkang J. Math. **32**(1), 45–49 (2001)
34. R. Pini, Invexity and generalized convexity. Optimization **22**, 513–525 (1991)
35. S.D. Purohit, S.L. Kalla, Certain inequalities related to the Chebyshev's functional involving Erdelyi-Kober operators. Scientia Ser. A Math. Sci. **25**, 53–63 (2014)

36. A. Rafiq, N.A. Mir, F. Ahmad, Weighted Čebyšev-Ostrowski type inequalities. Appl. Math. Mech. (English Edition) **28**(7), 901–906 (2007)
37. R.K. Raina, On generalized Wright's hypergeometric functions and fractional calculus operators. East Asian Math. J. **21**(2), 191–203 (2005)
38. M.Z. Sarikaya, On the Ostrowski type integral inequality. Acta Math. Univ. Comenianae **79**(1), 129–134 (2010)
39. M.Z. Sarikaya, F. Ertuğral, On the generalized Hermite-Hadamard inequalities (2017). https://www.researchgate.net/publication/321760443
40. M.Z. Sarikaya, H. Yildirim, On generalization of the Riesz potential. Indian J. Math. Math. Sci. **3**(2), 231–235 (2007)
41. E. Set, A. Gözpnar, J. Choi, Hermite-Hadamard type inequalities for twice differentiable m-convex functions via conformable fractional integrals. Far East J. Math. Sci. **101**(4), 873–891 (2017)
42. M. Tunç, Ostrowski type inequalities for functions whose derivatives are MT-convex. J. Comput. Anal. Appl. **17**(4), 691–696 (2014)
43. N. Ujević, Sharp inequalities of Simpson type and Ostrowski type. Comput. Math. Appl. **48**, 145–151 (2004)
44. T. Weir, B. Mond, Preinvex functions in multiple objective optimization. J. Math. Anal. Appl. **136**, 29–38 (1988)
45. X.M. Yang, X.Q. Yang, K.L. Teo, Generalized invexity and generalized invariant monotonicity. J. Optim. Theory Appl. **117**, 607–625 (2003)
46. Ç. Yildiz, M.E. Özdemir, M.Z. Sarikaya, New generalizations of Ostrowski-like type inequalities for fractional integrals. Kyungpook Math. J. **56**, 161–172 (2016)
47. L. Zhongxue, On sharp inequalities of Simpson type and Ostrowski type in two independent variables. Comput. Math. Appl. **56**, 2043–2047 (2008)

[illegible]

Some New Integral Inequalities via General Fractional Operators

Artion Kashuri, Themistocles M. Rassias, and Rozana Liko

Abstract Trapezoidal inequalities for functions of diverse natures are useful in numerical computations. The authors have proved an identity for a generalized integral operator via differentiable function. By applying the established identity, the generalized trapezoidal type integral inequalities have been discovered. It is pointed out that the results of this research provide integral inequalities for almost all fractional integrals discovered in recent past decades. Various special cases have been identified. Some applications of presented results to special means have been analyzed and new error estimates for the trapezoidal formula are provided as well. The ideas and techniques of this paper may stimulate further research.

1 Introduction

The following inequality, named Hermite–Hadamard inequality, is one of the most famous inequalities in the literature for convex functions.

Theorem 1 *Let $f : I \subseteq \mathbb{R} \longrightarrow \mathbb{R}$ be a convex function and $a_1, a_2 \in I$ with $a_1 < a_2$. Then the following inequality holds:*

$$f\left(\frac{a_1 + a_2}{2}\right) \leq \frac{1}{a_2 - a_1} \int_{a_1}^{a_2} f(x)dx \leq \frac{f(a_1) + f(a_2)}{2}. \tag{1}$$

This inequality (1) is also known as trapezium inequality.

A. Kashuri (✉) · R. Liko
Department of Mathematics, Faculty of Technical Science, University Ismail Qemali of Vlora, Vlorë, Albania

T. M. Rassias
Department of Mathematics, National Technical University of Athens, Athens, Greece
e-mail: trassias@math.ntua.gr

N. J. Daras, T. M. Rassias (eds.), *Computational Mathematics and Variational Analysis*, Springer Optimization and Its Applications 159,
https://doi.org/10.1007/978-3-030-44625-3_9

The trapezium inequality has remained an area of great interest due to its wide applications in the field of mathematical analysis. Authors of recent decades have studied (1) in the premises of newly invented definitions due to motivation of convex function. Interested readers see the references [1–16, 18, 19, 21, 22].

The aim of this paper is to establish trapezoidal type generalized integral inequalities for preinvex functions. Interestingly, the special cases of presented results are fractional integral inequalities. Therefore, it is important to summarize the study of fractional integrals. Let us recall some special functions and evoke some basic definitions as follows:

Definition 1 For $k \in \mathbb{R}^+$ and $x \in \mathbb{C}$, the k-gamma function is defined by

$$\Gamma_k(x) = \lim_{n \longrightarrow \infty} \frac{n!k^n(nk)^{\frac{x}{k}-1}}{(x)_{n,k}}. \tag{2}$$

Its integral representation is given by

$$\Gamma_k(\alpha) = \int_0^{\infty} t^{\alpha-1} e^{-\frac{t^k}{k}} dt. \tag{3}$$

One can note that

$$\Gamma_k(\alpha + k) = \alpha \Gamma_k(\alpha). \tag{4}$$

For $k = 1$, (3) gives integral representation of the gamma function.

Definition 2 ([13]) Let $f \in L[a_1, a_2]$. Then k-fractional integrals of order $\alpha, k > 0$ with $a_1 \geq 0$ are defined by

$$I_{a_1^+}^{\alpha,k} f(x) = \frac{1}{k\Gamma_k(\alpha)} \int_{a_1}^{x} (x-t)^{\frac{\alpha}{k}-1} f(t)dt, \quad x > a_1$$

and

$$I_{a_2^-}^{\alpha,k} f(x) = \frac{1}{k\Gamma_k(\alpha)} \int_{x}^{a_2} (t-x)^{\frac{\alpha}{k}-1} f(t)dt, \quad a_2 > x. \tag{5}$$

For $k = 1$, k-fractional integrals give Riemann–Liouville integrals. For $\alpha = k = 1$, k-fractional integrals give classical integrals.

Definition 3 ([20]) A set $S \subseteq \mathbb{R}^n$ is said to be an invex set with respect to the mapping $\eta : S \times S \longrightarrow \mathbb{R}^n$, if $x + t\eta(y, x) \in S$ for every $x, y \in S$ and $t \in [0, 1]$.

The invex set also termed as an η-connected set.

Definition 4 Let $S \subseteq \mathbb{R}^n$ be an invex set with respect to $\eta : S \times S \longrightarrow \mathbb{R}^n$. A function $f : S \longrightarrow [0, +\infty)$ is said to be preinvex with respect to η, if for every $x, y \in S$ and $t \in [0, 1]$,

$$f\big(x + t\eta(y, x)\big) \le (1 - t) f(x) + t f(y). \tag{6}$$

Also, let us define a function $\varphi : [0, \infty) \longrightarrow [0, \infty)$ satisfying the following conditions:

$$\int_0^1 \frac{\varphi(t)}{t} dt < \infty, \tag{7}$$

$$\frac{1}{A} \le \frac{\varphi(s)}{\varphi(r)} \le A \text{ for } \frac{1}{2} \le \frac{s}{r} \le 2 \tag{8}$$

$$\frac{\varphi(r)}{r^2} \le B \frac{\varphi(s)}{s^2} \text{ for } s \le r \tag{9}$$

$$\left| \frac{\varphi(r)}{r^2} - \frac{\varphi(s)}{s^2} \right| \le C|r - s| \frac{\varphi(r)}{r^2} \text{ for } \frac{1}{2} \le \frac{s}{r} \le 2, \tag{10}$$

where $A, B, C > 0$ are independent of $r, s > 0$. If $\varphi(r) r^\alpha$ is increasing for some $\alpha \ge 0$ and $\frac{\varphi(r)}{r^\beta}$ is decreasing for some $\beta \ge 0$, then φ satisfies (7)–(10), see [17]. Therefore, the left-sided and right-sided generalized integral operators are defined as follows:

$${}_{a_1^+}I_\varphi f(x) = \int_{a_1}^{x} \frac{\varphi(x - t)}{x - t} f(t) dt, \quad x > a_1, \tag{11}$$

$${}_{a_2^-}I_\varphi f(x) = \int_{x}^{a_2} \frac{\varphi(t - x)}{t - x} f(t) dt, \quad x < a_2. \tag{12}$$

The most important feature of generalized integrals is that; they produce Riemann–Liouville fractional integrals, k-Riemann–Liouville fractional integrals, Katugampola fractional integrals, conformable fractional integrals, Hadamard fractional integrals, etc.

Motivated by the above literatures, the main objective of this paper is to discover in Section 2, an interesting identity in order to study some new bounds regarding trapezoidal type integral inequalities. By using the established identity as an auxiliary result, some new estimates for trapezoidal type integral inequalities via generalized integrals are obtained. It is pointed out that some new fractional integral inequalities have been deduced from main results. In Section 3, some applications to special means and new error estimates for the trapezoidal formula are given. The ideas and techniques of this paper may stimulate further research in the field of integral inequalities.

2 Main Results

Let $a_1 < a_2$, $n \in \mathbb{N}^*$ and $m \in (0, 1]$ be a fixed number. Throughout this study, for brevity, we define

$$\Lambda_{m,n}(t) = \int_0^t \frac{\varphi\left(\frac{\eta(x,ma_1)}{n+1}u\right)}{u}du < \infty, \quad \eta(x, ma_1) > 0 \tag{13}$$

and

$$\Delta_{m,n}(t) = \int_0^t \frac{\varphi\left(\frac{\eta(a_2,mx)}{n+1}u\right)}{u}du < \infty, \quad \eta(a_2, mx) > 0 \tag{14}$$

for all $x \in P = [ma_1, a_2]$.

For establishing some new results regarding general fractional integrals we need to prove the following lemma.

Lemma 1 *Let $f : P \longrightarrow \mathbb{R}$ be a differentiable mapping on (ma_1, a_2). If $f' \in L(P)$, then the following identity for generalized fractional integrals holds:*

$$\frac{f(ma_1) + f(ma_1 + \eta(x, ma_1))}{2} - \frac{1}{2\Lambda_{m,n}(1)}$$

$$\times\Big[{}_{(ma_1)^+}I_\varphi f\left(ma_1+\tfrac{\eta(x,ma_1)}{n+1}\right) + {}_{(ma_1+\eta(x,ma_1))^-}I_\varphi f \left(ma_1+\tfrac{n}{n+1}\eta(x, ma_1)\right)\Big]$$

$$+\frac{f(mx) + f(mx + \eta(a_2, mx))}{2} - \frac{1}{2\Delta_{m,n}(1)}$$

$$\times\left[{}_{(mx)^+}I_\varphi f\left(mx + \frac{\eta(a_2, mx)}{n + 1}\right) + {}_{(mx+\eta(a_2,mx))^-}I_\varphi f\left(mx+\frac{n}{n+1}\eta(a_2, mx)\right)\right]$$

$$= \frac{\eta(x, ma_1)}{2(n + 1)\Lambda_{m,n}(1)} \tag{15}$$

$$\times\int_0^1 \Lambda_{m,n}(t)\left[f'\left(ma_1+\frac{(n + t)}{n + 1}\eta(x, ma_1)\right) - f'\left(ma_1+\frac{(1 - t)}{n + 1}\eta(x, ma_1)\right)\right]dt$$

$$+\frac{\eta(a_2, mx)}{2(n+1)\Delta_{m,n}(1)}$$

$$\times \int_0^1 \Delta_{m,n}(t)\left[f'\left(mx + \frac{(n+t)}{n+1}\eta(a_2, mx)\right) - f'\left(mx + \frac{(1-t)}{n+1}\eta(a_2, mx)\right)\right]dt.$$

We denote

$$I_{f,\Lambda_{m,n},\Delta_{m,n}}(x, a_1, a_2) := \frac{\eta(x, ma_1)}{2(n+1)\Lambda_{m,n}(1)} \tag{16}$$

$$\times \int_0^1 \Lambda_{m,n}(t)\left[f'\left(ma_1 + \frac{(n+t)}{n+1}\eta(x, ma_1)\right) - f'\left(ma_1 + \frac{(1-t)}{n+1}\eta(x, ma_1)\right)\right]dt$$

$$+\frac{\eta(a_2, mx)}{2(n+1)\Delta_{m,n}(1)}$$

$$\times \int_0^1 \Delta_{m,n}(t)\left[f'\left(mx + \frac{(n+t)}{n+1}\eta(a_2, mx)\right) - f'\left(mx + \frac{(1-t)}{n+1}\eta(a_2, mx)\right)\right]dt.$$

Proof Integrating by parts (16) and changing the variables of integration, we have

$$I_{f,\Lambda_{m,n},\Delta_{m,n}}(x, a_1, a_2) = \frac{\eta(x,ma_1)}{2(n+1)\Lambda_{m,n}(1)}$$

$$\times \left\{ \frac{(n+1)\Lambda_{m,n}(t) f\left(ma_1 + \frac{(n+t)}{n+1}\eta(x,ma_1)\right)}{\eta(x,ma_1)} \right|_0^1$$

$$-\frac{(n+1)}{\eta(x, ma_1)} \int_0^1 \frac{\varphi\left(\frac{\eta(x,ma_1)}{n+1}t\right)}{t} f\left(ma_1 + \frac{(n+t)}{n+1}\eta(x, ma_1)\right) dt$$

$$+\frac{(n+1)\Lambda_{m,n}(t) f\left(ma_1 + \frac{(1-t)}{n+1}\eta(x, ma_1)\right)}{\eta(x, ma_1)}\Bigg|_0^1$$

$$\left. -\frac{(n+1)}{\eta(x, ma_1)} \int_0^1 \frac{\varphi\left(\frac{\eta(x,ma_1)}{n+1}t\right)}{t} f\left(ma_1 + \frac{(1-t)}{n+1}\eta(x, ma_1)\right) dt \right\}$$

$$+\frac{\eta(a_2, mx)}{2(n+1)\Delta_{m,n}(1)} \times \left\{ \frac{(n+1)\Delta_{m,n}(t) f\left(mx + \frac{(n+t)}{n+1}\eta(a_2, mx)\right)}{\eta(a_2, mx)} \right|_0^1$$

$$-\frac{(n+1)}{\eta(a_2,mx)}\int_0^1 \frac{\varphi\left(\frac{\eta(a_2,mx)}{n+1}t\right)}{t} f\left(mx+\frac{(n+t)}{n+1}\eta(a_2,mx)\right)dt$$

$$+\frac{(n+1)\Delta_{m,n}(t)f\left(mx+\frac{(1-t)}{n+1}\eta(a_2,mx)\right)}{\eta(a_2,mx)}\Bigg|_0^1$$

$$\left.-\frac{(n+1)}{\eta(a_2,mx)}\int_0^1 \frac{\varphi\left(\frac{\eta(a_2,mx)}{n+1}t\right)}{t} f\left(mx+\frac{(1-t)}{n+1}\eta(a_2,mx)\right)dt\right\}$$

$$=\frac{\eta(x,ma_1)}{2(n+1)\Lambda_{m,n}(1)}\times\left\{\frac{(n+1)\Lambda_{m,n}(1)f\left(ma_1+\eta(x,ma_1)\right)}{\eta(x,ma_1)}\right.$$

$$-\frac{(n+1)}{\eta(x,ma_1)}\times {}_{(ma_1+\eta(x,ma_1))^-}I_\varphi f\left(ma_1+\frac{n}{n+1}\eta(x,ma_1)\right)$$

$$\left.+\frac{(n+1)\Lambda_{m,n}(1)f\left(ma_1\right)}{\eta(x,ma_1)}-\frac{(n+1)}{\eta(x,ma_1)}\times {}_{(ma_1)^+}I_\varphi f\left(ma_1+\frac{\eta(x,ma_1)}{n+1}\right)\right\}$$

$$+\frac{\eta(a_2,mx)}{2(n+1)\Delta_{m,n}(1)}\times\left\{\frac{(n+1)\Delta_{m,n}(1)f\left(mx+\eta(a_2,mx)\right)}{\eta(a_2,mx)}\right.$$

$$-\frac{(n+1)}{\eta(a_2,mx)}\times {}_{(mx+\eta(a_2,mx))^-}I_\varphi f\left(mx+\frac{n}{n+1}\eta(a_2,mx)\right)$$

$$\left.+\frac{(n+1)\Delta_{m,n}(1)f\left(mx\right)}{\eta(a_2,mx)}-\frac{(n+1)}{\eta(a_2,mx)}\times {}_{(mx)^+}I_\varphi f\left(mx+\frac{\eta(a_2,mx)}{n+1}\right)\right\}$$

$$=\frac{f(ma_1)+f(ma_1+\eta(x,ma_1))}{2}-\frac{1}{2\Lambda_{m,n}(1)}$$

$$\times\left[{}_{(ma_1)^+}I_\varphi f\left(ma_1+\frac{\eta(x,ma_1)}{n+1}\right)+{}_{(ma_1+\eta(x,ma_1))^-}I_\varphi f\left(ma_1+\frac{n}{n+1}\eta(x,ma_1)\right)\right]$$

$$+\frac{f(mx)+f(mx+\eta(a_2,mx))}{2}-\frac{1}{2\Delta_{m,n}(1)}$$

$$\times\left[{}_{(mx)^+}I_\varphi f\left(mx+\frac{\eta(a_2,mx)}{n+1}\right)+{}_{(mx+\eta(a_2,mx))^-}I_\varphi f\left(mx+\frac{n}{n+1}\eta(a_2,mx)\right)\right].$$

This completes the proof of the lemma.

Theorem 2 *Let $f : P \longrightarrow \mathbb{R}$ be a differentiable mapping on (ma_1, a_2). If $|f'|^q$ is preinvex on P for $q > 1$ and $p^{-1} + q^{-1} = 1$, then the following inequality for generalized fractional integrals holds:*

$$\left|I_{f,\Lambda_{m,n},\Delta_{m,n}}(x, a_1, a_2)\right| \leq \left(\frac{1}{2(n+1)}\right)^{\frac{1-q}{q}} \frac{\eta(x, ma_1)}{\Lambda_{m,n}(1)} \sqrt[p]{B_{\Lambda_{m,n}}(p)} \tag{17}$$

$$\times\left\{\sqrt[q]{|f'(ma_1)|^q + (2n+1)|f'(x)|^q} + \sqrt[q]{(2n+1)|f'(ma_1)|^q + |f'(x)|^q}\right\}$$

$$+\left(\frac{1}{2(n+1)}\right)^{\frac{1-q}{q}} \frac{\eta(a_2, mx)}{\Delta_{m,n}(1)} \sqrt[p]{C_{\Delta_{m,n}}(p)}$$

$$\times\left\{\sqrt[q]{|f'(mx)|^q + (2n+1)|f'(a_2)|^q} + \sqrt[q]{(2n+1)|f'(mx)|^q + |f'(a_2)|^q}\right\},$$

where

$$B_{\Lambda_{m,n}}(p) := \int_0^1 \left[\Lambda_{m,n}(t)\right]^p dt, \quad C_{\Delta_{m,n}}(p) := \int_0^1 \left[\Delta_{m,n}(t)\right]^p dt. \tag{18}$$

Proof From Lemma 1, preinvexity of $|f'|^q$, Hölder inequality, and properties of the modulus, we have

$$\left|I_{f,\Lambda_{m,n},\Delta_{m,n}}(x, a_1, a_2)\right| \leq \frac{\eta(x, ma_1)}{2(n+1)\Lambda_{m,n}(1)}$$

$$\times\left\{\int_0^1 \Lambda_{m,n}(t)\left[\left|f'\left(ma_1 + \frac{(n+t)}{n+1}\eta(x, ma_1)\right)\right.\right.\right.$$

$$\left.\left.\left.+\left|f'\left(ma_1 + \frac{(1-t)}{n+1}\eta(x, ma_1)\right)\right|\right]dt\right\} + \frac{\eta(a_2, mx)}{2(n+1)\Delta_{m,n}(1)}$$

$$\times\left\{\int_0^1 \Delta_{m,n}(t)\left[\left|f'\left(mx + \frac{(1-t)}{n+1}\eta(a_2, mx)\right)\right.\right.\right.$$

$$\left.\left.\left.+\left|f'\left(mx + \frac{(n+t)}{n+1}\eta(a_2, mx)\right)\right|\right]dt\right\}$$

$$\leq \frac{\eta(x,ma_1)}{2(n+1)\Lambda_{m,n}(1)}\left(\int_0^1\left[\Lambda_{m,n}(t)\right]^p dt\right)^{\frac{1}{p}}$$

$$\times\left\{\left(\int_0^1\left|f'\left(ma_1+\frac{(n+t)}{n+1}\eta(x,ma_1)\right)\right|^q dt\right)^{\frac{1}{q}}\right.$$

$$\left.+\left(\int_0^1\left|f'\left(ma_1+\frac{(1-t)}{n+1}\eta(x,ma_1)\right)\right|^q dt\right)^{\frac{1}{q}}\right\}$$

$$+\frac{\eta(a_2,mx)}{2(n+1)\Delta_{m,n}(1)}\left(\int_0^1\left[\Delta_{m,n}(t)\right]^p dt\right)^{\frac{1}{p}}$$

$$\times\left\{\left(\int_0^1\left|f'\left(mx+\frac{(1-t)}{n+1}\eta(a_2,mx)\right)\right|^q dt\right)^{\frac{1}{q}}\right.$$

$$\left.+\left(\int_0^1\left|f'\left(mx+\frac{(n+t)}{n+1}\eta(a_2,mx)\right)\right|^q dt\right)^{\frac{1}{q}}\right\}$$

$$\leq \frac{\eta(x,ma_1)}{2(n+1)\Lambda_{m,n}(1)}\sqrt[p]{B_{\Lambda_{m,n}}(p)}$$

$$\times\left\{\left(\int_0^1\left[\left(1-\frac{n+t}{n+1}\right)|f'(ma_1)|^q+\frac{(n+t)}{n+1}|f'(x)|^q\right]dt\right)^{\frac{1}{q}}\right.$$

$$\left.+\left(\int_0^1\left[\left(1-\frac{1-t}{n+1}\right)|f'(ma_1)|^q+\frac{(1-t)}{n+1}|f'(x)|^q\right]dt\right)^{\frac{1}{q}}\right\}$$

$$+\frac{\eta(a_2,mx)}{2(n+1)\Delta_{m,n}(1)}\sqrt[p]{C_{\Delta_{m,n}}(p)}$$

$$\times\left\{\left(\int_0^1\left[\left(1-\frac{1-t}{n+1}\right)|f'(mx)|^q+\frac{(1-t)}{n+1}|f'(a_2)|^q\right]dt\right)^{\frac{1}{q}}\right.$$

$$\left.+\left(\int_0^1\left[\left(1-\frac{n+t}{n+1}\right)|f'(mx)|^q+\frac{(n+t)}{n+1}|f'(a_2)|^q\right]dt\right)^{\frac{1}{q}}\right\}$$

$$= \left(\frac{1}{2(n+1)}\right)^{\frac{1-q}{q}} \frac{\eta(x, ma_1)}{\Lambda_{m,n}(1)} \sqrt[p]{B_{\Lambda_{m,n}}(p)}$$

$$\times\left\{\sqrt[q]{|f'(ma_1)|^q + (2n+1)|f'(x)|^q} + \sqrt[q]{(2n+1)|f'(ma_1)|^q + |f'(x)|^q}\right\}$$

$$+ \left(\frac{1}{2(n+1)}\right)^{\frac{1-q}{q}} \frac{\eta(a_2, mx)}{\Delta_{m,n}(1)} \sqrt[p]{C_{\Delta_{m,n}}(p)}$$

$$\times\left\{\sqrt[q]{|f'(mx)|^q + (2n+1)|f'(a_2)|^q} + \sqrt[q]{(2n+1)|f'(mx)|^q + |f'(a_2)|^q}\right\}.$$

The proof of this theorem is complete.

We point out some special cases of Theorem 2.

Corollary 1 *Taking $p = q = 2$ in Theorem 2, we get*

$$\left|I_{f,\Lambda_{m,n},\Delta_{m,n}}(x, a_1, a_2)\right| \leq \frac{\eta(x, ma_1)}{\Lambda_{m,n}(1)} \sqrt{2(n+1)B_{\Lambda_{m,n}}(2)} \tag{19}$$

$$\times\left\{\sqrt{|f'(ma_1)|^2 + (2n+1)|f'(x)|^2} + \sqrt{(2n+1)|f'(ma_1)|^2 + |f'(x)|^2}\right\}$$

$$+ \frac{\eta(a_2, mx)}{\Delta_{m,n}(1)} \sqrt{2(n+1)C_{\Delta_{m,n}}(2)}$$

$$\times\left\{\sqrt{|f'(mx)|^2 + (2n+1)|f'(a_2)|^2} + \sqrt{(2n+1)|f'(mx)|^2 + |f'(a_2)|^2}\right\}.$$

Corollary 2 *Taking $\varphi(t) = t$ in Theorem 2, we get*

$$\left|I_{f,\Lambda_{m,n},\Delta_{m,n}}(x, a_1, a_2)\right| \leq \left(\frac{1}{2(n+1)}\right)^{\frac{1-q}{q}} \frac{\eta(x, ma_1)}{\sqrt[p]{p+1}} \tag{20}$$

$$\times\left\{\sqrt[q]{|f'(ma_1)|^q + (2n+1)|f'(x)|^q} + \sqrt[q]{(2n+1)|f'(ma_1)|^q + |f'(x)|^q}\right\}$$

$$+ \left(\frac{1}{2(n+1)}\right)^{\frac{1-q}{q}} \frac{\eta(a_2, mx)}{\sqrt[p]{p+1}}$$

$$\times\left\{\sqrt[q]{|f'(mx)|^q + (2n+1)|f'(a_2)|^q} + \sqrt[q]{(2n+1)|f'(mx)|^q + |f'(a_2)|^q}\right\}.$$

Corollary 3 *Taking* $x = \frac{a_1+a_2}{2}$, $m = 1$, $\eta(x, ma_1) = x - ma_1$, *and* $\eta(a_2, mx) = a_2 - mx$ *in Corollary 2, we get*

$$\left|I_{f,\Lambda_{1,n},\Delta_{1,n}}\left(\frac{a_1+a_2}{2}, a_1, a_2\right)\right| \leq \left(\frac{1}{2(n+1)}\right)^{\frac{1-q}{q}} \frac{(a_2-a_1)}{2\sqrt[p]{p+1}} \tag{21}$$

$$\times\left\{\sqrt[q]{|f'(a_1)|^q+(2n+1)\left|f'\left(\frac{a_1+a_2}{2}\right)\right|^q}+\sqrt[q]{(2n+1)|f'(a_1)|^q+\left|f'\left(\frac{a_1+a_2}{2}\right)\right|^q}\right.$$

$$\left.+\sqrt[q]{\left|f'\left(\frac{a_1+a_2}{2}\right)\right|^q+(2n+1)|f'(a_2)|^q}+\sqrt[q]{(2n+1)\left|f'\left(\frac{a_1+a_2}{2}\right)\right|^q+|f'(a_2)|^q}\right\}.$$

Corollary 4 *Taking* $\varphi(t) = \frac{t^\alpha}{\Gamma(\alpha)}$ *in Theorem 2, we get*

$$\left|I_{f,\Lambda_{m,n},\Delta_{m,n}}(x, a_1, a_2)\right| \leq \left(\frac{1}{2(n+1)}\right)^{\frac{1-q}{q}} \frac{\eta(x, ma_1)}{\sqrt[p]{p\alpha+1}} \tag{22}$$

$$\times\left\{\sqrt[q]{|f'(ma_1)|^q+(2n+1)|f'(x)|^q}+\sqrt[q]{(2n+1)|f'(ma_1)|^q+|f'(x)|^q}\right\}$$

$$+\left(\frac{1}{2(n+1)}\right)^{\frac{1-q}{q}} \frac{\eta(a_2, mx)}{\sqrt[p]{p\alpha+1}}$$

$$\times\left\{\sqrt[q]{|f'(mx)|^q+(2n+1)|f'(a_2)|^q}+\sqrt[q]{(2n+1)|f'(mx)|^q+|f'(a_2)|^q}\right\}.$$

Corollary 5 *Taking* $\varphi(t) = \frac{t^{\frac{\alpha}{k}}}{k\Gamma_k(\alpha)}$ *in Theorem 2, we get*

$$\left|I_{f,\Lambda_{m,n},\Delta_{m,n}}(x, a_1, a_2)\right| \leq \left(\frac{1}{2(n+1)}\right)^{\frac{1-q}{q}} \frac{\eta(x, ma_1)}{\sqrt[p]{\frac{p\alpha}{k}+1}} \tag{23}$$

$$\times\left\{\sqrt[q]{|f'(ma_1)|^q+(2n+1)|f'(x)|^q}+\sqrt[q]{(2n+1)|f'(ma_1)|^q+|f'(x)|^q}\right\}$$

$$+\left(\frac{1}{2(n+1)}\right)^{\frac{1-q}{q}} \frac{\eta(a_2, mx)}{\sqrt[p]{\frac{p\alpha}{k}+1}}$$

$$\times\left\{\sqrt[q]{|f'(mx)|^q+(2n+1)|f'(a_2)|^q}+\sqrt[q]{(2n+1)|f'(mx)|^q+|f'(a_2)|^q}\right\}.$$

Corollary 6 *Taking $\varphi(t) = t(a_2 - t)^{\alpha-1}$ and $f(x)$ is symmetric to $x = \frac{ma_1+a_2}{2}$, in Theorem 2, we get*

$$\left| I_{f,\Lambda_{m,n},\Delta_{m,n}}\left(\frac{ma_1+a_2}{2}, a_1, a_2\right) \right| \tag{24}$$

$$\leq \left(\frac{1}{2(n+1)}\right)^{\frac{1-q}{q}} \times \frac{\alpha\eta\left(\frac{ma_1+a_2}{2}, ma_1\right)}{\left[a_2^\alpha - \left(a_2 - \frac{\eta\left(\frac{ma_1+a_2}{2}, ma_1\right)}{n+1}\right)^\alpha\right]} \sqrt[p]{B^*_{\Lambda_{m,n}}(p)}$$

$$\times \left\{ \sqrt[q]{|f'(ma_1)|^q + (2n+1)\left|f'\left(\frac{ma_1+a_2}{2}\right)\right|^q} + \sqrt[q]{(2n+1)|f'(ma_1)|^q + \left|f'\left(\frac{ma_1+a_2}{2}\right)\right|^q} \right\}$$

$$+ \left(\frac{1}{2(n+1)}\right)^{\frac{1-q}{q}} \times \frac{\alpha\eta\left(a_2, m\frac{(ma_1+a_2)}{2}\right)}{\left[a_2^\alpha - \left(a_2 - \frac{\eta\left(a_2, m\frac{(ma_1+a_2)}{2}\right)}{n+1}\right)^\alpha\right]} \sqrt[p]{C^*_{\Delta_{m,n}}(p)}$$

$$\times \left\{ \sqrt[q]{\left|f'\left(m\frac{(ma_1+a_2)}{2}\right)\right|^q + (2n+1)|f'(a_2)|^q} + \sqrt[q]{(2n+1)\left|f'\left(m\frac{(ma_1+a_2)}{2}\right)\right|^q + |f'(a_2)|^q} \right\},$$

where

$$B^*_{\Lambda_{m,n}}(p) := \frac{(n+1)}{\alpha^p\eta\left(\frac{ma_1+a_2}{2}, ma_1\right)} \int_{a_2 - \frac{\eta\left(\frac{ma_1+a_2}{2}, ma_1\right)}{n+1}}^{a_2} \left(a_2^\alpha - t^\alpha\right)^p dt \tag{25}$$

and

$$C^*_{\Delta_{m,n}}(p) := \frac{(n+1)}{\alpha^p\eta\left(a_2, m\frac{(ma_1+a_2)}{2}\right)} \int_{a_2 - \frac{\eta\left(a_2, m\frac{(ma_1+a_2)}{2}\right)}{n+1}}^{a_2} \left(a_2^\alpha - t^\alpha\right)^p dt. \tag{26}$$

Corollary 7 *Taking* $\varphi(t) = \frac{t}{\alpha}\exp\left[\left(-\frac{1-\alpha}{\alpha}\right)t\right]$ *for* $\alpha \in (0,1)$, *in Theorem 2, we get*

$$\left|I_{f,\Lambda_{m,n},\Delta_{m,n}}(x,a_1,a_2)\right| \tag{27}$$

$$\leq \left(\frac{1}{2(n+1)}\right)^{\frac{1-q}{q}} \times \frac{(\alpha-1)\eta(x,ma_1)}{\left\{\exp\left[\left(-\frac{1-\alpha}{\alpha}\right)\frac{\eta(x,ma_1)}{n+1}\right]-1\right\}} \sqrt[p]{B^{\diamond}_{\Lambda_{m,n}}(p)}$$

$$\times\left\{\sqrt[q]{|f'(ma_1)|^q + (2n+1)|f'(x)|^q} + \sqrt[q]{(2n+1)|f'(ma_1)|^q + |f'(x)|^q}\right\}$$

$$+\left(\frac{1}{2(n+1)}\right)^{\frac{1-q}{q}} \times \frac{(\alpha-1)\eta(a_2,mx)}{\left\{\exp\left[\left(-\frac{1-\alpha}{\alpha}\right)\frac{\eta(a_2,mx)}{n+1}\right]-1\right\}} \sqrt[p]{C^{\diamond}_{\Delta_{m,n}}(p)}$$

$$\times\left\{\sqrt[q]{|f'(mx)|^q + (2n+1)|f'(a_2)|^q} + \sqrt[q]{(2n+1)|f'(mx)|^q + |f'(a_2)|^q}\right\},$$

where

$$B^{\diamond}_{\Lambda_{m,n}}(p) := \frac{\alpha(n+1)}{(\alpha-1)^{p+1}\eta(x,ma_1)} \times \int_0^{\exp\left[\left(-\frac{1-\alpha}{\alpha}\right)\frac{\eta(x,ma_1)}{n+1}\right]} \frac{t^p}{t+1}dt \tag{28}$$

and

$$C^{\diamond}_{\Delta_{m,n}}(p) := \frac{\alpha(n+1)}{(\alpha-1)^{p+1}\eta(a_2,mx)} \times \int_0^{\exp\left[\left(-\frac{1-\alpha}{\alpha}\right)\frac{\eta(a_2,mx)}{n+1}\right]} \frac{t^p}{t+1}dt. \tag{29}$$

Theorem 3 *Let* $f : P \longrightarrow \mathbb{R}$ *be a differentiable mapping on* (ma_1, a_2). *If* $|f'|^q$ *is preinvex on* P *for* $q \geq 1$, *then the following inequality for generalized fractional integrals holds:*

$$\left|I_{f,\Lambda_{m,n},\Delta_{m,n}}(x,a_1,a_2)\right| \leq \left(\frac{1}{n+1}\right)^{\frac{1-q}{q}} \frac{\eta(x,ma_1)}{2\Lambda_{m,n}(1)}\left(B_{\Lambda_{m,n}}(1)\right)^{1-\frac{1}{q}} \tag{30}$$

$$\times\left\{\sqrt[q]{D_{\Lambda_{m,n}}|f'(ma_1)|^q + E_{\Lambda_{m,n}}(n)|f'(x)|^q}\right.$$

$$\left.+\sqrt[q]{E_{\Lambda_{m,n}}(n)|f'(ma_1)|^q + D_{\Lambda_{m,n}}|f'(x)|^q}\right\}$$

$$+\left(\frac{1}{n+1}\right)^{\frac{1-q}{q}} \frac{\eta(a_2, mx)}{2\Delta_{m,n}(1)} \left(C_{\Delta_{m,n}}(1)\right)^{1-\frac{1}{q}}$$

$$\times \left\{ \sqrt[q]{F_{\Delta_{m,n}} |f'(mx)|^q + G_{\Delta_{m,n}}(n)|f'(a_2)|^q} \right.$$

$$\left. + \sqrt[q]{G_{\Delta_{m,n}}(n)|f'(mx)|^q + F_{\Delta_{m,n}} |f'(a_2)|^q} \right\},$$

where

$$D_{\Lambda_{m,n}} := \int_0^1 (1-t)\Lambda_{m,n}(t)dt, \quad E_{\Lambda_{m,n}}(n) := \int_0^1 (n+t)\Lambda_{m,n}(t)dt, \tag{31}$$

$$F_{\Delta_{m,n}} := \int_0^1 (1-t)\Delta_{m,n}(t)dt, \quad G_{\Delta_{m,n}}(n) := \int_0^1 (n+t)\Delta_{m,n}(t)dt \tag{32}$$

and $B_{\Lambda_{m,n}}(1)$, $C_{\Delta_{m,n}}(1)$ *are defined as in Theorem 2.*

Proof From Lemma 1, preinvexity of $|f'|^q$, power mean inequality, and properties of the modulus, we have

$$\left|I_{f,\Lambda_{m,n},\Delta_{m,n}}(x, a_1, a_2)\right| \le \frac{\eta(x, ma_1)}{2(n+1)\Lambda_{m,n}(1)}$$

$$\times \left\{ \int_0^1 \Lambda_{m,n}(t) \left[\left| f'\left(ma_1 + \frac{(n+t)}{n+1}\eta(x, ma_1)\right) \right| + \left| f'\left(ma_1 + \frac{(1-t)}{n+1}\eta(x, ma_1)\right) \right| \right] dt \right\}$$

$$+ \frac{\eta(a_2, mx)}{2(n+1)\Delta_{m,n}(1)}$$

$$\times \left\{ \int_0^1 \Delta_{m,n}(t) \left[\left| f'\left(mx + \frac{(1-t)}{n+1}\eta(a_2, mx)\right) \right| + \left| f'\left(mx + \frac{(n+t)}{n+1}\eta(a_2, mx)\right) \right| \right] dt \right\}$$

$$\leq \frac{\eta(x, ma_1)}{2(n+1)\Lambda_{m,n}(1)} \left(\int_0^1 \Lambda_{m,n}(t)dt \right)^{1-\frac{1}{q}}$$

$$\times \left\{ \left(\int_0^1 \Lambda_{m,n}(t) \left| f'\left(ma_1 + \frac{(n+t)}{n+1}\eta(x, ma_1) \right) \right|^q dt \right)^{\frac{1}{q}} \right.$$

$$\left. + \left(\int_0^1 \Lambda_{m,n}(t) \left| f'\left(ma_1 + \frac{(1-t)}{n+1}\eta(x, ma_1) \right) \right|^q dt \right)^{\frac{1}{q}} \right\}$$

$$+ \frac{\eta(a_2, mx)}{2(n+1)\Delta_{m,n}(1)} \left(\int_0^1 \Delta_{m,n}(t)dt \right)^{1-\frac{1}{q}}$$

$$\times \left\{ \left(\int_0^1 \Delta_{m,n}(t) \left| f'\left(mx + \frac{(1-t)}{n+1}\eta(a_2, mx) \right) \right|^q dt \right)^{\frac{1}{q}} \right.$$

$$\left. + \left(\int_0^1 \Delta_{m,n}(t) \left| f'\left(mx + \frac{(n+t)}{n+1}\eta(a_2, mx) \right) \right|^q dt \right)^{\frac{1}{q}} \right\}$$

$$\leq \frac{\eta(x, ma_1)}{2(n+1)\Lambda_{m,n}(1)} \left(B_{\Lambda_{m,n}}(1) \right)^{1-\frac{1}{q}}$$

$$\times \left\{ \left(\int_0^1 \Lambda_{m,n}(t) \left[\left(1 - \frac{n+t}{n+1} \right) |f'(ma_1)|^q + \frac{(n+t)}{n+1} |f'(x)|^q \right] dt \right)^{\frac{1}{q}} \right.$$

$$\left. + \left(\int_0^1 \Lambda_{m,n}(t) \left[\left(1 - \frac{1-t}{n+1} \right) |f'(ma_1)|^q + \frac{(1-t)}{n+1} |f'(x)|^q \right] dt \right)^{\frac{1}{q}} \right\}$$

$$+ \frac{\eta(a_2, mx)}{2(n+1)\Delta_{m,n}(1)} \left(C_{\Delta_{m,n}}(1) \right)^{1-\frac{1}{q}}$$

$$\times \left\{ \left(\int_0^1 \Delta_{m,n}(t) \left[\left(1 - \frac{1-t}{n+1} \right) |f'(mx)|^q + \frac{(1-t)}{n+1} |f'(a_2)|^q \right] dt \right)^{\frac{1}{q}} \right.$$

$$\left. + \left(\int_0^1 \Delta_{m,n}(t) \left[\left(1 - \frac{n+t}{n+1} \right) |f'(mx)|^q + \frac{(n+t)}{n+1} |f'(a_2)|^q \right] dt \right)^{\frac{1}{q}} \right\}$$

$$= \left(\frac{1}{n+1}\right)^{\frac{1-q}{q}} \frac{\eta(x, ma_1)}{2\Lambda_{m,n}(1)} \left(B_{\Lambda_{m,n}}(1)\right)^{1-\frac{1}{q}}$$

$$\times \left\{ \sqrt[q]{D_{\Lambda_{m,n}}|f'(ma_1)|^q + E_{\Lambda_{m,n}}(n)|f'(x)|^q} \right.$$

$$\left. + \sqrt[q]{E_{\Lambda_{m,n}}(n)|f'(ma_1)|^q + D_{\Lambda_{m,n}}|f'(x)|^q} \right\}$$

$$+ \left(\frac{1}{n+1}\right)^{\frac{1-q}{q}} \frac{\eta(a_2, mx)}{2\Delta_{m,n}(1)} \left(C_{\Delta_{m,n}}(1)\right)^{1-\frac{1}{q}}$$

$$\times \left\{ \sqrt[q]{F_{\Delta_{m,n}}|f'(mx)|^q + G_{\Delta_{m,n}}(n)|f'(a_2)|^q} \right.$$

$$\left. + \sqrt[q]{G_{\Delta_{m,n}}(n)|f'(mx)|^q + F_{\Delta_{m,n}}|f'(a_2)|^q} \right\}.$$

The proof of this theorem is complete.

We point out some special cases of Theorem 3.

Corollary 8 *Taking $q = 1$ in Theorem 3, we get*

$$\left|I_{f,\Lambda_{m,n},\Delta_{m,n}}(x, a_1, a_2)\right| \leq \frac{\eta(x, ma_1)}{2\Lambda_{m,n}(1)} \times \left(D_{\Lambda_{m,n}} + E_{\Lambda_{m,n}}(n)\right) \left[|f'(ma_1)| + |f'(x)|\right] \tag{33}$$

$$+ \frac{\eta(a_2, mx)}{2\Delta_{m,n}(1)} \times \left(F_{\Delta_{m,n}} + G_{\Delta_{m,n}}(n)\right) \left[|f'(mx)| + |f'(a_2)|\right].$$

Corollary 9 *Taking $\varphi(t) = t$ in Theorem 3, we get*

$$\left|I_{f,\Lambda_{m,n},\Delta_{m,n}}(x, a_1, a_2)\right| \leq \frac{(n+1)}{4\sqrt[q]{3(n+1)}} \eta(x, ma_1) \tag{34}$$

$$\times \left\{ \sqrt[q]{|f'(ma_1)|^q + (3n+2)|f'(x)|^q} + \sqrt[q]{(3n+2)|f'(ma_1)|^q + |f'(x)|^q} \right\}$$

$$+ \frac{(n+1)}{4\sqrt[q]{3(n+1)}} \eta(a_2, mx)$$

$$\times \left\{ \sqrt[q]{|f'(mx)|^q + (3n+2)|f'(a_2)|^q} + \sqrt[q]{(3n+2)|f'(mx)|^q + |f'(a_2)|^q} \right\}.$$

Corollary 10 *Taking $x = \frac{a_1+a_2}{2}$, $m = 1$, $\eta(x, ma_1) = x - ma_1$, and $\eta(a_2, mx) = a_2 - mx$ in Corollary 9, we get*

$$\left|I_{f,\Lambda_{1,n},\Delta_{1,n}}\left(\frac{a_1+a_2}{2}, a_1, a_2\right)\right| \leq \frac{(n+1)(a_2-a_1)}{8\sqrt[q]{3(n+1)}} \tag{35}$$

$$\times\left\{\sqrt[q]{|f'(a_1)|^q+(3n+2)\left|f'\left(\frac{a_1+a_2}{2}\right)\right|^q}+\sqrt[q]{(3n+2)|f'(a_1)|^q+\left|f'\left(\frac{a_1+a_2}{2}\right)\right|^q}\right.$$

$$\left.+\sqrt[q]{\left|f'\left(\frac{a_1+a_2}{2}\right)\right|^q+(3n+2)|f'(a_2)|^q}+\sqrt[q]{(3n+2)\left|f'\left(\frac{a_1+a_2}{2}\right)\right|^q+|f'(a_2)|^q}\right\}.$$

Corollary 11 *Taking $\varphi(t) = \frac{t^\alpha}{\Gamma(\alpha)}$ in Theorem 3, we get*

$$\left|I_{f,\Lambda_{m,n},\Delta_{m,n}}(x, a_1, a_2)\right| \leq \left(\frac{1}{n+1}\right)^{\frac{1-q}{q}} \frac{\Gamma(\alpha+1)}{2\Gamma(\alpha+2)}\sqrt[q]{\frac{\Gamma(\alpha+2)}{\Gamma(\alpha+3)}}\eta(x, ma_1) \tag{36}$$

$$\times\left\{\sqrt[q]{|f'(ma_1)|^q+\left[n(\alpha+2)+(\alpha+1)\right]|f'(x)|^q}\right.$$

$$\left.+\sqrt[q]{\left[n(\alpha+2)+(\alpha+1)\right]|f'(ma_1)|^q+|f'(x)|^q}\right\}$$

$$+\left(\frac{1}{n+1}\right)^{\frac{1-q}{q}} \frac{\Gamma(\alpha+1)}{2\Gamma(\alpha+2)}\sqrt[q]{\frac{\Gamma(\alpha+2)}{\Gamma(\alpha+3)}}\eta(a_2, mx)$$

$$\times\left\{\sqrt[q]{|f'(mx)|^q+\left[n(\alpha+2)+(\alpha+1)\right]|f'(a_2)|^q}\right.$$

$$\left.+\sqrt[q]{\left[n(\alpha+2)+(\alpha+1)\right]|f'(mx)|^q+|f'(a_2)|^q}\right\}.$$

Corollary 12 *Taking $\varphi(t) = \frac{t^{\frac{\alpha}{k}}}{k\Gamma_k(\alpha)}$ in Theorem 3, we get*

$$\left|I_{f,\Lambda_{m,n},\Delta_{m,n}}(x, a_1, a_2)\right| \leq \left(\frac{1}{n+1}\right)^{\frac{1-q}{q}} \frac{\Gamma_k(\alpha+k)}{2\Gamma_k(\alpha+k+1)}\sqrt[q]{\frac{\Gamma_k(\alpha+k+1)}{\Gamma_k(\alpha+k+2)}}\eta(x, ma_1) \tag{37}$$

$$\times\left\{\sqrt[q]{|f'(ma_1)|^q+\left[n\left(\frac{\alpha}{k}+2\right)+\left(\frac{\alpha}{k}+1\right)\right]|f'(x)|^q}\right.$$

$$\left.+\sqrt[q]{\left[n\left(\frac{\alpha}{k}+2\right)+\left(\frac{\alpha}{k}+1\right)\right]|f'(ma_1)|^q+|f'(x)|^q}\right\}$$

$$+\left(\frac{1}{n+1}\right)^{\frac{1-q}{q}}\frac{\Gamma_k(\alpha+k)}{2\Gamma_k(\alpha+k+1)}\sqrt[q]{\frac{\Gamma_k(\alpha+k+1)}{\Gamma_k(\alpha+k+2)}}\eta(a_2,mx)$$

$$\times\left\{\sqrt[q]{|f'(mx)|^q+\left[n\left(\frac{\alpha}{k}+2\right)+\left(\frac{\alpha}{k}+1\right)\right]|f'(a_2)|^q}\right.$$

$$\left.+\sqrt[q]{\left[n\left(\frac{\alpha}{k}+2\right)+\left(\frac{\alpha}{k}+1\right)\right]|f'(mx)|^q+|f'(a_2)|^q}\right\}.$$

Remark 1 Applying our Theorems 2 and 3, for $n\in\mathbb{N}^*$ and appropriate choices of function $\varphi(t)=t$; $\varphi(t)=\frac{t^\alpha}{\Gamma(\alpha)}$; $\frac{t^{\frac{\alpha}{k}}}{k\Gamma_k(\alpha)}$; $\varphi(t)=t(a_2-t)^{\alpha-1}$, where $f(x)$ is symmetric to $x=\frac{ma_1+a_2}{2}$ and $m\in(0,1]$ is a fixed number; $\varphi(t)=\frac{t}{\alpha}\exp\left[\left(-\frac{1-\alpha}{\alpha}\right)t\right]$, for $\alpha\in(0,1)$ such that $\eta(x,ma_1)=x-ma_1$ and $\eta(a_2,mx)=a_2-mx$, $\forall x\in P$, we can deduce some new general fractional integral inequalities. The details are left to the interested reader.

3 Applications and New Error Estimates

Consider the following special means for different real numbers α, β and $\alpha\beta\neq 0$, as follows:

1. The arithmetic mean:

$$A:=A(\alpha,\beta)=\frac{\alpha+\beta}{2},$$

2. The harmonic mean:

$$H:=H(\alpha,\beta)=\frac{2}{\frac{1}{\alpha}+\frac{1}{\beta}},$$

3. The logarithmic mean:

$$L := L(\alpha, \beta) = \frac{\beta - \alpha}{\ln|\beta| - \ln|\alpha|},$$

4. The generalized log-mean:

$$L_r := L_r(\alpha, \beta) = \left[\frac{\beta^{r+1} - \alpha^{r+1}}{(r+1)(\beta - \alpha)}\right]^{\frac{1}{r}}; \quad r \in \mathbb{Z} \setminus \{-1, 0\}.$$

It is well-known that L_r is monotonic nondecreasing over $r \in \mathbb{Z}$ with $L_{-1} := L$. In particular, we have the following inequality $H \leq L \leq A$. Now, using the theory results in Section 2, we give some applications to special means for different real numbers.

Proposition 1 *Let $a_1, a_2 \in \mathbb{R} \setminus \{0\}$, where $a_1 < a_2$ and $x \in [a_1, a_2]$. Then for $r \in \{2, 3, \dots\}$, where $q > 1$ and $p^{-1} + q^{-1} = 1$, the following inequality holds:*

$$\left| A\left(a_1^r, \left(\frac{a_1 + a_2}{2}\right)^r\right) + A\left(\left(\frac{a_1 + a_2}{2}\right)^r, a_2^r\right) - 2L_r^r(a_1, a_2) \right|$$
$$\leq \frac{2r(a_2 - a_1)}{\sqrt[p]{p+1}} \tag{38}$$
$$\times \left\{ \sqrt[q]{A\left(|a_1|^{q(r-1)}, \left|\frac{a_1 + a_2}{2}\right|^{q(r-1)}\right)} + \sqrt[q]{A\left(\left|\frac{a_1 + a_2}{2}\right|^{q(r-1)}, |a_2|^{q(r-1)}\right)} \right\}.$$

Proof Applying Theorem 2 for $x = \frac{a_1+a_2}{2}$, $m = 1$, $n = 0$, $\eta(x, ma_1) = x - ma_1$, $\eta(a_2, mx) = a_2 - mx$, $f(x) = x^r$, and $\varphi(t) = t$, one can obtain the result immediately.

Proposition 2 *Let $a_1, a_2 \in \mathbb{R} \setminus \{0\}$, where $a_1 < a_2$ and $x \in [a_1, a_2]$. Then, for $q > 1$ and $p^{-1} + q^{-1} = 1$, the following inequality holds:*

$$\left| \frac{1}{H\left(a_1, \frac{a_1+a_2}{2}\right)} + \frac{1}{H\left(\frac{a_1+a_2}{2}, a_2\right)} - \frac{2}{L(a_1, a_2)} \right| \leq \frac{2(a_2 - a_1)}{\sqrt[p]{p+1}} \tag{39}$$
$$\times \left\{ \frac{1}{\sqrt[q]{H\left(|a_1|^{2q}, \left|\frac{a_1+a_2}{2}\right|^{2q}\right)}} + \frac{1}{\sqrt[q]{H\left(\left|\frac{a_1+a_2}{2}\right|^{2q}, |a_2|^{2q}\right)}} \right\}.$$

Proof Applying Theorem 2 for $x = \frac{a_1+a_2}{2}$, $m = 1$, $n = 0$, $\eta(x, ma_1) = x - ma_1$, $\eta(a_2, mx) = a_2 - mx$, $f(x) = \frac{1}{x}$ and $\varphi(t) = t$, one can obtain the result immediately.

Proposition 3 *Let $a_1, a_2 \in \mathbb{R} \setminus \{0\}$, where $a_1 < a_2$ and $x \in [a_1, a_2]$. Then, for $r \in \{2, 3, \ldots\}$ and $q \geq 1$, the following inequality holds:*

$$\left| A\left(a_1^r, \left(\frac{a_1+a_2}{2}\right)^r\right) + A\left(\left(\frac{a_1+a_2}{2}\right)^r, a_2^r\right) - 2L_r^r(a_1, a_2) \right|$$

$$\leq \sqrt[q]{\frac{2}{3}} \frac{r(a_2-a_1)}{8} \tag{40}$$

$$\times \left\{ \sqrt[q]{A\left(2|a_1|^{q(r-1)}, \left|\frac{a_1+a_2}{2}\right|^{q(r-1)}\right)} + \sqrt[q]{A\left(2\left|\frac{a_1+a_2}{2}\right|^{q(r-1)}, |a_1|^{q(r-1)}\right)} \right.$$

$$\left. + \sqrt[q]{A\left(2|a_2|^{q(r-1)}, \left|\frac{a_1+a_2}{2}\right|^{q(r-1)}\right)} + \sqrt[q]{A\left(2\left|\frac{a_1+a_2}{2}\right|^{q(r-1)}, |a_2|^{q(r-1)}\right)} \right\}.$$

Proof Applying Theorem 3 for $x = \frac{a_1+a_2}{2}$, $m = 1$, $n = 0$, $\eta(x, ma_1) = x - ma_1$, $\eta(a_2, mx) = a_2 - mx$, $f(x) = x^r$, and $\varphi(t) = t$, one can obtain the result immediately.

Proposition 4 *Let $a_1, a_2 \in \mathbb{R} \setminus \{0\}$, where $a_1 < a_2$ and $x \in [a_1, a_2]$. Then for $q \geq 1$, the following inequality holds:*

$$\left| \frac{1}{H\left(a_1, \frac{a_1+a_2}{2}\right)} + \frac{1}{H\left(\frac{a_1+a_2}{2}, a_2\right)} - \frac{2}{L(a_1, a_2)} \right| \leq \sqrt[q]{\frac{2}{3}} \frac{(a_2-a_1)}{8} \tag{41}$$

$$\times \left\{ \frac{1}{\sqrt[q]{H\left(2|a_1|^{2q}, \left|\frac{a_1+a_2}{2}\right|^{2q}\right)}} + \frac{1}{\sqrt[q]{H\left(2\left|\frac{a_1+a_2}{2}\right|^{2q}, |a_1|^{2q}\right)}} \right.$$

$$\left. + \frac{1}{\sqrt[q]{H\left(2|a_2|^{2q}, \left|\frac{a_1+a_2}{2}\right|^{2q}\right)}} + \frac{1}{\sqrt[q]{H\left(2\left|\frac{a_1+a_2}{2}\right|^{2q}, |a_2|^{2q}\right)}} \right\}.$$

Proof Applying Theorem 3 for $x = \frac{a_1+a_2}{2}$, $m = 1$, $n = 0$, $\eta(x, ma_1) = x - ma_1$, $\eta(a_2, mx) = a_2 - mx$, $f(x) = \dfrac{1}{x}$, and $\varphi(t) = t$, one can obtain the result immediately.

Remark 2 Applying our Theorems 2 and 3 for $x = \frac{a_1+a_2}{2}$, $m = 1$, $n = 0$, $\eta(x, ma_1) = x - ma_1$, $\eta(a_2, mx) = a_2 - mx$, and appropriate choices of function $\varphi(t) = \frac{t^\alpha}{\Gamma(\alpha)}$, $\frac{t^{\frac{\alpha}{k}}}{k\Gamma_k(\alpha)}$, $\varphi(t) = t(a_2 - t)^{\alpha-1}$, where $f(x)$ is symmetric to $x = \frac{a_1+a_2}{2}$, $\varphi(t) = \frac{t}{\alpha}\exp\left[\left(-\frac{1-\alpha}{\alpha}\right)t\right]$, for $\alpha \in (0, 1)$, such that $|f'|^q$ to be preinvex, we can deduce some new general fractional integral inequalities using above special means. The details are left to the interested reader.

Next, we provide some new error estimates for the trapezoidal formula.

Let Q be the partition of the points $a_1 = x_0 < x_1 < \ldots < x_k = a_2$ of the interval $[a_1, a_2]$. Let us consider the following quadrature formula:

$$\int_{a_1}^{a_2} f(x)dx = T(f, Q) + E(f, Q),$$

where

$$T(f, Q) = \sum_{i=0}^{k-1}\left[\frac{f(x_i) + f(x_{i+1})}{2} + f\left(\frac{x_i + x_{i+1}}{2}\right)\right](x_{i+1} - x_i)$$

is the trapezoidal version and $E(f, Q)$ denote their associated approximation error.

Proposition 5 *Let $f : [a_1, a_2] \longrightarrow \mathbb{R}$ be a differentiable function on (a_1, a_2), where $a_1 < a_2$. If $|f'|^q$ is convex on $[a_1, a_2]$ for $q > 1$ and $\frac{1}{p} + \frac{1}{q} = 1$, then the following inequality holds:*

$$\left|E(f, Q)\right| \le \left(\frac{1}{2}\right)^{\frac{1-q}{q}} \frac{1}{2\sqrt[p]{p+1}}\sum_{i=0}^{k-1}(x_{i+1} - x_i)^2 \tag{42}$$

$$\times\left\{\sqrt[q]{|f'(x_i)|^q + \left|f'\left(\frac{x_i + x_{i+1}}{2}\right)\right|^q} + \sqrt[q]{\left|f'\left(\frac{x_i + x_{i+1}}{2}\right)\right|^q + |f'(x_{i+1})|^q}\right\}.$$

Proof Applying Theorem 2 for $x = \frac{a_1+a_2}{2}$, $m = 1$, $n = 0$, $\eta(x, ma_1) = x - ma_1$, $\eta(a_2, mx) = a_2 - mx$, and $\varphi(t) = t$ on the subintervals $[x_i, x_{i+1}]$ $(i = 0, \ldots, k-1)$ of the partition Q, we have

$$\left|\frac{f(x_i) + f(x_{i+1})}{2} + f\left(\frac{x_i + x_{i+1}}{2}\right) - \frac{2}{x_{i+1} - x_i}\int_{x_i}^{x_{i+1}} f(x)dx\right|$$

$$\leq \left(\frac{1}{2}\right)^{\frac{1-q}{q}} \frac{(x_{i+1} - x_i)}{2\sqrt[p]{p+1}} \tag{43}$$

$$\times \left\{ \sqrt[q]{|f'(x_i)|^q + \left| f'\left(\frac{x_i + x_{i+1}}{2}\right)\right|^q} + \sqrt[q]{\left| f'\left(\frac{x_i + x_{i+1}}{2}\right)\right|^q + |f'(x_{i+1})|^q} \right\}.$$

Hence from (43), we get

$$\left|E(f, Q)\right| = \left| \int_{a_1}^{a_2} f(x)dx - T(f, Q)\right|$$

$$\leq \left| \sum_{i=0}^{k-1} \left\{ \int_{x_i}^{x_{i+1}} f(x)dx - \frac{1}{2}\left[\frac{f(x_i) + f(x_{i+1})}{2} + f\left(\frac{x_i + x_{i+1}}{2}\right)\right](x_{i+1} - x_i) \right\} \right|$$

$$\leq \sum_{i=0}^{k-1} \left| \left\{ \int_{x_i}^{x_{i+1}} f(x)dx - \frac{1}{2}\left[\frac{f(x_i) + f(x_{i+1})}{2} + f\left(\frac{x_i + x_{i+1}}{2}\right)\right](x_{i+1} - x_i) \right\} \right|$$

$$\leq \left(\frac{1}{2}\right)^{\frac{1-q}{q}} \frac{1}{2\sqrt[p]{p+1}} \sum_{i=0}^{k-1} (x_{i+1} - x_i)^2$$

$$\times \left\{ \sqrt[q]{|f'(x_i)|^q + \left| f'\left(\frac{x_i + x_{i+1}}{2}\right)\right|^q} + \sqrt[q]{\left| f'\left(\frac{x_i + x_{i+1}}{2}\right)\right|^q + |f'(x_{i+1})|^q} \right\}.$$

The proof of this proposition is complete.

Proposition 6 *Let $f : [a_1, a_2] \longrightarrow \mathbb{R}$ be a differentiable function on (a_1, a_2), where $a_1 < a_2$. If $|f'|^q$ is convex on $[a_1, a_2]$ for $q \geq 1$, then the following inequality holds:*

$$\left|E(f, Q)\right| \leq \frac{1}{8\sqrt[q]{3}} \sum_{i=0}^{k-1} (x_{i+1} - x_i)^2 \tag{44}$$

$$\times \left\{ \sqrt[q]{|f'(x_i)|^q + 2\left| f'\left(\frac{x_i + x_{i+1}}{2}\right)\right|^q} + \sqrt[q]{2|f'(x_i)|^q + \left| f'\left(\frac{x_i + x_{i+1}}{2}\right)\right|^q} \right.$$

$$\left. + \sqrt[q]{\left| f'\left(\frac{x_i + x_{i+1}}{2}\right)\right|^q + 2|f'(x_{i+1})|^q} + \sqrt[q]{2\left| f'\left(\frac{x_i + x_{i+1}}{2}\right)\right|^q + |f'(x_{i+1})|^q} \right\}.$$

Proof The proof is analogous as to that of Proposition 5 but use Theorem 3 for $x = \frac{a_1+a_2}{2}$, $m = 1$, $n = 0$, $\eta(x, ma_1) = x - ma_1$, $\eta(a_2, mx) = a_2 - mx$, and $\varphi(t) = t$.

Remark 3 Applying our Theorems 2 and 3, for $x = \frac{a_1+a_2}{2}$, $m = 1$, $n = 0$, $\eta(x, ma_1) = x - ma_1$, $\eta(a_2, mx) = a_2 - mx$, and appropriate choices of function $\varphi(t) = t$; $\varphi(t) = \frac{t^\alpha}{\Gamma(\alpha)}$; $\frac{t^{\frac{\alpha}{k}}}{k\Gamma_k(\alpha)}$; $\varphi(t) = t(a_2 - t)^{\alpha-1}$, where $f(x)$ is symmetric to $x = \frac{a_1+a_2}{2}$; $\varphi(t) = \frac{t}{\alpha} \exp\left[\left(-\frac{1-\alpha}{\alpha}\right) t\right]$, for $\alpha \in (0, 1)$, such that $|f'|^q$ to be convex, we can deduce some new general fractional integral inequalities using above ideas and techniques. The details are left to the interested reader.

References

1. S.M. Aslani, M.R. Delavar, S.M. Vaezpour, Inequalities of Fejér type related to generalized convex functions with applications. Int. J. Anal. Appl. **16**(1), 38–49 (2018)
2. F.X. Chen, S.H. Wu, Several complementary inequalities to inequalities of Hermite-Hadamard type for s-convex functions. J. Nonlinear Sci. Appl. **9**(2), 705–716 (2016)
3. Y.M. Chu, M.A. Khan, T.U. Khan, T. Ali, Generalizations of Hermite-Hadamard type inequalities for MT-convex functions. J. Nonlinear Sci. Appl. **9**(5), 4305–4316 (2016)
4. Z. Dahmani, On Minkowski and Hermite-Hadamard integral inequalities via fractional integration. Ann. Funct. Anal. **1**(1), 51–58 (2010)
5. M.R. Delavar, M. De La Sen, Some generalizations of Hermite-Hadamard type inequalities. SpringerPlus **5** (2016). Article no. 1661
6. M.R. Delavar, S.S. Dragomir, On η-convexity. Math. Inequal. Appl. **20**, 203–216 (2017)
7. S.S. Dragomir, R.P. Agarwal, Two inequalities for differentiable mappings and applications to special means of real numbers and trapezoidal formula. Appl. Math. Lett. **11**(5), 91–95 (1998)
8. M.A. Khan, Y. Khurshid, T. Ali, Hermite-Hadamard inequality for fractional integrals via η-convex functions. Acta Math. Univ. Comenianae **79**(1), 153–164 (2017)
9. M.A. Khan, Y.-M. Chu, A. Kashuri, R. Liko, G. Ali, New Hermite-Hadamard inequalities for conformable fractional integrals. J. Funct. Spaces (2018), 9. Article ID 6928130
10. W.J. Liu, Some Simpson type inequalities for h-convex and (α, m)-convex functions. J. Comput. Anal. Appl. **16**(5), 1005–1012 (2014)
11. W. Liu, W. Wen, J. Park, Hermite-Hadamard type inequalities for MT-convex functions via classical integrals and fractional integrals. J. Nonlinear Sci. Appl. **9**, 766–777 (2016)
12. M.V. Mihai, Some Hermite-Hadamard type inequalities via Riemann-Liouville fractional calculus. Tamkang J. Math, **44**(4), 411–416 (2013)
13. S. Mubeen, G.M. Habibullah, k-Fractional integrals and applications. Int. J. Contemp. Math. Sci. **7**, 89–94 (2012)
14. M.A. Noor, K.I. Noor, M.U. Awan, S. Khan, Hermite-Hadamard inequalities for s-Godunova-Levin preinvex functions. J. Adv. Math. Stud. **7**(2), 12–19 (2014)
15. O. Omotoyinbo, A. Mogbodemu, Some new Hermite-Hadamard integral inequalities for convex functions. Int. J. Sci. Innovation Tech. **1**(1), 1–12 (2014)
16. M.E. Özdemir, S.S. Dragomir, C. Yildiz, The Hadamard's inequality for convex function via fractional integrals. Acta Mathematica Scientia **33**(5), 153–164 (2013)
17. M.Z. Sarikaya, H. Yildirim, On generalization of the Riesz potential. Indian J. Math. Math. Sci. **3**(2), 231–235 (2007)
18. E. Set, M.A. Noor, M.U. Awan, A. Gözpinar, Generalized Hermite-Hadamard type inequalities involving fractional integral operators. J. Inequal. Appl. **169**, 1–10 (2017)

19. H. Wang, T.S. Du, Y. Zhang, k-fractional integral trapezium-like inequalities through (h, m)-convex and (α, m)-convex mappings. J. Inequal. Appl. **2017**(311), 20 (2017)
20. T. Weir, B. Mond, Preinvex functions in multiple objective optimization. J. Math. Anal. Appl. **136**, 29–38 (1988)
21. Y. Zhang, T.S. Du, H. Wang, Y.J. Shen, A. Kashuri, Extensions of different type parameterized inequalities for generalized (m,h)-preinvex mappings via k-fractional integrals. J. Inequal. Appl. **2018**(49), 30 (2018)
22. X.M. Zhang, Y.-M. Chu, X.H. Zhang, The Hermite-Hadamard type inequality of GA-convex functions and its applications. J. Inequal. Appl. 11 (2010). Article ID 507560

Asymptotic Statistical Results: Theory and Practice

Christos P. Kitsos and Amílcar Oliveira

Abstract The target of this paper is to discuss the existent difference of Asymptotic Theory in Statistics comparing to Mathematics. There is a need for a limiting distribution in Statistics, usually the Normal one. Adopting the sequential principle the first-order autoregression model and the stochastic approximation are referred for their particular interest for asymptotic results.

1 Introduction

Let $x_1, x_2, \ldots$ be a design variable with $y_1, y_2, \ldots$ being the corresponding response described by a probability density function $f(y|x, \theta)$ with $\theta \in \Theta \subseteq \mathbb{R}^p$ is the involved p-term parameter. That is we form a parametric model and we assume that the sequential principle of design has been adopted, [38], i.e. the x_{n+1}-th observation is of the form in the Linear case

$$x_{n+1} = x_{n+1}(y(n), x(n)),$$

while in the Non-Linear case is

$$x_{n+1} = x_{n+1}(y(n), x(n); \theta),$$

i.e. in the Non-Linear case the design points depend on the parameter we wish to estimate.

C. P. Kitsos
West Attica University, Athens, Greece
e-mail: xkitsos@uniwa.gr

A. Oliveira (✉)
Universidade Aberta and CEAUL, Lisbon, Portugal
e-mail: amilcar.oliveira@uab.pt

N. J. Daras, T. M. Rassias (eds.), *Computational Mathematics and Variational Analysis*, Springer Optimization and Its Applications 159,
https://doi.org/10.1007/978-3-030-44625-3_10

Moreover $x(n)$, $y(n)$ are the vectors

$$x(n) = (x_1, x_2, \dots, x_n)$$
$$y(n) = (y_1, y_2, \dots, y_n)$$

When the $(n+1)$-th observation is added we form

$$x(n+1) = (x(n), x_{n+1})$$
$$y(n+1) = (y(n), y_{n+1})$$

As it has been stated [13] the sequential nature of the design is irrelevant to any method of inference based on the strong likelihood principle. For the Linear case, I. Ford verified it in his unpublished Glasgow PhD Thesis, while C.P. Kitsos verified the Non-Linear case in his unpublished Glasgow PhD Thesis.

As the sample size n tends to infinite there are nice results, as far as the normality of the estimates is concerned. But to propose an experiment with infinite trials, might not be considered as practically applicable, especial in Biostatistics, or to nuclear experiments. Optimal experimental design theory offers a strong theoretical background to restrict the sample size, but still approximated results are needed for sensitive practical applications. We are working towards this line of thought at this paper.

In such a case the likelihood function, $Lik(\theta)$, can be evaluated through its definition and the fact that the prior knowledge for θ exists, and x_1 is specified. Thus

$$\begin{aligned} Lik(\theta) &= \prod_{i=1}^{n} f(y_i|x_i, \theta) \qquad (1)\\ &= \prod_{i=1}^{n} p(y_i|x(i), y(i-1), \theta) p(x_i|x(i-1), y(i-1), \theta)\\ &= \prod_{i=1}^{n} p(y_i|x(i), y(i-1), \theta) \end{aligned}$$

Example 1 Consider the log-likelihood on n observations, $l_n(\theta)$ say, when $p(y_i|x_i, \theta)$ is given. Then for the $n+1$ observations the log-likelihood $l_{n+1}(\theta)$ is

$$l_{n+1}(\theta) = \sum_{i=1}^{n+1} log[p(y_i|x_i, \theta)] = l_n(\theta) + logp(y_{n+1}|x_{n+1}, \theta)$$

Let $\hat{\theta}_n$, $\hat{\theta}_{n+1}$ be the LME obtained from $l_n(\theta)$, $l_{n+1}(\theta)$, respectively. Considering that

$$\frac{\partial l_{n+1}(\theta)}{\partial \theta} = \frac{\partial l_n(\theta)}{\partial(\theta)} + \frac{\partial logp(y_{n+1}|x_{n+1}, \theta)}{\delta \theta}$$

From the definition of MLE of $\hat{\theta}_{n+1}$ it holds:

$$0 = \frac{\partial l_n(\theta)}{\partial \theta} |_{\hat{\theta}_{n+1}} + S_c(y_{n+1}|x_{n+1}, \hat{\theta}_{n+1})$$

By $S(\cdot|\cdot)$ we denote the score function for a single observation. Considering the first-order Taylor expansion about $\hat{\theta}_n$ we obtain

$$0 = (\hat{\theta}_{n+1} - \hat{\theta}_n)\frac{\partial^2 l_n(\hat{\theta}_n)}{\partial \theta} + S_c(y_{n+1}|x_{n+1}, \hat{\theta}_n)$$

Thus

$$\hat{\theta}_{n+1} = \hat{\theta}_n - \frac{1}{\frac{\partial^2 l_n(\hat{\theta}_n)}{\partial \theta^2}} \times S_c(\cdot|\cdot)$$

In practice we approximate the second-order derivative (Hessian) with Fisher's information.

The above discussed example is associated with the iterative schemes, different than the one we shall discuss in Section 3. A typical asymptotic method is the Newton–Raphson adopted method for solving the maximum likelihood and obtain maximum likelihood estimators (MLE). Equation (3) can be solved, sometimes, only by iteration.

Example 2 Consider a random sample from the Γ distribution. The density of a single observation x_i is

$$f(x_i|\theta_1, \theta_2) = \frac{{\theta_1}^{\theta_2}}{(\theta_2 - 1)!} e^{-x_i\theta_1} {x_i}^{\theta_2 - 1}$$

Then, with $\theta = (\theta_1, \theta_2)$ the parameter vector the likelihood function is

$$Lik(x|\theta) = \prod_{i=1}^{n} f(x_i|\theta) = [\frac{{\theta_1}^{\theta_2}}{(\theta_2 - 1)!}]^n exp[-\theta_1 \sum_{i=1}^{n} x_i + (\theta_2 - 1) \sum_{i=1}^{n} lnx_i]$$

The log-Lik, $l(\theta)$, can be evaluated and then evaluated the MLE $\hat{\theta} = (\hat{\theta}_1.\hat{\theta}_2)$

Indeed

$$\frac{\partial(\theta)}{\partial \theta_1} = 0 \Rightarrow n(-\bar{x} + \frac{\hat{\theta}_2}{\hat{\theta}_1}) = 0$$

$$\frac{\partial(\theta)}{\partial \theta_2} = 0 \Rightarrow n[ln\hat{\theta}_1 + \bar{x} - \frac{d}{d\theta_2}[(\hat{\theta}_2 - 1)!]] = 0$$

with

$$\bar{x} = n^{-1} \sum_{i=1}^{n} x_i, \quad \bar{x}^* = n^{-1} \sum_{i=1}^{n} lnx_i$$

Since $\theta_2 \neq \theta_1$ we get from the second equation:

$$ln\hat{\theta}_2 - ln\hat{x} + \hat{x}^* - \frac{d}{d\theta_2} ln[(\hat{\theta}_2 - 1)!] = 0$$

which has to be solved by the Newton–Raphson iterative scheme. We adopt the approximation

$$\frac{d}{d\theta_2} ln[(\hat{\theta}_2 - 1)!] \approx ln\hat{\theta}_2 - \frac{1}{2\hat{\theta}_2}$$

and therefore the last equation provides

$$\hat{\theta}_2 = \frac{1}{2(ln\bar{x} - \bar{x}^*)} \tag{2}$$

This value can be the one, as $\hat{\theta}_2^{(0)}$, which feeds Newton–Raphson scheme to provide the MLE. The dispersion matrix, $Cov(\hat{\theta})$, is

$$Cov(\hat{\theta}) = Const \begin{pmatrix} \frac{d^2}{d\theta_2^2} ln[(\hat{\theta}_2 - 1)!] & \frac{1}{\hat{\theta}_1} \\ \frac{1}{\hat{\theta}_1} & \frac{\hat{\theta}_2}{\hat{\theta}_1^2} \end{pmatrix}$$

with $Const = \frac{\hat{\theta}_1^2}{n[\hat{\theta}_2 \frac{d^2}{d\theta_2^2}[ln(\theta_2 - 1)!] - 1]}$.

In the sequence the first-order autoregressive model is faced as a sequential design problem and therefore the asymptotic behavior of the sequence of estimators is discussed.

2 First-Order Autoregression Model

Consider the general linear model

$$y = X\theta + \varepsilon \tag{3}$$

with $y \in \Re^{n \times 1}$, $X \in \Re^{n \times p}$, with rank p, $\theta \in \Re^{p \times 1}$ the parameter vector and $\varepsilon \in \Re^{n \times 1}$. We assume that the noise ε comes a Normal distribution, i.e. $\varepsilon \sim N(0, \sigma^2 I)$. Then it is known, [16] that with $n \to \infty$

$$\hat{\theta} = (X^T X)^{-1} X^T y, \ \hat{\theta} \sim N(\theta, \sigma^2 (X^T X)^{-1})$$

Notice that

$$I = \sigma^{-2}(X^T X) \tag{4}$$

is the Fisher's information matrix. From model (3) with $p = 1$, $x_1 = 0$, and design procedure

$$x_i = y_{i-1}, \quad i = 2, \ldots, n \tag{5}$$

i.e. each new observation is the previous response it leads to what is known as the first-order autoregressive model, so useful in Econometrics as the AR(1) model without intercept, of the form

$$y_i = \theta y_{i-1} + \varepsilon_i, \quad i = 2, \ldots, n \tag{6}$$

In other words the first-order autoregressive model is a sequential procedure, [13, 22]. From (4) the estimate of θ at stage n is

$$\hat{\theta}_n = \frac{\sum_{i=2}^{n} y_i y_{i-1}}{\sum_{i=2}^{n} y_{i-1}^2} \tag{7}$$

The sample information M_n, evaluated through Fisher's information I_n at stage n is

$$M_n = \frac{I_n}{\sigma^2} = I_n^* = \sum_{i=1}^{n-1} y_i^2 \tag{8}$$

[28] that for $|\theta| < 1$ and $n \to \infty$ it holds:

$$I_n^{-1/2}(\hat{\theta}_n - \theta) \xrightarrow{\mathscr{L}} N(0, \sigma^2), \tag{9}$$

where $\mathscr{L}$ means in distribution. Moreover (9) it is not true for $|\theta| = 1$, for which particular case, an asymptotic normal distribution is suggested. To overpass the asymptotic nature [28] proved that for $|\theta| \leq 1$ if n_k is the first $n(\geq 2)$ for which $I_n \geq k$, then as $k \to \infty$ that results the asymptotically as a limiting result, i.e. $I_{n_k}^{-1/2}(\hat{\theta}_{n_k} - \theta)$ tends in distribution, uniformly in θ, to $N(0, \sigma^2)$. The particular case of $n = 2$ was discussed in [1]. Still the "sequential nature" of the input observations (which are not independent as the usual theory requests) can be evaluated at the nominal level a, as

$$\hat{\theta}_n \pm t_{n-1,1-\alpha/2} \sqrt{\frac{RSS}{n-1} I_n^{*-1}} \tag{10}$$

with, RSS, being the residual sum of squares, namely

$$RSS = \sum y_i^2 - \hat{\theta}_n \sum y_{i+1} y_i \tag{11}$$

see [22] for details and simulation results.

Relaxing the assumption on Normal errors as in [3] investigated an autoregressive model as in (6), the AR(1) process with $\theta \in [0, 1)$ and the new assumptions about the errors:

- ε_2: is a non-negative strict while noise with $E(\varepsilon_i^2) < \infty$.
- $F(x) = P[\varepsilon_i < x]$, assuming $P[\varepsilon_i = 0] < 1$

In such a case they proved that

$$\theta_n^* = min\{\frac{y_2}{y_1}, \frac{y_3}{y_2}, \ldots, \frac{y_n}{y_{n-1}}\} \tag{12}$$

is strongly consistent estimator, see [10] among others, of θ if and only if

$$F(k) - F(\lambda) < 1, \quad \forall k, \lambda \in \Re^+$$

In Econometrics [39] the OLS (ordinary least squares) estimate $\hat{\theta}_n$ it seems to be more applied estimate than other approaches. The limiting results might be theoretically nice, but the asymptotic distribution is very difficult to be approached in practice. Therefore simulation results investigate the limiting behavior, with truncation to accepted sample size n [22].

The sequential nature of the model does not influence the estimation. We are not losing neither in point estimation, as the average mean square error (MSE) is "small," nor in interval estimation, for different nominal levels [22]. Moreover the Normal character of the error distribution remains valid. All these results with $|\theta| < 1$. For values $\theta \pm 1.5$ the results are not valid as such cases, "far" from the interval $(-1, 1)$. Notice that when $\theta = 1$ in (6) we obtain a random walk process with $E(Y_n) = Y_0$, $Var(Y_n) = n\sigma^2$, and $Cov(Y_n, Y_{n-k}) = (n-k)\sigma^2$, see [20] among others, for detail. In the sequence we shall investigate the asymptotic nature of the stochastic approximation scheme.

3 Stochastic Approximation

The stochastic approximation (SA) method is, to our opinion, the most well-known and applicable sequential method in Statistics. It is the typical asymptotic statistical tool to solve an equation. Sometimes is refereed as the Newton–Raphson (NR), see [31], point of view in statistics.

If refers mainly to binary response problems, while in Section 2 we have been discussing the "continuous case," when in principle a non-linear model was assumed to fit the data. The AR(1) model without intercept discussed already although a linear one leads to a ratio estimate which has a non-linear feature, see [4].

We shall restrict our discussion to the binary response problems. For the outcome $y_i = 1$ or $0, i = 1, 2, \ldots, n$ linked with the input variables X through a probability model "T" in the sense that

$$P(Y_i = 1) = T(x_i) = 1 - P(Y_i = 0), \tag{13}$$

where x_i is the value of x going with observation y_i.

The stochastic approximation, is an iterative scheme, with target to provide the root of the equation

$$E\{y(x)\} = T(x) = p, \quad p \in \Re \tag{14}$$

We shall denote this unique root with θ, to emphasize that the parameter we want to estimate is the root, assuming that $T(x)$ and p are provided, as [32] in their pioneering paper imposed a number of assumptions mainly on the Borel measurable function T. We shall briefly review the method, provided emphasis on the asymptotic behavior of the method. The imposed assumptions try to provide the appropriate bounds, where, within these bounds, the procedure tries to evaluate θ, as it is working in the neighborhood of the root. The main assumptions are: (A) for the root θ.

$$(A) \quad (x - \theta)[T(x) - p] > 0, \quad Pr[[Y(x) - p] \leq k] = 1, \forall x, k \in \Re \tag{15}$$

and (B) for the function $T(x)$

$$(B) \quad inf|T(x) - p| = \delta > 0, \tag{16}$$

They eventually proved, that there exists a sequence of real numbers α_n, $n = 1, 2, \ldots$, with

$$\alpha_n > 0, \quad \sum \alpha_n = \infty, \quad \sum \alpha_n^2 < \infty$$

such that the sequence of stimuli, with x_1 arbitrary

$$x_{n+1} = x_n - \alpha_n(y_n - p), n = 1, 2, \ldots \tag{17}$$

converges to θ in mean square, i.e. as $n \to \infty$.

$$limE(x_n - \theta)^2 = 0 \tag{18}$$

In the early stages of developing the method, [21] weakened the assumption (A) and proved that as $n \to \infty$

$$x_n \to p \quad w.p1$$

Kiefer and Wolfowitz [21] modified the method to evaluate the extremes of a function and [17] worked on a real application and considered the sequence a_n, $n = 1, 2, \ldots$ as the weigh associated with trial n. It is clean that the "easiest" sequence is $a_n = \frac{1}{n}$. Letting $a = T(\theta)$, $a_n = \frac{c}{nb}$, $b = T'(\theta) = T'(x)|_{x=\theta}$ and assuming that

$$y_i = a + b(x_i - \theta) + \varepsilon_i \tag{19}$$

with $E(\varepsilon_i) = 0$, $Var(\varepsilon_i) = \sigma^2$ and ε_i are iid.

Then as $n \to \infty$ it can be proved the very useful (and nice!) [32] asymptotic results.

$$\begin{cases} lim E(x_n - \theta) = 0 \\ lim Var(x_n - \theta) = \frac{\sigma^2}{n} \frac{c^2}{b^2(2c-1)}, \; c > 0.5 \end{cases} \tag{20}$$

It was in [8, 33] and [18] who investigate the asymptotic normality, while [18] made the assumptions slightly weaker. The most interesting result, we believe, is the fact that for the sequence $a_n = \frac{c}{n}, n = 1, 2, \ldots$ as $n \to \infty$ the typical asymptotic result is still valid.

The sequence tends in distribution to the Normal distribution. Therefore the minimum variance can be obtained. Namely:

$$\begin{cases} \sqrt{n}(x_n - \theta) \xrightarrow{\mathscr{L}} N(0, \frac{\sigma^2 c^2}{2bc-1}), & bc > \frac{1}{2} \\ minVar \quad when, & c = c_{opt} = b^{-1} = [T'(\theta)]^{-1} \end{cases} \tag{21}$$

The SA scheme is an elegant example, to our opinion, of how a famous iterative scheme, such as the Newton–Raphson one, has been transferred to the statistical asymptotic theory providing limiting results in distribution. The multivariate case, although discussed in [34] has not been applied extensively.

It is very nice bounded the unknown function $T(x)$, we try to move near to the unknown solution, that is why x_1 can be arbitrary. Notice that Newton–Raphson heavily depends on the initial guess, and if it is not in the neighborhood of the root, the method diverges [31]. Moreover to form the iteration as a statistical asymptotic we need Normality. This is succeeded eventually due to (21), which also provides the way that the minimum variance will be achieved.

To our knowledge SA is the most "elegant transfer" of Mathematical knowledge to Statistical Theory: The iterative scheme is not enough in Asymptotic Statistics, the limiting results to Normality are the crucial point. We traced the pioneering papers to provide evidence of the development of this asymptotical result. As stochastic approximation is a sequential method and eventually a martingale (the next step depends on the current situation and not "on the history" of the sequence) the Measure Theory background is provided in Appendix.

We shall apply SA to real problems in the next section when interest is focused on the "best" estimation of percentiles.

4 Asymptotic Statistical Results for Percentiles

The estimation of the percentile of a cumulative distribution F, cdf F, was tackled by [9], in their pioneering paper. They were adopting Hermite orthogonal polynomial for estimating the $100p\%$ percentile, L_p, in the sense that $F(L_p) = p$. For the proof of the asymptotic normality of the sequence

$$\sqrt{\frac{n}{p(1-p)}} f(L_p)(Q_p - L_p) \longrightarrow N(0,1) \tag{22}$$

with $f(x)$ being the density function and Q_p the sample quantile of order p, see ([10], Theorem 8.5.3).

Neither the Hermite orthogonal polynomial nor the asymptotic limiting result in (22) does solve the problem of estimating L_p especially in the case of low probability level, so essential in experimental carcinogenesis, [25] and biological problems, [24].

The problem has been also faced by [7, 15]. Moreover an optimal design approach can be formed in a Bio-assay. It is rather appreciated an analysis on a small sample size as, for example, in cancer (Ca) problems there is not usually large samples available in practice.

One of the merits of the stochastic approximation discussed in Section 3 is the asymptotic Normality, see (21), which includes that the min of variance occurs at $C_{opt} = [T'(\theta)]^{-1}$, in principle. In the percentile problem we face the general form of the function $T(x)$ represents $F(L_p)$ as the parameter we wish to estimate is L_p and T can be approximated, [26, 40], by the cumulative distribution function F. As the equation to be solved is

$$F(L_p) - p = 0 \tag{23}$$

recall (14), the SA scheme can be applied.

For the class of functions $F(.)$ from the multistage model (MM), see [26, 27] where is proved the following theorem:

Theorem 1 ([26]) *Within the class of MM model with cdf $F(.)$ such that*

$$F(x) = 1 - exp[-\sum_{i=0}^{k} \theta_i x^i]$$

the iterative scheme

$$L_{p,n+1} = L_{p,n} - (n \sum_{i=1}^{k} \tau_i)^{-1}(y_n - p), n = 1, 2, \ldots \tag{24}$$

converges to the percentile point L_p in m.s, i.e.

$$L_{p,n+1} \longrightarrow L_p, p \in (0,1) \tag{25}$$

where

$$\tau_i = i\theta_i L_p^{i-1}(1-p),\, F(L_p) = p,\;\; p \in (0,1).$$

From [8, 22, 23], for the sequence $L_{p,n}$ it is derived that

$$n^{1/2}(L_{p,n} - L_p) \xrightarrow{\mathscr{L}} N(0, \sigma^2(F'(L_p))^{-2}), n \to \infty \tag{26}$$

As θ_i, $i = 1, 2, \ldots, k$ are not known, we can devote, [26] n_0 observations out of the total number of collected n observations at the first stage to estimate τ_i by $\hat{\tau}_i$ as

$$\hat{\tau}_i = \hat{\tau}_{i,n_0} = i\theta_{i,n_0} L_{p,n_0}^{i-1}(1-p). \tag{27}$$

With $\hat{\tau}_i$ feeding (24) the sequence converges asymptotically in m.s. to L_p, and minimizes, due to (26), the variance in the limit, recall also (21) and that the optimum step function is $a_n = n^{-1}C_{opt}$, with $C_{opt} = (F'(L_p))^{-1}$ so $\sigma^2_{min}(F'(L_p))^{-1}$. That is we know the minimum variance and the appropriate sequence to reach it.

Corollary 1 *The iterative scheme (24) provides a D-optimal estimators.*

Moreover:

Corollary 2 *The (optimum) iterative scheme (24) minimizes the entropy of the limiting design.*

Indeed:

It is known, ([2], pg 262) that for the multivariate Normal distribution $N(\mu, V)$ the corresponding entropy $H(.)$ equals $logdet V$, so in this case holds

$$H(L_{p,n}) = logdet(\sigma^2 F'(L_p)^{-2})$$

The $logdet(.)$ reaches the minimum, so does the entropy as

$$minH(L_{p,n}) = log[\sigma^2(F'(L_p))^{-2} = 2log\sigma - 2logF'(L_p) = 2[log\sigma - logC_{opt}^{-1}]$$
$$= 2[log\sigma + logC_{opt}]$$

Notice that we arrive at the same result due to the fact that $F'(L_p) = \sum_{i=0}^{1} \tau_i = C_{opt}^{-1}$.

For the maximum entropy sampling and design see [35, 36].

In [26] was considered from the family of the multistage models, [27], the one-hit model. He performed 1000 simulations with value of $p = 0.01, 0.02, 0.04$ and $n = 200, 500$.

The results provide evidence that even with $n = 200$ (unpublished results are valid also for $n = 50$) the proposed iterative scheme approaches the true value.

5 Discussion

The sequential procedure either as an (optimal) design approach or as a data augmentation and design in the first-order autoregression model appears to provide food for thought. First, the observations are not independent and therefore the likelihood might not obey to the existent results. This problem has been overpassed, [11] for the Linear case, and in [12, 22–24] for the Non-linear case. Second, problem could be in the asymptotic results. Due to [28, 30] the AR(1) with no intersection, [4] converges under the restriction $|\theta| < 1$. For the design point of view, the sequential principle, through stochastic approximation, see [23–25, 29], provides evidence that the asymptotic nature of the design obeys to the limiting theorems of the stochastic approximation. Moreover if the initial Design is D-optimal the limiting one will be also D-optimal, [23]. Moreover the limiting design provides a minimum Entropy one, as already discussed in the application of the SA in cancer carcinogenesis problems, see also [24–26]. Third, notice that the asymptotic results in Statistics are, in principle, different than in Mathematics. Statistics demand the evaluation of the limiting distribution in asymptotic results. In other words we need a distribution which "carries" the existing iterative scheme (sequential procedure) to the limit. Such a case is the SA which "carries" Newton–Raphson iterative scheme under the Normality! Fourth, essential point is that in Statistics the asymptotic results are strongly needed for the Non-Linear problems, [12–14, 22]. The existence of the estimators, either in continuous or in binary case is insured thanks to [19, 37]. The superiority of the Linear case is due to [41]; the sequence design measures also converge to the optimal design measure. Despite the excellent attempt, in [5, 6] to transfer this result in Non-Linear case, still there is not such a result in Non-Linear situation. So, in conclusion, the asymptotic results in Statistics offer to the experimentalist the solution to various estimators problems, and can be adopted especially now when the computation is simple.

Acknowledgments This research is based on the partial results of the Funded by FCT—Fundação para a Ciência e a Tecnologia, Portugal, through the project UID/MAT/00006/2019.

Appendix

Let $\{X_n\}$ be a stochastic process such that the joint distribution of $(X_1, X_2, \ldots, X_n)$ has a strictly positive continuous density. The sequence $\{X_n\}$ will be called absolute fair if for $n = 1, 2, \ldots$.

$$E(X_1) = 0, \quad E(X_{n+1}|X_1, X_2, \ldots, X_n) = 0$$

The sequence $\{Y_n\}$, with $Y_n = X_1 + X_2 + \ldots + X_n + c$ is a martingale if

$$E(Y_{n+1}, |Y_1, \ldots, Y_n) = Y_n \quad n = 1, 2, \ldots$$

provided that $\{X_n\}$ is absolutely fair.

Now,

Let $\mathscr{A}_n$ be the σ-algebra generated by $(Y_1, Y_2, \ldots, Y_n)$, as above, then

$$E(Y_{n+1}|\mathscr{A}_n) = Y_n$$

from the martingale definition. We replace σ-algebra $\mathscr{A}_n$ by a larger σ-algebra $\mathscr{B}_n$: generated by $(Y_1, Y_2, \ldots, Y_n$ and additional random variables depending on the past). The so create process $\mathscr{B}_n$ (containing the past history of the process) is an increasing sequence, i.e. $\mathscr{B}_1 \subset \mathscr{B}_2 \subset \ldots$. Then the sequence $\{Y_n\}$ is a martingale with respect to $\mathscr{B}_n$ if and only if

$$E(Y_{n+1}|\mathscr{B}_n) = Y_n$$

Since $\mathscr{B}_n \supset \mathscr{B}_{n-1}$

$$E(Y_{n+1}|\mathscr{B}_{n-1}) = Y_{n-1}$$

Now, a move general result helpful in sequential methods is

$$E(Y_n|\mathscr{B}_k) = Y_k, k = 1, 2, \ldots.$$

That is the condition on knowledge expressed on $\mathscr{B}_k$ provides knowledge for the martingale at the k-th stage.

The Martingale Convergence Theorem provides the mathematical insight for the limiting Normality of the stochastic approximation.

Let $\{Y_n\}$ be an (infinite) martingale with $E\{Y_n^2\} < c < \infty$, $\forall n$. Then exists a random variable Y, such that

$$Y_n \longrightarrow Y, \quad \forall n, w_{p1}.$$

Furthermore $E(Y_n) = E(Y), \forall n$.

References

1. D.F. Andrews, A.M. Herzbg, A simple method for constructing exact tests for sequentially design experiments. Biometrika **60**, 489–497 (1973)
2. Y. Bard, *Nonlinear Parametric Estimation* (Academic, Cambridge, 1974)
3. C.B. Bell, E.P. Smith, Inference for non-negative autoregressive schemes. Commun. Stat. Theory Methods **15**, 2267–2293 (1986)

4. G.E.P. Box, G.M. Jeckins, *Time Series Analysis Forecasting and Control* (Holden, San Francisco, 1970)
5. P. Chadhuri, P.A. Mykland, Nonlinear experiments: optimal design and inference based on likelihood. J. Am. Stat. Assoc. **88**, 538–546 (1995)
6. P. Chadhuri, P.A. Mykland, In efficient designing of nonlinear experiments. Satistica Sinica **5**, 421–440 (1995)
7. S.C. Choi, Interval estimation of the LD_50 based on an up-and-down experiment. Biometrics **46**, 485–492 (1990)
8. K.L. Chung, On a stochastic approximation method. Annal Math. Stat. **25**, 463 (1954)
9. E.A. Cornish, R.A. Fisher, Moment and cumulants in the specification of distribution. Rev. Int. Stat. Inst. **5**, 307–315 (1937)
10. J.E. Dudewicz, *Introduction to Statistic and Probability* (Holt, Rinehart and Winston, New York, 1976)
11. I. Ford, Optimal static and sequential design: a critical review. Ph.D. Thesis. University of Glasgow (1976)
12. I. Ford, S.D. Silvey, A sequentially constructed design for estimating a nonlinear parametric function. Biometrika **67**, 381–388 (1980)
13. I. Ford, D.M. Titterington, C.F.J. Wu, Inference and sequential design. Biometrika **72**, 545–551 (1985)
14. I. Ford, C.P. Kitsos, D.M. Titterington, Recent advances in non-linear experimental design. Technometrics **31**, 49–60 (1989)
15. H.L. Gray, S. Wang, A general method for approximating tail probabilities. J. Am. Stat. Assoc. **86**, 160–166 (1991)
16. F.A. Graybill, *Theory and Application of the Linear Model* (Duxbury Press, Massachusetts, 1976)
17. L. Guttman, R. Guttman, An illustration of the use of stochastic approximation. Biometrics **15**, 551–559 (1959)
18. J.L. Hodges, E.C. Lehman, Two approximations to the Robbins Monro process. In: *Proceedings of the 3rd Berkley Symposium on Mathematical Statistics and Probability* (University of California Press, Berkley, 1956), pp. 95–104
19. R.J. Jennrich, Asymptotic properties of non-linear Least Squares estimators. Ann. Math. Stat. **40**, 633–643 (1969)
20. S. Kavlin, H. Taylor, *A First Course in Stochastic Processes* (Academic, Cambridge, 1975)
21. J. Kiefer, J. Wolfowitz, Stochastic estimation of the maximum of a regression function. Ann. Math Stat. **23**, 462–466 (1952)
22. C.P. Kitsos, Design and inference for the nonlinear problems. Ph.D. Thesis. Glasgow (1986)
23. C.P. Kitsos, Fully-sequential procedures in nonlinear design problems. Comp. Stat. Data Anal. **8**, 13–19 (1989)
24. C.P. Kitsos, Adopting sequential procedures for biological experiments. In: *MODA3 (Model Oriented Data Analysis)*, ed. by W.G. Muller, H.P. Wynn, A.A. Zhigliavsky (Physica-Verlag, Heidelberg, 1992), pp. 3–9
25. C.P. Kitsos, Sequential assays for experimental cancinogenesis. In: *ISI 50th Session, Book 1*, Beijing (1995), pp. 625–626
26. C.P. Kitsos, Optimal designs for estimating the percentiles of the risk in multistage models in carcinogenesis. Biom. J. **41**, 33–43 (1999)
27. C.P. Kitsos, *Cancer Bioassays. A Statistical Approach* (Lambert Academic Publishing, Saabrücken, 2012)
28. T.L. Lai, W. Siegmund, Fixed of an autoregressive parameter. Ann. Stat. **11**, 478–485 (1983)
29. T.L. Lai, H. Robbins, Adaptive design and stochastic approximation. Ann. Stat. **7**(6), 1196–1221 (1979)
30. T.L. Lai, C.Z. Wei, Least Squares estimates in stochastic regression models with applications to identification and central control dynamic systems. Biometrika Ann. Stat. **10**, 154–166 (1982)
31. J.M. Ortega, W.C. Pheinfold, *Iterative Solution of Nonlinear Equation in Several Variables* (Academic Press, Cambridge, 1970)

32. H. Robbins, S. Monro, A stochastic approximation method. Ann. Math. Stat. **22**, 400–407 (1951)
33. J. Sacks, Asymptotic distribution of stochastic approximation procedure. Annal. Math. Stat. **29**, 373–405 (1958)
34. D.J. Sakrison, Efficient recursive estimation: application to estimating the parameters of a covariance function. Int. J. Eng. Sci. **3**, 461–483 (1965)
35. P. Sebastiany, H.P. Wynn, Maximum entropy sampling and Bayesian experimental designs. J. R. Stat. Soc. Ser. B Methodol. **62**, 145–157 (2000)
36. P. Shewry, H.P. Wynn, Maximum entropy sampling. Appl. Stat. **14**, 165–170 (1987)
37. M.J. Silvapulle, On the existence of the maximum likelihood estimators for the binomial response models. J. R. Stat. Soc. Ser. B Methodol. **43**, 310–313 (1981)
38. S.D. Silvey, *Optimal Design* (Chapman and Hall, London, 1980)
39. J. Stewart, L. Gill, *Econometrics* (Prentice Hall, Upper Saddle River, 1998)
40. C.F.I. Wu, Asymptotic inference from sequential design in a nonlinear situation. Biometrika **72**, 553–558 (1985)
41. C.F.J. Wu, H.P. Wynn, The convergence of general step – length algorithms for regular optimum design criteria. Ann. Stat. **6**, 1273–1285 (1978)

On the Computational Methods in Non-linear Design of Experiments

Christos P. Kitsos and Amílcar Oliveira

Abstract In this paper the non-linear problem is discussed, for point and interval computational estimation. For the interval estimation an adjusted formulation is discussed due to Beale's measure of non-linearity. The non-linear experimental design problem is regarded when the errors of observations are assumed i.i.d. and normally distributed as usually. The sequential approach is adopted. The average-per-observation information matrix is adopted to the developed theoretical approach. Different applications are discussed and we provide evidence that the sequential approach might be the panacea for solving a non-linear optimal experimental design problem.

1 Background

When an experiment is performed n times, within the experimental region $X \subseteq \Re^k$, known also as design space, the response Y can be considered either as a discrete variable or a continuous one. Thus we formulate the problem as following the set of all possible response outcomes Ψ, known as response space, that can be one of the two following cases:

- Case 1: Ψ is a finite set of integers, i.e. $\Psi = \{0, 1, 2, \ldots, \lambda\}$ with cardinal number $\nu = \lambda + 1$.

C. P. Kitsos
West Attica University, Athens, Greece
e-mail: xkitsos@uniwa.gr

A. Oliveira (✉)
Universidade Aberta and CEAUL, Lisbon, Portugal
e-mail: amilcar.oliveira@uab.pt

N. J. Daras, T. M. Rassias (eds.), *Computational Mathematics and Variational Analysis*, Springer Optimization and Its Applications 159,
https://doi.org/10.1007/978-3-030-44625-3_11

The most common in practical problems, usually in bioassays [6], is the case, $\nu = 2$ and corresponds to what is known as binary response problems, i.e. when $Y \in \Psi = \{0, 1\}$. When $n > 2$ we are referring to polytomous experiments. In some cases, like Poisson experiments, the set Ψ is countable infinite.

- Case 2: Ψ has the power of the continuum, i.e. cardinality c, as Ψ can be any interval in $\Re$.

In the former Case 1 and for binary problems the outcome $Y_i = 0$ or 1, $i = 1, 2, \ldots, n$ is linked with the covariate $x \in X$ and the parameter vector θ, from the parameter space $\Theta \subseteq \Re^p$, through a probability model T as

$$p(x) = p(Y_i = 1 \mid x) = T(x; \theta) = 1 - P(Y_i = 0 \mid x). \tag{1}$$

In bioassays the typical situation is to consider logit, probit, or exponential models (known also as "One Hit" in experimental carcinogenesis) [14], with the corresponding ("smooth" monotone function) T function to be

$$T_L(x, \theta) = log\{p(x)(1 - p(x))\}, T_P(x, \theta) = \Phi^{-1}\{p(x)\}, T_E(x, \theta) = exp(-\theta x),$$

with "log," "Φ," and "exp" having their trivial meaning. Now the latter (Case 1) situation is faced with the general regression model: we define the (assumed correct) deterministic portion $f(x, \theta)$ with f being continues, twice differentiable and the error is applied, as a stochastic portion additively, in the form

$$Y_i = f(x_i, \theta) + e_i, i = 1, 2, \ldots, n, \theta \in \Theta \subseteq \Re^p, x \in X \subseteq \Re^k. \tag{2}$$

When for the real function f there exists a real g continuous differentiable function such that $f(x_i, \theta) = g(x_i)^T \theta$, $i = 1, 2, \ldots, n$, then the problem is reduced to the linear regression problem. It is assumed that the errors $e_i, i = 1, 2, \ldots, n$ are independent identical distributed (i.i.d.) from the normal distribution $N(0, \sigma^2)$. In principle, $\sigma^2 = \sigma^2(x; \theta)$ for the non-linear situation and $\sigma^2 = \sigma^2(x)$ in linear. That is, in non-linear experimental models the variance depends also on the parameters we need to estimate [7]. This is a crucial difference between linear and non-linear situation and emerges crucial computational problems: we are heavily depended on the prior knowledge of the parameter vector we try to estimate! To emphasize this dependence, the optimality criteria in non-linear case are denoted with the θ in parentheses, i.e. $D(\theta)$, $C(\theta)$, etc. [7].

Below two are the main problems we have to face either from a theoretical point of view or in applications:

- **Problem 1:** The underlying model describing the physical phenomenon is non-linear, as in (2). The target then is estimate $\theta \in \Theta$ as well as possible.
- **Problem 2:** A non-linear function, known as general non-linear aspect, of the unknown parameter θ, $\varphi(\theta)$ say, is asked to be estimated as well as possible, even when the underlying model is assumed linear.

In both problems interest is focused on the assumption about the errors. In this paper it is assumed that e_i are i.i.d. from the normal distribution $N(0, \sigma^2)$. For a robust approach as in Problem 1, see [20] and [21], while Problem 1 has been extensively discussed in [7, 9, 25, 30].

In the sequence it would be assumed that the parameter space $\Theta \subseteq \Re^p$ is compact. This assumption is necessary, although not often imposed, when limiting results in the sequential approach [7, 10, 11] for the sequence of estimators are obtained. Under this assumption we insure that the limit will be an element of Θ.

As the non-linear optimal experiment design problem depends heavily on the parameters involved, as the average-per-observation information matrix depends on the parameters [4, 7] and [5]. Therefore the sequential principle of design is adopted to overcome this parameter-dependence [10]. In non-linear situation it is still a problem the optimality of the limiting design, while the development of locally optimal designs has been extensively discussed for a certain family of models [28, 31]. Sequences are the basis of iterations, iterations are needed to approach solutions. Therefore numerical analysis techniques are behind [26, 27] to solve the existing computational problems [24].

Still the most common applications from biological assays are based on D-optimality, as well as the experimental carcinogenesis, while in chemistry D-optimality appears an esthetic appeal [4, 5, 11]. In [12] a collection of non-linear models is presented and support points are provided, while in [18] the chemical inside of the models and the appropriate computational techniques are provided.

Different problems have been discussed under C-optimality: the well-known calibration problem [19] and the rhythmometry problem from the "biological time series" [22]. Both problems can be generalized under a robust approach [4] and [5].

Example 1 A typical, one variable non-linear problem is the Michaelis–Menten model [15], for which different optimality criteria have been discussed. For the chemical kinetics, as in models (2) different reaction models have been considered to describe a chemical process. Therefore [12] provides all the appropriate support points, while for the family of models as in (1), see [23] for details, the support points have been provided by [31].

Example 2 The following non-linear regression models are common in practice, and discussed in computer packages.

$$\begin{cases} \text{Competz model}: & G(u,\vartheta) = \vartheta_0 exp\{\vartheta_1 exp(\vartheta_2 u)\} \\ \text{Janoschek model}: & J(u,\vartheta) = \vartheta_0 + \vartheta_1\{exp(\vartheta_2 u^{\vartheta_3})\} \\ \text{Logistic model}: & L(u,\vartheta) = \frac{\vartheta_0}{\{1+\vartheta_1 exp(\vartheta_2 u)\}} \\ \text{Bertalanffy model}: & B(u,\vartheta) = \{\vartheta_0 + \vartheta_1 exp(\vartheta_2 u)\} \\ \text{Tanh model}: & T(u,\vartheta) = \vartheta_0 + \vartheta_1 tanh(\vartheta_2(u-\vartheta_3)) \\ \text{3–parameters tanh–model}: & T_3(u,\vartheta) = \frac{1}{2}\vartheta_0\{1+\frac{2}{\pi}arctan(\vartheta_1(u-\vartheta_2))\} \\ \text{4–parameters tanh–model}: & T_4(u,\vartheta) = \vartheta_0 + \frac{2}{\pi}\vartheta_1 arctan(\vartheta_2(u-\vartheta_3)) \end{cases}$$

2 Optimality Criteria

For both models (1) and (2) the expected value of the response Y will be denoted by $\eta = E(Y)$. Then it is

$$\eta = E(Y) = \begin{cases} g(x, \theta) \text{ models (2)} \\ T(x, \theta) \text{ models (1)} \end{cases} \tag{3}$$

Let $\nabla\eta$ denote the vector of partial derivatives of η with respect to $\theta \in \Theta \subseteq \Re^p$. Then for the exponential family of models, Fisher's information matrix is defined to be

$$I(\theta, x) = \sigma^{-2}(\nabla\eta)(\nabla\eta)^T. \tag{4}$$

Non-linearity is very important in statistics, as it usually creates computational problems. Moreover there is, with a strong theoretical inside, for the linear models. We would like to emphasize that the non-linear design theory does not offer, in principle, an extension of the linear case, but we try to "transfer" the problem to a linear one [1]. That is why measures of non-linearity, trying to measure how non-linear the model is, have been introduced [2]. If $\eta = \eta(\theta^T x)$, i.e. the covariates and the parameter appear together linearly, the model is called "intrinsic linear," then

$$\nabla\eta = [w(\theta^T x)]^{1/2} x \tag{5}$$

with $w(z) = [\frac{\partial\eta}{\partial z}]^2$, $z = \theta^T x$, the linear part.

Therefore Fisher's information matrix for this family of models (including logit and probit) for this particular subcase the information matrix is the product of a "weighting factor" and the norm of a vector—notice that by the definition in (4) Fisher's information can be considered as the norm of a scaled vector. Indeed:

$$I(\theta, x) = \sigma^{-2} w(\theta^T x) x x^T. \tag{6}$$

Example 3 Let $P(Y = 1) = T(\theta^T x)$ and $\theta_1 + \theta_2 x_1 = z = \theta^T x$ with $x = (1, x_1)$. Then $I(\theta, x)$ is evaluated as:

$$I(\theta, x) = \alpha(\theta) u u^T, \text{ with } \alpha(\theta) = T'^2 [T(1 - T)].$$

Notice that T can be either the logit or probit function. From the above discussion and (6) the following is true, see [9].

Proposition 1 *For the exponential family of models, Fisher's information is the measure of a vector, i.e. there exists a vector v such that*

$$I(\theta, x) = v v^T. \tag{7}$$

The concept of the average-per-observation information matrix plays an important role to our scenario for the definition of the optimality criteria, for the optimal experiment design problem. It is defined for ξ, the design measure [25], to be

$$M(\theta,\xi)=\begin{cases} n^{-1}\sum I(\theta,x_i) \ discretecase \\ \int_x I(\theta,x)\xi(dx) \ continuouscase \end{cases}. \tag{8}$$

On the basis of the experiment, the average-per-observation information matrix $M=(\theta,\xi)$ is obtained, which depends, in principle, on θ. That is to distinguish the linear from the non-linear optimality criterion the parameter θ will be denoted in non-linear case. The non-linear case despite the excellent work in [4] and [5] has not been equipped with a theorem similar to the Wu–Wynn convergence theorem in the linear case [34]. In [10] and the relevant applications appeared in [14] the limiting result concerns only the D-optimality with computational application only for D-optimality and minimum entropy.

Example 4 As far the logit or probit model concerns the $D(\theta)$-optimal design concentrated at two optimal design points, functions of the parameters we want to estimate, namely the points $x_1=\frac{x_0-\theta_1}{\theta_2}$, $x_2=\frac{-x_0-\theta_1}{\theta_2}$ with equal number of observations at each, i.e. for the design measure $\xi=\frac{1}{2}$. The value noticed as x_0 is different in both cases and is $x_0=1.54$ for the logit model and $x_0=1.14$ for the probit model, provided the design space $X=[\alpha,\beta]\subseteq\Re$ is symmetric about $-\frac{\theta_1}{\theta_2}$. The optimal design points $\frac{\pm x_0-\theta_1}{\theta_2}$ do not belong to X, then $x_1=\alpha$, $x_2=\beta$. Notice that the optimal design points depend on the parameters we want to estimate; therefore, initial guesses are necessary.

Unfortunately neither [7] nor [4] stated that, when the sample size is "small," the likelihood function might not provide maximum likelihood estimates. That occurs when only a batch of successes or failures are in the outcome. Simulation results in [10] for $T_E(x,\theta)$ model provided evidence that, in practice, even with $n=50$ two or three cases out of one thousand simulations might appear to be pathological. The situation is getting worse with smaller n. Roughly speaking the likelihood estimator exists when, at some design points, the response might be success for some experiments and failure for the rest of experiments, see [29] and [9]. The existence of estimators for models (4) has been discussed in [8], see also [9].

Example 5 To overpass this "θ-dependence," in non-linear case, the sequential approach was proposed. When the sequential principle of designing is adopted the experimenter selects n points and evaluates an estimate of θ, $\hat{\theta}_n$, say. Then the experimenter redesigns using the vector $\hat{\theta}_n$ and a new estimator is obtained. Three questions are arisen:

- How the initial design is chosen for the experimenter?
- How the sequence of points is created through an iterative scheme?
- Does the sequence of estimators converge (usually in mean square) to θ?

We briefly discuss these questions:

The initial design is build up on the locally optimal experiment design [7]. The problem is that the optimum design points in non-linear problems depend on the parameter under investigation. Therefore a prior knowledge of the neighborhood of θ is asked. In practice the experimenter can provide such information. The sequence of points $x(n+1) = (x(n), x_{n+1}) = (x_1, x_2, \ldots, x_n, x_{n+1})$ is created by choosing that point, as a next stage, which minimizes the determinant of the covariance matrix for a current estimate of θ, i.e. keeping a D-optimal design. Recall that $M(\theta, \xi)$ is asymptotically the covariance matrix C, namely $C^{-1} \simeq nM(\theta, \xi)$. Moreover, at each point it is allocated an equal batch of points. Different batch sizes can be proposed, some worked by adding one observation per stage, see Example 3, relation (12). We give the following definition [10].

Definition 1 A sequential procedure which allocates $n_0 = dim\theta$ observations per stage is called a fully sequential procedure.

Then, the following is true due to [10].

Proposition 2 *If the initial design is D-optimal the limiting design is then a D-optimal, if a fully sequential procedure, through stochastic approximation, is adopted. The sequence of estimators will converge in mean square to* θ, *if the stochastic approximation scheme is adopted for model (2).*

Now the following can be proved, involving the eigenvalues of $M(\theta, \xi)$.

Proposition 3 *For models (2) the iterative scheme*

$$\theta_{n+1} = \theta_n - (\delta s^2 n^{-1}) \Sigma_{i=1}^{p} \frac{\mu_i}{\lambda_i} V_i \tag{9}$$

converges to θ, with $\delta \in (0, 1)$, s^2 a suitable estimator of σ^2, $\theta_n \in \Theta \subseteq \Re^p$, λ_i, V_i, $i = 1, 2, \ldots, p$ the eigenvalues and the eigenvectors of $M = M(\theta, \xi)$, respectively, and μ_j, $j = 1 \ldots, p$ the eigenvalues of ∇f.

Indeed:

Apply Newton and Kantorovich Theorem [24], the iterative scheme

$$\theta_{n+1} = \theta_n - (\delta s^2 (M(\theta, \xi))^{-1} \frac{\partial f}{\partial \theta} \tag{10}$$

converges to θ, as $\sigma^2 nM(\theta, \xi)$ is the Hessian, from a numerical analysis point of view. Consider the eigenvalue–eigenvector decomposition of M at [17], while for biorhythms, see [35].

$$M = \sum \lambda_i V_i V_i^T, \ \frac{\partial f(x; \theta)}{\partial \theta} = \sum \mu_j V_j \tag{11}$$

as $V_i V_j^T = \delta_{ij}$, with $i, j = 1, 2, \ldots, p$ and δ Kronecker's, then (10) is reduced to (9). We comment that theoretically any δ in (0, 1) is appropriate. It depends on the collected data and the experimentalist what to choose. Usually the 0.5 value accelerates the iteration. In the following paragraph we discuss the Ca bioassay as a non-linear optimal design problem through the sequential approach.

3 A Computational Approach for the Cancer Bioassay

The mathematical and statistical insight of the cancer risk assessment has been discussed by [33], while the covariates offer another approach on the influence of a cancer bioassay [13]. The statistical modeling and a number of approaches can be traced at [17].

In order to perform an experiment, a "good" initial estimate of θ is needed. Thus, the experiment should be designed in an optimal sense, with an initial guess about θ [16]. The initial guess is used to redefine the design points, which in non-linear case depends on θ. The procedure is stopped either when a pre-defined accuracy of θ occurs, or when for a pre-defined accuracy ε positive occurs

$$|logdet M(\theta_n, \xi) - logdet M(\theta^*_{n+1}, \xi)| < \varepsilon.$$

Now, let us consider the recursive scheme to estimate the root, r say, of a real valued function Q defined on the real axis $\Re$, such as $Q(r) - p = 0$. Only observations of Q, the values $y_n = y_n(x)$ say exist, i.e. $y_n = Q(x_n) + e_n$. As usually e_n is declared the errors, coming from a distribution with zero mean and variance σ^2. This methods known as stochastic approximation originated by Robbins and Monro [26], see also [32]. Notice that when $Q(x)$ is a cumulative distribution function (cdf) and p is within [0, 1] the root r corresponds to the percentile point of the distribution. The iteration is defined through the sequence:

$$x_{n+1} = x_n - \alpha_n(y_n - p), \; n = 1, 2, \ldots$$

with x_1 arbitrary, $\alpha_n > 0$ a fixed sequence of real numbers (step sizes) and y_n are binary responses at stage n. The sequence x_n converges to r in mean square, where a typical step size of $\alpha_n = cn^{-\alpha}$, $\alpha < 1$, $n = 1, 2, \ldots$ is assumed. The minimum variance estimate is reached with $c = C_{opt} = (Q'(r))^{-1}$, and in the case of $\alpha = 1$ the minimum value of the variance is $\sigma^2_{min} = (Q'(r))^{-2}\sigma^2$.

Example 6 Under this line of thought when the Weibull model $F(x) = 1 - exp(-(\theta x)^k)$ is considered [16], then one easily obtains (notice that knowledge on both parameters θ and k is needed) $p - th$ percentile as:

$$L_p = (-\frac{1}{\theta^k} ln(1 - p))^{1/k}$$

see [14] and [16] for details.

Thus, the question is how we can evaluate the percentile, L_p, so essential in a number of applications, especially to low-dose extrapolation problems. Perform an experiment, get the estimates and an estimate of the percentile can be evaluated. In low-dose extrapolation problems p is small, such that $p = 0.01$. It can be proved [14] in relation to the optimal c value for the Robbins–Monro iterative scheme that

$$F'(L_p) = exp(-(\theta L_p)^k)\theta^k k L_p^{k-1} = (1-p)\theta^k k L_p^{k-1} = r_k = C_{opt}^{-1},$$

where the definition of r_k is obvious. In order to get the estimates of the Weibull model and to proceed, n_0 observations can be devoted to get an initial estimate. Therefore the iterative scheme

$$L_{p,n+1} = L_{p,n} - (nr_k)^{-1}(y_n - p),\ n = n_0 + 1, n_0 + 2, \ldots$$

converges to L_p.

It has been pointed out that the method works well, as far as the convergence concern and the size of experiments needed [14].

Example 7 For the carcinogenic experiments, it is considered the model $T_H(x, \theta) = 1 - exp\{(\theta_0 + \theta_1 x)\}$, known as One Hit model. For this model the $100p$ percentile, L_p, can be evaluated sequentially, through the iteration

$$L_{p,n+1} = L_{p,n} - (n\hat{\theta}_1(1-p))^{-1}(y_n - p) \tag{12}$$

see Kitsos ([11] and [12]). In such a case the sequence $L_{p,n+1}$ converges in mean square to L_p.

4 Measuring Non-linearity in Models

For the p-term model (2) the residual sum of square is defined to be

$$RSS_p = \sum_{i=1}^{n}(y_i - \hat{y}_i)^2 \tag{13}$$

with $\hat{y}_i$ being the fitted value. The RSS_p provides a measure of the discrepancy between fit and data and therefore is proportional to the estimate of σ^2:

$$\hat{\sigma}^2 = \frac{1}{n-p} RSS_p. \tag{14}$$

Now, the measure of instability of predictions is discussed. As a measure of the instability of predictions $\hat{y}_i, i = 1, \ldots, n$ the total variance of predictions (TVP) for the non-linear model (2) has to be evaluated. By definition, it is

$$TVP = \sum_{i=1}^{n} Var(\hat{y}_i). \tag{15}$$

We prove that TVP is proportional of $\sigma^2 p$.

Proposition 4 *For a non-linear model, as (2), the first-order approximation of the total variance of prediction is*

$$TVP \approx \sigma^2 p. \tag{16}$$

Proof Recall the non-linear model (2). A linear Taylor expansion in the neighborhood of true ϑ, say ϑ_t, is

$$f(\vartheta) \cong f(\vartheta_t) + F(\vartheta - \vartheta_t), \tag{17}$$

where the matrix F is

$$F = F(\vartheta) = \frac{\partial f(\vartheta)}{\partial \vartheta} = (\frac{\partial f_i(\vartheta)}{\partial \vartheta_j}), i = 1, 2, \ldots, n \quad j = 1, 2, \ldots, p = dim\Theta \tag{18}$$

and

$$f_i(\theta) = f(u_i, \theta); \; f(\theta) = (f_1(\theta), \ldots, f_n(\theta))'.$$

Hence the model in (17) can be expressed by the (approximated) linear model:

$$z = F\beta + e \tag{19}$$

with $z = f(u, \vartheta) - f(u, \vartheta_t)$, and e the introduced error due to the approximation.

Moreover the sum of squares from the non-linear model (2) is approximated by the sum of squares for the linear model (19) as follows:

$$SS(\vartheta) = \|y - f(\vartheta)\|_2^2 = \|y - f(\vartheta_t) - F(\vartheta - \vartheta_t)\|_2^2 = \|z - F\beta\|_2^2.$$

For the linear model (19) let us consider a future observation y_i which provides eventually the $F_i, i = 1, 2, \ldots$ vector of $\hat{y}_i$. Then the corresponding variance is

$$Var(\hat{y}_i) = Var(F_i\hat{\beta}) = F_i^{'} Var(\hat{\beta})F_i = \sigma^2 F_i^{'} M F_i = \sigma^2 W_i \tag{20}$$

with

$$W_i = F_i^{'}(F'F)^{-1}F_i, i = 1, 2, \ldots, \text{and } \; Var(\hat{\beta}) = M(\vartheta)\sigma^2 \simeq M(\theta, \xi)\sigma^2. \tag{21}$$

Therefore it can be evaluated:

$$TVP = \sum Var(\hat{y}_i) = \sigma^2 \sum_i^n F_i'(F'F)^{-1}F_i = \sigma^2 tr[F(F'F)^{-1}F']$$
$$= \sigma^2 tr[F'F(F'F)^{-1}] = \sigma^2 tr(Ip) = \sigma^2 p,$$

as it holds that $tr(AB) = tr(BA)$.

Moreover we try to offer more criteria for the non-linear model (2), see [1]; therefore, in statistics a measure of non-linearity, that is a measure of the deviation from linearity, has to be also calculated. The most well-known measure of non-linearity is Beale's measure [2].

The target is not only the point estimation of ϑ, but the construction of a confidence region for it. To avoid banana-shape confidence regions the empirical measures of non-linearity were suggested by Beale, see [2, 9] and [28], bias and non-linearity are connected, while in [1] is adopted a differential geometry approach to prove that Beale's intrinsic measure of curvature is the one quarter of the mean square intrinsic curvature, see also ([28], pg.158), ([9]), pg.80). Moreover replicating the experiment r times the curvature at any point in any direction is reduced by $r^{-\frac{1}{2}}$, while [15] suggested an upper bound approximation for Beale's measure, to avoid calculations and ϑ dependence.

Beale's intrinsic non-linearity has been proved in [1] to be

$$B_0 = \frac{\sigma^2}{p+2} \sum_{i=1}^{n} \{\frac{1}{4} tr[MH_i]^2 + \frac{1}{2} tr[MH_i MH_i]\} \tag{22}$$

with $M = (F'F)^{-1}$ with F as in 17 the Hessian H is defined as:

$$H = H(\vartheta) = \frac{\partial^2 SS(\vartheta)}{\partial\vartheta\partial\vartheta'}. \tag{23}$$

The Hessian associated with the first-order vector of derivatives $h = h(\vartheta) = \nabla SS(\vartheta)$ provides the second-order approximation:

$$SS(\vartheta) = SS(\vartheta_0) + h'(\vartheta_0)(\vartheta - \vartheta_0) + \frac{1}{2}(\vartheta - \vartheta_0)' H(\vartheta_0)(\vartheta - \vartheta_0), \tag{24}$$

which is minimized by Newton's method. So, estimates of ϑ are obtained. In practice and for computational efficiency the fact that

$$\frac{1}{2} E\{H(\vartheta)\} = F'(\vartheta)F(\vartheta) = M(\vartheta) \tag{25}$$

is used, see [10, 11]. The Beale's measure B equals

$$B = \begin{cases} 1 + \frac{n}{n-1} B_0 & if\ p = 1 \\ 1 + \frac{n(p+2)}{(n-p)p} B_0 & if\ p \geq 1 \end{cases}. \tag{26}$$

Therefore we state and prove (in Appendix 2) the following:

Proposition 5 *For a non-linear problem the confidence intervals for the parameter vector θ can be adjusted with the relation:*

$$SS(\vartheta_t) - SS(\hat{\vartheta}) \leq B_k p S^2 F(\alpha; p, n-p) \tag{27}$$

with

$$B_k = \begin{cases} 2.2 & with\ p = 1 \\ 1 + \frac{n}{n-2}\frac{1}{\sqrt{F_\alpha}} & with\ p = 2 \\ 1 + 0.41\frac{n}{n-3} & with\ p = 3, n > 9 \\ 1 + \frac{n(p+2)}{2(n-p)p}\frac{1}{\sqrt{F_\alpha}} & with\ p \geq 4 \end{cases} \tag{28}$$

and $F_\alpha = F(\alpha; p, n-p)$ the F distribution as usually, at α level with p and $n-p$ degrees of freedom (df).

So, in Proposition (5) as above, we provide an easy and compact form for the most non-linear models to help the experimenter, who has no knowledge on curvature models, to obtain a better approximation than the one in $B_k = 1$ that is used in practice.

There is a number of criteria C1–C4, see Appendix 1, as another attempt to choose the "best" model and moreover to approach the non-linear theory with the linear problem [1, 9, 11].

So, the proposed in [3], selection criterion for a non-linear model:

$$Bor = \frac{RSS}{s^2} - (n-p) + 3s^2 B$$

with B the Beale's measure, which is computationally tedious, can be replaced with the following criterion, we propose in this paper:

$$C^*5 : B_{max} = \frac{RSS}{s^2} - (n-p) + 3s^2 B_k.$$

Example 8 For the data set in Table 3.1 of ([28], pg.93) Fisher's information matrix has been calculated and it is

$$I(\hat{\theta}, \xi) = \begin{pmatrix} 43145 & -157.0 \\ -157.0 & 584.5 \end{pmatrix}$$

For this data set with $p = 2$ we evaluate $B_k = 1.04938$ and the approximated confidence region for the vector $\theta = (\theta_1, \theta_2)$ is the ellipsis [15]:

$$43.14\theta_1^2 + 0.58\theta_2^2 - 3.14\theta_1\theta_2 - 82.92\theta_1 - 0.92\theta_2 + 1.50 \leq 0.$$

Now, consider the artificial non-linear data (x_i, y_i), with $x_i = i = 1, \ldots, 8$ and $y_i = \{10, 20, 35, 45, 60, 65, 67, 70\}$. The linear model $\hat{y}_i = 5.97 + 9.05x_i$ provides a $R^2 = 0.93$ larger than the non-linear model $\hat{y}_i = 38.5 + 0.0135exp(x_i)$ with $R^2 = 0.37$. The corresponding standard deviations are $s = 6.16$ and $s = 19.56$. No doubt that B_k is larger in the non-linear case, and a simple plot of data provides evidence that the linear model is not the appropriate one. The value of $B_k = 1.5879$, recall (28) with $p = 2$, influences the confidence region, and it is for the non-linear model:

$$\beta_0^2 + 2620.5\beta_1^2 + 93\beta_0\beta_1 - 39.75\beta_0 - 3618.28\beta_1 + 1453.65 \leq 0.$$

5 Conclusions

There are a number of computational problems that are faced in the non-linear design theory of experiments. The Newton–Raphson method (notice that also the stochastic approximation can be considered as the statistical point of view of N-R) does not always converges, if the initial guess is not in the neighborhood of the root. That is why we propose to apply the bisection method for 2 or 3 steps, to get an "appropriate" initial value and feed the Newton–Raphson (N-R) which usually converges then very fast. The non-linear problem suffers on θ-dependence. Therefore not only the point estimation is needed, but the interval estimation appears essential. We adjusted the confidence intervals with the methods we proposed, adopting Beale's measure of non-linearity.

The confidence regions are certainly more accurate in the non-linear case if the proposed values of B_k are adopted to adjust the typical $ps^2F(\alpha; p, n-p)$ value. The non-linear problem deserves more attention as a number of applications are non-linear. Or to put it in other way, we consider the linear case because the non-linear is difficult to be faced, since the time of Gauss [9]. Consider Figures 1 and 2: although they are expressing different problems, their feature is quite different—the "fat" Figure 1 ensures big error on both parameters, while the adjusted "thin" confidence interval provides evidence that sort error. Moreover Figure 2, although from a non-linear problem, provides typical confidence interval, while usually in non-linear problems the confidence intervals are "banana shape," therefore problems exist on estimation. The evaluated area is related to D-optimality criterion—the one which minimizes the area (or the volume in more than 2 dimensions) of the corresponding confidence interval. Iterative schemes seem more appropriate and we adopted the sequential approach to evaluate percentiles, so useful to bioassay, especially to cancer problems $p \simeq 0.01$ known as low-dose extrapolation problem. The statistical insight in non-linear problems is different than in the strict mathematical way of thinking. But it still appears to be of highest importance with more computational difficulties than the linear problem.

Fig. 1 Ellipse with center (1.0981, 3.76554) and area enclosed = 30.2688

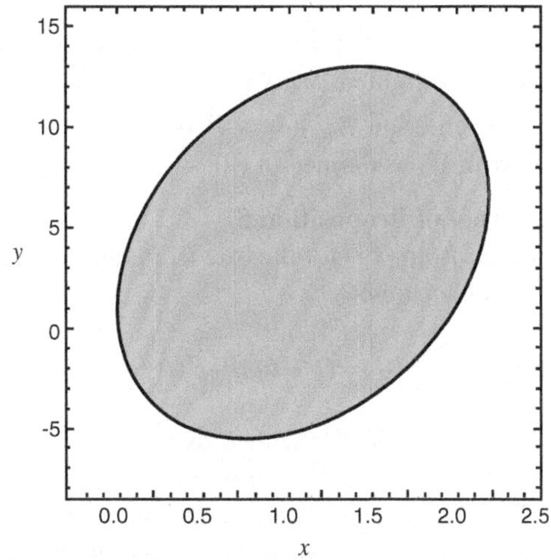

Fig. 2 Ellipse with center (−69.9238, 1.93116) and area enclosed = 95.443

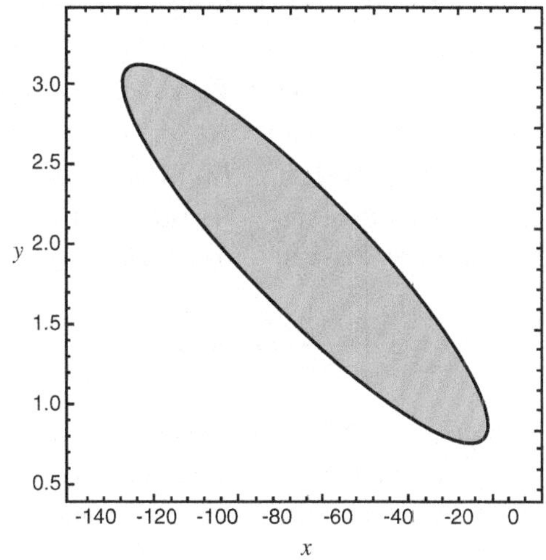

Acknowledgments This research is based on the partial results of the fund by FCT—Fundação para a Ciência e a Tecnologia, Portugal, through the project UID/MAT/00006/2019.

Appendix

1. Local Selection Criteria for NLM

C1: Mallow's C_p-statistic $C_p(\theta) = (\frac{RSS}{s^2}) - n + 2p$

C2: Akaike's criterion $AC(\theta) = nln(\frac{RSS}{n} + \frac{n(n+p)}{n-p-2}$

C3: Weighted $WRSS(\theta) = \sum_i [\frac{(y_i - \hat{y}_i)^2}{1-W_i}]$
C4: p—Weighted $WRSS_p(\theta) = \frac{WRSS(\theta)}{s^2} - n + 2p$
C5: B-criterion $B_{max} = \frac{RSS}{s^2} - (n-p) + 3s^2 B$
with W_i as defined in (9).

2. Proof of Proposition 5
Indeed from (19) for the true ϑ, ϑ_t and the estimate $\hat{\vartheta}$ it holds for the corresponding sum of squares:

$$SS(\vartheta_t) - SS(\hat{\vartheta}) = (\hat{\vartheta} - \vartheta_t)F'F(\hat{\vartheta} - \vartheta) = (\hat{\vartheta} - \vartheta_t)'M^{-1}(\hat{\vartheta} - \vartheta) \leq Bps^2 F(a;\, p, n-p) \tag{29}$$

from ([9], pg.81) with B as in (26).

But B is as in [23] and we extend it to (28). We approximate B_0 by the value of the intrinsic curvature of the model γ_{max}. This curvature provides a measure to "how good is the target-plane approximation," in other words "how far from linear" is the model. But there is an upper bound of γ_{max}, see ([28], pg.135), namely

$$B_0 \cong \gamma_{max} < \frac{1}{2}(F(a;\, p, n-p))^{-\frac{1}{2}} = \gamma_0. \tag{30}$$

So, with $p = 1$, it is $F(a; 1; n-p) = t^2_{n-p}$, i.e. $\gamma_0 = \frac{1}{2}t(a; n-1)$, that is, $B = 1 + \frac{n}{n+1}B_0 < 1 + \frac{n}{n-1}\frac{1}{2}t(a; n-1) \approx 2.2 := B_k$.

When $p = 2$, recall (26) and (30):

$$B = 1 + \frac{n(p+2)}{(n-p)p}B_0 \cong 1 + \frac{4n}{2(n-2)}\gamma_{max} < 1 + \frac{n}{n-2}\frac{1}{\sqrt{F}} = B_k.$$

With $p = 3$ similar calculations provide evidence that $B_k = 1 + 0.41\frac{n}{n-3}$ approximating $\sqrt{F}$ by a rough approximation 2 with $n > 9$ at $a = 0.05$.

Also with $p \geq 4$ considering (29) and (30) the last part of (28) is defined, q.e.d.

References

1. D.M. Bates, D.G. Watts, *Non-linear Regression Analysis and its Applications* (John Wiley and Sons, Hoboken, 1998)
2. E.M.L. Beale, Confidence regions in non-linear estimation. J. R. Stat. Soc. Ser. B Methodol. **22**, 41–88 (1960)
3. D.S. Borowiak, *Model Discrimination in Nonlinear Regression Models* (Marcel Dekker, New York, 1989)
4. P. Chadhuri, P.A. Mykland, Nonlinear experiments: optimal design and inference based on likelihood. J. Am. Stat. Assoc. **88**, 538–546 (1995)
5. P. Chadhuri, P.A. Mykland, In efficient designing of nonlinear experiments. Statistica Sinica **5**, 421–440 (1995)
6. D.J. Finney, *Statistical Methods in Biological Assay* (C. Griffin and Co. Ltd., Glasgow, 1978)

7. I. Ford, C.P. Kitsos, D.M. Titterington, Recent advances in non-linear experimental design. Technometrics **31**, 49–60 (1989)
8. R.J. Jennrich, Asymptotic properties of non-linear least squares estimators. Ann. Math. Stat. **40**, 633–643 (1969)
9. C.P. Kitsos, Design and Inference for the Nonlinear Problems. Ph.D. Thesis. Glasgow (1986)
10. C.P. Kitsos, Fully-sequential procedures in nonlinear design problems. Comp. Stat. Data Anal. **8**, 13–19 (1989)
11. C.P. Kitsos, Adopting sequential procedures for biological experiments. In: *MODA3 (Model Oriented Data Analysis)*, ed. by W.G. Muller, H.P. Wynn, A.A. Zhigliavsky (Physica-Verlag, Heidelberg, 1992), pp. 3–9
12. C.P. Kitsos, On the support points of D-optimal nonlinear experiment design for chemical kinetics. In: *Model Oriented Data Analysis*, ed. by C. Kitsos, W. Mueller (Physica-Verlag, Heidelberg, 1995), pp. 71–76
13. C.P. Kitsos, The role of covariates in experimental carcinogenesis. Biomed. Lett. **35**, 95–106 (1998)
14. C.P. Kitsos, Optimal designs for estimating the percentiles of the risk in multistage models in carcinogenesis. Biomet. J. **41**, 33–43 (1999)
15. C.P. Kitsos, Design aspects for the Michaelis-Menten model. Biomet Lett. **38**, 53–66 (2001)
16. C.P. Kitsos, Optimal design for bioassays in carcinogenesis. In: *Quantitative Methods for Cancer and Human Health Risk Assessment*, ed. by L. Edler, C. Kitsos (Wiley, Hoboken, 2005), pp. 267–279
17. C.P. Kitsos, L. Edler, Cancer risk assessment for mixtures. In: *Quantitative Methods for Cancer and Human Health Risk Assessment*, ed. by L. Edler, C. Kitsos (Wiley, Hoboken, 2005), pp. 283–298
18. C.P. Kitsos, K.G. Kolovos, A compilation of the D-optimal designs in chemical kinetics. Chem. Eng. Commun. **200**(2), 185–204 (2013). http://dx.doi.org/10.1080/00986445.2012.699481
19. C.P. Kitsos, K.G. Kolovos, *Optimal Calibration Procedures for Calibrating the pH Meters* (Lambert Academic Publishing, Saarbrücken, 2010). ISBN: 978-3-8433-5286-4
20. C.P. Kitsos, Ch.H. Muller, Robust estimation of non-linear aspects. In: *MODA4 (Model Oriented Data Analysis)*, ed. by C. Kitsos, W. Muller (Physica, Heidelberg, 1995), pp. 71–76
21. C.P. Kitsos, Ch.H. Muller, Robust linear calibration. Statistics **27**, 93–103 (1995)
22. C.P. Kitsos, D.M. Titterington, B. Torsney, An optimal design problem in rhythmometry. Biometrics **44**, 657–671 (1988)
23. P. McCallagh, J. Nelder, *Generalized Linear Models* (Chapman and Hall, London, 1989)
24. J.M. Ortega, W.C. Pheinfold, *Iterative Solution of Nonlinear Equation in Several Variables* (Academic, London, 1970)
25. F. Pukelsheim, *Optimal Design of Experiments* (John Wiley, Hoboken, 1993)
26. H. Robbins, S. Monro, A stochastic approximation method. Ann. Math. Stat. **22**, 400–407 (1951)
27. D.J. Sakrison, Efficient recursive estimation: application to estimating the parameters of a covariance function. Int. J. Eng. Sci. **3**, 461–483 (1965)
28. G.A.F. Seber, C.J. Wild, *Nonlinear Regression* (John Wiley and Sons, Hoboken, 1989)
29. M.J. Silvapulle, On the existence of the maximum likelihood estimators for the binomial response models. J. R. Stat. Soc. Ser. B Methodol. **43**, 310–313 (1981)
30. S.D. Silvey, *Optimal Design* (Chapman and Hall, London, 1980)
31. R.R. Sitter, B. Torsney, D-optimal designs for generalized linear models. In: *MODA4 (Model Oriented Data Analysis)*, ed. by C. Kitsos, W. Muller (Physica, Heidelberg, 1995), pp. 87–102
32. J. Wolfowitz, On the stochastic approximation method of Robbins and Monro. Ann. Math. Stat. **23**, 457–461 (1973)
33. W. Wosniok, C. Kitsos, K. Watanabe, Statistical issues in the application of multistage and biologically based models. In: *Prospective on Biologically Based Cancer Risk Assessment*. NATO Pilot Study Publication ed. by V.J. Cogliano, G.E. Luebeck, G.A. Zapponi (Plenum Publishing Co., New York, 1998)

34. C.F.J. Wu, H.P. Wynn, The convergence of general step – length algorithms for regular optimum design criteria. Ann. Stat. **6**, 1273–1285 (1978)
35. V. Zarikas, V. Gikas, C.P. Kitsos, Evaluation of the optimal design "cosinor model" for enhancing the potential of robotic theodolite kinematic observation. Measurement **43**(10), 1416–1424 (2010). http://dx.doi.org/10.1016/j.measurement.2010.08.006

Geometric Derivation and Analysis of Multi-Symplectic Numerical Schemes for Differential Equations

Odysseas Kosmas, Dimitrios Papadopoulos, and Dimitrios Vlachos

Abstract In the current work we present a class of numerical techniques for the solution of multi-symplectic PDEs arising at various physical problems. We first consider the advantages of discrete variational principles and how to use them in order to create multi-symplectic integrators. We then consider the nonstandard finite difference framework from which these integrators derive. The latter is now expressed at the appropriate discrete jet bundle, using triangle and square discretization. The preservation of the discrete multi-symplectic structure by the numerical schemes is shown for several one- and two-dimensional test cases, like the linear wave equation and the nonlinear Klein–Gordon equation.

1 Introduction and Motivation

In general, symplectic integrators are robust, efficient, and accurate in preserving the long time behavior of the solutions of Hamiltonian ordinary differential equations (ODEs) [1]. The basic feature of a symplectic integrator is that the numerical performance is designed to preserve a physical observable property, i.e., the symplectic form at each time step. Recently, it was shown that many conservative partial differential equations (PDEs) allow for description similar to the symplectic structure of Hamiltonian ODEs, called the multi-symplectic formulation (see, e.g., Refs. [2–5]). For example, in Ref. [2] the authors develop the multi-symplectic

O. Kosmas (✉)
Modelling and Simulation Centre, MACE, University of Manchester, Manchester, UK
e-mail: odysseas.kosmas@manchester.ac.uk

D. Papadopoulos
Delta Pi Systems Ltd., Thessaloniki, Greece
e-mail: dimitris@delta-pi-systems.eu

D. Vlachos
Department of Informatics & Telecommunications, University of Peloponnese, Tripoli, Greece
e-mail: dvlachos@uop.gr

N. J. Daras, T. M. Rassias (eds.), *Computational Mathematics and Variational Analysis*, Springer Optimization and Its Applications 159,
https://doi.org/10.1007/978-3-030-44625-3_12

structure of Hamiltonian PDEs from a Lagrangian formulation, using the variational principle. The wave equation and its multi-symplectic structure have been studied by [6–8] from the Hamiltonian viewpoint.

On the other hand, in the past decades, nonstandard finite difference schemes have been well established by Mickens [9–11] to compensate the weaknesses that may be caused by standard finite difference methods, such as the numerical instabilities. Regarding the positivity, the boundedness, and the monotonicity of solutions, nonstandard finite difference schemes have a better performance than standard ones, due to their flexibility to construct a nonstandard finite difference method. The latter can preserve certain properties and structures, which are obeyed by the original equations.

In the present paper, following our previous work [12] we pay special attention to the geometric structure of multi-symplectic integrators through the use of nonstandard finite difference schemes for variational partial differential equations (PDEs). The considered approach comes as a first step towards developing a Veselov type discretization for PDEs in variational form, e.g., [2, 4, 5] and combines it with nonstandard finite difference schemes of Mickens [9–11]. The resulting multisymplectic-momentum integrators have very good energy performance in the level of the conservation of a nearby Hamiltonian, under appropriate circumstances, up to exponentially small error [2].

In Section 2 we present a short overview of the standard numerical techniques relying on variational integrator schemes and their special case of exponential variational integrators in Section 3. Afterwards, nonstandard finite difference properties are employed for the derivation of nonstandard variational integrators by using a triangle discretization of the spacetime (Section 4.1). Then, in Sections 5 and 6, we demonstrate concrete applications of the proposed integrators, for the numerical solution of the linear wave equation, the Laplace equation, and the Poisson equation. In Section 7, we perform dispersion analysis and convergence experiments to further illustrate the numerical properties of the method. Finally, in Section 8, we summarize the main conclusions coming out of our study.

2 Review of Variational Integrators

The discrete Euler–Lagrange equations can be derived in correspondence to the steps of derivation of the Euler–Lagrange equations in the continuous formulation of Lagrangian dynamics [3]. Denoting the tangent bundle of the configuration manifold Q by TQ, the continuous Lagrangian $L : TQ \to \mathbb{R}$ can be defined. In the discrete setting, considering approximate configurations $q_k \approx q(t_k)$ and $q_{k+1} \approx q(t_{k+1})$ at the time nodes t_k, t_{k+1}, with $h = t_{k+1} - t_k$ being the fixed time step, a discrete Lagrangian $L_d : Q \times Q \to \mathbb{R}$ is defined to approximate the action integral along the curve segment between q_k and q_{k+1}, i.e.,

$$L_d(q_k, q_{k+1}) \approx \int_{t_k}^{t_{k+1}} L(q(t), \dot{q}(t))dt. \tag{1}$$

Defining the discrete trajectory $\gamma_d = (q_0, \ldots, q_N)$, $N \in \mathbb{N}$, one can obtain the action sum

$$S_d(\gamma_d) = \sum_{k=1}^{N-1} L_d(q_k, q_{k+1}). \tag{2}$$

The discrete Hamilton's principle states that a motion γ_d of the discrete mechanical system extremizes the action sum, i.e., $\delta S_d = 0$. Through differentiation and rearrangement of the terms, holding the end points q_0 and q_N fixed, the discrete Euler–Lagrange equations are obtained [3]

$$D_2 L_d(q_{k-1}, q_k) + D_1 L_d(q_k, q_{k+1}) = 0, \quad k = 1, \ldots, N-1, \tag{3}$$

where the notation $D_i L_d$ indicates derivative with respect to the i-th argument of L_d, see also [3, 12–16].

The definition of the discrete conjugate momentum at time steps k and $k+1$ reads

$$p_k = -D_1 L_d(q_k, q_{k+1}), \quad p_{k+1} = D_2 L_d(q_k, q_{k+1}), \quad k = 0, \ldots, N-1. \tag{4}$$

The above equations, also known as position–momentum form of a variational integrator, can be used when an initial condition (q_0, p_0) is known, to obtain (q_1, p_1).

To construct high order methods, we approximate the action integral along the curve segment between q_k and q_{k+1} using a discrete Lagrangian that depends only on the end points. We obtain expressions for configurations q_k^j and velocities $\dot{q}_k^j$ for $j = 0, \ldots, S-1$, $S \in \mathbb{N}$ at time $t_k^j \in [t_k, t_{k+1}]$ by expressing $t_k^j = t_k + C_k^j h$ for $C_k^j \in [0, 1]$ such that $C_k^0 = 0$, $C_k^{S-1} = 1$ using

$$q_k^j = g_1(t_k^j) q_k + g_2(t_k^j) q_{k+1}, \qquad \dot{q}_k^j = \dot{g}_1(t_k^j) q_k + \dot{g}_2(t_k^j) q_{k+1}, \tag{5}$$

where $h \in \mathbb{R}$ is the time step. We choose functions

$$g_1(t_k^j) = \sin\left(u - \frac{t_k^j - t_k}{h} u\right) (\sin u)^{-1}, \quad g_2(t_k^j) = \sin\left(\frac{t_k^j - t_k}{h} u\right) (\sin u)^{-1}, \tag{6}$$

to represent the oscillatory behavior of the solution, see [17, 18]. For continuity, $g_1(t_{k+1}) = g_2(t_k) = 0$ and $g_1(t_k) = g_2(t_{k+1}) = 1$ is required.

For any different choice of interpolation used, we define the discrete Lagrangian by the weighted sum

$$L_d(q_k, q_{k+1}) = h \sum_{j=0}^{S-1} w^j L(q(t_k^j), \dot{q}(t_k^j)), \tag{7}$$

where it can be easily proved that for maximal algebraic order

$$\sum_{j=0}^{S-1} w^j (C_k^j)^m = \frac{1}{m+1}, \tag{8}$$

where $m = 0, 1, \ldots, S-1$ and $k = 0, 1, \ldots, N-1$, see [17, 18].

Applying the above interpolation technique with the trigonometric expressions of (6), following the phase lag analysis of [13, 14, 17, 18], the parameter u can be chosen as $u = \omega h$. For problems that include a constant and known domain frequency ω (such as the harmonic oscillator) the parameter u can be easily computed. For the solution of orbital problems of the general N-body problem, where no unique frequency is given, a new parameter u must be defined by estimating the frequency of the motion of any moving point mass [16, 19–21].

3 Exponential Integrators

We now consider the Hamiltonian systems

$$\ddot{q} + \Omega q = g(q), \qquad g(q) = -\nabla U(q), \tag{9}$$

where Ω is a diagonal matrix (will contain diagonal entries ω with large modulus) and $U(q)$ is a smooth potential function. We are interested in the long time behavior of numerical solutions when ωh is not small.

Since $q_{n+1} - 2\cos(h\omega)q_n + q_{n-1} = 0$ is an exact discretization of (9) we can consider the numerical scheme

$$q_{n+1} - 2\cos(h\omega)q_n + q_{n-1} = h^2 \psi(\omega h) g(\phi(\omega h) q_n), \tag{10}$$

where the functions $\psi(\omega h)$ and $\phi(\omega h)$ are even, real-valued functions satisfying $\psi(0) = \phi(0) = 1$, see [1]. The resulting methods using the latter numerical scheme are known as exponential integrators (for some examples of those integrators, see the Appendix).

3.1 Exponential High Order Variational Integrators

If we now use the phase fitted variational integrator for the system (9) the result of the discrete Euler–Lagrange equations (3) will be

$$q_{n+1} + \Lambda(u, \omega, h, S)q_n + q_{n-1} = h^2\Psi(\omega h)g(\Phi(\omega h)q_n), \tag{11}$$

where

$$\Lambda(u, \omega, h, S) = \frac{\sum_{j=0}^{S-1} w^j \left[\dot{g}_1(t_k^j)^2 + \dot{g}_2(t_k^j)^2 - \omega^2\big(g_1(t_k^j)^2 + g_2(t_k^j)^2\big)\right]}{\sum_{j=0}^{S-1} w^j \left[\dot{g}_1(t_k^j)\dot{g}_2(t_k^j) - \omega^2 g_1(t_k^j)g_2(t_k^j)\right]}. \tag{12}$$

Using the above expressions, to obtain exponential variational integrators that use expressions for configurations q_k^j and velocities $\dot{q}_k^j$ taken from (5), we get

$$\Lambda(u, \omega, h, S) = -2\cos(\omega h). \tag{13}$$

In [16] we have proved (using the phase lag analysis of [22]) that exponentially fitted methods using phase fitted variational integrators can be derived when (13) holds. So phase fitted variational integrators using trigonometric interpolation can be considered as exponential integrators, i.e., when using phase fitted variational integrators, keeping the phase lag zero the resulting methods are exponentially fitted methods (exponential integrators). Those methods have been tested on several numerical results in [16].

3.2 Frequency Estimation for Mass Points Motion in Three Dimensions

In our previous work [16], we constructed adaptive time step variational integrators using phase fitting techniques and estimated the required frequency through the use of a harmonic oscillator with given frequency ω. Here, in solving the general N-body problem by using a constant time step, a new frequency estimation is necessary in order to find for each body i) the frequency at an initial time t_0 and ii) the frequency at time t_k for $k = 1, \ldots, N-1$.

It is now clear that, by applying the trigonometric interpolation (6), the parameter u can be chosen as $u = \omega h$. For problems for which the domain of frequency ω is fixed and known (such as the harmonic oscillator) the parameter u can be easily computed. For the solution of orbital problems involved in the general N-body problem, where no unique frequency is determined, the parameter u must be defined by estimating the frequency of the motion of any moving material point.

Towards this purpose, we consider the general case of N masses moving in three dimensions. If $q_i(t)$ $(i = 1, \ldots, N)$ denotes the trajectory of the i-th particle, its curvature can be computed from the known expression

$$k_i(t) = \frac{|\dot{q}_i(t) \times \ddot{q}_i(t)|}{|\dot{q}_i(t)|^3}, \tag{14}$$

where $\dot{q}_i(t)$ is the velocity of the i-th mass with magnitude $|\dot{q}_i(t)|$ at a point $q_i(t)$. After a short time h, the angular displacement of that mass is $h|\dot{q}_i(t) \times \ddot{q}_i(t)|/|\dot{q}_i(t)|^2$, which for each mass's actual frequency gives the expression

$$\omega_i(t) = \frac{|\dot{q}_i(t) \times \ddot{q}_i(t)|}{|\dot{q}_i(t)|^2}. \tag{15}$$

From (14) and (15) the well-known relation $\omega_i(t) = k_i(t)|\dot{q}_i(t)|$ holds (see also [16]).

For the specific case of many-particle physical problems, that can be described via a Lagrangian of the form $L(q, \dot{q}) = \frac{1}{2}\dot{q}^T M(q)\dot{q} - V(q)$, where $M(q)$ represents a symmetric positive definite mass matrix and V is a potential function, the continuous Euler–Lagrange equations are $M(q)\ddot{q} = -\nabla V(q)$. In this case, the expression for frequency estimation (15), referred to the i-th body at time t_k, $k = 1, \ldots, N-1$, takes the form

$$\omega_i(t_k) = h^{-1}\frac{\left|M^{-1}(q_k)p_k \times \left(M^{-1}(q_k)p_k - M^{-1}(q_{k-1})p_{k-1}\right)\right|}{\left|M^{-1}(q_k)p_k\right|^2}, \tag{16}$$

where the quantities on the right-hand side are the mass matrix, the configuration, and the momentum of the i-th body. Since the frequency $\omega_i(t_k)$ must be also known at an initial time instant t_0 (in which the initial positions are $\bar{q}_0$ and initial momenta are $\bar{p}_0$), using the continuous Euler–Lagrange equation at t_0 we obtain

$$\omega_i(t_0) = \frac{\left|M^{-1}(\bar{q}_0)\bar{p}_0 \times \left(-M^{-1}(\bar{q}_0)\nabla V(\bar{q}_0)\right)\right|}{\left|M^{-1}(\bar{q}_0)\bar{p}_0\right|^2}. \tag{17}$$

Equations (16) and (17) provide an "estimated frequency" for each mass in the general motion of the N-body problem. This allows us to derive high order variational integrator methods using trigonometric interpolation where the frequency is estimated at every time step of the integration procedure. These methods show better energy behavior, i.e., smaller total energy oscillation than other methods which employ constant frequency, see [14, 16].

Before closing this section, it should be mentioned that the linear stability of our method is comprehensively analyzed in our previous works [14, 16, 19].

4 Triangle and Square Discretization

In order to express the discrete Lagrangian and discrete Hamilton function, we will use the definition of the tangent bundle TQ and cotangent bundle T^*Q as in [2] to fields over the higher-dimensional manifold X. In this way, we also view fields over X as sections of some fiber bundle $B \to X$, with fiber Y, and then consider the first jet bundle J^1B and its dual (J^1B^*) as the appropriate analogs of the tangent and cotangent bundles.

It is then possible to use the generalization of the Veselov discretization [4, 5] to multi-symplectic field theory, by discretizing the spacetime X. For simplicity reasons we will restrict ourselves to the discrete analogue of $\dim X = 2$. Thus, we take $X = \mathbb{Z} \times \mathbb{Z} = (i, j)$ and the fiber bundle Y to be $X \times F$ for some smooth manifold F [2, 12].

4.1 Triangle Discretization

Assume that we have a uniform quadrangular mesh in the base space, with mesh lengths Δx and Δt. The nodes in this mesh are denoted by $(i, j), \in \mathbb{Z} \times \mathbb{Z}$, corresponding to the points $(x_i, t_j) := (i\Delta x, j\Delta t) \in \mathbb{R}^2$. We denote the value of the field u at the node (i, j) by u^i_j. We label the triangle at (i, j) with three ordered triple $((i, j), (i+1, j), (i, j+1))$ as Δ_{ij}, and we define X_Δ to be the set of all such triangles, see Figure 1.

Then, the discrete jet bundle is defined as follows [2]:

$$J^1_\Delta Y := \{(u^j_i, u^j_{i+1}, u^{j+1}_i) \in \mathbb{R}^3 : ((i, j), (i+1, j), (i, j+1)) \in X_\Delta\}, \tag{18}$$

which is equal to $X_\Delta \times \mathbb{R}^3$. The field u can be now defined by averaging the fields over all vertices of the triangle (see Figure 1a)

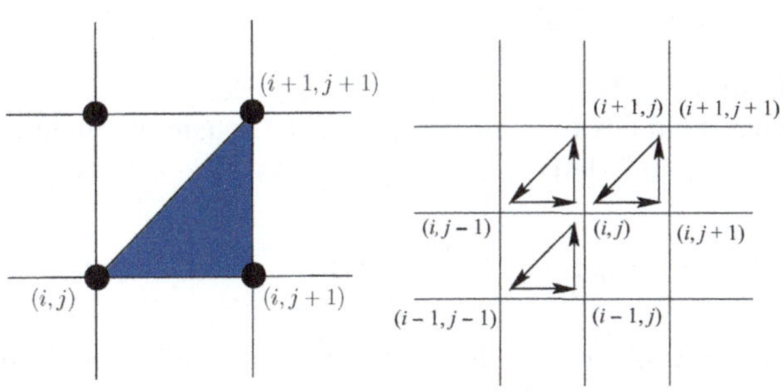

Fig. 1 The triangles which touch (i, j)

$$u \to \frac{u_i^j + u_i^{j+1} + u_{i+1}^{j+1}}{3}, \tag{19}$$

while the derivatives can be expressed using nonstandard finite differences [9–11]

$$\frac{du}{dt} \to \frac{u_i^{j+1} - u_i^j}{\phi(\Delta t)}, \qquad \frac{du}{dx} \to \frac{u_{i+1}^{j+1} - u_i^{j+1}}{\psi(\Delta x)}, \tag{20}$$

with [9, 10]

$$\phi(\Delta t) = 2\sin\left(\frac{\Delta t}{2}\right), \quad \psi(\Delta x) = 2\sin\left(\frac{\Delta x}{2}\right). \tag{21}$$

Using the latter expressions, we can obtain the discrete Lagrangian at any triangle, which depends on the edges of the triangle, i.e., $L_d(u_i^j, u_i^{j+1}, u_{i+1}^{j+1})$, while the discrete Euler–Lagrange field equations are

$$D_1 L_d(u_i^j, u_i^{j+1}, u_{i+1}^{j+1}) + D_2 L_d(u_i^{j-1}, u_i^j, u_{i+1}^j) + D_3 L_d(u_{i-1}^{j-1}, u_{i-1}^j, u_i^j) = 0, \tag{22}$$

see Figure 1 (right).

4.2 Square Discretization

For the cases where square discretization is used, and if we also denote a square at (i, j) with four ordered quaternion $((i, j), (i+1, j), (i+1, j+1), (i, j+1))$ by $\square_j^i$, we can consider $X_\square$ to be the set of all such squares, see Figure 1. Then, the discrete jet bundle is defined as (for more details see [2] and the references therein)

$$J_\square^1 Y := \Big\{(u_i^j, u_{i+1}^j, u_{i+1}^{j+1}, u_i^{j+1}) \in \mathbb{R}^4 : ((i, j), (i+1, j), (i+1, j+1), (i, j+1)) \in X_\square\Big\}, \tag{23}$$

which is equal to $X_\square \times \mathbb{R}^4$.

By averaging the fields over all vertices of the square, the field u can be now obtained as (see Figure 2 (left))

$$u \to \frac{u_i^j + u_{i+1}^j + u_i^{j+1} + u_{i+1}^{j+1}}{4}. \tag{24}$$

As above, the expressions for the derivatives can be taken from [9–11] for the discrete Lagrangian, which now depends on the edges of the square, i.e., $L_d(u_i^j, u_{i+1}^j, u_{i+1}^{j+1}, u_i^{j+1})$. As a result, the discrete Euler–Lagrange field equations are

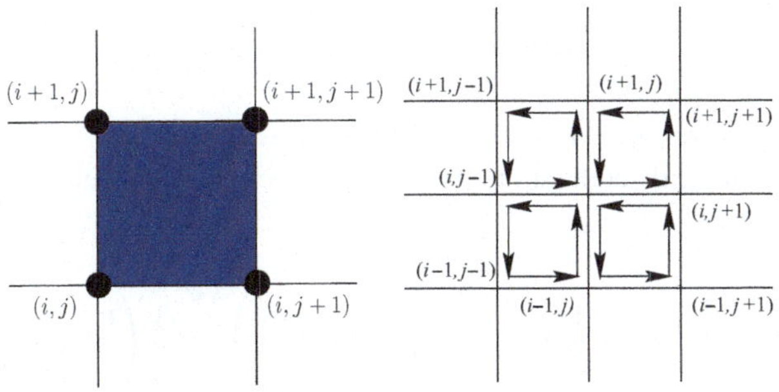

Fig. 2 The squares which touch (i, j)

$$D_1L_d(u_i^j, u_i^{j+1}, u_{i+1}^{j+1}, u_{i+1}^j) + D_2L_d(u_i^{j-1}, u_i^j, u_{i+1}^j, u_{i+1}^{j-1}) +$$
$$D_3L_d(u_{i-1}^{j-1}, u_{i-1}^j, u_i^j, u_i^{j-1}) + D_4L_d(u_{i-1}^j, u_{i-1}^{j+1}, u_i^{j+1}, u_i^j) = 0, \tag{25}$$

see Figure 2 (right).

5 Numerical Examples Using Triangle Discretization

To illustrate the proposed method, we consider the basic PDEs of three physical problems, i.e., the linear wave equation, the Laplace equation, and the Poisson equation (see [2] and [23, 24]). In the following subsections, for representation requirements, quadrilaterals have been used by interpolating the solution on triangles.

5.1 *Linear Wave Equation*

The linear wave equation contains second-order partial derivatives of the wave function $u(x, t)$ with respect to time and space, respectively, as (see, e.g., [23, 24])

$$\frac{\partial^2 u}{\partial t^2} + c\frac{\partial^2 u}{\partial x^2} = 0. \tag{26}$$

This equation may be considered for the description of the wave function, i.e., the amplitude of oscillation, that is created from a one-dimensional medium (e.g., a string extended in the x-direction). For the special case that the velocity of the wave, representing by the parameter c, is chosen as $c = -1$, the corresponding Lagrangian is [12]

$$L(u, u_t, u_x) = \frac{1}{2}u_t^2 - \frac{1}{2}u_x^2, \tag{27}$$

where the derivatives are $\partial u/\partial t = u_t$ and $\partial u/\partial x = u_x$.

If we use triangle discretization, described in Section 4.1, we end up with discrete Lagrangian

$$L_d\left(u_i^j, u_i^{j+1}, u_{i+1}^{j+1}\right) = \frac{1}{2}\Delta t \Delta x \left[\frac{1}{2}\left(\frac{u_i^{j+1} - u_i^j}{\phi(\Delta t)}\right)^2 - \frac{1}{2}\left(\frac{u_{i+1}^{j+1} - u_i^{j+1}}{\psi(\Delta x)}\right)^2\right], \tag{28}$$

where Δt and Δx are the mesh lengths for time and space, respectively. Applying the above discrete Lagrangian to the discrete Euler–Lagrange field equations (22), we get

$$\frac{u_i^{j+1} - 2u_i^j + u_i^{j-1}}{(\phi(\Delta t))^2} - \frac{u_{i+1}^j - 2u_i^j + u_{i-1}^j}{(\psi(\Delta x))^2} = 0. \tag{29}$$

The latter expression represents the variational integrator for the linear wave equation (26), resulting through the use of the proposed nonstandard finite difference schemes.

In Figure 3 the solution $u(x, t)$ of (29) is shown in a 3-D diagram. We have chosen as initial conditions $0 < x < 1$, $u(x, 0) = 0.5[1-\cos(2\pi x)]$, $u_t(x, 0) = 0.1$ and as boundary conditions $u(0, t) = u(1, t)$, $u_x(0, t) = u_x(1, t)$, the latter being periodic. The grid discretization has been taken to be $\Delta t = 0.01$ and $\Delta x = 0.01$. As seen, the time evolution of the solution $u(x = const., t)$ is a continuous function, while the periodicity is preserved.

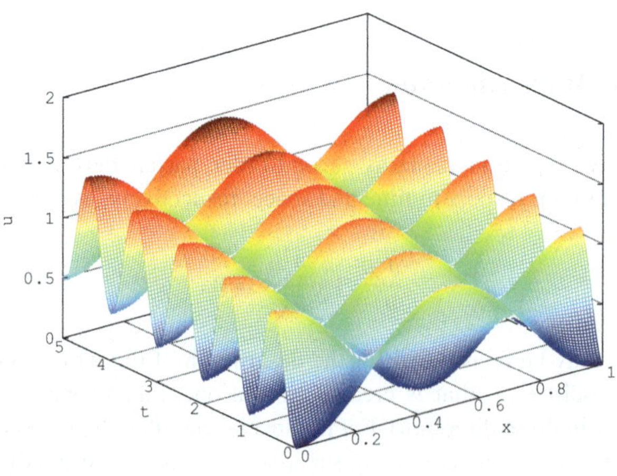

Fig. 3 The waveforms of linear wave equation (26)

5.2 Laplace Equation

As another physical example, we have chosen the Laplace equation over a 2-D scalar field $u(x, y)$. It is written as

$$u_{xx} + u_{yy} = 0. \tag{30}$$

The function $u(x, y)$ may describe a potential in a 2-D medium or a potential inside a 3-D medium, which does not depend on the third coordinate z. Thus, the two-dimensional second-order PDE (30) governs a variety of equilibrium physical phenomena such as temperature distribution in solids, electric field in electrostatics, inviscid and irrotational two-dimensional flow (potential flow), groundwater flow, etc.

The corresponding continuous Lagrangian of (30) takes the form

$$L(u, u_x, u_y) = \frac{1}{2}u_x^2 + \frac{1}{2}u_y^2. \tag{31}$$

By applying the triangle discretization of Section 4.1, the discrete Lagrangian can be written as

$$L_d\left(u_i^j, u_i^{j+1}, u_{i+1}^{j+1}\right) = \frac{1}{2}\Delta x \Delta y\left[\frac{1}{2}\left(\frac{u_i^{j+1} - u_i^j}{\phi(\Delta x)}\right)^2 + \frac{1}{2}\left(\frac{u_{i+1}^{j+1} - u_i^{j+1}}{\psi(\Delta y)}\right)^2\right]. \tag{32}$$

From the latter Lagrangian, working in a similar manner to that followed in Section 4.2 results the integrator from the proposed nonstandard finite difference schemes

$$\frac{u_i^{j+1} - 2u_i^j + u_i^{j-1}}{(\phi(\Delta x))^2} + \frac{u_{i+1}^j - 2u_i^j + u_{i-1}^j}{(\psi(\Delta y))^2} = 0. \tag{33}$$

The solution of the above equation, when considering the boundary conditions $u(x, 0) = 0$, $u(x, 1) = 1$ and $u(0, y) = u(1, y) = 0$, is plotted in Figure 4. The grid discretization has been chosen to be $\Delta x = 0.02$ and $\Delta y = 0.02$.

5.3 Poisson Equation

As a final application to illustrate the advantages of the proposed variational integrator relying on nonstandard finite difference schemes, we examine the Poisson equation, which is an elliptic PDE of the form

$$- u_{xx} - u_{yy} = f(x, y). \tag{34}$$

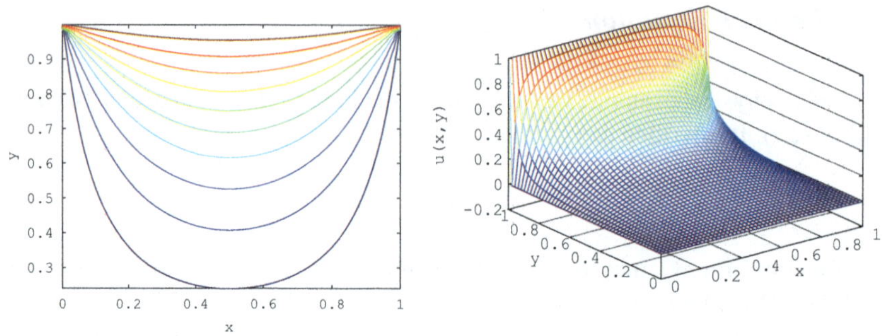

Fig. 4 Contour plot (left) and three-dimensional surface plot (right) of the solution of Laplace equation with boundary conditions $u(x, 0) = 0$, $u(x, 1) = 1$, $u(0, y) = u(1, y) = 0$, and discretization: $\Delta x = 0.02$, $\Delta y = 0.02$

Obviously, this equation in physical applications presents an additional complexity compared to the Laplace equation (30). Now the right-hand side is a nonzero function $f(x, y)$, which may be considered as a source (or a load) function defined on some two-dimensional domain denoted by $\Omega \subset \mathbb{R}^2$ (it could also be a general nonlinear function $f(u, x, y)$). A solution u satisfying (34) will also satisfy specific conditions on the boundaries of the domain Ω. For example, for the element $\partial\Omega$ the general condition holds

$$\alpha u + \beta \frac{\partial u}{\partial n} = g \qquad \text{on} \quad \partial\Omega, \tag{35}$$

where $\partial u/\partial n$ denotes the directional derivative in the direction normal to the boundary $\partial\Omega$ and α and β are constants [23, 24].

As it is well known, the system of (34) and (35) is referred to as a boundary value problem for the Poisson equation. If the constant β in Equation (35) is zero, then the boundary condition is of Dirichlet type, and the boundary value problem is referred to as the Dirichlet problem for the Poisson equation. Alternatively, if the constant α is zero, then we correspondingly have a Neumann boundary condition, and the problem is referred to as a Neumann problem. A third possibility exists when the Dirichlet conditions hold on a part of the boundary $\partial\Omega_D$, and Neumann conditions hold on the remainder $\partial\Omega \setminus \partial\Omega_D$ (or indeed mixed conditions where α and β are both nonzero), see [23, 24] and the references therein.

Equation (34) can also be obtained by starting from the Lagrangian

$$L(u, u_x, u_y) = \frac{1}{2}u_x^2 + \frac{1}{2}u_y^2 - fu. \tag{36}$$

The triangle discretization of Section 4.1 in the Poisson problem defines the discrete Lagrangian

$$L_d\left(u_i^j, u_i^{j+1}, u_{i+1}^{j+1}\right) = \frac{1}{2}\Delta x \Delta y \left[\frac{1}{2}\left(\frac{u_i^{j+1} - u_i^j}{\phi(\Delta x)}\right)^2 + \frac{1}{2}\left(\frac{u_{i+1}^{j+1} - u_i^{j+1}}{\psi(\Delta y)}\right)^2\right]$$
$$- \frac{f_i^j u_i^j + f_i^{j+1} u_i^{j+1} + f_{i+1}^{j+1} u_{i+1}^{j+1}}{3}. \quad (37)$$

By inserting the latter discrete Lagrangian into the discrete Euler–Lagrange field equations (22) and elaborating as done in [2], the resulting integrator from the proposed nonstandard finite difference schemes is

$$-\frac{u_i^{j+1} - 2u_i^j + u_i^{j-1}}{(\phi(\Delta x))^2} - \frac{u_{i+1}^j - 2u_i^j + u_{i-1}^j}{(\psi(\Delta y))^2} = f_i^j + \partial f_i^j / \partial u_i^j. \quad (38)$$

As a special case we chose the source term $f(x, y) \equiv 1$, so $\partial f_i^j / \partial u_i^j = 0$ in (38), and the boundary conditions $u(0, y) = u(1, y) = 0$ and $u(x, 0) = u(x, 1) = 0$. Figure 5 shows the numerical results obtained with the discretization $\Delta x = 0.02$ and $\Delta y = 0.02$.

6 Numerical Examples Using Square Discretization

To illustrate the behavior of the proposed method, we will consider the Klein–Gordon equation, which plays a significant role in many scientific applications such as solid state physics, nonlinear optics, and quantum field theory, see for example [25].

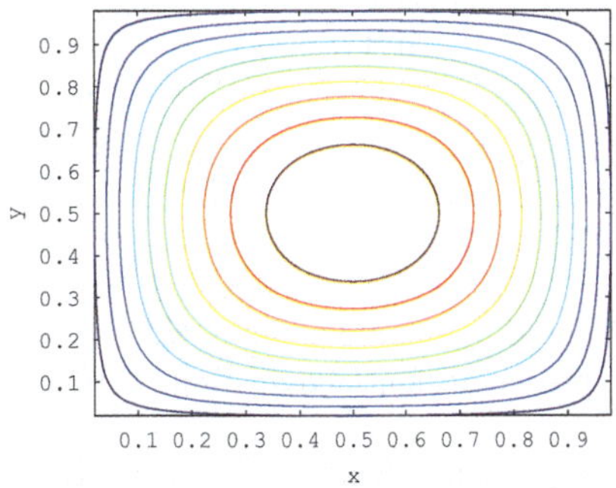

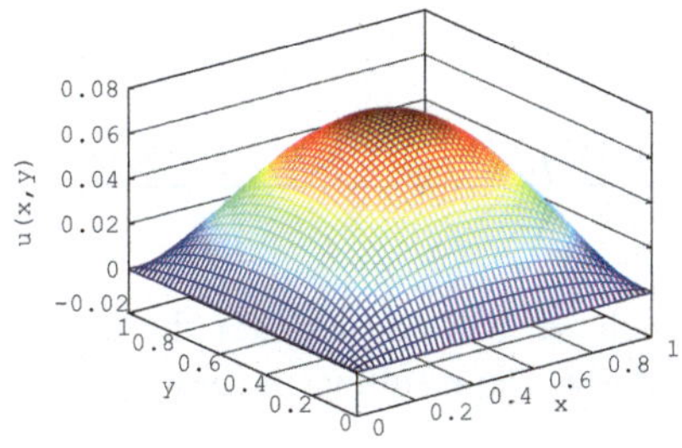

Fig. 5 Contour plot (left) and three-dimensional surface plot (right) of the solution of Poisson equation, using the variational integrator with nonstandard finite difference schemes. The source term was chosen $f(x, y) \equiv 1$, while the boundary conditions $u(0, y) = u(1, y) = 0$, $u(x, 0) = u(x, 1) = 0$ for discretization $\Delta x = 0.02$, $\Delta y = 0.02$

6.1 Klein–Gordon

For the general case, the initial-value problem of the one-dimensional nonlinear Klein–Gordon equation is given by

$$u_{tt} + \alpha u_{xx} + g(u) = f(x, t), \qquad (39)$$

where $u = u(x, t)$ represents the wave displacement at position x and time t, α is a known constant and $g(u)$ is the nonlinear force which in the physical applications has also other forms [25].

Here we will consider the special case that $\alpha = -1$, $g(u) = u^3 - u$, and $f(x, t) = 0$ resulting in

$$u_{tt} = u_{xx} - u^3 + u.$$

The above equation can be described using the Lagrangian

$$L(u, u_t, u_x) = \frac{1}{2}u_t^2 - \frac{1}{2}u_x^2 - \frac{1}{4}u^4 - \frac{1}{2}u^2.$$

Following Section 4.2 we can obtain the discrete Lagrangian that now uses square discretization as

$$\begin{aligned} L_d(u_i^j, u_{i+1}^j, u_{i+1}^{j+1}, u_i^{j+1}) = {} & \frac{\Delta t \Delta x}{2}\left(\frac{u_i^{j+1} - u_i^j}{2\phi(\Delta t)} + \frac{u_{i+1}^{j+1} - u_{i+1}^j}{2\phi(\Delta t)}\right)^2 - \\ & \frac{\Delta t \Delta x}{2}\left(\frac{u_{i+1}^{j+1} - u_i^{j+1}}{2\psi(\Delta x)} + \frac{u_{i+1}^j - u_i^j}{2\psi(\Delta x)}\right)^2 - \\ & -\frac{\Delta t \Delta x}{4} u_i^j u_{i+1}^j u_{i+1}^{j+1} u_i^{j+1} + \\ \frac{\Delta t \Delta x}{2}\left(\frac{u_i^j u_{i+1}^j + u_i^j u_{i+1}^{j+1} + u_i^j u_i^{j+1} + u_{i+1}^j u_{i+1}^{j+1} + u_{i+1}^{j+1} u_i^{j+1} + u_{i+1}^{j+1} u_i^{j+1}}{6}\right), \end{aligned}$$

which we will consider for the discrete Euler–Lagrange equations (25) in order to derive the resulting integrator from the proposed nonstandard finite difference schemes.

Figure 6 shows the numerical results obtained with the discretization $\Delta t = 0.05$ and $\Delta x = 0.05$. To that we have used initial conditions $u(x, 0) = A(1 + \cos(\frac{2\pi x}{L}))$, where $A = 5$ and $u_t(x, 0) = 0$, while the boundary conditions were $u(-1, t) = u(1, t)$ and $u_x(-1, t) = u_x(1, t)$.

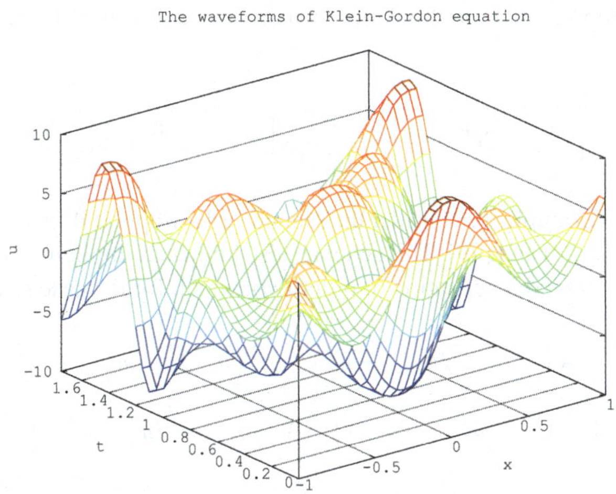

Fig. 6 Numerical solution of the Klein–Gordon equation (6.1) using square discretization of Section 4.2

7 Analysis of the Proposed Schemes

A dispersion analysis and mesh convergence experiments are performed in this section in order to show the numerical properties of the proposed method.

7.1 Dispersion Analysis

We will now turn our study to the dispersion–dissipation properties of the derived numerical schemes and compare them with the ones of [2]. To that end, similar to [26], we consider the discrete analog of the Fourier mode

$$u_i^j = \hat{u}e^{\mathbf{i}(ik\Delta x + j\omega\Delta t)}, \tag{40}$$

where $\mathbf{i}^2 = 1$. Using $\bar{k} = k\Delta x$ and $\bar{\omega} = k\Delta t$, the latter equation results in

$$u_i^j = \hat{u}e^{\mathbf{i}(i\bar{k} + j\bar{\omega})}. \tag{41}$$

Following the above, the multi-symplectic scheme of [2], also known as leapfrog algorithm, for the case of the linear wave (26) gives

$$\frac{u_i^{j+1} - 2u_i^j + u_i^{j-1}}{(\Delta t)^2} - \frac{u_{i+1}^j - 2u_i^j + u_{i-1}^j}{(\Delta x)^2} = 0. \tag{42}$$

When substituting (41) in the latter equation, we get the discrete dispersion relationship

$$\frac{e^{\mathbf{i}\bar{k}}}{(\Delta t)^2}\left[e^{2\mathbf{i}\bar{\omega}} - 2e^{\mathbf{i}\bar{\omega}} + 1\right] - \frac{e^{\mathbf{i}\bar{\omega}}}{(\Delta x)^2}\left[e^{2\mathbf{i}\bar{k}} - 2e^{\mathbf{i}\bar{k}} + 1\right] = 0. \tag{43}$$

As a second example we consider the second-order implicit Runge–Kutta scheme described in [27, 28] and [29]. This scheme, also known as implicit Crank–Nicolson, is a symplectic time discretization of order two, which for the case of (26) gives

$$\begin{aligned} &4\left(u_i^{j+2} - 2u_i^{j+1} + u_i^j\right) - \lambda^2\left(u_{i-1}^{j+2} - 2u_i^{j+2} + u_{i+1}^{j+2}\right) \\ &-2\lambda^2\left(u_{i-1}^{j+1} - 2u_i^{j+1} + u_{i+1}^{j+1}\right) - \lambda^2\left(u_{i-1}^{j} - 2u_i^{j} + u_{i+1}^{j}\right) = 0, \end{aligned} \tag{44}$$

where

$$\lambda^2 = \left(\frac{\Delta t}{\Delta x}\right)^2. \tag{45}$$

Substituting to the above integrator the form (41) we obtain the discrete dispersion relationship

$$\frac{4e^{\mathbf{i}\bar{k}}}{(\Delta t)^2}\left[e^{2\mathbf{i}\bar{\omega}} - 2e^{\mathbf{i}\bar{\omega}} + 1\right] - \frac{\left[e^{2\mathbf{i}\bar{k}} - 2e^{\mathbf{i}\bar{k}} + 1\right]}{(\Delta x)^2}\left[e^{2\mathbf{i}\bar{\omega}} - 2e^{\mathbf{i}\bar{\omega}} + 1\right] = 0. \tag{46}$$

For the case of the linear wave equation (26) the integrator with the proposed technique, i.e., (29) for u_i^j of (41) gives

$$(\cos\bar{\omega} - 1)(1 - \cos\Delta x) - \left(\cos\bar{k} - 1\right)(1 - \cos\Delta t) = 0. \tag{47}$$

For now we will restrict ourselves only to $\lambda \leq 1$, but due to symmetry, all other cases can be easily obtained. Figure 7 shows the discrete dispersion relationships for $\lambda = \{0.95, 0.9, 0.85, 0.8\}$. Specifically, to each subplot we can see the dispersion curve of the leapfrog scheme, i.e., equation (43), with blue line, the red line corresponds to the proposed method, described by (47), while the green line is the one for the implicit Runge–Kutta scheme, equation (46). For all the choices of λ tested the behavior of the method using nonstandard finite difference schemes is close to the excellent behavior of the leapfrog scheme, and much better than the implicit Runge–Kutta scheme.

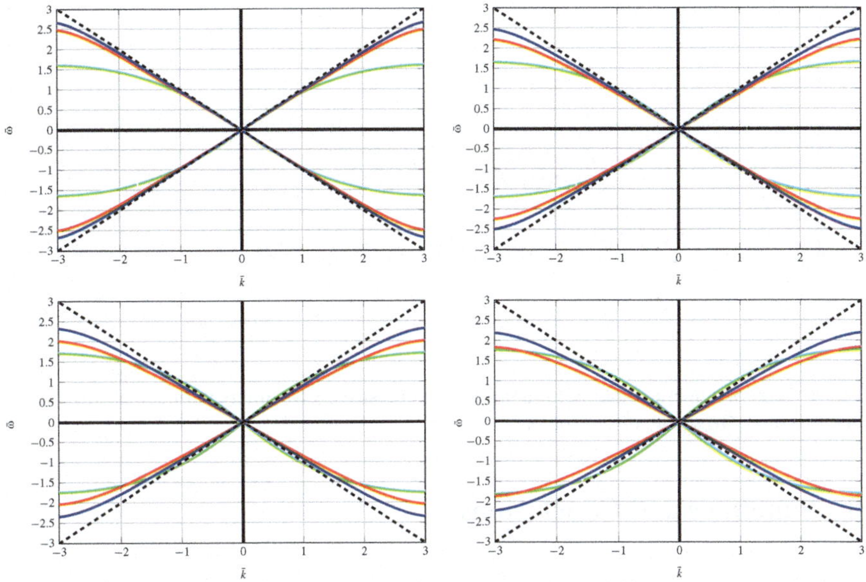

Fig. 7 Dispersion curves for the linear wave equation with the proposed method (red), the leapfrog scheme of [2] (blue), the implicit Runge–Kutta (green), and the analytic one (dashed black) for $\lambda = \{0.95, 0.9, 0.85, 0.8\}$

7.2 Convergence Experiments

In order to show the grid independence of the solution, following the finite element convention, the l^{∞}-norm error is calculated between the solutions on two successive grids according to

$$e_h = \max_i \{|u_i^f - u_i^c|, \ldots, |u_{n_{\text{el}}}^f - u_{n_{\text{el}}}^c|\}, \tag{48}$$

where u_i^f is the solution on the fine grid and u_i^c is the solution on a coarse grid interpolated on the fine one. Here, n_{el} are the total number of elements, where the elements of the mesh are either triangles or squares. A sample convergence of the calculations for the Klein–Gordon case is shown in Figure 8 in a logarithmic plot for triangle and square discretizations and for different time steps. It can be easily seen that by decreasing the space discretization the error is also decreased linearly in the log scale.

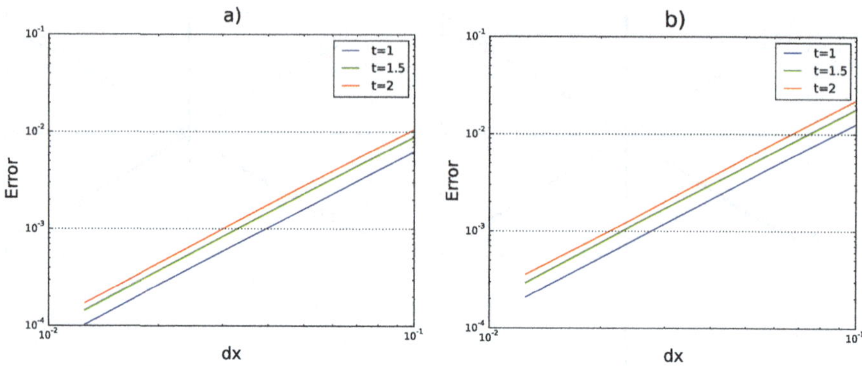

Fig. 8 Error of numerical solution as a function of grid size Δx for different time steps: (**a**) triangle discretization, (**b**) square discretization

8 Summary and Conclusions

The derivation of advantageous multi-symplectic numerical methods, relying on nonstandard finite difference schemes, is investigated. The numerical solution of the linear wave equation, the 2-D Laplace equation, and the 2-D Poisson equation, which are addressed in this study, shows a good energy behavior and the preservation of the discrete multi-symplectic structure of the proposed numerical schemes. Moreover, we showed with the help of dispersion analysis and mesh convergence experiments the numerical properties of the proposed method.

Future applications may include the field equation of incompressible fluid dynamics, like that of Cotter et al. [30] and Pavlov et al. [31], which could be of interest in investigating the properties of 3-D media. For partial differential equations arising in the field of fluid dynamics, dissipative terms should be taken into consideration. These dissipative perturbations necessitate application of techniques similar to [32, 33] but in the case of PDEs. Furthermore, a possible application in complex geometries, as they appear in real world problems, would necessitate the extension of this methodology to non-uniform grids.

The variational method presented in this work can be applied in a variety of physical problems, ranging from magnetic field simulations in NMR [34] to inverse problems that arise in geophysics [35] and others. Future work may include comparison with other numerical methods used for the solution of PDEs, such as the finite element method or the finite volume method.

Acknowledgments Dr. Odysseas Kosmas wishes to acknowledge the support of EPSRC via grand EP/N026136/1 "Geometric Mechanics of Solids."

Appendix

By denoting $\mathrm{sinc}(\xi) = \sin(\xi)/\xi$, special cases of the exponential integrators described using (10) can be obtained, i.e.,

- Gautschi type exponential integrators [36] for

$$\psi(\Omega h) = \mathrm{sinc}^2\left(\frac{\Omega h}{2}\right), \qquad \phi(\Omega h) = 1$$

- Deuflhard type exponential integrators [37] for

$$\psi(\Omega h) = \mathrm{sinc}(\Omega h), \qquad \phi(\Omega h) = 1$$

- García-Archilla et al. type exponential integrators [38] for

$$\psi(\Omega h) = \mathrm{sinc}^2(\Omega h), \qquad \phi(\Omega h) = \mathrm{sinc}(\Omega h)$$

Finally, in [1] a way to write the Störmer–Verlet algorithm as an exponential integrators is presented.

References

1. E. Hairer, C. Lubich, G. Wanner, Geometric numerical integration illustrated by the Störmer-Verlet method. Acta Numerica **12**, 399 (2003)
2. J.E. Marsden, G.W. Patrick, S. Shkoller, Multisymplectic geometry, variational integrators, and nonlinear PDEs. Commun. Math. Phys. **199**, 351 (1998).
3. J.E. Marsden, M. West, Discrete mechanics and variational integrators. Acta Numerica **10**, 357 (2001)
4. A.P. Veselov, Integrable discrete-time systems and difference operators. Funkts. Anal. Prilozhen. **22**, 1 (1988)
5. A.P. Veselov, Integrable Lagrangian correspondences and the factorization of matrix polynomials. Funkts. Anal. Prilozhen. **25**, 38 (1991)
6. T.J. Bridges, Multi-symplectic structures and wave propagation. Math. Proc. Camb. Philos. Soc. **121**, 1 (1997)
7. T.J. Bridges, S. Reich, Multi-symplectic integrators: numerical schemes for Hamiltonian PDEs that conserve symplecticity. Phys. Lett. A **284**, 4–5 (2001)
8. T.J. Bridges, S. Reich, Numerical methods for Hamiltonian PDEs. J. Phys. **39**, 19 (2006)
9. R.E. Mickens, *Applications of Nonstandard Finite Difference Schemes* (World Scientific Publishing, Singapore, 2000)
10. R.E. Mickens, Nonstandard finite difference schemes for differential equations. J. Differ. Equ. Appl. **8**, 823 (2002)
11. R.E. Mickens, Dynamic consistency: a fundamental principle for constructing nonstandard finite difference schemes for differential equations. J. Differ. Equ. Appl. **11**, 645 (2005)
12. O.T. Kosmas, D. Papadopoulos, Multisymplectic structure of numerical methods derived using nonstandard finite difference schemes. J. Phys. Conf. Ser. **490** (2014)

13. O.T. Kosmas, Charged particle in an electromagnetic field using variational integrators. Numer. Anal. Appl. Math. **1389**, 1927 (2011)
14. O.T. Kosmas, S. Leyendecker, Analysis of higher order phase fitted variational integrators. Adv. Comput. Math. **42**, 605 (2016)
15. O.T. Kosmas, D.S. Vlachos, Local path fitting: a new approach to variational integrators. J. Comput. Appl. Math. **236**, 2632 (2012)
16. O.T. Kosmas, S. Leyendecker, Variational integrators for orbital problems using frequency estimation. Adv. Comput. Math. **45**, 1–21 (2019)
17. O.T. Kosmas, D.S. Vlachos, Phase-fitted discrete Lagrangian integrators. Comput. Phys. Commun. **181**, 562–568 (2010)
18. O.T. Kosmas, S. Leyendecker, Phase lag analysis of variational integrators using interpolation techniques. Proc. Appl. Math. Mech. **12**, 677–678 (2012)
19. O.T. Kosmas, S. Leyendecker, Stability analysis of high order phase fitted variational integrators. Proceedings of WCCM XI – ECCM V – ECFD VI, vol. 1389 (2014), pp. 865–866
20. O.T. Kosmas, S. Leyendecker, Family of high order exponential variational integrators for split potential systems. J. Phys. Conf. Ser. **574** (2015)
21. O.T. Kosmas, D.S. Vlachos, A space-time geodesic approach for phase fitted variational integrators. J. Phys. Conf. Ser. **738** (2016)
22. L. Brusca, L. Nigro, A one-step method for direct integration of structural dynamic equations. Int. J. Numer. Methods Eng. **15**, 685–699 (1980)
23. L.C. Evans, *Partial Differential Equations* (American Mathematical Society, Providence, 1998)
24. V.I. Arnold, *Lectures on Partial Differential Equations* (Springer, Berlin, 2000)
25. H. Han, Z. Zhang, Split local absorbing conditions for one-dimensional nonlinear Klein-Gordon equation on unbounded domain. J. Comput. Phys. **227**, 8992 (2008)
26. J.W. Thomas, *Numerical Partial Differential Equations*, vol. 1. Finite Difference Methods (Springer, New York, 1995)
27. J.M. Sanz-Serna, M.P. Calvo, *Numerical Hamiltonian Problems* (Chapman & Hall, London, 1994)
28. J.M. Sanz-Serna, Solving numerically Hamiltonian systems. In: *Proceedings of the International Congress of Mathematicians* (Birkhäuser, Basel, 1995)
29. S. Reich, Multi-symplectic Runge-Kutta collocation methods for Hamiltonian wave equations. J. Comput. Phys. **157**, 473 (2000)
30. C.J. Cotter, D.D. Holm, P.E. Hydon, Multisymplectic formulation of fluid dynamics using the inverse map. Proc. R. Soc. A **463**, 2671 (2007)
31. D. Pavlov, P. Mullen, Y. Tong, E. Kanso, J.E. Marsden, M. Desbrun, Structure-preserving discretization of incompressible fluids. Physica D **240**, 443 (2011)
32. E. Hairer, C. Lubich, Invariant tori of dissipatively perturbed Hamiltonian systems under symplectic discretization. Appl. Numer. Math **29**, 57–71 (1999)
33. D. Stoffer, On the qualitative behaviour of symplectic integrators. III: Perturbed integrable systems. J. Math. Anal. Appl. **217**, 521–545 (1998)
34. D. Papadopoulos, M.A. Voda, S. Stapf, F. Casanova, M. Behr, B. Blümich, Magnetic field simulations in support of interdiffusion quantification with NMR. Chem. Eng. Sci. **63**, 4694 (2008)
35. D. Papadopoulos, M. Herty, V. Rath, M. Behr, Identification of uncertainties in the shape of geophysical objects with level sets and the adjoint method. Comput. Geosci. **15**, 737 (2011)
36. W. Gautschi, Numerical integration of ordinary differential equations based on trigonometric polynomials. Numer. Math. **3**, 1 (1961)
37. P. Deuflhard, A study of extrapolation methods based on multistep schemes without parasitic solutions. Z. Angew. Math. Phys. **30**, 2 (1979)
38. B. García-Archilla, M.J. Sanz-Serna, R.D. Skeel, Long-time-step methods for oscillatory differential equations. SIAM J. Sci. Comput. **20**, 3 (1999)

Additive (ρ_1, ρ_2)-Functional Inequalities in Complex Banach Spaces

Jung Rye Lee, Choonkil Park, and Themistocles M. Rassias

Abstract In this paper, we introduce and solve the following additive (ρ_1, ρ_2)-functional inequalities:

$$\begin{aligned} \|f(x+y) - f(x) - f(y)\| \geq \|\rho_1(f(x-y) - f(x) + f(y))\| \\ + \|\rho_2(f(y-x) - f(y) + f(x))\|, \end{aligned} \tag{1}$$

where ρ_1 and ρ_2 are fixed complex numbers with $|\rho_1| + |\rho_2| > 1$, and

$$\begin{aligned} \|f(x-y) - f(x) + f(y)\| \geq \|\rho_1(f(x+y) - f(x) - f(y))\| \\ + \|\rho_2(f(y-x) - f(y) + f(x))\|, \end{aligned} \tag{2}$$

where ρ_1 and ρ_2 are fixed complex numbers with $1 + |\rho_1| > |\rho_2| > 1$. Using the fixed point method and the direct method, we prove the Hyers–Ulam stability of the additive (ρ_1, ρ_2)-functional inequalities (1) and (2) in complex Banach spaces.

J. R. Lee
Department of Mathematics, Daejin University, Pocheon-si, Korea
e-mail: jrlee@daejin.ac.kr

C. Park
Department of Mathematics, Hanyang University, Seoul, South Korea
e-mail: baak@hanyang.ac.kr

T. M. Rassias (✉)
Department of Mathematics, National Technical University of Athens, Athens, Greece
e-mail: trassias@math.ntua.gr

N. J. Daras, T. M. Rassias (eds.), *Computational Mathematics and Variational Analysis*, Springer Optimization and Its Applications 159,
https://doi.org/10.1007/978-3-030-44625-3_13

1 Introduction and Preliminaries

The stability problem of functional equations originated from a question of Ulam [24] concerning the stability of group homomorphisms. Hyers [12] gave a first affirmative partial answer to the question of Ulam for Banach spaces. Hyers' Theorem was generalized by Aoki [1] for additive mappings and by Rassias [22] for linear mappings by considering an unbounded Cauchy difference. A generalization of the Rassias theorem was obtained by Găvruta [11] by replacing the unbounded Cauchy difference by a general control function in the spirit of Rassias' approach. The stability of quadratic functional equation was proved by Skof [23] for mappings $f : E_1 \to E_2$, where E_1 is a normed space and E_2 is a Banach space. Cholewa [7] noticed that the theorem of Skof is still true if the relevant domain E_1 is replaced by an abelian group.

Park [16, 17] defined additive ρ-functional inequalities and proved the Hyers–Ulam stability of the additive ρ-functional inequalities in Banach spaces and non-Archimedean Banach spaces. The stability problems of various functional equations have been extensively investigated by a number of authors (see [5, 6, 9, 10, 14, 25]).

We recall a fundamental result in fixed point theory.

Theorem 1 ([2, 8]) *Let (X, d) be a complete generalized metric space and let $J : X \to X$ be a strictly contractive mapping with Lipschitz constant $\alpha < 1$. Then for each given element $x \in X$, either*

$$d(J^n x, J^{n+1} x) = \infty$$

for all nonnegative integers n or there exists a positive integer n_0 such that

(1) $d(J^n x, J^{n+1} x) < \infty, \qquad \forall n \geq n_0;$
(2) *the sequence $\{J^n x\}$ converges to a fixed point y^* of J;*
(3) *y^* is the unique fixed point of J in the set $Y = \{y \in X \mid d(J^{n_0} x, y) < \infty\}$;*
(4) $d(y, y^*) \leq \frac{1}{1-\alpha} d(y, Jy)$ *for all $y \in Y$.*

In 1996, Isac and Rassias [13] were the first to provide applications of stability theory of functional equations for the proof of new fixed point theorems with applications. By using fixed point methods, the stability problems of several functional equations have been extensively investigated by a number of authors (see [3, 4, 18–21]).

This paper is organized as follows: In Sections 2 and 3, we solve the additive (ρ_1, ρ_2)-functional inequality (1) and prove the Hyers–Ulam stability of the additive (ρ_1, ρ_2)-functional inequality (1) in Banach spaces by using the fixed point method and by using the direct method, respectively. In Sections 4 and 5, we solve the additive (ρ_1, ρ_2)-functional inequality (1) and prove the Hyers–Ulam stability of the additive (ρ_1, ρ_2)-functional inequality (2) in Banach spaces by using the fixed point method and by using the direct method, respectively.

Throughout this paper, let X be a real or complex normed space and Y a complex Banach space.

2 Additive (ρ_1, ρ_2)-Functional Inequality (1): A Fixed Point Method

Assume that ρ_1 and ρ_2 are fixed complex numbers with $|\rho_1| + |\rho_2| > 1$.

In this section, we solve and investigate the additive (ρ_1, ρ_2)-functional inequality (1) in complex Banach spaces.

Lemma 1 *If a mapping $f : X \to Y$ satisfies*

$$\begin{aligned}\|f(x+y) - f(x) - f(y)\| &\geq \|\rho_1(f(x-y) - f(x) + f(y))\| \\ &\quad + \|\rho_2(f(y-x) - f(y) + f(x))\| \end{aligned} \tag{3}$$

for all $x, y \in X$, then $f : X \to Y$ is additive.

Proof Assume that $f : X \to Y$ satisfies (3).

Letting $x = y = 0$ in (3), we get $\|f(0)\| \geq (|\rho_1|+|\rho_2|)\|f(0)\|$ and so $f(0) = 0$, since $|\rho_1| + |\rho_2| > 1$.

Letting $y = 0$ in (3), we get

$$0 \geq |\rho_2|\|f(-x) + f(x)\|$$

for all $x \in X$. Thus $f(-x) = -f(x)$ for all $x \in X$. So

$$\|f(x+y) - f(x) - f(y)\| \geq (|\rho_1| + |\rho_2|)\|f(x-y) - f(x) + f(y)\| \tag{4}$$

and so

$$\|f(x-y) - f(x) + f(y)\| \geq (|\rho_1| + |\rho_2|)\|f(x+y) - f(x) - f(y)\| \tag{5}$$

for all $x, y \in X$. It follows from (4) and (5) that

$$\|f(x+y) - f(x) - f(y)\| \geq (|\rho_1| + |\rho_2|)^2\|f(x+y) - f(x) - f(y)\|$$

for all $x, y \in X$. So $f(x+y) = f(x)+f(y)$ for all $x, y \in X$, since $|\rho_1|+|\rho_2| > 1$. Thus f is additive.

Using the fixed point method, we prove the Hyers–Ulam stability of the additive (ρ_1, ρ_2)-functional inequality (3) in complex Banach spaces.

Theorem 2 *Let $\varphi : X^2 \to [0, \infty)$ be a function such that there exists an $L < 1$ with*

$$\varphi\left(\frac{x}{2}, \frac{y}{2}\right) \leq \frac{L}{2}\varphi(x, y) \tag{6}$$

for all $x, y \in X$. Let $f : X \to Y$ be an odd mapping satisfying

$$\|\rho_1(f(x-y) - f(x) + f(y))\| + \|\rho_2(f(y-x) - f(y) + f(x))\| \\ \leq \|f(x+y) - f(x) - f(y)\| + \varphi(x, y) \quad (7)$$

for all $x, y \in X$. Then there exists a unique additive mapping $A : X \to Y$ such that

$$\|f(x) - A(x)\| \leq \frac{L}{2(1-L)(|\rho_1| + |\rho_2|)}\varphi(x, -x)$$

for all $x \in X$.

Proof Letting $y = -x$ in (7), we get

$$\|f(2x) - 2f(x)\| \leq \frac{1}{|\rho_1| + |\rho_2|}\varphi(x, -x) \quad (8)$$

for all $x \in X$.

Consider the set

$$S := \{h : X \to Y, \ \ h(0) = 0\}$$

and introduce the generalized metric on S:

$$d(g, h) = \inf\{\mu \in \mathbf{R}_+ : \|g(x) - h(x)\| \leq \mu\varphi(x, -x), \ \ \forall x \in X\},$$

where, as usual, $\inf \phi = +\infty$. It is easy to show that (S, d) is complete (see [15]).

Now we consider the linear mapping $J : S \to S$ such that

$$Jg(x) := 2g\left(\frac{x}{2}\right)$$

for all $x \in X$.

Let $g, h \in S$ be given such that $d(g, h) = \varepsilon$. Then

$$\|g(x) - h(x)\| \leq \varepsilon\varphi(x, -x)$$

for all $x \in X$. Hence

$$\|Jg(x) - Jh(x)\| = \left\|2g\left(\frac{x}{2}\right) - 2h\left(\frac{x}{2}\right)\right\| \leq 2\varepsilon\varphi\left(\frac{x}{2}, -\frac{x}{2}\right) \\ \leq 2\varepsilon\frac{L}{2}\varphi(x, -x) = L\varepsilon\varphi(x, -x)$$

for all $x \in X$. So $d(g, h) = \varepsilon$ implies that $d(Jg, Jh) \leq L\varepsilon$. This means that

$$d(Jg, Jh) \leq Ld(g, h)$$

for all $g, h \in S$.

It follows from (8) that

$$\left\| f(x) - 2f\left(\frac{x}{2}\right)\right\| \leq \frac{1}{|\rho_1| + |\rho_2|}\varphi\left(\frac{x}{2}, -\frac{x}{2}\right) \leq \frac{L}{2(|\rho_1| + |\rho_2|)}\varphi(x, -x)$$

for all $x \in X$. So $d(f, Jf) \leq \frac{L}{2(|\rho_1|+|\rho_2|)}$.

By Theorem 1, there exists a mapping $A : X \to Y$ satisfying the following:

(1) A is a fixed point of J, i.e.,

$$A(x) = 2A\left(\frac{x}{2}\right) \tag{9}$$

for all $x \in X$. The mapping A is a unique fixed point of J in the set

$$M = \{g \in S : d(f, g) < \infty\}.$$

This implies that A is a unique mapping satisfying (9) such that there exists a $\mu \in (0, \infty)$ satisfying

$$\|f(x) - A(x)\| \leq \mu\varphi(x, -x)$$

for all $x \in X$;

(2) $d(J^l f, A) \to 0$ as $l \to \infty$. This implies the equality

$$\lim_{l\to\infty} 2^n f\left(\frac{x}{2^n}\right) = A(x)$$

for all $x \in X$;

(3) $d(f, A) \leq \frac{1}{1-L}d(f, Jf)$, which implies

$$\|f(x) - A(x)\| \leq \frac{L}{2(1 - L)(|\rho_1| + |\rho_2|)}\varphi(x, -x)$$

for all $x \in X$.

It follows from (6) and (7) that

$$\begin{aligned}
&\|A(x + y) - A(x) - A(y)\| \\
&= \lim_{n\to\infty} 2^n \left\| f\left(\frac{x + y}{2^n}\right) - f\left(\frac{x}{2^n}\right) - f\left(\frac{y}{2^n}\right)\right\| + \lim_{n\to\infty} 2^n\varphi\left(\frac{x}{2^n}, \frac{y}{2^n}\right) \\
&\geq \lim_{n\to\infty} 2^n|\rho_1| \left\| f\left(\frac{x - y}{2^n}\right) - f\left(\frac{x}{2^n}\right) + f\left(\frac{y}{2^n}\right)\right\|
\end{aligned}$$

$$+\lim_{n\to\infty} 2^n|\rho_2| \left\| f\left(\frac{y-x}{2^n}\right) - f\left(\frac{y}{2^n}\right) + f\left(\frac{x}{2^n}\right)\right\|$$
$$= \|\rho_1(A(x-y) - A(x) + A(y))\| + \|\rho_2\left(A(y-x) - A(y) + A(x)\right)\|$$

for all $x, y \in X$. So

$$\|A\,(x+y) - A(x) - A(y)\| \geq \|\rho_1(A(x-y) - A(x) + A(y))\| + \|\rho_2\left(A(y-x) - A(y) + A(x)\right)\|$$

for all $x, y \in X$. By Lemma 1, the mapping $A : X \to Y$ is additive.

Corollary 1 *Let $r > 1$ and θ be nonnegative real numbers, and let $f : X \to Y$ be an odd mapping satisfying*

$$\|\rho_1(f(x-y) - f(x) + f(y))\| + \|\rho_2\left(f(y-x) - f(y) + f(x)\right)\| \leq \|f\,(x+y) - f(x) - f(y)\| + \theta(\|x\|^r + \|y\|^r) \tag{10}$$

for all $x, y \in X$. Then there exists a unique additive mapping $A : X \to Y$ such that

$$\|f(x) - A(x)\| \leq \frac{2\theta}{(2^r - 2)(|\rho_1| + |\rho_2|)}\|x\|^r$$

for all $x \in X$.

Proof The proof follows from Theorem 2 by taking $\varphi(x, y) = \theta(\|x\|^r + \|y\|^r)$ for all $x, y \in X$. Choosing $L = 2^{1-r}$, we obtain the desired result.

Theorem 3 *Let $\varphi : X^2 \to [0, \infty)$ be a function such that there exists an $L < 1$ with*

$$\varphi\,(x, y) \leq 2L\varphi\left(\frac{x}{2}, \frac{y}{2}\right) \tag{11}$$

for all $x, y \in X$. Let $f : X \to Y$ be an odd mapping satisfying (7). *Then there exists a unique additive mapping $A : X \to Y$ such that*

$$\|f(x) - A(x)\| \leq \frac{1}{2(1-L)(|\rho_1| + |\rho_2|)}\varphi\,(x, -x)$$

for all $x \in X$.

Proof Let (S, d) be the generalized metric space defined in the proof of Theorem 2.
Now we consider the linear mapping $J : S \to S$ such that

$$Jg(x) := \frac{1}{2}g\,(2x)$$

for all $x \in X$.

It follows from (8) that

$$\left\| f(x) - \frac{1}{2} f(2x) \right\| \le \frac{1}{2(|\rho_1| + |\rho_2|)} \varphi(x, -x)$$

for all $x \in X$.

The rest of the proof is similar to the proof of Theorem 2.

Corollary 2 *Let $r < 1$ and θ be positive real numbers, and let $f : X \to Y$ be an odd mapping satisfying* (10)*. Then there exists a unique additive mapping $A : X \to Y$ such that*

$$\| f(x) - A(x) \| \le \frac{2\theta}{(2 - 2^r)(|\rho_1| + |\rho_2|)} \|x\|^r$$

for all $x \in X$.

Proof The proof follows from Theorem 3 by taking $\varphi(x, y) = \theta(\|x\|^r + \|y\|^r)$ for all $x, y \in X$. Choosing $L = 2^{r-1}$, we obtain the desired result.

Remark 1 If ρ_1 and ρ_2 are real numbers such that $|\rho_1| + |\rho_2| > 1$ and Y is a real Banach space, then all the assertions in this section remain valid.

3 Additive (ρ_1, ρ_2)-Functional Inequality (1): A Direct Method

In this section, we prove the Hyers–Ulam stability of the additive (ρ_1, ρ_2)-functional inequality (3) in complex Banach spaces by using the direct method.

Theorem 4 *Let $\varphi : X^2 \to [0, \infty)$ be a function such that*

$$\Psi(x, y) := \sum_{j=1}^{\infty} 2^j \varphi\left(\frac{x}{2^j}, \frac{y}{2^j}\right) < \infty \tag{12}$$

for all $x, y \in X$. Let $f : X \to Y$ be an odd mapping satisfying (7)*. Then there exists a unique additive mapping $A : X \to Y$ such that*

$$\| f(x) - A(x) \| \le \frac{1}{2(|\rho_1| + |\rho_2|)} \Psi(x, -x) \tag{13}$$

for all $x \in X$.

Proof Letting $y = -x$ in (7), we get

$$\|f(2x) - 2f(x)\| \le \frac{1}{|\rho_2| + |\rho_2|}\varphi(x, -x) \tag{14}$$

and so

$$\left\|f(x) - 2f\left(\frac{x}{2}\right)\right\| \le \frac{1}{|\rho_2| + |\rho_2|}\varphi\left(\frac{x}{2}, -\frac{x}{2}\right)$$

for all $x \in X$. Thus

$$\begin{aligned}\left\|2^l f\left(\frac{x}{2^l}\right) - 2^m f\left(\frac{x}{2^m}\right)\right\| &\le \sum_{j=l}^{m-1}\left\|2^j f\left(\frac{x}{2^j}\right) - 2^{j+1} f\left(\frac{x}{2^{j+1}}\right)\right\| \\ &\le \sum_{j=l}^{m-1}\frac{2^j}{|\rho_2| + |\rho_2|}\varphi\left(\frac{x}{2^{j+1}}, -\frac{x}{2^{j+1}}\right)\end{aligned} \tag{15}$$

for all nonnegative integers m and l with $m > l$ and all $x \in X$. It follows from (15) that the sequence $\{2^k f(\frac{x}{2^k})\}$ is Cauchy for all $x \in X$. Since Y is a Banach space, the sequence $\{2^k f(\frac{x}{2^k})\}$ converges. So one can define the mapping $A : X \to Y$ by

$$A(x) := \lim_{k\to\infty} 2^k f\left(\frac{x}{2^k}\right)$$

for all $x \in X$. Moreover, letting $l = 0$ and passing to the limit $m \to \infty$ in (15), we get (13).

It follows from (7) and (12) that

$$\begin{aligned}&\|A(x+y) - A(x) - A(y)\| \\ &= \lim_{n\to\infty} 2^n \left\|f\left(\frac{x+y}{2^n}\right) - f\left(\frac{x}{2^n}\right) - f\left(\frac{y}{2^n}\right)\right\| + \lim_{n\to\infty} 2^n \varphi\left(\frac{x}{2^n}, \frac{y}{2^n}\right) \\ &\ge \lim_{n\to\infty} 2^n |\rho_1| \left\|f\left(\frac{x-y}{2^n}\right) - f\left(\frac{x}{2^n}\right) + f\left(\frac{y}{2^n}\right)\right\| \\ &+ \lim_{n\to\infty} 2^n |\rho_2| \left\|f\left(\frac{y-x}{2^n}\right) - f\left(\frac{y}{2^n}\right) + f\left(\frac{x}{2^n}\right)\right\| \\ &= \|\rho_1(A(x-y) - A(x) + A(y))\| + \|\rho_2(A(y-x) - A(y) + A(x))\|\end{aligned}$$

for all $x, y \in X$. So

$$\begin{aligned}\|A(x+y) - A(x) - A(y)\| &\ge \|\rho_1(A(x-y) - A(x) + A(y))\| \\ &+ \|\rho_2(A(y-x) - A(y) + A(x))\|\end{aligned}$$

for all $x, y \in X$. By Lemma 1, the mapping $A : X \to Y$ is additive.

Now, let $T : X \to Y$ be another additive mapping satisfying (13). Then we have

$$\begin{aligned}\|A(x) - T(x)\| &= \left\|2^q A\left(\frac{x}{2^q}\right) - 2^q T\left(\frac{x}{2^q}\right)\right\| \\ &\leq \left\|2^q A\left(\frac{x}{2^q}\right) - 2^q f\left(\frac{x}{2^q}\right)\right\| + \left\|2^q T\left(\frac{x}{2^q}\right) - 2^q f\left(\frac{x}{2^q}\right)\right\| \\ &\leq \frac{2^q}{|\rho_1| + |\rho_2|}\Psi\left(\frac{x}{2^q}, -\frac{x}{2^q}\right),\end{aligned}$$

which tends to zero as $q \to \infty$ for all $x \in X$. So we can conclude that $A(x) = T(x)$ for all $x \in X$. This proves the uniqueness of A.

Corollary 3 *Let $r > 1$ and θ be nonnegative real numbers, and let $f : X \to Y$ be an odd mapping satisfying* (10). *Then there exists a unique additive mapping $A : X \to Y$ such that*

$$\|f(x) - A(x)\| \leq \frac{2\theta}{(2^r - 2)(|\rho_1| + |\rho_2|)}\|x\|^r$$

for all $x \in X$.

Theorem 5 *Let $\varphi : X^2 \to [0, \infty)$ be a function and let $f : X \to Y$ be an odd mapping satisfying* (7) *and*

$$\Psi(x, y) := \sum_{j=0}^{\infty} \frac{1}{2^j}\varphi(2^j x, 2^j y) < \infty \tag{16}$$

for all $x, y \in X$. Then there exists a unique additive mapping $A : X \to Y$ such that

$$\|f(x) - A(x)\| \leq \frac{1}{2(|\rho_1| + |\rho_2|)}\Psi(x, -x)$$

for all $x \in X$.

Proof It follows from (14) that

$$\left\|f(x) - \frac{1}{2}f(2x)\right\| \leq \frac{1}{2(|\rho_1| + |\rho_2|)}\varphi(x, -x)$$

for all $x \in X$. Hence

$$\begin{aligned}\left\|\frac{1}{2^l}f(2^l x) - \frac{1}{2^m}f(2^m x)\right\| &\leq \sum_{j=l}^{m-1}\left\|\frac{1}{2^j}f\left(2^j x\right) - \frac{1}{2^{j+1}}f\left(2^{j+1}x\right)\right\| \\ &\leq \sum_{j=l}^{m-1}\frac{1}{2^{j+1}(|\rho_1| + |\rho_2|)}\varphi(2^j x, -2^j x) \qquad (17)\end{aligned}$$

for all nonnegative integers m and l with $m > l$ and all $x \in X$. It follows from (17) that the sequence $\{\frac{1}{2^n} f(2^n x)\}$ is a Cauchy sequence for all $x \in X$. Since Y is complete, the sequence $\{\frac{1}{2^n} f(2^n x)\}$ converges. So one can define the mapping $A : X \to Y$ by

$$A(x) := \lim_{n\to\infty} \frac{1}{2^n} f(2^n x)$$

for all $x \in X$. Moreover, letting $l = 0$ and passing to the limit $m \to \infty$ in (17), we get (17).

The rest of the proof is similar to the proof of Theorem 4.

Corollary 4 *Let $r < 1$ and θ be positive real numbers, and let $f : X \to Y$ be an odd mapping satisfying* (10). *Then there exists a unique additive mapping $A : X \to Y$ such that*

$$\|f(x) - A(x)\| \leq \frac{2\theta}{(2-2^r)(|\rho_1| + |\rho_2|)} \|x\|^r$$

for all $x \in X$.

4 Additive (ρ_1, ρ_2)-Functional Inequality (2): A Fixed Point Method

From now on, assume that ρ_1 and ρ_2 are fixed complex numbers with $1 + |\rho_1| > |\rho_2| > 1$.

In this section, we solve and investigate the additive (ρ_1, ρ_2)-functional inequality (2) in complex Banach spaces.

Lemma 2 *If an odd mapping $f : X \to Y$ satisfies*

$$\begin{aligned} \|f(x-y) - f(x) + f(y)\| &\geq \|\rho_1(f(x+y) - f(x) - f(y))\| \\ &\quad + \|\rho_2(f(y-x) - f(y) + f(x))\| \end{aligned} \tag{18}$$

for all $x, y, z \in X$, then $f : X \to Y$ is additive.

Proof It follows from (18) and the oddness of f that

$$(1 - |\rho_2|)\|f(x-y) - f(x) + f(y)\| \geq |\rho_1|\|f(x+y) - f(x) - f(y)\| \tag{19}$$

and so

$$(1 - |\rho_2|)\|f(x+y) - f(x) - f(y)\| \geq |\rho_1|\|f(x-y) - f(x) + f(y)\| \tag{20}$$

for all $x, y \in X$. It follows from (19) and (20) that

$$\|f(x+y) - f(x) - f(y)\| \geq \frac{|\rho_1|^2}{(1-|\rho_2|)^2}\|f(x+y) - f(x) - f(y)\|$$

for all $x, y \in X$. So $f(x+y) = f(x) + f(y)$ for all $x, y \in X$, since $1 + |\rho_1| > |\rho_2| > 1$. Thus f is additive.

Using the fixed point method, we prove the Hyers–Ulam stability of the additive (ρ_1, ρ_2)-functional inequality (18) in complex Banach spaces.

Theorem 6 *Let $\varphi : X^2 \to [0, \infty)$ be a function such that there exists an $L < 1$ with*

$$\varphi\left(\frac{x}{2}, \frac{y}{2}\right) \leq \frac{L}{2}\varphi(x, y) \tag{21}$$

for all $x, y \in X$. Let $f : X \to Y$ be an odd mapping satisfying

$$\begin{aligned} &\|\rho_1(f(x+y) - f(x) - f(y))\| + \|\rho_2(f(y-x) - f(y) + f(x))\| \\ &\leq \|f(x-y) - f(x) + f(y)\| + \varphi(x, y) \end{aligned} \tag{22}$$

for all $x, y \in X$. Then there exists a unique additive mapping $A : X \to Y$ such that

$$\|f(x) - A(x)\| \leq \frac{L}{2(1-L)(|\rho_2| - 1)}\varphi(x, -x)$$

for all $x \in X$.

Proof Letting $y = -x$ in (22), we get

$$\|f(2x) - 2f(x)\| \leq \frac{1}{|\rho_2| - 1}\varphi(x, -x) \tag{23}$$

for all $x \in X$.

Consider the set

$$S := \{h : X \to Y,\ \ h(0) = 0\}$$

and introduce the generalized metric on S:

$$d(g, h) = \inf\{\mu \in \mathbf{R}_+ : \|g(x) - h(x)\| \leq \mu\varphi(x, -x),\ \ \forall x \in X\},$$

where, as usual, $\inf \phi = +\infty$. It is easy to show that (S, d) is complete (see [15]).

Now we consider the linear mapping $J : S \to S$ such that

$$Jg(x) := 2g\left(\frac{x}{2}\right)$$

for all $x \in X$.

Let $g, h \in S$ be given such that $d(g, h) = \varepsilon$. Then

$$\|g(x) - h(x)\| \le \varepsilon\varphi(x, -x)$$

for all $x \in X$. Hence

$$\begin{aligned}\|Jg(x) - Jh(x)\| &= \left\|2g\left(\frac{x}{2}\right) - 2h\left(\frac{x}{2}\right)\right\| \le 2\varepsilon\varphi\left(\frac{x}{2}, -\frac{x}{2}\right) \\ &\le 2\varepsilon\frac{L}{2}\varphi(x, -x) = L\varepsilon\varphi(x, -x)\end{aligned}$$

for all $x \in X$. So $d(g, h) = \varepsilon$ implies that $d(Jg, Jh) \le L\varepsilon$. This means that

$$d(Jg, Jh) \le Ld(g, h)$$

for all $g, h \in S$.

It follows from (23) that

$$\left\|f(x) - 2f\left(\frac{x}{2}\right)\right\| \le \frac{1}{|\rho_2| - 1}\varphi\left(\frac{x}{2}, -\frac{x}{2}\right) \le \frac{L}{2(|\rho_2| - 1)}\varphi(x, -x)$$

for all $x \in X$. So $d(f, Jf) \le \frac{L}{2(|\rho_2|-1)}$.

By Theorem 1, there exists a mapping $A : X \to Y$ satisfying the following:

(1) A is a fixed point of J, i.e.,

$$A(x) = 2A\left(\frac{x}{2}\right) \tag{24}$$

for all $x \in X$. The mapping A is a unique fixed point of J in the set

$$M = \{g \in S : d(f, g) < \infty\}.$$

This implies that A is a unique mapping satisfying (24) such that there exists a $\mu \in (0, \infty)$ satisfying

$$\|f(x) - A(x)\| \le \mu\varphi(x, -x)$$

for all $x \in X$;

(2) $d(J^l f, A) \to 0$ as $l \to \infty$. This implies the equality

$$\lim_{l\to\infty} 2^n f\left(\frac{x}{2^n}\right) = A(x)$$

for all $x \in X$;

(3) $d(f, A) \le \frac{1}{1-L} d(f, Jf)$, which implies

$$\|f(x) - A(x)\| \le \frac{L}{2(1-L)(|\rho_2|-1)} \varphi(x, -x)$$

for all $x \in X$.

It follows from (21) and (22) that

$$\begin{aligned}
&\|A(x-y) - A(x) + A(y)\| \\
&= \lim_{n\to\infty} 2^n \left\| f\left(\frac{x-y}{2^n}\right) - f\left(\frac{x}{2^n}\right) + f\left(\frac{y}{2^n}\right)\right\| + \lim_{n\to\infty} 2^n \varphi\left(\frac{x}{2^n}, \frac{y}{2^n}\right) \\
&\ge \lim_{n\to\infty} 2^n |\rho_1| \left\| f\left(\frac{x+y}{2^n}\right) - f\left(\frac{x}{2^n}\right) - f\left(\frac{y}{2^n}\right)\right\| \\
&+ \lim_{n\to\infty} 2^n |\rho_2| \left\| f\left(\frac{y-x}{2^n}\right) - f\left(\frac{y}{2^n}\right) + f\left(\frac{x}{2^n}\right)\right\| \\
&= \|\rho_1(A(x+y) - A(x) - A(y))\| + \|\rho_2(A(y-x) - A(y) + A(x))\|
\end{aligned}$$

for all $x, y \in X$. So

$$\begin{aligned}
\|A(x-y) - A(x) + A(y)\| &\ge \|\rho_1(A(x+y) - A(x) - A(y))\| \\
&\quad + \|\rho_2(A(y-x) - A(y) + A(x))\|
\end{aligned}$$

for all $x, y \in X$. By Lemma 2, the mapping $A : X \to Y$ is additive.

Corollary 5 *Let $r > 1$ and θ be nonnegative real numbers, and let $f : X \to Y$ be an odd mapping satisfying*

$$\begin{aligned}
&\|\rho_1(f(x+y) - f(x) - f(y))\| + \|\rho_2(f(y-x) - f(y) + f(x))\| \\
&\le \|f(x-y) - f(x) + f(y)\| + \theta(\|x\|^r + \|y\|^r) \qquad (25)
\end{aligned}$$

for all $x, y \in X$. Then there exists a unique additive mapping $A : X \to Y$ such that

$$\|f(x) - A(x)\| \le \frac{2\theta}{(2^r-2)(|\rho_2|-1)} \|x\|^r$$

for all $x \in X$.

Proof The proof follows from Theorem 6 by taking $\varphi(x, y) = \theta(\|x\|^r + \|y\|^r)$ for all $x, y \in X$. Choosing $L = 2^{1-r}$, we obtain the desired result.

Theorem 7 *Let $\varphi : X^2 \to [0, \infty)$ be a function such that there exists an $L < 1$ with*

$$\varphi(x, y) \le 2L\varphi\left(\frac{x}{2}, \frac{y}{2}\right) \tag{26}$$

for all $x, y \in X$. Let $f : X \to Y$ be an odd mapping satisfying (22). *Then there exists a unique additive mapping $A : X \to Y$ such that*

$$\|f(x) - A(x)\| \le \frac{1}{2(1-L)(|\rho_2| - 1)}\varphi(x, -x)$$

for all $x \in X$.

Proof Let (S, d) be the generalized metric space defined in the proof of Theorem 6.

Now we consider the linear mapping $J : S \to S$ such that

$$Jg(x) := \frac{1}{2}g(2x)$$

for all $x \in X$.

It follows from (23) that

$$\left\| f(x) - \frac{1}{2}f(2x) \right\| \le \frac{1}{2(|\rho_2| - 1)}\varphi(x, -x)$$

for all $x \in X$.

The rest of the proof is similar to the proof of Theorem 6.

Corollary 6 *Let $r < 1$ and θ be positive real numbers, and let $f : X \to Y$ be an odd mapping satisfying* (25). *Then there exists a unique additive mapping $A : X \to Y$ such that*

$$\|f(x) - A(x)\| \le \frac{2\theta}{(2 - 2^r)(|\rho_2| - 1)}\|x\|^r$$

for all $x \in X$.

Proof The proof follows from Theorem 7 by taking $\varphi(x, y) = \theta(\|x\|^r + \|y\|^r)$ for all $x, y \in X$. Choosing $L = 2^{r-1}$, we obtain the desired result.

Remark 2 If ρ is a real number such that $|\rho_1| + |\rho_2| < 1$ and Y is a real Banach space, then all the assertions in this section remain valid.

5 Additive (ρ_1, ρ_2)-Functional Inequality (2): A Direct Method

In this section, we prove the Hyers–Ulam stability of the additive (ρ_1, ρ_2)-functional inequality (18) in complex Banach spaces by using the direct method.

Theorem 8 *Let $\varphi : X^2 \to [0, \infty)$ be a function such that*

$$\Psi(x, y) := \sum_{j=1}^{\infty} 2^j \varphi\left(\frac{x}{2^j}, \frac{y}{2^j}\right) < \infty \tag{27}$$

for all $x, y \in X$. Let $f : X \to Y$ be an odd mapping satisfying (22)*. Then there exists a unique additive mapping $A : X \to Y$ such that*

$$\|f(x) - A(x)\| \le \frac{1}{2(|\rho_2| - 1)} \Psi(x, -x) \tag{28}$$

for all $x \in X$.

Proof Letting $y = -x$ in (22), we get

$$\|f(2x) - 2f(x)\| \le \frac{1}{|\rho_2| - 1} \varphi(x, -x) \tag{29}$$

and so

$$\left\|f(x) - 2f\left(\frac{x}{2}\right)\right\| \le \frac{1}{|\rho_2| - 1} \varphi\left(\frac{x}{2}, -\frac{x}{2}\right)$$

for all $x \in X$. Thus

$$\begin{aligned} \left\|2^l f\left(\frac{x}{2^l}\right) - 2^m f\left(\frac{x}{2^m}\right)\right\| &\le \sum_{j=l}^{m-1} \left\|2^j f\left(\frac{x}{2^j}\right) - 2^{j+1} f\left(\frac{x}{2^{j+1}}\right)\right\| \\ &\le \sum_{j=l}^{m-1} \frac{2^j}{|\rho_2| - 1} \varphi\left(\frac{x}{2^{j+1}}, -\frac{x}{2^{j+1}}\right) \end{aligned} \tag{30}$$

for all nonnegative integers m and l with $m > l$ and all $x \in X$. It follows from (30) that the sequence $\{2^k f(\frac{x}{2^k})\}$ is Cauchy for all $x \in X$. Since Y is a Banach space, the sequence $\{2^k f(\frac{x}{2^k})\}$ converges. So one can define the mapping $A : X \to Y$ by

$$A(x) := \lim_{k \to \infty} 2^k f\left(\frac{x}{2^k}\right)$$

for all $x \in X$. Moreover, letting $l = 0$ and passing to the limit $m \to \infty$ in (30), we get (28).

It follows from (30) and (27) that

$$\begin{aligned}
&\|A(x-y) - A(x) + A(y)\| \\
&= \lim_{n\to\infty} 2^n \left\| f\left(\frac{x-y}{2^n}\right) - f\left(\frac{x}{2^n}\right) + f\left(\frac{y}{2^n}\right)\right\| + \lim_{n\to\infty} 2^n \varphi\left(\frac{x}{2^n}, \frac{y}{2^n}\right) \\
&\geq \lim_{n\to\infty} 2^n |\rho_1| \left\| f\left(\frac{x+y}{2^n}\right) - f\left(\frac{x}{2^n}\right) - f\left(\frac{y}{2^n}\right)\right\| \\
&+ \lim_{n\to\infty} 2^n |\rho_2| \left\| f\left(\frac{y-x}{2^n}\right) - f\left(\frac{y}{2^n}\right) + f\left(\frac{x}{2^n}\right)\right\| \\
&= \|\rho_1(A(x+y) - A(x) - A(y))\| + \|\rho_2(A(y-x) - A(y) + A(x))\|
\end{aligned}$$

for all $x, y \in X$. So

$$\begin{aligned}
\|A(x-y) - A(x) + A(y)\| \geq \|\rho_1(A(x+y) - A(x) - A(y))\| \\
+ \|\rho_2(A(y-x) - A(y) + A(x))\|
\end{aligned}$$

for all $x, y \in X$. By Lemma 2, the mapping $A : X \to Y$ is additive.

Now, let $T : X \to Y$ be another additive mapping satisfying (28). Then we have

$$\begin{aligned}
\|A(x) - T(x)\| &= \left\|2^q A\left(\frac{x}{2^q}\right) - 2^q T\left(\frac{x}{2^q}\right)\right\| \\
&\leq \left\|2^q A\left(\frac{x}{2^q}\right) - 2^q f\left(\frac{x}{2^q}\right)\right\| + \left\|2^q T\left(\frac{x}{2^q}\right) - 2^q f\left(\frac{x}{2^q}\right)\right\| \\
&\leq \frac{2^q}{|\rho_2| - 1} \Psi\left(\frac{x}{2^q}, -\frac{x}{2^q}\right),
\end{aligned}$$

which tends to zero as $q \to \infty$ for all $x \in X$. So we can conclude that $A(x) = T(x)$ for all $x \in X$. This proves the uniqueness of A.

Corollary 7 *Let $r > 1$ and θ be nonnegative real numbers, and let $f : X \to Y$ be an odd mapping satisfying (25). Then there exists a unique additive mapping $A : X \to Y$ such that*

$$\|f(x) - A(x)\| \leq \frac{2\theta}{(2^r - 2)(|\rho_2| - 1)} \|x\|^r$$

for all $x \in X$.

Theorem 9 *Let $\varphi : X^2 \to [0, \infty)$ be a function and let $f : X \to Y$ be an odd mapping satisfying $f(0) = 0$, (22) and*

$$\Psi(x, y) := \sum_{j=0}^{\infty} \frac{1}{2^j}\varphi(2^j x, 2^j y) < \infty \tag{31}$$

for all $x, y \in X$. Then there exists a unique additive mapping $A : X \to Y$ such that

$$\|f(x) - A(x)\| \le \frac{1}{2(|\rho_2| - 1)}\Psi(x, -x)$$

for all $x \in X$.

Proof It follows from (29) that

$$\left\| f(x) - \frac{1}{2}f(2x)\right\| \le \frac{1}{2(|\rho_2| - 1)}\varphi(x, -x)$$

for all $x \in X$. Hence

$$\begin{aligned}\left\|\frac{1}{2^l}f(2^l x) - \frac{1}{2^m}f(2^m x)\right\| &\le \sum_{j=l}^{m-1}\left\|\frac{1}{2^j}f\left(2^j x\right) - \frac{1}{2^{j+1}}f\left(2^{j+1}x\right)\right\| \\ &\le \sum_{j=l}^{m-1}\frac{1}{2^{j+1}(|\rho_2| - 1)}\varphi(2^j x, -2^j x) \end{aligned}\tag{32}$$

for all nonnegative integers m and l with $m > l$ and all $x \in X$. It follows from (32) that the sequence $\{\frac{1}{2^n}f(2^n x)\}$ is a Cauchy sequence for all $x \in X$. Since Y is complete, the sequence $\{\frac{1}{2^n}f(2^n x)\}$ converges. So one can define the mapping $A : X \to Y$ by

$$A(x) := \lim_{n\to\infty}\frac{1}{2^n}f(2^n x)$$

for all $x \in X$. Moreover, letting $l = 0$ and passing to the limit $m \to \infty$ in (32), we get (32).

The rest of the proof is similar to the proof of Theorem 8.

Corollary 8 *Let $r < 1$ and θ be positive real numbers, and let $f : X \to Y$ be an odd mapping satisfying* (25). *Then there exists a unique additive mapping $A : X \to Y$ such that*

$$\|f(x) - A(x)\| \le \frac{2\theta}{(2 - 2^r)(|\rho_2| - 1)}\|x\|^r$$

for all $x \in X$.

Acknowledgments C. Park was supported by Basic Science Research Program through the National Research Foundation of Korea funded by the Ministry of Education, Science and Technology (NRF-2017R1D1A1B04032937).

References

1. T. Aoki, On the stability of the linear transformation in Banach spaces. J. Math. Soc. Japan **2**, 64–66 (1950)
2. L. Cădariu, V. Radu, Fixed points and the stability of Jensen's functional equation. J. Inequal. Pure Appl. Math. **4**(1), Art. ID 4 (2003)
3. L. Cădariu, V. Radu, On the stability of the Cauchy functional equation: a fixed point approach. Grazer Math. Ber. **346**, 43–52 (2004)
4. L. Cădariu, V. Radu, Fixed point methods for the generalized stability of functional equations in a single variable. J. Fixed Point Theory Appl. **2008**, Art. ID 749392 (2008)
5. L. Cădariu, L. Găvruta, P. Găvruta, On the stability of an affine functional equation. J. Nonlinear Sci. Appl. **6**, 60–67 (2013)
6. A. Chahbi, N. Bounader, On the generalized stability of d'Alembert functional equation. J. Nonlinear Sci. Appl. **6**, 198–204 (2013)
7. P.W. Cholewa, Remarks on the stability of functional equations. Aequationes Math. **27**, 76–86 (1984)
8. J. Diaz, B. Margolis, A fixed point theorem of the alternative for contractions on a generalized complete metric space. Bull. Am. Math. Soc. **74**, 305–309 (1968)
9. N. Eghbali, J.M. Rassias, M. Taheri, On the stability of a k-cubic functional equation in intuitionistic fuzzy n-normed spaces. Results Math. **70**, 233–248 (2016)
10. G.Z. Eskandani, P. Găvruta, Hyers-Ulam-Rassias stability of pexiderized Cauchy functional equation in 2-Banach spaces. J. Nonlinear Sci. Appl. **5**, 459–465 (2012)
11. P. Găvruta, A generalization of the Hyers-Ulam-Rassias stability of approximately additive mappings. J. Math. Anal. Appl. **184**, 431–436 (1994)
12. D.H. Hyers, On the stability of the linear functional equation. Proc. Nat. Acad. Sci. U.S.A. **27**, 222–224 (1941)
13. G. Isac, Th.M. Rassias, Stability of ψ-additive mappings: applications to nonlinear analysis. Int. J. Math. Math. Sci. **19**, 219–228 (1996)
14. H. Khodaei, On the stability of additive, quadratic, cubic and quartic set-valued functional equations. Results Math. **68**, 1–10 (2015)
15. D. Miheţ, V. Radu, On the stability of the additive Cauchy functional equation in random normed spaces. J. Math. Anal. Appl. **343**, 567–572 (2008)
16. C. Park, Additive ρ-functional inequalities and equations. J. Math. Inequal. **9**, 17–26 (2015)
17. C. Park, Additive ρ-functional inequalities in non-Archimedean normed spaces. J. Math. Inequal. **9**, 397–407 (2015)
18. C. Park, Fixed point method for set-valued functional equations. J. Fixed Point Theory Appl. **19**, 2297–2308 (2017)
19. C. Park, Set-valued additive ρ-functional inequalities. J. Fixed Point Theory Appl. **20**(2), 20–70 (2018)
20. C. Park, D. Shin, J. Lee, Fixed points and additive ρ-functional equations. J. Fixed Point Theory Appl. **18**, 569–586 (2016)
21. V. Radu, The fixed point alternative and the stability of functional equations. Fixed Point Theory **4**, 91–96 (2003)
22. Th.M. Rassias, On the stability of the linear mapping in Banach spaces. Proc. Am. Math. Soc. **72**, 297–300 (1978)

23. F. Skof, Propriet locali e approssimazione di operatori. Rend. Sem. Mat. Fis. Milano **53**, 113–129 (1983)
24. S.M. Ulam, *A Collection of the Mathematical Problems* (Interscience Publ., New York, 1960)
25. Z. Wang, Stability of two types of cubic fuzzy set-valued functional equations. RM **70** (2016), 1–14.

First Study for Ramp Secret Sharing Schemes Through Greatest Common Divisor of Polynomials

Gerasimos C. Meletiou, Dimitrios S. Triantafyllou, and Michael N. Vrahatis

Abstract A ramp secret sharing scheme through greatest common divisor of polynomials is presented. Verification and shelf correcting protocols are also developed. The proposed approach can be implemented in a hybrid way using numerical and symbolical arithmetic. Numerical examples illustrating the proposed sharing schemes are also given.

1 Introduction

A (t, n)-threshold secret sharing scheme is a method in which the dealer distributes the secret to n participants [2]. In this scheme any t participants, $1 \leqslant t \leqslant n$, can cooperate and retrieve the secret but any $t - 1$ participants cannot reconstruct the secret. An (s, t, n)-threshold ramp scheme is a generalization of a threshold secret sharing scheme using two parameters. Namely, the value s which determines the lower threshold and t which is the upper threshold. In a ramp scheme, any t (or more than t) of the n players can compute the secret (exactly as in a (t, n)-threshold scheme). It is also required that no subset of s (or less than s) players can determine any information about the secret. We note that a $(t-1, t, n)$-ramp scheme is exactly the same as a (t, n)-threshold scheme. The parameters of a ramp scheme satisfy the conditions $0 \leqslant s < t \leqslant n$. For more information see [3, 11]. A ramp scheme is

G. C. Meletiou (✉)
University of Ioannina, School of Agriculture, Arta, Greece
e-mail: gmelet@uoi.gr

D. S. Triantafyllou
Department of Mathematics and Engineering Sciences, Hellenic Military Academy, Vari, Greece
e-mail: dtriant@sse.gr

M. N. Vrahatis
Computational Intelligence Laboratory (CILab), Department of Mathematics, University of Patras, Patras, Greece
e-mail: vrahatis@math.upatras.gr

N. J. Daras, T. M. Rassias (eds.), *Computational Mathematics and Variational Analysis*, Springer Optimization and Its Applications 159,
https://doi.org/10.1007/978-3-030-44625-3_14

essentially a non-perfect secret sharing scheme. Ramp schemes are useful because they can achieve a high information rate.

In Section 2 the proposed ramp secret sharing scheme and two self-correcting protocols are presented based on the greatest common divisor (GCD) of polynomials. In Section 3 a numerical linear algebra technique for computing the GCD of polynomials through Sylvester matrices is presented. In Section 4 an example evaluating the proposed method is presented. The two self-correcting protocols are also evaluated. In Section 5 a synopsis and concluding remarks are given.

2 Ramp Secret Sharing Scheme

Throughout this paper, it is assumed that all participants are cooperating giving the real data to each other. Also, verification protocols are given to tackle the case where an error in a cooperation is appeared.

Cryptographic Scheme Let D be the dealer, $P_1,\ P_2, \ldots,\ P_n$ be n participants, $s(x) := \prod_{i=1}^{n} m_i(x)$ be the secret, where $m_i(x),\ i = 1, 2, \ldots, n$ are n polynomials of degree $d_{m_i},\ i = 1, 2, \ldots, n$ such that for $i \neq j,\ i, j = 1, 2, \ldots, n$ the $m_i(x),\ m_j(x)$ are coprime and $d(x)$ be a polynomial known by the dealer, where $d(x),\ m_i(x)$ are coprime for all $i = 1, 2, \ldots, n$. Each participant $P_i,\ i = 1, 2, \ldots, n$ receives the following information (share) from the dealer D:

$$
\begin{aligned}
p_1(x) &= d(x) \cdot m_2(x) \cdots m_n(x), \\
p_2(x) &= d(x) \cdot m_1(x) \cdot m_3(x) \cdots m_n(x), \\
&\vdots \\
p_i(x) &= d(x) \cdot m_1(x) \cdots m_{i-1}(x) \cdot m_{i+1}(x) \cdots m_n(x), \\
&\vdots \\
p_n(x) &= d(x) \cdot m_1(x) \cdots m_{n-1}(x).
\end{aligned}
\tag{1}
$$

Thus,

$$
p_i(x) = d(x) \cdot \prod_{\substack{j=1 \\ j \neq i}}^{n} m_i(x), \qquad i = 1, 2, \ldots, n.
$$

Remark 1 In a future work our scope is to study the polynomials over fields in such a way that the factorization will not be feasible.

Remark 2 The participant P_i knows the whole product $p_i(x)$, as a polynomial of degree d_{m_i} $i = 1, 2, \ldots, n$ but he does not know the following factorization:

$$
d(x) \cdot m_1(x) \cdots m_{i-1}(x) \cdot m_{i+1}(x) \cdots m_n(x).
$$

Remark 3 In general, a direct corollary of the fundamental theorem of algebra [13] states that a polynomial can be factorized over the complex domain into a product $a_n(x - r_1)(x - r_2)\cdots(x - r_n)$, where a_n is the leading coefficient and $r_1, r_2, \ldots, r_n$ are all of its n complex roots. On the other hand, it is well known that, for polynomials of degrees more than four, no general closed-form formulas for their roots exist. For these cases we can apply various root-finding algorithm for the approximation of the roots of a polynomial. This approximation is not an easy task and it is depended on the inner tolerance that it will be used during the floating point operations. Specifically, for a polynomial of degree n the required bit operations are $O\big(n^{12} + n^9(\log(|p|))^3\big)$, where $p = a_n x^n + a_{n-1}x^{n-1} + \cdots + a_0$ is the polynomial and $\left|\sum_{i=0}^{n} a_i\, x^i\right| = \left(\sum_{i=0}^{n} a_i^2\right)^{1/2}$ for a polynomial with real coefficients [6]. Victor Pan in 2002 [9] presented almost optimal algorithms for numerical factorization of univariate polynomials, while Sagraloff and Mehlhorn in 2016 [10] introduced a hybrid of the Descartes method and Newton iteration which is in comparable complexity with Pan's algorithms.

In our case, after approximating all the roots of $p_i(x)$ the participant P_i may retrieve the secret by computing all the combinations of the roots. The scheme can be improved and become more robust by giving the dealer to participants different inner tolerances in order the participants not to be able to approximate all the roots. In that case the scheme looses its self-correcting protocols.

Theorem 1 *If all participants $P_1,\ P_2, \ldots,\ P_n$ are cooperating, they can derive the secret $s(x)$.*

Proof Suppose that the two participants P_i and P_j, $1 \leqslant i,\ j \leqslant n,\ i \neq j$ are cooperating. Participant P_i knows the following information from the dealer:

$$p_i(x) = d(x) \cdot m_1(x) \cdots m_{i-1}(x) \cdot m_{i+1}(x) \cdots m_n(x),$$

while participant P_j knows the following information from the dealer:

$$p_j(x) = d(x) \cdot m_1(x) \cdots m_{j-1}(x) \cdot m_{j+1}(x) \cdots m_n(x).$$

The GCD of the previous two polynomials is given as follows:

$$\gcd\{p_i(x),\ p_j(x)\} := g_{i,j}(x) = d(x) \cdot \prod_{\substack{k=1 \\ k \neq i,j}}^{n} m_i(x).$$

Then,

i. participant P_i finds out the identity $m_j(x)$ of participant P_j by dividing $p_i(x)$ with the GCD: $g_{i,j}(x)$: $p_i(x)/g_{i,j}(x) = m_j(x)$ and
ii. participant P_j finds out the identity $m_i(x)$ of participant P_i by dividing $p_j(x)$ with the GCD: $g_{i,j}(x)$: $p_j(x)/g_{i,j}(x) = m_i(x)$.

In addition, participant P_i informs participant P_j about his/her identity m_j and vice versa. Thus, both participants P_i and P_j know $m_i(x), m_j(x)$ and the whole product $d(x) \cdot \prod_{j=1}^{n} m_i(x)$.

If all the n participants cooperate, then by revealing their identities they are able to derive the secret $s(x) = \prod_{i=1}^{n} m_i(x)$. □

Assume now that two participants, e.g., P_1 and P_2 decide to cooperate in order to compute a part of the secret. They can derive $m_1(x)$ and $m_2(x)$ and the Least Common Multiple (LCM) of them:

$$r(x) := \operatorname{lcm}\{p_1(x),\ p_2(x)\} = d(x) \cdot \prod_{i=1}^{n} m_i(x) = d(x) \cdot s(x),$$

as a polynomial. The secret $s(x)$ divides $r(x)$ and is divided by $m_1(x) \cdot m_2(x)$, thus

$$m_1(x) \cdot m_2(x)/s(x)/\operatorname{lcm}\{p_1(x),\ p_2(x)\}. \tag{2}$$

In other terms the shares $p_1(x)$ and $p_2(x)$ define a restriction for the secret $s(x)$. That means that from the shares of the participants P_1 and P_2, a partial information for the secret $s(x)$ can be derived (ramp scheme). In order to compute all the roots of their polynomials they have to use polynomial root-finding algorithms of complexity mentioned above.

More general, assume that k participants cooperate in order to compute a part of the secret, $1 < k < n$ and without loss of generality let $P_1, P_2, \ldots, P_k$ be the participants. They can compute $m_1(x), m_2(x), \ldots, m_k(x)$, $\operatorname{lcm}\{p_1(x),\ p_2(x), \ldots, p_k(x)\} = d(x) \cdot \prod_{i=1}^{n} m_i(x) = d(x) \cdot s(x)$ as polynomial and its roots and they can recognize the roots of $m_i(x),\ i = 1, 2, \ldots, k$ as roots of the secret. Therefore, we have the following condition:

$$\prod_{i=1}^{k} m_i(x)/s(x)/\operatorname{lcm}\{p_1(x),\ p_2(x), \ldots, p_k(x)\}. \tag{3}$$

The "division interval" becomes more narrow since $m_1(x) \cdot m_2(x)/\prod_{i=1}^{k} m_i(x)$. In the case where $k = n - 1$ the condition (3) becomes

$$\prod_{i=1}^{n-1} m_i(x)/s(x)/d(x) \cdot s(x). \tag{4}$$

Assume that $\hat{s}(x)$ satisfies (4) and let

$$\hat{d}(x) := \frac{\operatorname{lcm}\{p_1(x),\ p_2(x), \ldots, p_{n-1}(x)\}}{\hat{s}(x)} = \frac{d(x) \cdot s(x)}{\hat{s}(x)},$$

$$\hat{m}_n(x) := \frac{\hat{s}(x)}{\prod_{i=1}^{n-1} m_i(x)},$$

$$\hat{p}_n(x) := \hat{d}(x) \cdot \prod_{i=1}^{n-1} m_i(x).$$

It is easy to verify that $\hat{s}(x)$ is a pseudo secret for the shares $p_1(x), \ldots, p_{n-1}(x)$, $\hat{p}_n(x)$.

Self-Correcting Protocols In the following proposition, two self-correcting protocols are presenting, in the case where error information in cooperations is given.

Proposition 1 (Self-Correcting Protocols)

Protocol 1 Suppose that the participant P_i, $1 \leqslant i \leqslant n$ is pairwise cooperating with all the other participants. Then, from these cooperations, the following pieces of information can be derived:

$$\begin{array}{lll}
P_i,\ P_1: & \gcd\{p_i(x),\ p_1(x)\} \longrightarrow & m_1(x), m_i(x) \\
P_i,\ P_2: & \gcd\{p_i(x),\ p_2(x)\} \longrightarrow & m_2(x), m_i(x) \\
\vdots & \vdots & \vdots \\
P_i,\ P_{i-1}: & \gcd\{p_i(x),\ p_{i-1}(x)\} \longrightarrow & m_{i-1}(x), m_i(x) \\
P_i,\ P_{i+1}: & \gcd\{p_i(x),\ p_{i+1}(x)\} \longrightarrow & m_{i+1}(x), m_i(x) \\
\vdots & \vdots & \vdots \\
P_i,\ P_n: & \gcd\{p_i(x),\ p_n(x)\} \longrightarrow & m_n(x), m_i(x)
\end{array}$$

where $m_i(x), m_j(x)$, $j = 1, \ldots, n$, $j \neq i$ are obtained by dividing $p_j(x)$, $p_i(x)$ with $\gcd\{p_i(x), p_j(x)\}$ respectively. Also, $m_i(x)$ in all the previous computations should be the same. Thus, if there is anywhere a different result, then in the corresponding pair will be an error.

Protocol 2 Suppose that every participant is pairwise cooperating with two other participants. Without loss of generality assume that every participant is cooperating in a cycling order first with his/her previous and afterwards with his/her next participant. Let P_i, $1 \leqslant k \leqslant n$ be a participant that gave an error information. By the above assumption, P_i cooperates firstly with the P_{i-1} and next with the P_{i+1}. Obviously, in the case where $i = 1$ then $i - 1$ is considered n and if $i = n$, then $i + 1$ is considered 1. Then, from these cooperations, the following pieces of information can be derived:

$$\begin{array}{lll}
P_i,\ P_{i-1}: & \gcd\{p_i(x),\ p_{i-1}(x)\} \longrightarrow & m_{i-1}(x), m_i(x) \\
P_i,\ P_{i+1}: & \gcd\{p_i(x),\ p_{i+1}(x)\} \longrightarrow & m_{i+1}(x), m_i(x)
\end{array}$$

Thus, from these cooperations, P_{i-1} and P_{i+1} should reveal the same identity $m_i(x)$. In the case where it does not hold, then P_i has given an error information.

3 Computation of The Greatest Common Divisor of Polynomials

For completeness purposes, we shall allow us to briefly discuss a few basic concepts regarding a numerical method for computing the GCD of a set of polynomials. For more details we refer the interested reader to [1, 4, 5, 12, 14].

Definition 1 Let

$$p_i(x) = p_{i,d_{m_i}} x^{d_{m_i}} + p_{i,d_{m_i-1}} x^{d_{m_i}-1} + p_{i,d_{m_i-2}} x^{d_{m_i}-2} + \cdots + p_{i,0}, \qquad i=1,2,\ldots,n,$$

be n polynomials as defined in (2). Without loss of generality let $p_1(x)$ be the polynomial of maximal degree d_{m_1}. Let $p_2(x)$ be the polynomial with the second maximum degree. Consider the following matrices:

$$S_1 = \begin{bmatrix} p_{1,d_{m_1}} & p_{1,d_{m_1-1}} & p_{1,d_{m_1-2}} & \cdots & p_{1,d_0} & 0 & \ldots & 0 & 0 \\ 0 & p_{1,d_{m_1}} & p_{1,d_{m_1-1}} & \cdots & p_{1,d_1} & p_{1,d_0} & \cdots & 0 & 0 \\ \vdots & \vdots & \vdots & \ddots & \ddots & \ddots & \ddots & \vdots & \vdots \\ 0 & 0 & 0 & p_{1,d_{m_1}} & & \cdots & & p_{1,d_1} & p_{1,d_0} \end{bmatrix},$$

and

$$S_i = \begin{bmatrix} p_{i,d_{m_i}} & p_{i,d_{m_i-1}} & p_{i,d_{m_i-2}} & \cdots & p_{i,d_0} & 0 & \ldots & 0 & 0 \\ 0 & p_{i,d_{m_i}} & p_{i,d_{m_i-1}} & \cdots & p_{i,d_1} & p_{i,d_0} & \cdots & 0 & 0 \\ \vdots & \vdots & \vdots & \ddots & \ddots & \ddots & \ddots & \vdots & \vdots \\ 0 & 0 & 0 & p_{i,d_{m_i}} & & \cdots & & p_{i,d_1} & p_{i,d_0} \end{bmatrix}, \quad i = 2, \ldots, n,$$

where S_1 is an $d_{m_2} \times (d_{m_1} + d_{m_2})$ block matrix representing $p_1(x)$ and S_i is an $d_{m_1} \times (d_{m_1} + d_{m_2})$ matrix which represents $p_i(x)$, $i = 2, \ldots, n$. The classical Sylvester matrix is defined as the following $(d_{m_1} \cdot d_{m_2} + dm_2) \times (d_{m_1} + d_{m_2})$ matrix [1]:

$$S = \begin{bmatrix} S_1 \\ S_2 \\ \vdots \\ S_n \end{bmatrix}.$$

By collecting the first row of every block S_i, $i = 2, 3, \ldots, n$ as follows:

$$B = \begin{bmatrix} p_{2,d_{m_2}} & p_{2,d_{m_2}-1} & p_{2,d_{m_2}-2} & \cdots & p_{2,0} \\ p_{3,d_{m_3}} & p_{3,d_{m_3}-1} & p_{3,d_{m_3}-2} & \cdots & p_{3,0} \\ \vdots & \vdots & \vdots & \ddots & \vdots \\ p_{n,d_{m_n}} & p_{n,d_{m_n}-1} & p_{n,d_{m_n}-2} & \cdots & p_{n,0} \end{bmatrix},$$

and reconstructing the Sylvester matrix S as follows:

$$S^* = \begin{bmatrix} B & \theta & \theta & \Theta & & & \\ \theta & B & \theta & \Theta & & & \\ \theta & \theta & B & \Theta & & & \\ & & & \ddots & \ddots & & \\ & & & \Theta & & \theta & B \\ & & & S_0 & & & \end{bmatrix},$$

where Θ is a zero square matrix, θ is a zero column vector, and I is the identity matrix, we get a matrix with n same blocks called modified Sylvester matrix S^* [12].

Theorem 2 ([1, 12]) *Let S^* be the modified Sylvester matrix of n polynomials $p_1(x), p_2(x), \ldots, p_n(x)$ and $P \cdot S^* = L \cdot U$ the LU factorization with partial pivoting of S^*. Then the nonzero elements of the last nonzero row of U define the coefficients of GCD of the polynomials $p_1(x), p_2(x), \ldots, p_n(x)$.*

Taking advantage of the special form of S^* and zeroing and updating only specific entries of the first block at each step we get the following modified LU factorization algorithm (for more details we refer the interested reader to [12]):

Algorithm The modified LU factorization

STEP 1 : Construct the modified generalized Sylvester matrix S^*

STEP 2 : While number of same blocks > 3 do
 if number of same blocks of $(S^*)^{(i)}$ = odd
  move the last block after S_1
 endif
 Compute the upper triangular matrix U
 $U = LU(B)$, where B contains
 the two first same blocks of $(S^*)^{(i)}$

STEP 3 : Compute the upper triangular matrix U' :
 $U' = LU(S^*)^{(k)}$

Computational Complexity The required computational complexity of the previous algorithm is given as follows [12]:

$$O\left((d_{m_1}+d_{m_2})^3\left(2\log_2 2(d_{m_1})-\frac{1}{3}\right)+(d_{m_1}+d_{m_2})^2\left(2n\log_2 2(d_{m_1})+d_{m_2}\right)\right),$$

which is significant reduced in compare with the complexity of the classical LU factorization [4, 5] for a matrix of size of the generalized Sylvester.

Numerical Stability The modified LU factorization computes the exact factorization of a slightly perturbed initial matrix. For the modified LU factorization, it holds that

$$S^* + E = L \cdot U,$$

with

$$\| E \|_\infty \leqslant (d_{m_1}+d_{m_2})^2 \rho u \prod_{i=1}^{\log_2 2(d_{m_1})} \| L_i \|_\infty \| S^* \|_\infty,$$

where ρ is the growth factor and u the unit round off [12]. More details for the stability of the LU factorization with partial pivoting can be found in [14].

4 Hybrid Implementation

In this section we present an example evaluating the proposed method in a hybrid way. The computation of the GCD of polynomials is achieved using floating point arithmetic and the modified LU factorization algorithm and the divisions of polynomials symbolically.

Table 1 Polynomials of participants given by the dealer

Participant	Polynomial
P_1	$x^6 - 42x^5 + 685x^4 - 5460x^3 + 22084x^2 - 43008x + 31680$
P_2	$x^6 - 41x^5 + 645x^4 - 4855x^3 + 17834x^2 - 29424x + 15840$
P_3	$x^6 - 40x^5 + 607x^4 - 4324x^3 + 14572x^2 - 21376x + 10560$
P_4	$x^6 - 39x^5 + 571x^4 - 3861x^3 + 12100x^2 - 16692x + 7920$

Let us suppose that we have one dealer and four participants. Let $s(x) = x^4 - 10x^3 + 35x^2 - 50x + 24$ be the secret and $d(x) = x^3 - 33x^2 + 362x - 1320$ the polynomial chosen by the dealer for increasing the difficulty of breaking the secret key. The dealer gives to participants the polynomials shown in Table 1.

Let us assume that the participants P_1 and P_2 inform each other about their polynomials and compute their GCD. The information that they found is

$$\begin{aligned} g_{12}(x) &:= \gcd\{p_1(x),\ p_2(x)\} \\ &= x^5 - 40x^4 + 605x^3 - 4250x^2 + 13584x - 15840. \end{aligned}$$

The participant P_1 divides his/her polynomial $p_1(x)$ with $g_{12}(x)$ and finds the factor $m_2(x)$ of participant P_2:

$$\begin{aligned} \frac{p_1(x)}{g_{12}(x)} &= \frac{x^6 - 42x^5 + 685x^4 - 5460x^3 + 22084x^2 - 43008x + 31680}{x^5 - 40x^4 + 605x^3 - 4250x^2 + 13584x - 15840} \\ &= x - 2 \\ &= m_2(x). \end{aligned}$$

Similarly, participant P_2 divides his/her polynomial $p_2(x)$ with the GCD $g_{12}(x)$ and finds the factor $m_1(x)$ of participant P_1:

$$\begin{aligned} \frac{p_2(x)}{g_{12}(x)} &= \frac{x^6 - 41x^5 + 645x^4 - 4855x^3 + 17834x^2 - 29424x + 15840}{x^5 - 40x^4 + 605x^3 - 4250x^2 + 13584x - 15840} \\ &= x - 1 \\ &= m_1(x). \end{aligned}$$

Thus, participants P_1 and P_2 have found a part of the secret $s(x)$ but they have to cooperate with the other two participants as well in order to compute the secret $s(x)$, since they have to find out $m_3(x)$ and $m_4(x)$ in order to deconvolute $d(x)$ from their LCM of their polynomials. If all participants reveal their polynomials $p_i(x),\ i = 1, 2, 3, 4$ and compute the GCD $g(x)$, then:

$$g(x) := \gcd\{p_1(x),\ p_2(x),\ p_3(x),\ p_4(x)\}$$
$$= x^3 - 33x^2 + 362x - 1320.$$

Remark 4 The previous GCD was computed using the modified LU factorization algorithm with inner tolerance 10^{-16}. Since in numerical arithmetic different tolerances may lead to different results, the dealer may include the tolerance as information to the participants along with their polynomials.

Each participant divides his/her polynomial $p_i(x), i = 1, 2, 3, 4$ with the computed GCD $g(x)$ and obtain the results presented in Table 2.

Participant, e.g., P_2 can now cooperate with P_1 as shown before in order to find out $m_1(x)$ and inform P_1 about it. Thus P_1 can now multiply $g_1(x)$ with $m_1(x)$ in order to retrieve the secret $s(x)$:

$$\begin{aligned} g_1(x) \cdot m_1(x) &= \left(x^3 - 9x^2 + 26x - 24\right) \cdot \left(x - 1\right) \\ &= x^4 - 10x^3 + 35x^2 - 50x + 24 \\ &= s(x). \end{aligned}$$

The same result is also obtained if all participants cooperate in cyclic pairs as shown in Table 3. The secret is obtained by the product $\prod_{i=1}^{4} m_i(x)$:

$$s(x) = \prod_{i=1}^{4} m_i(x) = x^4 - 10x^3 + 35x^2 - 50x + 24.$$

Table 2 Polynomials of participants after division and the corresponding results

Participant	Polynomial division
P_1	$\frac{p_1(x)}{g(x)} = \frac{x^6 - 42x^5 + 685x^4 - 5460x^3 + 22084x^2 - 43008x + 31680}{x^3 - 33x^2 + 362x - 1320}$
P_2	$\frac{p_2(x)}{g(x)} = \frac{x^6 - 41x^5 + 645x^4 - 4855x^3 + 17834x^2 - 29424x + 15840}{x^3 - 33x^2 + 362x - 1320}$
P_3	$\frac{p_3(x)}{g(x)} = \frac{x^6 - 40x^5 + 607x^4 - 4324x^3 + 14572x^2 - 21376x + 10560}{x^3 - 33x^2 + 362x - 1320}$
P_4	$\frac{p_4(x)}{g(x)} = \frac{x^6 - 39x^5 + 571x^4 - 3861x^3 + 12100x^2 - 16692x + 7920}{x^3 - 33x^2 + 362x - 1320}$
Participant	Result$g_i(x)$
P_1	$g_1(x) = x^3 - 9x^2 + 26x - 24 = (x-2)(x-3)(x-4)$
P_2	$g_2(x) = x^3 - 8x^2 + 19x - 12 = (x-1)(x-3)(x-4)$
P_3	$g_3(x) = x^3 - 7x^2 + 14x - 8 = (x-1)(x-2)(x-4)$
P_4	$g_4(x) = x^3 - 6x^2 + 11x - 6 = (x-1)(x-2)(x-3)$

Table 3 Cooperation participant P_2 in pairs

Participants	GCD $g_i(x)$
P_1, P_2	$g_{12}(x) = \gcd\{p_1(x), p_2(x)\} = x^5 - 40x^4 + 605x^3 - 4250x^2 + 13584x - 15840$
P_2, P_3	$g_{23}(x) = \gcd\{p_2(x), p_3(x)\} = x^5 - 38x^4 + 531x^3 - 3262x^2 + 8048x - 5280$
P_3, P_4	$g_{34}(x) = \gcd\{p_3(x), p_4(x)\} = x^5 - 36x^4 + 463x^3 - 2472x^2 + 4684x - 2640$
P_4, P_1	$g_{41}(x) = \gcd\{p_4(x), p_1(x)\} = x^5 - 38x^4 + 533x^3 - 3328x^2 + 8772x - 7920$
Participants	Polynomial deconvolution $m_i(x)$
P_1, P_2	$m_2(x) = \frac{p_1(x)}{g_{12}(x)} = x - 2$
P_2, P_3	$m_3(x) = \frac{p_2(x)}{g_{23}(x)} = x - 3$
P_3, P_4	$m_4(x) = \frac{p_3(x)}{g_{34}(x)} = x - 4$
P_4, P_1	$m_1(x) = \frac{p_4(x)}{g_{41}(x)} = x - 1$

Table 4 Cooperation of participant P_2 and P_4 with P_1 in pairs

Participants	GCD $g_i(x)$
P_1, P_2	$g_1(x) = \gcd\{p_1(x), p_2(x)\} = x^5 - 40x^4 + 605x^3 - 4250x^2 + 13584x - 15840$
P_2, P_3	$g_3(x) = \gcd\{p_2(x), p_3(x)\} = x^4 - 34x^3 + 395x^2 - 1682x + 1320$
P_2, P_4	$g_4(x) = \gcd\{p_3(x), p_4(x)\} = x^5 - 37x^4 + 497x^3 - 2867x^2 + 6366x - 3960$
Participants	Polynomial deconvolution $m_i(x)$
P_1, P_2	$m_2(x) = \frac{p_1(x)}{g_1(x)} = x - 2$
	$m_1(x) = \frac{p_2(x)}{g_1(x)} = x - 1$
P_2, P_3	$\hat{m}_3(x) = \frac{p_2(x)}{g_3(x)} = x^2 - 7x + 12 = (x - 3) \cdot (x - 4)$
	$\tilde{m}_3(x) = \frac{\hat{p}_3(x)}{g_3(x)} = x^2 - 17x + 30 = (x - 2) \cdot (x - 15)$
P_2, P_4	$m_4(x) = \frac{p_2(x)}{g_4(x)} = x - 4$
	$m_2(x) = \frac{p_4(x)}{g_4(x)} = x - 2$

Let us suppose that one of the participants, e.g., P_3 gives wrong information and let $\hat{p}_3(x) = x^6 - 51x^5 + 1003x^4 - 9417x^3 + 41764x^2 - 72900x + 39600$ be the false instead of the real one $p_3(x) = x^6 - 40x^5 + 607x^4 - 4324x^3 + 14572x^2 - 21376x + 10560$. $\hat{p}_3(x)$ will be either coprime with the polynomials of the other participants or may have some common roots but not the right ones.

Protocol 1

Let participant P_2 cooperate per pair with all other ones. The results are summarized in Table 4. Every P_i, $i = 1, 3, 4$ through the cooperation with P_2 should give as result $m_2(x)$ to P_2. As it is shown in Table 4 only participant P_3 did not give the factor $m_2(x)$ as it was supposed to do. Thus P_3 had given wrong information.

Table 5 Cyclic cooperation in pairs

Participants	GCD $g_i(x)$
P_2, P_3	$g_{23}(x) = \gcd\{p_1(x), \hat{p}_3(x)\} = x^4 - 34x^3 + 395x^2 - 1682x + 1320$
P_3, P_4	$g_{34}(x) = \gcd\{\hat{p}_3(x), p_4(x)\} = x^5 - 36x^4 + 463x^3 - 2472x^2 + 4684x - 2640$
Participants	Polynomial deconvolution $m_i(x)$
P_2, P_3	$\hat{m}_2(x) = \dfrac{p_2(x)}{g_{23}(x)} = x^2 - 7x + 12 = (x-3)\cdot(x-4)$
P_3, P_4	$\hat{m}_4(x) = \dfrac{p_4(x)}{g_{34}(x)} = x - 3$

Protocol 2
Participant P_3 cooperates per pair with the previous participant P_2 and the next one P_4 as shown in Table 5. From the cooperation with P_3 the participants P_2 and P_4 should take as result the same polynomial $m_3(x) = \frac{p_2(x)}{g_{23}(x)} = x - 3$ and $m_3(x) = \frac{p_4(x)}{g_{34}(x)} = x - 3$ but they have taken different ones. Also $\hat{m}_4(x)$ should be a polynomial of larger degree since it should include $m_3(x)$ and $m_4(x)$ as a product. Also $\hat{m}_2(x)$ should include $m_2(x)$ and $m_3(x)$ as a product. The third participant gave wrong results with both the previous and next participant, thus he/she gave wrong information.

5 Synopsis and Concluding Remarks

In this paper a ramp secret sharing scheme is presented. The scheme is based on the computation of the greatest common divisor of polynomials. A subset of the participants can derive information about the secret and can approximate the roots of their shares, but the approximation of all real roots of a polynomial is, in general, a hard task. The dealer can make more difficult the approximation of the roots of the polynomials by selecting roots with many decimal digits. Two correcting protocols are also presented in order to recognize the participant that gave false information. The computation of the greatest common divisor of polynomials is implemented through stable numerical linear algebra methods in an efficient way. Polynomial divisions can be evaluated either numerically through Horner's algorithm which is proved that is optimal [7, 8] in respect of floating point operations or symbolically.

The proposed scheme will be improved in a future work by constructing, by the dealer a vector with different inner tolerances for the participants. In that way, any participant will compute all the roots of its polynomial but not all of them will be the same with that of the secret ones. Furthermore, we will use polynomials over fields of non-zero characteristic in order to significantly improve the robustness of the proposed method.

References

1. S. Barnett, Greatest common divisor of several polynomials. Linear Multilinear A **8**, 271–279 (1980)
2. G.R. Blakley, Safeguarding cryptographic keys, in *Proceedings AFIPS 1979 National Computer Conference* (1979), pp. 313–317
3. G.R. Blakley, C. Meadows, Security of ramp schemes, in *Advances in Cryptology*, ed. by G.R. Blakley, D. Chaum. CRYPTO 1984. Lecture Notes in Computer Science, (Springer, Berlin, Heidelberg, 1984), **196**, 242–268
4. R.L. Burden, J.D. Faires, *Numerical Analysis*, 6th edn. (Brooks/Cole Publishing Company, Pacific Grove, 1997)
5. B.N. Datta, *Numerical Linear Algebra and Applications*, 2nd edn. (SIAM, Philadelphia, 2010)
6. A.K. Lenstra, H.W. Lenstra, Jr., L. Lovász, Factoring polynomials with rational coefficients. Math. Ann. **261**, 515–534 (1982)
7. A.M. Ostrowski, *On Two Problems in Abstract Algebra Connected with Horner's Rule*. Studies in Mathematics and Mechanics (Academic, New York, 1954), pp. 40–48
8. V.Y. Pan, On means of calculating values of polynomials. Russ. Math. Surv. **21**, 105–136 (1966)
9. V. Pan, Univariate polynomials: nearly optimal algorithms for numerical factorization and root-finding. J. Symb. Comput. **33**, 701–733 (2002)
10. M. Sagraloff, K. Mehlhorn, Computing real roots of real polynomials. J. Symb. Comput. **73**, 46–86 (2016)
11. D.R. Stinson, Ideal ramp schemes and related combinatorial objects. Discrete Math. **341**, 299–307 (2018)
12. D. Triantafyllou, M. Mitrouli, On rank and null space computation of the generalized Sylvester matrix. Numer. Algorithms **54**, 297–324 (2010)
13. B.L. van der Waerden, *Algebra*, vol. I (Springer, New York, 1991)
14. J.H. Wilkinson, *The Algebraic Eigenvalue Problem* (Clarendon Press, Oxford, 1965)

From Representation Theorems to Variational Inequalities

Muhammad Aslam Noor and Khalida Inayat Noor

Abstract We start with an historical introduction and then give an overview of the most important representation theorems for the linear(nonlinear) continuous functionals by the bifunction. Then we extensively study representations theorems for nonlinear operators. These representation results contain the difference of two(more) monotone operators, complementarity problems, systems of the absolute values equations and difference of two convex functions as special cases. These problems are very important and significant, which provide a unified and general framework of studying a wide class of unrelated problems.

1 Introduction

Convexity theory describes a broad spectrum of very interesting developments involving a link among various fields of mathematics, physics, economics and engineering sciences. Some of these developments have made mutually enriching contacts with other fields. Ideas explaining these concepts led to the developments of new and powerful techniques to solve a wide class of linear and nonlinear problems. Convexity theory provides us with a unified framework to develop highly efficient and powerful numerical methods to solve nonlinear problems, see [1–17, 19–27].

It is well known that the minimum of differentiable convex functions on convex sets can be characterized by variational inequalities. This simple and deep result is due to Stampacchia [99] and Fichera [11]. Variational inequalities can be viewed as natural generalization of the variational principle, which have played the crucial part in the developments of modern fields of mathematical and engineering sciences. Closely related to the variational inequalities, we have the various representation theorems for linear and continuous functionals, the origin of which can be traced back to Riesz [94, 95] and Frechet [13]. Riesz [94] and Frechet [13] proved that

M. A. Noor (✉) · K. I. Noor
COMSATS University Islamabad, Islamabad, Pakistan

N. J. Daras, T. M. Rassias (eds.), *Computational Mathematics and Variational Analysis*, Springer Optimization and Its Applications 159,
https://doi.org/10.1007/978-3-030-44625-3_15

an arbitrary linear and continuous functional on the real Hilbert spaces can be represented by an inner product. This result plays an important part in the existence theory of differential equations. Since the inner product is a bifunction. Naturally the question arises whether a representation result holds for an arbitrary bifunction? The answer to this question is in affirmative. Lax and Milgram [25] proved that a linear and continuous functional on the Hilbert spaces can be represented by an arbitrary bifunctions. This result enables us to study the existence theory of a solution of the differential and also played an important part in obtaining the error bounds for the approximate solution of the boundary value problems, which is obtained by using the finite element method and its variant forms. Lions and Stampacchia [26] proved that linear continuous functions on the convex set can be represented by the variational inequalities. In fact, one can be easily obtained the Lax–Milgram Lemma and the Riesz–Frechet results as special cases of the variational inequalities.

Another important aspect of these results is the relationship with the variational principles, the origin of which can be traced back to Newton, Euler, Lagrange and Bernoulli's brothers. From the day of discovery of the representation theorems, many important contributions have been made in these directions. In every case, new approaches and techniques are applied to generalize and extend these results. Question arises whether such type of representations hold for nonlinear continuous operators? The answer is in affirmative. Motivated by these facts, Noor [39] and Noor et al. [76] have shown that the nonlinear operator can be represented by an arbitrary bifunction, which can be viewed as a novel generalization of the Lax–Milgram Lemma and Riesz–Frechet representation theorem. For the applications, error estimates, numerical results and other aspects of these nonlinear representation theorems, see Noor [36–41].

In recent years, several related problems are being considered and investigated extensively from practical and applications view point. The projection methods and its variant forms have been used to reformulate the variational inequalities as fixed point problems. These alternative fixed point formulations have been used to suggest and analyse various iterative methods for solving variational inequalities. This alternative formulation has also been used to study the sensitivity analysis, error bounds, dynamic systems and neural networking for variational inequalities. Shi [96] showed that the variational inequalities are equivalent to system of equations, which are known as Wiener–Hopf equations. This alternative formulation proved to more effective than the projection technique, see, for example, [14, 15, 18, 33, 49–51, 53–73, 77–86, 89, 90, 97, 98, 100–108, 110, 111, 113–115].

Lion and Stampacchia [26] used another method to consider the existence of a unique solution of the variational inequalities, which is known as the auxiliary principle technique. Glowinski et al. [26] used this technique to consider the existence of a solution the mixed variational inequalities. This technique is more flexible and unifying one. This approach have been used to suggest not only the iterative methods for solving variational inequalities, equilibrium problems and other type of complicated variational problems. In this article, our main focus is the applicability of auxiliary principle technique in considering various problems. This technique can be used to study the problems, where projection method, Weiner–

Hopf equations approach, resolvent methods cannot be applied. The main idea of this approach is to consider an auxiliary problem related to the original problem. In this way, one defines a mapping connecting the original and auxiliary problems. Then, one uses the fixed point technique, to show that the mapping connection these two problems have a fixed point, which is the solution of the original problem. We demonstrate the main features of this approach and show that this approach can be used to applied to discuss the representation theorems for linear(nonlinear) continuous functionals on the whole space. See [31, 45–48, 52, 87, 92, 95, 109, 112] for more information.

Our main motivation of this paper is manifold.

1. To give a brief summary of the basic theory of representation results and their connection with variational inequalities.
2. To point out the close relationship with other problems such as system of absolute values equation, complementarity problems, difference of two(more) monotone operators and minimization of the difference of convex functions. These problems can be viewed as special cases of mildly nonlinear variational inequalities, which were introduced and studied by Noor [36–40]. These special problems are being studied extensively in recent years.
3. Some recent iterative methods for solving mildly nonlinear variational inequalities using the auxiliary principle technique.
4. Some open problems are highlighted for further research.

2 Preliminary Results

Let K be a nonempty closed set in a real Hilbert space H. We denote by $\langle \cdot, \cdot \rangle$ and $\| \cdot \|$ be the inner product and norm, respectively. First of all, we recall some concepts from convex analysis. which are needed in the derivation of the main results.

Definition 1 ([6, 35, 91]) The set K in H is said to be a convex set, if

$$u + t(v - u) \in K, \qquad \forall u, v \in K, t \in [0, 1].$$

Definition 2 A function F is said to be convex function, if

$$F((1 - t)u + tv) \leq (1 - t)F(u) + tF(v), \quad \forall u, v \in K, \quad t \in [0, 1]. \tag{1}$$

For $t = \frac{1}{2}$, the convex function reduces to

$$F\Big(\frac{u + v}{2}\Big) \leq \frac{1}{2}\{F(u) + F(v)\}, \quad \forall u, v \in K,$$

which is known as the mid-convex (Jensen-convex) function. It is known that if the function is continuous on the interior of the convex set, then convex function and mid-convex are equivalent.

Convex functions are closely related to the integral inequalities and variational inequalities. These types of inequalities have played crucial part in developing fields such as Numerical Analysis, Operations Research, Transportation, Financial Mathematics, Structural Analysis, Dynamical Systems, Sensitivity Analysis, etc.

It is well known that a function F is a convex function, if and only if, it satisfies the inequality

$$F\left(\frac{a+b}{2}\right) \leq \frac{2}{b-a}\int_a^b F(x)dx \leq \frac{F(a)+F(b)}{2}, \quad \forall a, b \in I = [0, 1], \tag{2}$$

which is known as the Hermite–Hadamard type inequality. Such type of the inequalities provide us with the upper and lower bounds for the mean value integral.

If the convex function F is differentiable, then $u \in K$ is the minimum of the F, if and only if, $u \in K$ satisfies the inequality

$$\langle F'(u), v - u\rangle \geq 0, \quad \forall v \in K. \tag{3}$$

The inequalities of the type (3) are called the variational inequalities, which were introduced and studied by Stampacchia [99] in 1964. For the applications, formulation, sensitivity, dynamical systems, generalizations and other aspects of the variational inequalities, see [1–7, 9, 11–23, 27, 32, 34, 42, 57, 73–75] and the references therein.

We now recall the well-known representation theorems, which are mainly due to Riesz–Frechet [13, 94, 95] and Lax–Milgram [25].

Theorem 1 *Let f be a linear and continuous functional. Then there exists a unique $u \in H$ such that*

$$\langle u, v\rangle = \langle f, v\rangle, \quad \forall v \in H. \tag{4}$$

which is known as the Riesz–Frechet representation theorem see [13, 94, 95].

This result says that a linear continuous functional f on the whole space can be represented by an inner product.

One can show that $u \in H$ is the minimum of a differentiable convex functional $I[v]$, where

$$I[v] = \langle v, v\rangle - 2\langle f, v\rangle, \quad \forall v \in H, \tag{5}$$

if and only if, $u \in H$ is a solution of (4). Note that the functional $I[v]$ defined by (5) is a quadratic convex function, which is strongly convex function. Due to this fact the minimum of the functional $I[v]$ exists and is unique. This equivalence formulation can be used to discuss the existence of a unique solution of (4).

Since the inner product $\langle ., .\rangle$ is a bifunction. Naturally the question arises whether this result is also true for any arbitrary bifunction. The answer to this is in affirmative. In this direction, we have the following representation result.

Theorem 2 ([25]) *Let $a(.,.) : H \times H \to H$ be a continuous cocoercive bifunction and let f be a linear continuous function. Then there exists a unique $u \in H$ such that*

$$a(u, v) = \langle f, v\rangle, \quad \forall v \in H, \tag{6}$$

which is known as the Lax–Milgram Lemma [25].

It is obvious that Theorem 2 contains Theorem 1 as a special case. For the generalizations and applications of the Lax–Milgram Lemma, see [10, 24, 36–39, 76, 88] and the references therein.

If the bifunction $a(.,.)$ is bilinear, symmetric and positive, then one can show that $u \in H$ satisfies (6), if and only if, $u \in H$ is the minimum of the functional $I_1[v]$, where

$$I_1[v] = a(v, u) - 2\langle f, v\rangle, \quad \forall v \in H. \tag{7}$$

Now, we consider the problem of finding the functional $I[.]$ defined by (7) on the convex set in a real Hilbert space. This problem was first considered by Stampacchia [99] in 1964. He proved that the minimum of the functional (7) on the convex set K in the real Hilbert space can be characterized by an inequality

$$a(u, v - u) \geq \langle f, v - u\rangle, \quad \forall v \in K, \tag{8}$$

which is known as the variational inequality. It is amazing that the variational inequalities have important applications in almost every field of pure and applied sciences. Variational inequalities have appeared a novel and power technique to consider a class of related problems in a unified framework. Consequently this result says that every linear continuous function on the convex set can be represented by the variational inequality. Obviously, representation results can be viewed as special cases of the variational inequalities.

A question arises that whether similar representation results hold for the nonlinear operators? Motivated by this question, Noor [37–39] have shown that the nonlinear operators can be represented by an arbitrary bifunction $a(.,.)$. He considered the functional $J[v]$ defined as:

$$J[v] = a(v, v) - 2f(v), \quad \forall v \in H \tag{9}$$

proved that the minimum of the functional $J[v]$ on the convex set$K$ can be characterized by the inequality of the form:

$$a(u, v - u) \geq \langle f'(u), v - u\rangle \quad \forall v \in K, \tag{10}$$

where $f'(.)$ is the differential of the functional. The inequality (10) is called the mildly nonlinear variational inequality studied by Noor [37–39]. Note that if f

is a linear continuous function, then inequality (10) reduces to the variational inequality (8) and contains Lax–Milgram Lemma and Riesz–Frechet representation theorem as special cases.

Motivated by the above fact, Noor [36–39] considered more general variational inequalities of which (10) is a special case. To be more precise and to convey the main ideas. we include all the details.

For a given continuous bifunction $a(.,.)$ and a nonlinear operator A, find $u \in K$ such that

$$a(u, v-u) \geq \langle A(u), v-u\rangle, \forall v \in K. \tag{11}$$

The inequalities of the type (11) are known as the mildly nonlinear variational inequalities [40]. It have been shown that the obstacle mildly nonlinear boundary values can be studied in the framework of the mildly nonlinear variational inequalities. It is obvious that the nonlinear operator on the convex sets can be represented by an arbitrary bifunction, which is another aspect of the mildly nonlinear variational inequalities. The mildly nonlinear Variational inequality (11) implies that the nonlinear operators can be represented by an arbitrary bifunction on the convex set.

Noor [38] has considered and studied a more general variational inequality, which contains all the above problems as special cases.

For given nonlinear operators T, A, we consider the problem of finding $u \in K$, such that

$$\langle Tu, v-u\rangle \geq \langle A(u), v-u\rangle, \quad \forall v \in K, \tag{12}$$

which is called the strongly nonlinear variational inequality.

As special cases of (12), we now mention some very important and interesting problems, which are as follows:

I. Problem (12) can be interpreted as variational inequality involving difference of two monotone operators, which is itself a very difficult problem. This problem can be viewed as the problem of finding the minimum of two differences of convex functions, known DC-problem. Such type of problems have applications in optimization theory and imaging process in medical sciences and earthquake.

II. If $K^* = \{u \in H : \langle u, v\rangle \geq 0, \quad \forall v \in K$ is a polar (dual) cone of a cone K in H, then problem (12) is equivalent to finding $u \in K$ such that

$$Tu - A(u) \in K^* \quad \text{and} \quad \langle Tu - A(u), u\rangle = 0, \tag{13}$$

which is known as the strongly nonlinear complementarity problems, which were introduced and studied by Noor [42–44]. Obviously strongly complementarity problems include the complementarity problems, which were introduced by Lemke and Cottle in game theory and management sciences, [4] as special cases. We would like to emphasize that it was Karamardian who proved that variational inequalities and complementarity problems are equivalent, see [4]. This equivalent formulation has proved to be very fruitful from applications and numerical point of views.

III. Problem (12) includes the absolute value equations, which is being investigated extensively in recent years using quite different techniques and ideas, as a special case. To be more precise, take $K = H$, then one easily shows that problem (12) is equivalent to finding $u \in$ such that

$$Tu - A|u| = b, \tag{14}$$

which is called the absolute value equations, where b is a given data. This problem was rediscovered by Mangasarian [28, 29], Mangasarian and Meyer [30] and Noor et al. [87]. Clearly, system of absolute value equations is a special case of strongly complementarity problems.

We also recall the well-known approximation (projection) result, which plays a significant part in the discussion.

Lemma 1 *Let K be a closed and convex set in H. Then, for a given $z \in H$, $u \in K$ satisfies the inequality*

$$\langle u - z, v - u \rangle \geq 0, \quad \forall v \in K,$$

if and only if,

$$u = P_K z,$$

where P_K is the projection of H onto the closed and convex set K.

3 Main Results

In this section, we consider the auxiliary principle technique for the strongly nonlinear variational inequalities (12). This technique can be used to discuss the existence of a unique solution and to consider some iterative methods for solving the variational inequalities of type (12) and absolute value equations (14). This technique is mainly due to Lions and Stampacchia [26]. Noor [38–40] used this technique to study the existence of a unique solution of problem (11). Glowinski et al. [19] modified this technique and used the modified technique for solving mixed variational inequalities. This technique is a powerful and effective tool for solving various types of variational inequalities and equilibrium problems. We only give the main flavour of this technique. The following result can be viewed as a significant refinement of a result of Noor [38–40].

Theorem 3 *Let T be a strongly monotone with constant $\alpha > 0$ and Lipschitz continuous with constant $\beta > 0$, respectively, and the operator A be Lipschitz continuous with constant $\gamma >$. If there exists a parameter $\rho > 0$ such that*

$$0 \leq \rho \leq \frac{\alpha - \gamma}{\beta^2 - \gamma^2}, \quad \rho\gamma < 1, \quad \gamma < \alpha, \tag{15}$$

then there exists a unique solution $u \in K$ satisfying (12).

Proof

Uniqueness Let $u_1 \neq u_2$ be two solution of (12). Then

$$\langle Tu_1, v - u_1 \rangle \geq \langle A(u_1), v - u_1 \rangle, \quad \forall v \in K. \tag{16}$$

and

$$\langle Tu_2, v - u_2 \rangle \geq \langle A(u_2), v - u_2 \rangle, \quad \forall v \in K. \tag{17}$$

Taking $v = u_2$ in (16) and $v = u_1$ in (17), adding the resultants, we have

$$\langle Tu_1 - Tu_2, u_1 - u_2 \rangle \leq \langle A(u_1) - A(u_2), u_2 - u_1 \rangle. \tag{18}$$

Since the operator T is strongly monotone with constant $\alpha > 0$ and the operator A is Lipschitz continuous with constant $\gamma >$ such that

$$\begin{aligned} \alpha \|u_1 - u_2\|^2 &\leq \langle Tu_1 - Tu_2, u_1, -u_2 \rangle \\ &\leq \langle A(u_1) - A(u_2), u_2, -u_1 \rangle \\ &\leq \gamma \|u_1 - u_2\|^2, \end{aligned}$$

from which it follows that

$$(\alpha - \gamma)\|u_1 - u_2\|^2 \leq 0,$$

which implies that $u_1 = u_2$, the uniqueness of the solution.

Existence In order to prove the existence of a solution of problem 12, we use the auxiliary principle technique. To be more precise, for a given $u \in K$, satisfying (12), consider the problem of finding $w \in K$ such that

$$\langle \rho Tu + w - u, v - w \rangle \geq \langle \rho A(u), v - w \rangle, \quad \forall v \in K, \tag{19}$$

where $\rho > 0$ is a constant. The problem (19) is known as the auxiliary variational inequalities. It is clear that, if $w(u) = u$, then w is a solution of the problem (12). That is the problem (19) defines a mapping associated with the problem (12). Thus it is enough to show that the mapping $w(u)$ is a contraction mapping and consequently it has a fixed point $w(u) = u$, which is the solution of the original problem. For $w_1 \neq w_2$(corresponding to $u_1 \neq u_2) \in K$, we have

$$\langle \rho T u_1 + w_1 - u_1, v - w_1 \rangle \geq \langle \rho A(u_1), v - w_1 \rangle, \quad \forall v \in K, \tag{20}$$

and

$$\langle \rho T u_2 + w_2 - u_2, v - w_2 \rangle \geq \langle \rho A(u_2), v - w_2 \rangle, \quad \forall v \in K. \tag{21}$$

Taking $v = u_2$ in (20), $v = u_1$ in (21) and adding the resultants, we have

$$\begin{aligned}\|w_1-w_2\|^2 &\leq \langle u_1-u_2-\rho(Tu_1-Tu_2), w_1-w_2\rangle+\rho\langle(A(u_1)-A(u_2), w_1-w_2\rangle \\ &\leq \|u_1-u_2-\rho(Tu_1-Tu_2)\|\|w_1-w_2\|+\rho\|A(u_1)-A(u_2)\|\|w_1-w_2\|,\end{aligned}$$

from which it follows that

$$\begin{aligned}\|w_1 - w_2\| &\leq \|u_1 - u_2 - \rho(Tu_1 - Tu_2)\| + \rho\|A(u_1) - A(u_2)\|, \\ &\leq \|u_1 - u_2 - \rho(Tu_1 - Tu_2)\| + \rho\gamma\|u_1 - u_2\|, \end{aligned} \tag{22}$$

where we have used the fact that the operator A is Lipschitz continuous with constant $\gamma > 0$. Using the strongly monotonicity and Lipschitz continuity of the operator T with constants $\alpha > 0$ and $\beta > 0$, respectively, we have

$$\begin{aligned}\|u_1-u_2-\rho(Tu_1-Tu_2)\|^2 &= \langle u_1-u_2-\rho(Tu_1-Tu_2), u_1-u_2-\rho(Tu_1-Tu_2)\rangle \\ &\leq \|u_1 - u_2\|^2 - 2\alpha\rho\|u_1 - u_2\|^2 + \rho^2\beta^2\|u_1 - u_2\|^2 \\ &= 1 - 2\rho\alpha + \rho^2\beta^2)\|u_1 - u_2\|^2. \end{aligned} \tag{23}$$

From (22) and (23), we have

$$\begin{aligned}\|w_1 - w_2\| &\leq \{\rho\gamma + \sqrt{1 - 2\rho\alpha + \rho^2\beta^2}\}\|u_1 - u_2\| \\ &= \theta\|u_1 - u_2\|, \end{aligned} \tag{24}$$

where $\theta = \rho\gamma + \sqrt{1 - 2\rho\alpha + \beta^2\rho^2}$.

From (15), it follows that $\theta < 1$. This implies that the mapping w defined by problem (19) is a contraction mapping and consequently, it has a fixed point $w(u) = u \in K$ satisfying (12).

Remark 1 We would like to mention that the auxiliary variational inequality (19) is equivalent to the minimization problem of finding $u \in K$ of the functional

$$I_1[w] = \langle w - u, w - u\rangle - 2\rho\langle Tu - A(u), w - u\rangle, \tag{25}$$

which is quadratic differentiable convex function associated with the problem (19). It follows that the strongly nonlinear variational inequalities are equivalent to quadratic programming problems. This equivalent formulation can used to solve the variational inequalities using quadratic programming problems techniques. This aspect of variational inequality offers another direction for further research.

It is known that a function which can constitute an equivalent optimization problem is called a merit (gap) function. Merit functions turn out to be very useful in designing new globally convergent algorithms and in analysing the rate of convergence of some iterative methods. Various merit (gap) functions for variational inequalities and complementarity problems have been suggested and proposed. Merit functions can be used to obtain the error bounds for the approximate solution of the variational inequalities. Error bounds are functions which provide a measure of the distance between a solution set and an arbitrary point. Therefore, error bounds play an important role in the analysis of global or local convergence analysis of algorithms for solving variational inequalities. Using the technique of Noor [74], one can easily show that the functional $I_1[w]$ defined by (25) is a merit function for the strongly nonlinear variational inequalities. We can consider the merit function associated with (12) as:

$$M_\rho(u) = \max_{v\in K}\{\langle Tu - A(u), v - u\rangle - \frac{1}{2\rho}\|v - u\|^2\}, \quad \forall v \in K, \tag{26}$$

which is called the regularized merit function. This merit can be viewed as a natural extension of the merit function of variational inequalities, which was suggested and investigated by Fukushima [16] using different technique.

4 Iterative Methods

We now how that the auxiliary principle technique can be used to suggest some iterative methods for solving strongly general variational inequalities. Note that, if $w = u$, then w is a solution of the inequality (12). This observation allows us to suggest some iterative methods for solving variational inequalities (12).

Algorithm 1 For a given u_0, find the approximate solution u_{n+1} by the iterative scheme

$$\langle \rho Tu_n + u_{n+1} - u_n, v - u_{n+1}\rangle \geq \langle \rho A(u_n), v - u_{n+1}\rangle, \forall v \in K.$$

Using the auxiliary variational inequality (19), one can suggest and analyse the two-step and three-step iterative methods for solving the variational inequalities.

Algorithm 2 For a given $u_0 \in K$, compute the approximate solution u_{n+1} by the iterative schemes

$$\langle \rho Tu_n + y_n - u_n, v - y_n\rangle \geq \langle \rho A(u_n), v - y_n\rangle, \forall v \in K,$$

$$\langle \rho T y_n + w_n - y_n, v - w_n \rangle \geq \langle \rho A(y_n), v - w_n \rangle, \forall v \in K,$$

$$\langle T w_n + u_{n+1} - w_n, v - u_{n+1} \rangle \geq \langle \rho A(w_n), v - u_{n+1} \rangle, \forall v \in K.$$

Using Lemma 1, Algorithm 2 can be written in the equivalent form as:

Algorithm 3 For a given u_0, find the approximate solution u_{n+1} by the iterative scheme

$$y_n = (1 - \gamma_n)u_n + \gamma_n P_K[u_n - \rho(T u_n - A(u_n))]$$

$$w_n = (1 - \beta_n)u_n + \beta_n P_K[y_n - \rho(T y_n - A(y_n))]$$

$$u_{n+1} = (1 - \alpha_n)u_n + \alpha_n P_K[w_n - \rho(T w_n - A(w_n))], \quad n = 0, 1, 2, ..$$

where $0 \leq \alpha_n \leq 1, 0 \leq \alpha_n \leq 1, 0 \leq \alpha_n \leq 1$ are constants.

Algorithm 3 is known as Noor iterations. For $\gamma_n = 0$, and $\beta_n = 0$, we can obtain Mann iteration and Ishikawa two-step method. Using the technique of Noor [74], one studies the convergence criteria of Algorithm 3.

We now use the auxiliary principle technique to suggest proximal type iterative methods for solving variational inequalities.

For a given $u \in K$ satisfying (12), consider the problem of finding $w \in K$ such that

$$\rho T w + w - u, v - w \rangle \geq \langle \rho A(w), v - w \rangle, \quad \forall v \in K, \tag{27}$$

which is called the auxiliary variational inequality. Note that problems (12) and (27) are quite different from each other. We note that, if $w = u$, then w is a solution of the problem (12). This observation is used to suggest the following iterative methods for solving the variational inequalities.

Algorithm 4 For a given u_0, find the approximate solution u_{n+1} by the iterative scheme

$$\langle \rho T u_{n+1} + u_{n+1} - u_n, v - u_{n+1} \rangle \geq \langle \rho A(u_{n+1}), v - u_{n+1} \rangle, \forall v \in K,$$

which is known as the implicit or proximal point method.

We would like to mention that to implement the implicit methods, one has to use the predictor-corrector technique. Using this idea, Algorithm 4 is equivalent to the following two-step method.

Algorithm 5 For a given u_0, find the approximate solution u_{n+1} by the iterative scheme

$$\langle T u_n + y_n - u_n, v - y_n \rangle \geq \langle \rho A(u_n), v - u_n \rangle$$

$$\langle \rho T y_n + u_{n+1} - u_n, v - u_{n+1} \rangle \geq \langle \rho A(y_n), v - u_{n+1} \rangle, \forall v \in K.$$

Using the Lemma 1, Algorithm 5 is equivalent to the following iterative method.

Algorithm 6 For a given u_0, find the approximate solution u_{n+1} by the iterative scheme

$$y_n = P_K[u_n - \rho(Tu_n - A(u_n))]$$
$$u_{n+1} = P_K[u_n - \rho(Ty_n - A(y_n))].$$

Algorithm 6 can be viewed as a natural generalization of the extragradient method of Korpelevich [23]. The convergence of this method requires only that the monotone operator T and antimonotone operator A. We again consider the other auxiliary principle technique to suggest a more general implicit method.

For a given $u \in K$ satisfying (12), consider the problem of finding w such that

$$\langle \rho Tw+(1-\lambda)w+\lambda u-u, v-(1-\lambda)w+\lambda u\rangle \geq \langle \rho A(w), v(1-\lambda)w+\lambda u\rangle, \quad \forall v \in K, \tag{28}$$

where $\lambda \in [0, 1]$ is a constant. For appropriate and suitable choice of the parameter λ, one can obtain different auxiliary variational inequalities.

Using this auxiliary variational inequality, we suggest another iterative method.

Algorithm 7 For a given u_0, find the approximate solution u_{n+1} by the iterative scheme

$$\langle \rho Tu_{n+1} + (1-\lambda)u_{n+1} + \lambda u_n - u_n, v - (1-\lambda)u_{n+1} - \lambda u_n\rangle$$
$$\geq \langle \rho A(u_{n+1}), v - (1-\lambda)u_{n+1} - \lambda u_n\rangle, \forall v \in K.$$

Using Lemma 1, Algorithm 7 can be rewritten in the following equivalent form.

Algorithm 8 For a given u_0, find the approximate solution u_{n+1} by the iterative scheme

$$u_{n+1} = P_K[(1-\lambda)u_n + \lambda u_{n+1} - \rho(Tu_{n+1} - A(u_{n+1}))], n = 0, 1, 2, \ldots,$$

which is equivalent to the following iterative method.

Algorithm 9 For a given u_0, find the approximate solution u_{n+1} by the iterative scheme

$$\langle \rho Tu_{n+1} + (1-\lambda)(u_{n+1} - u_n), v - u_{n+1}\rangle \geq \langle \rho A(u_{n+1}), v - u_{n+1}\rangle, \forall v \in K.$$

We would like to point out that one can suggest and analyse several iterative methods for solving the strongly variational inequalities using the auxiliary principle technique. Implementation and comparison with other iterative methods is itself an interesting problem for future research.

5 Conclusion and Future Research

In this article, we have reviewed the applications of the auxiliary principle technique for solving the strongly nonlinear variational inequalities involving the difference of two(more) monotone operators. The study of this area is a fruitful field of intellectual endeavour. Our main aim was to show the effective of this powerful technique. It is expected that this method can be used to discuss the existence, uniqueness of the solutions of the absolute value equations, complementarity problems and representation theorems. It is true that each of these problems requires special considerations of the physical problems that model. However, many of these concepts we have discussed are fundamental to all these problems.

It is worth mentioning that the strongly nonlinear variational inequality theory (12) is applicable for studying free and moving mildly boundary value problems of even order. This theory can be extended for odd-order and nonsymmetric boundary value problems. The auxiliary principle technique can be extended for solving the general strongly nonlinear variational inequalities, which include the problem (12) and its variant forms as special cases.

For given nonlinear operators T, A, g, h we consider the problem of finding $u \in H : h(u) \in K$, such that

$$\langle Tu, g(v) - h(u) \rangle \geq \langle A(u), g(v) - h(u) \rangle, \quad \forall v \in H : g(v) \in K, \tag{29}$$

which is called the extended strongly nonlinear variational inequality. For different and appropriate choice of the operators T, A, g, h one find several new classes of variational inequalities, complementarity problems, absolute value problems and representation theorems as special cases of problem (29). We have given only a very brief introduction of this fast growing field. It is our hope that this introduction may motivate the readers to consider these fascinating problems from different aspects and find novel and innovative applications of these extended general variational inequalities. For some recent developments, see the references.

Acknowledgments The authors would like to thank the Rector, COMSATS University Islamabad, Pakistan, for providing excellent research and academic environments. Prof. Dr. Muhammad Aslam Noor would like to express his sincere gratitude to Prof. Dr. C. F. Schubert, Mathematics Department, Queens University, Kingston, Ontario, Canada, for introducing the most interesting and fascinating field of Variational Inequalities. Authors are grateful to Prof. Dr. Themistocles M. Rassias for his kind invitation and support.

References

1. C. Baiocchi, A. Capelo, *Variational and Quasi-Variational Inequalities* (Wiley, New York, 1984)
2. E. Blum, W. Oettli, From optimization and variational inequalities to equilibrium problems. Math. Student **63**(1994), 325–333
3. R.W. Cottle, Nonlinear programs with positively bounded Jacobians. SIAM J. Appl. Math. **14**, 147–158 (1966)

4. R.W. Cottle, J.S. Pang, R.E. Stone, *The Linear Complementarity Problem* (Academic Press, New York, 1992)
5. J. Crank, *Free and Moving Boundary Problems* (Clarendon Press, Oxford, 1984)
6. G. Cristescu, L. Lupsa, *Non-Connected Convexities and Applications* (Kluwer Academic Publishers, Dordrecht, 2002)
7. G. Duvaut, J.L. Lions, *Inequalities in Mechanics and Physics* (Springer, Berlin, 1976)
8. I. Ekland, R. Temam, *Convex Analysis and Variational Problems* (North-Holland, Amsterdam, 1976)
9. R. Eskandari, M. Frank, V. M. Manuilov, M.S. Moslehian, Extensions of the Lax-Milgram theorem to Hilbert C^*-modules (preprint, 2019)
10. W. Fechner, Functional inequalities motivated by the Lax-Milgram lemma. J. Math. Anal. Appl. **402**, 411–414 (2013)
11. G. Fichera, Problemi elastostatici con vincoli unilaterali: il problema di Signorini con ambique condizione al contorno. Atti. Acad. Naz. Lincei. Mem. Cl. Sci. Nat. Sez. Ia **7**(8), 91–140 (1963/1964)
12. V.M. Filippov, *Variational Principles for Nonpotential Operators*, vol. 77 (American Mathematical Society, Providence, 1989)
13. M. Frechet, Sur les ensembles des fonctions et les operations lineaires. C. R. Acad. Sci. Paris **144**, 1414–1416 (1907)
14. T.L. Friesz, D.H. Bernstein, N.J. Mehta, S. Ganjliazadeh, Day-to-day dynamic network disequilibria and idealized traveler information systems. Oper. Res. **42**, 1120–1136 (1994)
15. T.L. Friesz, D.H. Bernstein, R. Stough, Dynamic systems, variational inequalities and control theoretic models for predicting time-varying urban network flows. Trans. Sci. **30**, 14–31 (1996)
16. M. Fukushima, Equivalent differentiable optimization problems and descent methods for asymmetric variational inequality problems. Math. Program. **53**, 99–110 (1992)
17. F. Giannessi, A. Maugeri, *Variational Inequalities and Network Equilibrium Problems* (Plenum Press, New York, 1995)
18. F. Giannessi, A. Maugeri, P.M. Pardalos, *Equilibrium Problems: Nonsmooth Optimization and Variational Inequality Models* (Kluwer Academic Publishers, Dordrecht, 2001)
19. R. Glowinski, J.J. Lions, R. Tremolieres, *Numerical Analysis of Variational Inequalities* (North-Holland, Amsterdam, 1981)
20. P.T. Harker, J.S. Pang, Finite dimensional variational inequalities and nonlinear complementarity problems: a survey of theory, algorithms and applications. Math. Program. **48**, 161–220 (1990)
21. N. Kikuchi, J.T. Oden, *Contact problems in Elasticity* (SIAM, Philadelphia, 1988)
22. D. Kinderlehrer, G. Stampacchia, *An Introduction to Variational Inequalities and Their Applications* (SIAM, Philadelphia, 2000)
23. G.M. Korpelevich, The extragradient method for finding saddle points and other problems. Matecon **12**, 747–756 (1976)
24. H. Kozono, Lax-Milgram theorem in Banach spaces and its generalization to the elliptic system of boundary value problems. Manuscripta Math. **141**, 637–662 (2013)
25. P.D. Lax, A.N. Milgram, Parabolic equations. Ann. Math. Study **33**, 167–190 (1954)
26. J.L. Lions, G. Stampacchia, Variational inequalities. Commun. Pure Appl. Math. **20**, 493–512 (1967)
27. D.T. Luc, M.A. Noor, Local uniqueness of solutions of general variational inequalities. J. Optim. Theory Appl. **117**, 103–119 (2003)
28. O.L. Mangasarian, A generalized Newton method for absolute value equations. Optim. Lett. **3**, 101–108 (2009)
29. O. Mangasarian, Absolute value equation solution via dual complementarity. Optim. Lett. **7**, 625–630 (2013)
30. O.L. Mangasarian, R.R. Meyer, Absolute value equations. Linear Algebra Appl. **419**(2), 359–367 (2006)

31. B. Martinet, Regularization d'inequations variationnelles par approximations successive. Revue Fran. d'Informat. Rech. Oper. **4**, 154–159 (1970)
32. U. Mosco, *Implicit Variational Problems and Quasi Variational Inequalities*. Lecture Notes in Mathematics, vol. 543 (Springer, Berlin, 1976), pp. 83–126
33. A. Moudafi, M.A. Noor, Sensitivity analysis for variational inclusions by Wiener-Hopf equations technique. J. Appl. Math. Stoch. Anal. **12**, 223–232 (1999)
34. A. Nagurney, *Network Economics, A Variational Inequality Approach* (Kluwer Academics Publishers, Boston, 1999)
35. C.P. Niculescu, L.E. Persson, *Convex Functions and Their Applications* (Springer, New York, 2018)
36. M.A. Noor, The Riesz-Frechet theorem and monotonicity. M.Sc. Thesis, Queen's University, Kingston (1971)
37. M.A. Noor, Bilinear forms and convex set in Hilbert space. Boll. Union. Math. Ital. **5**, 241–244 (1972)
38. M.A. Noor, On variational inequalities, Ph.D. Thesis, Brunel University, London (1975)
39. M.A. Noor, Variational inequalities and approximations. Punjab Univer. J. Math. **8**, 25–40 (1975)
40. M.A. Noor, Mildly nonlinear variational inequalities. Math. Anal. Numer. Theory Approx. **24**, 99–110 (1982)
41. M.A. Noor, Strongly nonlinear variational inequalities. C. R. Math. Rep. **4**, 213–218 (1982)
42. M.A. Noor, On the nonlinear complementarity problems. J. Math. Anal. Appl. **123**, 455–460 (1987)
43. M.A. Noor, The quasi complementarity problem. J. Math. Anal. Appl. **130**, 344–353 (1988)
44. M.A. Noor, Fixed-point approach for complementarity problems. J. Math. Anal. Appl. **133**, 437–448 (1988)
45. M.A. Noor, General variational inequalities. Appl. Math. Lett. **1**, 119–121 (1988)
46. M.A. Noor, Quasi variational inequalities. Appl. Math. Lett. **1**, 367–370 (1988)
47. M.A. Noor, Wiener-Hopf equations and variational inequalities. J. Optim. Theory Appl. **79**, 197–206 (1993)
48. M.A. Noor, Variational inequalities in physical oceanography, in *Ocean Wave Engineering* ed. by M. Rahman. (Computer Mechanics Publications, Southampton, 1994), pp. 201–226
49. M.A. Noor, Sensitivity analysis for quasi variational inequalities. J. Optim. Theory Appl. **95**, 399–407 (1997)
50. M.A. Noor, Some recent advances in variational inequalities, Part I, basic concepts. New Zealand J. Math. **26**, 53–80 (1997)
51. M.A. Noor, Some recent advances in variational inequalities, Part II, other concepts. New Zealand J. Math. **26**, 229–255 (1997)
52. M.A. Noor, Some iterative techniques for variational inequalities. Optimization **46**, 391–401 (1999)
53. M.A. Noor, Generalized quasi variational inequalities and implicit Wiener-Hopf equations. Optimization **45**, 197–222 (1999)
54. M.A. Noor, A modified extragradient method for general monotone variational inequalities. Comput. Math. Appl. **38**, 19–24 (1999)
55. M.A. Noor, Some algorithms for general monotone mixed variational inequalities. Math. Comput. Model. **29**, 1–9 (1999)
56. M.A. Noor, Set-valued mixed quasi variational inequalities and implicit resolvent equations. Math. Comput. Model. **29**, 1–11 (1999)
57. M.A. Noor, New approximation schemes for general variational inequalities. J. Math. Anal. Appl. **251**, 217–229 (2000)
58. M.A. Noor, Three-step iterative algorithms for multivalued quasi variational inclusions. J. Math. Anal. Appl. **255**, 589–604 (2001)
59. M.A. Noor, Modified resolvent algorithms for general mixed variational inequalities. J. Comput. Appl. Math. **135**, 111–124 (2001)

60. M.A. Noor, A class of new iterative methods for general mixed variational inequalities. Math. Comput. Model. **31**(13), 11–19 (2001)
61. M.A. Noor, A predictor-corrector method for general variational inequalities. Appl. Math. Lett. **14**, 53–87 (2001)
62. M.A. Noor, Projection-splitting algorithms for general monotone variational inequalities. J. Comput. Anal. Appl. **4**, 47–61 (2002)
63. M.A. Noor, Proximal methods for mixed variational inequalities. J. Optim. Theory Appl. **115**, 447–451 (2002)
64. M.A. Noor, Implicit dynamical systems and quasi variational inequalities. Appl. Math. Comput. **134**, 69–81 (2002)
65. M.A. Noor, Implicit resolvent dynamical systems for quasi variational inclusions. J. Math. Anal. Appl. **269**, 216–226 (2002)
66. M.A. Noor, Sensitivity analysis framework for general quasi variational inequalities. Comput. Math. Appl. **44**, 1175–1181 (2002)
67. M.A. Noor, A Wiener-Hopf dynamical system for variational inequalities. New Zealand J. Math. **31**, 173–182 (2002)
68. M.A. Noor, New extragradient-type methods for general variational inequalities. J. Math. Anal. Appl. **277**, 379–395 (2003)
69. M.A. Noor, Extragradient method for pseudomonotone variational inequalities. J. Optim. Theory Appl. **117**, 475–488 (2003)
70. M.A. Noor, Well-posed variational inequalities. J. Appl. Math. Comput. **11**, 165–172 (2003)
71. M.A. Noor, Mixed quasi variational inequalities. Appl. Math. Comput. **146**, 553–578 (2003)
72. M.A. Noor, Multivalued general equilibrium problems. J. Math. Anal. Appl. **283**(1), 140–149 (2003)
73. M.A. Noor, Auxiliary principle technique for equilibrium problems. J. Optim. Theory Appl. **122**(2), 371–386 (2004)
74. M.A. Noor, Some dvelopments in general variational inequalites. Appl. Math. Comput. **152**, 199–277 (2004)
75. M.A. Noor, Extended general variational inequalities. Appl. Math. Lett. **22**(2), 182–185 (2009)
76. K.I. Noor, M.A. Noor, A generalization of the Lax-Milgram lemma. Can. Math. Bull. **23**(2), 179–184 (1980)
77. M.A. Noor, K.I. Noor, Multivalued variational inequalities and resolvent equations. Math. Comput. Model. **26**(4), 109–121 (1997)
78. M.A. Noor, K.I. Noor, Sensitivity analysis for quasi variational inclusions. J. Math. Anal. Appl. **236**, 290–299 (1999)
79. M.A. Noor, K.I. Noor, Self-adaptive projection algorithms for general variational inequalities. Appl. Math. Comput. **151**, 659–670 (2004)
80. M.A. Noor, W. Oettli, On general nonlinear complementarity problems and quasi equilibria. Le Matematiche **49**, 313–331 (1994)
81. M.A. Noor, E. Al-Said, Change of variable method for generalized complementarity problems. J. Optim. Theory Appl. **100**, 389–395 (1999)
82. M.A. Noor, E.A. Al-Said, Finite difference method for a system of third-order boundary value problems. J. Optim. Theory Appl. **112**, 627–637 (2002)
83. M.A. Noor, T.M. Rassias, A class of projection methods for general variational inequalities. J. Math. Anal. Appl. **268**, 334–343 (2002)
84. M.A. Noor, K.I. Noor, T.M. Rassias, Some aspects of variational inequalities. J. Comput. Appl. Math. **47**, 285–312 (1993)
85. M.A. Noor, K.I. Noor, T.M. Rassias, Set-valued resolvent equations and mixed variational inequalities. J. Math. Anal. Appl. **220**, 741–759 (1998)
86. M.A. Noor, Y.J. Wang, N.H. Xiu, Some projection iterative schemes for general variational inequalities. J. Inequal. Pure Appl. Math. **3**(3), 1–8 (2002)
87. M.A. Noor, K.I. Noor, S. Batool, On generalized absolute value equations. U. P. B. Sci. Bull. Ser. A **80**(4), 63–70 (2018)

88. N. Nyamoradi, M.R. Hamidi, An extension of the Lax-Milgram theorem and its application to fractional differential equations. Electron. J. Differ. Equ. **2015**, 95 (2015)
89. M. Pappalardo, M. Passacantando, Stability for equilibrium problems: from variational inequalities to dynamical systems. J. Optim. Theory Appl. **113**, 567–582 (2002)
90. M. Patriksson, *Nonlinear Programming and Variational Inequality Problems: A Unified Approach* (Kluwer Academic Publishers, Dordrecht, 1998)
91. J. Pecaric, F. Proschan, Y.L. Tong, *Convex Functions, Partial Orderings and Statistical Applications* (Academic Press, New York, 1992)
92. B.T. Polyak, *Introduction to Optimization* (Optimization Software, New York, 1987)
93. T.M. Rassias, M.A. Noor, K.I. Noor, Auxiliary principle technique for the general Lax-Milgram lemma. J. Nonlinear Funct. Anal. **2018**, 34 (2018)
94. F. Riesz, Sur une espace de geometric alytique des aystemes fonctions sommables. C. R. Acad. Sci. Paris **144**, 1409–1411 (1907)
95. F. Riesz, Zur theorie des Hilbertschen rauemes. Acta Sci. **7**, 34–38 (1934/1935)
96. P. Shi, Equivalence of variational inequalities with Wiener-Hopf equations. Proc. Am. Math. Soc. **111**, 339–346 (1991)
97. M. Sibony, Methodes iteratives pour les equations et inequations aux derivees partielles nonlineaires de type monotone. Calcolo **7**, 65–183 (1970)
98. M.V. Solodov, P. Tseng, Modified projection type methods for monotone variational inequalities. SIAM J. Control Optim. **34**, 1814–1830 (1996)
99. G. Stampacchia, Formes bilineaires coercitives sur les ensembles convexes. C. R. Acad. Paris **258**, 4413–4416 (1964)
100. E. Tonti, Variational formulation for every nonlinear problem. Int. J. Eng. Sci. **22**, 1343–1371 (1984)
101. R.L. Tobin, Sensitivity analysis for variational inequalities. J. Optim. Theory Appl. **48**, 191–204 (1986)
102. P. Tseng, A modified forward-backward splitting method for maximal monotone mappings. SIAM J. Control Optim. **38**, 431–446 (2000)
103. P. Tseng, On linear convergence of iterative methods for variational inequality problem. J. Comput. Appl. Math. **60**, 237–252 (1995)
104. Y.J. Wang, N.H. Xiu, C.Y. Wang, Unified framework of projection methods for pseudomonotone variational inequalities. J. Optim. Theory Appl. **111**, 643–658 (2001)
105. Y.J. Wang, N.H. Xiu, C.Y. Wang, A new version of extragradient projection method for variational inequalities. Comput. Math. Appl. **42**, 969–979 (2001)
106. Y.S. Xia, J. Wang, On the stability of globally projected dynamical systems. J. Optim. Theory Appl. **106**, 129–150 (2000)
107. N. Xiu, J. Zhang, M.A. Noor, Tangent projection equations and general variational equalities. J. Math. Anal. Appl. **258**, 755–762 (2001)
108. N.H. Xiu, J. Zhang, Some recent advances in projection-type methods for variational inequalities. J. Comput. Appl. Math. **152**, 559–585 (2003)
109. X.Q. Yang, G.Y. Chen, A class of nonconvex functions and variational inequalities. J. Math. Anal. Appl. **169**, 359–373 (1992)
110. N.D. Yen, Holder continuity of solutions to a parametric variational inequality. Appl. Math. Optim. **31**, 245–255 (1995)
111. N.D. Yen, G.M. Lee, Solution sensitivity of a class of variational inequalities. J. Math. Anal. Appl. **215**, 46–55 (1997)
112. E.A. Youness,E-convex sets, E-convex functions and E-convex programming. J. Optim. Theory Appl. **102**, 439–450 (1999)
113. D. Zhang, A. Nagurney, On the stability of the projected dynamical systems. J. Optim. Theory Appl. **85**, 97–124 (1995)
114. Y.B. Zhao, Extended projection methods for monotone variational inequalities. J. Optim. Theory Appl. **100**, 219–231 (1999)
115. D.L. Zhu, P. Marcotte, Co-coercivity and its role in the convergence of iterative schemes for solving variational inequalities. SIAM J. Optim. **6**, 714–726 (1996)

Unsupervised Stochastic Learning for User Profiles

Nikolaos K. Papadakis

Abstract An unsupervised learning method for user profiles is examined. A user profile is considered the set of all the queries a user issues against an information or a database system. The mechanism of the Markovian model is employed where probabilistic locality translates to semantic locality in ways that facilitate a hierarchical clustering with optimal properties.

AMS Subject Classification 91B70, 91G20, 60J20, 60J22

1 Introduction to Profiling and Contributions

Profile identification is the problem of discovering groups of users with similar interests. Data mining techniques are currently used for discovering usage patterns on web log data for user profile clustering. These techniques have drawbacks mainly due to nature of the web log data. High dimensionality and scarcity of the web log data are difficult to handle with traditional statistical analysis and machine learning methods. Another difficulty is the identification of sessions and users from the web access data due to the incompleteness of the available data.

Dourish and Chalmers discussed three types of metaphors for the design and use of an information space, namely spatial navigation, semantic navigation, and social navigation [1]. Jasper is another information filtering and sharing system [2]. It maintains a collection of summaries and annotated reference links to documents on the World-Wide Web (WWW). Generalised Similarity Analysis (GSA) is a framework for structuring and visualising hyper-media networks [3]. It unifies a number of similarity-based models and visualises salient structural patterns of interconnected documents.

N. K. Papadakis (✉)
Hellenic Military Academy, Vari Attikis, Greece
e-mail: npapadakis@sse.gr

N. J. Daras, T. M. Rassias (eds.), *Computational Mathematics and Variational Analysis*, Springer Optimization and Its Applications 159,
https://doi.org/10.1007/978-3-030-44625-3_16

Existing approaches to web personalisation often rely heavily on human input for determining the personalisation actions. For example, the use of static profiles obtained through user registration is the predominant techniques used by many web-based companies. Collaborative filtering techniques (e.g., GroupLens [4]) usually involve explicit ratings of objects by users. Content-based filtering techniques such as those used by WebWatcher [5] and Letizia [6] rely on client-side personal profiles and content similarity of web documents to these profiles. This type of input is often a subjective description of the users by the users themselves, and thus prone to biases. Furthermore, the profile is static, and its performance degrades over time as the profile ages. Using content similarity alone as a way to obtain aggregate profiles may result in missing important relationships among web object based on their usage. A number of approaches have been developed dealing with specific aspects of web usage mining for the purpose of automatically discovering user profiles. For example, Perkowitz and Etzioni [7] proposed the idea of optimising the structure of profile identification is the problem of discovering groups of users with similar interests. Data mining techniques are currently used for discovering usage patterns on web log data for user profile clustering. These techniques have drawbacks mainly due to nature of the web log data. High dimensionality and scarcity of the web log data are difficult to handle with traditional statistical analysis and machine learning methods. Another difficulty is the identification of sessions and users from the web access data due to the incompleteness of the available data. Schechter et al. [8] have developed techniques for using path profiles of users to predict future HTTP requests, which can be used for network and proxy caching. Spiliopoulou et al. [9] and Buchner and Mulvenna [10] have applied data mining techniques to extract usage patterns from web logs, for the purpose of deriving marketing intelligence. Shahabi et al. [11], Yan et al. [12], and Nasraoui et al. [13] have proposed clustering of user sessions to predict future user behaviour. B. Mobasher et al. proposed a technique for capturing common user profiles based on association-rule discovery and usage-based clustering. The clustering technique employed called Association-Rule Hypergraph Partitioning is capable of clustering URL references directly, hence eliminating the need for discovering clusters of user transactions as the starting point for obtaining aggregate profiles.

Log file mining on the other hand is the discovery of user access patterns from web servers. Chen et al. pioneered work in log file mining [14] and several similar approaches followed, but most techniques make assumptions such as the insignificance of backward references and the need to handle noise that are not applicable to product development.

2 Unsupervised Learning with Stochastic Hierarchical Clustering

A scheme that employs stochastic techniques on query logs, rather on the web log data is presented here. Users' query logs are the list of queries asked by users, each query considered a list of keywords. Query logs present richer semantics and less sparsity than web access logs. The keyword sequence is viewed as a stochastic process assuming semantic dependency for keywords in the same query.

The goal of the proposed approach is to cluster the stochastic state space in a simple and general way, while keeping the calculations feasible even for very large state spaces. This clustering of the state space is translated to a semantic clustering of the keywords. To achieve this clustering to be hierarchical and to the desired semantic granularity, an effective algorithm is presented where a stochastic process is started from all possible initial conditions. Geometric properties of convergence are identified towards an unsupervised learning scheme with optimal properties. In this section we present some basic concepts of the Markovian process theory, some notation is also introduced. We start with preliminaries on the Markovian process and the specific method will be introduced thereafter. Throughout this paper bold letters are vectors, capital letters are matrices, and standard letters are scalars. A bold letter with subscript, for example, is a column vector where its place to a matrix is implied through the subscript. All vectors are column vectors except the equilibrium vector.

2.1 *Preliminaries: The Markovian Process and its Transient Behaviour*

Let $P = \{p_{ij}\}$ the one step transition probability matrix of a Markovian process of N states. A *chain* is a kind of collective trapping state, a set of states the process never leaves once enters that set. A transient state cannot be a member of a chain but can be associated with one if it is possible for the process to enter that chain from that transient state. All states in a chain are therefor recurrent. The n-step transition probability matrix we denote $\Phi(n) = \{\phi_{ij}(n)\}$ and it is $\Phi(n) = P^n = \underbrace{P * P * \ldots\ldots * P}_{\text{n } times}$. The ith row of $\Phi(n)$ corresponds to the probability vector of the system being in each of the states after n steps (transient response n steps into the process), if started from state i, or equivalently from the initial condition $\boldsymbol{e}_i$, where $\boldsymbol{e}_i$ the ith, $N-$ dimensional orthocanonical vector which has 1 at the ith place and zeros elsewhere (an example for three dimensions is $\boldsymbol{e}_1 = (1, 0, 0)$, $\boldsymbol{e}_2 = (0, 1, 0)$, $\boldsymbol{e}_3 = (0, 0, 1)$). As n goes to infinity the matrix $\Phi(n)$ approaches the limit matrix $P^{\infty} = \Phi$. All states belonging to the same chain will have identical rows in the matrix Φ, each chain therefor can itself be thought of as a monodesmic

Fig. 1 A Markovian process showing two non-recurrent states with identical future

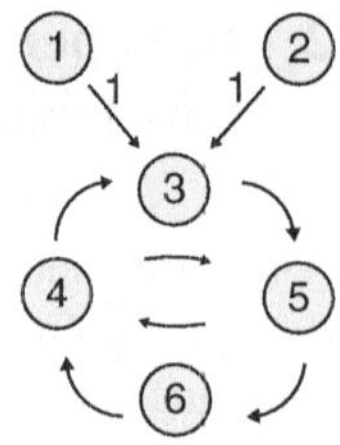

Markov process[1] having its own equilibrium probability vector, the identical rows of its states. The transient states associated with more than one chain will have a row in the matrix Φ that is the sum of the rows corresponding to each chain with which they are associated, weighted by the probabilities of entering each of these chains. In the general case conclusively, a Markovian process can be thought of a collection of its chains together with any possible transient states associated with them Fig. 1. A Markovian process of more than one chain will not converge to a single equilibrium probability vector. Depending on the initial condition the equilibrium vector will be the one of the corresponding chain the initial condition is associated with. In this case the nodes associated with any particular chain could be identified by the corresponding group of identical rows of the limiting matrix Φ. The convergence process in a monodesmic process has a single equilibrium point. This is the point where all transient state series from all initial conditions meet eventually. The rate of this convergence is proportional to the product of the eigenvalues of the transition probability matrix, since the rate of convergence in each dimension equals the corresponding eigenvalue. An eigenvalue of 1 corresponds to no convergence along the corresponding direction, the number of eigenvalues that are equal to 1 equals the number of chains in the process. The matrix $T(n) = \Phi(n) - \Phi$, the transient matrix, contains the transient components of the Markov process. The sum of these matrices up to time n is the transient n-sum matrix $T_{\Sigma}(n) = \sum_{k=0}^{n} T(k)$. The z-transform of $T(n)$ evaluated at 1 is the transient sum matrix $T_g(1)$ that contains the infinite sum of the transient components of the Markovian process and can be calculated from

$$T_g(1) = \sum_{n=0}^{\infty}(P^n - \Phi) = [I - P + \Phi]^{-1} - \Phi \tag{1}$$

It is clear that $T_g(1)$ is the limit of $T_{\Sigma}(n)$. The rows of $T_{\Sigma}(n)$, $T(n)$, and $T_g(1)$ will be denoted $\boldsymbol{\tau}_i(n), \boldsymbol{t}_i(n)$, and $\boldsymbol{t}_g(i)$ and their values with $\tau_{ij}(n), t_{ij}(n)$, and $t_g(ij)$, respectively. The expected number of occupancies of the nodes at time n if started from i is

[1] A Markovian process of a single chain, having a limiting matrix with identical rows, duodesmic if the process consists of two chains, tridesmic of three chains, etc.

$$\bar{\boldsymbol{\eta}}_i(n) = \sum_{k=0}^{n} \boldsymbol{\phi}_i(k) \tag{2}$$

The expected fractional occupancies from i is

$$\boldsymbol{f}_i(n) = \frac{\bar{\boldsymbol{\eta}}_i(n)}{n+1} = x_{i1}\boldsymbol{v}_1 + \frac{\boldsymbol{\tau}_i(n)}{n+1} \tag{3}$$

and the difference of the fractional occupancies of each node at step n due to starting the system from i rather than j is then:

$$\boldsymbol{f}_i(n) - \boldsymbol{f}_j(n) = \frac{\boldsymbol{\tau}_i(n) - \boldsymbol{\tau}_j(n)}{n+1} \tag{4}$$

The vector of fractional occupancies $\boldsymbol{f}_i(n)$ converges to the equilibrium vector $x_{i1}\boldsymbol{v}_1$ as was expected, the difference of fractional occupancies due to different initial condition converges to zero.

2.2 *Markovian State Space Partition: The Convergence Process on the Eigenvectors of the One Step Transition Probability Matrix*

In this section we realise that the convergence from an initial condition to the respective equilibrium state happens on the direction of the eigenvectors of the one step transition probability matrix and the rate on each direction equals the respective eigenvalue. The l_1 normalised eigenvector that corresponds to an eigenvalue of 1 is the equilibrium state for the respective initial condition since the process converges with rate 1 on this direction. Let i, j nodes of some chain and $\boldsymbol{e}_i = \boldsymbol{\phi}_i(0)$, $\boldsymbol{e}_j = \boldsymbol{\phi}_j(0)$ the respective initial conditions. Let also λ_i, $i = 1, 2, .., N$ be the eigenvalues of P, $|\lambda_i| \leq 1, \forall i$ and $\boldsymbol{v}_i$ the respective normalised[2] eigenvectors. Let us also suppose that $|\lambda_i| \geq |\lambda_j|$ when $i < j$. The initial conditions can be expressed as linear combinations of the normalised eigenvectors:

$$\begin{aligned} \boldsymbol{e}_i &= \boldsymbol{\phi}_i(0) = x_{i1}\boldsymbol{v}_1 + x_{i2}\boldsymbol{v}_2 + \cdots + x_{iN}\boldsymbol{v}_N \\ \boldsymbol{e}_j &= \boldsymbol{\phi}_j(0) = x_{j1}\boldsymbol{v}_1 + x_{j2}\boldsymbol{v}_2 + \cdots + x_{jN}\boldsymbol{v}_N \end{aligned} \tag{5}$$

The process if started from $\boldsymbol{e}_i$ or $\boldsymbol{e}_j$ converges to some equilibrium eigenvector (respective eigenvalue equal to 1), lets say to $\boldsymbol{v}_1$. We can rearrange the indices in

[2]Normalised with respect to the l_1 norm, the sum of the absolute coordinates equals 1. All probability vectors are normalised with respect to l_1.

equation (5) so that $\boldsymbol{v}_1$ is the equilibrium eigenvector of the chain, $|\lambda_1| = 1$ and $|\lambda_i| < 1, \forall i = 2, .., N$, where N is reduced to the cardinality of the chain the node i belongs to. The system's response n-steps into the process when started from i and j can be expressed as:

$$\begin{aligned}\boldsymbol{\phi}_i(n) &= \boldsymbol{v}_1 + \lambda_2^n x_{i2}\boldsymbol{v}_2 + \cdots + \lambda_N^n x_{iN}\boldsymbol{v}_N = \boldsymbol{v}_1 + \boldsymbol{t}_i(n)\\ \boldsymbol{\phi}_j(n) &= \boldsymbol{v}_1 + \lambda_2^n x_{j2}\boldsymbol{v}_2 + \cdots + \lambda_N^n x_{jN}\boldsymbol{v}_N = \boldsymbol{v}_1 + \boldsymbol{t}_j(n)\end{aligned} \tag{6}$$

The vector of the expected number of occupancies, n steps into the process that started from node i, is then:

$$\begin{aligned}\bar{\boldsymbol{\eta}}_i(n) &= \sum_{k=0}^{n}\boldsymbol{\phi}_i(k) = (n+1)\boldsymbol{v}_1 + x_{i2}\boldsymbol{v}_2\sum_{k=0}^{n}\lambda_2^k + \ldots\ldots + x_{iN}\boldsymbol{v}_N\sum_{k=0}^{n}\lambda_N^k\\ &= (n+1)\boldsymbol{v}_1 + \sum_{k=0}^{n}\boldsymbol{t}_i(k) = (n+1)\boldsymbol{v}_1 + \boldsymbol{\tau}_i(n)\end{aligned} \tag{7}$$

and the expected fractional occupancies at step n if started from i is then:

$$\boldsymbol{f}_i(n) = \frac{\bar{\boldsymbol{\eta}}_i(n)}{n+1} = \frac{1}{n+1}\sum_{p=1}^{N}\frac{1-\lambda_p^n}{1-\lambda_p}x_{ip}\boldsymbol{v}_p \tag{8}$$

The difference of the expected occupancies of each node at step n due to starting the system from i rather than j is then:

$$\bar{\boldsymbol{\eta}}_i(n) - \bar{\boldsymbol{\eta}}_j(n) = \sum_{k=0}^{n}\boldsymbol{\phi}_i(k) - \sum_{k=0}^{n}\boldsymbol{\phi}_j(k) = \sum_{k=0}^{n}(\boldsymbol{\phi}_i(k) - \boldsymbol{\phi}_j(k)) \tag{9}$$

which from (6) and (7) becomes:

$$\bar{\boldsymbol{\eta}}_i(n)-\bar{\boldsymbol{\eta}}_j(n)=\sum_{p=2}^{N}\sum_{k=0}^{n}\lambda_p^k(x_{ip}-x_{jp})\boldsymbol{v}_p=\sum_{p=2}^{N}\frac{1-\lambda_p^n}{1-\lambda_p}(x_{ip}-x_{jp})\boldsymbol{v}_p=\boldsymbol{\tau}_i(n)-\boldsymbol{\tau}_j(n) \tag{10}$$

and the difference of fractional occupancies due to the different initial condition is then:

$$\boldsymbol{f}_i(n) - \boldsymbol{f}_j(n) = \frac{\bar{\boldsymbol{\eta}}_i(n) - \bar{\eta}_j(n)}{n+1} = \frac{1}{n+1}\sum_{p=2}^{N}\frac{1-\lambda_p^n}{1-\lambda_p}(x_{ip} - x_{jp})\boldsymbol{v}_p \tag{11}$$

which is a linear combination of the eigenvectors.

To see how the first passage times relate to the coordinates on the eigenvectors we let n go to infinity and (10) becomes:

$$\bar{\boldsymbol{\eta}}_i(\infty) - \bar{\boldsymbol{\eta}}_j(\infty) = \sum_{p=2}^{N} \frac{x_{ip} - x_{jp}}{1 - \lambda_p} \boldsymbol{v}_p = \boldsymbol{t}_g(i) - \boldsymbol{t}_g(j) \tag{12}$$

The mean first passage time $\bar{\theta}_{ji}$ from node j to node i is given by $\frac{t_g(ii) - t_g(ji)}{\pi_i}$. Equation (12) shows that the mean first passage time from j to i is related to the coefficients of the initial conditions $\boldsymbol{e}_i$ and $\boldsymbol{e}_j$ in their expression as a linear combination of the eigenvectors of P. In particular from (12) we get:

$$\pi_i \bar{\theta}_{ji} = \frac{x_{i2} - x_{j2}}{1 - \lambda_2} v_{2i} + \cdots + \frac{x_{iN} - x_{jN}}{1 - \lambda_N} v_{Ni} \tag{13}$$

in words this means that the first passage time from j to i depends on the difference of the coordinates of the respective initial conditions on the eigenvectors of P amplified by the respective eigenvalue. While the first passage time is a measure of the state connectivity, it is not easy to compute and it is not symmetric. To measure the distance between the states we will use another quantity instead that we show is related to the first passage times but is also symmetric and quantifies the decomposability of random Markovian processes (Figure 2).

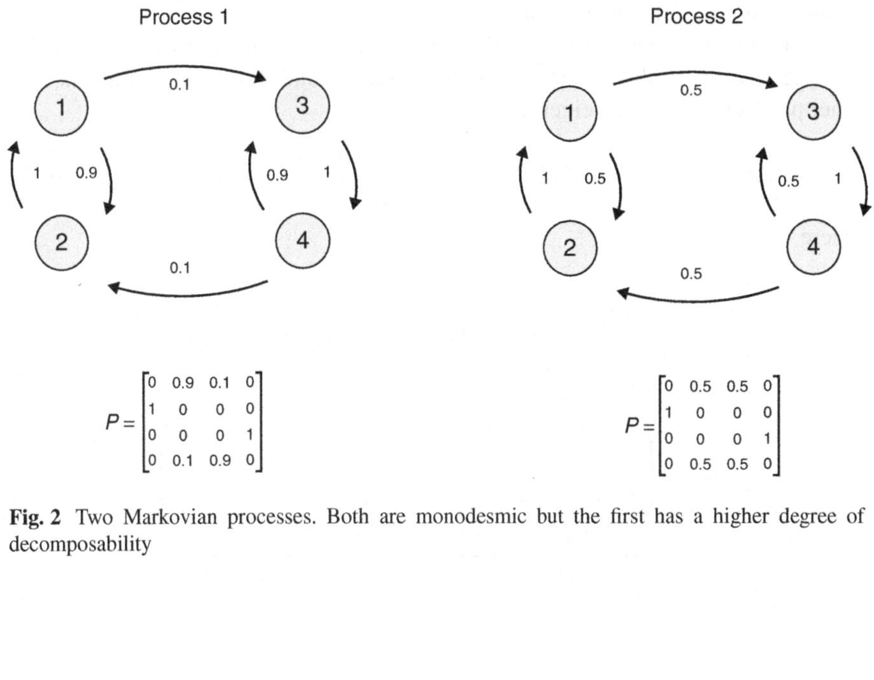

Fig. 2 Two Markovian processes. Both are monodesmic but the first has a higher degree of decomposability

2.3 The Initial Conditions as Random Variables and Their Covariance

Starting the convergence process of a Markovian network from a specific node is, due to the Markovian property, equivalent to passing from this node at any given moment. The transient response of the system n steps into the process is therefor what we should expect to happen n steps into the future after each time the system visits the specific node. For some nodes, the transient responses of the system after passing/starting from these nodes are more similar than for others. We want to measure this similarity of the nodes as initial conditions or transient points and to use this to cluster the states. Here we will realise that the covariance of the system's transient fractional occupancies from different initial conditions is an indicator of the similarity of the initial conditions (states) in respect to the system's transient response after passing from these states.

Lets see i and j as random variables with the sense of initial conditions. The values the random variable i and j take at time n are the coordinates of the process's transient fractional occupancy vector $\boldsymbol{f}_i(n)$ and $\boldsymbol{f}_j(n)$ at time n. The mean value therefor for i and j is $\frac{1}{N}$. From the definition of covariance, the fact that $\bar{i} = \bar{j} = \frac{1}{N}$ and the fact that $\forall n, \|\boldsymbol{f}_i(n)\|_1 = \|\boldsymbol{f}_j(n)\|_1 = 1$ we get:

$$cov_n(i, j) = \frac{1}{N}\langle \boldsymbol{f}_i(n), \boldsymbol{f}_j(n)\rangle - \frac{1}{N^2} \tag{14}$$

where $\langle\rangle$ signifies the inner product.

From (11) we have an expression of $\boldsymbol{f}_i(n)$ and $\boldsymbol{f}_j(n)$ as linear combination of the eigenvectors of the one step probability matrix and thus we can express their inner product through the eigenvectors as follows:

$$\langle \boldsymbol{f}_i(n), \boldsymbol{f}_j(n)\rangle = (\boldsymbol{\lambda}(n)\Box \boldsymbol{x}_i) \cdot U \cdot (\boldsymbol{x}_j^T \Box \boldsymbol{\lambda}(n)^T) = \boldsymbol{\lambda}(n) \cdot (X_i \Box U \Box X_j^T) \cdot \boldsymbol{\lambda}(n)^T \tag{15}$$

where

$$\boldsymbol{\lambda}(n) = \frac{1}{n+1}(\frac{1-\lambda_1^n}{1-\lambda_1}, \ldots, \frac{1-\lambda_N^n}{1-\lambda_N}),$$

$$X_i = \underbrace{\begin{pmatrix} x_{i1} & x_{i1} & \ldots & x_{i1} \\ x_{i2} & x_{i2} & \ldots & x_{i2} \\ \ldots & \ldots & \ldots & \ldots \\ x_{iN} & x_{iN} & \ldots & x_{iN} \end{pmatrix}}_{N times}, X_j = \underbrace{\begin{pmatrix} x_{j1} & x_{j1} & \ldots & x_{j1} \\ x_{j2} & x_{j2} & \ldots & x_{j2} \\ \ldots & \ldots & \ldots & \ldots \\ x_{jN} & x_{jN} & \ldots & x_{jN} \end{pmatrix}}_{N times},$$

$$U = \begin{pmatrix} \langle \boldsymbol{v}_1, \boldsymbol{v}_1 \rangle & \dots & \langle \boldsymbol{v}_1, \boldsymbol{v}_N \rangle \\ \langle \boldsymbol{v}_2, \boldsymbol{v}_1 \rangle & \dots & \langle \boldsymbol{v}_2, \boldsymbol{v}_N \rangle \\ \dots & \dots & \dots \\ \langle \boldsymbol{v}_N, \boldsymbol{v}_1 \rangle & \dots & \langle \boldsymbol{v}_N, \boldsymbol{v}_N \rangle \end{pmatrix}$$

where $\boldsymbol{v}_i$ the eigenvectors of the one step transition probability matrix and $\Box$ the pointwise or congruent matrix multiplication of commensurate matrices (matrices of the same dimensions). The properties of the $\Box$ operation can be found in [15].

To show how the covariance of i and j is related to the first passage times from i and j we start from another expression of their covariance:

$$2cov_n(i, j) = var_n(i) + var_n(j) - var_n(i - j) \tag{16}$$

while the variance of the difference $(i - j)$ is

$$var_n(i - j) = \frac{\|\boldsymbol{f}_i(n) - \boldsymbol{f}_j(n)\|^2}{N} \tag{17}$$

From (11) we can get an expression of the vector $\boldsymbol{f}_i(n) - \boldsymbol{f}_j(n)$ as a linear combination of the eigenvectors and therefor we can express its norm in respect to the eigenbase:

$$\begin{aligned} \|\boldsymbol{f}_i(n) - \boldsymbol{f}_j(n)\|^2 &= \langle \boldsymbol{f}_i(n) - \boldsymbol{f}_j(n), \boldsymbol{f}_i(n) - \boldsymbol{f}_j(n) \rangle \\ &= (\boldsymbol{\lambda}(n)\Box(\boldsymbol{x}_i - \boldsymbol{x}_j)) \cdot U \cdot ((\boldsymbol{x}_i - \boldsymbol{x}_j)^T \Box \boldsymbol{\lambda}(n)^T) \\ &= \boldsymbol{\lambda}(n) \cdot (A \Box U \Box A^T) \cdot \boldsymbol{\lambda}(n)^T \end{aligned} \tag{18}$$

where

$$\boldsymbol{\lambda}(n) = \frac{1}{n+1}(\frac{1-\lambda_1^n}{1-\lambda_1}, \dots, \frac{1-\lambda_N^n}{1-\lambda_N}),$$

$$A = \underbrace{\begin{pmatrix} x_{i1} - x_{j1} & x_{i1} - x_{j1} & \dots & x_{i1} - x_{j1} \\ x_{i2} - x_{j2} & x_{i2} - x_{j2} & \dots & x_{i2} - x_{j2} \\ \dots & \dots & \dots & \dots \\ x_{iN} - x_{jN} & x_{iN} - x_{jN} & \dots & x_{iN} - x_{jN} \end{pmatrix}}_{Ntimes},$$

$$U = \begin{pmatrix} \langle \boldsymbol{v}_1, \boldsymbol{v}_1 \rangle & \dots & \langle \boldsymbol{v}_1, \boldsymbol{v}_N \rangle \\ \langle \boldsymbol{v}_2, \boldsymbol{v}_1 \rangle & \dots & \langle \boldsymbol{v}_2, \boldsymbol{v}_N \rangle \\ \dots & \dots & \dots \\ \langle \boldsymbol{v}_N, \boldsymbol{v}_1 \rangle & \dots & \langle \boldsymbol{v}_N, \boldsymbol{v}_N \rangle \end{pmatrix}$$

From (9), (18), (17), and (13) we get that the variance of $i - j$ is an indication of the difference of all the state occupancies due to starting the process from i rather than j and similarly to the first passage time is calculated from the differences of the respective coordinates of the initial conditions on the eigenbase. From the previous and (16) we get that the covariance is decreasing(increasing) as the variance increases(decreases) thus is an indication of similarity of the initial conditions (random variables) i and j.

Furthermore if i, j are nodes in different chains then from (14) we get that $cov_n(i, j)$ takes its minimum value equal to $-\frac{1}{N^2}$. This holds for all n, since if i, j in different chains, then $\boldsymbol{f}_i(n)$ is normal to $\boldsymbol{f}_j(n)$, while their mean is constant and equals $\frac{1}{N}$ for all n. In all other cases we get that $cov_n(i, j) > -\frac{1}{N^2}$. This is true because $\boldsymbol{f}_i(n)$ and $\boldsymbol{f}_j(n)$ are both stochastic and since in this case not normal to each other, their inner product is always positive. The maximum value of $cov_n(i, j)$ is when $\boldsymbol{f}_i = \boldsymbol{f}_j = \boldsymbol{e}_k$ for some k and this value equals $\frac{N-1}{N^2}$. We thus conclude that $-\frac{1}{N^2} \leq cov_n(i, j) \leq \frac{N-1}{N^2}$ and the covariance of i and j as initial condition random variables presents an adequate measure of the similarity of nodes i and j as initial conditions since the covariance takes its minimum value when the two nodes belong to different chains and in all other cases it increases as the dissimilarity of i, j, expressed by the variance of the difference of the system's transient fractional occupancies due to starting the system from one of these nodes rather from the other, decreases (Figure 3).

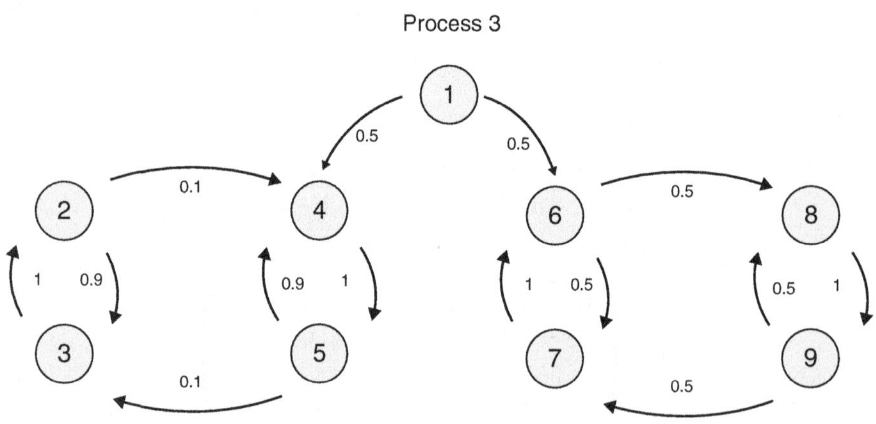

$$
P_3 = \begin{bmatrix}
0 & 0 & 0 & 0.5 & 0 & 0.5 & 0 & 0 & 0 \\
0 & 0 & 0.9 & 0.1 & 0 & 0 & 0 & 0 & 0 \\
0 & 1 & 0 & 0 & 0 & 0 & 0 & 0 & 0 \\
0 & 0 & 0 & 0 & 1 & 0 & 0 & 0 & 0 \\
0 & 0 & 0.1 & 0.9 & 0 & 0 & 0 & 0 & 0 \\
0 & 0 & 0 & 0 & 0 & 0 & 0.5 & 0.5 & 0 \\
0 & 0 & 0 & 0 & 0 & 1 & 0 & 0 & 0 \\
0 & 0 & 0 & 0 & 0 & 0 & 0 & 0 & 1 \\
0 & 0 & 0 & 0 & 0 & 0 & 0.5 & 0.5 & 0
\end{bmatrix}
$$

Fig. 3 A Markovian process showing a non-recurrent state leading to two different futures

3 Unsupervised Learning of User Profiles

A model is presented that captures behaviours and measures their difference numerically. A behaviour will be constructed from the queries that a user asks, the content of the queries and the order asked. The content of a query is the list of keywords it contains. The behaviour of a user as well as also the behaviour of all users collectively will be modelled through a state transition diagram. In the proposed model each user is represented by a Markovian process, constructed from the user's queries. The aggregate behaviour of all users is captured in another Markovian process. The aggregate process is then used to weight the distance between processes of different users and classify any other profile as well. The Markovian process for each user and the aggregate process are constructed simultaneously in a similar manner. Each keyword corresponds to a state. One step transition probabilities between the states are assigned modelling a query this way. Each time a keyword appears in a query, its state counter is advanced, if a state transition has occurred, the interstate link counter is also advanced and the destination state's counter as well. We measure this way both the occurrences of the keyword but also the sequencing of these occurrences. We batch process G keywords, we advance the counters, we then update the probabilities as above. Before we process the next set of G keywords, the counters are cleared. If user u asks keyword k_j immediately following keyword k_i and this occurred k times, then the one step transition probability p_{ij} is being updated as follows: If p_{ij} is the current probability (before the update) based on M queries, then the new probability (based on $M + G$ queries) is

$$p_{ij} = \frac{Mp_{ij} + k}{G + M}$$

The merits of modelling a query as a transition between states and the user profile as a Markovian process will only be clear once we introduce the distance implied by the aggregate Markovian process. A semantic distance between the keywords is implied by the aggregate Markovian process and used to weight the differences of the occurrences of these keywords in the queries of the different users. Users that have no keyword in common may end up having 'close' profiles if their keywords were asked together by many other uses, thus the general behaviour signifies the two users semantically close.

3.1 The Distance Between User Profiles

In this section we will use the previous results to construct a distance function between different user profiles when each profile is represented by a Markovian process. Given a general user behaviour in the form of a Markovian process with state space S and one step transition probability matrix P and two user profiles represented by two Markovian processes with state spaces $S_1 \subset S$, $S_2 \subset S$ and one step transition probability matrices P_1, P_2 we want to calculate the distance between the two user profiles in the context of the general behaviour.

Suppose that the general behaviour is captured in the process of Figure 4a. In the same figure (b)–(f) and (g) are six other processes representing six users. Users 1 and 2 (u_1, u_2) use the same keywords but in a different way, whereas user 1 and user 3 (u_3) use different keywords but the keywords they use are pairwise close to each other as is shown in the process representing the general behaviour. User 1 and user 3 should be closer to each other than user 4 and user 5 (u_4 and u_5) are since users 1 and 3 use keywords that are closer with respect to the general behaviour than the keywords the other two users use.

The distance measure between users has the form $d(u_i, u_j) = x^T Ax$ and it is well defined since A is a positive definite symmetric matrix that holds the similarities of keywords in the initial Markovian model of keyword usage. The similarity of keywords is expressed in terms of their covariance in the feature space of the transient fractional occupancy matrix and the distance between users is introduced in the form of a generalised Euclidean distance.

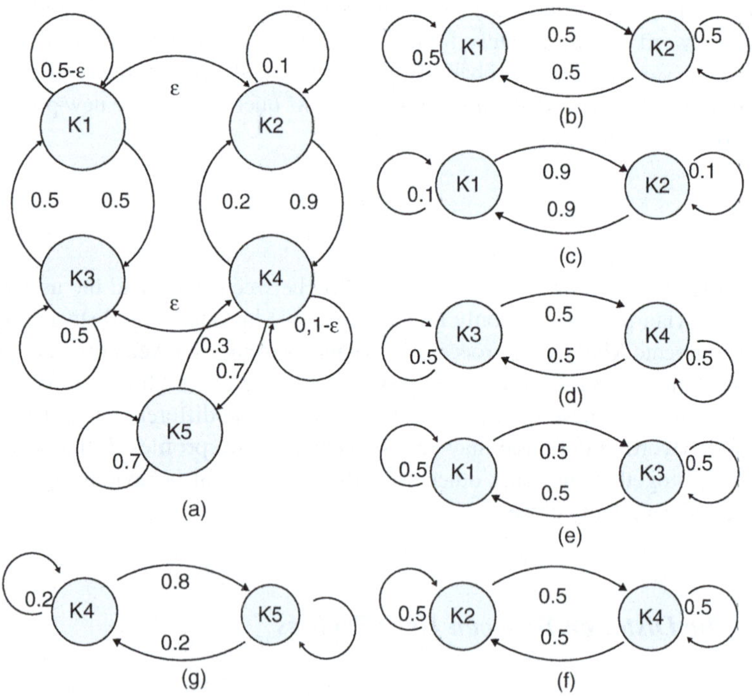

Fig. 4 A process representing the general behaviour and five other processes representing the behaviour of five users. (**a**) General behaviour. (**b**) Behaviour of user 1. (**c**) Behaviour of user 2. (**d**) Behaviour of user 3. (**e**) Behaviour of user 4. (**f**) Behaviour of user 5. (**g**) Behaviour of user 6

The construction of the distance between user profiles is based on a direct comparison (out of context) between them and an adjustment of the result of this comparison according to the general behaviour (bringing the result into context). In all circumstances from now on when we refer to a profile we will refer to the equilibrium vector that corresponds to the Markovian process assigned to this profile denoted by π_i for user i.

In the general case then the *out of context* distance between users u_i and u_j represented by π_i and π_j, respectively, equals:

$$\delta_{i,j} = \|\pi_i - \pi_j\|^2 \tag{19}$$

Users are now compared and their difference is in the form of a vector. The next step in constructing the distance between users is bringing this result into context by applying a non-uniform scaling to the coordinates of this vector before calculating its length. This non-uniform scaling will be according to the directions where the general behaviour projects with maximum variance. The treatment here is the typical PCA for cancelling out the dependencies in the feature space. To form keyword clusters in the general behaviour we proceed in calculating the matrix of expected fractional occupancies at time n, denoted by $F_G(n)$. It is $F_G(n) = \sum_{k=0}^{n} P^k$ where P is the one step transition probability matrix of the general behaviour. A termination condition is needed, an n to stop the series at the point where the slow convergence has taken over but not before the rapid convergence has finished. Since the rapid convergence will form clusters in the rows of $F_G(n)$ with respect to their Euclidean distance and since the slow convergence will not significantly change this forming in one iteration, our criterion for terminating the series is the comparison of the consecutive vector variances of the rows of $F_G(n)$. If the change in consecutive variances is below a threshold, we stop the calculation and in any case we stop the iteration if n exceed the dimension of N. From now on we will use just F_G meaning $F_G(n)$ calculated at the desired n. The rows of matrix F_G will act as observations, with a covariance matrix $\Sigma(F_G)$, the elements of which are the covariances of pairs of nodes, as was explained above. To bring the difference of two users into context we perform a non-uniform scaling along the axes of the orthogonal coordinate system of the observations maximum variance directions. This will penalise the length of δ_{ij} according to the amount of its projection on the direction of the eigenvectors of $\Sigma(F_G)$. In Markovian terms the projection of δ_{ij} on a direction of a slow shrinking direction of a slow shrinking eigenvector will be more penalised(increased)than the projection on a direction of a fast eigenvector. The more δ_{ij} projects on a slow shrinking eigenvector the more unrelated (with respect to the general behaviour) are the keywords used by the two users under comparison.

We now define the distance of u_i and u_j and we denote d_{ij} to be:

$$d_{ij} = \boldsymbol{\delta}_{ij}^T \Sigma(F_G) \boldsymbol{\delta}_{ij} \tag{20}$$

where the dimensionality of $\boldsymbol{\delta}_{ij}$ is extended to the dimensionality of $\Sigma(F_G)$ by filling in with zeros the respective coordinates. The proposed distance is obviously well defined since it is defined as a generalised Euclidean distance function using a covariance matrix which is always positive definite and achieves to adjust the direct difference of the two users according to the general behaviour. If the general behaviour indicates that k_1 is closer to k_3 than it is to k_2 then the same direct difference between two users will count more if it is between the keywords k_1 and k_2 rather than between k_1 and k_3. This distance can be used with any standard clustering method to produce clusters with high abstraction from the raw data since the semantic locality between two users is defined through the general behaviour of all the other users.

3.2 *Hierarchical Clustering of User Profiles*

The most widely used hierarchical clustering methods are the agglomerative family of single and complete link, group average, weighted group average, weighted and unweighted centroid, and the Ward's method. A dendrogram tree can be constructed in all of them that shows the relationship between clusters at each level of clustering and between levels. The distance measure is always an Euclidean distance. The above methods measure distance between clusters but they do it differently. In practice this means the shapes of clusters they can naturally identify is also different. The distance just introduced permits the clustering of the user profiles using any of the above methods. In our case nevertheless, the data, at least in the interesting cases, are over correlated leading to well separated clusters. Single and complete link methods are more sensitive to outliers therefor in the example of the following section we choose the average link, mainly due to its simplicity over the other group based methods.

3.3 *Working Example of Unsupervised Learning Through Hierarchical Clustering*

Consider the nine state Markovian network shown in Figure 5. The one step transition probability matrix is

$$P = \begin{bmatrix} 0.25 & 0 & 0 & 0.75 & 0 & 0 & 0 & 0 & 0 \\ 0 & 0.15 & 0.55 & 0.3 & 0 & 0 & 0 & 0 & 0 \\ 0 & 0.25 & 0.75 & 0 & 0 & 0 & 0 & 0 & 0 \\ 0.55 & 0.3 & 0 & 0.15 & 0 & 0 & 0 & 0 & 0 \\ 0 & 0 & 0 & 0 & 1 & 0 & 0 & 0 & 0 \\ 0 & 0 & 0 & 0 & 0 & 0.65 & 0.1 & 0 & 0.25 \\ 0 & 0 & 0 & 0 & 0 & 0.1 & 0.65 & 0.25 & 0 \\ 0 & 0 & 0 & 0 & 0 & 0 & 0.75 & 0.25 & 0 \\ 0 & 0 & 0 & 0 & 0 & 0.25 & 0 & 0 & 0.75 \end{bmatrix}$$

illustrating three chains, $(1, 2, 3, 4)$, 5 and $(6, 7, 8, 9)$. We will form an average link hierarchical clustering to a set of 15 user profiles with the distance implied by the network in Figure 5 and defined in Equation (20). The set of user profiles is given in the following matrix:

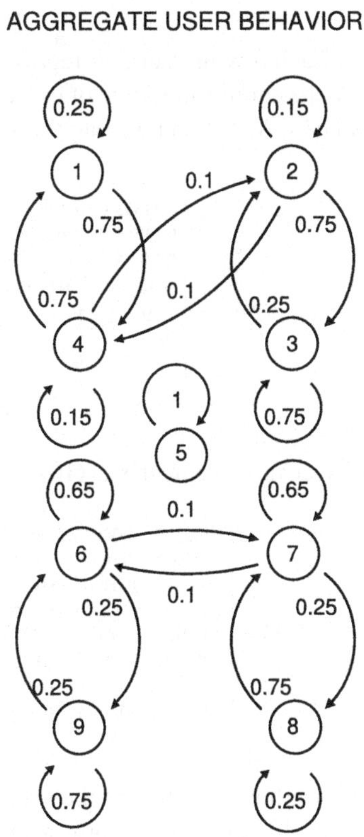

Fig. 5 A tridesmic Markovian network of 9 states. There are two chains of four states and one isolated node

$$B = \begin{bmatrix} 0.25 & 0 & 0 & 0.75 & 0 & 0 & 0 & 0 & 0 \\ 0 & 0.75 & 0.25 & 0 & 0 & 0 & 0 & 0 & 0 \\ 0 & 0.25 & 0.75 & 0 & 0 & 0 & 0 & 0 & 0 \\ 0.75 & 0 & 0 & 0.25 & 0 & 0 & 0 & 0 & 0 \\ 0 & 0 & 0 & 0 & 1 & 0 & 0 & 0 & 0 \\ 0 & 0 & 0 & 0 & 0 & 0.75 & 0 & 0 & 0.25 \\ 0 & 0 & 0 & 0 & 0 & 0 & 0.75 & 0.25 & 0 \\ 0 & 0 & 0 & 0 & 0 & 0 & 0.25 & 0.75 & 0 \\ 0 & 0 & 0 & 0 & 0 & 0.25 & 0 & 0 & 0.75 \\ 0.5 & 0 & 0 & 0.5 & 0 & 0 & 0 & 0 & 0 \\ 0 & 0 & 0 & 0 & 0.5 & 0.5 & 0 & 0 & 0 \\ 0 & 0.5 & 0.5 & 0 & 0 & 0 & 0 & 0 & 0 \\ 0 & 0 & 0 & 0 & 0 & 0 & 0.5 & 0.5 & 0 \\ 0 & 0 & 0 & 0 & 0 & 0.5 & 0 & 0 & 0.5 \\ 0.1 & 0.1 & 0.1 & 0.1 & 0.1 & 0.1 & 0.1 & 0.1 & 0.2 \end{bmatrix}$$

Each row in matrix B represents a user. The fractional occupancies in each row are the equilibrium state of the respective user's state transition diagram. We choose n equal $N+1$ and we find the sample $S(10)$ to be:

$$S(10) = \begin{bmatrix} 0.33603 & 0.10966 & 0.063645 & 0.23724 & 0 & 0 & 0 & 0 & 0 \\ 0.14953 & 0.27602 & 0.20055 & 0.17311 & 0 & 0 & 0 & 0 & 0 \\ 0.19094 & 0.44121 & 0.6281 & 0.23695 & 0 & 0 & 0 & 0 & 0 \\ 0.3235 & 0.17311 & 0.1077 & 0.35271 & 0 & 0 & 0 & 0 & 0 \\ 0 & 0 & 0 & 0 & 1 & 0 & 0 & 0 & 0 \\ 0 & 0 & 0 & 0 & 0 & 0.45761 & 0.1663 & 0.13804 & 0.33008 \\ 0 & 0 & 0 & 0 & 0 & 0.1663 & 0.56798 & 0.51549 & 0.093898 \\ 0 & 0 & 0 & 0 & 0 & 0.046015 & 0.17183 & 0.27464 & 0.023941 \\ 0 & 0 & 0 & 0 & 0 & 0.33008 & 0.093898 & 0.071824 & 0.55208 \end{bmatrix}$$

The covariance matrix $\Sigma(10)$ of $S(10)$ is

$$\Sigma(10) = \begin{bmatrix} 0.022 & 0.0097 & 0.00637 & 0.0192 & -0.0125 & -0.0125 & -0.0125 & -0.0125 & -0.0125 \\ 0.00967 & 0.023 & 0.0290 & 0.0127 & -0.0125 & -0.0125 & -0.0125 & -0.0125 & -0.0125 \\ 0.00637 & 0.0290 & 0.0436 & 0.0108 & -0.0125 & -0.0125 & -0.0125 & -0.0125 & -0.0125 \\ 0.019213 & 0.012719 & 0.0108 & 0.0192 & -0.0125 & -0.0125 & -0.0125 & -0.0125 & -0.0125 \\ -0.0125 & -0.0125 & -0.0125 & -0.0125 & 0.100 & -0.0125 & -0.0125 & -0.0125 & -0.0125 \\ -0.0125 & -0.0125 & -0.0125 & -0.0125 & -0.0125 & 0.0297 & 0.0089 & 0.00547 & 0.027 \\ -0.0125 & -0.0125 & -0.0125 & -0.0125 & -0.0125 & 0.0089 & 0.0359 & 0.0323 & 0.00228 \\ -0.0125 & -0.0125 & -0.0125 & -0.0125 & -0.0125 & 0.00547 & 0.0324 & 0.0319 & -0.000751 \\ -0.0125 & -0.0125 & -0.0125 & -0.0125 & -0.0125 & 0.02797 & 0.00228 & -0.000751 & 0.0400 \end{bmatrix}$$

To see how d_{ij} clusters the users in B we multiply B with the square root matrix of $\Sigma(10)$ and we perform a hierarchical clustering with the standard Euclidean distance on the resulting user matrix. This is because for any symmetric and positive definite matrix A and a vector u it holds:

$$\|Au\| = (Au)'Au = u'A'Au = u'A^2u$$

therefor for two vectors u_1 and u_2 it is

$$\|A(u_1 - u_2)\| = (u_1 - u_2)'A^2(u_1 - u_2)$$

The root matrix of $\Sigma(10)$ is

$$\sqrt{\Sigma(10)} =$$

$$\begin{bmatrix}
0.10584 & 0.011068 & 0.00061105 & 0.06869 & -0.034435 & -0.0332 & -0.026487 & -0.039359 & -0.03418 \\
0.011068 & 0.095801 & 0.089844 & 0.03011 & -0.032004 & -0.030925 & -0.02391 & -0.03738 & -0.031799 \\
0.00061105 & 0.089844 & 0.1796 & 0.015566 & -0.025143 & -0.024426 & -0.023307 & -0.02531 & -0.024864 \\
0.06869 & 0.03011 & 0.015566 & 0.092387 & -0.031853 & -0.030728 & -0.026882 & -0.034196 & -0.031557 \\
-0.034435 & -0.032004 & -0.025143 & -0.031853 & 0.30497 & -0.030426 & -0.025629 & -0.034779 & -0.03123 \\
-0.0332 & -0.030925 & -0.024426 & -0.030728 & -0.030426 & 0.13837 & 0.019157 & -0.00098687 & 0.075535 \\
-0.026487 & -0.02391 & -0.023307 & -0.026882 & -0.025629 & 0.019157 & 0.1463 & 0.10495 & -0.0032024 \\
-0.039359 & -0.03738 & -0.02531 & -0.034196 & -0.034779 & -0.00098687 & 0.10495 & 0.12067 & -0.01944 \\
-0.03418 & -0.031799 & -0.024864 & -0.031557 & -0.03123 & 0.075535 & -0.0032024 & -0.01944 & 0.17097
\end{bmatrix}$$

multiplied by B yields

$$B1 = B\sqrt{\Sigma(10)} =$$

$$\begin{bmatrix}
0.077978 & 0.02535 & 0.011827 & 0.086463 & -0.032499 & -0.031346 & -0.0267 & -0.035487 & -0.0322 \\
0.008454 & 0.094312 & 0.11228 & 0.026474 & -0.030289 & -0.029301 & -0.0237 & -0.034362 & -0.0300 \\
0.0032254 & 0.091333 & 0.15716 & 0.019202 & -0.026859 & -0.026051 & -0.0234 & -0.028327 & -0.0265 \\
0.096555 & 0.015829 & 0.0043498 & 0.074614 & -0.03379 & -0.032582 & -0.0265 & -0.038069 & -0.0335 \\
-0.034435 & -0.032004 & -0.025143 & -0.031853 & 0.30497 & -0.030426 & -0.0256 & -0.034779 & -0.031 \\
-0.033445 & -0.031144 & -0.024536 & -0.030936 & -0.030627 & 0.12266 & 0.0135 & -0.0056002 & 0.0993 \\
-0.029705 & -0.027278 & -0.023807 & -0.028711 & -0.027916 & 0.014121 & 0.135 & 0.10888 & -0.00726 \\
-0.036141 & -0.034012 & -0.024809 & -0.032368 & -0.032491 & 0.0040491 & 0.115 & 0.11674 & -0.0153 \\
-0.033935 & -0.03158 & -0.024754 & -0.03135 & -0.031029 & 0.091245 & 0.00238 & -0.014827 & 0.147 \\
0.087266 & 0.020589 & 0.00808 & 0.0805 & -0.033144 & -0.031964 & -0.0266 & -0.036778 & -0.0328 \\
-0.033818 & -0.031465 & -0.0247 & -0.0312 & 0.13727 & 0.053974 & -0.00323 & -0.017883 & 0.0221 \\
0.0058397 & 0.092823 & 0.13472 & 0.0228 & -0.028574 & -0.027676 & -0.0236 & -0.031345 & -0.0283 \\
-0.032923 & -0.030645 & -0.0243 & -0.0305 & -0.030204 & 0.0090852 & 0.12562 & 0.11281 & -0.0113 \\
-0.03369 & -0.031362 & -0.0246 & -0.0311 & -0.0308 & 0.10695 & 0.00797 & -0.010214 & 0.123 \\
-0.0015628 & 0.00390 & 0.0137 & 0.0019 & 0.00282 & 0.0157 & 0.0137 & 0.0014732 & 0.024
\end{bmatrix}$$

We now perform a hierarchical clustering (average link) to the rows of $B1$ with the Euclidean metric. The result is shown in Figure 6. The clustering of the nodes according to their covariances as initial conditions is shown in Figure 7.

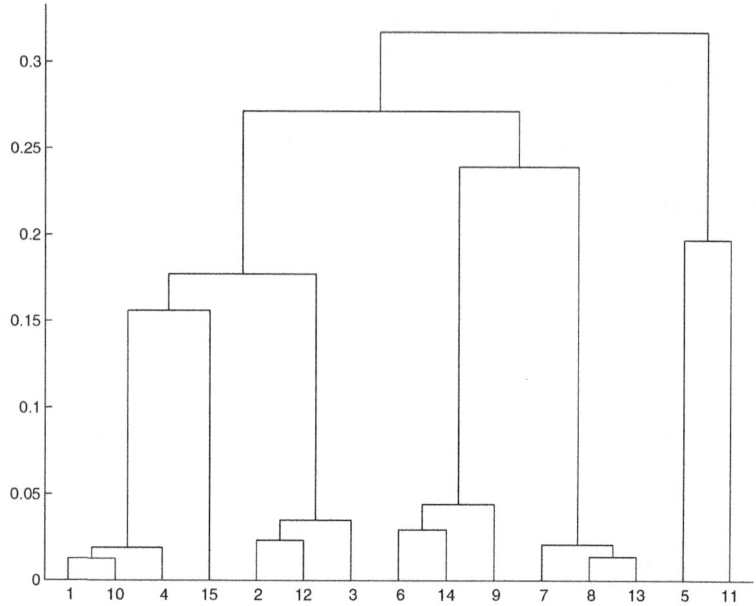

Fig. 6 Hierarchical clustering of the 15 user profiles in B with the distance implied by the network in (5)

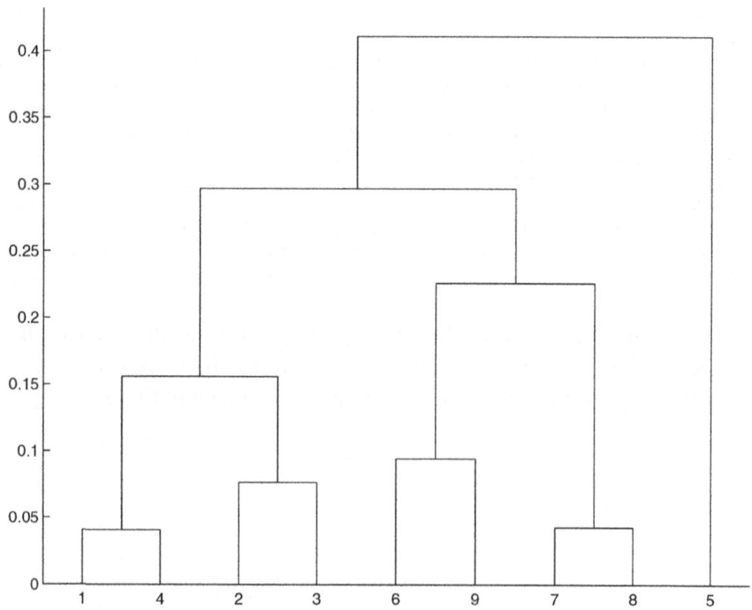

Fig. 7 Hierarchical clustering of the nodes of the network

References

1. T. Erickson, From interface to interplace: the spatial environment as a medium for interaction, in *Proceedings of HCI'94* (1994)
2. C. Chen, Structuring and visualising the WWW with generalised similarity analysis, in *Proceedings of the 8th ACM Conference on Hypertext (Hypertext'97), Southampton* (1997), pp. 177–186
3. C. Chen, Generalised similarity analysis and pathfinder network scaling. Interact. Comput. **10**(2), 107–128 (1998)
4. J. Herlocker, J. Konstan, A. Borchers, J. Riedl, An algorithmic framework for performing collaborative filtering, in *Proceedings of the 1999 Conference on Research and Development in Information Retrieval* (1999)
5. T. Joachims, D. Freitag, T. Mitchell, WebWatcher: a tour guide for the world wide web, in *Proceedings of the International Joint Conference in AI (IJCAI97)* (1997)
6. H. Lieberman, Letizia: an agent that assists web browsing, in *Proceedings of the 14th International Joint Conference in AI (IJCAI95)* (AAAI Press, Menlo Park, 1995)
7. M. Perkowitz, O. Etzioni, Adaptive web sites: automatically synthesising web pages, in *Proceedings of Fifteenth National Conference on Artificial Intelligence, Madison* (1998)
8. S. Schechter, M. Krishnan, M.D. Smith, Using path profiles to predict HTTP requests, in *Proceedings of 7th International World Wide Web Conference, Brisbane* (1998)
9. M. Spiliopoulou, L.C. Faulstich, WUM: a web utilisation miner, in *Proceedings of EDBT Workshop WebDB98, Valencia*. Lecture Notes in Computer Science, vol. 1590 (Springer, Berlin, 1999)
10. A. Buchner, M.D. Mulvenna, Discovering Internet Marketing Intelligence through Online Analytical Web Usage Mining. SIGMOD Rec. 27(4), 54–61 (1998)
11. C. Shahabi, A.M. Zarkesh, J. Adibi, V. Shah, Knowledge discovery from users Web-page navigation, in *Proceedings of Workshop on Research Issues in Data Engineering, Birmingham* (1997)
12. T. Yan, M. Jacobsen, H. Garcia-Molina, U. Dayal, From user access patterns to dynamic hypertext linking, in *Proceedings of the 5th International World Wide Web Conference, Paris* (1996)
13. O. Nasraoui, H. Frigui, A. Joshi, R. Krishnapuram, Mining Web access logs using relational competitive fuzzy clustering, in *Proceedings of the Eight International Fuzzy Systems Association World Congress* (1999)
14. M.S. Chen, J.S. Park, P.S. Yu, Data mining for path traversal patterns in a web environment, in *Proceedings of 16th International Conference on Distributed Computing Systems, Hong Kong* (1996), pp. 385–392
15. R.A. Howard, *Dynamic Probabilistic Systems* (Wiley, New York, 1971)

On the Solution of Boundary Value Problems for Ordinary Differential Equations of Order n and $2n$ with General Boundary Conditions

I. N. Parasidis, E. Providas, and S. Zaoutsos

Abstract We present a method for examining the existence and uniqueness and obtaining the exact solution to boundary value problems consisting of the differential equation $Au = f$, where A is a linear ordinary differential operator of order n, and multipoint and integral boundary conditions. We also derive a formula for computing the exact solution to even order boundary value problems encompassing the differential equation $A^2u = f$ subject to $2n$ general boundary conditions. The method is based on the correct extensions of operators in Banach spaces.

1 Introduction

Let A be the general linear ordinary differential operator of order n,

$$A = a_0(x)\frac{d^n}{dx^n} + a_1(x)\frac{d^{n-1}}{dx^{n-1}} + \cdots + a_{n-1}(x)\frac{d}{dx} + a_n(x), \tag{1}$$

where the coefficients $a_i(x),\ i = 0, \ldots, n$, are all continuous complex functions on $[0, 1]$ and $a_0(x) \neq 0$.

We are concerned primarily with the exact solution of the boundary value problem consisting of the linear ordinary differential equation of order n,

$$Au(x) = f(x), \quad x \in (0, 1), \tag{2}$$

I. N. Parasidis · E. Providas (✉) · S. Zaoutsos
University of Thessaly, Gaiopolis, Larissa, Greece
e-mail: paras@teilar.gr; providas@uth.gr; szaoutsos@teilar.gr

N. J. Daras, T. M. Rassias (eds.), *Computational Mathematics and Variational Analysis*, Springer Optimization and Its Applications 159,
https://doi.org/10.1007/978-3-030-44625-3_17

and the n general boundary conditions

$$\sum_{j=0}^{m_1}\sum_{k=0}^{n-1} \nu_{ij}^k u^{(k)}(x_j) + \sum_{k=0}^{n-1}\int_0^1 h_{ik}(t)u^{(k)}(t)dt = \beta_i, \quad i = 1, \dots, n, \tag{3}$$

where the nonhomogeneous term $f(x)$ is a continuous complex functions on $[0, 1]$, the functions $h_{ik}(x)$, $i = 1, \dots, n$, $k = 0, \dots, n-1$, are continuous on the interval $[0, 1]$, ν_{ij}^k, $i = 1, \dots, n$, $j = 0, \dots, m_1$, $k = 0, \dots, n-1$, and the nonhomogeneous terms β_i, $i = 1, \dots, n$, are real constants, and lastly $0 \leq x_0 < x_1 < \cdots < x_{m_1} \leq 1$ are ordered boundary points; $u(x)$ is the solution function with n continuous derivatives on $[0, 1]$.

We also deal with a particular class of boundary value problems made up of the even order ordinary differential equation,

$$A^2u(x) = f(x), \quad x \in (0, 1), \tag{4}$$

and the $2n$ general boundary conditions

$$\sum_{j=0}^{m_1}\sum_{k=0}^{n-1} \nu_{ij}^k u^{(k)}(x_j) + \sum_{k=0}^{n-1}\int_0^1 h_{ik}(t)u^{(k)}(t)dt = 0, \quad i = 1, \dots, n, \tag{5}$$

$$\sum_{j=0}^{m_1}\sum_{k=0}^{n-1} \nu_{ij}^k (Au)^{(k)}(x_j) + \sum_{k=0}^{n-1}\int_0^1 h_{ik}(t)(Au)^{(k)}(t)dt = 0, \quad i = 1, \dots, n, \tag{6}$$

where A^2 stands for the composite product $A(A)$ and $u(x)$ is the solution function with $2n$ continuous derivatives on $[0, 1]$.

In all cases, we state the requirements and establish criteria for the existence and uniqueness and derive formulae for computing the exact solutions. The background of our technique is the theory of the extensions of operators in Banach spaces, see, for example, [7, 8, 11, 13, 14, 16] and the recent articles [2, 3].

For further studies on the general boundary value problems of the above types one may look at the works, among others, [1, 4–6, 9, 10, 12, 15, 17] and the references therein.

The outline of the paper is as follows. In Section 2, we give some definitions and explain the notation. In Section 3, we deal with boundary value problems of order n with homogeneous conditions and provide a convenient solution formula. Boundary value problems of order n with nonhomogeneous conditions are studied in Section 4 and two solution formulae are derived. Section 5 is devoted to boundary value problems of order $2n$. The implementation of the solution process is elucidated by considering several illustrative examples in Section 6. Finally some concluding remarks are given in Section 7.

2 Preliminaries and Definitions

Let $\mathbf{C}$ be the set of all complex numbers. Let $X = C^n[0, 1]$, where $C^n[0, 1]$ denotes the space of all complex valued functions on $[0, 1]$ with continuous derivatives of order n, $X^2 = C^{2n}[0, 1]$ and $Y = C[0, 1]$.

Organize the boundary conditions (3) in the form

$$\Phi_i(u) - \sum_{j=1}^{m} n_{ij}\Psi_j(u) = \beta_i, \quad i = 1, \ldots, n, \tag{7}$$

where $0 \leq m \leq nm_1 + 2n - 1$, $\Phi_i : X \to \mathbf{C},\ i = 1, \ldots, n$ and $\Psi_j : X \to \mathbf{C},\ j = 1, \ldots, m$ are linear and bounded complex valued functionals, i.e. $\Phi_i \in X^*,\ i = 1, \ldots, n$ and $\Psi_j \in X^*,\ j = 1, \ldots, m$, where X^* denotes the adjoint space of X, and the coefficients $n_{ij},\ i = 1, \ldots, n,\ j = 1, \ldots, m$, are constants. Equation (7) can be written conveniently using matrix notations, namely

$$\Phi(u) - N\Psi(u) = \mathbf{b}, \tag{8}$$

where $\Phi = \operatorname{col}(\Phi_1, \ldots, \Phi_n) \in X_n^*$, $\Psi = \operatorname{col}(\Psi_1, \ldots, \Psi_m) \in X_m^*$, N is an $n \times m$ constant matrix, and $\mathbf{b} = \operatorname{col}(\beta_1, \ldots, \beta_n)$ is a vector of real constants. The functionals $\Phi_1, \ldots, \Phi_n$ are chosen such as they satisfy the relation (9) below. Precisely, they describe initial or boundary conditions of a simpler problem whose exact solution is known; usually they are associated to a Cauchy problem.

Let $A : X \overset{on}{\to} Y$ be the linear operator in (1) with $D(A) = X$ and let the operator $\widehat{A} : X \to Y$ be a correct restriction of A, in symbol $\widehat{A} \subset A$, defined by

$$\begin{aligned} \widehat{A}u &= Au, \\ D(\widehat{A}) &= \{u : u \in D(A),\ \Phi(u) = \mathbf{0}\}. \end{aligned} \tag{9}$$

We recall that an operator $\widehat{A} : X \to Y$ is called *correct* if $R(\widehat{A}) = Y$ and the inverse operator $\widehat{A}^{-1}$ exists and is continuous on Y.

Let $\dim \ker A = n$ and the vector $z = (z_1, \ldots, z_n) \in [\ker A]_n$ be a basis of $\ker A$, i.e. $z_1, \ldots, z_n$ are fundamental solutions of the homogeneous equation $Au = 0$, such that $\Phi_1, \ldots, \Phi_n$ are biorthogonal to elements $z_1, \ldots, z_n$, namely

$$Az = (Az_1 \ \ldots \ Az_n) = \mathbf{0}, \quad \Phi(z) = \begin{bmatrix} \Phi_1(z_1) & \cdots & \Phi_1(z_n) \\ \vdots & \ldots & \vdots \\ \Phi_n(z_1) & \cdots & \Phi_n(z_n) \end{bmatrix} = I_n, \tag{10}$$

where the element $\Phi_i(z_j)$ of the $n \times n$ matrix $\Phi(z)$ is the value of the functional Φ_i on the element z_j and $\Phi_i(z_j) = \delta_{ij}$, with δ_{ij} being the Kronecker delta.

Finally, it is easy to verify that

$$\Phi(zN) = \Phi(z)N = N, \tag{11}$$

a relation which we will use several times in the sequel.

It is understood that, the inverse operator of the operator $\widehat{A}$ is denoted by $\widehat{A}^{-1}$ and $\widehat{A}^{-2}$ is the composite product $\widehat{A}^{-1}(\widehat{A}^{-1})$. Also, unless otherwise is declared or deduced, lower case Latin letters and brackets designate vectors, while capital Latin letters and square brackets symbolize matrices. The $n \times n$ unit and zero matrices are denoted by I_n and 0_n, respectively, and the zero column vector by $\mathbf{0}$.

3 BVPs of Order *n* with Homogeneous Conditions

We first study general boundary value problems of the type (2), (3) with homogeneous boundary conditions, i.e. $\beta_i \equiv 0,\ i = 1, \ldots, n$. We prove the following theorem for examining the existence and uniqueness of the solution and obtaining it in closed form.

Theorem 1 *Let the operator $A_1 : X \to Y$ be a restriction of A to a subset $D(A_1) \subset D(A)$, specifically*

$$A_1 u = Au,$$
$$D(A_1) = \{u : u \in D(A),\ \ \Phi(u) - N\Psi(u) = \mathbf{0}\}. \tag{12}$$

Then:

(i) The operator A_1 is injective if and only if

$$\det V = \det[I_m - \Psi(z)N] \neq 0. \tag{13}$$

(ii) If (i) is true, then the operator A_1 is correct and the unique solution to the general boundary value problem

$$A_1 u = f, \quad \forall f \in Y, \tag{14}$$

is given by

$$\begin{aligned} u &= A_1^{-1} f \\ &= \widehat{A}^{-1} f + zNV^{-1}\Psi(\widehat{A}^{-1} f). \end{aligned} \tag{15}$$

Proof

(i) Let $\det V \neq 0$ and will prove that $\ker A_1 = \{0\}$. Assume there exists an element $u_0 \in \ker A_1$. Then from (12), we have

$$Au_0 = 0, \quad \Phi(u_0) - N\Psi(u_0) = \mathbf{0}. \tag{16}$$

The second equation by making use of the biorthogonality condition (10) can be written as

$$\Phi(u_0) - \Phi(z)N\Psi(u_0) = \Phi(u_0) - \Phi(zN\Psi(u_0)) = \Phi\left(u_0 - zN\Psi(u_0)\right) = \mathbf{0}. \tag{17}$$

From this and (9) it is implied that $u_0 - zN\Psi(u_0) \in D(\widehat{A}) \subset D(A)$ and hence

$$\widehat{A}\left(u_0 - zN\Psi(u_0)\right) = A\left(u_0 - zN\Psi(u_0)\right) = 0, \tag{18}$$

by the first equation of (16) and the definition of z in (10). Moreover, $\widehat{A}$ is correct and therefore

$$u_0 - zN\Psi(u_0) = 0. \tag{19}$$

Acting now by the vector Ψ and taking like terms, we obtain

$$[I_m - \Psi(z)N]\Psi(u_0) = V\Psi(u_0) = \mathbf{0}. \tag{20}$$

Since by hypothesis $\det V \neq 0$, it is implied that $\Psi(u_0) = \mathbf{0}$ and thus $u_0 = 0$ by (19). That is $\ker A_1 = \{0\}$ and so A_1 is an injective operator.

Conversely. We will prove that if A_1 is injective, then $\det V \neq 0$, or equivalently if $\det V = 0$, then A_1 is not injective. Let $\det V = 0$. Then there exists a vector of constants not all equal to zero $\mathbf{c} = \mathrm{col}(c_1, \ldots, c_m) \neq \mathbf{0}$ such that $V\mathbf{c} = \mathbf{0}$. Consider the element $u_0 = zN\mathbf{c}$; notice that $u_0 \neq 0$ because otherwise $N\mathbf{c} = \mathbf{0}$, since the components of z are linearly independent, and $V\mathbf{c} = [I_m - \Psi(z)N]\mathbf{c} = \mathbf{0}$ implies $\mathbf{c} = \mathbf{0}$ which contradicts the hypothesis $\mathbf{c} \neq \mathbf{0}$. Additionally, $u_0 \in D(A_1)$, since $\Phi(u_0) = N\mathbf{c}$, $\Psi(u_0) = \Psi(z)N\mathbf{c}$ and

$$\Phi(u_0) - N\Psi(u_0) = N\mathbf{c} - N\Psi(z)N\mathbf{c} = N[I_m - \Psi(z)N]\mathbf{c} = NV\mathbf{c} = \mathbf{0}. \tag{21}$$

Further, $u_0 \in \ker A_1$ since $A_1u_0 = Au_0 = AzN\mathbf{c} = 0$. Hence A_1 is not injective.

(ii) Consider the nonlocal boundary value problem (14) for any $f \in Y$, specifically

$$Au = f, \quad \Phi(u) - N\Psi(u) = \mathbf{0}. \tag{22}$$

By the same arguments as above $\Phi(u - zN\Psi(u)) = \mathbf{0}$ and hence $u - zN\Psi(u) \in D(\widehat{A})$ and

$$\widehat{A}\left(u - zN\Psi(u)\right) = A\left(u - zN\Psi(u)\right) = Au = f, \quad \forall f \in Y. \tag{23}$$

Since the operator $\widehat{A}$ is correct, then for every $u \in D(A_1)$ and $f \in Y$, we have

$$u - zN\Psi(u) = \widehat{A}^{-1}f. \tag{24}$$

Acting by the vector Ψ on both sides of (24) and taking like terms, we obtain successively

$$\begin{aligned}\Psi(u) - \Psi(z)N\Psi(u) &= \Psi(\widehat{A}^{-1}f),\\ [I_m - \Psi(z)N]\Psi(u) &= \Psi(\widehat{A}^{-1}f),\\ V\Psi(u) &= \Psi(\widehat{A}^{-1}f).\end{aligned} \tag{25}$$

By the statement (i) $\det V \neq 0$ and therefore

$$\Psi(u) = V^{-1}\Psi(\widehat{A}^{-1}f). \tag{26}$$

Substituting (26) into (24), we get

$$u = \widehat{A}^{-1}f + zNV^{-1}\Psi(\widehat{A}^{-1}f). \tag{27}$$

From this and the fact that the operator A_1 is injective, we obtain the unique solution (15) to the problem (14).

It remains to prove that the operator A_1 is correct. Because f in (22) is taken for any $f \in Y$ it is implied that $R(A_1) = Y$. Moreover, from (15) follows the boundedness of A_1^{-1}, since the operator $\widehat{A}^{-1}$ and the functionals $\Psi_1, \ldots, \Psi_m$ are bounded. Hence, the operator A_1 is correct if and only if the condition (13) holds and in this case the unique solution of (14) is given by (15). The theorem is proved. □

4 BVPs of Order n with Nonhomogeneous Conditions

In this section we consider general boundary value problems of order n of the type (2) and (3) with nonhomogeneous boundary conditions. We start by stating first the following theorem.

Theorem 2 *The operator $B_1 : X \to Y$ defined by*

$$B_1u = Au,$$
$$D(B_1) = \{u : u \in D(A),\ \ \Phi(u) - N\Psi(u) = \mathbf{b}\}$$

is injective if and only if

$$\det V = \det[I_m - \Psi(z)N] \neq 0. \tag{28}$$

Proof Let $\det V \neq 0$. Assume there exist $u_1,\ u_2 \in D(B_1)$ and $B_1u_1 = B_1u_2$. Then,

$$Au_1 = Au_2, \quad \Phi(u_1) - N\Psi(u_1) = \mathbf{b}, \quad \Phi(u_2) - N\Psi(u_2) = \mathbf{b}. \tag{29}$$

By subtracting and since the operator A and the functionals $\{\Phi_i\},\ \{\Psi_i\}$ are linear, we have

$$A(u_1 - u_2) = 0, \quad \Phi(u_1 - u_2) - N\Psi(u_1 - u_2) = \mathbf{0}. \tag{30}$$

Setting $v = u_1 - u_2$, we get

$$Av = 0, \quad \Phi(v) - N\Psi(v) = \mathbf{0}, \tag{31}$$

and by using the definition (12),

$$A_1 v = 0. \tag{32}$$

From the assumption that $\det V \neq 0$ and Theorem 1 follows that $v = 0$ and hence $u_1 = u_2$, which proves that the operator B_1 is injective.

Conversely. We will show that if B_1 is injective, then $\det V = 0$, or equivalently if $\det V = 0$, then B_1 is not injective. Let $\det V = 0$. Suppose $u_1,\ u_2 \in D(B_1)$, and $Bu_1 = Bu_2$. Working as above, we get (32). Since $\det V = 0$, Theorem 1 implies that the operator A_1 is not injective. This means that there exists at least a $v_0 \neq 0$ that solves (32). Then $v_0 = u_1 - u_2 \neq 0$ or $u_1 \neq u_2$ and hence B_1 is not injective. The proof is completed. □

Next, and before we consider the completely nonhomogeneous boundary value problem $B_1u = f$, we look at the boundary value problem $B_1u = 0$. For this we show the following theorem.

Theorem 3 *If the operator B_1 is injective, then the unique solution to the boundary value problem,*

$$B_1 u = 0, \tag{33}$$

is given by

$$u = zNV^{-1}\Psi(z)\mathbf{b} + z\mathbf{b}. \tag{34}$$

Proof From (33), we have

$$B_1u = Au = 0, \quad \Phi(u) - N\Psi(u) = \mathbf{b}. \tag{35}$$

We recall that the complementary solution to the homogeneous equation $Au = 0$ is given by

$$u = z\mathbf{c}, \tag{36}$$

where $\mathbf{c} = \text{col}(c_1, \ldots, c_n)$ is a vector of arbitrary constants. To determine these constants we substitute (36) into the second equation of (35), viz.

$$\Phi(z\mathbf{c}) - N\Psi(z\mathbf{c}) = \mathbf{b}. \tag{37}$$

By applying the orthogonality relation, we have

$$\mathbf{c} - N\Psi(z\mathbf{c}) = \mathbf{b}. \tag{38}$$

Multiplying by the vector z and then acting by the vector Ψ, we get

$$\Psi\,(z\mathbf{c} - zN\Psi(z\mathbf{c})) = [I_m - \Psi(z)N]\,\Psi(z\mathbf{c}) = V\Psi(z\mathbf{c}) = \Psi(z\mathbf{b}). \tag{39}$$

From Theorem 2, $\det V \neq 0$ and hence

$$\Psi(z\mathbf{c}) = V^{-1}\Psi(z\mathbf{b}). \tag{40}$$

Substitution of (40) into (38) and then into (36) yields the solution in (34). □

Now by the help of the two previous theorems we derive a condition for checking the existence and uniqueness and a formula for obtaining the solution of the completely nonhomogeneous boundary value problem $B_1 u = f$.

Theorem 4 *If the operator B_1 is injective, then the operator B_1 is correct and the unique solution to the boundary value problem,*

$$B_1 u = f, \tag{41}$$

for any $f \in Y$ is given by

$$\begin{aligned} u &= B_1^{-1} f \\ &= \widehat{A}^{-1} f + zNV^{-1}\left[\Psi(\widehat{A}^{-1} f) + \Psi(z)\mathbf{b}\right] + z\mathbf{b}. \end{aligned} \tag{42}$$

Proof By the principle of superposition the solution of the completely nonhomogeneous boundary value problem (41) can be constructed as the sum $u = v + w$ where v and w are solutions of the boundary value problems

$$Av = f, \quad \Phi(v) - N\Psi(v) = \mathbf{0}, \quad \text{or} \quad A_1 v = f, \tag{43}$$

and

$$Aw = 0, \quad \Phi(w) - N\Psi(w) = \mathbf{b}, \quad \text{or} \quad B_1 w = 0. \tag{44}$$

respectively. Thus, from Theorems 1 and 3 and in particular the relations (15) and (34), we obtain

$$\begin{aligned} u &= v + w \\ &= \left(\widehat{A}^{-1} f + zNV^{-1}\Psi(\widehat{A}^{-1} f)\right) + \left(zNV^{-1}\Psi(z)\mathbf{b} + z\mathbf{b}\right) \\ &= \widehat{A}^{-1} f + zNV^{-1}\left[\Psi(\widehat{A}^{-1} f) + \Psi(z)\mathbf{b}\right] + z\mathbf{b}, \end{aligned} \tag{45}$$

which is the solution (42) to the boundary value problem (41).

Since (43) holds for any $f \in Y$ it is implied that $R(B_1) = Y$. Also, because the operator $\widehat{A}^{-1}$ and the functionals $\Psi_1, \ldots, \Psi_m$ in (42) are bounded it is concluded that the operator B_1 is bounded. Hence B_1 is correct. □

5 BVPs of Order 2*n* of Special Form

In this section we deal with the 2*n*th order boundary value problem (4)–(6). This problem, of course, can be attacked directly as in the previous section but it is better to take advance of the symmetric form that it has and to develop a solution formula that requires only *n* fundamental solutions and hence it is easier to manage.

By using the notation in Section 2, we write conveniently the boundary equations (5) and (6) in the matrix form

$$\Phi(u) - N\Psi(u) = \mathbf{0}, \quad \Phi(Au) - N\Psi(Au) = \mathbf{0}, \tag{46}$$

and prove the next theorem.

Theorem 5 *Let the operator $A_2 : X^2 \to Y$ be a restriction of A^2 to a subset $D(A_2) \subset D(A^2)$, namely*

$$\begin{aligned} A_2 u &= A^2 u = A(Au), \\ D(A_2) &= \left\{u : u \in C^{2n}[0,1],\ \Phi(u) - N\Psi(u) = \mathbf{0},\ \Phi(Au) - N\Psi(Au) = \mathbf{0}\right\}. \end{aligned} \tag{47}$$

Then:

(i) The operator A_2 is injective if and only if

$$\det V = \det[I_m - \Psi(z)N] \neq 0. \tag{48}$$

(ii) If (i) is true, then the operator A_2 is correct and the unique solution to the boundary value problem,

$$A_2 u = f, \quad \forall f \in Y, \tag{49}$$

is obtained by the formula

$$u = A_2^{-1} f = \widehat{A}^{-2} f + \left(zN \ \widehat{A}^{-1} zN\right) \begin{bmatrix} V & -\Psi(\widehat{A}^{-1}z)N \\ 0_m & V \end{bmatrix}^{-1} \begin{pmatrix} \Psi(\widehat{A}^{-2} f) \\ \Psi(\widehat{A}^{-1} f) \end{pmatrix}. \quad (50)$$

Proof From the definition (47) we have $A_2 u = A^2 u = A(Au) = f$. By setting $Au = y$, the $2n$th order boundary value problem (49) can be factorized to the following two boundary value problems of order n,

$$Au = y, \quad u \in C^{2n}[0, 1], \quad \Phi(u) - N\Psi(u) = \mathbf{0}, \quad (51)$$

and

$$Ay = f, \quad y \in C^n[0, 1], \quad \Phi(y) - N\Psi(y) = \mathbf{0}. \quad (52)$$

From (48) and Theorem 1 it follows that the problem (52) is correct and has a unique solution, specifically

$$y = \widehat{A}^{-1} f + zNV^{-1}\Psi(\widehat{A}^{-1} f). \quad (53)$$

Consequently, by applying Theorem 1 once more, it is implied that the problem (51) is correct and possesses a unique solution given by

$$u = \widehat{A}^{-1} y + zNV^{-1}\Psi(\widehat{A}^{-1} y). \quad (54)$$

Substituting (53) into (54) and by using Remark 1, we obtain the solution (50) to the problem (49).

The correctness of the operator A_2 follows from Theorem 1. □

Remark 1 The inverse matrix in (50) can be written explicitly as

$$\begin{bmatrix} V & -\Psi(\widehat{A}^{-1}z)N \\ 0_m & V \end{bmatrix}^{-1} = \begin{bmatrix} V^{-1} & V^{-1}\Psi(\widehat{A}^{-1}z)NV^{-1} \\ 0_n & V^{-1} \end{bmatrix}. \quad (55)$$

6 Example Problems

In this section, we select three example problems to solve, both to demonstrate the implementation of the method presented in the previous sections as well as to reveal its efficiency.

Problem 1 Let the second order boundary value problem

$$\begin{aligned} &u'' - u = e^{2x}, \quad x \in (0, 1), \\ &u(0) - 2u'(0) = 0, \\ &u'(0) + eu(1) - 3\int_0^1 e^t u(t)dt = 0. \end{aligned} \tag{56}$$

To solve this problem we apply Theorem 1. We take

$$Au = u'' - u, \quad D(A) = C^2[0, 1], \tag{57}$$

and write the boundary conditions in the form

$$\Phi(u) - N\Psi(u) = \Phi(u) - \begin{bmatrix} 2 & 0 & 0 \\ 0 & -e & 3 \end{bmatrix} \Psi(u) = \mathbf{0}, \tag{58}$$

where

$$\Phi(u) = \begin{pmatrix} \Phi_1(u) \\ \Phi_2(u) \end{pmatrix} = \begin{pmatrix} u(0) \\ u'(0) \end{pmatrix}, \quad \Psi(u) = \begin{pmatrix} \Psi_1(u) \\ \Psi_2(u) \\ \Psi_3(u) \end{pmatrix} = \begin{pmatrix} u'(0) \\ u(1) \\ \int_0^1 e^t u(t)dt \end{pmatrix}. \tag{59}$$

It is easy to verify that

$$z_1 = \frac{e^x + e^{-x}}{2}, \quad z_2 = \frac{e^x - e^{-x}}{2}, \tag{60}$$

are fundamental solutions of the homogeneous equation $Au = 0$ and biorthogonal to boundary functionals Φ_1, Φ_2, i.e. $\Phi_i(z_j) = \delta_{ij}$. For the vector $z = (z_1, z_2)$, we compute the matrix

$$\Psi(z) = \begin{bmatrix} \Psi_1(z_1) & \Psi_1(z_2) \\ \Psi_2(z_1) & \Psi_2(z_2) \\ \Psi_3(z_1) & \Psi_3(z_2) \end{bmatrix} = \begin{bmatrix} 0 & 1 \\ \frac{e+e^{-1}}{2} & \frac{e-e^{-1}}{2} \\ \frac{e^2+1}{4} & \frac{e^2-3}{4} \end{bmatrix}, \tag{61}$$

and then the determinant

$$\det V = \det[I_3 - \Psi(z)N] = -\frac{3(e^2 - 3)}{4} \neq 0. \tag{62}$$

This ensures that the boundary value problem (56) has a unique solution.

To find the solution we need the exact solution of the correct problem,

$$\begin{aligned} &\widehat{A}u = u'' - u = f, \quad x \in (0, 1), \\ &D(\widehat{A}) = \{u : u \in D(A), \ u(0) = 0, \ u'(0) = 0\}, \end{aligned} \tag{63}$$

which for $f(x) = e^{2x}$ is given by

$$\widehat{A}^{-1} f = \frac{1}{3} e^{2x} - \frac{1}{2} e^{x} + \frac{1}{6} e^{-x}. \tag{64}$$

From (59) and (64), we get

$$\Psi(\widehat{A}^{-1} f) = \begin{pmatrix} \Psi_1(\widehat{A}^{-1} f) \\ \Psi_2(\widehat{A}^{-1} f) \\ \Psi_3(\widehat{A}^{-1} f) \end{pmatrix} = \begin{pmatrix} 0 \\ \frac{2e^2 - 3e + e^{-1}}{6} \\ \frac{4e^3 - 9e^2 + 11}{36} \end{pmatrix}. \tag{65}$$

Hence, by substituting the matrices N and V, the vectors z and $\Psi(\widehat{A}^{-1} f)$, and $\widehat{A}^{-1} f$ from above into (15), we get the solution to the boundary value problem (56), namely

$$u(x) = \frac{1}{3} \left(e^{2x} + e^{-x} \right). \tag{66}$$

Problem 2 Consider the boundary value problem

$$\begin{aligned} &u''(x) = f(x), \quad 0 < x < 1, \\ &u(0) - \int_0^1 h(x) u(x) dx = 0, \quad u'(0) - \int_0^1 h(x) u'(x) dx = 0, \end{aligned} \tag{67}$$

where $f(x)$, $h(x)$ are continuous on $[0, 1]$.

Here, Theorem 5 is appropriate. We take

$$Au = u'(x), \quad D(A) = C^1[0, 1], \tag{68}$$

and

$$\begin{aligned} &\Phi(u) = u(0), \quad \Psi(u) = \int_0^1 h(x) u(x) dx, \\ &\Phi(Au) = u'(0), \quad \Psi(Au) = \int_0^1 h(x) u'(x) dx. \end{aligned} \tag{69}$$

and $N = [1]$. Additionally, we consider the operator $\widehat{A}$ defined by

$$\widehat{A} u = u'(x), \quad u(0) = 0, \tag{70}$$

Notice that $z = 1$ is a solution of the homogeneous problem $Au = 0$ and $\Phi(z) = 1$, i.e. Φ is biorthogonal to z. Thus, if

$$\det V = \det[1 - \Psi(z)] = 1 - \Psi(z) = 1 - \int_0^1 h(x)dx \neq 0, \tag{71}$$

then the boundary value problem (67) has a unique solution which can be obtained by (50) where

$$\begin{aligned}
\widehat{A}^{-1} f &= \int_0^x f(t)dt, \\
\widehat{A}^{-2} f &= \int_0^x \left(\int_0^t f(s)ds \right) dt, \\
\widehat{A}^{-1} z &= x.
\end{aligned} \tag{72}$$

and

$$\begin{aligned}
\Psi(\widehat{A}^{-1} f) &= \int_0^1 h(t)\widehat{A}^{-1} f dt, \\
\Psi(\widehat{A}^{-2} f) &= \int_0^1 h(t)\widehat{A}^{-2} f dt, \\
\Psi(\widehat{A}^{-1} z) &= \int_0^1 th(t)dt.
\end{aligned} \tag{73}$$

Problem 3 Finally, consider the nonlocal three point boundary value problem

$$\begin{aligned}
&u''(x) + 2p(x)u'(x) + [p'(x) + p^2(x)]u(x) = f(x), \quad x \in (0, 1), \\
&u(0) - k_1 u(\frac{1}{2}) - k_2 u(1) = 0, \\
&u'(0) - k_1 u'(\frac{1}{2}) - k_2 u'(1) = 0,
\end{aligned} \tag{74}$$

where $p(x) \in C^1[0, 1]$, $p(0) = p(\frac{1}{2}) = p(1)$ and k_1, k_2 are real constants.

Observe that the problem (74) can be written as follows:

$$\left[u'(x) + p(x)u(x)\right]' + p(x)\left[u'(x) + p(x)u(x)\right] = f(x), \quad x \in (0, 1), \tag{75}$$

$$u(0) - k_1 u(\frac{1}{2}) - k_2 u(1) = 0,$$

$$u'(0)+p(0)u(0)-k_1 \left[u'(\frac{1}{2})+p(\frac{1}{2})u(\frac{1}{2})\right] -k_2 \left[u'(1) + p(1)u(1)\right] =0. \tag{76}$$

Thus, we can take

$$Au = u'(x) + p(x)u(x), \quad D(A) = C^1[0, 1], \tag{77}$$

and

$$\Phi(u) = u(0), \quad \Psi_1 = u(\frac{1}{2}), \quad \Psi_2 = u(1), \quad N = [k_1 \ k_2]. \tag{78}$$

It is easy to verify that

$$z(x) = e^{-\int_0^x p(s)ds} \tag{79}$$

is a fundamental solution of the homogeneous problem $Au = 0$ and biorthogonal to Φ, i.e. $\Phi(z) = 1$. Furthermore,

$$\Psi(z) = \begin{pmatrix} \Psi_1(z) \\ \Psi_2(z) \end{pmatrix} = \begin{pmatrix} z(\frac{1}{2}) \\ z(1) \end{pmatrix} = \begin{pmatrix} e^{-\int_0^{1/2} p(s)ds} \\ e^{-\int_0^1 p(s)ds} \end{pmatrix}. \tag{80}$$

The operators $\widehat{A}$, A_1, and A_2 are defined accordingly to (9), (12), and (47), respectively, and the problem (74) is recast in the operator form (49).

If the condition

$$\begin{aligned} \det V &= \det[I_2 - \Psi(z)N] \\ &= \det \begin{bmatrix} 1 - k_1\Psi_1(z) & -k_2\Psi_1(z) \\ -k_1\Psi_2(z) & 1 - k_2\Psi_2(z) \end{bmatrix} \\ &= 1 - (k_1\Psi_1(z) + k_2\Psi_2(z)) \neq 0 \end{aligned} \tag{81}$$

is satisfied, then the problem (74) possesses a unique solution. Note that the operator $\widehat{A}$ defined by

$$\widehat{A}u = u'(x) + p(x)u(x) = f(x), \quad u(0) = 0, \tag{82}$$

since $p(x)$, $f(x)$ are continuous functions on [0, 1], has the inverse operator

$$\widehat{A}^{-1} f(x) = e^{-\int_0^x p(t)dt} \int_0^x f(t) e^{\int_0^t p(s)ds} dt. \tag{83}$$

The unique solution of the problem (74) can be now obtained by substituting into (50).

7 Concluding Remarks

We have scrutinized the solution of nonhomogeneous boundary value problems incorporating a general linear ordinary differential operator of order n and general boundary conditions, including multipoint and integral conditions. We have established an easy to compute solvability criterion which when is satisfied ensures the existence and uniqueness of the solution of a general boundary value problem. Also, we have presented a technique for constructing the exact solution of a general boundary value problem when the exact solution of a simpler associated Cauchy or boundary value problem is known. Finally, we have demonstrated how a $2n$th order differential boundary value problem with special symmetric boundary conditions can be solved exactly by considering merely a corresponding nth order boundary value problem.

The solution process can be easily implemented to a Computer Algebra System (CAS). It will be of help to practitioners, scientists, and students who are interested in obtaining the solution of a boundary value problem without to have to rummage through the vast literature on differential equations and advanced specialized mathematics.

References

1. K.R. Aida-zade, V.M. Abdullaev, On the solution of boundary value problems with nonseparated multipoint and integral conditions. Differ. Equ. **49**, 1114–1125 (2013). https://doi.org/10.1134/S0012266113090061
2. B.N. Biyarov, Normal extensions of linear operators. Eurasian Math. J. **7**(3), 17–32 (2016)
3. B.N. Biyarov, G.K. Abdrasheva, Relatively bounded perturbations of correct restrictions and extensions of linear operators, in *Functional Analysis in Interdisciplinary Applications (FAIA 2017)*, ed. by T. Kalmenov, E. Nursultanov, M. Ruzhansky, M. Sadybekov (Springer, Cham, 2017). https://doi.org/10.1007/978-3-319-67053-9
4. J. Chamberlain, L. Kong, Q. Kong, Nodal solutions of nonlocal boundary value problems. Math. Model. Anal. **14**(4), 435–450 (2009). https://doi.org/10.3846/1392-6292.2009.14.435-450
5. M. Denche, A. Kourta, Boundary value problem for second-order differential operators with nonregular integral boundary conditions. Rocky Mountain J. Math. **36**(3), 893–913 (2006). https://doi.org/10.1216/rmjm/1181069435
6. J.M. Gallardo, Second-order differential operators with integral boundary conditions and generation of analytic semigroups. Rocky Mountain J. Math. **30**(4), 1265–1291 (2000). https://doi.org/10.1216/rmjm/1021477351
7. G. Grubb, A characterization of the non-local boundary value problems associated with an elliptic operator. Ann. Scuola Norm. Sup. Pisa **22**, 425–513 (1968)
8. B.K. Kokebaev, M. Otelbaev, A.N. Shynibekov, About restrictions and extensions of operators (in Russian). D.A.N. SSSR. **271**(6), 1307–1310 (1983)
9. A.M. Krall, The development of general differential and general differential-boundary systems. Rocky Mountain J. Math. **5**(4), 493–542 (1975) https://doi.org/10.1216/RMJ-1975-5-4-493
10. L. Liu, X. Hao, Y. Wu, Positive solutions for singular second order differential equations with integral boundary conditions. Math.Comput. Model. **57**, 836–847 (2013)
11. R.O. Oinarov, I.N. Parasidis, Correct extensions of operators with finite defect in Banach spaces (in Russian). Izv. Akad. Kaz. SSR. **5**, 42–46 (1988)
12. I.N. Parasidis, E. Providas, Exact solutions to problems with perturbed differential and boundary operators, in *Analysis and Operator Theory*, ed. by T. Rassias, V. Zagrebnov (Springer, Cham, 2019). https://doi.org/10.1007/978-3-030-12661-2

13. I.N. Parasidis, P. Tsekrekos, Correct self-adjoint and positive extensions of nondensely defined minimal symmetric operators. Abstr. Appl. Anal. **7**, 767–790 (2005)
14. I.N. Parasidis, P. Tsekrekos, Some quadratic correct extensions of minimal operators in Banach spaces. Operators Matrices **4**, 225–243 (2010)
15. M.A. Sadybekov, N.S. Imanbaev, A regular differential operator with perturbed boundary condition. Math. Notes **101**(5), 878–887 (2017). https://doi.org/10.1134/S0001434617050133
16. M.I. Vishik, On general boundary value problems for elliptic differential equations. Tr. Moskv. Mat. Obšv. **1**, 187–246 (1952). Translated in AMS Transl. **24**, 107–172 (1963)
17. G.D. Zhang, H.R. Sun, Multiple solutions for a fourth-order difference boundary value problem with parameter via variational approach. Appl. Math. Model. **36**(9), 4385–4392 (2012). https://doi.org/10.1016/j.apm.2011.11.064

Additive-Quadratic Functional Inequalities

Choonkil Park and Themistocles M. Rassias

Abstract In this paper, we introduce and solve the following additive-quadratic s-functional inequalities:

$$\begin{aligned} &\|f(x+y)+f(x-y)-2f(x)-f(y)-f(-y)\| \\ &\leq \left\| s\left(2f\left(\frac{x+y}{2}\right)-f(x)-f(y)+f\left(\frac{x-y}{2}\right)+f\left(\frac{y-x}{2}\right)\right)\right\|, \end{aligned} \tag{1}$$

where s is a fixed nonzero complex number with $|s| < \sqrt{2}$, and

$$\begin{aligned} &\left\|2f\left(\frac{x+y}{2}\right)-f(x)-f(y)+f\left(\frac{x-y}{2}\right)+f\left(\frac{y-x}{2}\right)\right\| \\ &\leq \|s(f(x+y)+f(x-y)-2f(x)-f(y)-f(-y))\|, \end{aligned} \tag{2}$$

where s is a fixed nonzero complex number with $|s| < \frac{1}{2}$. Using the direct method and the fixed point method, we prove the Hyers–Ulam stability of the additive-quadratic s-functional inequalities (1) and (2) in complex Banach spaces.

1 Introduction and Preliminaries

The stability problem of functional equations originated from a question of Ulam [21] concerning the stability of group homomorphisms. Hyers [12] gave a first affirmative partial answer to the question of Ulam for Banach spaces. Hyers'

C. Park
Department of Mathematics, Hanyang University, Seoul, South Korea
e-mail: baak@hanyang.ac.kr

T. M. Rassias (✉)
Department of Mathematics, National Technical University of Athens, Athens, Greece
e-mail: trassias@math.ntua.gr

N. J. Daras, T. M. Rassias (eds.), *Computational Mathematics and Variational Analysis*, Springer Optimization and Its Applications 159,
https://doi.org/10.1007/978-3-030-44625-3_18

Theorem was generalized by Aoki [1] for additive mappings and by Rassias [19] for linear mappings by considering an unbounded Cauchy difference. A generalization of the Rassias theorem was obtained by Găvruta [11] by replacing the unbounded Cauchy difference by a general control function in the spirit of Rassias' approach. The stability of quadratic functional equation was proved by Skof [20] for mappings $f : E_1 \to E_2$, where E_1 is a normed space and E_2 is a Banach space. Cholewa [7] noticed that the theorem of Skof is still true if the relevant domain E_1 is replaced by an Abelian group.

Park [16, 17] defined additive ρ-functional inequalities and proved the Hyers–Ulam stability of the additive ρ-functional inequalities in Banach spaces and non-Archimedean Banach spaces. The stability problems of various functional equations have been extensively investigated by a number of authors (see [5, 6, 9, 10, 14, 22]).

We recall a fundamental result in fixed point theory.

Theorem 1 ([2, 8]) *Let (X, d) be a complete generalized metric space and let $J : X \to X$ be a strictly contractive mapping with Lipschitz constant $\alpha < 1$. Then for each given element $x \in X$, either*

$$d(J^n x, J^{n+1} x) = \infty$$

for all nonnegative integers n or there exists a positive integer n_0 such that

(1) $d(J^n x, J^{n+1} x) < \infty, \qquad \forall n \geq n_0$;

(2) the sequence $\{J^n x\}$ converges to a fixed point y^* of J;

(3) y^* is the unique fixed point of J in the set $Y = \{y \in X \mid d(J^{n_0} x, y) < \infty\}$;

(4) $d(y, y^*) \leq \frac{1}{1-\alpha} d(y, Jy)$ for all $y \in Y$.

In 1996, Isac and Rassias [13] were the first to provide applications of stability theory of functional equations for the proof of new fixed point theorems with applications. By using fixed point methods, the stability problems of several functional equations have been extensively investigated by a number of authors (see [3, 4, 18]).

This paper is organized as follows: In Sections 2 and 3, we solve the additive-quadratic s-functional inequalities (1) and (2), and prove the Hyers–Ulam stability of the additive-quadratic s-functional inequalities (1) and (2) in Banach spaces by using the direct method. In Sections 4 and 5, we prove the Hyers–Ulam stability of the additive-quadratic s-functional inequalities (1) and (2) in Banach spaces by using the fixed point method.

Throughout this paper, let X be a real or complex normed space and Y a complex Banach space.

2 Stability of the Additive-Quadratic s-Functional Inequality (1): A Direct Method

Throughout this section, assume that s is a fixed nonzero complex number with $|s| < \sqrt{2}$.

In this section, we solve and investigate the additive-quadratic s-functional inequality (1) in complex Banach spaces.

Lemma 1 *Assume that a mapping $f : X \to Y$ satisfies $f(0) = 0$ and*

$$\begin{aligned}&\|f(x+y) + f(x-y) - 2f(x) - f(y) - f(-y)\| \\ &\quad \le \left\| s\left(2f\left(\frac{x+y}{2}\right) - f(x) - f(y) + f\left(\frac{x-y}{2}\right) + f\left(\frac{y-x}{2}\right)\right)\right\| \end{aligned} \tag{3}$$

for all $x, y \in X$. If f is odd, then $f : X \to Y$ is additive. If f is even, then $f : X \to Y$ is quadratic.

Proof Assume that $f : X \to Y$ satisfies (3) and that $f : X \to Y$ is odd.

It follows from (3) that

$$\|f(x+y) + f(x-y) - 2f(x)\| \le \left\| s\left(2f\left(\frac{x+y}{2}\right) - f(x) - f(y)\right)\right\| \tag{4}$$

for all $x, y \in X$.

Letting $x = y$ in (4), we get $\|f(2x) - 2f(x)\| \le 0$ and so $f(2x) = 2f(x)$ for all $x \in X$. Thus

$$\|f(x+y) + f(x-y) - 2f(x)\| \le \|s(f(x+y) - f(x) - f(y))\| \tag{5}$$

for all $x, y \in X$. Letting $z = x + y$ and $w = x - y$ in (5), we get

$$\begin{aligned}\|f(z) + f(w) - f(z+w)\| &= \left\| f(z) + f(w) - 2f\left(\frac{z+w}{2}\right)\right\| \\ \le \left\| s\left(f(z) - f\left(\frac{z+w}{2}\right) - f\left(\frac{z-w}{2}\right)\right)\right\| &= \left\| \frac{s}{2}(2f(z) - f(z+w) - f(z-w))\right\| \end{aligned} \tag{6}$$

for all $z, w \in X$. It follows from (5) and (6) that

$$\|f(z) + f(w) - f(z+w)\| \le \left\| \frac{s^2}{2}(f(z) + f(w) - f(z+w))\right\|$$

and so $f(z+w) = f(z) + f(w)$ for all $z, w \in X$, since $|s| < \sqrt{2}$. Thus $f : X \to Y$ is additive.

Assume that $f : X \to Y$ satisfies (3) and that $f : X \to Y$ is even.

It follows from (3) that

$$\begin{aligned}&\|f(x+y)+f(x-y)-2f(x)-2f(y)\|\\&\quad\le\left\|s\left(2f\left(\frac{x+y}{2}\right)+2f\left(\frac{x-y}{2}\right)-f(x)-f(y)\right)\right\|\end{aligned}\tag{7}$$

for all $x, y \in X$.

Letting $x = y$ in (7), we get $\|f(2x) - 4f(x)\| \le 0$ and so $f(2x) = 4f(x)$ for all $x \in X$. Thus

$$\begin{aligned}&\|f(x+y)+f(x-y)-2f(x)-2f(y)\|\\&\quad\le\left\|\frac{s}{2}(f(x+y)+f(x-y)-2f(x)-2f(y))\right\|\end{aligned}$$

and so $f(x + y) + f(x - y) = 2f(x) + 2f(y)$ for all $x, y \in X$, since $|s| < \sqrt{2}$. Thus $f : X \to Y$ is quadratic.

Now, we prove the Hyers–Ulam stability of the additive-quadratic s-functional inequality (1) in complex Banach spaces by using the direct method.

Theorem 2 *Let $\varphi : X^2 \to [0, \infty)$ be a function such that*

$$\Psi(x, y) := \sum_{j=1}^{\infty} 4^j \varphi\left(\frac{x}{2^j}, \frac{y}{2^j}\right) < \infty \tag{8}$$

for all $x, y \in X$. Let $f : X \to Y$ be a mapping satisfying $f(0) = 0$ and

$$\begin{aligned}&\|f(x+y)+f(x-y)-2f(x)-f(y)-f(-y)\| \qquad (9)\\&\le\left\|s\left(2f\left(\frac{x+y}{2}\right)-f(x)-f(y)+f\left(\frac{x-y}{2}\right)+f\left(\frac{y-x}{2}\right)\right)\right\|+\varphi(x,y)\end{aligned}$$

for all $x, y \in X$. If the mapping $f : X \to Y$ is odd, then there exists a unique additive mapping $A : X \to Y$ such that

$$\|f(x) - A(x)\| \le \frac{1}{2}\Phi(x, x) \tag{10}$$

for all $x \in X$, where $\Phi(x, y) := \sum_{j=1}^{\infty} 2^j \varphi\left(\frac{x}{2^j}, \frac{y}{2^j}\right)$ for all $x, y \in X$. If the mapping $f : X \to Y$ is even, then there exists a unique quadratic mapping $Q : X \to Y$ such that

$$\|f(x) - Q(x)\| \le \frac{1}{4}\Psi(x, x) \tag{11}$$

for all $x \in X$.

Proof Assume that $f : X \to Y$ satisfies (9) and is an odd mapping. Letting $y = x$ in (9), we get

$$\|f(2x) - 2f(x)\| \le \varphi(x, x) \tag{12}$$

and so

$$\left\|f(x) - 2f\left(\frac{x}{2}\right)\right\| \le \varphi\left(\frac{x}{2}, \frac{x}{2}\right)$$

for all $x \in X$. Thus

$$\begin{aligned}\left\|2^l f\left(\frac{x}{2^l}\right) - 2^m f\left(\frac{x}{2^m}\right)\right\| &\le \sum_{j=l}^{m-1} \left\|2^j f\left(\frac{x}{2^j}\right) - 2^{j+1} f\left(\frac{x}{2^{j+1}}\right)\right\| \\ &\le \frac{1}{2} \sum_{j=l+1}^{m} 2^j \varphi\left(\frac{x}{2^j}, \frac{x}{2^j}\right)\end{aligned} \tag{13}$$

for all nonnegative integers m and l with $m > l$ and all $x \in X$. It follows from (13) that the sequence $\{2^k f(\frac{x}{2^k})\}$ is Cauchy for all $x \in X$. Since Y is a Banach space, the sequence $\{2^k f(\frac{x}{2^k})\}$ converges. So one can define the mapping $A : X \to Y$ by

$$A(x) := \lim_{k\to\infty} 2^k f\left(\frac{x}{2^k}\right)$$

for all $x \in X$. Moreover, letting $l = 0$ and passing to the limit $m \to \infty$ in (13), we get (10).

It follows from (8) and (9) that

$$\begin{aligned}&\|A(x+y) + A(x-y) - 2A(x) - A(y) - A(-y)\| \\ &= \lim_{n\to\infty} 2^n \left\|f\left(\frac{x+y}{2^n}\right) + f\left(\frac{x-y}{2^n}\right) - 2f\left(\frac{x}{2^n}\right) - f\left(\frac{y}{2^n}\right) - f\left(\frac{-y}{2^n}\right)\right\| \\ &\le \lim_{n\to\infty} 2^n \left\|s\left(2f\left(\frac{x+y}{2^{n+1}}\right) - f\left(\frac{x}{2^n}\right) - f\left(\frac{y}{2^n}\right) + f\left(\frac{x-y}{2^{n+1}}\right) + f\left(\frac{y-x}{2^{n+1}}\right)\right)\right\| \\ &\quad + \lim_{n\to\infty} 2^n \varphi\left(\frac{x}{2^n}, \frac{y}{2^n}\right) \\ &= \left\|s\left(2A\left(\frac{x+y}{2}\right) - A(x) - A(y) + A\left(\frac{x-y}{2}\right) + A\left(\frac{y-x}{2}\right)\right)\right\|\end{aligned}$$

for all $x, y \in X$. So

$$\begin{aligned}&\|A(x+y) + A(x-y) - 2A(x) - A(y) - A(-y)\| \\ &\le \left\|s\left(2A\left(\frac{x+y}{2}\right) - A(x) - A(y) + A\left(\frac{x-y}{2}\right) + A\left(\frac{y-x}{2}\right)\right)\right\|\end{aligned}$$

for all $x, y \in X$. By Lemma 1, the mapping $A : X \to Y$ is additive, since $A : X \to Y$ is an odd mapping.

Now, let $T : X \to Y$ be another additive mapping satisfying (10). Then we have

$$\begin{aligned}\|A(x) - T(x)\| &= \left\|2^q A\left(\frac{x}{2^q}\right) - 2^q T\left(\frac{x}{2^q}\right)\right\| \\ &\leq \left\|2^q A\left(\frac{x}{2^q}\right) - 2^q f\left(\frac{x}{2^q}\right)\right\| + \left\|2^q T\left(\frac{x}{2^q}\right) - 2^q f\left(\frac{x}{2^q}\right)\right\| \\ &\leq 2^q \Phi\left(\frac{x}{2^q}, \frac{x}{2^q}\right),\end{aligned}$$

which tends to zero as $q \to \infty$ for all $x \in X$. So we can conclude that $A(x) = T(x)$ for all $x \in X$. This proves the uniqueness of A.

Assume that $f : X \to Y$ satisfies (9) and is an even mapping. Letting $y = x$ in (9), we get

$$\|f(2x) - 4f(x)\| \leq \varphi(x, x) \tag{14}$$

and so

$$\left\|f(x) - 4f\left(\frac{x}{2}\right)\right\| \leq \varphi\left(\frac{x}{2}, \frac{x}{2}\right)$$

for all $x \in X$. Thus

$$\begin{aligned}\left\|4^l f\left(\frac{x}{2^l}\right) - 4^m f\left(\frac{x}{2^m}\right)\right\| &\leq \sum_{j=l}^{m-1} \left\|4^j f\left(\frac{x}{2^j}\right) - 4^{j+1} f\left(\frac{x}{2^{j+1}}\right)\right\| \\ &\leq \frac{1}{4} \sum_{j=l+1}^{m} 4^j \varphi\left(\frac{x}{2^j}, \frac{x}{2^j}\right)\end{aligned} \tag{15}$$

for all nonnegative integers m and l with $m > l$ and all $x \in X$. It follows from (15) that the sequence $\{4^k f(\frac{x}{2^k})\}$ is Cauchy for all $x \in X$. Since Y is a Banach space, the sequence $\{4^k f(\frac{x}{2^k})\}$ converges. So one can define the mapping $Q : X \to Y$ by

$$Q(x) := \lim_{k\to\infty} 4^k f\left(\frac{x}{2^k}\right)$$

for all $x \in X$. Moreover, letting $l = 0$ and passing to the limit $m \to \infty$ in (15), we get (11).

It follows from (8) and (9) that

$$\|Q(x + y) + Q(x - y) - 2Q(x) - Q(y) - Q(-y)\|$$

$$
\begin{aligned}
&= \lim_{n\to\infty} 4^n \left\| f\left(\frac{x+y}{2^n}\right) + f\left(\frac{x-y}{2^n}\right) - 2f\left(\frac{x}{2^n}\right) - f\left(\frac{y}{2^n}\right) - f\left(\frac{-y}{2^n}\right) \right\| \\
&\le \lim_{n\to\infty} 4^n \left\| s\left(2f\left(\frac{x+y}{2^{n+1}}\right) - f\left(\frac{x}{2^n}\right) - f\left(\frac{y}{2^n}\right) + f\left(\frac{x-y}{2^{n+1}}\right) + f\left(\frac{y-x}{2^{n+1}}\right)\right) \right\| \\
&\qquad + \lim_{n\to\infty} 4^n \varphi\left(\frac{x}{2^n}, \frac{y}{2^n}\right) \\
&= \left\| s\left(2Q\left(\frac{x+y}{2}\right) - Q(x) - Q(y) + Q\left(\frac{x-y}{2}\right) + Q\left(\frac{y-x}{2}\right)\right) \right\|
\end{aligned}
$$

for all $x, y \in X$. So

$$
\begin{aligned}
&\|Q(x+y) + Q(x-y) - 2Q(x) - Q(y) - Q(-y)\| \\
&\le \left\| s\left(2Q\left(\frac{x+y}{2}\right) - Q(x) - Q(y) + Q\left(\frac{x-y}{2}\right) + Q\left(\frac{y-x}{2}\right)\right) \right\|
\end{aligned}
$$

for all $x, y \in X$. By Lemma 1, the mapping $Q : X \to Y$ is quadratic, since $Q : X \to Y$ is an even mapping.

Now, let $T : X \to Y$ be another quadratic mapping satisfying (11). Then we have

$$
\begin{aligned}
\|Q(x) - T(x)\| &= \left\| 4^q Q\left(\frac{x}{2^q}\right) - 4^q T\left(\frac{x}{2^q}\right) \right\| \\
&\le \left\| 4^q Q\left(\frac{x}{2^q}\right) - 4^q f\left(\frac{x}{2^q}\right) \right\| + \left\| 4^q T\left(\frac{x}{2^q}\right) - 4^q f\left(\frac{x}{2^q}\right) \right\| \\
&\le \frac{4^q}{2} \Phi\left(\frac{x}{2^q}, \frac{x}{2^q}\right),
\end{aligned}
$$

which tends to zero as $q \to \infty$ for all $x \in X$. So we can conclude that $Q(x) = T(x)$ for all $x \in X$. This proves the uniqueness of Q.

Corollary 1 *Let $r > 2$ and θ be nonnegative real numbers and $f : X \to Y$ be a mapping satisfying $f(0) = 0$ and*

$$
\begin{aligned}
&\|f(x+y) + f(x-y) - 2f(x) - f(y) - f(-y)\| \qquad (16) \\
&\le \left\| s\left(2f\left(\frac{x+y}{2}\right) - f(x) - f(y) + f\left(\frac{x-y}{2}\right) + f\left(\frac{y-x}{2}\right)\right) \right\| + \theta(\|x\|^r + \|y\|^r)
\end{aligned}
$$

for all $x, y \in X$. If the mapping $f : X \to Y$ is odd, then there exists a unique additive mapping $A : X \to Y$ such that

$$
\|f(x) - A(x)\| \le \frac{2\theta}{2^r - 2}\|x\|^r
$$

for all $x, y \in X$. If the mapping $f : X \to Y$ is even, then there exists a unique quadratic mapping $Q : X \to Y$ such that

$$\|f(x) - Q(x)\| \le \frac{2\theta}{2^r - 4}\|x\|^r$$

for all $x \in X$.

Proof The proof follows from Theorem 2 by taking $\varphi(x, y) = \theta(\|x\|^r + \|y\|^r)$ for all $x, y \in X$.

Theorem 3 *Let $\varphi : X^2 \to [0, \infty)$ be a function and let $f : X \to Y$ be a mapping satisfying $f(0) = 0$, (9) and*

$$\Psi(x, y) := \sum_{j=0}^{\infty} \frac{1}{2^j}\varphi(2^j x, 2^j y, 2^j z) < \infty \tag{17}$$

for all $x, y, z \in X$. If the mapping $f : X \to Y$ is odd, then there exists a unique additive mapping $A : X \to Y$ such that

$$\|f(x) - A(x)\| \le \frac{1}{2}\Psi(x, x) \tag{18}$$

for all $x \in X$. If the mapping $f : X \to Y$ is even, then there exists a unique quadratic mapping $Q : X \to Y$ such that

$$\|f(x) - Q(x)\| \le \frac{1}{4}\Phi(x, x) \tag{19}$$

for all $x \in X$, where $\Phi(x, y) := \sum_{j=0}^{\infty} \frac{1}{4^j}\varphi\left(2^j x, 2^j y\right)$ for all $x, y \in X$.

Proof Assume that $f : X \to Y$ satisfies (9) and is an odd mapping.

It follows from (12) that

$$\left\|f(x) - \frac{1}{2}f(2x)\right\| \le \frac{1}{2}\varphi(x, x)$$

for all $x \in X$. Thus

$$\begin{aligned}\left\|\frac{1}{2^l}f\left(2^l x\right) - \frac{1}{2^m}f\left(2^m x\right)\right\| &\le \sum_{j=l}^{m-1}\left\|\frac{1}{2^j}f\left(2^j x\right) - \frac{1}{2^{j+1}}f\left(2^{j+1}x\right)\right\| \\ &\le \frac{1}{2}\sum_{j=l}^{m-1}\frac{1}{2^j}\varphi\left(2^j x, 2^j x\right)\end{aligned} \tag{20}$$

for all nonnegative integers m and l with $m > l$ and all $x \in X$. It follows from (20) that the sequence $\{\frac{1}{2^k} f(2^k x)\}$ is Cauchy for all $x \in X$. Since Y is a Banach space, the sequence $\{\frac{1}{2^k} f(2^k x)\}$ converges. So one can define the mapping $A : X \to Y$ by

$$A(x) := \lim_{k\to\infty} \frac{1}{2^k} f\left(2^k x\right)$$

for all $x \in X$. Moreover, letting $l = 0$ and passing to the limit $m \to \infty$ in (20), we get (18).

Assume that $f : X \to Y$ satisfies (9) and is an even mapping. It follows from (14) that

$$\left\| f(x) - \frac{1}{4} f(2x) \right\| \le \frac{1}{4}\varphi(x, x)$$

for all $x \in X$. Thus

$$\begin{aligned} \left\| \frac{1}{4^l} f\left(2^l x\right) - \frac{1}{4^m} f\left(2^m x\right) \right\| &\le \sum_{j=l}^{m-1} \left\| \frac{1}{4^j} f\left(2^j x\right) - \frac{1}{4^{j+1}} f\left(2^{j+1} x\right) \right\| \qquad (21) \\ &\le \frac{1}{4} \sum_{j=l}^{m-1} \frac{1}{4^j} \varphi\left(2^j x, 2^j x\right) \end{aligned}$$

for all nonnegative integers m and l with $m > l$ and all $x \in X$. It follows from (21) that the sequence $\{\frac{1}{4^k} f(2^k x)\}$ is Cauchy for all $x \in X$. Since Y is a Banach space, the sequence $\{\frac{1}{4^k} f(2^k x)\}$ converges. So one can define the mapping $Q : X \to Y$ by

$$Q(x) := \lim_{k\to\infty} \frac{1}{4^k} f\left(2^k x\right)$$

for all $x \in X$. Moreover, letting $l = 0$ and passing to the limit $m \to \infty$ in (21), we get (19).

The rest of the proof is similar to the proof of Theorem 2.

Corollary 2 *Let $r < 1$ and θ be positive real numbers and $f : X \to Y$ be a mapping satisfying $f(0) = 0$ and (16). If the mapping $f : X \to Y$ is odd, then there exists a unique additive mapping $A : X \to Y$ such that*

$$\|f(x) - A(x)\| \le \frac{2\theta}{2 - 2^r} \|x\|^r$$

for all $x, y \in X$. If the mapping $f : X \to Y$ is even, then there exists a unique quadratic mapping $Q : X \to Y$ such that

$$\|f(x) - Q(x)\| \le \frac{2\theta}{4 - 2^r} \|x\|^r$$

for all $x \in X$.

Proof The proof follows from Theorem 3 by taking $\varphi(x, y) = \theta(\|x\|^r + \|y\|^r)$ for all $x, y \in X$.

3 Stability of the Additive-Quadratic s-Functional Inequality (2): A Direct Method

Throughout this section, assume that s is a fixed nonzero complex number with $|s| < \frac{1}{2}$.

In this section, we solve and investigate the additive-quadratic s-functional inequality (2) in complex Banach spaces.

Lemma 2 *Assume that a mapping $f : X \to Y$ satisfies $f(0) = 0$ and*

$$\left\|2f\left(\frac{x+y}{2}\right) - f(x) - f(y) + f\left(\frac{x-y}{2}\right) + f\left(\frac{y-x}{2}\right)\right\| \qquad (22)$$
$$\leq \|s\left(f\left(x+y\right) + f(x-y) - 2f(x) - f(y) - f(-y)\right)\|$$

for all $x, y \in X$. If f is odd, then $f : X \to Y$ is additive. If f is even, then $f : X \to Y$ is quadratic.

Proof Assume that $f : X \to Y$ satisfies (22) and that $f : X \to Y$ is odd.

It follows from (22) that

$$\left\|2f\left(\frac{x+y}{2}\right) - f(x) - f(y)\right\| \leq \|s\left(f\left(x+y\right) + f(x-y) - 2f(x)\right)\| \qquad (23)$$

for all $x, y \in X$.

Letting $y = 0$ in (23), we get $\left\|2f\left(\frac{x}{2}\right) - f(x)\right\| \leq 0$ and so $f(2x) = 2f(x)$ for all $x \in X$. Thus

$$\|f\left(x+y\right) - f(x) - f(y)\| \leq \|s\left(f\left(x+y\right) + f(x-y) - 2f(x)\right)\| \qquad (24)$$

for all $x, y \in X$. Letting $z = x + y$ and $w = x - y$ in (24), we get

$$\left\|\frac{1}{2}(2f\left(z\right) - f\left(z+w\right) - f\left(z-w\right))\right\| = \left\|f\left(z\right) - f\left(\frac{z+w}{2}\right) - f\left(\frac{z-w}{2}\right)\right\| \qquad (25)$$
$$\leq \left\|s\left(f\left(z\right) + f(w) - 2f\left(\frac{z+w}{2}\right)\right)\right\| = \|s\left(f\left(z\right) + f(w) - f\left(z+w\right)\right)\|$$

for all $z, w \in X$. It follows from (24) and (25) that

$$\|f(x+y) - f(x) - f(y)\| \leq \left\|2s^2\left(f(x+y) - f(x) - f(y)\right)\right\|$$

and so $f(x+y) = f(x) + f(y)$ for all $x, y \in X$, since $|s| < \frac{1}{2}$. Thus $f : X \to Y$ is additive.

Assume that $f : X \to Y$ satisfies (22) and that $f : X \to Y$ is even.

It follows from (22) that

$$\left\| 2f\left(\frac{x+y}{2}\right) + 2f\left(\frac{x-y}{2}\right) - f(x) - f(y) \right\| \\ \leq \| s\,(f\,(x+y) + f(x-y) - 2f(x) - 2f(y)) \| \tag{26}$$

for all $x, y \in X$.

Letting $y = 0$ in (26), we get $\left\| 4f\left(\frac{x}{2}\right) - f(x) \right\| \leq 0$ and so $f(2x) = 4f(x)$ for all $x \in X$. Thus

$$\frac{1}{2} \| f\,(x+y) + f(x-y) - 2f(x) - 2f(y) \| \\ \leq \| s\,(f\,(x+y) + f\,(x-y) - 2f(x) - 2f(y)) \|$$

and so $f\,(x+y) + f(x-y) = 2f(x) + 2f(y)$ for all $x, y \in X$, since $|s| < \frac{1}{2}$. Thus $f : X \to Y$ is quadratic.

Now, we prove the Hyers–Ulam stability of the additive-quadratic s-functional inequality (2) in complex Banach spaces by using the direct method.

Theorem 4 *Let $\varphi : X^2 \to [0, \infty)$ be a function such that*

$$\Psi(x, y) := \sum_{j=0}^{\infty} 4^j \varphi\left(\frac{x}{2^j}, \frac{y}{2^j}\right) < \infty \tag{27}$$

for all $x, y \in X$. Let $f : X \to Y$ be a mapping satisfying $f(0) = 0$ and

$$\left\| 2f\left(\frac{x+y}{2}\right) - f(x) - f(y) + f\left(\frac{x-y}{2}\right) + f\left(\frac{y-x}{2}\right) \right\| \\ \leq \| s\,(f\,(x+y) + f(x-y) - 2f(x) - f(y) - f(-y)) \| + \varphi(x, y) \tag{28}$$

for all $x, y \in X$. If the mapping $f : X \to Y$ is odd, then there exists a unique additive mapping $A : X \to Y$ such that

$$\| f(x) - A(x) \| \leq \Phi(x, 0) \tag{29}$$

for all $x \in X$, where $\Phi(x, y) := \sum_{j=0}^{\infty} 2^j \varphi\left(\frac{x}{2^j}, \frac{y}{2^j}\right)$ for all $x, y \in X$. If the mapping $f : X \to Y$ is even, then there exists a unique quadratic mapping $Q : X \to Y$ such that

$$\| f(x) - Q(x) \| \leq \Psi(x, 0) \tag{30}$$

for all $x \in X$.

Proof Assume that $f : X \to Y$ satisfies (28) and is an odd mapping. Letting $y = 0$ in (28), we get

$$\left\|2f\left(\frac{x}{2}\right) - f(x)\right\| \le \varphi(x, 0) \tag{31}$$

for all $x \in X$. Thus

$$\begin{aligned} \left\|2^l f\left(\frac{x}{2^l}\right) - 2^m f\left(\frac{x}{2^m}\right)\right\| &\le \sum_{j=l}^{m-1} \left\|2^j f\left(\frac{x}{2^j}\right) - 2^{j+1} f\left(\frac{x}{2^{j+1}}\right)\right\| \\ &\le \sum_{j=l}^{m-1} 2^j \varphi\left(\frac{x}{2^j}, 0\right) \end{aligned} \tag{32}$$

for all nonnegative integers m and l with $m > l$ and all $x \in X$. It follows from (32) that the sequence $\{2^k f(\frac{x}{2^k})\}$ is Cauchy for all $x \in X$. Since Y is a Banach space, the sequence $\{2^k f(\frac{x}{2^k})\}$ converges. So one can define the mapping $A : X \to Y$ by

$$A(x) := \lim_{k\to\infty} 2^k f\left(\frac{x}{2^k}\right)$$

for all $x \in X$. Moreover, letting $l = 0$ and passing to the limit $m \to \infty$ in (32), we get (29).

It follows from (27) and (28) that

$$\begin{aligned} &\left\|2A\left(\frac{x+y}{2}\right) - A(x) - A(y) + A\left(\frac{x-y}{2}\right) + A\left(\frac{y-x}{2}\right)\right\| \\ &= \lim_{n\to\infty} 2^n \left\|2f\left(\frac{x+y}{2^{n+1}}\right) - f\left(\frac{x}{2^n}\right) - f\left(\frac{y}{2^n}\right) + f\left(\frac{x-y}{2^{n+1}}\right) + f\left(\frac{y-x}{2^{n+1}}\right)\right\| \\ &\le \lim_{n\to\infty} 2^n \left\|s\left(f\left(\frac{x+y}{2^n}\right) + f\left(\frac{x-y}{2^n}\right) - 2f\left(\frac{x}{2^n}\right) - f\left(\frac{y}{2^n}\right) - f\left(\frac{-y}{2^n}\right)\right)\right\| \\ &\qquad + \lim_{n\to\infty} 2^n \varphi\left(\frac{x}{2^n}, \frac{y}{2^n}\right) \\ &= \|s\left(A\left(x+y\right) + A(x-y) - 2A(x) - A(y) - A(-y)\right)\| \end{aligned}$$

for all $x, y \in X$. So

$$\begin{aligned} &\left\|2A\left(\frac{x+y}{2}\right) - A(x) - A(y) + A\left(\frac{x-y}{2}\right) + A\left(\frac{y-x}{2}\right)\right\| \\ &\le \|s\left(A\left(x+y\right) + A(x-y) - 2A(x) - A(y) - A(-y)\right)\| \end{aligned}$$

for all $x, y \in X$. By Lemma 2, the mapping $A : X \to Y$ is additive, since $A : X \to Y$ is an odd mapping.

Now, let $T : X \to Y$ be another additive mapping satisfying (29). Then we have

$$\begin{aligned}\|A(x) - T(x)\| &= \left\|2^q A\left(\frac{x}{2^q}\right) - 2^q T\left(\frac{x}{2^q}\right)\right\| \\ &\le \left\|2^q A\left(\frac{x}{2^q}\right) - 2^q f\left(\frac{x}{2^q}\right)\right\| + \left\|2^q T\left(\frac{x}{2^q}\right) - 2^q f\left(\frac{x}{2^q}\right)\right\| \\ &\le 2^{q+1}\Phi\left(\frac{x}{2^q}, 0\right),\end{aligned}$$

which tends to zero as $q \to \infty$ for all $x \in X$. So we can conclude that $A(x) = T(x)$ for all $x \in X$. This proves the uniqueness of A.

Assume that $f : X \to Y$ satisfies (28) and is an even mapping. Letting $y = 0$ in (28), we get

$$\left\|4f\left(\frac{x}{2}\right) - f(x)\right\| \le \varphi(x, 0) \tag{33}$$

for all $x \in X$. Thus

$$\begin{aligned}\left\|4^l f\left(\frac{x}{2^l}\right) - 4^m f\left(\frac{x}{2^m}\right)\right\| &\le \sum_{j=l}^{m-1}\left\|4^j f\left(\frac{x}{2^j}\right) - 4^{j+1} f\left(\frac{x}{2^{j+1}}\right)\right\| \\ &\le \sum_{j=l}^{m-1} 4^j \varphi\left(\frac{x}{2^j}, \frac{x}{2^j}\right)\end{aligned} \tag{34}$$

for all nonnegative integers m and l with $m > l$ and all $x \in X$. It follows from (34) that the sequence $\{4^k f(\frac{x}{2^k})\}$ is Cauchy for all $x \in X$. Since Y is a Banach space, the sequence $\{4^k f(\frac{x}{2^k})\}$ converges. So one can define the mapping $Q : X \to Y$ by

$$Q(x) := \lim_{k\to\infty} 4^k f\left(\frac{x}{2^k}\right)$$

for all $x \in X$. Moreover, letting $l = 0$ and passing to the limit $m \to \infty$ in (34), we get (30).

It follows from (27) and (28) that

$$\begin{aligned}&\left\|2Q\left(\frac{x+y}{2}\right) - Q(x) - Q(y) + Q\left(\frac{x-y}{2}\right) + Q\left(\frac{y-x}{2}\right)\right\| \\ &= \lim_{n\to\infty} 4^n \left\|2f\left(\frac{x+y}{2^{n+1}}\right) - f\left(\frac{x}{2^n}\right) - f\left(\frac{y}{2^n}\right) + f\left(\frac{x-y}{2^{n+1}}\right) + f\left(\frac{y-x}{2^{n+1}}\right)\right\|\end{aligned}$$

$$
\begin{aligned}
&\le \lim_{n\to\infty} 4^n \left\| s\left(f\left(\frac{x+y}{2^n}\right) + f\left(\frac{x-y}{2^n}\right) - 2f\left(\frac{x}{2^n}\right) - f\left(\frac{y}{2^n}\right) - f\left(\frac{-y}{2^n}\right)\right)\right\| \\
&\qquad + \lim_{n\to\infty} 4^n \varphi\left(\frac{x}{2^n}, \frac{y}{2^n}\right) \\
&= \| s\left(Q\left(x+y\right) + Q(x-y) - 2Q(x) - Q(y) - Q(-y)\right)\|
\end{aligned}
$$

for all $x, y \in X$. So

$$
\begin{aligned}
&\left\| 2Q\left(\frac{x+y}{2}\right) - Q(x) - Q(y) + Q\left(\frac{x-y}{2}\right) + Q\left(\frac{y-x}{2}\right)\right\| \\
&\le \| s\left(Q\left(x+y\right) + Q(x-y) - 2Q(x) - Q(y) - Q(-y)\right)\|
\end{aligned}
$$

for all $x, y \in X$. By Lemma 2, the mapping $Q : X \to Y$ is quadratic, since $Q : X \to Y$ is an even mapping.

Now, let $T : X \to Y$ be another quadratic mapping satisfying (30). Then we have

$$
\begin{aligned}
\|Q(x) - T(x)\| &= \left\| 4^q Q\left(\frac{x}{2^q}\right) - 4^q T\left(\frac{x}{2^q}\right)\right\| \\
&\le \left\| 4^q Q\left(\frac{x}{2^q}\right) - 4^q f\left(\frac{x}{2^q}\right)\right\| + \left\| 4^q T\left(\frac{x}{2^q}\right) - 4^q f\left(\frac{x}{2^q}\right)\right\| \\
&\le 2\cdot 4^q \Phi\left(\frac{x}{2^q}, 0\right),
\end{aligned}
$$

which tends to zero as $q \to \infty$ for all $x \in X$. So we can conclude that $Q(x) = T(x)$ for all $x \in X$. This proves the uniqueness of Q.

Corollary 3 *Let $r > 2$ and θ be nonnegative real numbers and $f : X \to Y$ be a mapping satisfying $f(0) = 0$ and*

$$
\begin{aligned}
&\left\| 2f\left(\frac{x+y}{2}\right) - f(x) - f(y) + f\left(\frac{x-y}{2}\right) + f\left(\frac{y-x}{2}\right)\right\| \qquad (35)\\
&\le \| s\left(f\left(x+y\right) + f(x-y) - 2f(x) - f(y) - f(-y)\right)\| + \theta(\|x\|^r + \|y\|^r)
\end{aligned}
$$

for all $x, y \in X$. If the mapping $f : X \to Y$ is odd, then there exists a unique additive mapping $A : X \to Y$ such that

$$
\|f(x) - A(x)\| \le \frac{2^r\theta}{2^r - 2}\|x\|^r
$$

for all $x, y \in X$. If the mapping $f : X \to Y$ is even, then there exists a unique quadratic mapping $Q : X \to Y$ such that

$$
\|f(x) - Q(x)\| \le \frac{2^r\theta}{2^r - 4}\|x\|^r
$$

for all $x \in X$.

Proof The proof follows from Theorem 4 by taking $\varphi(x, y) = \theta(\|x\|^r + \|y\|^r)$ for all $x, y \in X$.

Theorem 5 *Let $\varphi : X^2 \to [0, \infty)$ be a function and let $f : X \to Y$ be a mapping satisfying $f(0) = 0$, (28) and*

$$\Psi(x, y) := \sum_{j=1}^{\infty} \frac{1}{2^j}\varphi(2^j x, 2^j y, 2^j z) < \infty \tag{36}$$

for all $x, y, z \in X$. If the mapping $f : X \to Y$ is odd, then there exists a unique additive mapping $A : X \to Y$ such that

$$\|f(x) - A(x)\| \leq \Psi(x, 0) \tag{37}$$

for all $x \in X$. If the mapping $f : X \to Y$ is even, then there exists a unique quadratic mapping $Q : X \to Y$ such that

$$\|f(x) - Q(x)\| \leq \Phi(x, 0) \tag{38}$$

for all $x \in X$, where $\Phi(x, y) := \sum_{j=1}^{\infty} \frac{1}{4^j}\varphi\left(2^j x, 2^j y\right)$ for all $x, y \in X$.

Proof Assume that $f : X \to Y$ satisfies (28) and is an odd mapping.

It follows from (31) that

$$\left\| f(x) - \frac{1}{2} f(2x) \right\| \leq \frac{1}{2}\varphi(2x, 0)$$

for all $x \in X$. Thus

$$\begin{aligned} \left\| \frac{1}{2^l} f\left(2^l x\right) - \frac{1}{2^m} f\left(2^m x\right) \right\| &\leq \sum_{j=l}^{m-1} \left\| \frac{1}{2^j} f\left(2^j x\right) - \frac{1}{2^{j+1}} f\left(2^{j+1} x\right) \right\| \\ &\leq \frac{1}{2} \sum_{j=l+1}^{m} \frac{1}{2^j}\varphi\left(2^j x, 0\right) \end{aligned} \tag{39}$$

for all nonnegative integers m and l with $m > l$ and all $x \in X$. It follows from (20) that the sequence $\{\frac{1}{2^k} f(2^k x)\}$ is Cauchy for all $x \in X$. Since Y is a Banach space, the sequence $\{\frac{1}{2^k} f(2^k x)\}$ converges. So one can define the mapping $A : X \to Y$ by

$$A(x) := \lim_{k\to\infty} \frac{1}{2^k} f\left(2^k x\right)$$

for all $x \in X$. Moreover, letting $l = 0$ and passing to the limit $m \to \infty$ in (39), we get (37).

Assume that $f : X \to Y$ satisfies (28) and is an even mapping. It follows from (33) that

$$\left\| f(x) - \frac{1}{4} f(2x) \right\| \leq \frac{1}{4} \varphi(2x, 0)$$

for all $x \in X$. Thus

$$\begin{aligned} \left\| \frac{1}{4^l} f\left(2^l x\right) - \frac{1}{4^m} f\left(2^m x\right) \right\| &\leq \sum_{j=l}^{m-1} \left\| \frac{1}{4^j} f\left(2^j x\right) - \frac{1}{4^{j+1}} f\left(2^{j+1} x\right) \right\| \\ &\leq \frac{1}{4} \sum_{j=l+1}^{m} \frac{1}{4^j} \varphi\left(2^j x, 0\right) \end{aligned} \tag{40}$$

for all nonnegative integers m and l with $m > l$ and all $x \in X$. It follows from (40) that the sequence $\{\frac{1}{4^k} f(2^k x)\}$ is Cauchy for all $x \in X$. Since Y is a Banach space, the sequence $\{\frac{1}{4^k} f(2^k x)\}$ converges. So one can define the mapping $Q : X \to Y$ by

$$Q(x) := \lim_{k \to \infty} \frac{1}{4^k} f\left(2^k x\right)$$

for all $x \in X$. Moreover, letting $l = 0$ and passing to the limit $m \to \infty$ in (40), we get (38).

The rest of the proof is similar to the proof of Theorem 2.

Corollary 4 *Let $r < 1$ and θ be positive real numbers and $f : X \to Y$ be a mapping satisfying $f(0) = 0$ and (35). If the mapping $f : X \to Y$ is odd, then there exists a unique additive mapping $A : X \to Y$ such that*

$$\|f(x) - A(x)\| \leq \frac{2^r \theta}{2 - 2^r} \|x\|^r$$

for all $x, y \in X$. If the mapping $f : X \to Y$ is even, then there exists a unique quadratic mapping $Q : X \to Y$ such that

$$\|f(x) - Q(x)\| \leq \frac{2^r \theta}{4 - 2^r} \|x\|^r$$

for all $x \in X$.

Proof The proof follows from Theorem 5 by taking $\varphi(x, y) = \theta(\|x\|^r + \|y\|^r)$ for all $x, y \in X$.

4 Stability of the Additive-Quadratic s-Functional Inequality (1): A Fixed Point Method

Using the fixed point method, we prove the Hyers–Ulam stability of the additive-quadratic functional equation (1) in complex Banach spaces.

Theorem 6 *Let $\varphi : X^2 \to [0, \infty)$ be a function such that there exists an $L < 1$ with*

$$\varphi\left(\frac{x}{2}, \frac{y}{2}\right) \leq \frac{L}{4}\varphi(x, y) \leq \frac{L}{2}\varphi(x, y) \tag{41}$$

for all $x, y \in X$. Let $f : X \to Y$ be a mapping satisfying (9) and $f(0) = 0$. If the mapping $f : X \to Y$ is odd, then there exists a unique additive mapping $A : X \to Y$ such that

$$\|f(x) - A(x)\| \leq \frac{L}{2(1-L)}\varphi(x, x) \tag{42}$$

for all $x \in X$. If the mapping $f : X \to Y$ is even, then there exists a unique quadratic mapping $Q : X \to Y$ such that

$$\|f(x) - Q(x)\| \leq \frac{L}{4(1-L)}\varphi(x, x) \tag{43}$$

for all $x \in X$.

Proof Assume that $f : X \to Y$ satisfies (9) and is an odd mapping.

Consider the set

$$S := \{h : X^2 \to Y, \ \ h(x) = 0 \ \forall x \in X\}$$

and introduce the generalized metric on S:

$$d(g, h) = \inf\{\mu \in \mathbf{R}_+ : \|g(x) - h(x)\| \leq \mu\varphi(x, x), \ \ \forall x \in X\},$$

where, as usual, $\inf \phi = +\infty$. It is easy to show that (S, d) is complete (see [15]).

Now we consider the linear mapping $J : S \to S$ such that

$$Jg(x) := 2g\left(\frac{x}{2}\right)$$

for all $x \in X$.

Let $g, h \in S$ be given such that $d(g, h) = \varepsilon$. Then

$$\|g(x) - h(x)\| \leq \varepsilon\varphi(x, x)$$

for all $x \in X$. Hence

$$\|Jg(x) - Jh(x)\| = \left\|2g\left(\frac{x}{2}\right) - 2h\left(\frac{x}{2}\right)\right\| \le 2\varepsilon\varphi\left(\frac{x}{2}, \frac{x}{2}\right)$$
$$\le 2\varepsilon\frac{L}{2}\varphi(x, x) = L\varepsilon\varphi(x, x)$$

for all $x \in X$. So $d(g, h) = \varepsilon$ implies that $d(Jg, Jh) \le L\varepsilon$. This means that

$$d(Jg, Jh) \le Ld(g, h)$$

for all $g, h \in S$.

It follows from (12) that

$$\left\|f(x) - 2f\left(\frac{x}{2}\right)\right\| \le \varphi\left(\frac{x}{2}, \frac{x}{2}\right) \le \frac{L}{2}\varphi(x, x)$$

for all $x \in X$. So $d(f, Jf) \le \frac{L}{2}$.

By Theorem 1, there exists a mapping $A : X \to Y$ satisfying the following:

(1) A is a fixed point of J, i.e.,

$$A(x) = 2A\left(\frac{x}{2}\right) \tag{44}$$

for all $x \in X$. The mapping A is a unique fixed point of J. This implies that A is a unique mapping satisfying (44) such that there exists a $\mu \in (0, \infty)$ satisfying

$$\|f(x) - A(x)\| \le \mu\varphi(x, x)$$

for all $x \in X$;

(2) $d(J^l f, A) \to 0$ as $l \to \infty$. This implies the equality

$$\lim_{l\to\infty} 2^n f\left(\frac{x}{2^n}, z\right) = A(x)$$

for all $x \in X$;

(3) $d(f, A) \le \frac{1}{1-L}d(f, Jf)$, which implies

$$\|f(x) - A(x)\| \le \frac{L}{2(1-L)}\varphi(x, x)$$

for all $x \in X$.

Therefore, there is a unique additive mapping $A : X \to Y$ satisfying (47).

Assume that $f : X \to Y$ satisfies (9) and is an even mapping.

Now we consider the linear mapping $J' : S \to S$ such that

$$J'g(x) := 4g\left(\frac{x}{2}\right)$$

for all $x, z \in X$.

Let $g, h \in S$ be given such that $d'(g, h) = \varepsilon$. Then

$$\|g(x) - h(x)\| \le \varepsilon\varphi(x, x)$$

for all $x \in X$. Hence

$$\begin{aligned}\|J'g(x) - J'h(x)\| &= \left\|4g\left(\frac{x}{2}\right) - 4h\left(\frac{x}{2}\right)\right\| \le 4\varepsilon\varphi\left(\frac{x}{2}, \frac{x}{2}\right) \\ &\le 4\varepsilon\frac{L}{4}\varphi(x, x) = L\varepsilon\varphi(x, x)\end{aligned}$$

for all $x \in X$. So $d'(g, h) = \varepsilon$ implies that $d'(J'g, J'h) \le L\varepsilon$. This means that

$$d'(J'g, J'h) \le Ld'(g, h)$$

for all $g, h \in S$.

It follows from (14) that

$$\left\|f(x) - 4f\left(\frac{x}{2}\right)\right\| \le \varphi\left(\frac{x}{2}, \frac{x}{2}\right) \le \frac{L}{4}\varphi(x, x)$$

for all $x \in X$. So $d'(f, J'f) \le \frac{L}{4}$.

By Theorem 1, there exists a mapping $Q : X \to Y$ satisfying the following:

(1) Q is a fixed point of J', i.e.,

$$Q(x) = 4Q\left(\frac{x}{2}\right) \tag{45}$$

for all $x \in X$. The mapping Q is a unique fixed point of J'. This implies that Q is a unique mapping satisfying (45) such that there exists a $\mu \in (0, \infty)$ satisfying

$$\|f(x) - Q(x)\| \le \mu\varphi(x, x)$$

for all $x \in X$;

(2) $d(J'^l f, Q) \to 0$ as $l \to \infty$. This implies the equality

$$\lim_{l\to\infty} 4^n f\left(\frac{x}{2^n}\right) = Q(x)$$

for all $x \in X$;

(3) $d(f, Q) \leq \frac{1}{1-L} d(f, J'f)$, which implies

$$\|f(x) - Q(x)\| \leq \frac{L}{4(1-L)} \varphi(x, x)$$

for all $x \in X$.

Therefore, there is a unique quadratic mapping $Q : X \to Y$ satisfying (48).

Corollary 5 *Let $r > 2$ and θ be nonnegative real numbers and $f : X \to Y$ be a mapping satisfying* (16) *and $f(0) = 0$. If the mapping $f : X \to Y$ is odd, then there exists a unique additive mapping $A : X \to Y$ such that*

$$\|f(x) - A(x)\| \leq \frac{4\theta}{2^r - 4} \|x\|^r$$

for all $x, y \in X$. If the mapping $f : X \to Y$ is even, then there exists a unique quadratic mapping $Q : X \to Y$ such that

$$\|f(x) - Q(x)\| \leq \frac{2\theta}{2^r - 4} \|x\|^r$$

for all $x \in X$.

Proof The proof follows from Theorem 6 by taking $L = 2^{2-r}$ and $\varphi(x, y) = \theta(\|x\|^r + \|y\|^r)$ for all $x, y \in X$.

Theorem 7 *Let $\varphi : X^2 \to [0, \infty)$ be a function such that there exists an $L < 1$ with*

$$\varphi(x, y) \leq 2L\varphi\left(\frac{x}{2}, \frac{y}{2}\right) \leq 4L\varphi\left(\frac{x}{2}, \frac{y}{2}\right) \tag{46}$$

for all $x, y \in X$. Let $f : X \to Y$ be a mapping satisfying (9) *and $f(0) = 0$. If the mapping $f : X \to Y$ is odd, then there exists a unique additive mapping $A : X \to Y$ such that*

$$\|f(x) - A(x)\| \leq \frac{1}{2(1-L)} \varphi(x, x) \tag{47}$$

for all $x \in X$. If the mapping $f : X \to Y$ is even, then there exists a unique quadratic mapping $Q : X \to Y$ such that

$$\|f(x) - Q(x)\| \leq \frac{1}{4(1-L)} \varphi(x, x) \tag{48}$$

for all $x \in X$.

Proof Consider the complete metric space (S, d) given in the proof of Theorem 6.

Assume that $f : X \to Y$ satisfies (9) and is an odd mapping.

Now we consider the linear mapping $J : S \to S$ such that

$$Jg(x) := \frac{1}{2} g\,(2x)$$

for all $x \in X$.

It follows from (12) that

$$\left\| f(x) - \frac{1}{2} f\,(2x) \right\| \leq \frac{1}{2} \varphi(x, x)$$

for all $x \in X$. So $d(f, Jf) \leq \frac{1}{2}$.

By the same reasoning as in the proof of Theorem 6, one can show that there exists a unique additive mapping $A : X \to Y$ such that

$$\|f(x) - A(x)\| \leq \frac{1}{2(1 - L)} \varphi\,(x, x)$$

for all $x \in X$.

Assume that $f : X \to Y$ satisfies (9) and is an odd mapping.

Now we consider the linear mapping $J' : S \to S$ such that

$$J'g(x) := 4g\left(\frac{x}{2}\right)$$

for all $x \in X$.

It follows from (14) that

$$\left\| f(x) - \frac{1}{4} f\,(2x) \right\| \frac{1}{4} \varphi(x, x)$$

for all $x \in X$. So $d'(f, J'f) \leq \frac{1}{4}$.

By the same reasoning as in the proof of Theorem 6, one can show that there exists a unique quadratic mapping $Q : X \to Y$ such that

$$\|f(x) - Q(x)\| \leq \frac{1}{4(1 - L)} \varphi\,(x, x)$$

for all $x \in X$.

Corollary 6 *Let $r < 1$ and θ be nonnegative real numbers and $f : X \to Y$ be a mapping satisfying* (16) *and* $f(0) = 0$. *If the mapping $f : X \to Y$ is odd, then there exists a unique additive mapping $A : X \to Y$ such that*

$$\|f(x) - A(x)\| \le \frac{2\theta}{2^r - 2}\|x\|^r$$

for all $x, y \in X$. *If the mapping* $f : X \to Y$ *is even, then there exists a unique quadratic mapping* $Q : X \to Y$ *such that*

$$\|f(x) - Q(x)\| \le \frac{\theta}{2^r - 4}\|x\|^r$$

for all $x \in X$.

Proof The proof follows from Theorem 7 by taking $L = 2^{r-1}$ and $\varphi(x, y) = \theta(\|x\|^r + \|y\|^r)$ for all $x, y \in X$.

5 Stability of the Additive-Quadratic *s*-Functional Inequality (2): A Fixed Point Method

Using the fixed point method, we prove the Hyers–Ulam stability of the additive-quadratic functional equation (2) in complex Banach spaces.

Theorem 8 *Let* $\varphi : X^2 \to [0, \infty)$ *be a function such that there exists an* $L < 1$ *with*

$$\varphi\left(\frac{x}{2}, \frac{y}{2}\right) \le \frac{L}{4}\varphi(x, y) \le \frac{L}{2}\varphi(x, y) \tag{49}$$

for all $x, y \in X$. *Let* $f : X \to Y$ *be a mapping satisfying* (28) *and* $f(0) = 0$. *If the mapping* $f : X \to Y$ *is odd, then there exists a unique additive mapping* $A : X \to Y$ *such that*

$$\|f(x) - A(x)\| \le \frac{1}{1 - L}\varphi(x, 0) \tag{50}$$

for all $x \in X$. *If the mapping* $f : X \to Y$ *is even, then there exists a unique quadratic mapping* $Q : X \to Y$ *such that*

$$\|f(x) - Q(x)\| \le \frac{1}{1 - L}\varphi(x, 0) \tag{51}$$

for all $x \in X$.

Proof Assume that $f : X \to Y$ satisfies (28) and is an odd mapping.
Consider the set

$$S := \{h : X^2 \to Y, \; h(x) = 0 \; \forall x \in X\}$$

and introduce the generalized metric on S:

$$d(g,h)=\inf\{\mu\in\mathbf{R}_+ : \|g(x)-h(x)\|\le\mu\varphi(x,0),\ \ \forall x\in X\},$$

where, as usual, $\inf\phi=+\infty$. It is easy to show that (S,d) is complete (see [15]).

Now we consider the linear mapping $J:S\to S$ such that

$$Jg(x):=2g\left(\frac{x}{2}\right)$$

for all $x\in X$.

It follows from (31) that

$$\left\|f(x)-2f\left(\frac{x}{2}\right)\right\|\le\varphi(x,x)$$

for all $x\in X$. So $d(f,Jf)\le 1$.

By Theorem 1, there exists a mapping $A:X\to Y$ satisfying the following:

(1) A is a fixed point of J, i.e.,

$$A(x)=2A\left(\frac{x}{2}\right) \tag{52}$$

for all $x\in X$. The mapping A is a unique fixed point of J. This implies that A is a unique mapping satisfying (52) such that there exists a $\mu\in(0,\infty)$ satisfying

$$\|f(x)-A(x)\|\le\mu\varphi(x,0)$$

for all $x\in X$;

(2) $d(J^l f,A)\to 0$ as $l\to\infty$. This implies the equality

$$\lim_{l\to\infty}2^n f\left(\frac{x}{2^n},z\right)=A(x)$$

for all $x\in X$;

(3) $d(f,A)\le\frac{1}{1-L}d(f,Jf)$, which implies

$$\|f(x)-A(x)\|\le\frac{1}{1-L}\varphi(x,0)$$

for all $x\in X$.

Therefore, there is a unique additive mapping $A:X\to Y$ satisfying (50).

Assume that $f:X\to Y$ satisfies (28) and is an even mapping.

Now we consider the linear mapping $J':S\to S$ such that

$$J'g(x) := 4g\left(\frac{x}{2}\right)$$

for all $x \in X$.

It follows from (33) that

$$\left\| f(x) - 4f\left(\frac{x}{2}\right)\right\| \le \varphi(x, 0)$$

for all $x \in X$. So $d'(f, J'f) \le 1$.

By Theorem 1, there exists a mapping $Q : X \to Y$ satisfying the following:

(1) Q is a fixed point of J', i.e.,

$$Q(x) = 4Q\left(\frac{x}{2}\right) \tag{53}$$

for all $x \in X$. The mapping Q is a unique fixed point of J'. This implies that Q is a unique mapping satisfying (53) such that there exists a $\mu \in (0, \infty)$ satisfying

$$\|f(x) - Q(x)\| \le \mu\varphi(x, 0)$$

for all $x \in X$;

(2) $d(J'^l f, Q) \to 0$ as $l \to \infty$. This implies the equality

$$\lim_{l\to\infty} 4^n f\left(\frac{x}{2^n}\right) = Q(x)$$

for all $x \in X$;

(3) $d(f, Q) \le \frac{1}{1-L} d(f, J'f)$, which implies

$$\|f(x) - Q(x)\| \le \frac{1}{1-L}\varphi(x, 0)$$

for all $x \in X$.

Therefore, there is a unique quadratic mapping $Q : X \to Y$ satisfying (51).

Corollary 7 *Let $r > 2$ and θ be nonnegative real numbers and $f : X \to Y$ be a mapping satisfying* (35) *and $f(0) = 0$. If the mapping $f : X \to Y$ is odd, then there exists a unique additive mapping $A : X \to Y$ such that*

$$\|f(x) - A(x)\| \le \frac{2^r\theta}{2^r - 4}\|x\|^r$$

for all $x, y \in X$. If the mapping $f : X \to Y$ is even, then there exists a unique quadratic mapping $Q : X \to Y$ such that

$$\|f(x) - Q(x)\| \leq \frac{2^r\theta}{2^r - 4}\|x\|^r$$

for all $x \in X$.

Proof The proof follows from Theorem 8 by taking $L = 2^{2-r}$ and $\varphi(x, y) = \theta(\|x\|^r + \|y\|^r)$ for all $x, y \in X$.

Theorem 9 *Let* $\varphi : X^2 \to [0, \infty)$ *be a function such that there exists an* $L < 1$ *with*

$$\varphi(x, y) \leq 2L\varphi\left(\frac{x}{2}, \frac{y}{2}\right) \leq 4L\varphi\left(\frac{x}{2}, \frac{y}{2}\right) \tag{54}$$

for all $x, y \in X$. *Let* $f : X \to Y$ *be a mapping satisfying* (28) *and* $f(0) = 0$. *If the mapping* $f : X \to Y$ *is odd, then there exists a unique additive mapping* $A : X \to Y$ *such that*

$$\|f(x) - A(x)\| \leq \frac{L}{1 - L}\varphi(x, x) \tag{55}$$

for all $x \in X$. *If the mapping* $f : X \to Y$ *is even, then there exists a unique quadratic mapping* $Q : X \to Y$ *such that*

$$\|f(x) - Q(x)\| \leq \frac{L}{1 - L}\varphi(x, x) \tag{56}$$

for all $x \in X$.

Proof Consider the complete metric space (S, d) given in the proof of Theorem 8.

Assume that $f : X \to Y$ satisfies (28) and is an odd mapping.

Now we consider the linear mapping $J : S \to S$ such that

$$Jg(x) := \frac{1}{2}g(2x)$$

for all $x \in X$.

It follows from (31) that

$$\left\| f(x) - \frac{1}{2}f(2x)) \right\| \leq \frac{1}{2}\varphi(2x, 0) \leq \frac{2L}{2}\varphi(x, 0)$$

for all $x \in X$. So $d(f, Jf) \leq L$.

By the same reasoning as in the proof of Theorem 8, one can show that there exists a unique additive mapping $A : X \to Y$ such that

$$\|f(x) - A(x)\| \leq \frac{L}{1-L}\varphi(x, x)$$

for all $x \in X$.

Assume that $f : X \to Y$ satisfies (28) and is an odd mapping.

Now we consider the linear mapping $J' : S \to S$ such that

$$J'g(x) := \frac{1}{4}g(2x)$$

for all $x \in X$.

It follows from (33) that

$$\left\| f(x) - \frac{1}{4}f(2x) \right\| \leq \frac{1}{4}\varphi(2x, 0) \leq \frac{4L}{4}\varphi(x, 0)$$

for all $x \in X$. So $d'(f, J'f) \leq L$.

By the same reasoning as in the proof of Theorem 6, one can show that there exists a unique quadratic mapping $Q : X \to Y$ such that

$$\|f(x) - Q(x)\| \leq \frac{L}{1-L}\varphi(x, x)$$

for all $x \in X$.

Corollary 8 *Let $r < 1$ and θ be nonnegative real numbers and $f : X \to Y$ be a mapping satisfying* (35) *and $f(0) = 0$. If the mapping $f : X \to Y$ is odd, then there exists a unique additive mapping $A : X \to Y$ such that*

$$\|f(x) - A(x)\| \leq \frac{2^r\theta}{2-2^r}\|x\|^r$$

for all $x, y \in X$. If the mapping $f : X \to Y$ is even, then there exists a unique quadratic mapping $Q : X \to Y$ such that

$$\|f(x) - Q(x)\| \leq \frac{2^r\theta}{2-2^r}\|x\|^r$$

for all $x \in X$.

Proof The proof follows from Theorem 9 by taking $L = 2^{r-1}$ and $\varphi(x, y) = \theta(\|x\|^r + \|y\|^r)$ for all $x, y \in X$.

Acknowledgments C. Park was supported by Basic Science Research Program through the National Research Foundation of Korea funded by the Ministry of Education, Science and Technology (NRF-2017R1D1A1B04032937).

References

1. T. Aoki, On the stability of the linear transformation in Banach spaces. J. Math. Soc. Japan **2**, 64–66 (1950)
2. L. Cădariu, V. Radu, Fixed points and the stability of Jensen's functional equation. J. Inequal. Pure Appl. Math. **4**(1), 4 (2003)
3. L. Cădariu, V. Radu, On the stability of the Cauchy functional equation: a fixed point approach. Grazer Math. Ber. **346**, 43–52 (2004)
4. L. Cădariu, V. Radu, Fixed point methods for the generalized stability of functional equations in a single variable. Fixed Point Theory Appl. **2008**, 749392 (2008)
5. L. Cădariu, L. Găvruta, P. Găvruta, On the stability of an affine functional equation. J. Nonlinear Sci. Appl. **6**, 60–67 (2013)
6. A. Chahbi, N. Bounader, On the generalized stability of d'Alembert functional equation. J. Nonlinear Sci. Appl. **6**, 198–204 (2013)
7. P.W. Cholewa, Remarks on the stability of functional equations. Aequationes Math. **27**, 76–86 (1984)
8. J. Diaz, B. Margolis, A fixed point theorem of the alternative for contractions on a generalized complete metric space. Bull. Am. Math. Soc. **74**, 305–309 (1968)
9. N. Eghbali, J.M. Rassias, M. Taheri, On the stability of a k-cubic functional equation in intuitionistic fuzzy n-normed spaces. Results Math. **70**, 233–248 (2016)
10. G.Z. Eskandani, P. Găvruta, Hyers-Ulam-Rassias stability of pexiderized Cauchy functional equation in 2-Banach spaces. J. Nonlinear Sci. Appl. **5**, 459–465 (2012)
11. P. Găvruta, A generalization of the Hyers-Ulam-Rassias stability of approximately additive mappings. J. Math. Anal. Appl. **184**, 431–436 (1994)
12. D.H. Hyers, On the stability of the linear functional equation. Proc. Natl. Acad. Sci. U.S.A. **27**, 222–224 (1941)
13. G. Isac, T.M. Rassias, Stability of ψ-additive mappings: applications to nonlinear analysis. Int. J. Math. Math. Sci. **19**, 219–228 (1996)
14. H. Khodaei, On the stability of additive, quadratic, cubic and quartic set-valued functional equations. Results Math. **68**, 1–10 (2015)
15. D. Miheţ, V. Radu, On the stability of the additive Cauchy functional equation in random normed spaces. J. Math. Anal. Appl. **343**, 567–572 (2008)
16. C. Park, Additive ρ-functional inequalities and equations. J. Math. Inequal. **9**, 17–26 (2015)
17. C. Park, Additive ρ-functional inequalities in non-Archimedean normed spaces. J. Math. Inequal. **9**, 397–407 (2015)
18. V. Radu, The fixed point alternative and the stability of functional equations. Fixed Point Theory **4**, 91–96 (2003)
19. T.M. Rassias, On the stability of the linear mapping in Banach spaces. Proc. Am. Math. Soc. **72**, 297–300 (1978)
20. F. Skof, Propriet locali e approssimazione di operatori. Rend. Sem. Mat. Fis. Milano **53**, 113–129 (1983)
21. S.M. Ulam, *A Collection of the Mathematical Problems* (Interscience Publ. New York, 1960)
22. Z. Wang, Stability of two types of cubic fuzzy set-valued functional equations. Results Math. **70**, 1–14 (2016)

Time-Delay Multi-Agent Systems for a Movable Cloud

Rabha W. Ibrahim

Abstract Adaptable splitting of calculations between movable devices and cloud is a significant and exciting research subject for movable cloud computing systems. Current work's emphasis on the single-agent, neighbor-agent, or multi-agent systems, which aim to optimize the request achievement time for one particular agent. Due to the monopolistic competition for cloud resources among a large number of agents, the divested computations may be performed with a definite scheduling delay on the cloud. In this effort, we introduce three distributed dynamical systems, based on fractional calculus, describing the delay for single-agent, neighbor-agent, and multi-agent systems. We impose a new class of mixed-index time fractional differential equations. In addition, we investigate the asymptotic stability properties by using the fractional Cauchy's method. We discuss the case of time-delay systems. A simulation is illustrated.

1 Introduction

Fractional calculus (see [1–3]) is a major branch of the non-linearity studies. It is the main tool of the generalized differential and integral operators depending on the rational power. These operators play significant roles not only in mathematical science, but also in all sciences. This theory carries good technical descriptions in many subjects containing computer science and dynamics in financial markets. Lately, the fractional calculus is a profit technique to unfolding dynamic processes in chaotic or complex systems such as relaxation or dielectric conduct in control systems.

Recently, there has been a rising attention in synchronized control of multi-agent systems due to industrial advances in computing, communication, and extensive

R. W. Ibrahim (✉)
Informetrics Research Group, Ton Duc Thang University, Ho Chi Minh City, Vietnam

Faculty of Mathematics and Statistics, Ton Duc Thang University, Ho Chi Minh City, Vietnam
e-mail: rabhaibrahim@tdtu.edu.vn

N. J. Daras, T. M. Rassias (eds.), *Computational Mathematics and Variational Analysis*, Springer Optimization and Its Applications 159,
https://doi.org/10.1007/978-3-030-44625-3_19

applications. There are various possible benefits of such systems over a single-agent, involving more elasticity, robustness, and reliability. The control system has established significant attention. It aims to scheme suitable procedures and algorithms such that the agents can attain and preserve a predefined topologically. These systems are widely acted in a moving cloud. Moreover, these systems are related to multi-agent systems with a large number of agents (see [4, 5]). The class of dynamic systems with delay (time-varying) expected in reality, therefore, it is necessary to investigate the solutions. One of these systems is the distributed multi-agent system. Note that, the joint value is the ordinary of the initial positions of all the agents, the consensus is called static system; when the common value is the ordinary of the time-delay reference inputs of all the agents, it is called dynamic system. It has been assumed that the time delays were continuous in the setting of the cloud. Newly, the author introduced a definition of the fractional cloud computing systems by utilizing various types of fractional calculus (see [6–11]). Time-delay problem in the cloud system appeared in the second-order multi-agent system, where the time delays were uniform and occurred only in the transmission of position information between neighbors. For the second-order multi-agent systems, the data switched between neighbors may contain velocity evidence as well as position statistics, and more significantly, the time-delay may be multiple in many practical locations.

In this work, we deliver three distributed dynamical systems, based on fractional calculus, describing the delay for single-agent, neighbor-agent, and multi-agent systems. We impose a new class of mixed-index time fractional differential equations. In addition, we investigate the asymptotic stability properties by using the fractional Cauchy's method. We discuss the case of time-delay systems. A simulation is illustrated.

The remainder of this article prepared as follows: Section 2 deals with some recent works and studies in this direction. Section 3 treats with some notions of fractional calculus theory and define the problem to be investigated. Section 4 devotes to conversing the outcomes of the fractional dynamic problem under direct time-delay properties. Section 5 addresses the outcomes by two numerical examples. Section 6 includes some concluding remarks and a discussion on future works.

2 Literature

Shiraz et al.[12] reviewed some recent works, including the distribution of cloud computing systems. They adapted that the mobile cloud computing system (MCCS) is the modern applied resolution for improving the capacity by covering the services and resources of computing clouds to smart mobile devices on demand source. Hwang et al. [13] introduced and converged modern distributed computing systems to describe the technology and information involving computer clusters and virtualization. Dinh et al. [14] imposed a survey paper on an MCCS. The study involves definitions, architecture, and applications as well as issues, existing outcomes, and methods are presented. Magalhães et al. [15] suggested an MCCS for resource usage analysis and simulation in a cloud environment. Kampas et al. [16] created a novel method and systems for integrating multi-agent MCCS and developed the architectures by utilizing service-oriented orchestration. Sung et al. [17] introduced a new model for key distribution based on data re-encryption. They presented a comprehensive MCCS which yields an effective and secure CCS (cloud computing system) on mobile devices. Li et al. [18] used a randomly distributed dynamic of MCCS.

3 Algorithms

In this section, the requests are demonstrated as a sequence of processing agents. The agent characterizes a type of operation onto the information sources. The directed boundaries epitomize the colony between the agents. It incomes that an agent starts to run whenever the pattern unit completes. Each agent is permitted to run either locally on the mobile device or remotely on the cloud. Furthermore, the input information of the request is invented to be from the devices of the mobility scheme, and the output information should be supplied back to the mobile device. The presentation metric is the implementation time of the request. Equally the finishing time signifies the reactive delay for the request; we basically utilize the term delay, in a fractional differential operator (FDO). The delay is the summation of the calculation time of all the agents and the information communication time between the agents.

3.1 *Single-Agent System*

In this section, we establish a dynamical system of a single-agent. The system describes the movement of the agent i, where $i = 1, \ldots, n$. Moreover, by utilizing the definition of the FDO, the delay is minimized by the minimization of FDO. Let the finishing time of the agent i denoted by ς_i if it is discharged onto the cloud $(1 \leq i \leq n)$; else, it is denoted by ω_i, where $\omega_i > \varsigma_i$. Let χ_i be the request information of the agent i from the movable side or the cloud side (binary decision variable), then we formulate the system as follows:

$$D^{\wp} \chi_i^t = \sum_{i=1}^{n} \omega_i D^{\wp_i} \chi_i^t + \varsigma_i \chi_i^t + u_i^t, \quad (1)$$

$$\Big(\wp_1 < \wp_2 < \ldots < \wp_n < \wp \in (0, 1],\ i = 1, \ldots, n\Big),$$

where u_i^t is the controller, and $D^{\wp}$ is the Riemann–Liouville fractional differential operator, given by

$$D^{\wp}\chi_i^t = \frac{1}{\Gamma(1-\wp)}\frac{d}{dt}\int_0^t (t-\varsigma)^{-\wp}\chi_i^{\varsigma}\,d\varsigma, \quad 0<\varsigma\le t<\infty,$$

with the average of the time-delay $t-\varsigma$ for the system and

$$D^{\wp_i}\chi_i^t = \frac{1}{\Gamma(1-\wp_i)}\frac{d}{dt}\int_0^t (t-\varsigma_i)^{-\wp_i}\chi_i^{\varsigma_i}\,d\varsigma_i, \quad 0<\varsigma_i\le t<\infty,$$

with the average of the time-delay $t-\varsigma_i$ for the agent i. The fractional integral operator is defined by

$$I^{\wp}\chi_i^t = \frac{1}{\Gamma(\wp)}\int_0^t (t-\varsigma)^{\wp-1}\chi_i^{\varsigma}\,d\varsigma.$$

In term of matrices, System (1) can be converted into

$$D^{\wp}\chi^t = \Upsilon\chi^t + U^t, \tag{2}$$

where $\chi = (\chi_1,\ldots,\chi_n)^{\top}$, $U = (u_1,\ldots,u_n)^{\top}$, $\wp = (\wp_1,\ldots,\wp_n)$, and $\Upsilon = (\lambda_{ii})_{i=1}^n$. Note that λ depends on $\wp$, ω, and ς.

3.2 Neighbor-Agent System

In this section, we discuss the neighbor-agent system. The agent runs the applications in the cloud for computation clearing. Presume the applications involve of a sequence of n agents. Each agent can apply on the movable cloud side or on the fixed cloud side. The finishing time of the agent i is ς_i if it is discharged onto cloud $(1\le i\le n)$; else, it is ω_i, where $\omega_i > \varsigma_i$. If two neighboring agents i and $i+1$ run on various borders, then the information communication time is τ_i; else the information communication time between i and $i+1$ becomes nil when they run on the same border. To systematization the input/output information, the application should be from/to the mobile device. In this case, we increase two virtual agents 0 and $n+1$ as the access and departure agents. Then we have the following neighbor-agent system, describing the request dynamic χ_i, $i=1,\ldots,n$ of the agent i:

$$D^{\wp}\chi_i^t = \sum_{i=1}^n \omega_{i,i+1}D^{\wp_i}\chi_i^t + \tau_i\left|\chi_i^t - \chi_{i+1}^t\right| + u_i^t. \tag{3}$$

If there is no connection between χ_i and χ_{i+1}, then System (3) reduces to (1), where $\tau_i = \varsigma_i$. In term of the matrix, we have the following system:

$$D^{\wp}\chi^t = \Upsilon\chi^t + F^t + U^t, \tag{4}$$

where $\chi = (\chi_1, \ldots, \chi_n)^\top$, $U = (u_1, \ldots, u_n)^\top, \wp = (\wp_1, \ldots, \wp_n)$, $F = (f_1, \ldots, f_n)^\top$, $f_i = \tau_i \left|\chi_i^t - \chi_{i+1}^t\right|$, and $\Upsilon = \left(\lambda_{i,i+1}\right)_{i=1}^n$. Our aim is to minimize the delay in System (4).

3.3 Multi-Agent System

In this section, we demonstrate the system model of the multiple agents working out dividing. The system involves two measures: the cloud and a mobile user. For a mobile user, we require a set of agents that refer requests to the cloud for dividing implementation of the request. The display manager of the user's middleware accumulates data on the wireless channel and device environments. This data is referred with the requests to the cloud. In our system, we accept that the customers are requesting for dividing implementation of the same request. Nevertheless, we may cover the system by seeing that the customers request for several requests. For this purpose, we deal with the following system:

$$D^{\wp}\chi_i^t = \sum_{j\in J}\omega_{ij}D^{\wp_i}\left(\chi_i^t - \chi_j^t\right) + u_i^t, \tag{5}$$

where

$$\left(\omega_{ij} \neq 0,\ J := \{1, \ldots, n\}\right).$$

In term of the matrix, we have the following system:

$$D^{\wp}\chi^t = \Upsilon\chi^t + U^t, \tag{6}$$

where $\chi = (\chi_1, \ldots, \chi_n)^\top$, $U = (u_1, \ldots, u_n)^\top, \wp = (\wp_1, \ldots, \wp_n)$, and $\Upsilon = \left(\lambda_{ij}\right)_{i,j=1}^n$. Our aim is to minimize the delay in System (6). We assume that the initial state of each agent is calculated by $\chi^0 = \chi(0)$ such that the total initial state of the system is

$$\mathcal{T}^0 = \sum_{i\in J}\frac{\chi_i^0}{n}.$$

We essentially have two adjusting categories to discharge the incomes which have been employed at the critical point, (movement and waiting). Movement incomes to modify the implementation place of the module, while waiting incomes to only delay the implementation of the agents at the cloud adjacent.The above systems are described the fractional velocity of the request in the mobile cloud including implicitly the delay for receiving this request. The environment of the moving cloud is completely designed by the above systems, depending on its connection. There are various methods to minimize the velocity (numerically, algebraically, and analytically). All of these methods are constructed a long list of constraints and conditions to minimize the system. Our investigation basically depends on a method which avoids using any constraint. This method is called the Cauchy's method [19]. We first modify this method in order to consider the above systems by utilizing the definition of the fractional gradient. Consequently, we employ the modified Cauchy's method by suggestion the discrete data.

4 Results

To construct our outcomes, we introduce the following generalized concepts:

Definition 4.1 The fractional gradient can be expressed by

$$\begin{aligned} \nabla_i^{\wp}\chi^t &= \sum_{i=1}^{n}\left(D^{\wp_i}\chi_i^t\right)e_i, \\ \nabla_{i+1}^{\wp}\chi^t &= \sum_{i=1}^{n-1}\left(D^{\wp_i}\chi_i^t\right)e_i, \\ &\vdots, \\ \nabla_{i+k}^{\wp}\chi^t &= \sum_{i=1}^{n-k}\left(D^{\wp_i}\chi_i^t\right)e_i, \quad k \le n, \end{aligned} \tag{7}$$

where $\chi = (\chi_1, \ldots, \chi_n)^{\top}$ and e_i is the coordinate vector corresponding to χ_i, $i = 1, \ldots, n.$

Note that the above definition is a generalization for the usual gradient (see [20]).

Definition 4.2 The fractional gradient is called a fractional $\wp$−Hölder continuous if and only if

$$\|\nabla_i^{\wp}\chi^t - \nabla_i^{\wp}\chi^{\varsigma}\| \le \ell\|t_i - \varsigma_i\|^{\wp}, \quad \wp \in (0, 1].$$

We denote by χ^* the minimum value of χ^t. The above inequality is a generalization of the usual Lipschitz condition ($\wp = 1$), where Lipschitz continuous $\subseteq$ $\wp$-Hölder continuous.

Definition 4.3 The fractional Cauchy's method can be defined on the set

$$S_\alpha^{\wp} := \left\{ t_\alpha : t_\alpha = t - \alpha \nabla^{\wp}, \quad \alpha \in (0, \frac{\ell}{2\Gamma(\wp)}) \right\}$$

as follows: for a sequence $\{t_i\}_{i=1}^n$, where

$$\begin{aligned} t_{i+1} &= t_i - \alpha_i \nabla_i^{\wp} \chi^t, \\ t_{i+2} &= t_{i+1} - \alpha_{i+1} \nabla_{i+1}^{\wp} \chi^t, \\ &\vdots, \\ t_{i+k} &= t_{i+k-1} - \alpha_{i+k-1} \nabla_{i+k-1}^{\wp} \chi^t. \end{aligned} \tag{8}$$

Proposition 4.4 *If* $t_{i+k+1} = t_{i+k} - \alpha_{i+k} \nabla_{i+k}^{\wp} \chi^t$, *with* $\alpha_{i+k} \in (0, \frac{\ell}{2(k+1)\Gamma(\wp)})$, *then the sequence* $\{t_i\}_{i=1}^n$ *converges to the point* t^* *which minimizes* χ_i, $i = 1, \ldots, n$.

Proof From (8), we have

$$\begin{aligned} \|t_{i+1} - t_i\| &\leq \alpha \|\nabla_i^{\wp} \chi^t\| \leq \alpha \|\nabla_i^{\wp} \chi^t - \nabla_i^{\wp} \chi^0\| + \alpha \|\nabla_i^{\wp} \chi^0\|, \quad \alpha := \max_i \alpha_i \\ &\leq \alpha \ell \|t_i\|^{\wp} + c_0, \quad c_0 = \alpha \|\nabla_i^{\wp} \chi^0\|. \end{aligned}$$

Hence, $\{t_i\}_{i=1}^n$ is bounded and so it has cluster points. Suppose that the maximum one is t^*, but all cluster points are in $S_\alpha^{\wp}$ then the sequence $\{t_i\}_{i=1}^n$ converges to the point t^* which minimizes χ_i, $i = 1, \ldots, n$. This completes the proof.

The aim of this section is to establish the lower bound of solution, for a convex and Lipschitzian gradient, full convergence of the series $\{t_i\}_{i=1}^n$ to minimize (2), (4), and (6), with the first approach without any suggestion on its level sets. In the sequel, we let $t \in [0, 1]$ then we have the following outcomes:

Theorem 4.5 *Assume that all the functions in Systems* (2), (4) *or* (6) *are Lipschitzian continuous. If*

$$\frac{\overline{\lambda}}{\Gamma(\wp + 1)} < 1, \quad \overline{\lambda} = \max \lambda_i,$$

then the sequence $\{t_i\}_{i=1}^n$ *generated by* (8) *converges to the minimizer* χ^*.

Proof By Proposition 4.4, $\lim_i t_i = t^* \in S_\alpha^{\wp}$. Therefore, it is enough to show that Systems (2), (4), and (6) have a unique solution χ, which is equal to χ^*. Suppose that there exists a constant $\mu > 0$ and $\varphi > 0$ such that

$$\|U^t - U^\varsigma\| \le \mu \|t - \varsigma\|, \quad t, \varsigma \in [0, 1]$$

and

$$\|F^t - F^\varsigma\| \le \varphi \|t - \varsigma\|, \quad t, \varsigma \in [0, 1].$$

For Systems (2) and (6), we have the following conclusion:

$$|\chi^t| \le |\chi^0| + \frac{\overline{\lambda}}{\Gamma(\wp + 1)} |\chi^t| + \frac{\mu}{\Gamma(\wp + 1)} |U^t|,$$

which implies that

$$\|\chi\| \le \frac{|\chi^0| + \frac{\mu}{\Gamma(\wp+1)} \|U\|}{1 - \frac{\overline{\lambda}}{\Gamma(\wp+1)}}.$$

Thus, the set that χ^t can be minimized is

$$M := \left\{ \chi : \|\chi\| \le \frac{|\chi^0| + \frac{\mu}{\Gamma(\wp+1)} \|U\|}{1 - \frac{\overline{\lambda}}{\Gamma(\wp+1)}} \right\}.$$

Moreover, we have

$$|\chi^t - \chi^\varsigma| \le \frac{\overline{\lambda}}{\Gamma(\wp + 1)} |\chi^t - \chi^\varsigma| + \frac{1}{\Gamma(\wp + 1)} |U^t - U^\varsigma|,$$

consequently, we attain

$$\|\chi^t - \chi^\varsigma\| \le \frac{\mu}{\Gamma(\wp + 1) - \overline{\lambda}} \|t - \varsigma\|.$$

Since μ is arbitrary, then one can select it such that $\frac{\mu}{\Gamma(\wp+1) - \overline{\lambda}} < 1$; thus, χ^t is a contraction mapping, which implies that System (2) has a unique solution (Banach fixed point theorem) satisfying $\chi^t \to \chi^* \in M$. Similarly, we obtain a unique solution for System (4) in the set

$$\Omega := \left\{ \chi : \|\chi\| \le \frac{|\chi^0| + \frac{\mu}{\Gamma(\wp+1)} \|U\| + \frac{\varphi}{\Gamma(\wp+1)} \|F\|}{1 - \frac{\overline{\lambda}}{\Gamma(\wp+1)}} \right\}.$$

This completes the proof.

5 Simulation

In this section, we illustrate a simulation. The basic technique that we shall follow is applying the method (8) on an objective function involving χ_i^t. Consider a 3-agent fractional system of nonlinear differential equations:

$$\begin{cases} D^{\wp}\, \chi_1^t = 3\chi_1^t - \cos(\chi_2^t\chi_3^t) - \frac{3}{2} \\ \\ D^{\wp}\, \chi_2^t = 4(\chi_1^t)^2 - 625(\chi_2^t)^2 + 2\chi_2^t - 1 \\ \\ D^{\wp}\, \chi_3^t = \exp(-\chi_1^t\chi_2^t) + 20\chi_3^t + (10\pi - 3)/3. \end{cases} \tag{9}$$

To minimize System (9), we let $\left(D^{\wp}\, \chi_1^t, D^{\wp}\, \chi_2^t, D^{\wp}\, \chi_3^t\right)^{\top} = (0, 0, 0)^{\top}$.

In the matrix form, we define the function

$$\Theta(\boldsymbol{\chi}^{\mathbf{t}}) = \begin{bmatrix} 3\chi_1^t - \cos(\chi_2^t\chi_3^t) - \frac{3}{2} \\ \\ 4(\chi_1^t)^2 - 625(\chi_2^t)^2 + 2\chi_2^t - 1 \\ \\ \exp(-\chi_1^t\chi_2^t) + 20\chi_3^t + (10\pi - 3)/3 \end{bmatrix},$$

where

$$\boldsymbol{\chi}^{\mathbf{t}} = \begin{bmatrix} \chi_1^t \\ \chi_2^t \\ \chi_3^t \end{bmatrix}$$

and the objective function

$$\begin{aligned} \Psi(\boldsymbol{\chi}^{\mathbf{t}}) &= \tfrac{1}{2}\Theta^{\top}(\boldsymbol{\chi}^{\mathbf{t}})\Theta(\boldsymbol{\chi}^{\mathbf{t}}) \\ &= \tfrac{1}{2}\Big(\left(3\chi_1^t - \cos(\chi_2^t\chi_3^t) - \tfrac{3}{2}\right)^2 + \left(4(\chi_1^t)^2 - 625(\chi_2^t)^2 + 2\chi_2^t - 1\right)^2 \\ &+ \left(\exp(-\chi_1^t\chi_2^t) + 20\chi_3^t + \tfrac{1}{3}(10\pi - 3)\right)^2\Big). \end{aligned}$$

With the initial data

$$\mathbf{x}^{(0)} = \begin{bmatrix} \chi_1^t \\ \chi_2^t \\ \chi_3^t \end{bmatrix} = \begin{bmatrix} 0 \\ 0 \\ 0 \end{bmatrix}.$$

Obviously,

$$\chi^{t^{(1)}} = \chi^{t^{(0)}} - \alpha_0 \nabla^{\wp} \Psi(\chi^{t^{(0)}})$$

, where

$$\nabla^{\wp} \Psi(\chi^{t^{(0)}}) = J_{\Theta}^{\wp}(\chi^{t^{(0)}})^{\mathrm{T}} \Theta(\chi^{t^{(0)}}).$$

The fractional Jacobian matrix of order $\wp \in (0, 1]$, is defined by $J_{\Theta}^{\wp}(\chi^{t^{(0)}})$

$$J_{\Theta}^{\wp} \approx \wp! \begin{bmatrix} 3 & \sin(\chi_2^t \chi_3^t)\chi_3^t & \sin(\chi_2^t \chi_3^t)\chi_2^t \\ 8\chi_1^t & -1250\chi_2^t + 2 & 0 \\ -\chi_2^t \exp(-\chi_1^t \chi_2^t) & -\chi_1^t \exp(-\chi_1^t \chi_2^t) & 20 \end{bmatrix}, \quad \wp! = \Gamma(\wp + 1).$$

Then evaluating these terms at $\chi^{t^{(0)}}$ and $\wp = 1$,

$$J_{\Theta}^{\wp}\left(\chi^{t^{(0)}}\right) = \begin{bmatrix} 3 & 0 & 0 \\ 0 & 2 & 0 \\ 0 & 0 & 20 \end{bmatrix}, \qquad \Theta(\chi^{t^{(0)}}) = \begin{bmatrix} -2.5 \\ -1 \\ 10.472 \end{bmatrix}$$

. So that

$$\chi^{t^{(1)}} = 0 - \alpha_0 \begin{bmatrix} -7.5 \\ -2 \\ 209.44 \end{bmatrix}$$

and

$$\Psi\left(\chi^{t^{(0)}}\right) = 0.5\left((-2.5)^2 + (-1)^2 + (10.472)^2\right) = 58.456.$$

For a selected value of α_0, such that $\Psi(\chi^{t^{(1)}}) \leq \Psi(\chi^{t^{(0)}})$. This can be done with any of a variety of line search algorithms. One might also obtain $\alpha_0 = 0.001$ which yields

$$\chi^{t^{(1)}} = \begin{bmatrix} 0.0075 \\ 0.002 \\ -0.20944 \end{bmatrix},$$

evaluating at this value,

$$\Psi\left(\chi^{t^{(1)}}\right) = 0.5\left((-2.48)^2 + (-1.00)^2 + (6.28)^2\right) = 23.306, \quad \wp = 1.$$

The decrease from $\Psi(\chi^{t^{(0)}}) = 58.456$ to the next step's value of $\Psi(\chi^{t^{(1)}}) = 23.306$ is a sizable decrease in the objective function. Extra steps should reduce its value until an outcome to the system was delivered.

Case 1 If $\wp = 0.75$, then we conclude that

$$J_{\Theta}^{0.75}\left(\chi^{t^{(0)}}\right) = \begin{bmatrix} 3.675 & 0 & 0 \\ 0 & 2.45 & 0 \\ 0 & 0 & 24.5 \end{bmatrix}, \qquad \Theta(\chi^{t^{(0)}}) = \begin{bmatrix} -2.5 \\ -1 \\ 10.472 \end{bmatrix}.$$

Thus, we obtain

$$\chi^{t^{(1)}} = 0 - \alpha_0 \begin{bmatrix} -9.187 \\ -2.45 \\ 265.564 \end{bmatrix}$$

and

$$\Psi\left(\chi^{t^{(0)}}\right) = 0.5\left((-2.5)^2 + (-1)^2 + (10.472)^2\right) = 58.456.$$

For a suitable value of α_0, such that $\Psi(\chi^{t^{(1)}}) \leq \Psi(\chi^{t^{(0)}})$. This can be done with any of a variety of line search algorithms. One might also select $\alpha_0 = 0.001$, which yields

$$\chi^{t^{(1)}} = \begin{bmatrix} 0.0091 \\ 0.0024 \\ -0.2655 \end{bmatrix}, \qquad \Theta(\chi^{t^{(1)}}) = \begin{bmatrix} -2.47 \\ -1 \\ 5.166 \end{bmatrix}$$

calculating at this value,

$$\Psi\left(\chi^{t^{(1)}}\right) = 0.5\left((-2.47)^2 + (-1.00)^2 + (5.166)^2\right) = 16.894, \quad \wp = 0.75.$$

The decrease from $\Psi(\chi^{t^{(0)}}) = 58.456$ to the next step's value of $\Psi(\chi^{t^{(1)}}) = 16.894$ is a sizable decrease in the objective function. The fractional value $\wp = 0.75$ makes the method converging faster than $\wp = 1$.

Case 2 If $\wp = 0.5$, then we attain

$$J_{\Theta}^{0.5}\left(\chi^{t^{(0)}}\right) = \begin{bmatrix} 5.31 & 0 & 0 \\ 0 & 3.54 & 0 \\ 0 & 0 & 35.4 \end{bmatrix}, \qquad \Theta(\chi^{t^{(0)}}) = \begin{bmatrix} -2.5 \\ -1 \\ 10.472 \end{bmatrix}.$$

Therefore, we have

$$\chi^{\mathbf{t}^{(1)}} = 0 - \alpha_0 \begin{bmatrix} -13.275 \\ -3.54 \\ 370.7 \end{bmatrix}$$

with

$$\Psi\left(\chi^{\mathbf{t}^{(0)}}\right) = 0.5\left((-2.5)^2 + (-1)^2 + (10.472)^2\right) = 58.456.$$

Now let $\alpha_0 = 0.001$, this implies

$$\chi^{\mathbf{t}^{(1)}} = \begin{bmatrix} 0.0132 \\ 0.0035 \\ -0.3707 \end{bmatrix}, \qquad \Theta(\chi^{\mathbf{t}^{(1)}}) = \begin{bmatrix} -2.46 \\ -1 \\ 3.062 \end{bmatrix}$$

calculating at this value,

$$\Psi\left(\chi^{\mathbf{t}^{(1)}}\right) = 0.5\left((-2.46)^2 + (-1.00)^2 + (3.062)^2\right) = 8.213, \quad \wp = 0.5.$$

The decrease from $\Psi(\chi^{\mathbf{t}^{(0)}}) = 58.456$ to the next step's value of $\Psi(\chi^{\mathbf{t}^{(1)}}) = 8.213$ is a sizable decrease in the objective function. The fractional value $\wp = 0.5$ makes the method converging faster than $\wp = 1$ and $\wp = 0.75$. Hence, we conclude that the fractional system leads to faster converge than the ordinary system. Note that, when $\wp = 0.25$ we have

$$\Psi\left(\chi^{\mathbf{t}^{(1)}}\right) = 0.5\left((-2.48)^2 + (-0.867)^2 + (6.7)^2\right) = 25.9, \quad \wp = 0.25.$$

Thus, a good converge of the method is given when $\wp \in (0.25, 1]$, where the conditions of the Theorem 4.5 are satisfied. Figure 1 shows the convergence of the method.

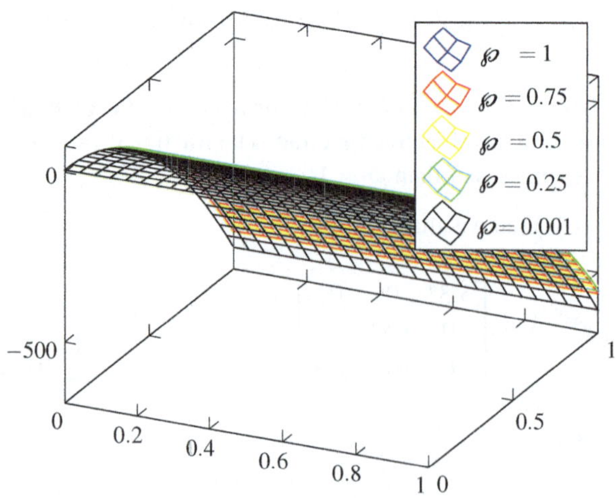

Fig. 1 System (9), with different values of $\wp$ and $\chi_3 = 0$

6 Conclusion

We developed three different agent systems in the movable cloud computing, single-agent, neighbor-agent, and multi-agent systems. It was aimed to optimize the request achievement time for one particular agent. Due to the monopolistic competition for cloud resources among a large number of agents, the divested computations may be performed with a definite scheduling delay on the cloud. The delay in the systems is computed implicitly (the kernel of this operator involves the fractional delay $(t - \varsigma)^{\wp}$). We employed the definition of fractional derivative to study the delay in the fractional systems. Moreover, we modified some recent and active methods to investigate the asymptotic stability properties. We have utilized the fractional Cauchy's method. We concluded that this generalized method is converged faster than the normal method.

References

1. A. Kilbas, H. Srivastava, J.J. Trujillo, *Theory and Applications of Fractional Differential Equations*, vol. 204 (Elsevier, Amsterdam, 2006)
2. J. Sabatier, O.P. Agrawal, J.A. Tenreiro Machado. *Advances in Fractional Calculus*, vol. 4, no. 9 (Springer, Dordrecht, 2007)
3. E. Tarasov, *Fractional Dynamics: Applications of Fractional Calculus to Dynamics of Particles, Fields and Media* (Springer, Berlin, 2011)
4. H. Qi, A. Gan, Research on mobile cloud computing: review, trend and perspectives, in *2012 Second International Conference on Digital Information and Communication Technology and it's Applications (DICTAP)* (IEEE, Piscataway, 2012)
5. W. Ren, Y.C. Cao, *Distributed Coordination of Multi-Agent Networks* (Springer, London, 2011)
6. R.W. Ibrahim, A. Gani, A new algorithm in cloud computing of multi-agent fractional differential economical system. Computing **98**(11), 1061–1074 (2016)
7. R.W. Ibrahim, H.A. Jalab, A. Gani, Cloud entropy management system involving a fractional power. Entropy **18**(1), 14 (2015)
8. R.W. Ibrahim, H.A. Jalab, A. Gani, Perturbation of fractional multi-agent systems in cloud entropy computing. Entropy **18**(1), 31 (2016)
9. R.W. Ibrahim, A. Gani, Hybrid cloud entropy systems based on Wiener process. Kybernetes **45**(7) , 1072–1083 (2016)
10. R.W. Ibrahim, H.A. Jalab, A. Gani, Entropy solution of fractional dynamic cloud computing system associated with finite boundary condition. Boundary Value Problems **94**, 1–12 (2016)
11. R.W. Ibrahim, A. Gani, A mathematical model of cloud computing in the economic fractional dynamic system. Iranian J. Sci. Technol. Trans. A Sci. **42**(1), 65–72 (2018)
12. S. Muhammad, et al., A review on distributed application processing frameworks in smart mobile devices for mobile cloud computing. IEEE Commun. Surv. Tutorials **15**(3), 1294–1313 (2013)
13. H. Kai, J. Dongarra, C. Geoffrey, *Distributed and Cloud Computing: From Parallel Processing to the Internet of Things* (Morgan Kaufmann, Los Altos, 2013)
14. T. Dinh, et al., A survey of mobile cloud computing: architecture, applications, and approaches. Wirel. Commun. Mob. Comput. **13**(18), 1587–1611 (2013)
15. D. Magalhäes, et al., Workload modeling for resource usage analysis and simulation in cloud computing. Comput. Electr. Eng. **47**, 69–81 (2015)

16. R. Kampas, et al., System and method for cloud enterprise services. U.S. Patent No. 9,235,442. 12 Jan. 2016
17. S. Sung, et al., A distributed mobile cloud computing model for secure big data, in *2016 International Conference on Information Networking (ICOIN)* (IEEE, Piscataway, 2016)
18. T. Li, et al., A convergence of key-value storage systems from clouds to supercomputers. Concurrency Comput. Practice Exp. **28**(1), 44–69 (2016)
19. A.A.Goldstein, Cauchy's method of minimization. Numer. Math. **4**(1), 146–150 (1962)
20. V.E.Tarasov, Fractional vector calculus and fractional Maxwell's equations. Ann. Phys. **323**(11), 2756–2778 (2008)

The Global-Local Transformation

Konstantinos A. Raftopoulos

Abstract The Global-Local transformation, a shape representation technique for manifolds of multiple dimensions is presented in this chapter. Useful properties of the transform space are examined and through experiments, unique advantages of the GL-transform are revealed. Applications in shape matching and radar shadow analysis demonstrate its effectiveness in real life scenarios.

AMS Subject Classification 60E10, 60G35, 41A35

1 Prologue

The Global-Local transformation, a shape representation technique with unique advantages, is formally presented in this chapter. It is also demonstrated with experiments that the presented theory is effective in application. Parts of this manuscript have appeared elsewhere; however, it is the first systematic approach to unify the theory in a single chapter, where a theoretical foundation is in the main focus.

2 The Global-Local Transformation in a Single Dimension

Let $(0, \lambda] \subset \mathbb{R}$ and $\alpha : (0, \lambda] \to \mathbb{R}^n$ a continuous 1-1 infinite times differentiable *curve* embedded in $\mathbb{R}^n$. Let also α be a *closed* curve, therefore isomorphic to a mapping from the unit circle S^1 into $\mathbb{R}^n$. Since α is continuous, 1-1 and onto

---

K. A. Raftopoulos (✉)
Hellenic Military Academy, Vari, Attika, Greece

University of Western Attika, Athens, Greece

National Technical University of Athens, Athens, Greece

University of California at Los Angeles, Los Angeles, USA
e-mail: raftop@cs.ucla.edu

N. J. Daras, T. M. Rassias (eds.), *Computational Mathematics and Variational Analysis*, Springer Optimization and Its Applications 159,
https://doi.org/10.1007/978-3-030-44625-3_20

$\alpha((0, \lambda]) \subset \mathbb{R}^n$ is an isomorphism and imposes the total ordering of $(0, \lambda] \subset \mathbb{R}$ into $\mathrm{Im}_\alpha((0, \lambda]) \subset \mathbb{R}^n$. Consider $\mathbb{R}^n$ endowed with the usual metric derived by the usual norm $\|.\|_n$, $\mathrm{Im}_\alpha((0, \lambda]) := \mathrm{I}_\alpha$ is the trajectory of the closed curve α and its diameter in $\mathbb{R}^n$ is

$$\Delta = \sup_{x,y \in \mathrm{I}_\alpha} \|x - y\|_n.$$

Consider also the trajectory endowed with the usual metric derived by the usual norm $\|.\|_1 = |.|$ (the absolute value) in its isomorphic $(0, \lambda]$. The metric derived from $\|.\|_1 := d_1$ is a one-dimensional metric embedded into $\mathbb{R}^n$ and can be derived from the metric defined from $\|.\|_n := d_n$ as an infinite sum of infinitesimal d_n distances. Therefore, from the triangle inequality of the metric: $\|y - x\|_n \leq \|y - x\|_1 = |\alpha^{-1}(y) - \alpha^{-1}(x)|, \forall x, y \in \mathrm{I}_\alpha$.

Now let x_0 be a point in the trajectory of α and consider the function v'_{x_0} as : $v'_{x_0} : I_\alpha \to (0, \Delta], x \mapsto v_{x_0}(x) := \|x - x_0\|_n$. v'_{x_0} is d_n restricted to the trajectory of α, fixed to x_0 and continuous as such. Define v_{x_0}, the composition of α and v'_{x_0} as follows:

Definition 1 $v_{x_0} := (v'_{x_0} \circ \alpha)$, therefore,

$$v_{x_0} : (0, \lambda] \to (0, \Delta], x \mapsto v_{x_0}(x) := v'_{x_0}(\alpha(x))$$

Call v_{x_0} the *view of the trajectory I_α from x_0*.

v_{x_0} is well defined on the choice of x_0. It is also continuous as a composition of the continuous curve α and the continuous metric v'_{x_0}.

Proposition 1 *For any choice of x_0, v_{x_0} satisfies a Lipschitz condition*

Proof Indeed let x_0 be a choice in the trajectory, v_{x_0} the respective view, $t_1, t_2 \in (0, \lambda)$ and $t_0 = \alpha^{-1}(x_0)$. It is $v_{x_0}(t_1) = \|\alpha(t_1) - x_0\|_n = \|\alpha(t_1) - \alpha(t_0)\|_n$, $v_{x_0}(t_2) = \|\alpha(t_2) - x_0\|_n = \|\alpha(t_2) - \alpha(t_0)\|_n$. We have now $|v_{x_0}(t_2) - v_{x_0}(t_1)| = |\|\alpha(t_2) - \alpha(t_0)\|_n - \|\alpha(t_1) - \alpha(t_0)\|_n| \leq \|\alpha(t_2) - \alpha(t_1)\|_n \leq \|\alpha(t_2) - \alpha(t_1)\|_1 = |\alpha^{-1}(\alpha(t_2)) - \alpha^{-1}(\alpha(t_2))| = |t_2 - t_1|$. Therefore, $\forall t_1, t_2 \in (0, \lambda]$ we have

$$\frac{|v_{x_0}(t_2) - v_{x_0}(t_1)|}{|t_2 - t_1|} \leq 1$$

and the proof is complete. □

Let $t \in (0, \lambda]$ and $\alpha(t) = x$.

Definition 2 Let ϕ a function defined on $(0, \lambda]$ and taking values in $C^\infty(0, \lambda]$ (the continuous infinite times differentiable real functions defined on $(0, \lambda]$) as follows:

$$\phi : (0, \lambda] \to C^\infty(0, \lambda] : t \mapsto \phi(t) := v_{\alpha(t)} = v_x \in C^\infty(0, \lambda].$$

If for each $t_1 \in (0, \lambda]$ one considers the graph of the view functions $v_{\alpha(t_1)}$ as the pairs of its coordinates, $(t_2, v_{\alpha(t_1)}(t_2))$, $t_2 \in (0, \lambda]$, then a geometric representation of ϕ being the surface $S_\alpha(x, y, z) = (t_1, t_2, v_{\alpha(t_1)}(t_2))$, $t_1, t_2 \in (0, \lambda]$ is possible.

Definition 3 The surface S_α is called the *Global-Local surface* of the closed curve α and it is obvious that

$$S_\alpha \subset (0, \lambda] \times (0, \lambda] \times (0, \Delta].$$

Consider $C^\infty(0, \lambda]$ endowed with the metric d_f defined by the norm $\|.\|_f$ (norm of the operator) as:

$$\|.\|_f : C^\infty(0, \lambda] \to \mathbb{R},\ f \mapsto \|f\|_f := \int_0^\lambda |f(t)|dt.$$

$C^\infty(0, \lambda]$ becomes a metric space with reference to the metric of the operator just introduced and ϕ is a function between metric spaces. It is trivial to show that for a given closed curve α, ϕ is continuous 1-1 and onto the Global-Local surface S_α.

Definition 4 With the help of ϕ one can now define an equivalence relationship $\sim$ in the set of the infinite times differentiable closed curves, $C_c^\infty(\mathbb{R}, \mathbb{R}^m)$ such that

$$a_1 \sim a_2 \iff \phi(a_1) = \phi(a_2) \iff S_{a_1} = S_{a_2}.$$

Definition 5 The Global-Local transformation of curve α is now defined and denoted by Φ as an isomorphism between $\frac{C_c^\infty(\mathbb{R},\mathbb{R}^m)}{\sim}$ and $\mathbb{R}^3$ as follows:

$$\Phi : \frac{C_c^\infty(\mathbb{R}, \mathbb{R}^m)}{\sim} \to \mathbb{R}^3,\ C_\alpha \mapsto \Phi(C_\alpha) \equiv S_\alpha,$$

where C_α is the class of closed curves being represented by α and S_α is the Global-Local surface of the closed curve α.

Some properties of the just introduced GL-transformation will be examined now. For easier exposition one can start by examining plane curves. Let α be a closed curve of length λ, embedded in the $2D$ plane. A direction has been agreed on the boundary α. In the following only this direction of traversing α is allowed. α is GL transformed to the surface S_α. If x, y are arbitrary chosen points on the curve α, denote with $(\overline{(x, y)}, (x, y), (y, x))$ the line segment that connects x and y on the plane, the boundary segment that connects x and y starting from x and the boundary segment that connects y and x starting form y, respectively. Denote their respective lengths with $(|\overline{(x, y)}|, |(x, y)|, |(y, x)|)$.

One can now demonstrated that ϕ is invertible, the inverse ϕ^{-1} maps each v_x (view of α from x) to the point x on the curve. For this suppose the closed curve α is GL transformed to the surface S_α. Choose v_x and find $x \in \alpha$ such that $\phi(x) = v_x$. Since the transformation is invariant to the coordinate system, to specify this x on the boundary of the curve one will have to reconstruct the whole curve starting from the surface S_α. This way x will be specified through the relative locations of the rest of the points. Due to the independence from the external coordinate system, any two points that have distance (shortest) less than the global maximum of the surface S_α can be considered the first two points on the curve. Their placement will determine the orientation and scaling of the curve. Choose then x to be any random point on the 2D plane and consider another point y as having short boundary distance from x equal to $d_1(x, y) = t_1$, where t_1 another random value in $(0, \lambda)$. The pair (x, y) defines a partition of the curve α into two pieces, (x, y) with length t_1 and (y, x) with length $\lambda - t_1$. The distance on the plane between x and y equals $d_2(x, y) = v_x(t_1)$.

Place y randomly anywhere on the plane. The orientation will be imposed by the direction of the axis that connects x and y and the scaling will be determined by the fact that the length of this axis will be assumed to be $v_x(t_1)$. Given the pair (x, y) one can proceed in finding at least two other points on the same curve α. The idea is to locate the points where the perpendicular bisector of $\overline{(x, y)}$ intercepts α. One can prove that these points are always (for any pair) at least two at number, due to the fact that the curve is closed. Furthermore, their location can be easily determined by the fact that each one of them lies at equal distances from x and y. These points are different from x and y. Call them *generated from* x, y and call x and y their *generators*.

An illustration of this procedure is shown in Figure 1. One can demonstrate that the above procedure, if applied repeatedly, starting from the already generated pairs, will eventually (to the limit of infinite repetitions) reconstruct the whole curve. For this purpose some propositions will be needed.

Proposition 2 *Let x, y two arbitrarily chosen points on a curve α that is GL transformed to the surface S_α. If $\overline{(x, y)}$, (x, y), and (y, x) as defined above, then the perpendicular bisector of $\overline{(x, y)}$ intercepts α at two points at least, one on each of the two curve segments above.*

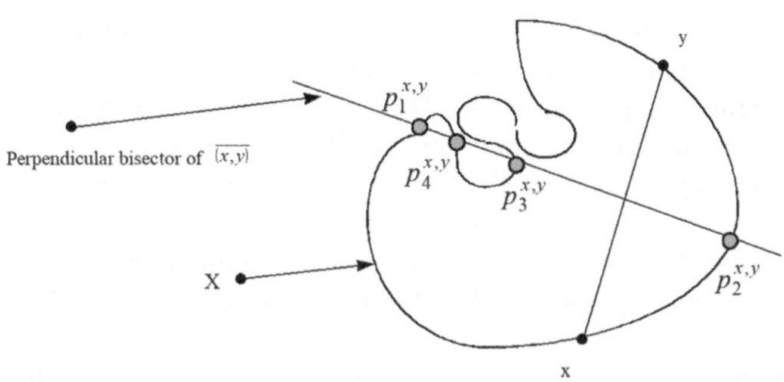

Fig. 1 x and y random points on the curve. The perpendicular bisector of $\overline{(x, y)}$ always cuts curve X in at least two points one on each of the boundary segments $\overset{\frown}{x, y}$ and $\overset{\frown}{y, x}$. See Proposition 2 for more

Proof Consider the function, $\delta_{x,t_1}(t) \equiv \phi_x(t_1+t) - \phi_{x+t_1}(t), t \in [0, \lambda - t_1]$. Function d_{x,t_1} is well defined as a point-wise difference and notice that $\delta_{x,t_1}(0) = \phi_x(t_1)$ and $\delta_{x,t_1}(\lambda - t_1) = -\phi_{x+t_1}(\lambda - t_1) = -\phi_x(t_1)$, therefore $\delta_{x,t_1}(0)\,\delta_{x,t_1}(\lambda - t_1) \le 0$.

Also δ_{x,t_1} is continuous in $[0, \lambda - t_1]$, as the difference of the continuous in $[0, \lambda]$ functions ϕ_x and ϕ_{x+t_1}. From a Fermat's theorem there exists $t_0 \in [0, \lambda - t_1]$ such that $\delta_{x,t_1}(t_0) = 0 \Leftrightarrow \phi_x(t_1+t_0) - \phi_{x+t_1}(t_0) = 0 \Leftrightarrow \phi_x(t_1+t_0) = \phi_{x+t_1}(t_0)$, but $t_1 + t_0$ and t_0, are boundary distances from x and y resp. that correspond to the same point z_{t_0} on (y, x), which means that z_{t_0} with boundary distance t_0 from y, has equal plane distance from x and y, therefore it has to be on the perpendicular bisector of $\overline{(x, y)}$ at a distance $v_y(t_0)$ from y and x and on (y, x).

Now consider the function $\delta'_{x,t_1}(t) \equiv \phi_x(t) - \phi_{x+t_1}(\lambda - (t_1 - t)), t \in [0, t_1]$. One can notice in a similar manner that $\delta'_{x,t_1}(0) = -\phi_{x+t_1}(\lambda - t_1) = -\phi_x(t_1)$ and $\delta'_{x,t_1}(t_1) = \phi_x(t_1)$, therefore $\delta'_{x,t_1}(0)\,\delta'_{x,t_1}(t_1) < 0$. Also δ'_{x,t_1} is continuous in $[0, t_1]$, as the difference of the continuous in $[0, \lambda]$ functions ϕ_x and ϕ_{x+t_1}. Then from the same Fermat's theorem there exists t'_0 in $[0, t_1]$ such that $\delta'_{x,t_1}\left(t'_0\right) = 0 \Leftrightarrow$ $\phi_x\left(t'_0\right) - \phi_{x+t_1}\left(\lambda - \left(t_1 - t'_0\right)\right) = 0 \Leftrightarrow \phi_x\left(t'_0\right) = \phi_{x+t_1}\left(\lambda - t_1 + t'_0\right)$, but t'_0 and $\lambda - t_1 + t'_0$, are boundary distances from x and y resp. that correspond to the same point $z_{t'_0}$ on (x, y), which means that point $z_{t'_0}$ with boundary distance t'_0 from x, has equal plane distances from x and y. Therefore, it has to be on the perpendicular bisector of $\overline{(x, y)}$ at a distance $v_x(t'_0)$ from y and x and on (x, y). □

It is now proved that there exist at least two different points z_{t_0} and $z_{t'_0}$ on the boundary of α and on the two different boundary segments produced by the pair (x, y). Lets continue with the construction of α from S_α, and thus with the invertibility of ϕ. The first two points that are generated from x and y, are on the bisector of $\overline{(x, y)}$. In order to be completely defined and placed on the plane we still need their relative position on the bisector. This is because there are two possible placements for each of the points on the perpendicular bisector that has the same equal distance from x and y. In general, generated points will be placed by means of their distance from other already generated points. In the case of z_{t_0} and $z_{t'_0}$, however, since there is no other generated point, except the generators x and y, their relative placement on the bisector is only restricted by their in-between distance, which can be easily found from the transformation surface S_α to be $\phi_{y+t_0}\left(\lambda - t_0 - t_1 + t'_0\right) = \phi_{x+t'_0}\left(t_1 - t'_0 + t_0\right)$. The choice of how these points are placed on the bisector will determine the mirror-wise direction of the produced curve, thus the invariance of GL-transform to translation, scaling-rotation, and mirroring, corresponds to our freedom of placing x, y, and z_{t_0}, respectively. This same procedure of generating points, if applied recursively to all the possible pairs, will eventually (at the limit) reconstruct the whole curve α. This is expressed in the following:

Proposition 3 *Let p_0, x, y arbitrarily chosen points on a closed curve α. There exists a sequence of generated points that starts from (x, y) and converges to p_0.*

Proof Without loss of generality assume that p_0 lies on segment $\overset{\frown}{(y, x)}$. In the opposite case the proof is similar. Starting from (x, y) generate points $p_{1,i}$ $i = 1(1)k_1, k_1 \geq 2$, (1 denotes the first step of generation, i is an index on the number of generated points for this step). According to Proposition 2, at least one of them, say $p_{1,1}$, lies on (y, x) and defines a partition of (y, x) into $\left(y, p_{1,1}\right)$ and $\left(p_{1,1}, x\right)$. Choose the boundary segment that contains p_0 (suppose it is $\left(p_{1,1}, x\right)$) and use its end points as generators for the next step of generation. At this second step the points $p_{1,1}$ and x will generate points $p_{2,i}, i = 1(1)k_2, k_2 \geq 2$, of which again from the same proposition at least one, assume $p_{2,1}$, is on the chosen segment $\left(p_{1,1}, x\right)$. Continuing this process of partitioning in two pieces the always smaller segments that contain p_0 and choosing the one that still contains p_0 as the one that provides new generators, one constructs a series of generated points $p_{n,1}$, on the curve, (p_n in the rest, since the chosen point is always labeled with index 1), that for each step (n) they come closer to p_0 on the boundary. In Figure 2 the first four of these points are shown.

At step n, point p_n is generated from p_{n-1} and one of p_{n-1}'s generators denoted by p_n^*. So p_n^* is a generator of p_{n-1} at step $n - 1$ but also a generator of p_n at step n. This means that the boundary segment between p_n^* and p_n is contained in the boundary segment between p_n^* and p_{n-1}, thus the length of the segment between p_n^* and p_n approaches zero as n tends to infinity and since all these segments contain p_0 it has been shown that

$$\lim_{n \to \infty} (p_n) = p_0.$$

□

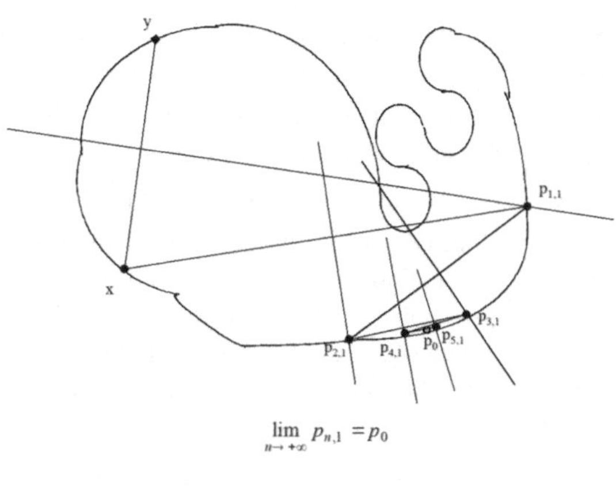

Fig. 2 The process of reaching any random point p_0 from any pair of points (x, y) by generating a sequence of points that converges to p_0

It has been proved that infinite steps of generations as above will converge to any random point p_0, starting from any random point x and scale-orientation choice of a second point y and therefore all points of α can be generated this way and justifies the intuition that intrinsic morphometry of shapes is independent of rigid transformations. We have then just proved the following:

Theorem 1 *A class of the quotient set* $\frac{C_c^{\infty}(\mathbb{R},\mathbb{R}^m)}{\sim}$ *contains all these closed curves that are related through rigid transformations (translation, scaling, rotation, mirroring).*

3 The Generalized Global-Local Transformation in Multiple Dimensions

The extension of the GL-transformation to the general case of closed n-dimensional hyper-objects embedded in $\mathbb{R}_m$ is described in this section.

Definition 6 Let $h : D^n \to \mathbb{R}^m$ a continuous infinite times differentiable mapping defined on the hyper-object D^n with local isomorphic mappings in $\mathbb{R}^n$, embedded into $\mathbb{R}^m$. Let also D^n be a *closed* hypersurface, that is h is equivalent to a continuous mapping of the unit hypersphere S^n into $\mathbb{R}^m$. The value set of h is denoted by I_h, thus $I_h \equiv h(D^n)$.

I_h is isomorphic to D^n through h but at the same time is also a subset of $\mathbb{R}^m$. Two metrics can therefore be defined on I_h, one induced by the usual metric

$$\|.\|_n \overset{def}{\Longleftrightarrow} d_n$$

defined in $\mathbb{R}^n$:

$$d_n(y_1, y_2) \overset{def}{\Longleftrightarrow} d_n(h^{-1}(y_1), h^{-1}(y_2)), \forall y_1, y_2 \in I_h \subset \mathbb{R}^m$$

and the other by the restriction of the usual metric

$$(\|.\|_m \overset{def}{\Longleftrightarrow} d_m),$$

defined in $\mathbb{R}^m$.

Through d_n and a choice $x_0 \in D^n$, an equivalence relation $\overset{x_0}{\sim}$ in D_n (and in I_h) is defined as follows:

$$x_1 \overset{x_0}{\sim} x_2 \overset{def}{\Longleftrightarrow} d_n(x_0, x_1) = d_n(x_0, x_2)$$

and denoted by $C_{(x_0,x)}$, the class centered at x_0 and represented by x. The function:

$$v'_{x_0} : \frac{D^n}{\underset{\sim}{x_0}} \to \mathbb{R} : C_{(x_0,x)} \mapsto d_n(x_0, x), x \in C_{(x_0,x)},$$

is well defined on the choice of $x_0 \in D^n$, 1-1,continuous (with the obvious modification of the metric d_n to the classes $C_{(x_0,x)}$) and therefore an isomorphism between $\frac{D^n}{\underset{\sim}{x_0}}$ and $\mathbb{R}$ and it is essentially the mapping from the boundaries of the d_n-balls centered at $h(x_0)$ in I_h to their real radius. Denote the inverse function by ϕ_{x_0}.

$$\phi_{x_0} : \mathbb{R} \to \frac{D^n}{\underset{\sim}{x_0}} : t_x \mapsto \phi_{x_0}(t_x) \equiv C_{(x_0,x)} \equiv h^{-1}(C_{(y_0,y)}).$$

It is rather obvious that the members of the quotient set $\frac{D^n}{\underset{\sim}{x_0}}$ correspond to d_n-isosurfaces of $I_h \subset \mathbb{R}^m$. Each $C_{(x_0,x)}$ is a sub-manifold of h, of dimension n embedded into R^m.

$$C_{(y_0,y)} : A^n \to h(A^n) \subset I_h \subset \mathbb{R}^m.$$

Each $C_{(x_0,x)}$ also corresponds to a d_n-isosurface of $I_h \subset \mathbb{R}^m$. It can be seen therefore as a *closed* munifold of dimension $n-1$ embedded into $\mathbb{R}^n$.

$$C_{(x_0,x)} : D^{n-1} \subset \mathbb{R}^{n-1} \to A^n \subset D^n,$$

Combining the two results above, a class of radius $r \in \mathbb{R}$ is defined as:

$$C_{(x_0,r)} \overset{def}{\Longleftrightarrow} C_{(y_0,y)} \circ C_{(x_0,x)} : D^{n-1} \subset R^{n-1} \to I_h \subset \mathbb{R}^m,$$

a closed manifold of dimension $n-1$. Now see again ϕ_{x_0} as:

$$\phi_{x_0} : D \subset \mathbb{R} \to C^\infty(D^{n-1}, \mathbb{R}^m) : r \mapsto \phi_{x_0}(r) \equiv C_{(x_0,r)}$$

and since x_0 is a random choice in D^n on can proceed with a definition:

Definition 7

$$\phi : D^n \to C^\infty(D, C^\infty(D^{n-1}, \mathbb{R}^m)) : x \mapsto \phi(x) \equiv \phi_x \in C^\infty(D, C^\infty(D^{n-1}, \mathbb{R}^m)),$$

where all the necessary metric adjustments are assumed.

ϕ can be recursively applied to the produced classes to yield a multidimensional hypersurface that corresponds to manifold h, and will describe the morphometry of h. To see this start from ϕ applied on D^n, the domain of h, producing the pairs: $(x, \phi_x), x \in D^n$. ϕ_x can be seen as the set of its points thus:

$$(x, \phi_x) \equiv (x, r, C_{(x,r)} \equiv h_1), x \in D^n \subset \mathbb{R}^n, r \in D \subset \mathbb{R}.$$

The third element is recognized as a closed manifold of dimension $n-1$, as was derived above. It can therefore be the subject of the same analysis and in particular one can apply ϕ on its domain D^{n-1}:

$$(x, r, x_1, r_1, h_2), x \in D^n, x_1 \in D^{n-1}, r \in D \subset \mathbb{R}, r_1 \in D_1 \subset \mathbb{R},$$

where h_2 is a closed manifold of dimension $n-2$ defined from the rest of the coordinates. Applying recursively ϕ on the domain D^{n-i} of the new manifold h_i, reducing its dimension every time, one will finally, after $n-1$ steps conclude as:

$$(x, r, x_1, r_1, \ldots, x_{n-2}, r_{n-2}, h_{n-1}), x_i \in D^{n-i}, r_i \in D_i \subset \mathbb{R}, i \in T_{n-2},$$

where h_{n-1} is a one-dimensional manifold, a closed curve in $\mathbb{R}^m$ transformed to the GL surface $S_{h_{n-1}}$, as was illustrated in the first paragraph of this paper. Putting it all together and expanding all x_i above with their coordinates and normalizing all D^i and D_i, one can have a final representation of h as a hypersurface embedded in the $\frac{(n+1)(n+2)}{2}$ dimensional hypercube. Call this representation, the *GL-hypersurface* of h and denote S_h.

Definition 8 With the help of ϕ one can define an equivalence relationship $\sim$ in the set of infinite times differentiable closed hyper-objects, $C_c^\infty(\mathbb{R}^n, \mathbb{R}^m)$ such that

$$h_1 \sim h_2 \iff \phi(h_1) = \phi(h_2) \iff S_{h_1} = S_{h_2}.$$

Proposition 4 *A class of the quotient set* $\frac{C_c^\infty(\mathbb{R}^n, \mathbb{R}^m)}{\sim}$ *contains all those closed hyper-objects that are related through rigid transformations (translation, scaling, rotation, mirroring).*

Proof Here a hyper-object has to be constructed back from its GL-hyper-surface, demonstrating invariance of rigid transformations through degrees of freedom in placing the generated points. The construction is a direct extension to the 1D case above, adjusting for the degrees of freedom when placing generated points. □

Definition 9 The *Global-Local* transformation of the hyper-object h is denoted by Φ and defined as an isomorphism between $\frac{C_c^\infty(\mathbb{R}^n, \mathbb{R}^m)}{\sim}$ and $\mathbb{R}^{\frac{(n+1)(n+2)}{2}}$ as follows:

$$\Phi : \frac{C_c^\infty(\mathbb{R}^n, \mathbb{R}^m)}{\sim} \to \mathbb{R}^{\frac{(n+1)(n+2)}{2}}, C_h \mapsto \Phi(C_h) \equiv S_h,$$

where C_h is the class of closed hyper-objects being represented by h and S_h the *Global-Local* surface of h.

4 GL-Landmarks on 1D Manifolds

In this section some properties of curves are discussed and related concepts are introduced. Let $\alpha \in C_c^\infty((0,\lambda], \mathbb{R}^m)$, a closed, continuous, infinite times differentiable curve embedded into $\mathbb{R}^m$. Let also $t_1, t_2, t_3 \in (0,\lambda]$ and $x_1 = \alpha(t_1), x_2 = \alpha(t_2), x_3 = \alpha(t_3)$.

Definition 10 The point x_1 is called *interesting from the point x_2, or x_2-int* iff

$$\frac{dv_{x_2}}{dt}(t_1) = 0,$$

the point x_1 is called *interesting from the boundary segment* $(x_2, x_3) \equiv (\alpha(t_2), \alpha(t_3))$ *or* (x_2, x_3)-int iff it is x-int for all $x = \alpha(t), t \in (t_2, t_3)$.

The above definition is based on the first derivative of the view functions. Further discrimination of the *interesting* points can be achieved based on the second derivative of the view functions.

Definition 11 An *interesting* point $x_1 = \alpha(t_1), t_1 \in (0,\lambda]$ is a $+x_2$-interesting point or $+x_2$-int iff in addition to

$$\frac{dv_{x_2}}{dt}(t_1) = 0,$$

it is also

$$\frac{d^2 v_{x_2}}{dt^2}(t_1) < 0.$$

If it is

$$\frac{d^2 v_{x_2}}{dt^2}(t_1) > 0,$$

say that x_1 is an $-x_2$-interesting point. Refer to + or − as the *sign* of the point's *interest.*

Let $t_1, t_2 \in (0,\lambda], x_1 = \alpha(t_1), x_2 = \alpha(t_2) \in I_\alpha$. Then the *length* of the boundary segment (x_1, x_2) is $\|(x_1, x_2)\|_1$ and equals $|t_1 - t_2|$.

Definition 12 Let $x = \alpha(t) \in I_a$. Lets call *strength* of the *interesting* point x and denote s_x, the algebraic summation of the lengths of the boundary segments x is *interesting* from. In this summation, each segment's length is added with the same sign as the *interest* of the point it corresponds to:

$$s_x \stackrel{def}{=} \sum_{s^+ \in S^+} \|s^+\|_1 - \sum_{s^- \in S^-} \|s^-\|_1,$$

for all s^-, s^+, such that x is $+s^+$-int and $-s^-$-int.

Each point on the contour is *interesting* from at least one other point on the same contour, therefore it only make sense to refer to the point's *strength*.

Definition 13 A point with positive non-zero strength is called *convex point*. A point with negative non-zero strength is called *concave point*. A point with zero strength is called *transitive point*.

Due to continuity and smoothness properties of the GL surface, concave, convex, and transitive points appear in contiguous contour segments of increasing strength. This way they form convex, concave, and transitive regions, respectively. These formations decompose a contour in a natural way. They will be called the *contour's morphological regions*.

Definition 14 The point in a convex/concave region that has the largest strength in this region will be called *convex/concave landmark* point, respectively.

From the definition of the landmark points and the continuity properties of the view functions it is easily shown that between two convex landmarks there is a concave landmark and between two concave landmarks there is a convex landmark. This contour's decomposition in interleaved convex-concave landmark points has important applications as is shown below.

4.1 The Global-Local Properties

The GL-transformation just introduced processes significant properties for shape representation. As a metric embedding is independent of the coordinate system therefore invariant of geometric transformations. Furthermore, local measures of the transformed curve become global measures on the GL surface, while local measures on the GL surface correspond to global shape descriptors for the curve. Local curvature on the transformed curve can be estimated by area integrals in the transform space, whereas points of local extrema on the GL surface correspond to *restrictions* and *dilations* of the curve. This correspondence of local and global descriptors through the GL-transformation is significant. In relation to curvature, it can be used for better resistance to noise, in relation to morphometry, restrictions, and dilations present a meaningful encoding of the shape's boundary. Indicative applications and examples follow.

4.2 *Encoding Curvature into Global Descriptors of the GL Surface*

Lets consider $C^\infty(0, \lambda]$ (the infinite times differentiable real functions defined on $(0, \lambda]$) endowed with the metric d_f defined by the norm $\|.\|_f$ (norm of the operator) defined as:

$$\|.\|_f : C^\infty(0, \lambda] \to \mathbb{R},\ f \mapsto \|f\|_f := \int_0^\lambda |f(t)|dt.$$

The space $C^\infty(0, \lambda]$ becomes a metric space with reference to the metric of the operator just introduced. Let also f a function defined on $(0, \lambda]$ and taking values in $\mathbb{R}$ as follows:

$$f : (0, \lambda] \to \mathbb{R}: t \mapsto f(t) := \|v_{\alpha(t)}\|_f = \|v_x\|_f.$$

Function f maps a point $x = \alpha(t)$ on the curve to the area that is enclosed by the view function $v_{\alpha(t)}$ and the $x'x$ axis. From the Lipschitz condition satisfied by $v_{\alpha(t)}$ and the strong continuity of f follows that the curvature of $f(t)$ is a bounded estimation of the curvature of $\alpha(t)$. This result is significant since in the presence of noise, where the curvature of the initial curve cannot be estimated without smoothing operations, f can be used to estimate the curvature of the noisy curve. The resulting from f representation is called the *View-Area Representation* or just *VAR*. In Figure 3 the initial shape (a) together with the view-area representation (d) is shown. Despite the noisy shape (a) its curvature can be estimated through the curvature of the representation in (d).

4.3 *Encoding Global Curve Descriptors into Curvature*

Lets now show how local measures on the GL surface identify morphological regions, namely *dilations* and *restrictions* on the corresponding curve. A way to measure these morphological formations is provided, by identifying regions around landmark points.

Let $\alpha \in C_c^\infty((0, \lambda], \mathbb{R}^m)$, a closed, continuous, infinite times differentiable curve in $\mathbb{R}^m$, $v_x(t)$ the view functions, $S_\alpha(x, y)$ the GL surface and $H_\alpha = \begin{vmatrix} \frac{\partial S_\alpha{}^2}{\partial x^2} & \frac{\partial S_\alpha{}^2}{\partial x \partial y} \\ \frac{\partial S_\alpha{}^2}{\partial y \partial x} & \frac{\partial S_\alpha{}^2}{\partial y^2} \end{vmatrix}$ the Hessian of S_α at (x, y). Let also $t_1, t_2 \in (0, \lambda]$ and $x_1 = \alpha(t_1), x_2 = \alpha(t_2))$. In extension to above definitions:

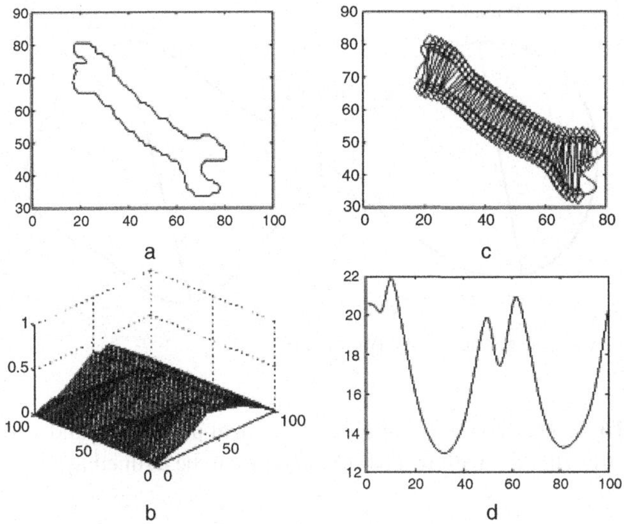

Fig. 3 A shape from the Kimia database (**a**), the GL surface (**b**), the GL-critical point outline (**c**), and the corresponding view-area representation (**d**)

Definition 15 A point (t_1, t_2) is called ϵ-*critical* iff

$$|\frac{\partial S_\alpha}{\partial t_1}(t_1, t_2)| \leq \epsilon, \epsilon \geq 0$$

and

$$|\frac{\partial S_\alpha}{\partial t_2}(t_1, t_2)| \leq \epsilon, \epsilon \geq 0$$

and

$$H_\alpha(t_1, t_2) > 0.$$

Here relaxed conditions are used, necessary for points of local extrema on the GL surface. Under this definition, critical points are points of local extrema together with points in a range around them on the GL surface, the range is according to ϵ. Now if (t_1, t_2) is ϵ critical, then the pair (x_1, x_2) with $x_1 = \alpha(t_1)$ and $x_2 = \alpha(t_2)$ are also ϵ critical defining a morphological region on the curve, namely a *dilation* or a *restriction*. In Figure 4 a shape with one restriction and two dilations is shown.

Local measures on the GL surface define global shape properties for the curve. Measurements of the morphological regions will now be performed.

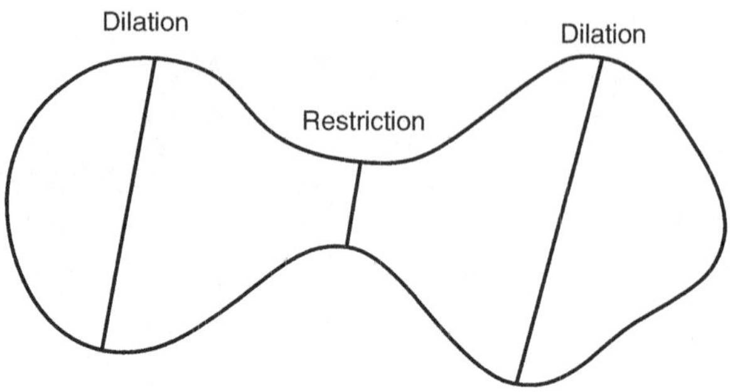

Fig. 4 A shape with one restriction and two dilations

Definition 16 Let $C_\epsilon(\alpha)$ the set of all the ϵ-critical points of α and x a point on the curve. The strength of x with respect to curve α can be defined as:

$$s_\alpha(x) = \int_{(x,y)\in C_\epsilon(\alpha)} k(x,y)\|x-y\|_m dy,$$

where $k(x,y)$ a kernel function ensuring that $\|x-y\|_m$ counts in the calculation of the point's *strength* only if the line connecting x and y lies in the interior of α. This is necessary because the *strength* has to convey perceptual compatibility by measuring local convex formations of the boundary segments. The significance of the point's *strength* lies in the fact that global morphometric properties of the contour are being measured and assigned locally. In the next section the effectiveness of the *GL-transform* and the resulting measure of *strength* are demonstrated in matching benchmark shapes.

5 Performance Evaluation: Matching Benchmark Shapes with the GL-Transform

In this section dynamic programming is employed, to measure distance between shapes from the Kimia silhouette database using each point's *strength* for calculating pairwise correspondences. Let α_1 and α_2 two curves. Given a correspondence c between the points of α_1 and α_2 in a way that $c(\alpha_1(t_1)) = \alpha_2(t_2)$ and $\mathscr{C}$ the set of all such correspondences the distance between the two curves is measured by minimizing the energy functional:

$$E(\alpha_1, \alpha_2, c) = \int_{x\in\alpha_1} |s_{\alpha_1}(x) - s_{\alpha_2}(c(x))| dx$$

with respect to the correspondence c. The distance therefore between the two curves based on points *strengths* is given and denoted by

$$\|\alpha_1 - \alpha_2\|_s = \min_{c \in \mathscr{C}} \{E(\alpha_1, \alpha_2, c)\}.$$

Curves are *discretized* by sampling 100 equally spaced points on each. An initial correspondence between a pair of points on both curves is then established by estimating the diameter of their enclosed area. An exhaustive search for the best initial correspondence is also possible if the results through the diameter are not accurate. All the possible correspondences between pairs of points on the two curves are then examined, the total ordering of the points on both curves keeps the complexity feasible as the method finds the optimum correspondence. In the table of Figure 5 18 shapes from the Kimia benchmark database are shown, including shapes from the classes of fish, mammals, rabbits, men, hands, planes, tools, and sea creatures. For each shape the three best matches are shown labeled accordingly. These are the three shapes among all the 99 shapes in the database that have the smallest distance from the initial image. For all images the three best matches are always correct. Since the presented shapes have different characteristics, this experiment demonstrates that GL-transform based methods achieve significant results in general shape recognition.

6 Targeted Application: SAR Radar Shadow Analysis

Synthetic aperture radar (SAR) is an active military radar technology for obtaining high resolution images even under adverse weather conditions. Due to its high success and wide use, jamming methods have been also developed, mainly by transmitting fake SAR signals reproducing nonexistent objects and scenes. However, SAR technology has improved diffuse scattering characteristics, thus images produced by SAR radars have profound radar shadows that cannot be recreated by fake signals. Unlike airphotos, there is no information within the radar shadow area, it is black, therefore deceptive jammers cannot reproduce shadow areas. On the other hand, radar shadows occur only in the cross-track dimension, therefore, the orientation of shadows in a radar image provides information about the look direction and the location of the near- and far-range, introducing *meaningful* noise into SAR images.

Radar shadows are therefore both discriminating of the obscuring objects and impossible to be reproduced by jammers at the same time. These facts make radar shadows ideal candidates for modern radar detection. However, there is an inherent difficulty in recognizing SAR shadow shapes. They convey *meaningful* noise, that is noise that should not be removed, otherwise the discriminating information is lost. Any system based on radar shadow detection therefor will have to demonstrate strong shape recognition performance in real time without removing noise by

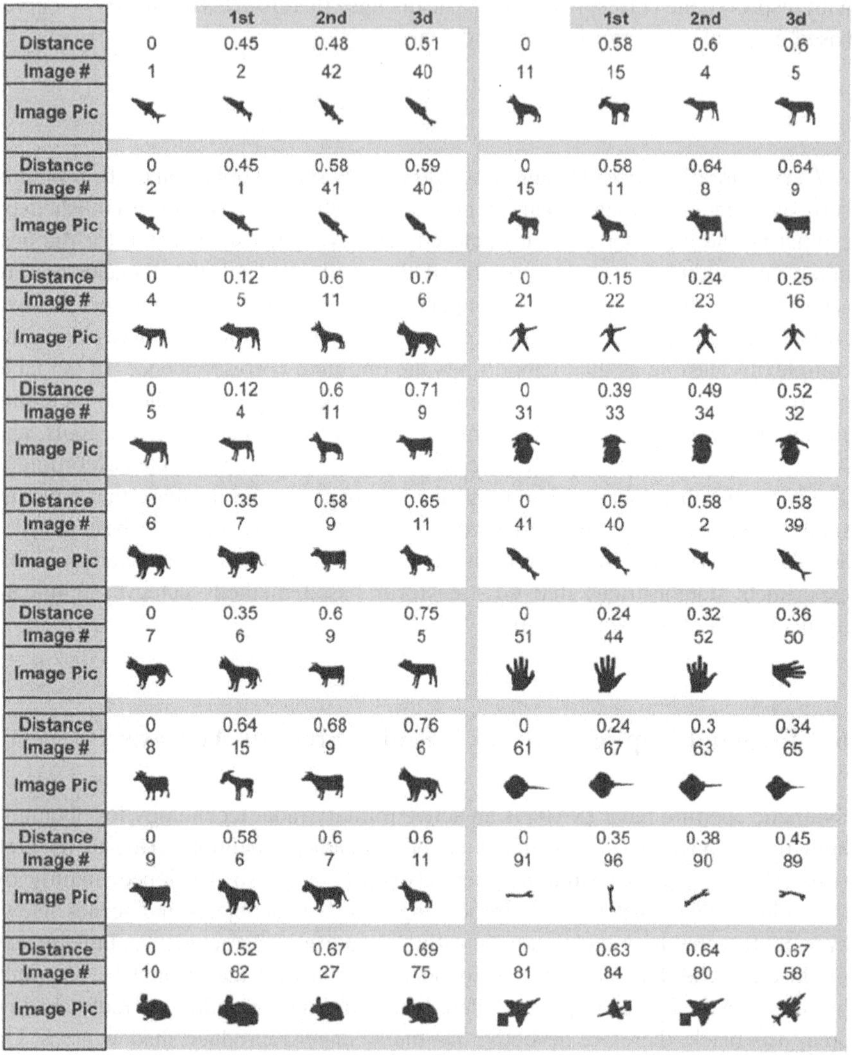

Fig. 5 Shapes from the Kimia database arranged in the order of their three closest counterparts

smoothing. We present in this section experimental results showing that the GL-transformation possesses all these properties that are necessary for this task and in fact outperforms the best alternatives currently in the literature, proving also that the theory above is effective in application. In Section 6.1 a set of experiments comparing GL-transform based methods to Local Area Integral Invariant (LAII) [1] is presented on benchmark shapes. LAII is a state-of-the-art method in terms of robustness to noise and low complexity without learning. The comparison is done:

- In noise resistance, Section 6.2.
- In matching scores for noisy shapes without smoothing using the KIMIA benchmark dataset, Section 6.3.

GL-transform is implemented in a discrete form through the use of VAR (Section 4.2) as an intuitive shape descriptor. More relevant experiments can be found in [2].

6.1 VAR vs LAII

LAII [1] is a representation of low complexity that generalizes the concept of curvature over the noisy segments of curves. A circle of certain radius is used, centered at each point, and curvature is measured as the ratio of the area of this circle that lies in the interior of the closed contour. In the case of zero curvature, e.g. a noisy straight line, half of the disk will lie in the interior of the shape, whereas in case of infinite curvature this portion will tend to zero or to one depending on the sign of curvature at this point. LAII therefore uses a *relaxed* locality to measure curvature but its essentially a local descriptor since only curvature is calculated but not any global features. VAR on the other hand is a global descriptor that uses the whole shape and as already demonstrated, novel hybrid global and local shape features are also captured in VAR.

6.2 VAR vs LAII as Noise Resistant Shape Descriptors

In Figure 6, a comparison between VAR and LAII, with respect to noise resistance has been quantified. Noisy versions of a given shape for increasing values of Gaussian noise are presented together with the VAR and LAII representations for each shape. To measure the effect of noise on the representation itself the point-wise distance between the noisy and the initial representations as a fraction of the initial value (without noise) is used. If $r_\delta(i), i = 1, 2, \ldots, n$ the representation for a particular value of noise δ and $r_0(i), i = 1, 2, \ldots, n$ the respective representation of the original (without induced noise) shape, then the number above each graph is calculated by

$$\sum_i \frac{|r_\delta(i) - r_0(i)|}{r_0(i)} \tag{1}$$

this number therefore shows the overall diversion of the noisy representation measured as a fraction of the initial representation. The results in Figure 6 reveal that VAR has increased resistance to Gaussian perturbations on the boundary.

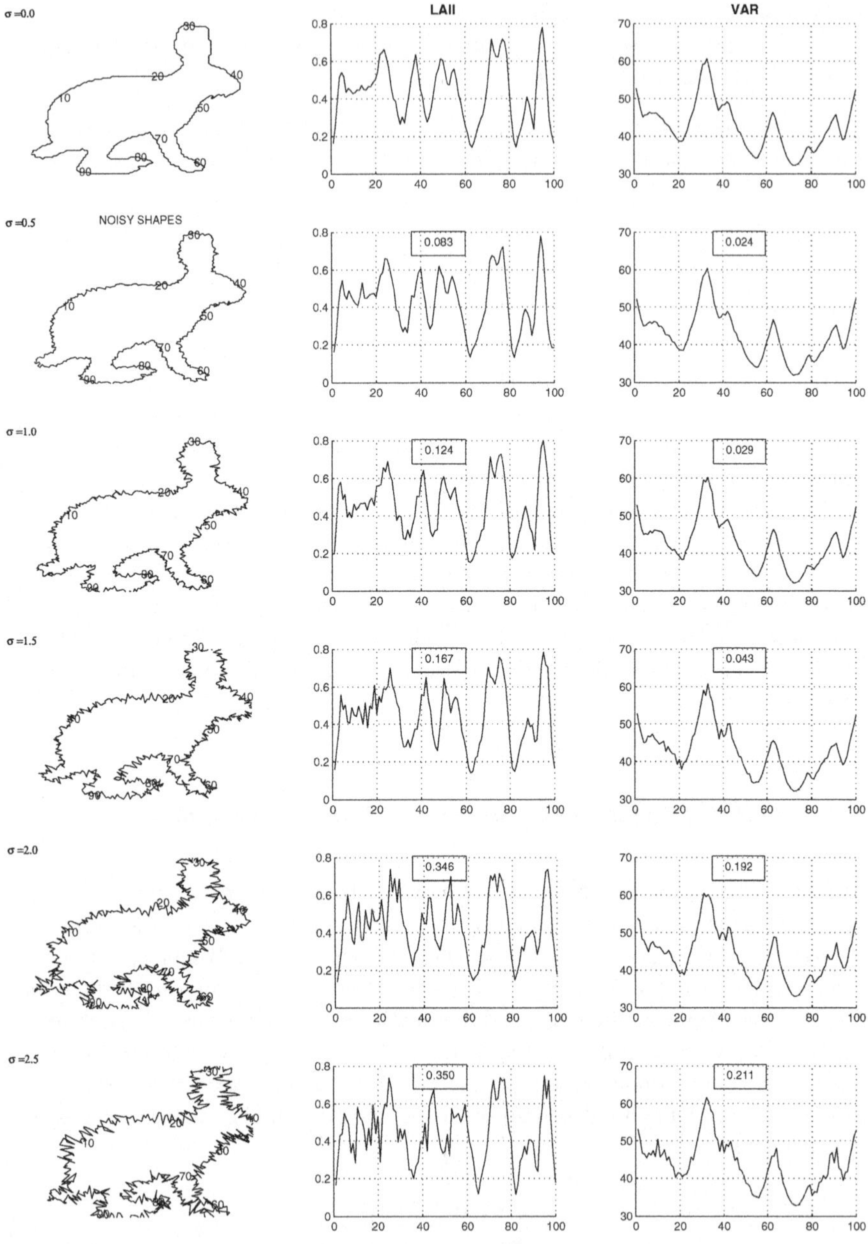

Fig. 6 Noisy versions of a given shape for increasing values of Gaussian perturbation and the corresponding Local Area Integral Invariant (LAII) and View-Area Representations (VAR). The number above each graph is used to measure the effect of noise. It is the sum of the point-wise absolute differences between the graph that corresponds to the noisy shape and the graph that corresponds to the original shape. Each point to point difference is considered as a fraction of the original value. The higher this number is the more distorted is the corresponding noisy graph compared to the corresponding original graph. The numbers justify the visual impression that VAR is more resisting to noise than LAII

6.3 *VAR vs LAII in Matching Noisy Shapes*

A dynamic programming approach for shape matching, [1, 3, 4], will be employed in this section to measure the distance between shapes from the KIMIA benchmark database that are altered by noise. The methodology of matching curves, under the dynamic programming framework is as follows: Let $\boldsymbol{\alpha_1}$ and $\boldsymbol{\alpha_2}$ denote two curves with lengths normalized to 1 and c a correspondence between the points of the two curves such that $c(\boldsymbol{\alpha_1}(s_1)) = \boldsymbol{\alpha_2}(s_2)$ and $\mathscr{C}$ the set of all such correspondences. The distance between the two curves is measured by minimizing the energy functional:

$$E(\boldsymbol{\alpha_1}, \boldsymbol{\alpha_2}, c) = \int_{s\in[0,1]} |\phi_{\alpha_1}(s) - \phi_{\alpha_2}(c(s))|ds \tag{2}$$

with respect to the correspondence c. The shape distance, therefore, between the two curves $\boldsymbol{\alpha_1}$ and $\boldsymbol{\alpha_2}$ will be given by

$$\min_{c\in\mathscr{C}}\{E(\boldsymbol{\alpha_1}, \boldsymbol{\alpha_2}, c)\}\,. \tag{3}$$

This methodology is widely used in shape matching, details can be found in the references above. Research under development on the comparison of curves with respect to the minimization of functionals defined on matchings can be found in [5–8]. Two experiments are presented.

The first experiment measures the success of both methods by counting the number of correct matches from the 6 closets matches for all the 24 noisy shapes. 24 shapes from the KIMIA silhouette database, [9] are used, with a noise perturbation of $\sigma = 0.5$, on the boundary. For the encoding phase LAII was implemented as follows. Starting from a binary image of the shape to be encoded, first extract the contour. Then using a circular kernel (constructed as a binary image of a circle of radius 15, as is suggested in [1]) convolve the filter with the shape image only at the contour points. The values of the convolution at each of the contour points are the values of LAII at these points. Notice that if initially only the contour is given (as a list of points) and not the binary image of the shape, then the binary image has to be constructed before the LAII values can be calculated. This is mentioned because it plays a role in the complexity of the method.

Curves are discretized by sampling 100 equally spaced points on each. The choice of the initial correspondence between a pair of points on the two curves is a critical issue for dynamic programming. An exhaustive search for all the possible combinations for this initial correspondence raises the matching complexity from $O(n^2)$ to $O(n^3)$, n the number of contour points. An alternative technique was used; to use VAR for restricting the search of initial correspondences to the intuitive points. For this purpose two implementations of dynamic programming were examined. In the first implementation the local extrema of VAR are used as the only possibilities for an initial correspondence between the two curves and therefore all the combinations of these extreme points on the two curves are examined. The extreme points are chosen due to their intuitive interpretation with respect to the original shape. It has been shown that the first and second derivatives of VAR quantify other intuitive interpretations which can be used to selectively reduce the possible choices of points that will be used in dynamic programming as initial correspondences at the matching phase.

In the second implementation all the combinations for every 5th point on the two curves are examined, starting from a random position. This implementation has been used by other authors (for every 10th or 15th points in contours of 100 points)[4]. VAR and LAII are compared using the two implementations of dynamic programming with identical parameters for the matching phase for both methods. VAR/VAR in the experiment means using VAR as the shape descriptor and also VAR for choosing the initial correspondence for dynamic programming (first implementation), VAR/5P means using VAR as the shape descriptor but the initial correspondence for dynamic programming is examined for every 5th point on both curves (second implementation). LAII/VAR and LAII/5P mean using LAII as the shape descriptor with the same two implementations for choosing the possible starting points for dynamic programming.

In the table of Figure 7, columns are indexed by noisy versions of the KIMIA small dataset of 24 benchmark shapes. Noisy versions were produced by applying a normal to the tangent at each point perturbation to the boundary of the original shapes, the magnitude of which was drawn from a Gaussian distribution with mean 1 and variance 0.5. The KIMIA small silhouette database contains 6 silhouettes for each of the 4 classes of fish, hands, planes, and rabbits. For each noisy shape the 6 best matches (1–6 is the rank of the match) against the original 24 shapes in the database are shown as 6 quartets of rows. Each quartet represents the result for both methods under the two different implementations. The total number of correct matches at each rank, method, and implementation are arranged in the third column. The actual shape distance is the small number just above each shape.

VAR produces better results than LAII regardless of implementation but it is also confirmed that the second implementation does not improve the recognition of either method. The first implementation has order of complexity $O(\kappa n^2) = O(n^2)$, since κ depends on the intuitive characteristics of the shape (captured in the local extrema of VAR) and not directly on the number of points. The second implementation on the other hand has a complexity of $O(n/5 \times n^2) = O(n^3)$, therefore the first implementation is faster especially in cases of simple shapes that have a large number of points. The second implementation covers a larger search space since 20 candidate points (assuming contours of 100 points) must be examined for each curve in a total of $20 \times 20 = 400$ combinations for matching each pair of curves, where the average number of local extrema of each VAR is around 5 in a total of 25 combinations. Comparing the approaches it is confirmed that restricting the choices of initial correspondences through the VAR leads to better execution time and better recognition results at the same time. Compared to context independent heuristic searches (e.g., uniform sampling), this is because the search space has been reduced in an intuitive way, reducing the probability of false positive matches.

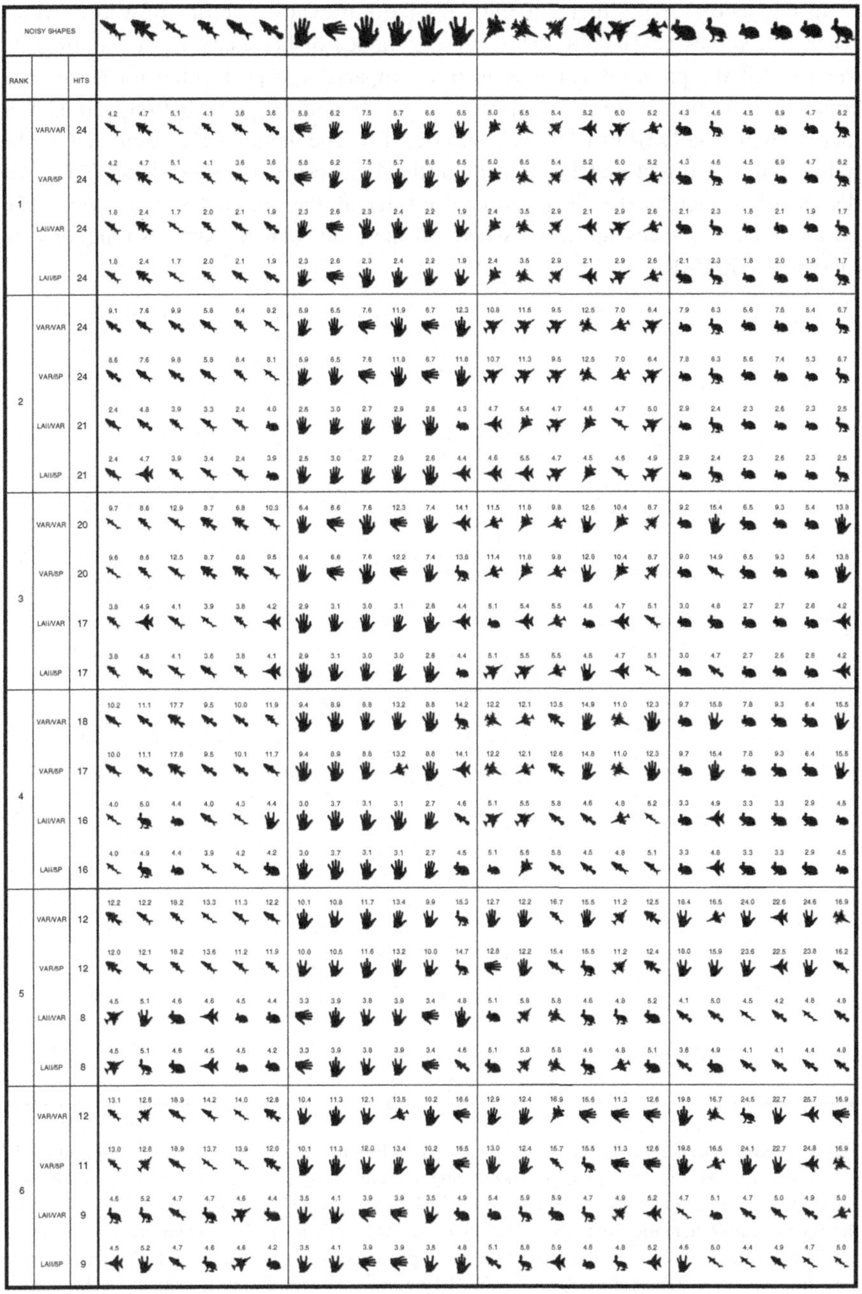

Fig. 7 Six closest matches to the noisy version of the KIMIA small data set of 24 shapes for two implementations of two different methods are shown. The actual shape distance is the number just above each shape. The first row holds the noisy images (query images) for Gaussian boundary perturbation of $\sigma = 0.5$. Each of the other rows holds the respective to the rank's index

In the second experiment of Figure 8 the retrieval capability of VAR/VAR is compared versus LAII/VAR for increasing values of boundary noise variance. In Figure 7 of the previous experiment the comparison is performed for a variance equal to 0.5. In the second experiment the boundary perturbation variance is increased gradually from 1.5 to 4.5 with step 1.0. The methods are compared by the number of the correct matches in each of the 4 closest matching orders. VAR/VAR demonstrates increased resistance to noise since all the first matches are correct for all the noise variances up to 3.5 where all the other matches are also higher than these of LAII/VAR.

Fig. 7 (continued) (first column) closest match to each of the noisy shapes for two different methods, VAR and LAII in two implementations using dynamic programming. The first implementation uses VAR to choose the initial correspondence for both methods (VAR/VAR and LAII/VAR) while the second implementation examines the correspondence exhaustively at each 5th contour point for both methods (VAR/5P and LAII/5P). VAR presents better results than LAII since all the first and second matches are correct for VAR (the noisy shapes match to shapes in the correct class), where three of the second matches are not correct for the Local Area Integral Invariant. The results for the other matches are also better for VAR compared to those of LAII. By comparing the implementations for both methods one can see that choosing the initial correspondence through VAR improves the performance but also the recognition accuracy since VAR selects context relevant starting points for dynamic programming

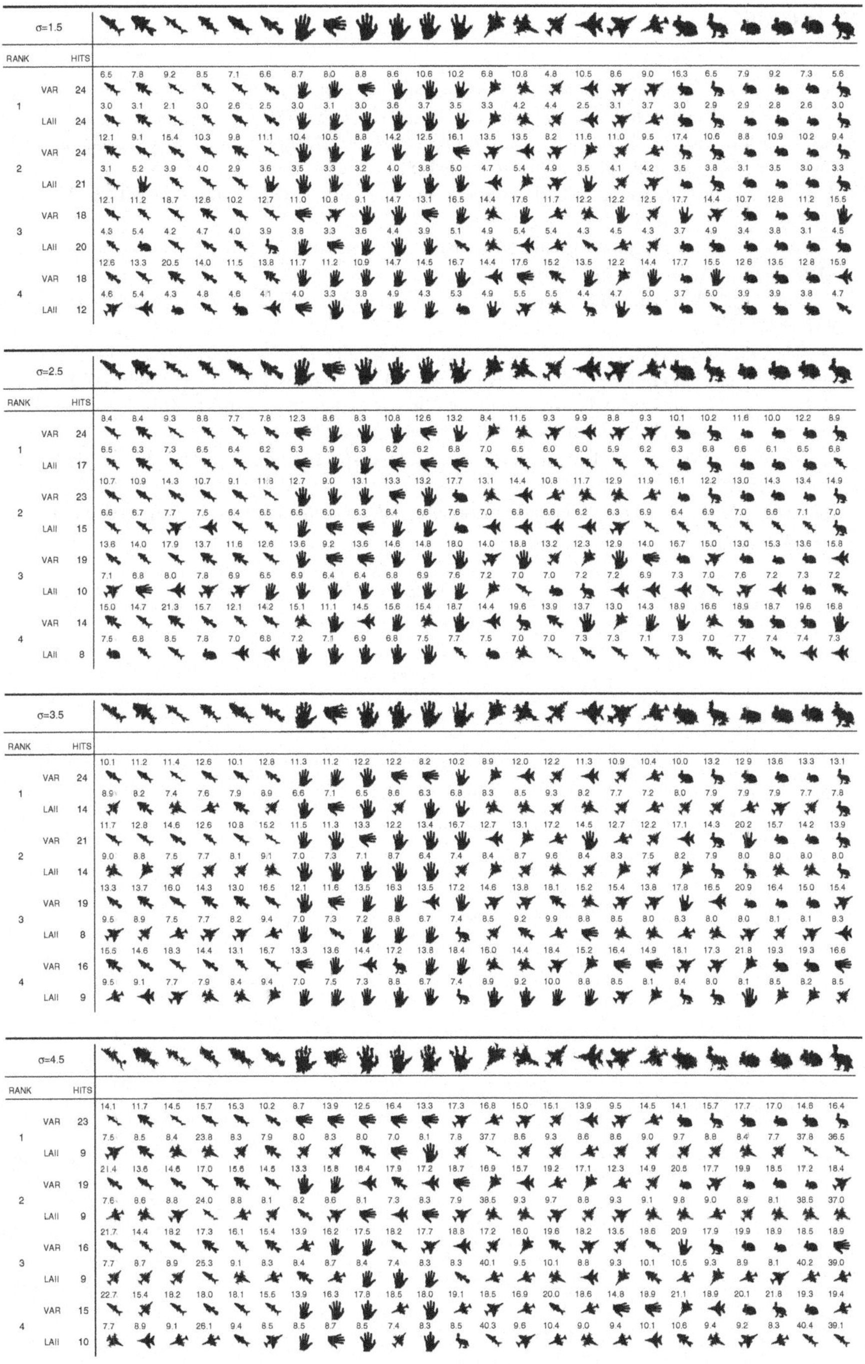

Fig. 8 Distance tables for the four best matches for increasing boundary noise. VAR/VAR (just VAR in the figure) and LAII/VAR (just LAII in the figure) are compared. The matching scores for LAII/VAR degrade with a faster pace compared to the scores of VAR/VAR which demonstrates increased resistance to noise

References

1. S. Manay, D. Cremers, B.-W. Hong, A. Yezzi, S. Soatto, Integral invariants for shape matching. IEEE Trans. Pattern Anal. Mach. Intell. **28**(10), 1602–1618 (2006)
2. K.A. Raftopoulos, S.D. Kollias, The Global-Local transformation for noise resistant shape representation. Comput. Vis. Image Underst. **115**(8), 1170–1186 (2011)
3. T.B. Sebastian, P.N. Klein, B.B. Kimia, On aligning curves. IEEE Trans. PAMI **25**, 116–125 (2003)
4. H. Ling, D. Jacobs, Shape classification using the inner-distance. IEEE Trans. Pattern Anal. Mach. Intell. **29**(2), 286–299 (2007)
5. P. Donatini, P. Frosini, Natural pseudo-distances between closed curves. Forum Math. **21**(6), 981–999 (2009)
6. D. Groisser, Certain optimal correspondences between plane curves II. Existence, local uniqueness, regularity, and other properties. Trans. Am. Math. Soc. **361**(6), 3001–3030 (2009)
7. D. Groisser, Certain optimal correspondences between plane curves I. Manifolds of shapes and bimorphisms. Trans. Am. Math. Soc. **361**(6), 2959–3000 (2009)
8. P.W. Michor, D. Mumford, Riemannian geometries on spaces of plane curves. J. Eur. Math. Soc. **8**(1), 1–48 (2006)
9. [Online]. Available: http://vision.lems.brown.edu/content/available-software-and-databases#Datasets-Shape

On Algorithms for Difference of Monotone Operators

Maede Ramazannejad, Mohsen Alimohammady, and Carlo Cattani

Abstract This review proposes a proximal algorithm for difference of two monotone operators in finite dimensional real Hilbert space. Our route begins with reviewing some properties of DC (difference of convex functions) programming and DCA (DC algorithms). Next, we recall some main results about a proximal point algorithm for DC programming.

1 Introduction

An important area of nonlinear analysis that emerged in the early 1960s [20–22, 40] is the theory of monotone operators. During nearly six decades, this field has reached a high level of maturity. Application of maximal monotone operators in some branches such as optimization, variational analysis, algorithms, mathematical economics is the reason why it has grown dramatically.

One of the most challenging issues in the theory of monotone operators is the problem of difference of monotone operators, because difference of two monotone operators is not always monotone. For this reason finding a zero of difference of two monotone operators has not been studied extensively. You can see the presence of this problem in DC programming, signal processing, machine learning, tomography, molecular biology, and optimization [7, 10, 25, 30, 32, 48, 57, 63].

M. Ramazannejad
Young Researchers and Elite Club, Islamic Azad University, Qaemshahr, Iran

M. Alimohammady
Department of Mathematics, Faculty of Mathematical Sciences, University of Mazandaran, Babolsar, Iran
e-mail: amohsen@umz.ac.ir

C. Cattani (✉)
Engineering School (DEIM), University of Tuscia, Viterbo, Italy
e-mail: cattani@unitus.it

N. J. Daras, T. M. Rassias (eds.), *Computational Mathematics and Variational Analysis*, Springer Optimization and Its Applications 159,
https://doi.org/10.1007/978-3-030-44625-3_21

The main object in this study is to offer a proximal algorithm for difference of two maximal monotone operators. It is the final section in our study but in the first step we try to review the features of the problem of minimizing difference of convex functions and the presented algorithms for this problem because finding the critical points of DC functions is a special case for finding the zeros of difference of two maximal monotone operators. The second step is about recalling a kind of DC programming and recommended proximal point algorithm for this.

2 DC Programming and DCA

This section is a small part of efforts of Pham Dinh Tao and Le Thi Hoai An in the way of development of DC programming and its applications in [48]. It is noteworthy that Pham Dinh Tao introduced DC programming and DCA in 1985 in their preliminary form.

The problem

$$\inf\{g(x) - h(x) : x \in \mathbb{R}^n\},$$

where g and h are convex functions, is called a *DC program*.

The importance of this problem to formulation of all most real-world optimization problems as DC programs cannot be exaggerated. The authors believe the main idea in convex analysis approach to DC programming is the use of the DC duality, which has been first studied by Toland in 1979 [61]. Though DC algorithms converge to a local solution generally, it is found from many literatures [3–6, 8–10, 47, 49] that DC algorithms converge quite often to a global solution. We see the different equivalent DC forms for primal and dual problems, because there is such an awareness that decompositions of the DC objective function may have an important influence on the qualities (robustness, stability, rate of convergence, and globality of sought solutions). In this regard, an interesting thing for guarantee of the convergence of DCA to a global solution is to check conditions on the choice of the DC decomposition and the initial point. We refer the reader to [3–6, 8–10, 47, 49] on applications of DCA in the study of many large-scale DC optimization problems and observing that DCA is more robust and efficient than related standard methods. Applying conjointly appropriate DC decompositions of convex functions and proximal regularization techniques [3, 50, 51] leads to proximal point algorithms [37, 53].

In this section, we want to express three outcomes of the main purposes in the paper of D.T. Pham and L.T.H. An [48]. To start, we will say the relationship between primal and dual solutions. We will continue argument by presenting of duality and local optimality conditions for DC optimization. Finally, the description of DCA will be reviewed.

We are now in a position to introduce some concepts and results about DC program, DCA, and their features.

Let $X = \mathbb{R}^n$, so the space X^* that is continuous dual of X can be identified with X. We assume X with norm $\|.\| = \langle ., . \rangle^{\frac{1}{2}}$ and that X and X^* are paired by $\langle ., . \rangle$. For an extended real value function $f : X \to \mathbb{R} \cup \{+\infty\}$, the domain of f is the set

$$\mathrm{dom} f = \{x \in \mathbb{R}^n : f(x) < +\infty\}.$$

The function f is said to be proper if its domain is nonempty.

If $B \subseteq X$, then the interior of B will be denoted by $\mathrm{int}B$ and if B is convex, then the relative interior of B will be denoted by $\mathrm{ri}B$.

Let $\Gamma_0(X)$ denote the set of all proper lower semicontinuous convex functions on X. The *indicator function* of subset C of X, written as $\chi_C(x)$, is defined at $x \in X$ by

$$\chi_C(x) = \begin{cases} 0 & x \in C, \\ +\infty & \text{otherwise.} \end{cases}$$

For a given function $f \in \Gamma_0(X)$,

$$f^*(x^*) = \sup\{\langle x, x^* \rangle - f(x) : x \in X\}$$

is *conjugate of the function* f which is a member of $\Gamma_0(X^*)$. Given a proper convex function f on X, $f : X \to \mathbb{R} \cup \{+\infty\}$, the *subdifferential* of f at x is given by

$$\partial f(x) = \{u \in X^* : f(y) \geq f(x) + \langle u, y - x \rangle \text{ for all } y \in X\}.$$

Also, $\mathrm{dom}\partial f = \{x \in X : \partial f(x) \neq \emptyset\}$ and $\mathrm{range}\partial f = \cup\{\partial f(x) : x \in \mathrm{dom}\partial f\}$. The *ε-subdifferential* of f was introduced by Brøndsted and Rockafellar in [23]. It is defined as:

$$\partial_\varepsilon f(x) := \{v \in X^* : f(y) \geq f(x) + \langle v, y - x \rangle - \varepsilon \text{ for all } y \in X\},$$

for any $\varepsilon \geq 0$, $x \in X$. Note that $\partial f(x) \subseteq \partial_\varepsilon f(x)$ for any $\varepsilon \geq 0$, $x \in X$. Here the convention $+\infty - (+\infty) = +\infty$ is followed.

Now we investigate the DC program

$$\text{(P)} \qquad \alpha = \inf\{f(x) := g(x) - h(x) : x \in X\},$$

where g and h belong to $\Gamma_0(X)$. Such a function f is called DC function on X and g, h are called its DC components. If g and h are in addition finite on all of X, then we say that $f = g - h$ is a finite DC function on X.

In [62], we observe that

$$\inf\{g(x) - h(x) : x \in X\} = \inf\{h^*(x^*) - g^*(x^*) : x^* \in X^*\}.$$

With such an equality in hand we are ready to start the main argument. The problem

$$\text{(D)} \qquad \alpha = \inf\{h^*(x^*) - g^*(x^*) : x^* \in X^*\}$$

is called *dual problem*. The attentive reader can see the perfect symmetry between primal and dual programs (P) and (D). Equivalence between problems (P) and (D) makes the reader relaxed to solve the easier problem.

Notice that since α is finite:

$$\text{dom}g \subset \text{dom}h \quad \text{and} \quad \text{dom}h^* \subset \text{dom}g^*. \tag{1}$$

We consider assumption (1) throughout this section.

A point $\bar{x}$ is called *local minimizer* of $g-h$ if $g(\bar{x})-h(\bar{x})$ is finite, in other words $\bar{x} \in \text{dom}g \cap \text{dom}h$, and there exists a neighborhood U of $\bar{x}$ such that

$$g(\bar{x}) - h(\bar{x}) \le g(x) - h(x), \quad \forall x \in U. \tag{2}$$

Under the convention $+\infty - (+\infty) = +\infty$, the property (2) is equivalent to

$$g(\bar{x}) - h(\bar{x}) \le g(x) - h(x), \quad \forall x \in U \cap \text{dom}g.$$

We say that $\bar{x}$ is a critical point of $g - h$ if $\partial g(\bar{x}) \cap \partial h(\bar{x}) \neq \emptyset$.

Here $\mathscr{P}$ and $\mathscr{D}$ denote the solution sets of problems (P) and (D), respectively. Let

$$\mathscr{P}_l = \{x \in X : \partial h(x) \subset \partial g(x)\},$$
$$\mathscr{D}_l = \{x^* \in X^* : \partial g^*(x^*) \subset \partial h^*(x^*)\}.$$

The first result to present is the following.

Theorem 2.1 *[3, 31, 46, 50]*

(i) $x \in \mathscr{P}$ *if and only if* $\partial_\varepsilon h(x) \subset \partial_\varepsilon g(x) \quad \forall \varepsilon > 0$.
(ii) *Dually,* $x^* \in \mathscr{D}$ *if and only if* $\partial_\varepsilon g^*(x^*) \subset \partial_\varepsilon h^*(x^*) \quad \forall \varepsilon > 0$.
(iii) $\cup\{\partial h(x) : x \in \mathscr{P}\} \subset \mathscr{D} \subset \textit{dom}h^*$.
The first inclusion becomes equality if g^* *is subdifferentiable in* $\mathscr{D}$ *(in particular if* $\mathscr{D} \subset ri(\textit{dom}g^*)$ *or if* g^* *is subdifferentiable in dom* h^**).*
In this case $\mathscr{D} \subset (\textit{dom}\partial g^* \cap \textit{dom}\partial h^*)$.
(iv) $\cup\{\partial g^*(x^*) : x^* \in \mathscr{D}\} \subset \mathscr{P} \subset \textit{dom}g$.

The proof of the properties (i) and (ii) appears in J.B. Hiriart-Urruty's paper [31], based on the behavior of the ε-directional derivative of a convex function as a function of the parameters ε. Also the ideas behind the proofs of properties (iii) and (iv) due to D.T. Pham and L.T.H. An. These proofs based on the theory of subdifferential for convex functions [3, 46, 50].

The property (i) is a difficult way to reach the global solution to problem (P). You can analyze the meaningful relationships between primal and dual solutions in (iii) and (iv). One noteworthy conclusion to be drawn from (ii) and (iv) above is the fact that solving the problem (P) leads to solving the problem (D) and vice versa.

The fundamental results linking duality and local optimality conditions for DC program are the following; proofs will not be presented. These beautiful results are cited of D.T. Pham and L.T.H. An [48].

Theorem 2.2 *[3, 31, 46, 61]*

(i) *If $\bar{x}$ is a local minimizer of $g - h$, then $\bar{x} \in \mathscr{P}_l$.*

(ii) *Let $\bar{x}$ be a critical point of $g - h$ and $\bar{y}^* \in \partial g(\bar{x}) \cap \partial h(\bar{x})$. Let U be a neighborhood of $\bar{x}$ such that $U \cap domg \subset dom\partial h$. If for any $x \in U \cap domg$ there is $y^* \in \partial h(x)$ such that*

$$h^*(y^*) - g^*(y^*) \geq h^*(\bar{y}^*) - g^*(\bar{y}^*),$$

then $\bar{x}$ is a local minimizer of $g - h$. More precisely,

$$g(x) - h(x) \geq g(\bar{x}) - h(\bar{x}), \quad \forall x \in U \cap domg.$$

Corollary 1 (Sufficient Local Optimality) *Let $\bar{x}$ be a point that admits a neighborhood U such that $\partial h(x) \cap \partial g(\bar{x}) \neq \emptyset$, $\forall x \in U \cap domg$. Then $\bar{x}$ is a local minimizer of $g - h$. More precisely,*

$$g(x) - h(x) \geq g(\bar{x}) - h(\bar{x}), \quad \forall x \in U \cap domg.$$

Corollary 2 (Sufficient Strict Local Optimality) *If $\bar{x} \in int(dom\, h)$ verifies $\partial h(\bar{x}) \subset int(\partial g(\bar{x}))$, then $\bar{x}$ is a strict local minimizer of $g - h$.*

Corollary 3 (DC Duality Transportation of a Local Minimizer) *Let $\bar{x} \in dom\partial h$ be a local minimizer of $g - h$ and let $\bar{y}^* \in \partial h(\bar{x})$, i.e. $\partial h(\bar{x})$ is nonempty and $\bar{x}$ admits a neighborhood U such that*

$$g(x) - h(x) \geq g(\bar{x}) - h(\bar{x}), \quad \forall x \in U \cap domg.$$

If

$$\bar{y}^* \in int(domg^*) \quad and \quad \partial g^*(\bar{y}^*) \subset U, \tag{3}$$

((3) holds if g^* is differentiable at $\bar{y}^*$), then $\bar{y}^*$ is a local minimizer of $h^* - g^*$.

Now we are ready to describe DCA for general DC programs. How? Well consider the problem

$$(S(x)) \quad \inf\{h^*(y^*) - g^*(y^*) : \; y^* \in \partial h(x)\},$$

for any fixed $x \in X$. By the conjugate definition, this problem is equivalent to

$$\inf\{\langle x, y^* \rangle - g^*(y^*) : \; y^* \in \partial h(x)\}. \tag{4}$$

Also consider the problem

$$(T(y^*)) \qquad \inf\{g(x) - h(x) : x \in \partial g^*(y^*)\},$$

for any fixed $y^* \in X^*$. For the same reason as (4), above problem is equivalent to

$$\inf\{\langle x, y^*\rangle - h(x) : x \in \partial g^*(y^*)\}. \tag{5}$$

Suppose $\mathscr{S}(x)$ and $\mathscr{T}(y^*)$ denote the solution sets of problems $(S(x))$ and $(T(y^*))$, respectively.

There is a face of DCA based on problems $(S(x))$ and $(T(y^*))$ which with an initial point $x_0 \in \mathrm{dom} g$ makes two sequences $\{x_k\}$ and $\{y_k^*\}$ such that

$$y_k^* \in \mathscr{S}(x_k), \quad x_{k+1} \in \mathscr{T}(y_k^*). \tag{6}$$

However the problems $S(x_k)$ and $(T(y_k^*))$ are simpler than (P) and (D), they are practically hard to solve. Therefore authors in [48] offered the simplified form of DCA as follows:

Start from any point $x_0 \in \mathrm{dom} g$ and consider the recursive process

$$y_k^* \in \partial h(x_k); \quad x_{k+1} \in \partial g^*(y_k^*).$$

This method aims at providing two sequences $\{x_k\}$ and $\{y_k^*\}$ which are easy to calculate and satisfy

- the sequences $(g - h)(x_k)$ and $(h^* - g^*)(y_k^*)$ are decreasing;
- every limit point x (resp. y^*) of the sequence $\{x_k\}$ (resp. $\{y_k^*\}$) is a critical point of $g - h$ (resp. $h^* - g^*$).

Whenever DCA produces the sequences $\{x_k\}$ and $\{y_k^*\}$ as described above, so

$$\{x_k\} \subset \mathrm{range}\partial g^* = \mathrm{dom}\partial g \quad \text{and} \quad \{y_k^*\} \subset \mathrm{range}\partial h = \mathrm{dom}\partial h^*.$$

Hence obviously sequences $\{x_k\}$ and $\{y_k\}$ in DCA are well defined if and only if

$$\mathrm{dom}\partial g \subset \mathrm{dom}\partial h \quad \text{and} \quad \mathrm{dom}\partial h^* \subset \mathrm{dom}\partial g^*.$$

You can find the convergence of DCA for general DC programs in [48, Theorem 3].

3 Proximal Point Algorithm for DC Program

The heart and soul of this section are each devoted to recalling one kind of proximal point algorithm for minimizing the difference of a nonconvex function and a convex function introduced by Nguyen Thai An and Nguyen Mau Nam [11]. We quote several of their results and urge the reader to carefully study their original paper.

The structure of the problem of this section is flexible enough to include the problem of minimizing a smooth function on a closed set or minimizing a DC function. On doing so, we will study the convergence results of this algorithm.

The first work regarding the proximal minimization algorithm is due to Martinet [38, 39]. Proximal minimization was extended to the general proximal point algorithm for finding the zero of an arbitrary maximal monotone operator by Rockafellar [53]. There are many papers on applying different proximal algorithms to special problems, such as loss minimization in machine learning [15, 19, 29, 33], optimal control [44], energy management [34], and signal processing [28].

Neal Parikh and Stephen Boyd in their monograph [45] mentioned some reasons of many reasons to study proximal algorithms. Here, we recall them. First, these algorithms work under extremely general conditions, including cases where the functions are nonsmooth and extended real valued. Second, they can be fast, since there can be simple proximal operators for functions that are otherwise challenging to handle in an optimization problem. Third, they are amenable to distributed optimization, so they can be used to solve very large-scale problems. Finally, they are often conceptually and mathematically simple, so they are easy to understand, derive, and implement for a particular problem. Indeed, many proximal algorithms can be interpreted as generalizations of other well known and widely used algorithms.

Consider concepts and notations as described in previous section but some notations, special to the present discussion, ought to be introduced.

Let f be a function from X into an extended real line $\mathbb{R} \cup \{+\infty\}$, finite at x. A set

$$\partial^F f(x) = \{x^* \in X^* : \liminf_{u \to x} \frac{f(u) - f(x) - \langle x^*, u - x\rangle}{\|u - x\|} \geq 0\} \tag{7}$$

is called *Fréchet subdifferential* of f at x. Its elements are sometimes referred to as Fréchet subgradients. If $x \notin \operatorname{dom} f$, then we set $\partial^F f(x) = \emptyset$.

The set (7) is closed and convex, but Fréchet subdifferential mapping does not have a closed graph.

Employing a limiting "robust regularization" procedure over the subgradient mapping $\partial^F f(.)$ leads us to the subdifferential of f at x defined by

$$\begin{aligned}\partial^L f(x) := \limsup_{x_k \xrightarrow{f} x} \partial^F f(x) = \{x^* \in X^* : \exists x_k \xrightarrow{f} x \text{ and } x_k^* \in \partial^F f(x_k) \\ \text{with } x_k^* \to x\},\end{aligned} \tag{8}$$

where $x_k \xrightarrow{f} x$ means that $x_k \to x$ and $f(x_k) \to f(x)$. Set (8) which is closed is called *limiting/Mordukhovich subdifferential* of f at x.

It is easy to see that $\partial^F f(x) \subset \partial^L f(x)$ for every $x \in X$. If the function f is differentiable, then the Fréchet subdifferential reduces to the derivative. The limiting subdifferential of f at x reduces to $\{\nabla f(x)\}$ if f is continuously differentiable on a neighborhood of x. Also the Fréchet and the limiting subdifferential coincide with the subdifferential in the sense of convex analysis if f is convex.

It needs be remarked that for a nonempty subset K of X, the normal cone or normality operator N_K of K is defined as

$$N_K(x) = \begin{cases} \{x^* \in X^* : \langle x^*, u - x\rangle \leq 0 \ \forall u \in K\} & \text{if } x \in K, \\ \emptyset & \text{otherwise.} \end{cases}$$

We can find in [24, Proposition 3.6.2], for a nonempty, closed, and convex subset $K \subset X$, the following properties hold:

(a) $N_K = \partial \chi_K$.
(b) $N_K(x)$ is closed and convex for all $x \in X$.
(c) $N_K(x)$ is a cone for all $x \in K$.

For this K now let $P_K : X \to K$ the *orthogonal projection* onto K. It is easy to check that

$$P_K(x) = (I + N_K)^{-1}(x) = \{u \in K : \|x - u\| = \text{dist}(x; K)\},$$

where I is an identity map and dist$(x; K)$ is used to denote the distance from x to K, i.e. $\text{dist}(x; K) = \inf_{x \in K} \|x - u\|$.

Clarke, in [26], showed that for locally Lipschitz functions, the *Clarke subdifferential* admits the simple presentation

$$\partial^C f(x) = co\, \partial^L f(x),$$

where $co\ K$ is the convex hull of an arbitrary set $K \subset X$. Some important consequences of this is the following.

Proposition 3.1 ([56, Exercise 8.8]) *Let $f = g + h$, where g is lower semicontinuous and h is continuously differentiable on a neighborhood of x. Then*

$$\partial^F f(x) = \partial^F g(x) + \nabla h(x) \quad \textit{and} \quad \partial^L f(x) = \partial^L g(x) + \nabla h(x).$$

Proposition 3.2 ([56, Theorem 10.1]) *If a lower semicontinuous function $f : X \to \mathbb{R} \cup \{+\infty\}$ has a local minimum at $\bar{x} \in dom f$, then $0 \in \partial^F f(\bar{x}) \subset \partial^L f(\bar{x})$. In the convex case, this condition is not only necessary for a local minimum but also sufficient for a global minimum.*

We state the definition of the *Kurdyka–Łojasiewicz property* from [12, 13].

The lower semicontinuous function $f : X \to \mathbb{R} \cup \{\infty\}$ has the Kurdyka–Łojasiewicz property at $\bar{x} \in \text{dom}\partial^L f$, if there exist $\eta \in (0, \infty]$, a neighborhood

U of $\bar{x}$ and a continuous concave function $\varphi : [0, \eta) \to \mathbb{R}_+$ such that $\varphi(0) = 0$, $\varphi \in C^1(0, \eta)$, for all $s \in (0, \eta)$ it is $\varphi'(s) > 0$, and for all $x \in U \cap [f(\bar{x}) < f < f(\bar{x}) + \eta]$ the Kurdyka–Łojasiewicz inequality holds, i.e.

$$\varphi'(f(x) - f(\bar{x}))\, \mathrm{dist}(0, \partial^L f(x)) \geq 1.$$

We cannot present the results of [11] without recalling the concept of proximal mapping [56]. Let $g : X \to \mathbb{R} \cup \{+\infty\}$ be a proper lower semicontinuous function. The *Moreau proximal mapping* with regularization parameter $t > 0$, $\mathrm{prox}_t^g(x)$, is defined by

$$\mathrm{prox}_t^g = \mathrm{argmin}_{u \in X}\{g(u) + \frac{t}{2}\|u - x\|^2\}.$$

When g is the indicator function χ_K, where K is a closed nonempty convex set, the proximal mapping of g reduces to projection mapping.

The following proposition shows us some conditions under which prox_t^g is well defined.

Proposition 3.3 ([18]) *If $g : X \to \mathbb{R} \cup \{+\infty\}$ be a proper lower semicontinuous function with* $\inf_{x \in X} g(x) > -\infty$. *Then, for every $t \in (0, +\infty)$, the set $prox_t^g(x)$ is nonempty and compact for every $x \in X$.*

Authors in [11] concentrated their attentions on the convergence analysis of a proximal point algorithm for solving nonconvex optimization problems of the following type

$$\min\{f(x) = g_1(x) + g_2(x) - h(x) : x \in X\}, \tag{9}$$

where $g_1 : X \to \mathbb{R} \cup \{+\infty\}$ is proper and lower semicontinuous, $g_2 : X \to \mathbb{R}$ is differentiable with L-Lipschitz gradient, and $h : X \to \mathbb{R}$ is convex. The problem (9) is so flexible which it can be changed to the DC problem

$$\min\{f(x) = g(x) - h(x) : x \in X\}, \tag{10}$$

where $g \in \Gamma_0(X)$ and $h : X \to \mathbb{R}$ is convex. Also the problem (9) includes the following problem on a closed constraint set,

$$\min\{g(x) : x \in K\}. \tag{11}$$

There exists a meaningful relationship between the Moreau proximal operator $\mathrm{prox}_{\frac{1}{\lambda}}^g$ and the subdifferential operator ∂g as follows:

$$\mathrm{prox}_{\frac{1}{\lambda}}^g = (I + \lambda \partial g)^{-1}. \tag{12}$$

The right-hand side of (12) is called *resolvent operator* for ∂f with parameter $\lambda > 0$. You see that the proximal operator is the resolvent of the subdifferential operator. Regarding the problem (11), the *proximal point algorithm* which is also called *proximal minimization algorithm* or *proximal iteration* is

$$x_{k+1} := \operatorname{prox}^{g}_{\frac{1}{\lambda}}(x_k), \tag{13}$$

where k is the iteration counter and x_k is the kth iterate of the algorithm.

We are ready for recalling a necessary optimality condition for minimizing the differences of functions in the nonconvex setting.

Proposition 3.4 ([41, Proposition 4.1]) *Consider the difference function $f = g - h$, where $g : X \to \mathbb{R} \cup \{+\infty\}$ and $h : X \to \mathbb{R}$ are lower semicontinuous functions. If $\bar{x} \in \operatorname{dom} f$ is a local minimizer of f, then we have the inclusion*

$$\partial^F h(\bar{x}) \subset \partial^F g(\bar{x}).$$

If in addition, h is convex, then $\partial h(\bar{x}) \subset \partial^L g(\bar{x})$.

In sequel, you see the optimality condition associated with the problem (9).

Proposition 3.5 *If $\bar{x} \in \operatorname{dom} f$ is a local minimizer of the function f considered in* (9)*, then*

$$\partial h(\bar{x}) \subset \partial^L g_1(\bar{x}) + \nabla g_2(\bar{x}). \tag{14}$$

A stationary point is any point like $\bar{x} \in \operatorname{dom} f$ that satisfies (14). Since the way to achieve condition (14) is tough, it can be relaxed to

$$[\partial^L g_1(\bar{x}) + \nabla g_2(\bar{x})] \cap \partial h(\bar{x}) \neq \emptyset. \tag{15}$$

The point $\bar{x}$ at the above condition is called *critical point*, so it is clear that every stationary point is a critical point. Therefore, if $0 \in \partial^L f(\bar{x})$, then $\bar{x}$ is a critical point of f, because in accordance with [41, Corollary 3.4] at any point $\bar{x}$ such that $g_1(\bar{x}) < +\infty$, we have

$$\partial^L (g_1 + g_2 - h)(\bar{x}) \subset \partial^L g_1(\bar{x}) + \nabla g_2(\bar{x}) - \partial h(\bar{x}).$$

It is time for recalling the main aim of this section. Authors in [11] introduced the generalized proximal point algorithm (GPPA) below to solve the problem (9):

(1) Initialization: Choose $x_0 \in \operatorname{dom} g_1$ and a tolerance $\varepsilon > 0$. Fix any $t > L$.
(2) Find

$$y_k \in \partial h(x_k).$$

(3) Find x_{k+1} as follows

$$x_{k+1} \in \text{prox}_t^{g_1}(x_k - \frac{\nabla g_2(x_k) - y_k}{t}). \tag{16}$$

(4) If $\|x_k - x_{k+1}\| \leq \varepsilon$, then exit. Otherwise, increase k by 1 and go back to step 2.

The definition of proximal mapping tells us (16) is equivalent to

$$x_{k+1} \in \text{argmin}_{x \in X}\{g_1(x) - \langle y_k - \nabla g_2(x_k), x - x_k\rangle + \frac{t}{2}\|x - x_k\|^2\}. \tag{17}$$

In the following a few of the most attractive conditions assuring the convergence of GPPA are presented without proof.

Theorem 3.3 *Consider the GPPA for solving* (9) *in which* $g_1(x) : X \to \mathbb{R} \cup \{+\infty\}$ *is proper and lower semicontinuous with* $\inf_{x\in X} g_1(x) > -\infty$, $g_2(x) : X \to \mathbb{R}$ *is differentiable with L-Lipschitz gradient, and* $h : X \to \mathbb{R}$ *is convex. Then:*

(i) For any $k \geq 1$, *we have*

$$f(x_k) - f(x_{k+1}) \geq \frac{t - L}{2}\|x_k - x_{k+1}\|^2. \tag{18}$$

(ii) If $\alpha = \inf_{x\in X} f(x) > -\infty$, *then* $\lim_{k\to+\infty} f(x_k) = \bar{l} \geq \alpha$ *and* $\lim_{k\to+\infty}\|x_k - x_{k+1}\| = 0$.

(iii) If $\alpha = \inf_{x\in X} f(x) > -\infty$ *and* $\{x_k\}$ *is bounded, then every cluster point of* $\{x_k\}$ *is a critical point of* f.

Proposition 3.6 *Suppose that* $\inf_{x\in X} f(x) > -\infty$, f *is proper and lower semicontinuous. If the GPPA sequence* $\{x_k\}$ *has a cluster point* $\bar{x}$, *then* $\lim_{k\to+\infty} f(x_k) = f(\bar{x})$. *Thus,* f *has the same value at all cluster points of* $\{x_k\}$.

A few further remarks on the convergence of GPPA are in order.

- If $h(x) = 0$, then the GPPA coincides with the proximal forward–backward algorithm for minimizing $f = g_1 + g_2$ in [12]. When $h(x) = 0$ and g_1 is the indicator function χ_K, where K is a nonempty closed set, then the GPPA turns into the projected gradient method (PGM) for minimizing the smooth function g_2 on a closed constraint set K:

$$x_{k+1} = P_K(x_k - \frac{1}{t}\nabla g_2(x_k)).$$

- Whenever $g_2 = 0$, the GPPA reduces to the PPA with a constant step size suggested in [58, 59]. Authors by setting $t = \frac{L}{2}$ could recover the convergence result of the primal sequence $\{x_k\}$ generated by the DCA for minimizing the DC function $f = g_1 - h$ in [47, Theorem 3.7].

In the following, authors presented some sufficient conditions that warrant the convergence of the sequence $\{x_k\}$ generated by the GPPA. Denote by Ω the set of cluster points of the sequence $\{x_k\}$.

Theorem 3.4 *Suppose that* $\inf_{x \in X} f(x) > -\infty$, *and* f *is lower semicontinuous. Suppose further that* ∇h *is L(h)-Lipschitz continuous and* f *has the Kurdyka–Łojasiewicz property at any point* $x \in dom f$. *If* $\Omega \neq \emptyset$, *then the GPPA sequence* $\{x_k\}$ *converges to a critical point of* f.

In the following theorem, you see some another sufficient conditions that warrant the convergence of the sequence $\{x_k\}$ generated by the GPPA.

Theorem 3.5 *Consider the difference of function* $f = g - h$ *with* $\inf_{x \in X} f(x) > -\infty$. *Suppose that* g *is differentiable and* ∇g *is L-Lipschitz continuous,* f *has the strong Kurdyka-Łojasiewicz property at any point* $x \in dom f$, *and* h *is a finite convex function. If* $\Omega \neq \emptyset$, *then the GPPA sequence* $\{x_k\}$ *converges to a critical point of* f.

Authors in [36] investigated the convergence of DC algorithm for DC programming with subanalytic data. From [16, Theorem 3.1] and [17, Corollary 16], it is obtained that a lower semicontinuous subanalytic function satisfies Kurdyka–Łojasiewicz property with a specific form of the function φ. From this you can find results in Theorems 3.4 and 3.5 as an extension of [36, Theorem 3.1].

Finally in the following proposition, authors exhibited some sufficient conditions for the set Ω to be nonempty.

Proposition 3.7 *Consider the function* $f = g - h$, *where* $g = g_1 + g_2$ *in* (9). *Let* $\{x_k\}$ *be the sequence generated by the GPPA for solving* (10). *The set of critical points* Ω *of* $\{x_k\}$ *is nonempty if one of the following conditions is satisfied:*

(i) *For any* α, *the lower level set* $L_{\leq \alpha} := \{x \in X : \ f(x) \leq \alpha\}$ *is bounded.*
(ii) $\liminf_{\|x\| \to +\infty} h(x) = +\infty$ *and* $\liminf_{\|x\| \to +\infty} \frac{g(x)}{h(x)} > 1$.

4 Inertial Proximal Algorithm for Difference of Two Maximal Monotone Operators

This section deals with presenting a proximal algorithm for difference of two maximal monotone operators. Before studying we investigate a bite of mathematical sociology.

Here our world will be extended to Hilbert space H. The notation $\langle .,. \rangle$ will be used for inner product in $H \times H$ and $\|.\|$ for the corresponding norm. A set valued operator $T : H \to 2^H$ is said to be *monotone* if

$$\langle x^* - y^*, x - y \rangle \geq 0, \quad \forall (x, x^*), \ (y, y^*) \in G(T),$$

wherein $G(T) := \{(x, y) \in H \times H;\ y \in Tx\}$ is the graph of T. The domain of T is $D(T) := \{x \in H;\ T(x) \neq \emptyset\}$.

A monotone operator T is called *maximal monotone* if its graph is maximal in the sense of inclusion.

Associated with a given monotone operator T, the resolvent operator for T and parameter $\lambda > 0$ is $J_\lambda^T := (I + \lambda T)^{-1}$. The resolvent J_λ^T of a monotone operator T is a single-valued nonexpansive map from $Im(I + \lambda T)$ to H [14, Proposition 3.5.3]. Moreover, the resolvent has full domain precisely when T is maximal monotone. For any $x \in H$, $\lim_{\lambda \to 0} J_\lambda^T(x) = Proj_{\overline{D(T)}} x$, wherein $Proj_{\overline{D(T)}}$ is the orthogonal projection on the closure of the domain of T. One of the best known approaches in the theory of optimization that is related to resolvent operators is *Yosida approximate* $T_\lambda := \frac{(I - J_\lambda^T)}{\lambda}$ of a maximal monotone operator T which satisfies in:

(i) For all $x \in H$, $T_\lambda(x) \in T(J_\lambda^T(x))$,
(ii) T_λ is Lipschitz with constant $\frac{1}{\lambda}$ and maximal monotone,
(iii) $T_\lambda(x)$ converges strongly to $T(x)$ as $\lambda \to 0$, for $x \in D(T)$,
(iv) $\|T_\lambda(x)\| \leq \|T^0(x)\|$ for every $x \in D(T)$, $\lambda > 0$, where T^0 is *minimal selection*

$$T^0(x) := \{y \in T(x);\ \|y\| = \min_{z \in T(x)} \|z\|\}, \qquad x \in D(T).$$

Consider the problem

$$\text{find } x \in H \text{ such that } 0 \in T(x) - S(x), \tag{19}$$

where $T, S : H \to 2^H$ are two maximal monotone operators on finite dimensional real Hilbert space H and it is equivalent to the problem

$$\text{find } x \in H \text{ such that } T(x) \cap S(x) \neq \emptyset. \tag{20}$$

Regarding the importance of this problem as mentioned, finding the critical points of the difference of two convex functions is the special case of finding the zeros of difference of two maximal monotone operators. Actually, an algorithm for difference of two maximal monotone operators plays a central role in the study of DC programming. The latter studies are limited to Moudafi [42, 43]. By [43], a regularization of the problem (19) is

$$\text{find } x \in H \text{ such that } 0 \in T(x) - S_\lambda(x). \tag{21}$$

For finding a solution of (19) Moudafi [43] suggested a sequence $\{x_n\}$ by

$$x_{n+1} = J_{\mu_n}^T (x_n + \mu_n S_{\lambda_n} x_n) \quad \forall n \in \mathbb{N}, \tag{22}$$

where $\mu_n > 0$ and x_0 is an initial point.

Here, the problem (19) is studied via the generalization of Moudafi's algorithm in [43] as the following:

$$x_{k+1} = J^T_{\beta_k}(x_k + \alpha_k(x_k - x_{k-1}) + \beta_k S_{\mu_k} x_k) \qquad \forall k \in \mathbb{N}, \tag{23}$$

with starting points $x_0, x_1 \in H$ and sequences $\{\mu_k\}$, $\{\alpha_k\}$, and $\{\beta_k\} \subset [0, +\infty)$ such that

(a) $\lim_{k\to+\infty} \mu_k = 0$;
(b) $\sum_{k=1}^{+\infty} \frac{\beta_k}{\mu_k} < +\infty$;
(c) $\lim_{k\to+\infty} \frac{\alpha_k}{\beta_k} = 0$;

also we suppose that

(d) $\sum_{k=1}^{+\infty} \alpha_k \|x_k - x_{k-1}\| < +\infty$;
(e) $\lim_{k\to+\infty} \frac{\|x_{k+1}-x_k\|}{\beta_k} = 0$.

We note that (23) is emanated from the evolution equation

$$x''(t) + \gamma x'(t) + \nabla f(x(t)) - \nabla g(x(t)) = 0, \tag{24}$$

where $\gamma > 0$ and algorithm (22) can be inspired from

$$x'(t) + \nabla f(x(t)) - \nabla g(x(t)) = 0, \tag{25}$$

in which both $f, g : H \to \mathbb{R}$ are differentiable convex functions and $\nabla f(x(t))$ and $\nabla g(x(t))$ are operators T and S in (19), respectively.
If $\nabla g(x(t)) = 0$, then (24) is *heavy ball with friction* system or (HBF) and (23) is equivalent to the standard gradient descent iteration (22) with an additional *inertia term* or *momentum term* $\alpha_k(x_k - x_{k-1})$. By the inertia term, the convergence of the solution trajectories of the (HBF) system to a stationary point of f can be faster than those of the first-order system (25) when $\nabla g(x(t)) = 0$ [52].
Another important advantage of algorithm (23) over algorithm (22) is using the condition of the local boundedness of S instead of the boundedness in (22).
In this section, we present different conditions under which (23) converges to a solution of (19). Now, we recall some required results and definitions.

A set valued operator $T : H \to 2^H$ is *locally bounded* at $\bar{x}$ if there exists a neighborhood U of $\bar{x}$ such that the set $T(U)$ is bounded.

Lemma 1 ([55]) *A maximal monotone operator T is locally bounded at a point $\bar{x} \in D(T)$ if and only if $\bar{x}$ belongs to the interior of $D(T)$.*

A set valued operator $T : H \to 2^H$ is upper semicontinuous at $\bar{x}$ if for any positive $\varepsilon > 0$ there exists $\delta > 0$ such that

$$\|x - \bar{x}\| \le \delta \Rightarrow T(x) \subseteq T(\bar{x}) + B(0, \varepsilon). \tag{26}$$

Lemma 2 ([1]) *Suppose that E is a Banach space. The maximal monotone operator $T : E \to 2^{E^*}$ where E^* is the dual space of E is demiclosed, i.e. the following conditions hold.*

(1) If $\{x_k\} \subset E$ converges strongly to x_0 and $\{u_k \in T(x_k)\}$ converges weak to u_0 in E^*, then $u_0 \in T(x_0)$.*
(2) If $\{x_k\} \subset E$ converges weakly to x_0 and $\{u_k \in T(x_k)\}$ converges strongly to u_0 in E^*, then $u_0 \in T(x_0)$.

Lemma 3 ([35]) *Suppose that $\{a_n\}$, $\{b_n\}$, and $\{c_n\}$ are three sequences of nonnegative numbers such that*

$$a_{n+1} \leq (1 + b_n)a_n + c_n \quad for\ all\ n \geq 1.$$

If $\sum_{n=1}^{\infty} b_n < +\infty$ and $\sum_{n=1}^{\infty} c_n < +\infty$, then $\lim_{n\to\infty} a_n$ exists.

In the following, we improve the conditions of Theorem 2.1 in [43].

Theorem 4.6 *Assume that S is locally bounded on $\overline{D(S)}$ and the solution set Ω of problem* (19) *is nonempty. If the conditions $(a), \ldots, (e)$ satisfy and $D(T) \subset D(S)$, then the sequence $\{x_k\}$ generated by* (23) *converges to a solution of* (19).

Proof Take $x^* \in \Omega$. According to (20), there exists $y^* \in T(x^*) \cap S(x^*)$ and from (23), $x^* = J_{\beta_k}^T(x^* + \beta_k y^*)$. From the triangular inequality, (iv), nonexpansivity of $J_{\beta_k}^T$ and the fact that S_{μ_k} is also nonexpansive with constant $\frac{1}{\mu_k}$, one quickly deduces that

$$\begin{aligned}
\|x_{k+1} - x^*\| &= \|J_{\beta_k}^T(x_k + \alpha_k(x_k - x_{k-1}) + \beta_k S_{\mu_k} x_k) - J_{\beta_k}^T(x^* + \beta_k y^*)\| \\
&\leq \|x_k + \alpha_k(x_k - x_{k-1}) + \beta_k S_{\mu_k}(x_k) - x^* - \beta_k y^*\| \\
&\leq \|x_k - x^*\| + \alpha_k\|x_k - x_{k-1}\| + \beta_k\|S_{\mu_k}(x_k) - y^*\| \\
&\leq \|x_k - x^*\| + \alpha_k\|x_k - x_{k-1}\| + \beta_k(\|S_{\mu_k}(x_k) - S_{\mu_k}(x^*)\| + \|S_{\mu_k}(x^*) - y^*\|) \\
&\leq (1 + \frac{\beta_k}{\mu_k})\|x_k - x^*\| + \alpha_k\|x_k - x_{k-1}\| + \beta_k(\|S^0 x^*\| + \|y^*\|).
\end{aligned}$$

Applying (a) and (b), $\sum_{k=0}^{\infty} \beta_k < \infty$. Also by (d) and Lemma 3, we have $\lim_{k\to+\infty} \|x_k - x^*\|$ exists. Hence, $\{x_k\}$ is bounded. Notice that there exist $\widetilde{x}$ and a subsequence $\{x_{k_\nu}\}$ such that $\lim_{\nu\to\infty} x_{k_\nu} = \widetilde{x}$, since H is a finite dimensional space. We see $J_{\mu_{k_\nu}}^S x_{k_\nu}$ tends to $\widetilde{x}$, because

$$\begin{aligned}
\|J_{\mu_{k_\nu}}^S x_{k_\nu} - \widetilde{x}\| &\leq \|J_{\mu_{k_\nu}}^S x_{k_\nu} - J_{\mu_{k_\nu}}^S \widetilde{x}\| + \|J_{\mu_{k_\nu}}^S \widetilde{x} - \widetilde{x}\| \\
&\leq \|x_{k_\nu} - \widetilde{x}\| + \|J_{\mu_{k_\nu}}^S \widetilde{x} - \widetilde{x}\|,
\end{aligned}$$

and $\lim_{\nu\to+\infty} J^S_{\mu_{k_\nu}}\widetilde{x} = Proj_{\overline{D(S)}}\widetilde{x} = \widetilde{x}$. This fact and the local boundedness of S imply that

$$\{S_{\mu_{k_\nu}} x_{k_\nu}\} \subseteq S(\{J^S_{\mu_{k_\nu}} x_{k_\nu}\}) \subseteq B, \tag{27}$$

where B is a bounded set. Therefore, $\{S_{\mu_{k_\nu}} x_{k_\nu}\}$ is bounded and there exist $\tilde{y}$ and a subsequence $\{S_{\mu_{k_{\nu'}}} x_{k_{\nu'}}\}$ such that $\lim_{\nu'\to\infty} S_{\mu_{k_{\nu'}}} x_{k_{\nu'}} = \tilde{y}$. Then $\tilde{y} \in S(\tilde{x})$ follows from

$$S_{\mu_{k_{\nu'}}} x_{k_{\nu'}} \in S(J^S_{\mu_{k_{\nu'}}} x_{k_{\nu'}}), \tag{28}$$

and Lemma 2. In sequel by (23), we have

$$S_{\mu_{k_{\nu'}}} x_{k_{\nu'}} - (\frac{x_{k_{\nu'}+1} - x_{k_{\nu'}}}{\beta_{k_{\nu'}}}) + \frac{\alpha_{k_{\nu'}}}{\beta_{k_{\nu'}}}(x_{k_{\nu'}} - x_{k_{\nu'}-1}) \in T x_{k_{\nu'}+1}, \tag{29}$$

tending ν' to $+\infty$ in (29) and using conditions (c), (e), the boundedness of $\{x_k\}$ and Lemma 2, it is obtained that $\widetilde{y} \in T\widetilde{x}$. By similar procedure in the proof of Theorem 2.1 in [43], $\widetilde{x}$ is unique. Now proof is complete.

Example 1 The best example of Theorem 4.6 can be seen in digital halftoning which is a procedure for producing a sample of pixels when a limited number of colors are available with a binary system so that it is a continuous-tone image. In this context Teuber et al. [60] minimized difference of two functions that one is corresponding to attraction of the dots by the image gray values and the other corresponds to the repulsion between the dots. They signified black pixel with 0 and white pixel with 1 and investigated images $u : G \to [0, 1]$ on an integer grid $G := \{1, \ldots, n_x\} \times \{1, \ldots, n_y\}$. If m be the number of black pixels generated by the dithering procedure and $p := (p_k)_{k=1}^m = ((p_{k,x}, p_{k,y})^T)_{k=1}^m \in \mathbb{R}^{2m}$ be their position vector, then $|p_k| := \sqrt{p_{k,x}^2 + p_{k,y}^2}$ is the Euclidean norm of the position of the k-th black pixel.

In [60], it is detected minimizer $\hat{p}$ of the functional

$$E(p) = \underbrace{\sum_{k=1}^{m} \sum_{(i,j)\in G} w(i,j)|p_k - \binom{i}{j}|}_{F(p)} - \underbrace{\lambda \sum_{k=1}^{m} \sum_{l=k+1}^{m} |p_k - p_l|}_{G(p)}, \tag{30}$$

where $w := 1 - u$ is the corresponding weight distribution and $\lambda := \frac{1}{m}\sum_{(i,j)\in G} w(i,j)$.
Given two functions $F(p)$ and $G(p)$, which are continuous and convex, since ∂F and ∂G are maximal monotone operators [54] and ∂G is locally bounded on $\mathbb{R}^{2m}$ [27], the problem of finding a minimizer of (30) is a special case of (19).

If conditions $(a),\ldots,(e)$ satisfy and $D(\partial F) \subset D(\partial G)$, then by Theorem 4.6 the generated sequence $\{x_k\}$ of (23) converges to a minimizer of (30).

In next result, the condition of the local boundedness of S in Theorem 4.6 is eliminated and domain of it will be entire H.

Corollary 4 *Assume that the solution set Ω of problem* (19) *is nonempty, conditions $(a), \ldots, (e)$ satisfy, and $D(S) = H$, then the sequence $\{x_k\}$ generated by* (23) *converges to a solution of* (21).

Proof *Since $D(S)$ is open, using Lemma 1 the operator S is locally bounded at any point of $D(S)$. The rest of proof is similar to Theorem 4.6.*

Remark 1 If $D(S) = H$ and $T-S$ is a monotone operator, then by [2, Theorem 2.1], $T - S$ is maximal monotone. Hence, (19) reduces to find a zero point of maximal monotone operator $T-S$ and iteration algorithm (23) changes to $x_{k+1} = J_{\beta_k}^{T-S}(x_k + \alpha_k(x_k - x_{k-1}))$.

Corollary 5 *Assume that S is a bounded value (i.e., for all $x \in H$, Sx is a bounded set) and upper semicontinuous at any point of $\overline{D(S)}$ and the solution set Ω of problem* (19) *is nonempty. If the conditions $(a), \ldots, (e)$ satisfy and $D(T) \subset D(S)$, then the sequence $\{x_k\}$ generated by* (23) *converges to a solution of* (19).

Proof Since S is a bounded value and upper semicontinuous at any point of $\overline{D(S)}$, so it is locally bounded. The rest of proof is similar to Theorem 4.6.

Two types of interesting particular instances of (19) are:

$$\text{find} \quad x^* \in H \quad \text{such that} \quad y^* \in T(x^*), \tag{31}$$

and

$$\text{find} \quad x^* \in H \quad \text{such that} \quad x^* \in T(x^*). \tag{32}$$

It is assumed that $G(S) := H \times \{y\}$ for an arbitrary point $y \in H$ in (31) and $G(S) := \{(x, x); x \in H\}$ for any point $x \in H$ in (32).

In the following, we present the results of these types of problems.

Corollary 6 *Assume that the operator $S : H \to H$ is continuous and the solution set Ω of problem* (19) *is nonempty. If the conditions $(a),\ldots,(e)$ satisfy and $D(T) \subset D(S)$, then the sequence $\{x_k\}$ generated by* (23) *converges to a solution of* (19).

Proof It is easy to check that the sequence $\{x_k\}$ is bounded and there exist $\widetilde{x}$ and a subsequence $\{x_{k_\nu}\}$ such that $\lim_{\nu\to\infty} x_{k_\nu} = \widetilde{x}$. In the proof of Theorem 4.6 it has been shown that $\lim_{\nu\to\infty} J_{\mu_{k_\nu}} x_{k_\nu} = \widetilde{x}$. Consequently, from

$$S_{\mu_{k_\nu}} x_{k_\nu} - (\frac{x_{k_\nu+1} - x_{k_\nu}}{\beta_{k_\nu}}) + \frac{\alpha_{k_\nu}}{\beta_{k_\nu}}(x_{k_\nu} - x_{k_\nu-1}) \in T x_{k_\nu+1}, \quad (33)$$

$S_{\mu_{k_\nu}}(x_{k_\nu}) = S(J^S_{\mu_{k_\nu}}(x_{k_\nu}))$, continuity of S and by passing to a subsequence, we can arrange that left side of (33) converges to $S(\widetilde{x})$. By Lemma 2, we see that $S(\widetilde{x}) \in T(\widetilde{x})$, i.e. $0 \in T(\widetilde{x}) - S(\widetilde{x})$.

Corollary 7 *Assume that $S : H \to H$ is Lipschitz continuous, the solution set Ω of problem* (19) *is nonempty and $D(T) \subset D(S)$. If conditions (c),...,(e) satisfy and if one replaces condition $\sum_{k=1}^{\infty} \beta_k < \infty$ with (a) and (b), then the generated sequence $\{x_k\}$ of the method*

$$x_{k+1} = J^T_{\beta_k}(x_k + \alpha_k(x_k - x_{k-1}) + \beta_k S(x_k))$$

converges to a solution of problem (19).

Remark 2 All results of this section has derived from Lemma 2. In an infinite dimensional real Hilbert space, the boundedness of sequence $\{x_k A\}$ in Theorem 4.6 implies that there exist subsequence $\{x_{k_\nu}\}$ and $\tilde{x} \in H$ such that $\{x_{k_\nu}\}$ converges weakly to $\tilde{x}$. The fundamental difficulties in proving $\tilde{y} \in S(\tilde{x})$ and $\tilde{y} \in T(\tilde{x})$ are showing strongly convergence of either $\{J^S_{\mu_{k_\nu}} x_{k_\nu}\}$ to $\tilde{x}$ or $\{S_{\mu_{k_\nu}} x_{k_\nu}\}$ to $\tilde{y}$ and the left side of (29) to $\tilde{y}$.

References

1. Y. Alber, I. Ryazantseva, *Nonlinear Ill-Posed Problems of Monotone Type*. Springer, New York, (2006)
2. M. Alimohammady, M. Ramazannejad, M. Roohi, Notes on the difference of two monotone operators. Optim. Lett. **8**(1), 81–84 (2014)
3. L.T.H. An, Analyse numérique des algorithmes de l'optimisation d.c. Approches locales et globales. Code et simulations numériques en grande dimension. Applications. Thése de Doctorat de l'Université de Rouen, (1994)
4. L.T.H. An, D.T. Pham, DCA with escaping procedure using a trust region algorithm for globally solving nonconvex quadratic programs. Technical Report, LMI-CNRS URA 1378, INSA-Rouen (1996)
5. L.T.H. An, D.T. Pham, Solving a class of linearly constrained indefinite quadratic problems by DC Algorithms. J. Global Optim. **11**(3), 253–285 (1997)
6. L.T.H. An, D.T. Pham, A Branch-and-Bound method via D.C. optimization algorithm and ellipsoidal technique for box constrained nonconvex quadratic programming problems. J. Global Optim. **13**(2), 171–206 (1998)
7. L.T.H. An, D.T. Pham, D.C. programming approach for large-scale molecular optimization via the general distance geometry problem. in Optimization in Computational Chemistry and Molecular Biology, Part of the Nonconvex Optimization and Its Applications book series (NOIA, volume 40), (Springer, Boston, 2000), 301–339
8. L.T.H. An, D.T. Pham, A continuous approach for globally solving linearly constrained quadratic. Optim. J. Math. Program. Oper. Res. **50**, 93–120 (2001)

9. L.T.H. An, D.T. Pham, D.C. programming approach to the multidimensional scaling problem, in *From Local to Global Optimization*. Nonconvex Optimization and Its Applications (NOIA), vol. 53 (Kluwer Academic Publishers, Dordrecht, 2001), pp. 231–276
10. L.T.H. An, D.T. Pham, The DC programming and DCA revisited with DC models of real world nonconvex optimization problems. Ann. Oper. Res. **133**, 25–46 (2005)
11. N.T. An, N.M. Nam, Convergence analysis of a proximal point algorithm for minimizing difference of functions. Optimization **66**(1), 129–147 (2017)
12. H. Attouch, J. Bolte, B. Svaiter, Convergence of descent methods for semi-algebraic and tame problems: proximal algorithms, forward-backward splitting, and regularized Gauss–Seidel methods. Math. Program Ser. A **137**, 91–124 (2011)
13. H. Attouch, J. Bolte, P. Redont, et al., Proximal alternating minimization and projection methods for nonconvex problems: an approach based on the Kurdyka-Łojasiewicz inequality. Math. Oper. Res. **35**, 438–457 (2010)
14. J.P. Aubin, H. Frankowska, Set-valued analysis. Reprint of the 1990 edition
15. F. Bach, R. Jenatton, J. Mairal, G. Obozinski, Optimization with sparsity-inducing penalties. Found. Trends Mach. Learn. **4**(1), 1–106 (2011)
16. J. Bolte, A. Daniilidis, A.S. Lewis, Lojasiewicz inequality for nonsmooth subanalytic functions with applications to subgradient dynamic systems. SIAM Optim. **17**, 1205–1223 (2007)
17. J. Bolte, A. Daniilidis, A.S. Lewis, et al., Clarke subgradients of stratifiable functions. SIAM J. Optim. **18**, 556–572 (2007)
18. J. Bolte, S. Sabach, M. Teboulle, Proximal alternating linearized minimization for nonconvex and nonsmooth problems. Math. Program. Ser. A **146**, 459–494 (2014)
19. S. Boyd, N. Parikh, E. Chu, B. Peleato, J. Eckstein, Distributed optimization and statistical learning via the alternating direction method of multipliers. Found. Trends Mach. Learn. **3**(1), 1–122 (2011)
20. F.E. Browder, Nonlinear maximal monotone mappings in Banach spaces. Math. Ann. **175**, 89–113 (1968)
21. F.E. Browder, Multi-valued monotone nonlinear mappings. Trans. Am. Math. Soc. **118**, 338–551 (1965)
22. F.E. Browder, The fixed point theory of multi-valued mappings in topological vector spaces. Math. Ann. **177**, 283–301 (1968)
23. A. Brøndsted, R.T. Rockafellar, On the subdifferentiability of convex functions. Proc. Am. Math. Soc. **16**, 605–611 (1965)
24. R.S. Burachik, A.N. Iusem, Set-Valued Mappings and Enlargements of Monotone Operators. Optimization and Its Applications (Springer, New York, 2008). ISBN: 978-0-387-69757-4
25. S. Chandra, Strong pseudo-convex programming. Indian J. Pure Appl. Math. **3**, 278–282 (1972).
26. F.H. Clarke, Generalized gradients and applications. Trans. Am. Math. Soc. **205**, 247–262 (1975). Fixed-Point Algorithms
27. F.H. Clarke, R.J. Stern, G. Sabidussi, Nonlinear Analysis, Differential Equations and Control. NATO Sciences Series C: Mathematical and Physical Sciences, vol. 528 (Kluwer Academic Publishers, Dordrecht, 1999)
28. P. Combettes, J.-C. Pesquet, Proximal splitting methods in signal processing, in *Fixed-Point Algorithms for Inverse Problems in Science and Engineering* (Springer, New York, 2011), pp. 185–212
29. C. Do, Q. Le, C. Foo, Proximal regularization for online and batch learning, in *International Conference on Machine Learning* (2009), pp. 257–264
30. G. Gasso, A. Rakotomamonji, S. Canu, Recovering sparse signals with non-convex penalties and DC programming. IEEE Trans. Signal Process. **57**(12), 4686–4698 (2009)
31. J.B. Hiriart-Urruty, From convex optimization to non convex optimization. Part I: Necessary and sufficent conditions for global optimality, in *Nonsmooth Optimization and Related Topics*. Ettore Majorana International Sciences, vol. 43 (Plenum Press, New York, 1988)
32. S. Huda, R. Mukerjee, Minimax second-order designs over cuboidal regions for the difference between two estimated responses. Indian J. Pure Appl. Math. **41**(1), 303–312 (2010)

33. R. Jenatton, J. Mairal, G. Obozinski, F. Bach, Proximal methods for sparse hierarchical dictionary learning, in *International Conference on Machine Learning* (2010)
34. M. Kraning, E. Chu, J. Lavaei, S. Boyd, Dynamic network energy management via proximal message passing. Foundations and Trends in Optimization **1**(2), 70–122 (2014)
35. D. Lei, L. Shenghong, Ishikawa iteration process with errors for nonexpansive mappings in uniformly convex Banach spaces. Int. J. Math. Math. Sci. **24**(1), 49–53 (2000)
36. H.A. Le Thi, V.N. Huynh, D.T. Pham, Convergence analysis of DC algorithm for DC programming with subanalytic data. Annals of Operations Research. Technical Report. Rouen: LMI, INSA-Rouen (2013)
37. P. Mahey, D.T. Pham, Proximal decomposition of the graph of maximal monotone operator. SIAM J. Optim. **5**, 454–468 (1995)
38. B. Martinet, Détermination approchée d'un point fixe d'une application pseudo-contractante. Rev C.R. Acad. Sci. Paris **274A**, 163–165 (1972)
39. B. Martinet, Régularisation d'iné quations variationnelles par approximations successives. Revue Française de Informatique et Recherche Opérationelle (1970)
40. G.J. Minty, Monotone networks. Proc. R. Soc. Lond. **257**, 194–212 (1960)
41. B.S. Mordukhovich, N.M. Nam, N.D. Yen, Fréchet subdifferential calculus and optimality conditions in nondifferentiable programming. Optimization **55**, 685–708 (2006)
42. A. Moudafi, On the difference of two maximal monotone operators: regularization and algorithmic approaches. Appl. Math. Comput. **202**, 446–452 (2008)
43. A. Moudafi, On critical points of the difference of two maximal monotone operators. Afr. Mat. **26**(3–4), 457–463 (2015)
44. B. O'Donoghue, G. Stathopoulos, S. Boyd, A splitting method for optimal control. IEEE Trans. Control Syst. Technol. **21**(6), 2432–2442 (2013)
45. N. Parikh, S. Boyd, Proximal algorithms. Found. Trends Optim. **1**(3), 123–231 (2013)
46. D.T. Pham, Duality in D.C. (difference of convex functions) optimization. Subgradient methods, in *Trends in Mathematical Optimization*. International Series of Numerical Mathematics, vol. 84. Birkhäuser, Basel (1988), pp. 277–293
47. D.T. Pham, L.T.H. An, A D.C. optimization algorithms for solving the trust region subproblem. SIAM J. Optim. **8**(2), 476–505 (1998)
48. D.T. Pham, L.T.H. An, Convex analysis approach to D.C. programming: theory, algorithms and applications. Acta Math. Vietnamica **22**(1), 289–355 (1997)
49. D.T. Pham, L.T.H. An, Difference of convex functions optimization algorithms (DCA) for globally minimizing nonconvex quadratic forms on Euclidean balls and spheres. Oper. Res. Lett. **19**(5), 207–216 (1996)
50. D.T. Pham, L.T.H. An, Optimisation d.c. (différence de deux fonctions convexes). Dualité et Stabilité. Optimalités locale et globale. Algorithmes de l'optimisation d.c. (DCA). Technical Report. LMI-CNRS URA 1378, INSA Rouen (1994)
51. D.T. Pham, L.T.H. An, Polyhedral d.c. optimization. Theory, algorithms and applications. Technical Report. LMI-CNRS URA 1378, INSA-Rouen (1994)
52. N. Qian, On the momentum term in gradient descent learning algorithms. Neural Netw. **12**(1), 145–151 (1999)
53. R.T. Rockafellar, Monotone operators and the proximal point algorithm. SIAM J. Control Optim. **14**(5), 877–898 (1976)
54. R.T. Rockafellar, On the maximal monotonicity of subdifferential mappings. Pacific J. Math. **33**, 209–216 (1970)
55. R.T. Rockafellar, Local boundedness of nonlinear monotone operators. Michigan Math. J. **16**, 397–407 (1969)
56. R.T. Rockafellar, R. Wets, *Variational analysis. Grundlehren der Mathematischen Wissenschaften Variational analysis*. Grundlehren der Mathematischen Wissenschaften [Variational analysis. Fundamental Principles of Mathematical Sciences], vol. 317 (Springer, Berlin, 1998)
57. T. Schüle, C. Schnörr, S. Weber, J. Hornegger, Discrete tomography by convex–concave regularization and D.C. programming. Discrete Appl. Math. **151**, 229–243 (2005)

58. J.C. Souza, P.R. Oliveira, A proximal point algorithm for DC functions on Hadamard manifolds. J Global Optim. **63**, 797–810 (2015)
59. W. Sun, R.J.B. Sampaio, M.A.B. Candido, Proximal point algorithm for minimization of DC functions. J. Comput. Math. **21**, 451–462 (2003)
60. T. Teuber, G. Steidl, P. Gwosdek, Ch. Schmaltz, J. Weickert, Dithering by differences of convex functions. SIAM J. Imag. Sci. **4**(1), 79–108 (2011)
61. J.F. Toland, On subdifferential calculus and duality in nonconvex optimization. Bull. Soc. Math. France Mémoire **60**, 173–180 (1979)
62. J.F. Toland, Duality in nonconvex optimization. J. Math. Anal. Appl. **66**, 399–415 (1978)
63. Z. Zhang, J.T. Kwok, D.-Y. Yeung, Surrogate maximization/minimization algorithms. Mach. Learn. **69**, 1–33 (2007)

A Mathematical Model for Simulation of Intergranular μ-Capacitance as a Function of Neck Growth in Ceramic Sintering

Branislav M. Randjelović and Zoran S. Nikolić

Abstract In this paper we will define a new mathematical model for predicting an evolution of an equivalent intergranular μ-capacitance during ceramic sintering. The contact between two adjacent grains will be defined as a structure that forms a μ-capacitor recognized as an intergranular μ-capacitor unit. It will be assumed that its μ-capacitance changes as the neck grows by diffusion. Diffusion mechanisms responsible for transport matter from the grain boundary to the neck are the volume diffusion and grain boundary diffusion. Such model does not need special geometric assumptions because the microstructural development can be simulated by a set of simple local rules and overall neck growth law which can be arbitrarily chosen. To find the total capacitance we will identify μ-capacitors in series and in parallel. More complicated connections of μ-capacitors will be transformed into simpler structure using delta to star transformation and/or star to delta transformation. In this way some μ-capacitors will be step-by-step replaced by their equivalent μ-capacitors. The developed model can be applied for the prediction of an evolution of the intergranular capacitance during ceramic sintering of $BaTiO_3$ system with spherical particle distributions.

1 Introduction

Control of ceramics processing involves the raw materials used and the way in which they are formed and heat-treated. Sintering is widely used for consolidation of ceramic materials due to low sintering temperatures, relatively fast densification and homogenization, and high final densities which are the main advantages of this method. Densification is based mostly on rearrangement and shape change of solid grains. Fast homogenization is of particular importance as a result of diffusion through the solid and liquid phase. Most of these phenomena are influenced by the

B. M. Randjelović · Z. S. Nikolić (✉)
University of Niš, Faculty of Electronic Engineering, Niš, Serbia
e-mail: zoran.nikolic@ni.ac.rs

N. J. Daras, T. M. Rassias (eds.), *Computational Mathematics and Variational Analysis*, Springer Optimization and Its Applications 159,
https://doi.org/10.1007/978-3-030-44625-3_22

presence of additives that control the microstructural and dimensional development during sintering, where additives may be present as a solid phase only, as a solid solution within the host material, as segregates at interfaces within the host material, or as a liquid phase.

As a result of a better understanding of ceramics and their properties, new high-performance ceramic materials have been developed in the last 10–20 years. Their processing depends on an improved understanding of essential basic mechanisms, on their microstructure evolution, and their final properties. Ceramic materials are of particular interest because of their unique and outstanding properties such as special electrical properties, superior mechanical properties, greater chemical stability, etc. Hence there is a growing interest in the electronic industry for high-performance electronic ceramic components, including for computer, signal processing, telecommunications, power transmission, and control system use. These materials are highly specialized, being prepared from specifically formulated compositions, processed under closely controlled conditions, and fabricated into complex shapes with specific electrical properties.

It may be of particular interest to gain better understanding of common basic (diffusion) sintering mechanisms and how they affect the production of ceramics with optimal properties. One way is by computer simulation using appropriate numerical approach. Process modeling by simulation enables the evolution of parameters to be followed throughout the processing time and offers a framework within which experimental observations can be assessed.

For electronic materials design, material structure is very important. Ceramics' microstructures show that the contact region between two interconnected interacting grains can be defined as impedance with capacitance dominant [1]. Two contacted grains make a structure that forms an intergranular capacitor, where both intergranular structure and electrical properties of ceramics depend on diffusion processes. In the papers [2, 3] the authors proposed the equivalent intergranular impedance models for an aggregate of grains. The equivalent circuit was envisaged as a series of grouped elements, i.e. several resistance and capacitance elements in parallel, each corresponding to a different process occurring at the electrode interface, at the grain boundaries, or within the grains themselves [4]. These approaches were based on the planar μ-capacitor that forms in the contact region between the grains. The lack of these models was the over-estimated intergranular capacitance due to the time-dependent distance between capacitor plates, i.e. the dielectric thickness.

The purpose of the present paper is to determine new relationship between electrical, geometrical, and diffusion parameters important for improved μ-capacitance model definition. In that sense, we will propose a new computational method for predicting an evolution of the intergranular capacitance during sintering ceramics. Microstructural development will be simulated by a set of simple local rules and overall neck growth law which can be arbitrarily chosen. This approach will be extended to the multi-grain model and used to enable the establishment of interrelation between structural and electrical parameters of $BaTiO_3$ ceramics. Mentioned computer-simulation model can be used to assist in the creation, modification, analysis, and optimization of new high-performance electronic ceramics materials.

2 Model System Topology

2.1 Domain Structure

For simulation of sintering it is convenient to use the multi-grain model with regular grain shape because it needs to store only the position, orientation, and size of each grain. The model does not need other geometric assumptions because the microstructural development can be simulated by a set of simple local rules and overall neck growth law.

We will assume that a point in three-dimensional (3D) space R^3 is characterized by three Cartesian coordinates (x, y, z), where these coordinates uniquely define a position-vector $\mathbf{r} = (x, y, z)$. The initial microstructure consisting of N spherical grains will be now represented by 3D domains of regular shape, i.e.

$$G^m = (\mathbf{r}_0^m, R_m),\ m = \overline{1, N}\,,$$

where $\mathbf{r}_0 = (x_0, y_0, z_0)$ is the position-vector of the center of the mass of the m-th grain of radius R_m. Thus, for two domains G^a and G^b, the center-to-center distance will be defined as the length of the line segment between their centers, i.e. as the Euclidean distance function [5]

$$\text{dist}(\mathbf{r}_0^a, \mathbf{r}_0^b) = \sqrt{(x_0^b - x_0^a)^2 + (y_0^b - y_0^a)^2 + (z_0^b - z_0^a)^2}\,. \tag{1}$$

2.2 Domain Displacement

In our approach, domain displacement will be possible if and only if a new position of m-th domain undergoing displacement, G_{T}^m, is not already occupied by other domains, i.e.

$$G_{\mathrm{T}}^m \cap G^k = \emptyset\ \ (k = \overline{1, N}, m \neq k).$$

Then the domain displacement will be modeled as the domain translation by a distance-vector $\mathbf{q} = (q_x, q_y, q_z)$ accomplished through [6]

$$\mathbf{r}_0^m = (x_0^m, y_0^m, z_0^m)\ \mapsto\ \mathbf{r}_{\mathrm{T}}^m = (x_0^m + q_x, y_0^m + q_y, z_0^m + q_z)\,,$$

where $\mathbf{r}_{\mathrm{T}}^m$ is the position-vector of the center of domain G^m after its displacement.

2.3 Skeleton Structure

Sintered materials are in general characterized by a connected microstructure composed of solid-phase grains in contact with one another in a solidified matrix phase. It is convenient to assume that if loose powder grains are brought into contact, inter-atomic forces cause small circles-of-contact (necks) to form between them. Such an interconnected structure is a solid skeleton structure, which always forms during sintering, so that inter-grain contacts exist in all liquid phase sintered microstructures. Even more, German and Liu's theoretical analysis [7] has shown that such a connected grain structure is favored.

Solid skeleton may be generally defined as a series of connected 3D domains arranged in long chain (interconnected chain of grains) [8], i.e.

$$SS^{m}_{K(l)} = \bigcup_{k=1}^{K(m)} G^{v(m,k)}\,, \tag{2}$$

where $K(m)$ is the vector of the number of domains included in m-th skeleton and $v(m, k)$ is the vector of their ordinal numbers.

Initial microstructure (the so-called green density) consisting of N domains can be now completely described with N isolated solid skeletons of unit length using the definition (2), i.e $SS^{s}_{1} = G^{s}\;\;(s = \overline{1, N})$. Now, if two domains $G^{\mathrm{a}}(\equiv SS^{\mathrm{a}}_{1})$ and $G^{\mathrm{b}}(\equiv SS^{\mathrm{b}}_{1})$ are connected, then they will form a *solid skeleton* unit consisting of two domains with an equilibrium contact area (the neck) between them, where part of their boundaries replaced by domain boundary, i.e. [9]

$$SS^{\mathrm{a}}_{2} = G^{\mathrm{a}} \cup G^{\mathrm{b}} \text{ and } SS^{\mathrm{b}}_{0} = \emptyset.$$

The geometry and topological aspects of the multi-grain model can be most easily described by the network in which grain (domain) centers are identified by vertices and a link (center-to-center distance) joins a pair of vertices $\mathbf{r}^{i}_{0}$ and $\mathbf{r}^{j}_{0}$, where the length of the link $\mathbf{D}_{ij}$ can be computed by Equation (1) as the distance dist$(\mathbf{r}^{i}_{0}, \mathbf{r}^{j}_{0})$. The network corresponding to a model of connected grains (domains) is thus made up of a unique, interconnected set of closed polygons.

It is easy to understand that domains in solid skeleton, according to their role in a set of connected domains, can be categorized into three kinds of domains: an *End domain* with exactly one neck, a *Link domain* with exactly two necks, and a *Junction domain* with more than two necks. Thus, the skeleton network may be defined as a set of vertices (i.e., centers of mass of connected domains), some of which have exactly one nearest-neighbor (*End points*), where the remaining network vertices are either points with exactly two nearest-neighbors (*Link points*) or vertices with more than two nearest-neighbors (*Junction or nodal points*). Rigorous topological characterization of skeletal structure requires reducing the solid-phase space into a node network, which emphasizes connectivity. Note that the skeleton network, given as a system of functions of some topological parameters, changes monotonically with time.

3 Process Modeling

During charge transport through a polycrystalline solid different processes take place, such as bulk (intra-grain) conduction, conduction across the grain boundary, etc. Each of these processes usually may be represented by independent R-C combinations. We will investigate the phenomena occurring in the grain-boundary regions, i.e. the case where the grain boundary capacitance is the dominant impedance in the sample. We will define the method for analyzing the intergranular capacitance by separating the contributions of grain boundaries and grains and by introducing mathematical formalisms.

We will use an idealized microstructure of two touching grains to describe neck formation during the initial stage of the sintering. The neck between the grains forms as soon as the sintering begins. The neck growth continues until it is about 40–50% of the grain size [10]. Kang assumed the maximum neck size to be about 20% of the grain size [11].

3.1 Intergranular μ-Capacitance

The kinetics of the sintering can be obtained by equating the mass transported from the materials source and change in the neck volume (shown in Figure 1a and taking into account the grain boundary diffusion only). The stress gradient between neck center and surface causes a flux of atoms into the neck resulting in a reduction of aggregate surface area [10, 12] and the primary grain centers approach each other.

Ceramics' microstructures show that the contact region between two interacting grains can be defined as impedance with capacitance dominant [1]: approximately two contacted grains make a structure that forms a μ-capacitor (the intergranular μ-capacitor unit), where its intergranular structure depends on diffusion processes.

In view of that, we will assume that the contact region defined by the neck radius x can be approximately treated as μ-capacitor having dielectric thicknesses d (Figure 1b), whose μ-capacitance changes as the neck grows and the grains approaching each other, by diffusion. Although that diffusing atoms penetrate the

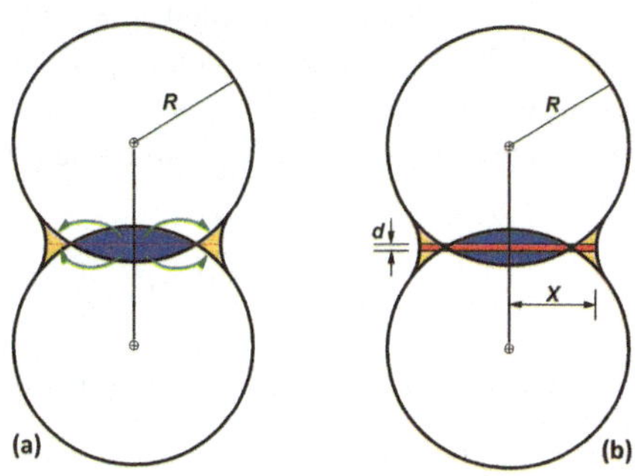

Fig. 1 Two-dimensional representation of the sintering of the two-grain model (equal sized grains): (**a**) the scheme of material transport driven by the difference in surface curvature between sources and sinks, (**b**) the scheme of μ-capacitor having dielectric thicknesses d

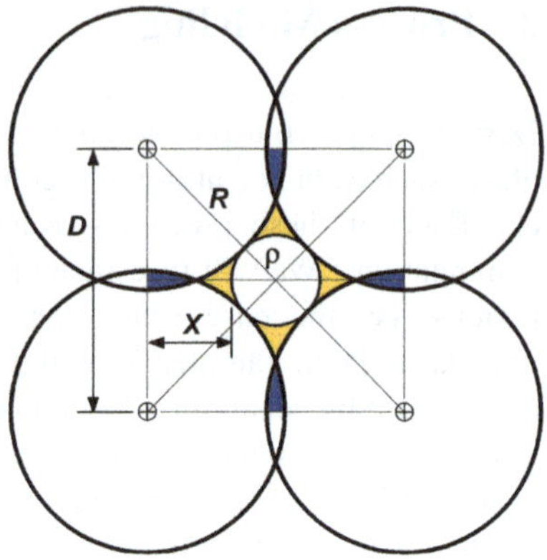

Fig. 2 Two-dimensional representation of the sintering of the four-grain model (with equal sized grains) with the circular pore of the radius ρ in between grains

neighboring grains and degrade the grain boundaries between contacting grains, we will assume the contact surface as the circle of the area πx^2 . Thus the intergranular μ-capacitance per grain contact (an intergranular μ-capacitance unit) can be defined as

$$\mu C_{IG} = \epsilon_0 \epsilon_r \frac{\pi x^2}{d} , \tag{3}$$

where ϵ_0 and ϵ_r are the dielectric constants of vacuum and the ceramic material, respectively.

Generally speaking, the sintering of a close-packed aggregate of grains is of more practical interest due to its complexity, although its initial regular pore geometry (defined by the pore radius ρ), becomes during sintering no longer circular due to the diffusion field between grains ongoing sintering becomes highly asymmetric. Taking into account the model geometry of porous system shown in Figure 2, Equation (3) can be replaced by new one, i.e.

$$\mu C_{IG} = \epsilon_0 \epsilon_r \frac{\pi x^2}{d} \left(\sqrt{(R+\rho)^2 - (D/2)^2} - \rho \right) . \tag{4}$$

In the absence of pressure, the neck growth is driven by the difference in surface curvature between sources and sinks, where the driving force depends on the system configuration and model geometry. As a matter of fact, the exact calculation of the curvature differences involves the solution of time-temperature dependent diffusion equations. Then Equation (3) must be rewritten as [2, 3]

$$\mu C_{IG}(t, T) = \epsilon_0 \epsilon_r \frac{\pi x^2(t, T)}{d} ,$$

where t is the sintering time and T is the sintering temperature.

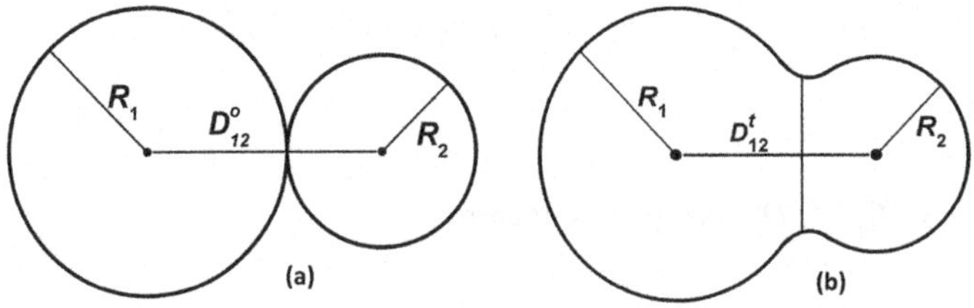

Fig. 3 Two-dimensional representation of the two-grain sintering kinetic model: (**a**) initial point contact structure; (**b**) after neck growth during a sintering time t, in which the initial center-to-center distance D_{12}^0 is reduced by ΔD_{12}^t. Grain boundary is assumed to be flat

If we now assume that the coordination number N_C has been conceived as the number of nearest-neighbors of a grain in the grain structure concerned [13], then the total intergranular μ-capacitance per grain will be estimated as

$$\mu C_{IG}^{\sum}(t, T) = N_C \cdot \mu C_{IG}(t, T) .$$

3.2 Neck Growth

We will define the basic sintering kinetics using the idealized two-grain sintering model (shown in Figure 3a) and the equation

$$\frac{\Delta D}{\Delta t} = f(D) , \tag{5}$$

where ΔD is the decrease in center-to-center distance (1) for a given time step Δt and $f(D)$ is the particular neck growth law. Assuming that at time $t + \Delta t$ the decrease of the link length is

$$\Delta D_{ki}^{t+\Delta t} = D_{ki}^t - D_{ki}^{t+\Delta t}ł, \tag{6}$$

then by substituting Equation (6) into Equation (5) and by assuming flat grain boundary (Figure 1b), each network link D_{ki}^t between a pair of vertices $\mathbf{r}_0^k$ and $\mathbf{r}_0^i$ can be updated by the iterative procedure (the sintering transformation) [8]

$$D_{ki}^{t+\Delta t} \mapsto D_{ki}^t - f(D_{ki}^t) \cdot \Delta t . \tag{7}$$

Taking into account the multi-grain sintering model, the simulation of neck growth will be based on the concept that the sintering law $f(\bullet)$ and the transformation (7) will be applied to each pair of contacting grains within the multi-grain model. The update of the state will be defined by the new topology of the skeleton structure (2), which will be accomplished by updating the position of i-th grain in 3D assuming the fixed position of k-th grain, i.e.

$$(x_0^i, y_0^i, z_0^i) \mapsto (F_{ki}x_0^i + (1 - F_{ki})x_0^k, F_{ki}y_0^i + (1 - F_{ki})y_0^k, F_{ki}z_0^i + (1 - F_{ki})z_0^k)$$

$$(i = \overline{1, N},\ i \neq k),$$

where $F_{ki} = D_{ki}^{t+\Delta t}/D_{ki}^t$ is the scale factor.

3.3 Time-Dependent Intergranular μ-Capacitance

The computational model considered here provides an effective methodology for computation of the time-dependent intergranular μ-capacitance (4). In our approach the grain boundary as a transition region in which some atoms are not exactly aligned with either grain will be viewed as capacitors' dielectric. It needs to be pointed out here that grain boundaries in real materials, which are usually considered to be two-dimensional, have a finite thickness (a few lattice parameters in very pure metals to a few hundred angstroms in ceramics, ~0.3–0.5 nm).

Having this in mind, we will apply the simulation method for sintering of $BaTiO_3$-ceramics assuming nearly constant dielectric thickness, $d = 0.5$ nm and constant neck growth rate between grains (relative neck size to grain size of about 0.4 after 1h at the sintering temperature, similar to the calculated value from conventional sintering models [14]). Figure 4 shows a typical time-dependent intergranular μ-capacitance obtained for the two-grain model with grains of sizes 10 μm and 5 μm. Due to the neck growth, densification manifested by decrease in center-to-center distance occurs, followed by a relatively small grain rearrangement.

This phenomenon will be more visible in the sintering of the multi-grain model in which the evolution of the solid skeleton network (intergranular contacts network)

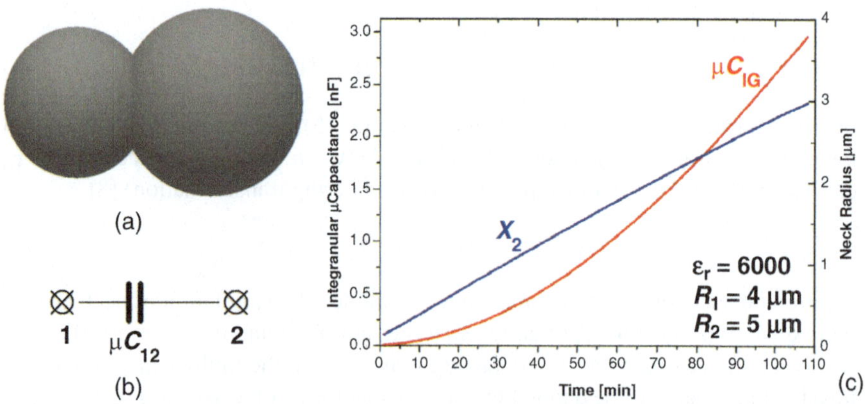

Fig. 4 (**a**) 3D representation of the two-grain model (two interconnected grains). (**b**) Appropriate electrical scheme of the intergranular μ-capacitor. (**c**) Time-dependent intergranular μ-capacitance for the two-grain model for $BaTiO_3$-ceramics (X_2 is the neck radius)

will have to be taken into account. In the next section we will propose a new computational method for predicting an evolution of the intergranular capacitance during ceramic sintering. The model does not need special geometric assumptions because the microstructural development can be simulated by a set of simple local rules and overall neck growth law which can be arbitrarily chosen. Computer-based approach will be extended to the multi-grain model and used to enable the establishment of interrelation between structural and electrical parameters.

4 Process Simulation

4.1 μ-Capacitor Connection Models

For simulation of the capacitance of a ceramic sample we will apply the multi-grain model in which interconnected grains may be replaced by μ-capacitors. Thus, multiple connections of μ-capacitors act like a single equivalent μ-capacitor. The total capacitance of this equivalent single capacitor depends both on the individual capacitors and how they are connected.

There are two simple and common types of connections, called in series and in parallel, for which we can easily calculate the total capacitance. Certain more complicated connections can also be related to combinations of in series and in parallel.

According to their position within the sample characterized by complex μ-capacitor network (identical to solid skeleton of connected grains) there are several characteristic μ-capacitor sub-networks: two simple such as μ-capacitors in series and μ-capacitors in parallel, and two more complicated μ-capacitors in delta connection and μ-capacitors in star connection.

μ-Capacitors in Series

μ-Capacitors are said to be connected together in series when they are effectively chained together in a single line (Figure 5a). Thus, the equivalent μ-capacitance will be calculated as equals the reciprocal of the sum of the reciprocals of the individual μ-capacitances, i.e.

$$\frac{1}{\mu C_t} = \frac{1}{\mu C_1} + \frac{1}{\mu C_2} + \cdots + \frac{1}{\mu C_n} = \sum_k \frac{1}{\mu C_k}. \tag{8}$$

The overall μ-capacitance decreases by added together μ-capacitors in series, i.e. this type of connection produces a total capacitance that is less than that of any of the individual μ-capacitance.

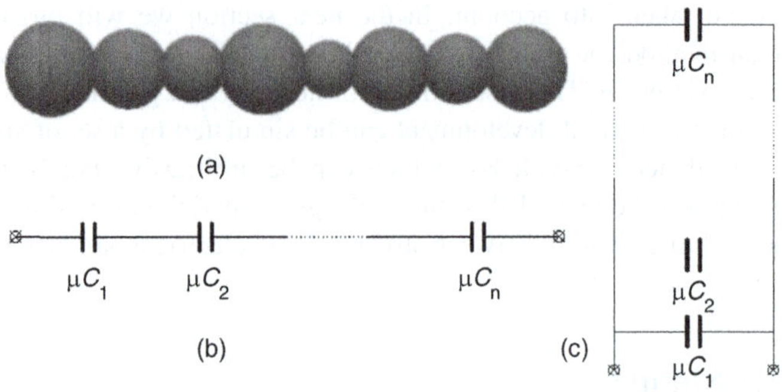

Fig. 5 Two simple μ-capacitor connection models. n μ-capacitors connected in series (the so-called straight chain connected structure) represented by (**a**) 3D model and (**b**) appropriate electrical scheme. (**c**) Electrical scheme for n μ-capacitors connected in parallel

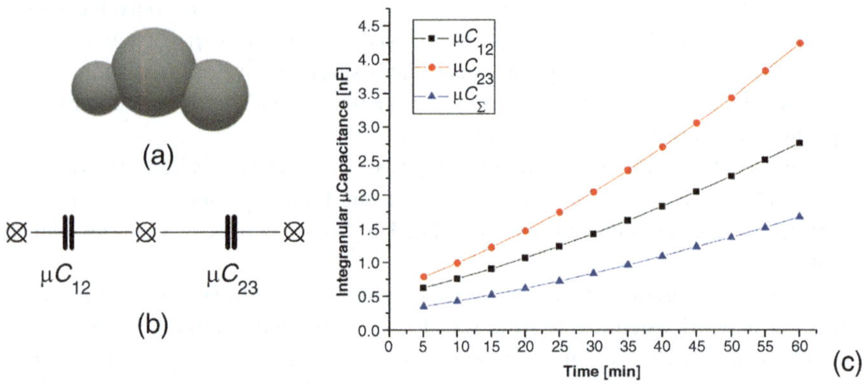

Fig. 6 Three-grain model $(R_1, R_2, R_3) = (3, 5, 4)$ μm . (**a**) 3D model and (**b**) appropriate electrical scheme. (**c**) Time-dependent intergranular μ-capacitance for sintering of $BaTiO_3$-ceramics

Figure 6 shows the time-dependent intergranular capacitance for three μ-capacitors connected in series. As it can be seen, μ-capacitances between grains numbered 1 and 2, as well as between grains numbered 2 and 3 show similar tendency during the sintering time. However, according to Equation (8) the total intergranular capacitance is less than that of two individual μ-capacitances.

μ-Capacitors in Parallel

μ-capacitors are said to be connected together in parallel combination when both of their terminals are respectively connected to each terminal of the other μ-capacitor or μ-capacitors (Figure 5c). The equivalent μ-capacitance as being the sum of all the individual μ-capacitance's added together, i.e.

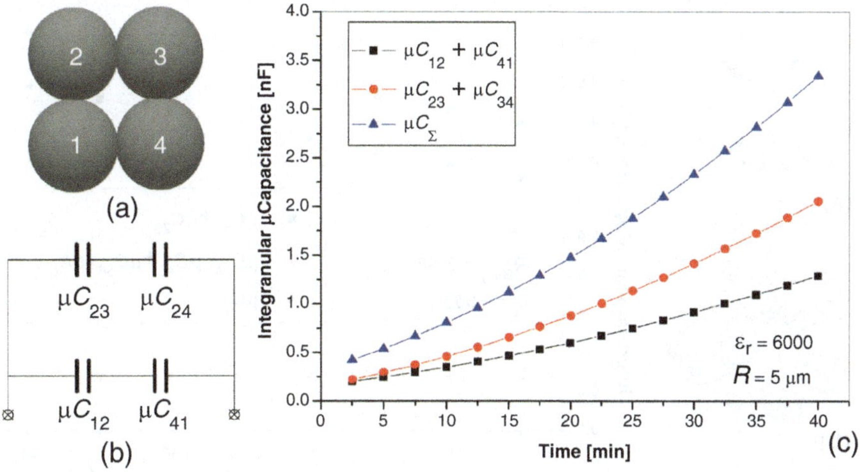

Fig. 7 Four-grain model. (**a**) 3D model and (**b**) appropriate electrical scheme. (**c**) Time-dependent intergranular μ-capacitance for $BaTiO_3$-ceramics

$$\mu C_t = \mu C_1 + \mu C_2 + \cdots + \mu C_n = \sum_k \mu C_k .$$

The equivalent μ-capacitance increases by added together μ-capacitors in parallel, so we create larger μ-capacitances than is possible using a single μ-capacitor.

As an example, Figure 7 shows the time-dependent μ-capacitance for the four-grain model in which individual μ-capacitors μC_{12} and μC_{41}, as well as μC_{23} and μC_{34} are connected in series but their equivalent μ-capacitors are connected in parallel.

Figure 8a shows the circular six-grain model with a large pore in between the grains, which grain structure is very possible in real sintered systems. The capacitance of this grain model may usually be computed taking into account that μ-capacitors are connected in parallel. If we assume that the points A and B (two terminals for equivalent μ-capacitance) are located as it is shown in Figure 8b, then two μ-capacitors connected in series will be in parallel branch with four μ-capacitors also connected in series. Figure 8c shows the equivalent μ-capacitance as a function of the sintering time. However, if the points A and B are moved to new location as it is shown in Figure 9a, then three μ-capacitors connected in series are in parallel branch with another three μ-capacitors also connected in series. Figure 9b shows the equivalent time-dependent μ-capacitance, whose dependence is different from the previous one.

For simulation of integral μ-capacitance and computation of equivalent electrical schemes (and transformation from one type of connection to another one) of some parts of the interconnected grain network another two μ-capacitor connections may be very important: *μ-capacitors in delta connection* (shown in Figure 13) and μ-

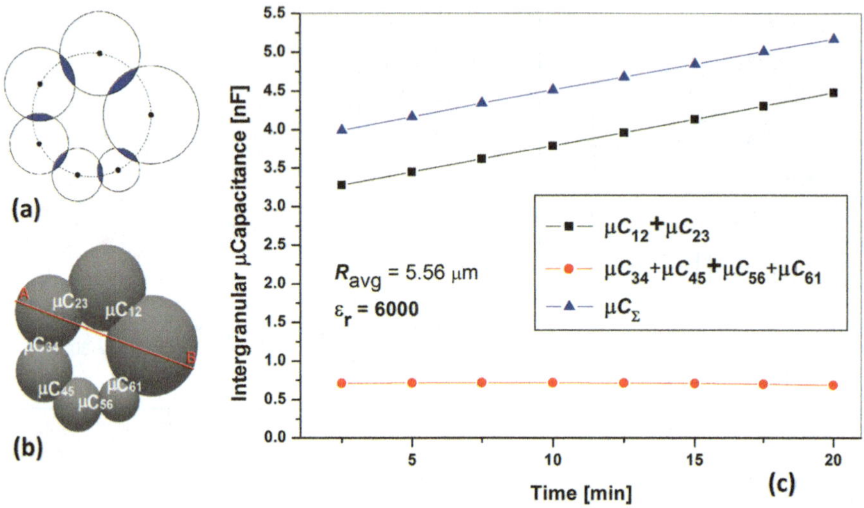

Fig. 8 Six-grain model with average grain radius 5.56 μm. (**a**) Circular model in 2D. (**b**) 3D model in which red line defines μ-capacitors in parallel (2+4 model). (**c**) Time-dependent intergranular μ-capacitance for sintering of $BaTiO_3$-ceramics

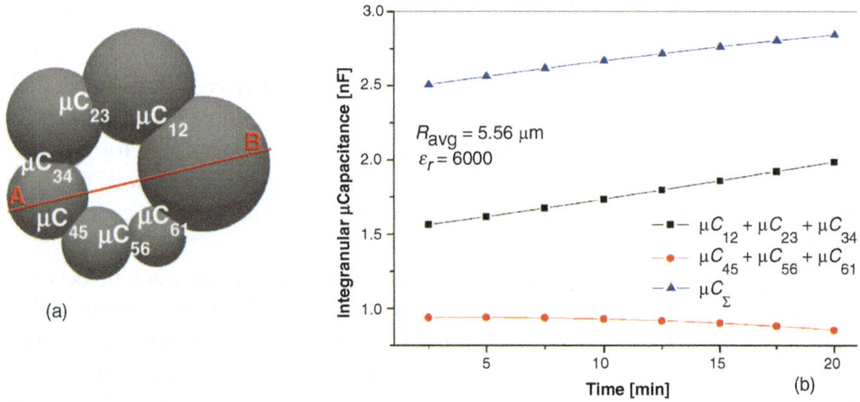

Fig. 9 Six-grain model with average grain radius 5.56 μm. (**a**) 3D model in which red line defines μ-capacitors in parallel (3+3 model). (**b**) Time-dependent intergranular μ-capacitance for sintering of $BaTiO_3$-ceramics

capacitors in star connection (shown in Figure 14). These two models require knowledge of the corresponding transformations of one model into the other and vice versa, defined in Appendices 2 and 3.

4.2 Time-Dependent Intergranular Capacitance

The defined and described intergranular μ-capacitor units, as well as specific μ-capacitors connections (aggregates of μ-capacitors) can be used to analyze the methodology for simulation and computation of (only time-dependent) intergranular capacitance of the sample and to gain insight of aggregate of μ-capacitor morphology.

In general, the multi-grain model represents (more or less) complex structure of μ-capacitors which occur in investigated sintered samples. It is, therefore, useful to use a set of rules for finding the equivalent capacitance of some real arrangement of μ-capacitors. It turns out that we can always find the equivalent capacitance by repeating application of simple rules. These rules are related to μ-capacitors connected in series, in parallel, in delta, and in star.

For intergranular capacitance computation, we need as the first step replace the solid skeleton network with the μ-capacitor network which is more appropriate and directly responsible for it. To find the total capacitance we identify μ-capacitors in series and in parallel. More complicated connections of μ-capacitors may be transformed into simpler structure using delta to star transformation (9)–(11) or star to delta transformations (12)–(14). In this way some μ-capacitors will be replaced by their equivalent μ-capacitors. At the end of this procedure we will find the total intergranular capacitance. Figure 10b demonstrates the evolution of the μ-capacitor network of 20 μ-capacitor units within the multi-grain model with thirteen grains. As we can see, this model initially contains both in series and in parallel connections of μ-capacitors. Figure 11b shows the overall time-dependent intergranular capacitance for sintering of $BaTiO_3$-ceramics multi-grain model. The computed result shows that the electrical capacitance of the equivalent μ-capacitor slowly increases according to the time-dependent neck growth within the μ-capacitor network. This process is characterized with appropriate densification manifested by decrease in center-to-center distance, followed by a relatively small grain rearrangement.

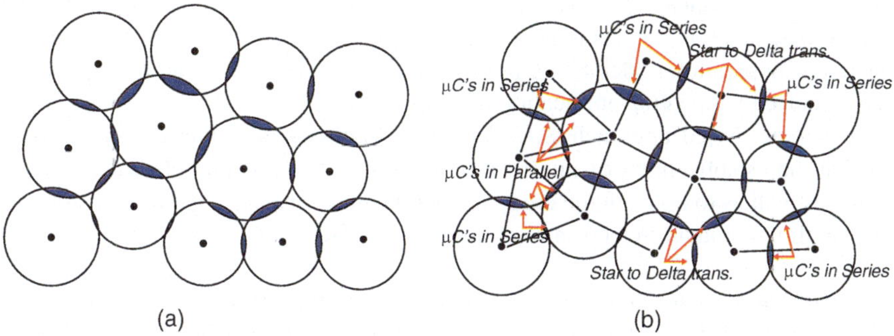

Fig. 10 Thirteen-grain model. (**a**) 2D model with 20 μ-capacitors. (**b**) Iterative steps in simplification of the complex intergranular geometry towards some of intergranular μ-capacitor units

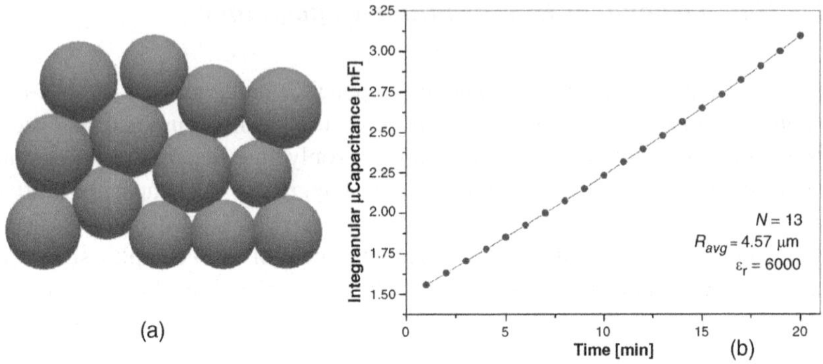

Fig. 11 Thirteen-grain model. (**a**) 3D model and (**b**) time-dependent intergranular μ-capacitance for sintering of $BaTiO_3$-ceramics

5 Conclusion

The paper presents an attempt to recognize, model, and establish an intergranular μ-capacitance unit which was used for definition of appropriate mathematical model for computation of an equivalent time-dependent intergranular capacitance.

The contact between two adjacent grains was defined as a structure that forms a μ-capacitor (the intergranular μ-capacitor unit) whose μ-capacitance changes as the neck grows and the grains approaching each other, by diffusion. It was assumed that diffusion mechanisms responsible for transport matter from the grain boundary to the neck are the volume diffusion and grain boundary diffusion. The model does not need special geometric assumptions because the microstructural development can be simulated by a set of simple local rules and overall neck growth law which can be arbitrarily chosen. The developed model can be applied for the prediction of an evolution of the intergranular capacitance during ceramic sintering of real systems with spherical particle distributions. Since the multi-grain model represents complex structure of μ-capacitors that occur in investigated sintered samples, we defined a set of rules (an algorithm) for finding the equivalent capacitance of some real arrangement of μ-capacitors. These rules are related to μ-capacitors connected in series, in parallel, in delta, and in star, so that we can always find the equivalent capacitance by repeating application of simple rules.

Computer-based approach extended to the multi-grain model can be used to enable the establishment of interrelation between structural and electrical parameters, as well as to assist in the creation, modification, analysis, and optimization of new high-performance electronic ceramics materials.

Acknowledgments This work and research is supported by the Ministry of Education, Science and Technological Development of the Republic of Serbia, under projects TR32012 and III43007.

Appendix 1: Neck Radius Computation

In computation of the time-dependent neck radius we will use the densification model of sintering based on Coble's two-grain model [15] and the extended model [16] in which diffusion mechanisms responsible for transport matter from the grain boundary to the neck are the volume diffusion and grain boundary diffusion (Figure 1a). Next assumptions will be made: volume conservation, center-to-center approach, and neck geometry simplified by straight line. For simulation purposes increasing the grain radius, induced by overlapping (defined by radius X_1) and maintaining the volume conservation of the two-grain model, will be neglected. All mentioned geometrical simplifications and notations are shown in Figure 12: the volume of the spherical cap

$$V_{\text{SphCap}} = \frac{\pi h}{6}(3x_1^2 + h^2) \quad \left(h = R - \frac{D}{2}\right),$$

the volume of the ring

$$V_{\text{Ring}} = V_{\text{Cyl}} - V_{\text{SphSeg}},$$

where the volume of the cylinder is defined as

$$V_{\text{Cyl}} = \pi x_2^2 y_2,$$

and the volume of the spherical segment as

$$V_{\text{SphSeg}} = \frac{1}{6}\pi y_2(3x_1^2 + 3x_2^2 + y_2^2)\,.$$

Fig. 12 Two-grain model for computation of the time-dependent neck radius

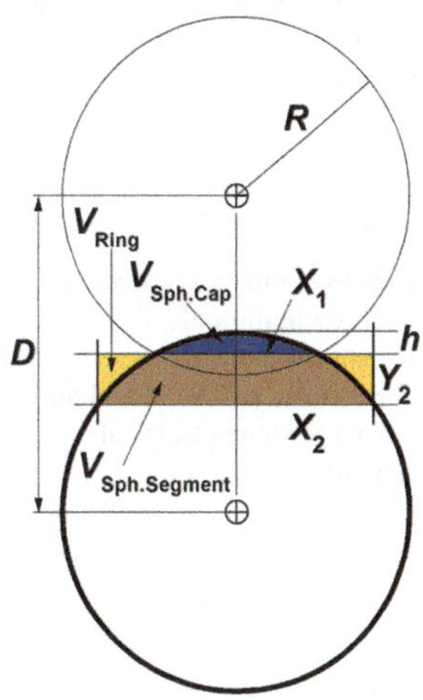

Taking into account distributing intersected volume to the neck width (at constant grain volume), the next algorithm (for numerical approaching, where ϵ is small enough positive number) will be applied for computation of the neck radius X_2:

$x_2 \mapsto x_1$
While $\left|V_{\text{Ring}} - V_{\text{SphCap}}\right| > \epsilon$
$\quad x_2 \mapsto x_2 + \Delta x$
Wend

Appendix 2: Delta to Star Transformation

Structure in Figure 13b can be transformed to the structure in Figure 14b as follows [17]:

$$\mu C_1 = \frac{\mu C_{12} \cdot \mu C_{13} + \mu C_{12} \cdot \mu C_{23} + \mu C_{13} \cdot \mu C_{23}}{\mu C_{23}}, \tag{9}$$

$$\mu C_2 = \frac{\mu C_{12} \cdot \mu C_{13} + \mu C_{12} \cdot \mu C_{23} + \mu C_{13} \cdot \mu C_{23}}{\mu C_{13}}, \tag{10}$$

$$\mu C_3 = \frac{\mu C_{12} \cdot \mu C_{13} + \mu C_{12} \cdot \mu C_{23} + \mu C_{13} \cdot \mu C_{23}}{\mu C_{12}}. \tag{11}$$

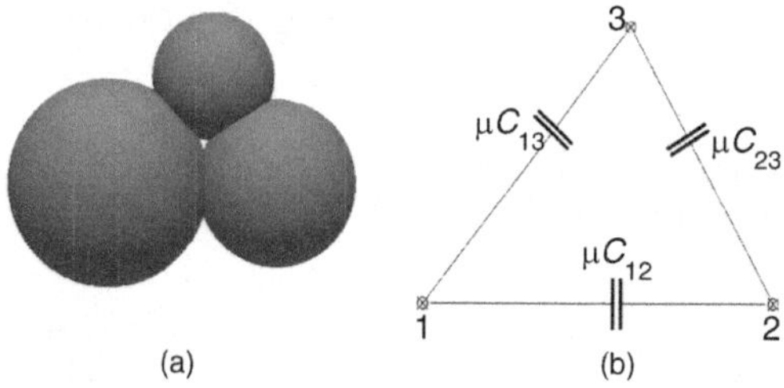

Fig. 13 Three-grain model as a representation of μ-capacitors in delta connection. (**a**) 3D model and (**b**) appropriate electrical scheme

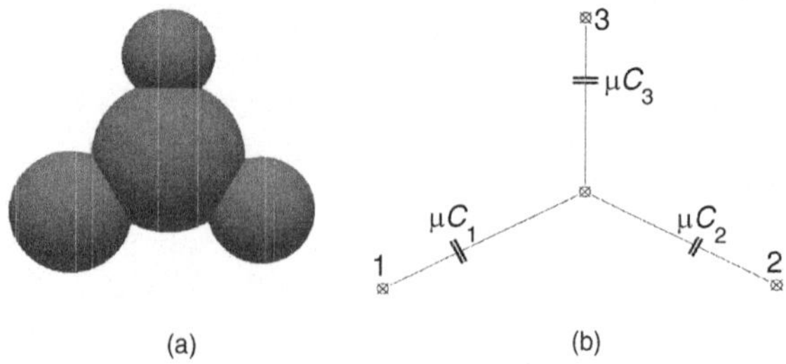

Fig. 14 Four-grain model as a representation of μ-capacitors in star connection. (**a**) 3D model and (**b**) appropriate electrical scheme

Appendix 3: Star to Delta Transformation

Structure in Figure 14b can be transformed to the structure in Figure 13b as follows [17]:

$$\mu C_{12} = \frac{\mu C_1 \cdot \mu C_2}{\mu C_1 + \mu C_2 + \mu C_3}, \tag{12}$$

$$\mu C_{13} = \frac{\mu C_1 \cdot \mu C_3}{\mu C_1 + \mu C_2 + \mu C_3}, \tag{13}$$

$$\mu C_{23} = \frac{\mu C_2 \cdot \mu C_3}{\mu C_1 + \mu C_2 + \mu C_3}. \tag{14}$$

References

1. J. Daniels, K.H. Hardtl, R. Wernicke, The PTC effect of barium titanate. Philips Tech. Rev. **38**(3), 73–82 (1978/1979)
2. Z.S. Nikolic, V.V. Mitic, I.Z. Mitrovic, Modeling of intergranular impedance as a function of consolidation parameters, in *Proceedings of 4th International Conference on Telecommunications in Modern Satellite, Cable and Broadcasting Services TELSIKS 99*, vol. 2, 673–676 (1999)
3. Z.S. Nikolic, Simulation of intergranular impedance as a function of diffusion processes. J. Mater. Sci. Mater. Electron. **13**, 743–749 (2002)
4. N.S. Hari, P. Padmini, T.R.N. Kutty, Complex impedance analyses of n-$BaTiO_3$ ceramics showing positive temperature coefficient of resistance. J. Mater. Sci.: Mater. Electron. **8**, 15–22 (1997)
5. Z. Liang, M.A. Ioannidis, I. Chatzis, Geometric and topological analysis of three-dimensional porous media: pore space partitioning based on morphological skeletonization. J Colloid Interface Sci. **221**, 13–24 (2000)
6. Z.S. Nikolic, Computer simulation of liquid phase sintering: gravity induced skeletal structure evolution - a review. Mater. Sci. Forum **624**, 19–42 (2009)
7. R.M. German, Y. Liu, Grain agglomeration in liquid phase sintering. J. Mater. Synth. Process **4**(1), 23–34 (1996)
8. Z.S. Nikolic, Theoretical study of skeletal structure evolution under topological constraints during sintering. Math. Comput. Modell. **57**, 1060–1069 (2013)
9. Z.S. Nikolic, Numerical method for computer study of liquid phase sintering: densification due to gravity induced skeletal settling, in *Approximation and Computation*. Springer Optimization and Its Applications, vol. 42, ed. by W. Gautschi, G. Mastroianni, T.M. Rassias, vol. XXII, 1st edn. (2010), pp. 409–424
10. M.N. Rahman, *Sintering of Ceramics* (CRC Press, New York, 2007)
11. S.-J.L. Kang, *Sintering - Densification, Grain Growth and Microstructure* (Elsevier, Burlington, MA, 2005)
12. R.L. Coble, Initial sintering of alumina and hematite. J. Am. Ceram. Soc. **41**(2), 55–62 (1958)
13. R. German, Coordination number changes during powder densification. Powder Technol. **253**, 368–376 (2014)
14. R.M. German, *Powder Metallurgy Science* (Metal Powder Industry Federation, Princeton, 1994)

15. R.L. Coble, Effects of particle size distribution in initial stage sintering. J. Am. Ceram. Soc. **56**, 461–466 (1973)
16. J.H. Chen, P.F. Johnson, Computer simulation of initial stage sintering in two-dimensional particulate systems, in *Microbeam Analysis*, ed. by P.E. Russell (1989), pp. 405–409
17. R. Fitzpatrick, Electromagnetism and Optics - An Introductory Course, The University of Texas at Austin, 2007

Variational Inequalities in Semi-inner Product Spaces

Nabin K. Sahu, Ouayl Chadli, and R. N. Mohapatra

Abstract Variational Inequalities play an important role in solving many outstanding problems ranging from Mechanics, Physics, Engineering, and Economics. The work of the Italian and French mathematicians laid a solid mathematical foundation and today, it is an interesting area of considerable research activity. The variational inequalities were first considered in Hilbert spaces and subsequently to Banach spaces. In 1961, Lumer introduced the theory of semi-inner product spaces. This was followed by the work of Giles and many other researchers. In this paper, we have mentioned most of the results in variational inequalities in semi-inner product spaces. The new results proved in this paper are for a system of variational inequalities in a semi-inner product space. These results throw a light into the structural study of variational inequalities in uniformly smooth Banach spaces.

MSC 2010 47H04, 47J20, 49J40, 47H10

1 Introduction

The theory of variational inequalities plays an important role in the study of both the qualitative and numerical analysis of a variety of subjects, ranging from mechanics, physics, and engineering to economics. This theory is a relatively young mathematical discipline, although its study dates back to the origins of the calculus of variations, where variational inequalities arose formally as "Euler equations" for constrained inequality problems. The basis of its development began in the

N. K. Sahu (✉)
Dhirubhai Ambani Institute of Information and Communication Technology, Gandhinagar, India

O. Chadli
Dept. of Economics, Ibn Zohr University, Agadir, Morocco

R. N. Mohapatra
Dept. of Mathematics, University of Central Florida, Orlando, FL, USA

N. J. Daras, T. M. Rassias (eds.), *Computational Mathematics and Variational Analysis*, Springer Optimization and Its Applications 159,
https://doi.org/10.1007/978-3-030-44625-3_23

sixties with the work of Fichera [12], who coined the term "Variational Inequality" in his paper on the solution of the frictionless contact problem between a linear elastic body and a rigid foundation posed by Signorini [37]. The foundation of the mathematical theory of variational inequalities was developed in the sixties and seventies especially by the Italian and French mathematical schools: Stampacchia [39, 40], Hartman and Stampacchia [20], Lions and Stampacchia [26], Lions [25], Fichera [13]. See also the well-known classical monographs by Kinderlehrer and Stampacchia [23], Freidman [14], Duvaut and Lions [10], Glowinski, Lions and Timolières [16], Baiocchi and Capelo [1], Kikuchi and Oden [22], where variational inequalities and their applications to problems of mathematical physics are broadly discussed in infinite-dimensional spaces. The book of Goeleven et al. [17] gives a good overview of the various mathematical approaches for coercive and noncoercive variational inequalities. Evolutionary variational inequalities have been treated initially, by Brézis [2], who also connected the notion of variational inequality to convex sub-differentials and maximal monotone operators.

Finite-dimensional variational inequalities, which is a generalization of nonlinear complementarity problems, also began in the middle of the 1960s but followed a different path. During five decades, they became a very fruitful discipline in the field of mathematical programming. The breakthrough in the theory of finite-dimensional variational inequalities occurred in 1980 when Dafermos [9] recognized that the traffic network equilibrium conditions as stated by Smith [38] had the structure of an inequality. This unveiled this methodology for the study of problems in economics, management science, operations research, and also in engineering, with a focus on transportation. Finite-dimensional variational inequality theory provides us with a tool for: formulating a variety of equilibrium problems; qualitatively analyzing the problems in terms of existence and uniqueness of solutions, stability and sensitivity analysis, and providing us with algorithms with accompanying convergence analysis for computational purposes, see the survey paper by Harker and Pang [19], and the books by Facchinei and Pang [11]. It contains, as special cases, such well-known problems in mathematical programming as: systems of nonlinear equations, optimization problems, complementarity problems, and is also related to fixed point problems. Observe that many of the applications explored to-date on the study of the equilibrium state of competitive systems arising in different disciplines have been formulated, studied, and solved as finite-dimensional variational inequality problems. Examples of equilibrium problems include: markets in which firms compete to determine their profit maximizing production outputs, spatial economic systems in which the optimal commodity production, consumption, and interregional trade patterns are to be computed, congested urban transportation systems in which users seek to determine their cost-minimizing routes of travel, general economic equilibrium problems in which all the commodity prices are to be determined, and general financial equilibrium problems in which the optimal composition of instruments in each sector's portfolio and the instrument prices are sought, see the books by Nagurney [28]. The complexity and often large-scale nature of such systems have stimulated the development of a variety of mathematical methodologies for their analysis and computation. Foremost of the methodologies has been the theory of finite-dimensional variational inequalities, which has yielded a powerful tool for both the qualitative analysis of equilibria governed by entirely distinct equilibrium concepts, as well as, theoretically rigorous computational procedures, see the books by Nagurney and Qiang [30], Nagurney and Li [29].

The finite-dimensional variational inequality problem, VI(F, K), is to determine a vector $\bar{x} \in K \subset \mathbb{R}^n$, such that

$$F(\bar{x})^\top \cdot (x - \bar{x}) \geq 0, \quad \text{for all } x \in K, \tag{1}$$

or, equivalently,

$$\langle F(\bar{x}), x - \bar{x} \rangle \geq 0, \quad \text{for all } x \in K,$$

where F is a given continuous function from K to $\mathbb{R}^n$, K is a given closed convex set, and $\langle \cdot, \cdot \rangle$ denotes the inner product in n-dimensional Euclidean space, as does "$\cdot$". In geometric terms, the variational inequality (1) states that $F(\bar{x})^\top$ is "orthogonal" to the feasible set K at the point $\bar{x}$. In order to show how finite-dimensional variational inequalities are used in the formulation of many economic equilibrium problems, we give the following examples applicable to equilibrium analysis:

- **Systems of Equations**. Many classical economic equilibrium problems have been formulated as systems of equations, since market clearing conditions necessarily equate the total supply with the total demand. In terms of a variational inequality problem, the formulation of a system of equations is as follows: when $K = \mathbb{R}^n$, it is easy to see that a vector $\bar{x}$ solves VI$(F, \mathbb{R}^n)$ if and only if $F(\bar{x}) = \mathbf{0}$. As an illustration, we present the following example on Market Equilibrium with equalities: Consider m consumers, with a typical consumer denoted by j, and n commodities, with a typical commodity denoted by i. Let p denote the n-dimensional vector of the commodity prices with components: $\{p_1, \cdots, p_n\}$. Assume that the demand for a commodity i, d_i, may, in general, depend upon the prices of all the commodities, that is, $d_i(p) = \sum_{j=1}^{m} d_i^j(p)$, where $d_i^j(p)$ denotes the demand for commodity i by consumer j at the price vector p. Similarly, the supply of a commodity i, s_i, may, in general, depend upon the prices of all the commodities, that is, $s_i(p) = \sum_{j=1}^{m} s_i^j(p)$, where $s_i^j(p)$ denotes the supply of commodity i of consumer j at the price vector p. We group the aggregate demands for the commodities into the n-dimensional column vector d with components: $\{d_1, \cdots, d_n\}$ and the aggregate supplies of the commodities into the n-dimensional column vector s with components: $\{s_1, \cdots, s_n\}$. The market equilibrium conditions that require that the supply of each commodity must be equal to the demand for each commodity at the equilibrium price vector $\bar{p}$, are equivalent to the following system of equations: $s(\bar{p}) - d(\bar{p}) = \mathbf{0}$. Clearly, this expression can be put into the standard nonlinear equation form, if we define the vectors $x \equiv p$ and $F(x) = s(p) - d(p)$.

- **Complementarity Problems**. The finite-dimensional variational inequality problem also contains the complementarity problem as a special case. Complementarity problems are defined on the nonnegative orthant. Let $\mathbb{R}^n_+$ denote the nonnegative orthant in $\mathbb{R}^n$, and let $F : \mathbb{R}^n \to \mathbb{R}^n$ be a nonlinear mapping. The nonlinear complementarity problem over $\mathbb{R}^n_+$ is a system of equations and inequalities stated as: Find $\bar{x} \geq \mathbf{0}$ such that

$$F(\bar{x}) \geq 0 \quad \text{and} \quad F(\bar{x})^\top \cdot \bar{x} = 0. \tag{2}$$

Whenever the operator F is affine, that is, whenever $F(x) = Mx + b$, where M is an $n \times n$ matrix and b an $n \times 1$ vector, problem (2) is then known as the linear complementarity problem. The relationship between the complementarity problem and the variational inequality problem is as follows: $\mathrm{VI}(F, \mathbb{R}^n_+)$ and (2) have precisely the same solutions, if any.

As an example, we present a nonlinear complementarity formulation of market equilibrium problem with equalities and inequalities as constraints: Assume that the prices must now be nonnegative in the market equilibrium example presented earlier. Hence, we consider the following situation, in which the demand functions are given as previously as are the supply functions, but now, instead of the market equilibrium conditions, which are represented by a system of equations, we have the following equilibrium conditions. For each commodity i $(i = 1, \cdots, n)$:

$$s_i(\bar{p}) - d_i(\bar{p}) \begin{cases} = 0, & \text{if } \bar{p}_i > 0; \\ \geq 0, & \text{if } \bar{p}_i = 0. \end{cases}$$

These equilibrium conditions state that if the price of a commodity is positive in equilibrium, then the supply of that commodity must be equal to the demand for that commodity. On the other hand, if the price of a commodity at equilibrium is zero, then there may be an excess supply of that commodity at equilibrium, that is, si $s_i(\bar{p}) - d_i(\bar{p}) > 0$, or the market clears. Furthermore, this system of equalities and inequalities guarantees that the prices of the instruments do not take on negative values, which may occur in the system of equations expressing the market clearing conditions. The nonlinear complementarity formulation of this problem is as the following: Find a vector price $\bar{p} \in \mathbb{R}^n_+$ such that

$$s(\bar{p}) - d(\bar{p}) \geq \mathbf{0} \quad \text{and} \quad \langle s(\bar{p}) - d(\bar{p}), \bar{p} \rangle = 0.$$

Since a nonlinear complementarity problem is a special case of a variational inequality problem, we may rewrite the nonlinear complementarity formulation of the market equilibrium problem above as the following variational inequality problem: find a vector price $\bar{p} \in \mathbb{R}^n_+$ such that

$$\langle s(\bar{p}) - d(\bar{p}), p - \bar{p} \rangle \geq 0, \quad \text{for all } p \in \mathbb{R}^n_+.$$

- **Fixed Point Problems**. Fixed point theory has been used to formulate, analyze, and compute solutions to economic equilibrium problems. The relationship between the variational inequality problem and a fixed point problem can be made through the use of a projection operator. The relationship between a variational inequality and a fixed point problem is as follows: Assume that K is closed and convex subset of $\mathbb{R}^n$. Then $\bar{x} \in K$ is a solution of the variational inequality problem VI(F, K) if and only if for any $\alpha > 0$, $\bar{x}$ is a fixed point of the map $P_K(I - \alpha F) : K \to K$, that is $\bar{x} = P_K(\bar{x} - \alpha F(\bar{x}))$, where $P_K(\cdot)$ is the projection operator on K.

2 Variational Inequalities in Banach Space

The theory of variational inequalities has been extended to Banach spaces (see Browder [3], Yao [42], Chang [7]). The classical variational inequality problem in Banach space can be traced back to Browder [3]. Let X be a Banach space with dual X^*. Let K be a nonempty closed convex subset of X, and $T : K \to X^*$ be a mapping. The classical variational inequality problem in a Banach space is to find $x \in K$ such that

$$(Tx, y - x) \geq 0, \quad \forall y \in K, \tag{3}$$

where $(.,.)$ denotes the duality pairing between X^* and X.

The above problem has been discussed by Browder [3] in a reflexive Banach space. He proved that

Theorem 2.1 ([3]) *Let E be a reflexive Banach space with dual E^*, and K be a nonempty closed convex subset of E, with $0 \in K^*$. Let $T : K \to E^*$ be a monotone and hemi-continuous map. Then the variational inequality problem*

$$(Tx, y - x) \geq 0, \quad \forall y \in K \tag{4}$$

has a solution in K.

Browder [3] also proved the result for a more general space. He proved the following result:

Theorem 2.2 ([3]) *Let K be a nonempty compact convex subset of a locally convex topological vector space E. let $A : K \to E^*$ be a continuous mapping. Then the variational inequality problem*

$$(Ax, y - x) \geq 0, \quad \forall y \in K \tag{5}$$

has a solution in K.

Definition 2.1 Let E be a normed space, $A : E \to E^*$ be a mapping, and $x_0 \in E$. The map A is said to be demi-continuous at x_0 if for any given $y \in E$, $A(x_0 + t_n y)$ weak* converges $A(x_0)$ when $t_n \to 0,\ t_n \geq 0$.

Chang [6] proved the following result for a monotone demi-continuous mapping in a reflexive Banach space.

Theorem 2.3 ([6]) *Let E be a reflexive Banach space and K be a nonempty closed convex bounded subset of E. Let $A : K \to E^*$ be a monotone demi-continuous mapping. Then*

(i) The variational inequality (5) has a solution.
(ii) The set of solutions of (5) is closed and convex.
(iii) If A is strictly monotone, then the problem (5) has a unique solution.

Plubtieng and Sombut [34] established the following result for a system of two variational inequalities in a reflexive Banach space.

Theorem 2.4 ([34]) *Let E be a reflexive Banach space and K be a compact convex subset of E. Let $A, B : K \to E^*$ be two continuous mappings. Then the system of variational inequalities*

$$(Ax, z - y) \geq 0,\ \forall z \in K$$
$$(By, z - x) \geq 0,\ \forall z \in K$$

has a solution $(x, y) \in K \times K$, and the set of solutions is closed.

Li [24] considered the following variational inequality problem:
Find $x \in K$ such that

$$(Tx - \xi, y - x) \geq 0,\ \ \forall y \in K, \tag{6}$$

where $\xi \in X^*$ and $(.,.)$ denotes the duality pairing between X^* and X.

The author has used the generalized projection in Banach space to solve the problem (6). The existence of solution was proved by using the well-known Fan-KKM Theorem. The problem was first converted into a fixed point type of problem using generalized projection operator then one iterative algorithm is constructed to approximate the solution.

There are many generalizations of the problem (3) and (6) in Banach spaces, and various solution techniques have been proposed. One may refer to [5, 8, 36], and the references there in. Most recently, Cai et al. [4] solved the variational inequality problem (3) in a Banach space using a double projection method. They have combined the Halpen's technique and the subgradient extragradient idea and proposed a new algorithm to solve the problem.

3 Introduction to Semi-inner Product

Studies of variational inequalities in Hilbert space has been carried out by many researchers. Subsequently, the research on variational inequalities were extended to Banach spaces and topological spaces. In 1961, Lumer [27] introduced the theory of semi-inner product spaces. We are going to demonstrate how variational inequalities in uniformly smooth Banach spaces can be studied with the help of the semi-inner product introduced by Lumer.

With a view to establishing simple operator theoretic results and to build Hilbert space like theory in general Banach spaces, G. Lumer [27] introduced the notion of semi-inner product on a Banach space X. Semi-inner product is an inner product like map $[.,.] : X \times X \to F(\mathbb{C}$ or $\mathbb{R})$ which is linear in the first argument, positive definite, and satisfies the Schwarz inequality. Precisely, the definition of the semi-inner product is as follows:

Definition 3.1 ([27]) A mapping $[.,.] : X \times X \to F(\mathbb{C}$ or $\mathbb{R})$ is called a semi-inner product if it satisfies the following:

1. $[x + y, z] = [x, z] + [y, z]$
   and $[\alpha x, y] = \alpha[x, y], \quad \forall x, y, z \in X$ and $\alpha \in F$;
2. $[x, x] > 0, \quad \forall x \neq 0$ and $x \in X$;
3. $|[x, y]|^2 \leq [x, x][y, y], \quad \forall x, y \in X.$

The vector space equipped with such a semi-inner product structure $[.,.]$ is called a semi-inner product space. Lumer discovered that the semi-inner product $[.,.]$ induces a norm in the natural way by defining $[x, x]^{\frac{1}{2}} = \|x\|$. Thus every semi-inner product space is a normed space endowed by this norm $\|x\| = [x, x]^{\frac{1}{2}}$. Lumer also proved that every normed linear space has at least one semi-inner product that is compatible with the norm. In fact every normed space has infinitely many semi-inner products.

Giles [15] extended the work of Lumer and proved that conjugate homogeneity property in the second argument holds true in a semi-inner product space, that is $[x, \lambda y] = \overline{\lambda}[x, y], \quad \forall x, y \in X, \ \lambda \in \mathbb{C}$. Giles also established the following Riesz representation type result in a uniformly convex smooth Banach space.

Theorem 3.1 *In a continuous semi-inner product space X which is uniformly convex and complete in its norm, to every continuous linear functional $f \in X^*$ there exists a unique vector $y \in X$ such that $f(x) = [x, y]$ for all $x \in X$.*

This theorem of Giles made the semi-inner product theory more applicable because in a uniformly convex smooth Banach space there exists a unique semi-inner product with the continuity property.

The following are examples of uniformly convex smooth Banach spaces where the semi-inner product is defined uniquely.

Example 3.1 The real sequence space l^p for $1 < p < \infty$ is a uniformly convex smooth Banach space with a unique semi-inner product defined by

$$[x, y] = \frac{1}{\|y\|_p^{p-2}} \sum_i x_i y_i |y_i|^{p-2}, \quad x, y \in l^p.$$

Example 3.2 (Giles [15]) The functions space $L^p(X, \mu)$ for $1 < p < \infty$ is a uniformly convex smooth Banach space with a unique semi-inner product defined by

$$[f, g] = \frac{1}{\|g\|_p^{p-2}} \int_X f(x)|g(x)|^{p-1}\, sgn(g(x))d\mu, \quad f, g \in L^p.$$

Definition 3.2 (Xu [41]) Let X be a real Banach space. The modulus of smoothness of X is defined as

$$\rho_X(t) = \sup\left\{\frac{\|x+y\| + \|x-y\|}{2} - 1 : \|x\| = 1, \|y\| = t, t > 0\right\}.$$

X is said to be uniformly smooth if $\lim\limits_{t\to 0} \frac{\rho_X(t)}{t} = 0$.

X is said to be p-uniformly smooth if there exists a positive real constant c such that $\rho_X(t) \le ct^p,\ p > 1$.

X is said to be 2-uniformly smooth if there exists a positive real constant c such that $\rho_X(t) \le ct^2$.

Lemma 3.1 (Xu [41]) *Let X be a smooth Banach space. Then the following statements are equivalent:*

(i) X is 2-uniformly smooth.
(ii) There is a constant $c > 0$ such that for every $x, y \in X$, the following inequality holds

$$\|x + y\|^2 \le \|x\|^2 + 2\langle y, f_x\rangle + c\|y\|^2, \tag{7}$$

where $f_x \in J(x)$ and $J(x) = \left\{x^* \in X^* : \langle x, x^*\rangle = \|x\|^2 \text{ and } \|x\|^* = \|x\|\right\}$ is the normalized duality mapping, where X^* denotes the dual space of X and $\langle x, x^*\rangle$ denotes the value of the functional x^* at x, that is $x^*(x)$.

Remark 3.1 Every normed linear space is a semi-inner product space (see Lumer [27]). In fact by Hahn–Banach theorem, for each $x \in X$, there exists at least one functional $f_x \in X^*$ such that $\langle x, f_x\rangle = \|x\|^2$. Given any such mapping f from X into X^*, it has been verified that $[y, x] = \langle y, f_x\rangle$ defines a semi-inner product. Hence we can write the inequality (7) as

$$\|x + y\|^2 \le \|x\|^2 + 2[y, x] + c\|y\|^2, \forall x, y \in X. \tag{8}$$

The constant c is chosen with the best possible minimum value. We call c, as the constant of smoothness of X.

Example 3.3 The functions space L^p is 2-uniformly smooth for $p \geq 2$ and it is p-uniformly smooth for $1 < p < 2$.

If $2 \leq p < \infty$, then we have for all $x, y \in L^p$,

$$\|x + y\|^2 \leq \|x\|^2 + 2[y, x] + (p - 1)\|y\|^2.$$

Here the constant of smoothness is $p - 1$.

4 Variational Inequalities in Banach Spaces in the Frame Work of Semi-inner Product

The objective of this section is to study the variational inequalities in the frame work of the semi-inner product. Semi-inner product is an alternative substitute for arbitrary bounded linear functionals. It helps to establish comparatively simpler theoretical results, and one can construct concrete examples by using the semi-inner product structure in the Banach space. As discussed in the previous section if the space X is uniformly convex and smooth, then there is a unique semi-inner product associated with it. Thus variational inequality problem (3) can be written as
Find $x \in K$ such that

$$[Tx, y - x] \geq 0, \quad \forall y \in K, \tag{9}$$

where $T : X \to X$ is a map, and $[.,.]$ is a unique semi-inner product associated with X. Other generalizations of variational inequality problems can be written accordingly.

It is well known that under certain conditions variational inequality problem is equivalent to complementarity problem in the sense of finding the solution. The complementarity problem in the Banach spaces in the frame work of the semi-inner product was studied by Nanda [31], Nath et al. [32], Khan [21], Sahu et al. [35].
Nanda [31] discussed the following complementarity problem:
Find $x \in K$ such that

$$[Tx, x] = 0, \tag{10}$$

where K be a closed convex cone with vertex 0 in a uniformly convex and strongly smooth Banach space X with the semi-inner product $[.,.]$, and $T : K \to X$ be a mapping. He proved that if T is Lipschitz continuous with Lipschitz constant b, T is strongly monotone with monotonicity constant c, and $b^2 < 2c < b^2 + 1$, then the problem (10) has a unique solution.
Nath et al. [32] considered the following complementarity problem:
Find $x \in K$ such that

$$[Tx - Cx + 2x, 2x] = 0, \tag{11}$$

where K be a closed convex cone with vertex 0 in a uniformly convex and strongly smooth Banach space X with the semi-inner product $[.,.]$, and $T, C : K \to X$ are some mappings. They proved that if T is contractive and C is non expansive, then there is a unique solution to the problem (11).
Khan [21] considered a more general problem and proved the following theorem:

Theorem 4.1 *Let X be a uniformly convex and strongly smooth Banach space with the semi-inner product $[.,.]$, and K be a closed convex cone in X. Let $T : K \to X$ be a contractive mapping, $S : K \to X$ be a non expansive mapping, and $A : X \to X$ be Lipschitz continuous with constant $b > 0$. Then there is a unique $x_0 \in K$ such that*

$$[Tx_0 - Sx_0 + (2+b)x_0, (2+b)x_0] = 0. \tag{12}$$

Sahu et al. [35] considered the following F-implicit complementarity and F-implicit variational inequality problems:

Definition 4.1 Extended F-implicit complementarity problem (Extended (F-ICP)): Let X be a continuous real semi-inner product space and K be a non empty closed convex subset of X. Let $F, f : K \to \mathbb{R}$ be two real valued functions on K, $g : K \to K$ be another function, and c be a real positive constant. The extended (F-ICP) is to find $\overline{x} \in K$ such that

$$[g(\overline{x}), \overline{x}] + F(g(\overline{x})) + cf(\overline{x}) = 0, \quad [y, \overline{x}] + F(y) \geq 0, \quad \forall y \in K. \tag{13}$$

Definition 4.2 Extended F-implicit variational inequality problem (Extended (F-IVIP)): Let X be a continuous real semi-inner product space and K be a non empty closed convex subset of X. Let $F, f : K \to \mathbb{R}$ be two real valued functions on K, $g : K \to K$ be another function, and c be a real positive constant. The extended (F-IVIP) is to find $\overline{x} \in K$ such that

$$[y - g(\overline{x}), \overline{x}] \geq F(g(\overline{x})) - F(y) + cf(\overline{x}), \quad \forall y \in K. \tag{14}$$

Sahu et al. [35] proved the following equivalence theorem between the above two problems in terms of their existence of solution.

Theorem 4.2 ([35]) *If $\overline{x}$ is a solution of the extended (F-ICP), then $\overline{x}$ is also a solution of the extended (F-IVIP). Conversely, if $F : K \to \mathbb{R}$ is a positive homogeneous and convex function, $f : K \to \mathbb{R}$ is positive and $\overline{x}$ solves the extended (F-IVIP), then $\overline{x}$ solves the extended (F-ICP).*

They also proved the following theorem for existence of solution of extended (F-IVIP) using the well-known KKM-lemma.

Theorem 4.3 ([35]) *Let X be a continuous real semi-inner product space and K be a nonempty closed convex subset of X. Assume that*

(a) $F, f : K \to \mathbb{R}$ *are lower semi-continuous functions,* $g : K \to K$ *is continuous, and* c *is a positive real constant;*
(b) *there exists a function* $h : K \times K \to \mathbb{R}$ *such that*

(i) $h(x, x) \geq 0, \;\; \forall x \in K$
(ii) $h(x, y) - [y - g(x), x] \leq F(y) - F(g(x)) - cf(x), \;\; \forall x, y \in K$
(iii) *the set* $\{y \in K : h(x, y) < 0\}$ *is convex for all* $x \in K$*;*

(c) *there exists a non empty, compact, and convex subset* C *of* K *such that for all* $x \in K \setminus C$ *(complement of* C*), there exists* $y \in C$ *such that* $[y - g(x), x] < F(g(x)) - F(y) + cf(x)$.

Then the extended (F-IVIP) has a solution. Moreover, the solution set of the extended (F-IVIP) is compact.

5 System of Multivariate Variational Inequalities in Uniformly Convex Banach Spaces in the Frame Work of Semi-inner Product

In this section, we introduce the following system of multivariate variational inequality problems: Find $(x_1, x_2, \ldots., x_N) \in K^N$ such that

$$
\begin{aligned}
&[A_1(x_1, x_2, \ldots., x_N), \; y_1 - x_1] \geq 0, \;\; \forall y_1 \in K; \\
&[A_2(x_1, x_2, \ldots., x_N), \; y_2 - x_2] \geq 0, \;\; \forall y_2 \in K; \\
&-------------------- \\
&-------------------- \\
&[A_N(x_1, x_2, \ldots., x_N), \; y_N - x_N] \geq 0, \;\; \forall y_N \in K, \qquad (15)
\end{aligned}
$$

where E is a uniformly convex smooth Banach space with a unique semi-inner product $[.,.]$, K is a nonempty closed convex bounded subset of E, and $A_1, A_2, \ldots.., A_N$ are N-variables monotone demi-continuous mappings from K^N into E.
To discuss the existence solution of the above system of multivariate variational inequalities, we need the following preliminaries.

We quote some results proved by Guan et al. [18] that are useful for proving our main result.

Lemma 5.1 ([18]) *Let* E *be a Banach space with the norm* $\|.\|$. *Consider the space* $E^N = E \times E \times \ldots \times E$, *Cartesian product of* E, N *times. Then the functional* $\|.\|_N : E^N \to \mathbb{R}$ *defined by*

$$\|x\|_N = \sqrt{\sum_{i=1}^{N} \|x_i\|^2}, \;\; \forall x = (x_1, x_2, \ldots., x_N) \in E^N$$

is a norm on E^N*. Also* $(E^N, \|.\|_N)$ *is a Banach space.*

Lemma 5.2 ([18]) *The following holds true for the space* E^N*:*

$$(E^N, \|.\|_N)^* = ((E, \|.\|)^*)^N.$$

Lemma 5.3 ([18]) *Let E be a reflexive Banach space with the norm* $\|.\|$*, and* $E^N = E \times E \times \ldots \times E$*, Cartesian product of E, N times. Then* $\|x\|_N = \sqrt{\sum_{i=1}^{N} \|x_i\|^2}$*,* $\forall x = (x_1, x_2, \ldots, x_N) \in E^N$ *is a norm for* E^N*, and* $(E^N, \|.\|_N)$ *is a reflexive Banach space.*

From now onwards we assume that E is a uniformly convex Banach space. Since every uniformly convex Banach space is reflexive, due to the Milman–Pettis theorem [33], all the above mentioned lemmas are also true for uniformly convex Banach spaces. By Milman–Pettis theorem [33], $(E^N, \|.\|_N)$ is a uniformly convex Banach space. Giles [15] had proved that in a uniformly convex smooth Banach space, it is possible to define a semi-inner product, uniquely. If the unique semi-inner product of E is $[.,.]$, then it is possible to define a semi-inner product in E^N, uniquely.
Let us define the functional $[.,.]_N : E^N \times E^N \to \mathbb{C}$ by

$$[X, Y]_N = \sum_{i=1}^{N} [x_i, y_i],$$

where $X = (x_1, x_2, \ldots, x_N),\ Y = (y_1, y_2, \ldots, y_N) \in E^N$.
In the following theorem, we verify that $[X, Y]_N$ is a semi-inner product on E^N.

Theorem 5.1 E^N *is a semi-inner product space with the semi-inner product* $[.,.]_N$

Proof $X = (x_1, x_2, \ldots, x_N),\ Y = (y_1, y_2, \ldots, y_N),\ Z = (z_1, z_2, \ldots, z_N) \in E^N$, and $\alpha \in \mathbb{C}$.

$$(i)\ [X+Y, Z]_N = [(x_1+y_1, x_2+y_2, \ldots, x_N+y_N), (z_1, z_2, \ldots, z_N)]_N$$
$$= \sum_{i=1}^{N} [x_i+y_i, z_i] = \sum_{i=1}^{N} [x_i, z_i] + \sum_{i=1}^{N} [y_i, z_i] = [X, Z]_N + [Y, Z]_N.$$

$$\begin{aligned}(ii)\ [\alpha X, Y]_N &= [\alpha(x_1, x_2, \ldots, x_N), (y_1, y_2, \ldots, y_N)]_N \\ &= [(\alpha x_1, \alpha x_2, \ldots, \alpha x_N), (y_1, y_2, \ldots, y_N)]_N \\ &= \sum_{i=1}^{N} [\alpha x_i, y_i] = \alpha \sum_{i=1}^{N} [x_i, y_i] = \alpha [X, Y]_N.\end{aligned}$$

$$(iii) \quad [X, X]_N = \sum_{i=1}^{N} [x_i, x_i] = \sum_{i=1}^{N} \|x_i\|^2 \geq 0$$

$$\text{and} \quad [X, X]_N = \sum_{i=1}^{N} \|x_i\|^2 = 0 \text{ if and only if } X = 0.$$

$$(iv) \quad |[X, Y]_N|^2 = |\sum_{i=1}^{N} [x_i, y_i]|^2 \leq \Big(\sum_{i=1}^{N} |[x_i, y_i]|\Big)^2 \leq \Big(\sum_{i=1}^{N} \|x_i\| \|y_i\|\Big)^2$$
$$\leq \Big(\sum_{i=1}^{N} \|x_i\|\Big)^2 \Big(\sum_{i=1}^{N} \|y_i\|\Big)^2$$
$$= [X, X]_N [Y, Y]_N.$$

From the above we saw that $[.,.]_N$ satisfies all the criterions of a semi-inner product, and hence E^N is a semi-inner product space with the semi-inner product $[.,.]_N$.

It is proved by Guan et al. [18] that $\|x\|_N = \sqrt{\sum_{i=1}^{N} \|x_i\|^2}$ is a norm for E^N. We see that $[X, X]_N = \sqrt{\sum_{i=1}^{N} \|x_i\|^2}$. Hence the semi-inner product generates the norm on E^N, that is $[X, X]_N = \|X\|_N^2$ for all $X \in E^N$.

5.1 Main Results

We assume that E is a uniformly convex Banach space with norm $\|.\|$ and semi-inner product $[.,.]$. Hence so is E^N with norm $\|X\|_N = \sqrt{\sum_{i=1}^{N} \|x_i\|^2}$ and semi-inner product $[X, Y]_N = \sum_{i=1}^{N} [x_i, y_i]$, where $X = (x_1, x_2, .., x_N)$, $Y = (y_1, y_2, .., y_N)$. We let K^N be nonempty closed convex bounded subset of E^N, and $A_i : K^N \to E$ be N-variables mappings for all $i = 1, 2, .., N$.

Definition 5.1 Let K be a nonempty subset of a Banach space E. An N-variable mapping $T : K^N \to E$ is said to be

(i) monotone if

$$[T(x_1, x_2, .., x_N) - T(y_1, y_2, .., y_N), x_i - y_i] \geq 0,$$
$$\forall i = 1, 2, .., N \text{ and for all } (x_1, x_2, .., x_N), (y_1, y_2, .., y_N) \in K^N.$$

(ii) strictly monotone if

$$[T(x_1, x_2, .., x_N) - T(y_1, y_2, .., y_N), x_i - y_i] = 0, \ \forall i = 1, 2, .., N$$
$$\Rightarrow (x_1, x_2, .., x_N) = (y_1, y_2, .., y_N).$$

Now we consider the variational inequality problem (15) and prove the existence of its solution.

Theorem 5.2 *Let K be a nonempty closed convex and bounded subset of a uniformly convex Banach space E. Let $A_i : K^N \to E$ be N-variables monotone demi-continuous mappings for all $i = 1, 2, .., N$. Then*

(i) The system of variational inequalities (15) has a solution $(x_1^*, x_2^*, .., x_N^*) \in K^N$.
(ii) The solutions set of (15) is closed and convex in K^N.
(iii) If $A_i's$ are strictly monotone for all $i = 1, 2, .., N$, then the system (15) has a unique solution.

Proof

(i) Let us define a map $A : K^N \to E^N$ by

$$A(X) = (A_1(X), A_2(X), .., A_N(X)),$$

where $X = (x_1, x_2, .., x_N)$ and $A_i's$ for all $i = 1, 2, .., N$ are N-variables monotone mappings.
We claim that A is also a monotone mapping. Since $A_i's$ are monotone for all $i = 1, 2, .., N$, we have

$$\begin{aligned} & [A(X) - A(Y), X - Y]_N \\ &= [\big(A_1(X) - A_1(Y), A_2(X) - A_2(Y), .., A_N(X) - A_N(Y)\big), X - Y]_N \\ &= \sum_{i=1}^{N} [A_i(X) - A_i(Y), x_i - y_i] \\ &\geq 0, \ \text{for all } X, Y \in K^N. \end{aligned}$$

Next, we prove that A is a demi-continuous map.

If for any given $X_0 \in K^N$ and $Y \in K^N$, then we have $X_0 + t_n Y \in K^N$. Since $A_i's$ are demi-continuous mappings, we see that

$$\begin{aligned} [A(X_0 + t_n Y), X]_N &= \sum_{i=1}^{N} [A_i(X_0 + t_n Y), x_i] \\ &\to \sum_{i=1}^{N} [A_i(X_0), x_i] \quad \text{as } t_n \to 0 \\ &= [A(X_0), X]_N, \ \ \forall X \in E^N. \end{aligned}$$

This proves that A is demi-continuous on K^N.

Since K is nonempty closed convex and bounded subset of E, K^N is a nonempty closed convex and bounded subset of E^N. Hence by Theorem 2.1, the following variational inequality

$$[A(X), Y - X]_N \geq 0, \ \ \forall Y = (y_1, y_2, .., y_N) \in K^N \tag{16}$$

has solution $X^* = (x_1^*, x_2^*, .., x_N^*) \in K^N$.

That is

$$[A(X^*), Y - X^*]_N \geq 0, \ \ \forall Y = (y_1, y_2, .., y_N) \in K^N. \tag{17}$$

We can write the above inequality (17) as

$$\sum_{i=1}^{N}[A_i(x_1^*, x_2^*, .., x_N^*), y_i - x_i^*] \geq 0, \ \ \forall Y = (y_1, y_2, .., y_N) \in K^N. \tag{18}$$

For any $y \in K$, if we take $Y = (y, x_2^*, x_3^*, .., x_N^*)$ in (17), then we get

$$[A_1(x_1^*, x_2^*, .., x_N^*), y - x_1^*] \geq 0, \ \ \forall y \in K. \tag{19}$$

For any $y \in K$, if we take $Y = (x_1^*, x_2^*, .., x_{j-1}^*, y, x_{j+1}^*, .., x_N^*)$ in (17), then we get

$$[A_j(x_1^*, x_2^*, .., x_N^*), y - x_j^*] \geq 0, \ \ \forall y \in K. \tag{20}$$

The above inequality (20) is true for all $j = 2, 3, .., N - 1$.
Similarly, for any $y \in K$, if we take $Y = (x_1^*, x_2^*, .., x_{N-1}^*, y)$ in (17), then we get

$$[A_N(x_1^*, x_2^*, .., x_N^*), y - x_N^*] \geq 0, \ \ \forall y \in K. \tag{21}$$

Therefore, from (19), (20), and (21) it is clear that $X^* = (x_1^*, x_2^*, .., x_N^*)$ is a solution of the system of multivariate variational inequality problem (15). Thus (i) is proved.
(ii) Let $X = (x_1, x_2, .., x_N)$ be an arbitrary solution of (15), that is

$$[A_1(x_1, x_2, ..., x_N), \ y_1 - x_1] \geq 0, \ \ \forall y_1 \in K;$$

$$[A_2(x_1, x_2, ..., x_N), \ y_2 - x_2] \geq 0, \ \ \forall y_2 \in K;$$

– –

– –

$$[A_N(x_1, x_2, ..., x_N), \ y_N - x_N] \geq 0, \ \ \forall y_N \in K.$$

Thus we have

$$[A(X), Y - X]_N = [(A_1(X), A_2(X), .., A_N(X)), Y - X]_N$$
$$= \sum_{i=1}^{N}[A_i(x_1, x_2, \dots, x_N), y_i - x_i] \geq 0, \quad \forall Y \in K^N.$$

Thus the system of variational inequality problems (15) is equivalent to the variational inequality problem (16). Since every uniformly convex Banach space is reflexive, and by using Theorem 2.3, we conclude that the solutions set of the variational inequality problem (16) is closed and convex. As a result the solutions set of the system of variational inequality problems (15) is also closed and convex.
(iii) We have

$$[A(X) - A(Y), X - Y]_N$$
$$= [\big(A_1(X) - A_1(Y), A_2(X) - A_2(Y), .., A_N(X) - A_N(Y)\big), X - Y]_N$$
$$= \sum_{i=1}^{N}[A_i(X) - A_i(Y), x_i - y_i]$$

Since $A's$ are strictly monotone we claim from the above equality that A is also strictly monotone. Now again by using Theorem 2.3, we arrive at the result that the variational inequality problem (16) has a unique solution. Therefore, the system of variational inequalities (15) has also a unique solution.

If $A_i = A$ for all $i = 1, 2, .., N$ in Theorem 5.2, then we have the following immediate corollary.

Corollary 5.1 *Let E be a uniformly convex Banach space, and K be a nonempty closed convex subset of E. Let $A : K^N \to E$ be an N-variables monotone demi-continuous mapping. Then*

(i) The system of multivariate variational inequalities

$$[A(x_1, x_2, .., x_N), y - x_i] \geq 0, \quad i = 1, 2, .., N$$

has a solution $(x_1^*, x_2^*, .., x_N^*) \in K^N$ for all $y \in K$.

Now we prove the following theorem:

Theorem 5.3 *Let E be a uniformly convex Banach space, and K be a nonempty compact convex subset of E. Let $A_1, A_2 : K \to E$ be two continuous mappings. Then the system of variational inequalities*

$$[A_1(x), z - y] \geq 0, \quad \forall z \in K$$
$$[A_2(y), z - x] \geq 0, \quad \forall z \in K \tag{22}$$

has a solution $(x, y) \in K \times K$. *Also the set of solutions of the above system of variational inequalities is closed.*

Proof Let us write $A_1(x, y) = A_1(y)$ and $A_2(x, y) = A_2(x)$ for all $(x, y) \in K \times K$. Then the system of variational inequalities (22) can be written as

$$[A_1(x, y), z - y] \geq 0, \quad \forall z \in K$$
$$[A_2(x, y), z - x] \geq 0, \quad \forall z \in K. \tag{23}$$

Let $A : K \times K \to E \times E$ be defined by

$$A(x, y) = \big(A_1(x, y), A_2(x, y)\big), \quad \forall (x, y) \in K \times K.$$

We denote the semi-inner product on $E \times E$ by $[.,.]_2$.
Since A_1 and A_2 are continuous on K, A is also continuous on $K \times K$. By Theorem 2.2, there exists a solution $(x^*, y^*) \in K \times K$, that is there exists $(x^*, y^*) \in K \times K$ such that

$$[A(x^*, y^*), (z_1, z_2) - (x^*, y^*)]_2 \geq 0, \quad \forall (z_1, z_2) \in K \times K$$
$$\Rightarrow [\big(A_1(x^*, y^*), A_2(x^*, y^*)\big), (z_1 - x^*, z_2 - y^*)]_2 \geq 0, \quad \forall (z_1, z_2) \in K \times K$$
$$\Rightarrow [\big(A_1(y^*), A_2(x^*)\big), (z_1 - x^*, z_2 - y^*)]_2 \geq 0, \quad \forall (z_1, z_2) \in K \times K$$
$$\Rightarrow [A_1(y^*), z_1 - x^*] + [A_2(x^*), z_2 - y^*] \geq 0, \quad \forall (z_1, z_2) \in K \times K. \tag{24}$$

If we take $z_2 = y^*$ and $z_1 = x^*$, separately in (24), we get

$$[A_1(y^*), z_1 - x^*] \geq 0, \forall z_1 \in K$$
$$\text{and} \quad [A_2(x^*), z_2 - y^*] \geq 0, \forall z_2 \in K.$$

Thus $(x^*, y^*) \in K \times K$ is a solution of the system of variational inequalities (22). Again, since A_1 and A_2 are continuous, so the set of solutions of (22) is closed.

References

1. C. Baiocchi, A. Capelo, *Variational and Quasivariational Inequalities: Applications to Free Boundary Problems* (Wiley, New York, 1984)
2. H. Brézis, Problèmes unilatéraux. J. Math. Pures Appl. **51**, 1–168 (1972)
3. F.E. Browder, Nonlinear monotone operators and convex sets in Banach spaces. Bull. Am. Math. Soc. **71**, 780–785 (1965)
4. G. Cai, A. Gibali, O.S. Iyiola , Y. Shehu, A new double projection method for solving variational inequalities in Banach spaces. J. Optim. Theory Appl. **178**, 219–239 (2018)
5. L.C. Ceng , Q.H. Ansari , J.C. Yao, On relaxed viscosity iterative methods for variational inequalities in Banach spaces. J. Comput. Appl. Math. **230**, 813–822 (2009)

6. S. Chang, Variational inequality and complementarity problem, in *Theory and Applications* (Shanghai Science and Technology Press, Shanghai, 1991)
7. S.S. Chang, The Mann and Ishikawa iterative approximation of solutions to variational inclusions with accretive type mappings. Comput. Math. Appl. **37**, 17–24 (1999)
8. S.S. Chang, Viscosity approximation methods for a finite family of nonexpansive mappings in Banach spaces. J. Math. Anal. Appl. **323**, 1402–1416 (2006)
9. S. Dafermos, Traffic equilibria and variational inequalities. Transport. Science **14**, 42–54 (1980)
10. G. Duvaut, J. L. Lions, *Inequalities in Mechanics and Physics*. (Springer-Verlag, Berlin, 1976)
11. F. Facchinei, J.S. Pang, *Finite-Dimensional Variational Inequalities and Complementarity Problems*. Springer Series in Operations Research, vols. I and II (Springer, New York, 2003)
12. G. Fichera, Problemi elastostatici con vincoli unilaterali, il problema di Signorini con ambigue condizioni al contorno. Mem. Accad. Naz. Lincei **8**(7), 91–140 (1964)
13. G. Fichera, Boundary value problems in elasticity with unilateral constraints, in *Encyclopedia of Physics*, vol. VI a/2, ed. by C. Truesdell. Mechanics of Solids II (Springer, Berlin, 1972), pp. 391–424
14. A. Friedman, *Variational Principles and Free Boundary Problems* (Wiley, New York, 1982)
15. J.R. Giles, Classes of semi-inner product spaces. Trans. Am. Math. Soc. **129**, 436–446 (1967)
16. R. Glowinski, J.-L. Lions, R. Trimolières, *Numerical Analysis of Variational Inequalities*. Studies in Applied Mathematics (North-Holland, Amsterdam, 1981)
17. D. Goeleven, D. Motreanu, Y. Dumont, M. Rochdi., *Variational and Hemivariational Inequalities: Theory, Methods and Applications. Volume I: Unilateral Analysis and Unilateral Mechanics*. Nonconvex Optimization and its Applications (Kluwer Academic Publishers, Dordrecht, 2003)
18. J. Guan, Y. Tang, Y. Xu, Y. Su, System of N fixed point operator equations with N-pseudo-contractive mapping reflexive Banach spaces. J. Nonlinear Sci. Appl. **10**, 2457–2470 (2017)
19. P.T. Harker, J.S. Pang, Finite dimensional variational inequality and nonlinear complementarity problems: a survey of theory, algorithms and applications. Math. Program. Ser. B **48**, 161–220 (1990)
20. P. Hartman, G. Stampacchia, On some nonlinear elliptic differential functional equations. Acta Math. **15**, 271–310 (1966)
21. M.S. Khan, An existence theorem for extended mildly nonlinear complementarity problem in semi-inner product spaces. Comment. Math. Univ. Carolin. **36**(1), 25–31 (1995)
22. N. Kikuchi, J.T. Oden, *Contact Problems in Elasticity: A Study of Variational Inequalities and Finite Element Methods*. SIAM Studies in Applied Mathematics (SIAM, Philadelphia, PA, 1988)
23. D. Kinderlehrer, G. Stampacchia, *An Introduction to Variational Inequalities and Their Applications* (Academic, New York, 1980); Classics in Applied Mathematics, vol. 31 (SIAM, Philadelphia, 2000)
24. J. Li, On the existence of solutions of variational inequalities in Banach spaces. J. Math. Anal. Appl. **295**, 115–126 (2004)
25. J.-L. Lions, *Quelques Méthodes de Résolution des Problèmes aux Limites non Linéaires* (Dunod, Paris, 1969)
26. J.-L. Lions, G. Stampacchia, Variational inequalities. Commun. Pure Appl. Anal. **20**, 493–519 (1967)
27. G. Lumer, Semi-inner product spaces. Trans. Am. Math. Soc. **100**, 29–43 (1961)
28. A. Nagurney, *Network Economics: A Variational Inequality Approach* (Kluwer Academic Publishers, Dordrecht, 1999)
29. A. Nagurney, D. Li, *Competing on Supply Chain Quality: A Network Economics Perspective* (Springer, Heidelberg, 2016)
30. A. Nagurney, Q. Qiang, *Fragile Networks, Identifying Vulnerabilities and Synergies in an Uncertain World* (Wiley, Hoboken, 2009)
31. S. Nanda, A nonlinear complementarity problem in semi-inner product spaces. Rendiconti di Mahematica **1**, 167–171 (1982)

32. B. Nath, S.N. Lal, R.N. Mukerjee, A generalized nonlinear complementarity problem in semi-inner product space. Indian J. Pure Appl. Math. **21**(2), 140–143 (1990)
33. B.J. Pettis, A proof that every uniformly convex space is reflexive. Duke Math J. **5**, 249–253 (1939)
34. S. Plubtieng, K. Sombut, Existence results for system of variational inequality problems with semimonotone operators. J. Inequal. Appl. (2010). https://doi.org/10.1155/2010/251510
35. N.K. Sahu , C. Nahak, S. Nanda, The extended F-implicit complementarity and variational inequality problems in semi-inner product spaces. Acta Universitatis Apulensis **35**, 111–123 (2013)
36. N. Shahzad, A. Udomene, Fixed point solutions of variational inequalities for asymptotically nonexpansive mappings in Banach spaces. Nonlinear Anal. **64**, 558–567 (2006)
37. A. Signorini, Sopra alcune questioni di elastostatica. Atti della Società Italiana per il Progresso delle Scienze (1933)
38. M.J. Smith, The existence, uniqueness and stability of traffic equilibria. Transport. Res. **13B**, 295–304 (1979)
39. G. Stampacchia, Formes bilineaires coercitives sur les ensembles convexes. C. R. Acad. Sci. Paris **258**, 4413–4416 (1964)
40. G. Stampacchia, Le probleme de Dirichlet pour les équations elliptiques du second ordre à coefficients discontinus. Ann. Inst. Fourier **258**, 189–258 (1965)
41. H.K. Xu, Inequalities in Banach spaces with applications. Nonlinear Anal. Theory Methods Appl. **16**(12), 1127–1138 (1991)
42. J.C. Yao, General variational inequalities in Banach spaces. Appl. Math. Lett. **5**(1), 51–54 (1992)

Results Concerning Certain Linear Positive Operators

Danyal Soybaş and Neha Malik

Abstract The present paper deals with some convergence results concerning Gupta-type operators. Varied sequences of linear positive operators have been discussed and their approximation properties been studied in literature. This work is a collection of these operators introduced over the past three decades.

1 Preliminaries

Approximation theory deals with the behaviour of functions to examine how they can be effectively approximated by simpler functions. This field of mathematical research gained significance in the mid-twentieth century. Researchers have studied the convergence behaviour by applying diverse methods for exactness. To accomplish this precision, adequate and voluminous analysis is involved. The objective of the prevailing estimation techniques concerning linear positive operators (abbrev. l.p.o.) is to deal with minimal errors resulting into effective approximation of functions. In general, there is no analytical approximation technique for a given form. The prime way is to begin with a suitable sequence of l.p.o. and then transform it to fulfil the desired requirements.

In the year 1963, Timan [61] realized that, instead of uniform approximation by algebraic polynomials, if we consider the point-wise approximation on the given segment, then conclusions could be reached on the whole domain, rather than on the contracting intervals. Moreover, 3 years later, when Lorentz [45] gave the global saturation theorem of Bernstein polynomials, work in this direction picked up momentum. Approximation theory concerning l.p.o. also largely relies

D. Soybaş
Department of Mathematics Education, Faculty of Education, Erciyes University, Kayseri, Turkey
e-mail: danyal@erciyes.edu.tr

N. Malik (✉)
Statistics and Mathematics Unit, Indian Statistical Institute, Bangalore, India

N. J. Daras, T. M. Rassias (eds.), *Computational Mathematics and Variational Analysis*, Springer Optimization and Its Applications 159,
https://doi.org/10.1007/978-3-030-44625-3_24

on the Weierstrass approximation theorem, which pertains to the convergence and approximation characteristics of l.p.o. on spaces of continuous functions forming a long and rich history.

The present paper deals with the convergence estimates for certain l.p.o. introduced by Prof. Vijay Gupta. In this, generalized versions of some recent operators introduced can also be seen and their approximation properties being investigated.

2 Ordinary and Simultaneous Approximation

This section is devoted to briefly allude ordinary and simultaneous approximation results of certain l.p.o. It involves investigation of direct results, which include point-wise convergence, asymptotic formula, error estimation in ordinary and simultaneous approximation, which are necessary for the convergence point of view. Direct results ascertain the order of approximation for the functions with specified smoothness, whereas the aspect of approximation of derivatives of the functions by the corresponding order derivatives of the operators taking place is referred to as simultaneous approximation.

We mention some direct approximation theorems in ordinary and simultaneous approximation for a few summation-integral type linear positive operators studied over the past three decades.

Dating back to the works of Durrmeyer [9], the following integral modification of Bernstein polynomials in the year 1967 was given in order to approximate Lebesgue integrable functions on [0, 1]:

$$M_n(f,x) = (n+1)\sum_{\nu=0}^{n} p_{n,\nu}(x) \int_0^1 p_{n,\nu}(t)\, f(t)\, dt, \quad f \in L_1(0,1),$$

where $p_{n,\nu}(x) = \binom{n}{\nu} x^{\nu}\,(1-x)^{n-\nu}$.

About a decade later, May [51] and Rathore [56] described a method for forming linear combinations of l.p.o. to improve the order of approximation. In the year 1989, Agrawal and Gupta [3] estimated simultaneous approximation results for these operators, which are as follows:

Theorem 2.1 *If $f \in L_B(0,1)$, the class of bounded and Lebesgue integrable functions on $[0,1]$ admits a derivative of order $2k+r+2$ at a point $x \in [0,1]$, then*

$$\lim_{n\to\infty} n^{k+1}\,[M^{(r)}(f,k,x) - f^{(r)}(x)] = \sum_{i=r}^{2k+r+2} f^{(i)}(x)\, Q(i,k,r,x)$$

and

$$\lim_{n\to\infty} n^{k+1}\,[M^{(r)}(f,k+1,x) - f^{(r)}(x)] = 0,$$

where $Q(i, k, r, x)$ are certain polynomials in x. Also, above two equations hold uniformly on $[0, 1]$ *if* $f^{(2k+r+2)} \in C[0, 1]$.

Theorem 2.2 *Let* $1 \leqslant p \leqslant 2k+2$, $f \in L_B(0, 1)$ *and* $n > 0$ *be arbitrary. If* $f^{(p+r)}$ *exists and is continuous on* $(a-\eta, b+\eta) \subset (0, 1)$ *having the modulus of continuity* $\omega_{f^{(p+r)}}(\delta)$ *on* $(a-\eta, b+\eta)$, $\eta > 0$, *then for n sufficiently large,*

$$||L_n^{(r)}(f, \cdot) - f^{(r)}|| \leqslant C_1\, n^{-1}\Big(||f^{(r)}|| + ||f^{(r+1)}||\Big) + C_2\, n^{-1/2}\, \omega_f(r+1)^{(n^{-1/2})} + \mathcal{O}(n^{-m}),$$

where $C_1 = C_1(k, p, r)$, $C_2 = C_2(k, p, r, f)$ *and* $||\cdot||$ *stands for sup-norm on* $[a, b] \subset (0, 1)$.

In the year 1994, a new kind of l.p.o. [15] was defined in order to provide better approximation than the modified Baskakov operators. By taking the weight function of Beta operators on $L_1[0, \infty)$, these operators were given in the following way:

$$L_n(f, x) = \sum_{k=0}^{\infty} p_{n,k}(x) \int_0^{\infty} b_{n,k}(t)\, f(t)\, dt, \quad x \in [0, \infty),$$

where $p_{n,k}(x) = \binom{n+k-1}{k} x^k\, (1+x)^{n+k}$ and $b_{n,k}(t) = \frac{t^k}{B(k+1,n)\,(1+t)^{n+k+1}}$, $B(k+1, n) = \frac{k!\,(n-1)!}{(n+k)!}$ being the Beta function.

Simultaneous approximation by these operators was studied and the following was proved.

Theorem 2.3 *Let* $f \in H[0, \infty)$, *the class of all measurable functions defined on* $[0, \infty)$ *satisfying* $\int_0^{\infty} \frac{|f(t)|}{(1+t)^{n+1}}\, dt < \infty$ *for some positive integer n. If f is bounded on every finite subinterval of* $[0, \infty)$, $f^{(r+2)}$ *exists at a fixed point* $x \in (0, \infty)$ *and* $f(t) = \mathcal{O}(t^{\alpha})$ *as* $t \to \infty$ *for some* $\alpha > 0$; *then*

$$\lim_{n\to\infty} n\,[L_n^{(r)}(f, x) - f^{(r)}(x)] = r^2\, f^{(r)}(x) + [(1+r) + x(1+2r)]\, f^{(r+1)}(x) + x(1+x)\, f^{(r+2)}(x).$$

Theorem 2.4 *Let* $f \in H[0, \infty)$ *be bounded on every finite subinterval of* $[0, \infty)$ *and* $f(t) = \mathcal{O}(t^{\alpha})$ *as* $t \to \infty$ *for some* $\alpha > 0$. *If* $f^{(r+1)}$ *exists and is continuous on* $(a-\eta, b+\eta) \subset (0, \infty)$, $\eta > 0$, *then for n sufficiently large,*

$$||L_n^{(r)}(f, \cdot) - f^{(r)}|| \leqslant C_1\, n^{-1}\Big(||f^{(r)}|| + ||f^{(r+1)}||\Big) + C_2\, n^{-1/2}\, \omega_f(r+1)^{(n^{-1/2})} + \mathcal{O}(n^{-m}),$$

for any $m > 0$; C_1 *and* C_2 *are constants independent of* f *and* n; $\omega_f(\delta)$ *is the modulus of continuity of* f *on* $(a-\eta, b+\eta)$ *and* $||\cdot||$ *denotes the sup-norm on* $[a, b]$.

In the year 2012, for $x \in [0, \infty)$, Gupta and Yadav [37] introduced the Baskakov–Beta–Stancu operators, viz., BBS operators as

$$L_{n,\alpha,\beta}(f,x)=n\int_0^\infty f\left(\frac{nt+\alpha}{n+\beta}\right)\frac{1+x}{(1+x+t)^{n+1}}\,{}_2F_1\left(n+1, 1-n; 1; \frac{-xt}{1+x+t}\right)dt,$$

where $0 \leqslant \alpha \leqslant \beta$ and ${}_2F_1\left(n+1, 1-n; 1; \frac{-xt}{1+x+t}\right)$ is the Gauss hypergeometric function.

They studied some direct results including a Voronovskaya type asymptotic formula and an estimation of error in simultaneous approximation for these BBS operators:

Theorem 2.5 *Let* $f \in C_\gamma[0,\infty) = \{f \in C[0,\infty) : f(t) = \mathcal{O}(t^\gamma), \gamma > 0\}$ *be bounded on every finite subinterval of* $[0,\infty)$ *admitting the derivative of order* $(r+2)$ *at a fixed* $x \in (0,\infty)$. *Let* $f(t) = \mathcal{O}(t^\gamma)$ *as* $t \to \infty$ *for some* $\gamma > 0$, *then*

$$\lim_{n\to\infty} n\,[L^{(r)}_{n,\alpha,\beta}(f,x) - f^{(r)}(x)] = r(r-\beta)\,f^{(r)}(x) + [(1+r+\alpha) + x(1+2r-\beta)]$$
$$\times\, f^{(r+1)}(x) + \,x(1+x)\,f^{(r+2)}(x).$$

Theorem 2.6 *Let* $f \in C_\gamma[0,\infty)$ *for some* $\gamma > 0$ *and* $r \leqslant m \leqslant r+2$. *If* $f^{(m)}$ *exists and is continuous on* $(a-\eta, b+\eta) \subset (0,\infty)$, $\eta > 0$, *then for* n *sufficiently large*

$$||L^{(r)}_{n,\alpha,\beta}(f,x) - f^{(r)}(x)||_{C[a,b]} \leqslant C_1\,n^{-1}\sum_{i=r}^{m}||f^{(i)}||_{C[a,b]} + C_2$$
$$n^{-1/2}\,\omega(f^{(m)}, n^{-1/2}) + \mathcal{O}(n^{-2}),$$

where C_1, C_2 *are constants independent of* f *and* n, $\omega(f,\delta)$ *is the modulus of continuity of* f *on* $(a-\eta, b+\eta)$ *and* $||\cdot||_{C[a,b]}$ *denotes the sup-norm on* $[a,b]$.

In order to generalize the well-known Szász–Mirakyan operators, Jain [42] introduced a class of discretely defined operators and estimated some direct theorems in ordinary approximation in the year 1972. These interesting operators led to the proposition of a new general sequence of summation-integral operators in 2013, which for special values reduce to the Phillips operators and Szász–Beta operators. These sequence of operators were defined as

$$D^{(\beta)}_{n,c}(f,x) = n\sum_{k=1}^{\infty} l^{(\beta)}_{n,k}(x)\int_0^\infty p_{n+c,k-1}(t,c)\,f(t)\,dt + l^{(\beta)}_{n,0}(x)\,f(0),$$

where $l_{n,k}^{(\beta)}(x) = \frac{nx(nx+k\beta)^{k-1}}{k!} e^{-(nx+k\beta)}$, $p_{n,k}(t,c) = \frac{(-t)^k}{k!} \phi_{n,c}^{(k)}(t)$ with $c \geqslant 0$ and two special cases: $\phi_{n,0}(t) = e^{-nt}$ and $\phi_{n,1}(t) = (1+t)^{-n}$.
Approximation properties of the above mentioned operators including point-wise convergence, asymptotic formula, rate of approximation in terms of modulus of continuity and weighted approximation have been investigated.
The point-wise convergence is given by

Theorem 2.7 *Let $f \in C[0,\infty)$ and $\beta \to 0$ as $n \to \infty$, then the sequence $\{D_{n,c}^{(\beta)}(f,x)\}$ converges uniformly to $f(x)$ in $[a,b]$, where $0 \leqslant a < b < \infty$.*

Theorem 2.8 *Let f be bounded and integrable on $[0,\infty)$ and has a second derivative at a point $x \in [0,\infty)$ with $\beta \to 0$ as $n \to \infty$, then*

$$\lim_{n\to\infty} n\,[D_{n,c}^{(\beta)}(f,x) - f(x)] = cx\, f'(x) + \frac{x(cx+2)}{2}\, f''(x).$$

Theorem 2.9 *Let $f \in C_B[0,\infty)$ and $0 \leqslant \beta \leqslant \beta'/n < 1$, then*

$$|D_{n,c}^{(\beta)}(f,x) - f(x)| \leqslant C\,\omega_2(f,\sqrt{\delta_n}) + \omega\left(f, \left|\frac{\beta' x}{n(1-\beta)} + \frac{cx}{(n-c)(1-\beta)}\right|\right),$$

where C is a positive constant and $\delta_n = \left(D_{n,c}^{(\beta)}(t-x,x)\right)^2 + D_{n,c}^{(\beta)}((t-x)^2,x)$.

The next result proved was the weighted approximation theorem, where the approximation formula holds on the interval $[0,\infty)$. Let $B_{x^2}[0,\infty) = \{f : for\ every\ x \in [0,\infty), |f(x)| \leqslant M_f(1+x^2), M_f\ being\ a\ constant\ depending\ on\ f\}$. $C_{x^2}^*[0,\infty)$ denotes the subspace of all continuous functions belonging to $B_{x^2}[0,\infty)$ and satisfying the condition $\lim\limits_{x\to\infty} \frac{f(x)}{1+x^2}$ is finite. The norm on $C_{x^2}^*[0,\infty)$ is $||f||_{x^2} = \sup\limits_{x\in[0,\infty)} \frac{f(x)}{1+x^2}$.

Theorem 2.10 *Let $\beta \to 0$ as $n \to \infty$. Then, for each $f \in C_{x^2}^*[0,\infty)$ and $n > 2c$,*

$$\lim_{n\to\infty} ||D_{n,c}^{(\beta)}(f) - f||_{x^2} = 0.$$

Govil et al. [14] studied some convergence estimates for the following Durrmeyer type integral modification of the Lupaş operators with the weights of Szász basis function for $f \in L_1[0,\infty)$.

$$D_n(f,x) = n\sum_{k=1}^{\infty} l_{n,k}(x) \int_0^{\infty} s_{n,k-1}(t)\, f(t)\, dt + l_{n,0}(x)\, f(0), \quad x \in [0,\infty),$$

where $l_{n,k}(x) = 2^{-nx} \frac{(nx)_k}{2^k\, k!}$ and $s_{n,k}(t) = e^{-nt} \frac{(nt)^k}{k!}$. Some direct theorems were studied by Gupta and collaborators for these operators.

Theorem 2.11 *Let f be a continuous function on $[0, \infty)$ for $n \to \infty$, then the sequence $\{D_n(f, x)\}$ converges uniformly to $f(x)$ in $[a, b] \subset [0, \infty)$.*

Theorem 2.12 *For $x \in [0, \infty)$ and $f \in C_B[0, \infty)$, there exists a constant $C > 0$, such that*

$$|D_n(f, x) - f(x)| \leqslant C\, \omega_2\left(f, \sqrt{\frac{3x}{n}}\right).$$

Theorem 2.13 *Let f be bounded and integrable on $[0, \infty)$ and has a second derivative at a point $x \in [0, \infty)$, then*

$$\lim_{n\to\infty} n\,[D_n(f, x) - f(x)] = \frac{3x}{2}\, f''(x).$$

Theorem 2.14 *For each $f \in C^*_{x^2}[0, \infty)$,*

$$\lim_{n\to\infty} ||D_n(f) - f||_{x^2} = 0.$$

In the year 2015, for $\rho > 0$ and $c = c_{n,x} > \beta$ $(n = 0, 1, 2, \ldots)$ for certain constant $\beta > 0$, Gupta [20] proposed the following generalized family of the hybrid integral operators, which include Lupaş–Szász type operators, Phillips operators and the Baskakov–Szász operators as particular cases.

$$P_n^{\rho,c}(f, x) = \sum_{k=1}^{\infty} l^c_{n,k}(x) \int_0^{\infty} s^{\rho}_{n,k}(t)\, f(t)\, dt + l^c_{n,0}(x)\, f(0), \quad x \geqslant 0,$$

where $l^c_{n,k}(x) = \left(\frac{c}{1+c}\right)^{ncx} \frac{(ncx)_k}{(1+c)^k\, k!}$ and $s^{\rho}_{n,k}(t) = n\rho\, e^{-n\rho t} \frac{(n\rho t)^{k\rho-1}}{\Gamma k\rho}$.
Some direct estimates are as follows.

Theorem 2.15 *Let f be a continuous function on $[0, \infty)$ for $n \to \infty$, then the sequence $\{P_n^{\rho,c}(f, x)\}$ converges uniformly to $f(x)$ in $[a, b] \subset [0, \infty)$.*

Theorem 2.16 *For $x \in [0, \infty)$ and $f \in C_B[0, \infty)$, there exists a constant $C > 0$, such that*

$$|P_n^{\rho,c}(f, x) - f(x)| \leqslant C\, \omega_2\left(f, \sqrt{\frac{[(1+c)\rho + c]x}{nc\rho}}\right).$$

Theorem 2.17 *Let f be bounded and integrable on $[0, \infty)$ and has a second derivative at a point $x \in [0, \infty)$, then*

$$\lim_{n\to\infty} n\,[P_n^{\rho,c}(f, x) - f(x)] = \frac{[(1+c)\rho + c]x}{2c\rho} f''(x).$$

Theorem 2.18 *For each $f \in C^*_{x^2}[0, \infty)$,*

$$\lim_{n\to\infty} ||P_n^{\rho,c}(f) - f||_{x^2} = 0.$$

Very recently, Gupta [36] gave a general family of the Srivastava–Gupta operators containing some well-known operators as special cases. These preserve not only the constant functions, but also linear functions. Several approximation properties can be investigated on the subject of these l.p.o.

3 Rate of Convergence

The rate of convergence measures how fast of a sequence of l.p.o. converges. Varied researchers have studied the rate of convergence for numerous l.p.o. for functions of bounded variation. Throughout this section, $V_a^b(g_x)$ will denote the total variation of g_x on $[a, b]$, where the auxiliary function $g_x(t)$ is defined by

$$g_x(t) = \begin{cases} f(t) - f(x-), & 0 \leqslant t < x \\ 0, & t = x \\ f(t) - f(x+), & x < t < \infty. \end{cases}$$

In the year 2002, the Bézier variant of the well-known Baskakov–Beta operators was introduced [16]. For $f \in H[0, \infty) = \{f : f \textit{ is locally bounded on } (0, \infty) \textit{ and } |f(t)| \leqslant M\,(1+t)^\beta,\ M > 0,\ \beta \in \mathbb{N}_0\}$, $\alpha \geqslant 1$, these operators were given as

$$B_{n,\alpha}(f, x) = \sum_{k=0}^{\infty} Q_{n,k}^{(\alpha)}(x) \int_0^\infty b_{n,k}(t)\, f(t)\, dt, \quad x \in [0, \infty),$$

where $Q_{n,k}^{(\alpha)}(x) = J_{n,k}^\alpha(x) - J_{n,k+1}^\alpha(x)$, $\sum\limits_{j=k}^{\infty} p_{n,j}(x) = J_{n,k}(x)$ is the Baskakov basis function and $b_{n,k}(t) = \frac{t^k}{B(k+1,n)\,(1+t)^{n+k+1}}$, $B(k+1, n) = \frac{k!\,(n-1)!}{(n+k)!}$. The rate of convergence of these operators $B_{n,\alpha}(f, x)$ for functions of bounded variation was estimated as follows:

Theorem 3.19 *Let $f \in H[0,\infty)$ and let at a fixed point $x \in (0,\infty)$, the one-sided limits $f(x\pm)$ exist. Then, for $\alpha \geqslant 1$, $\lambda > 2$, $x \in (0,\infty)$ and for $n > max\{1+\beta, \mathbb{N}(\lambda, x)\}$,*

$$\left|B_{n,\alpha}(f,x) - \left[\frac{1}{\alpha+1}f(x+) + \frac{\alpha}{\alpha+1}f(x-)\right]\right|$$

$$\leqslant |f(x+)-f(x-)| \cdot \frac{\alpha\sqrt{1+x}}{\sqrt{2enx}} + \frac{\alpha[3\lambda+(1+3\lambda)x]}{nx}\sum_{k=1}^{n} V_{x-x/\sqrt{k}}^{x+x/\sqrt{k}}(g_x)$$

$$+M\,\alpha(2^{\beta}-1)\,\frac{(1+x)^{\beta}}{x^{2\beta}}\,\mathscr{O}(n^{-\beta}) + \frac{2M\alpha\lambda(1+x)^{\beta+1}}{nx}.$$

A year later, for $f \in H_{\alpha}(0,\infty)$, the class of all locally integrable functions defined on $(0,\infty)$ and satisfying the growth condition $|f(t)| \leqslant M\,t^{\alpha}$, $(M > 0, \alpha \geqslant 0, t \to \infty)$, Srivastava and Gupta [60] introduced and investigated the following general family of l.p.o.

$$G_{n,c}(f,x) = n\sum_{k=1}^{\infty} p_{n,k}(x,c)\int_{0}^{\infty} p_{n+c,k-1}(t,c)f(t)dt + p_{n,0}(x,c)\,f(0),\; x \in [0,\infty),$$

where $p_{n,k}(x,c) = \frac{(-x)^k}{k!}\,\phi_{n,c}^{(k)}(x)$ and $\phi_{n,c}(x) = \begin{cases} e^{-nx} & c = 0, \\ (1+cx)^{-n/c} & c \in \mathbb{N}. \end{cases}$

Following is the rate of convergence for the operators $G_{n,c}(f,x)$.

Theorem 3.20 *Let $f \in H_{\alpha}(0,\infty)$ and suppose that the one-sided limits $f(x+)$ and $f(x-)$ exist for some fixed point $x \in (0,\infty)$. Then, for $r \in \mathbb{N}$, $c \in \mathbb{N}_0$ and $\lambda > 2$, there exists a positive constant $M > 0$, independent of n, such that, for n sufficiently large,*

$$\left|G_{n,c}(f,x) - \frac{1}{2}[f(x+)+f(x-)]\right|$$

$$\leqslant |f(x+)-f(x-)|\,A_{n,c}(x) + \frac{2\lambda(1+cx)+x]}{nx}\sum_{k=1}^{n} V_{x-x/\sqrt{k}}^{x+x/\sqrt{k}}(g_x) + M\,n^{-r},$$

where

$$A_{n,c}(x) = \begin{cases} \sqrt{\frac{1+cx}{8enx}}, & c \in \mathbb{N} \\ \frac{1}{2\sqrt{\pi nx}}, & c = 0. \end{cases}$$

Another sequence of l.p.o. defined for $f \in B_{\alpha}[0,\infty)$, $(\alpha > 0)$, the class of all measurable complex valued functions f satisfying the growth condition $|f(t)| \leqslant$

$M(1+t)^{\alpha}$ for all $t \in [0,\infty)$ and some $M > 0$, was given in [17]

$$M_n(f,x) = (n-1)\sum_{k=1}^{\infty} p_{n,k}(x) \int_0^{\infty} p_{n,k-1}(t)\, f(t)\, dt + (1+x)^{-n}\, f(0),$$

where $p_{n,k}(x) = \binom{n+k-1}{k} x^k (1+x)^{-(n+k)}$, $x \in [0,\infty)$. The rate of point-wise approximation by the operators $M_n(f,x)$ for functions of bounded variation was established as below.

Theorem 3.21 *Let $f \in B_{\alpha}[0,\infty)$ be a function of bounded variation on every finite subinterval of $[0,\infty)$. Then, for a fixed point $x \in (0,\infty)$, $\lambda > 2$ and $n \geqslant max\{1+\alpha, N(\lambda,x)\}$,*

$$\left| M_n(f,x) - \frac{1}{2}[f(x+)+f(x-)] \right|$$

$$\leqslant |f(x+)-f(x-)| \frac{(27x+25)}{4\sqrt{nx(1+x)}} + \frac{3\lambda+(3\lambda+1)x}{nx} \sum_{k=1}^{n} V_{x-x/\sqrt{k}}^{x+x/\sqrt{k}}(g_x)$$

$$+M\,(2^{\alpha}-1)\,\frac{(1+x)^{\alpha}}{x^{2\alpha}}\,\mathcal{O}(n^{-\alpha}) + \frac{2M\lambda(1+x)^{\alpha+1}}{nx}.$$

Gupta also estimated the rate of convergence for the operators $M_n(f,x)$ in terms of Chanturiya's modulus of variation. The Chanturiya's modulus of variation of jth order for the function g, bounded on a finite or infinite interval Y contained in I is denoted by $\nu_j(g,Y)$ and defined as the upper bound of the set of all numbers $\sum_{k=1}^{j} |g(b_k)-g(a_k)|$ over all systems of j non-overlapping intervals (a_k,b_k), $k = 1,2,3,\ldots,j$ contained in Y. In particular, if $j=0$, we have $\nu_0(g,Y)=0$, the sequence $\{\nu_j(g,Y)\}_{j=0}^{\infty}$ is called the modulus of variation.
This form for the operators $M_n(f,x)$ was given in the following way.

Theorem 3.22 *Let $f \in B_{\alpha}[0,\infty)$ be a function of bounded variation on every finite subinterval of $[0,\infty)$ and let the one-sided limits $f(x\pm)$ exist at a fixed point. Then, for $n \geqslant max\{1+\alpha, 4, N(\lambda,x)\}$,*

$$\left| M_n(f,x) - \frac{1}{2}[f(x+)+f(x-)] \right|$$

$$\leqslant |f(x+)-f(x-)| \frac{(27x+25)}{4\sqrt{nx(1+x)}} + M_1\,\frac{(1+x)^{\alpha}}{(nx^2)^{\alpha}} + \frac{2M\lambda(1+x)^{\alpha}}{nx}$$

$$+Q(x)\left\{\sum_{i=1}^{m-1} \frac{\nu_i(g;x-ix/\sqrt{n},x)+\nu_i(g_x;x,x+ix/\sqrt{n})}{i^3} + \frac{\nu_m(g_x;0,x)+\nu_m(g_x;x,2x)}{m^2}\right\},$$

where $Q(x) = 1 + 8\lambda(1+x)/x,\ m = [\sqrt{m}\,]$ *and* M, M_1 *are positive constants.*

In the year 2004, for $\alpha \geqslant 1$, the Durrmeyer variant of the Baskakov–Bézier operators $B_{n,\alpha}(f,x)$ were introduced by Gupta [18] (defined earlier). An estimate on the rate of convergence of $B_{n,\alpha}(f,x)$ for functions of bounded variation in terms of Chanturiya's modulus of variation was also found out.

Theorem 3.23 *Let* $f \in H[0,\infty)$ *be a function of bounded variation on every finite subinterval of* $[0,\infty)$ *and let the one-sided limits* $f(x\pm)$ *exist at a fixed point. Then, for* $\lambda > 2,\ \alpha \geqslant 1$ *and* $n \geqslant max\{4, \beta+1, N(\lambda,x)\}$,

$$\left|B_{n,\alpha}(f,x) - \left[\frac{1}{\alpha+1}f(x+) + \frac{\alpha}{\alpha+1}f(x-)\right]\right|$$

$$\leqslant |f(x+) - f(x-)| \cdot \frac{\alpha\sqrt{1+x}}{\sqrt{2enx}} + M_0\frac{(1+x)^\beta}{(nx^2)^\beta} + M_1\frac{(1+x)^{\beta+1}}{nx}$$

$$+\left(1 + \frac{8\alpha\lambda(1+x)}{x}\right)\left\{\sum_{j=1}^{m-1}\frac{\nu_j(g;x - jx/\sqrt{n}, x) + \nu_j(g_x;x, x + jx/\sqrt{n})}{j^3}\right.$$

$$\left.+\frac{\nu_m(g_x;0,x) + \nu_m(g_x;x,2x)}{m^3}\right\},$$

where M_0 *and* M_1 *are certain constants depending on* $\alpha,\ \lambda$ *and* β.

In the year 2006, Gupta et al. [38] studied a certain Durrmeyer type integral modification of Bernstein polynomials. They investigated simultaneous approximation and estimated the rate of convergence in simultaneous approximation. In the same year, Govil and Gupta [13] studied the convergence estimates in simultaneous approximation for the Bézier variant of the Baskakov–Beta operators by using the decomposition technique of functions of bounded variation.

Two years later, Gupta and Ivan [31] dealt with the rate of approximation for the Bézier variant of certain operators. They estimated the rate of convergence in simultaneous approximation.

In the year 2013, Govil et al. [14] estimated the rate of convergence for functions having bounded derivatives for certain Durrmeyer type generalization of Jain and Pethe operators [43].

4 Quantum Calculus

Into the early twenty-first century, application of quantum calculus emerged in the theory of approximation. Quantum calculus or q-calculus is a methodology comparable to the usual study of calculus but which is centred on the idea of deriving q-analogous results without the use of limits. The main tool is the q-

derivative. In approximation theory, Lupaş [46] was the first to introduce the concept of q-calculus. He introduced the q-variant of Bernstein polynomials. After this, numerous researchers have studied the approximation properties of different operators based on q-integers, for instance, the integral modifications of Bernstein operators using q-Beta and q-Gamma functions, q-Meyer–König–Zeller–Durrmeyer operators, q-Bernstein–Kantorovich operators, q-Phillips operators and others (cf. [8, 44] and references therein).

In the year 2008, Gupta and Heping [30] introduced the following q-Durrmeyer operators for $f \in C[0,1]$ and $0 < q < 1$:

$$M_{n,q}(f,x) = [n+1]_q \sum_{k=1}^{n} q^{1-k}\, p_{n,k}(q,x) \int_0^1 p_{n,k-1}(q,qt)\, f(t)\, d_q t$$
$$+p_{n,0}(q,x)\, f(0),$$

where $p_{n,k}(q,x) = \left[\begin{smallmatrix} n \\ k \end{smallmatrix}\right]_q x^k \prod_{s=0}^{n-k-1} (1-q^s x)$.
They studied the following approximation properties of $M_{n,q}(f,x)$.

Theorem 4.24 *Let $q_n \in (0,1)$. Then, the sequence $\{M_{n,q_n}(f)\}$ converges to f uniformly on $[0,1]$ for each $f \in C[0,1]$ if and only if $\lim_{n\to\infty} q_n = 1$.*

They set $p_{\infty,k}(q,x) := \frac{x^k}{(1-q)^k\,[k]_q!} \prod_{s=0}^{\infty} (1-q^s x)$. And for fixed $q \in (0,1)$, they defined $M_{\infty,q}(f,1) = f(1)$ and for $x \in [0,1)$,

$$M_{\infty,q}(f,x) = \frac{1}{1-q} \sum_{k=1}^{\infty} q^{1-k}\, p_{\infty,k}(q,x) \int_0^1 p_{\infty,k-1}(q,qt)\, f(t)\, d_q t + p_{\infty,0}(q,x)\, f(0).$$

Theorem 4.25 *Let $q \in (0,1)$. Then, for each $f \in C[0,1]$, the sequence $\{M_{n,q}(f,x)\}$ converges to $M_{\infty,q}(f,x)$ uniformly on $[0,1]$. Furthermore,*

$$||M_{n,q}(f) - M_{\infty,q}(f)|| \leqslant \left(2 + \frac{9}{1-q}\right) \omega(f, q^n).$$

Theorem 4.26 *Let $q \in (0,1)$ be fixed and let $f \in C[0,1]$. Then, $M_{\infty,q}(f,x) = f(x)$ for all $x \in [0,1]$ if and only if f is linear.*

Theorem 4.27 *For any $f \in C[0,1]$, $\{M_{\infty,q}(f)\}$ converges to f uniformly on $[0,1]$ as $q \to 1-$.*

In the same year 2008, for $f \in C[0,1]$, Gupta [19] proposed a similar q-analogue of Durrmeyer operators, given by

$$D_{n,q}(f,x) = [n+1]_q \sum_{k=0}^{n} q^{-k}\, p_{n,k}(q,x) \int_0^1 p_{n,k}(q,qt)\, f(t)\, d_q t.$$

The rate of convergence for these operators $\{D_{n,q}(f)\}$ and direct results in terms of $\omega(f,\cdot)$ have been estimated. A year later, Finta and Gupta [11] extended the above studies and investigated some local and global direct results for the q-Durrmeyer type operators. They also establish a simultaneous approximation theorem for $D_{n,q}f$, where f is a polynomial. The local theorem proved is as follows.

Theorem 4.28 *Let $n > 3$ be a natural number and let $q_0 = q_0(n) \in (0,1)$ be the least number such that $q^{n+2} - q^{n+1} - 2q^n - 2q^{n-1} - \ldots - 2q^3 - q^2 + q + 2 < 0$ for every $q \in (q_0, 1)$. Then, there exists an absolute constant $C > 0$, such that*

$$|D_{n,q}(f,x) - f(x)| \leqslant C\, \omega_2\left(f, [n+2]_q^{-1/2}\, \delta_n(x)\right) + \omega\left(f, \frac{1-x}{[n+2]_q}\right),$$

where $f \in C[0,1]$, $\delta_n^2(x) = x(1-x) + \frac{1}{[n]_q}$, $x \in [0,1]$ and $q \in (q_0, 1)$.

The global approximation result is as follows:

Theorem 4.29 *Let $n > 3$ be a natural number and let $q_0 = q_0(n) \in (0,1)$ be the least number such that $q^{n+2} - q^{n+1} - 2q^n - 2q^{n-1} - \ldots - 2q^3 - q^2 + q + 2 < 0$ for every $q \in (q_0, 1)$. Then, there exists an absolute constant $C > 0$, such that*

$$||D_{n,q}f - f|| \leqslant C\, \omega_2^{\varphi}(f, [n]_q^{-1/2}) + \vec{\omega}_{\psi}(f, [n+2]_q^{-1}),$$

where $f \in C[0,1]$, $\varphi^2(x) = x(1-x)$, $\psi(x) = 1 - x$ and $x \in [0,1]$.

Theorem on simultaneous approximation:

Theorem 4.30 *Let s be a fixed natural number and let $q = q(n) \in (0,1)$, such that $(q(n))^n \to 1$ and $3^n \left(\frac{n(n-1)\ldots(n-s+1)}{[n]_q\,[n-1]_q\,\ldots\,[n-s+1]_q} - 1\right) \to 0$ as $n \to \infty$. Then*

$$\lim_{n\to\infty} D_{n,q}^{(s)}(f,x) = f^{(s)}(x),$$

where $x \in [0,1]$ and f is a polynomial of degree s.

Two years later, Aral and Gupta [4] proposed a generalization of the Baskakov operators, based on q integers. For $f \in C[0,\infty)$, $q > 0$ and each positive integer n, new q-Baskakov operators were defined as

$$\mathscr{B}_{n,q}(f,x) = \sum_{k=0}^{\infty} \begin{bmatrix} n+k-1 \\ k \end{bmatrix}_q q^{\frac{k(k-1)}{2}} x^k\, (-x,q)_{n+k}^{-1}\, f\left(\frac{[k]_q}{q^{k-1}\,[n]_q}\right).$$

They estimated the rate of convergence in the weighted norm. Let $B_2(\mathbb{R}_+) := \{f : |f(x)| \leqslant B_f(1+x^2)\}$, where B_f is a constant depending on f, endowed with the norm $||f||_2 := \sup\limits_{x\geqslant 0} \frac{|f(x)|}{1+x^2}$.

Theorem 4.31 *Let $q = q_n$ satisfies $q_n > 0$ and let $q_1 \to 1$ as $n \to \infty$. For every $f \in B_2(\mathbb{R}_+)$,*

$$\lim_{n\to\infty} \sup_{x\geqslant 0} \frac{|B_{n,q_n}(f,x) - f(x)|}{(1+x^2)^3} = 0.$$

Furthermore,

$$|B_{n,q_n}(f,x) - f(x)| \leqslant M\,\omega_2\left(f, \sqrt{\frac{x}{[n]_q}\left(1+\frac{1}{q}x\right)}\right),$$

where $\omega_2(f,\delta)$ is the classical second order modulus of smoothness of f and f is bounded uniformly continuous function on $\mathbb{R}_+$.

Thus, Aral and Gupta proved that the rate of convergence of $B_{n,q_n}(f)$ to f in any closed subinterval of $\mathbb{R}_+$ is $\frac{1}{\sqrt{[n]_{q_n}}}$, which is at least as fast as $\frac{1}{\sqrt{n}}$, which is the rate of convergence of classical Baskakov operators. They also studied extensively the shape preserving and monotonicity properties for these q-Baskakov operators.
Next year, in 2012, Gupta et al. [39] proposed the q-analogue of the well-known Szász–Mirakyan–Baskakov operators:

$$G_n^q(f,x) = [n-1]_q \sum_{k=0}^{\infty} q^k\, s_{n,k}^q(x) \int_0^{\infty/A} p_{n,k}^q(t)\, f(t)\, d_q t, \quad x \in [0,\infty),$$

for every $n \in \mathbb{N}$, $q \in (0,1)$ and $s_{n,k}^q(x) = \frac{([n]_q\, x)^k}{[k]_q!}\, q^{k(k-1)/2}\, \frac{1}{E_q([n]_q\, x)}$, $p_{n,k}^q(t) = \begin{bmatrix} n+k-1 \\ k \end{bmatrix}_q q^{k(k-1)/2}\, \frac{t^k}{(1+t)_q^{n+k}}$.
They estimated some direct approximation results for these operators, which are as follows.

Theorem 4.32 *Let $f \in C_B[0,\infty)$. Then, for every $x \in [0,\infty)$, there exists a constant $L > 0$, such that*

$$\begin{aligned} &|G_n^q(f,x) - f(x)| \\ &\leqslant L\,\omega_2\left(f, \sqrt{\left(\left(G_n^q(t-x,x)\right)^2 + G_n^q((t-x)^2,x)\right)}\right) + \omega(f, G_n^q(t-x,x)). \end{aligned}$$

Theorem 4.33 *Let* $0 < \alpha \leqslant 1$ *and* $f \in C_B[0,\infty)$. *Then, if* $f \in Lip_M(\alpha)$, *i.e., the condition* $|f(y) - f(x)| \leqslant M\, |y-x|^{\alpha}$, $(x, y \in [0,\infty))$ *holds, then, for each* $x \in [0,\infty)$,

$$|G_n^q(f,x) - f(x)| \leqslant M\, \left(G_n^q((t-x)^2, x)\right)^{\alpha/2},$$

where M *is a constant depending on* α *and* f.

Theorem 4.34 *Let* f *be bounded and integrable on the interval* $[0,\infty)$, *second derivative of* f *exists at a fixed point* $x \in [0,\infty)$ *and* $q = q_n \in (0,1)$ *such that* $q_n \to 1$ *as* $n \to \infty$, *then*

$$\lim_{n\to\infty} [n]_{q_n}\, [G_n^q(f,x) - f(x)] = (1+2x)f'(x) + \left(\frac{x^2}{2} + x\right) f''(x).$$

The next result proved by Gupta et al. gives the rate of convergence of the operators $G_n^q(f,x)$ to $f(x)$, for all $f \in C[0,\infty)$, such that $\lim\limits_{x\to\infty} \frac{f(x)}{1+x^2} < \infty$.

Theorem 4.35 *Let* $f \in C[0,\infty)$ *and let* $\omega_{b+1}(f,\delta)$, $(b > 0)$ *be its modulus of continuity on the finite interval* $[0, b+1] \subset [0,\infty)$. *Then, for fixed* $q \in (0,1)$,

$$||G_n^q(f,x) - f(x)||_{C[0,b]} \leqslant N_f\,(1+b^2)\, G_n^q((t-b)^2, b)$$
$$+2\,\omega_{b+1}(f, \sqrt{G_n^q((t-b)^2, b)}),$$

where N_f *is a constant depending on* f.

5 Post-quantum Calculus

Few years ago, further extension of q-calculus, viz. post-quantum calculus ((p,q)-calculus) emerged in the theory of approximation. Lately, there has been a significant discussion amongst researchers in the area of (p,q)-calculus. In the year 2007, Vivek Sahai and Sarasvati Yadav [57] derived a link between the (p,q)-variant of special functions and two parameter quantum algebras. The delineation of these quantum algebras is with regard to the (p,q)-derivative operators. The first (p,q)-variant of Bernstein polynomials was proposed by Mohammad Mursaleen et al. [53], where they studied some direct estimates. After which, several other operators were appropriately modified in (p,q) setting. This led to the extension to (p,q)-analogues of several eminent operators. Various classes of discrete operators have been considered and studied extensively applying (p,q)-calculus (cf. [1, 5, 10, 41, 50, 52, 54]).

In the year 2016, for $x \in [0,1]$ and $0 < q < p \leqslant 1$, Gupta [21] proposed the following (p,q)-variant of genuine Bernstein–Durrmeyer operators:

$$D_n^{p,q}(f,x) = [n-1]_{p,q} \sum_{k=1}^{n-1} p^{-[n^2-k^2-n+k-2]/2}\, b_{n,k}^{p,q}(1,x)$$

$$\times \int_0^1 b_{n-2,k-1}^{p,q}(p,pqt)\, f(pt)\, d_{p,q}t$$

$$+ b_{n,0}^{p,q}(1,x)\, f(0) + b_{n,n}^{p,q}(1,x)\, f(1),$$

where $b_{n,k}^{p,q}(p,pqt) = \left[\begin{smallmatrix} n \\ k \end{smallmatrix}\right]_{p,q} (pt)^k (p \ominus pqt)^{n-k}$.

The (p,q)-Beta function was defined in this paper along with a relation between (p,q)-Beta and (p,q)-Gamma functions. Using some identities of (p,q) calculus, some direct estimates for the (p,q)-Bernstein–Durrmeyer operators were estimated, which are given as follows.

Theorem 5.36 *Let $f \in C[0,1]$ and $0 < q < p \leqslant 1$. Then, there exists an absolute constant $C > 0$, such that*

$$|D_n^{p,q}(f,x) - f(x)| \leqslant C\, \omega_2 \left(f, \sqrt{\frac{x(1-x)}{[n+1]_{p,q}}} \right).$$

The global direct approximation theorem was given as

Theorem 5.37 *Let $f \in C[0,1]$ and $0 < q < p \leqslant 1$. Then, there exists an absolute constant $C > 0$, such that*

$$||D_n^{p,q} f - f|| \leqslant C\, \omega_2^{\varphi} \left(f, [n+1]_{p,q}^{-1/2} \right),$$

where $\varphi = \sqrt{x(1-x)}$.

In the same year, Milovanović et al. [52] proposed the integral modification of the generalized Bernstein polynomials using the (p,q)-analogue of the Beta operators. For $x \in [0,1]$ and $0 < q < p \leqslant 1$, the following (p,q)-variant of Bernstein–Durrmeyer operators were introduced by them:

$$M_n^{p,q}(f,x) = [n+1]_{p,q} \sum_{k=1}^{n} p^{-(n-k+1)(n+k)/2}\, b_{n,k}^{p,q}(1,x) \int_0^1 b_{n,k-1}^{p,q}(t)\, f(t)\, d_{p,q}t$$

$$+ b_{n,0}^{p,q}(1,x)\, f(0),$$

where $b_{n,k}^{p,q}(t) = \left[\begin{smallmatrix} n \\ k \end{smallmatrix}\right]_{p,q} t^k (1 \ominus qt)_{p,q}^{n-k}$.
They established some direct results on local and global approximation and also, illustrated some graphs for the convergence of such operators. The main results proved can be seen as follows.

Theorem 5.38 *Let $n > 3$ be a natural number and let $0 < q < p \leqslant 1$ and $q_0 = q_0(n) \in (0, p)$. Then, there exists an absolute constant $C > 0$, such that*

$$|M_n^{p,q}(f, x) - f(x)| \leqslant C\, \omega_2\left(f, [n+2]_{p,q}^{-1/2}\, \delta_n(x)\right) + \omega\left(f, \frac{2x}{[n+2]_{p,q}}\right),$$

where $f \in C[0, 1]$, $\delta_n^2(x) = \varphi^2(x) + \frac{1}{[n+2]_{p,q}}$, $\varphi^2(x) = x(1-x)$, $x \in [0, 1]$ and $q \in (q_0, 1)$.

Theorem 5.39 *Let $n > 3$ be a natural number and let $0 < q < p \leqslant 1$ and $q_0 = q_0(n) \in (0, p)$. Then, there exists an absolute constant $C > 0$, such that*

$$||M_n^{p,q} f - f|| \leqslant C\, \omega_2^{\varphi}\left(f, [n+2]_{p,q}^{-1/2}\right) + \omega_{\psi}\left(f, [n+2]_{p,q}^{-1}\right),$$

where $f \in C[0, 1]$, $q \in (q_0, 1)$ and $\psi(x) = x$, $x \in [0, 1]$.

Further, to improve the approximation by preserving the quadratic functions, they used the King's technique and gave the following modification:

$$M_{n,p,q}^*(f, x) = [n+1]_{p,q} \sum_{k=1}^{n} p^{-(n-k+1)(n+k)/2}\, b_{n,k}^{p,q}(1, r_n(x))$$

$$\int_0^1 b_{n,k-1}^{p,q}(t)\, f(t)\, d_{p,q}t + b_{n,0}^{p,q}(1, r_n(x))\, f(0),$$

where $r_n(x) = \frac{x\,[n+2]_{p,q}}{p\,[n]_{p,q}}$ and $x \in \left[0, \frac{[n+2]_{p,q}}{p\,[n]_{p,q}}\right]$.

The following estimate was established for these modified operators.

Theorem 5.40 *Let $n > 3$ be a natural number and let $0 < q < p \leqslant 1$ and $q_0 = q_0(n) \in (0, p)$. Then, there exists an absolute constant $C > 0$, such that*

$$|M_{n,p,q}^*(f, x) - f(x)| \leqslant C\, \omega_2\left(f, \sqrt{\delta_n^{p,q}(x)}\right),$$

where $x \in \left[0, \frac{[n+2]_{p,q}}{p\,[n]_{p,q}}\right]$, $q \in (q_0, 1)$ and $\delta_n^{p,q}(x) = M_{n,p,q}^*((t-x)^2, x)$.

After a few months, Gupta and Aral [29] gave the (p, q)-analogue of Bernstein operators for $x \in [0, 1]$ and $0 < q < p \leqslant 1$ as

$$G_n^{p,q}(f, x) = [n+1]_{p,q} \sum_{k=0}^{n} p^{-[n^2+3n-k^2-k]/2}\, b_{n,k}^{p,q}(1, x) \int_0^1 b_{n,k}^{p,q}(p, pqt)\, f(t)\, d_{p,q}t,$$

where $b_{n,k}^{p,q}(p, pqt) = \left[{n \atop k}\right]_{p,q} (pt)^k (p \ominus pqt)_{p,q}^{n-k}$.

They established some local and global direct theorems, which were given as

Theorem 5.41 *Let $n > 3$ be a natural number and let $0 < q < p \leqslant 1$ and $q_0 = q_0(n) \in (0, p)$. Then, there exists an absolute constant $C > 0$, such that*

$$|G_n^{p,q}(f, x) - f(x)| \leqslant C\, \omega_2 \left(f, [n+2]_{p,q}^{-1/2}\, \delta_n(x) \right) + \omega \left(f, \frac{1-x}{[n+2]_{p,q}} \right),$$

where $f \in C[0, 1]$, $\delta_n^2(x) = \varphi^2(x) + \frac{1}{[n+3]_{p,q}}$, $\varphi^2(x) = x(1-x)$, $x \in [0, 1]$ and $q \in (q_0, 1)$.

Theorem 5.42 *Let $n > 3$ be a natural number and let $0 < q < p \leqslant 1$ and $q_0 = q_0(n) \in (0, p)$. Then, there exists an absolute constant $C > 0$, such that*

$$||G_n^{p,q} f - f|| \leqslant C\, \omega_2^{\varphi} \left(f, [n+2]_{p,q}^{-1/2} \right) + \vec{\omega}_{\psi} \left(f, [n+2]_{p,q}^{-1} \right),$$

where $f \in C[0, 1]$, $q \in (q_0, 1)$ and $\psi(x) = 1 - x$, $x \in [0, 1]$.

Moreover, Aral and Gupta [5] also introduced the (p, q)-variant of beta function of second kind and established a relation between the generalized beta and gamma functions using some identities of (p, q) calculus. For $x \in [0, \infty)$, $0 < q < p \leqslant 1$, the (p, q)-analogue of Baskakov–Durrmeyer operators defined by them were proposed as

$$H_n^{p,q}(f, x) = [n-1]_{p,q} \sum_{k=0}^{\infty} p^{(k+1)(k=2)/2}\, q^{[k(k+1)-2]/2}\, b_{n,k}^{p,q}(x)$$

$$\times \int_0^{\infty} \left[{n+k-1 \atop k}\right]_{p,q} \frac{t^k}{(1 \oplus pt)^{k+n}}_{p,q} f(p^k t)\, d_{p,q}t + b_{n,0}^{p,q}(1, x)\, f(0),$$

where $b_{n,k}^{p,q}(x) = \left[{n+k-1 \atop k}\right]_{p,q} p^{k+n(n-1)/2}\, q^{k(k-1)/2} \frac{x^k}{(1 \oplus x)_{p,q}^{n+k}}$.

They discussed the following weighted approximation theorem, where the approximation formula holds true on the interval $[0, \infty)$.

Theorem 5.43 *Let $p = p_n$, $q = q_n$ satisfying $0 < q_n < p_n \leqslant 1$ and for n sufficiently large $p_n \to 1$, $q_n \to 1$, $p_n^n \to 1$ and $q_n^n \to 1$. For each $f \in C[0, \infty)$,*

$$\lim_{n \to \infty} ||H_n^{p_n, q_n}(f) - f||_{x^2} = 0.$$

They also gave the following result to approximate all functions in $C_{x^2}^*[0, \infty)$.

Theorem 5.44 *Let $p = p_n$, $q = q_n$ satisfying $0 < q_n < p_n \leqslant 1$ and for n sufficiently large $p_n \to 1$, $q_n \to 1$, $p_n^n \to 1$ and $q_n^n \to 1$. For each $f \in C^*_{x^2}[0,\infty)$,*

$$\lim_{n\to\infty} \sup_{x\in[0,\infty)} \frac{|H_n^{p_n,q_n}(f,x) - f(x)|}{(1+x^2)^{1+\alpha}} = 0.$$

Following quantitative approximation result was also estimated:

Theorem 5.45 *Let $q \in (0,1)$ and $p \in (q,1]$. The operator $H_n^{p,q}$ maps space C_B into C_B and*

$$||H_n^{p,q}(f)||_{C_B} \leqslant ||f||_{C_B}.$$

After a few months, Gupta [22] introduced the (p,q)-Baskakov–Kantorovich operators:

$$K_n^{p,q}(f,x) = [n]_{p,q} \sum_{k=0}^{\infty} b_{n,k}^{p,q}(x)\; p^{-k}\; q^k \int_{[k]_{p,q}/q^{k-1}\,[n]_{p,q}}^{[k+1]_{p,q}/q^k\,[n]_{p,q}} f(t)\, d_{p,q}t,$$

where $b_{n,k}^{p,q}(x) = \begin{bmatrix} n+k-1 \\ k \end{bmatrix}_{p,q} p^{k+n(n-1)/2}\, q^{k(k-1)/2} \dfrac{x^k}{(1\oplus x)_{p,q}^{n+k}}$.
Some direct results are estimated by using linear approximating methods, viz., Steklov mean and K-functionals.

Theorem 5.46 *Let $q \in (0,1)$ and $p \in (q,1]$. The operator $K_n^{p,q}$ maps space C_B into C_B and*

$$||K_n^{p,q}(f)||_{C_B} \leqslant ||f||_{C_B}.$$

Theorem 5.47 *Let $f \in C_B[0,\infty)$. Then, for all $n \in \mathbb{N}$, there exists an absolute constant $C > 0$, such that*

$$\begin{aligned} &|K_n^{p,q}(f,x) - f(x)| \\ &\leqslant C\,\omega_2\Big(f, \{K_n^{p,q}((t-x)^2,x) + \big(K_n^{p,q}((t-x),x)\big)^2\}^{1/2}\Big) \\ &\quad + \omega\left(f, \left|\frac{1}{[2]_{p,q}\,[n]_{p,q}} + \left(\frac{1}{q\,p^{n-1}} - 1\right)x\right|\right). \end{aligned}$$

Further, they discussed the weighted approximation theorem, where the approximation formula holds true on the interval $[0,\infty)$.

Theorem 5.48 *Let $p = p_n$ and $q = q_n$ satisfies $0 < q_n < p_n \leqslant 1$ and for n sufficiently large, $p_n \to 1$, $q_n \to 1$ and $p_n^n \to 1$, $q_n^n \to 1$. For each $f \in C_{x^2}^*[0, \infty)$,*

$$\lim_{n\to\infty} ||K_n^{p_n,q_n}(f) - f||_{x^2} = 0.$$

In 2016, Gupta and Ali Aral [29] introduced the (p, q)-Durrmeyer type operators and estimated some approximation results. For $x \in [0, 1]$ and $0 < q < p \leqslant 1$, they defined the following (p, q)-analogue of Bernstein–Durrmeyer operators:

$$J_n^{p,q}(f, x)=[n+1]_{p,q} \sum_{k=0}^{n} p^{-[n^2+3n-k^2-k]/2}\, b_{n,k}^{p,q}(1, x) \int_0^1 b_{n,k}^{p,q}(p, pqt)\, f(t)\, d_{p,q}t,$$

where $b_{n,k}^{p,q}(p, pqt) = \begin{bmatrix} n \\ k \end{bmatrix}_{p,q} (pt)^k\, (p \ominus pqt)_{p,q}^{n-k}$.
For $p = 1$, these operators will not reduce to the q-Durrmeyer operators, though for $p = 1 = q$, these will reduce to the Durrmeyer operators. Gupta and Aral even studied the rate of convergence of these operators $J_n^{p,q}$ and compared the results with a graphical representation. We mention just the local and global theorems, which are given below.

Theorem 5.49 *Let $n > 3$ be a natural number and let $0 < q < p \leqslant 1$, $q_0 = q_0(n) \in (0, p)$. Then there exists an absolute constant $C > 0$, such that*

$$|J_n^{p,q}(f, x) - f(x)| \leqslant C\, \omega_2\left(f, [n+2]_{p,q}^{-1/2}\, \delta_n(x)\right) + \omega\left(f, \frac{1-x}{[n+2]_{p,q}}\right), \quad f \in C[0, 1],$$

where $\delta_n(x) = x(1-x) + \frac{1}{[n+3]_{p,q}}$, $x \in [0, 1]$ and $q \in (q_0, 1)$.

Theorem 5.50 *Let $n > 3$ be a natural number and let $0 < q < p \leqslant 1$, $q_0 = q_0(n) \in (0, p)$. Then there exists an absolute constant $C > 0$, such that*

$$||J_n^{p,q}(f) - f|| \leqslant C\, \omega_2^{\phi}\left(f, [n+2]_{p,q}^{-1/2}\right) + \omega_{\psi}\left(f, [n+2]_{p,q}^{-1}\right), \quad f \in C[0, 1],$$

where $\phi(x) = \sqrt{x(1-x)}$, $\psi(x) = 1 - x$, $x \in [0, 1]$.

After a year, Aral and Gupta [6] introduced the (p, q)-analogue of the Szász–Beta operators. Using (p, q)-variant of Beta function of second kind, they proposed the following for $x \in [0, \infty)$ and $0 < q < p \leqslant 1$:

$$F_n^{p,q}(f,x)=\sum_{k=1}^{\infty}s_{n,k}^{p,q}(x)\,\frac{1}{B_{p,q}(k,n+1)}\int_0^{\infty}\frac{t^{k-1}}{(1\oplus pt)_{p,q}^{k+n+1}}\,f(p^{k+1}\,q\,t)\,d_{p,q}t$$
$$+\frac{f(0)}{E_{p,q}([n]_{p,q}\,x)},$$

where $s_{n,k}^{p,q}(x)=\frac{1}{E_{p,q}([n]_{p,q}\,x)}\,\frac{q^{k(k-1)/2}}{[k]_{p,q}!}\,\left([n]_{p,q}\,x\right)^k$.
They established direct theorem in weighted spaces in terms of suitable weighted modulus of smoothness, a Voronovskaya type theorem and Grüss-type inequality.

Theorem 5.51 *Let $p=p_n$ and $q=q_n$ satisfy $0<q_n<p_n\leqslant 1$ and for n sufficiently large, $p_n\to 1$, $q_n\to 1$ and $p_n^n\to 1$, $q_n^n\to 1$. For each $f\in C_{x^2}^*[0,\infty)$,*

$$\lim_{n\to\infty}||F_n^{p_n,q_n}(f)-f||_{x^2}=0.$$

The functions satisfying $|f(t)|\leqslant M\,(1+t)^m$, for some $M>0$ are considered. For $m>0$, the weight $\rho(x)=(1+x)^{-m}$, $x\in I=[0,\infty)$. The polynomial weighted space associated with this weight is defined by $C_\rho(I)=\{f\in C(I):||f||_\rho<\infty\}$, where $||f||_\rho=\sup_{x\in I}\rho(x)\,|f(x)|$.

Theorem 5.52 *Set $\rho(x)=(1+x)^2$. For any $f\in C_\rho[0,\infty)$, $x\geqslant 0$ and $n\in\mathbb{N}$,*

$$\rho(x)\,|f(x)-F_n^{p,q}(f,x)|$$
$$\leqslant C\,\omega^2\left(f,\sqrt{x(x+1)\left(-3+6p^2+\frac{p^{12}}{q^{12}}-4\frac{p^6}{q^9}+\mathcal{O}\left(\frac{1}{p\,[n-1]_{p,q}}\right)\right)}\right)_\rho.$$

For $n>2\sqrt{3}$,

$$||f-F_n^{p,q}(f)||_\rho\leqslant\omega_\phi^2\left(f,\left(-3+6p^2+\frac{p^{12}}{q^{12}}-4\frac{p^6}{q^9}+\mathcal{O}\left(\frac{1}{p\,[n-1]_{p,q}}\right)\right)\right)_\rho,$$

where $\omega_\phi^2(f,t)_\rho=\sup_{h\in(0,t]}\sup_{x\in I(\phi,h)}|\rho(x)\,\Delta_{h\phi(x)}f(x)|$ and $I(\phi,h)=\{x>0,h\phi(x)\leqslant x\}$.

For Grüss type inequality, they considered two functions $f,g\in C_\rho[0,\infty)$ and defined the positive bilinear functional:

$$F_n(f,g,x)=F_n^{p,q}(fg,x)-F_n^{p,q}(f,x)\,F_n^{p,q}(g,x).$$

They measured the rate of convergence of this positive bilinear functional on weighted spaces as

Theorem 5.53 *For any* $f \in C_\rho[0, \infty)$, $x \geqslant 0$ *and* $n \in \mathbb{N}$,

$$||F_n(f, g)||_{\rho^2} \leqslant \sqrt{C(f)}\,\sqrt{C(g)},$$

where

$$C(f) = \omega_\phi^2 \left(f^2, \left(-3 + 6p^2 + \frac{p^{12}}{q^{12}} - 4\frac{p^6}{q^9} + \mathcal{O}\left(\frac{1}{p\,[n-1]_{p,q}} \right) \right)^{1/2} \right)$$
$$+ \left\{ 5 + \frac{3}{p\,[n-1]_{p,q}} \right\} ||f||_\rho$$
$$\times\, \omega_\phi^2 \left(f, \left(-3 + 6p^2 + \frac{p^{12}}{q^{12}} - 4\frac{p^6}{q^9} + \mathcal{O}\left(\frac{1}{p\,[n-1]_{p,q}} \right) \right)^{1/2} \right).$$

Next, they estimated the Voronovskaya type theorem, for which, they made the following assumptions:

$$\lim_{n\to\infty} [n]_{p_n,q_n}(p_n - 1) = \alpha$$

and

$$\lim_{n\to\infty} \left(-3 + 6p_n^2 + \frac{p_n^{12}}{q_n^{12}} - 4\frac{p_n^6}{q_n^9} \right) = \gamma.$$

Theorem 5.54 *Let* $f \in C(\mathbb{R}^+)$. *If* $x \in \mathbb{R}^+$, f *is two times differentiable in* x *and* f'' *is continuous in* x, $p = p_n$ *and* $q = q_n$ *satisfy* $0 < q_n < p_n \leqslant 1$ *and for* n *sufficiently large,* $p_n \to 1$, $q_n \to 1$ *and* $p_n^n \to 1$, $q_n^n \to 1$. *Then, the following holds true:*

$$\lim_{n\to\infty} [n]_{p_n,q_n}[F_n^{p_n,q_n}(f, x) - f(x)] = x(1 + \alpha x) f''(x).$$

In the same year, Finta and Gupta [12] proved the existence of the limit operator of the slight modification of the sequence of (p, q)-Bernstein–Durrmeyer operators. They also establish the rate of convergence of this limit operator.

Aral and Gupta [7] defined a (p, q)-analogue of Gamma function. Along with, they proposed (p, q)-Szász–Durrmeyer operators and obtained some direct results.

Malik and Gupta [49] considered the (p, q)-analogue of Baskakov–Beta operators and using it, they estimated some approximation theorems and graphically represented the convergence of these operators.

Acu et al. [2] discussed the local and global approximation results for certain (p, q)-Durrmeyer type operators. They also used King's technique to obtain optimal convergence and illustrated the comparisons graphically for different values of parameters p and q.

Very recently, Gupta [24] introduced the (p, q)-Szász–Mirakyan–Baskakov operators. The moments are estimated and some direct theorems including weighted approximation and approximation in terms of modulus of continuity by linear approximating method using Steklov mean are obtained.

Remark 1 Diverse linear positive operators have been studied in literature by many researchers. We refer readers some of these recent studies (cf. [23, 25–28, 32–35, 40, 47, 48, 55, 58, 59]).

Acknowledgments This work was supported by Erciyes University when second author visited Turkey during September 2019.

References

1. T. Acar, Generalization of Szász-Mirakyan operators. Math. Methods Appl. Sci. **39**(10), 2685–2695 (2016)
2. A.M. Acu, V. Gupta, N. Malik, Local and global approximation for certain (p, q)-Durrmeyer type operators. Complex Anal. Oper. Theory **12**(8), 1973–1989 (2018)
3. P.N. Agrawal, V. Gupta, Simultaneous approximation by linear combination of modified Bernstein polynomials. Bull. Greek Math. Soc. **30**, 21–29 (1989)
4. A. Aral, V. Gupta, Generalized q-Baskakov operators. Math. Slovaca **61**(4), 619–634 (2011)
5. A. Aral, V. Gupta, (p, q)-type beta functions of second kind. Adv. Oper. Theory **1**(1), 134–146 (2016)
6. A. Aral, V. Gupta, (p, q)-variant of Szász-Beta operators, Rev. R. Acad. Cienc. Exactas Fs. Nat., Ser. A Mat. **111**(3), 719–733 (2017)
7. A. Aral, V. Gupta, Applications of (p, q)-Gamma function to Szász Durrmeyer operators. Publ. Inst. Math., Nouv. Sér. **102**(116), 211–220 (2017)
8. A. Aral, V. Gupta, R.P. Agarwal, *Applications of q-Calculus in Operator Theory* (Springer, New York, 2013)
9. J.L. Durrmeyer, Une formule d' inversion de la Transformee de Laplace: Applications a la Theorie des Moments, These de 3e Cycle, Faculte des Sciences de l' Universite de Paris (1967)
10. Z. Finta, Approximation properties of (p, q)-Bernstein type operators. Acta Univ. Sapientiae, Math. **8**(2), 222–232 (2016)
11. Z. Finta, V. Gupta, Approximation by q-Durrmeyer operators. J. Appl. Math. Comput. **29**, 401–415 (2009)
12. Z. Finta, V. Gupta, Approximation theorems for limit (p, q)-Bernstein-Durrmeyer operator, Facta Univ., Ser. Math. Inf. **32**(2), 195–207 (2017)
13. N. K. Govil, V. Gupta, Simultaneous approximation for the Bézier variant of Baskakov-Beta operators. Math. Comput. Modell. **44**, 1153–1159 (2006)
14. N.K. Govil, V. Gupta, D. Soybaş, Certain new classes of Durrmeyer type operators. Appl. Math. Comput. **225**, 195–203 (2013)
15. V. Gupta, A note on modified Baskakov type operators. Approximation Theory Appl. **10**(3), 74–78 (1994)
16. V. Gupta, Rate of convergence on Baskakov-Beta-Bézier operators for bounded variation functions. Int. J. Math. Math. Sci. **32**(8), 471–479 (2002)
17. V. Gupta, Rate of approximation by a new sequence of linear positive operators. Comput. Math. Appl. **45**, 1895–1904 (2003)

18. V. Gupta, Rate of convergence of Durrmeyer type Baskakov-Bézier operators for locally bounded functions. Turk. J. Math. **28**, 271–280 (2004)
19. V. Gupta, Some approximation properties of q-Durrmeyer operators. Appl. Math. Comput. **197**, 172–178 (2008)
20. V. Gupta, Direct estimates for a new general family of Durrmeyer type operators. Boll. Unione Mat. Ital. **7**, 279–288 (2015)
21. V. Gupta, (p, q)-Genuine Bernstein Durrmeyer operators. Boll. Unione Mat. Ital. **9**(3), 399–409 (2016)
22. V. Gupta, (p, q)-Baskakov-Kantorovich operators. Appl. Math. Inf. Sci. **10**(4), 1551–1556 (2016)
23. V. Gupta, Some examples of genuine approximation operators. Gen. Math. **26**(1–2), 3–9 (2018)
24. V. Gupta, (p, q)-Szász-Mirakyan-Baskakov operators. Complex Anal. Oper. Theory **12**(1), 17–25 (2018)
25. V. Gupta, A large family of linear positive operators. Rend. Circ. Mat. Palermo (2) (2019). https://doi.org/10.1007/s12215-019-00430-3
26. V. Gupta, A note on the general family of operators preserving linear functions. Rev. R. Acad. Cienc. Exactas Fs. Nat., Ser. A Mat. **113**(4), 3717–3725 (2019)
27. V. Gupta, Estimate for the difference of operators having different basis functions. Rend. Circ. Mat. Palermo (2) (2019). https://doi.org/10.1007/s12215-019-00451-y
28. V. Gupta, R.P. Agarwal, *Convergence Estimates in Approximation Theory* (Springer, New York, 2014)
29. V. Gupta, A. Aral, Bernstein Durrmeyer operators based on two parameters. Facta Univ., Ser. Math. Inf. **31**(1), 79–95 (2016)
30. V. Gupta, W. Heping, The rate of convergence of q-Durrmeyer operators for $0 < q < 1$. Math. Methods Appl. Sci. **31**, 1946–1955 (2008)
31. V. Gupta, M. Ivan, Rate of simultaneous approximation for the Bézier variant of certain operators. Appl. Math. Comput. **199**, 392–395 (2008)
32. V. Gupta, N. Malik, Approximation of functions by complex genuine Pólya-Durrmeyer operators. Comput. Methods Funct. Theory **17**(1), 3–17 (2017)
33. V. Gupta, Th. M. Rassias, *Moments of Linear Positive Operators and Approximation* (Springer, Cham, 2019)
34. V. Gupta, D. Soybaş, Approximation by complex genuine hybrid operators. Appl. Math. Comput. **244**, 526–532 (2014)
35. V. Gupta, D. Soybaş, Convergence of integral operators based on different distributions. Filomat **30**(8), 2277–2287 (2016)
36. V. Gupta, H.M. Srivastava, A general family of the Srivastava-Gupta operators preserving linear functions. Eur. J. Pure Appl. Math. **11**(3), 575–579 (2018)
37. V. Gupta, R. Yadav, Direct estimates in simultaneous approximation for BBS operators. Appl. Math. Comput. **218**, 11290–11296 (2012)
38. V. Gupta, T. Shervashidze, M. Craciun, Rate of approximation for certain Durrmeyer operators. Georgian Math. J. **13**(2), 277–284 (2006)
39. V. Gupta, A. Aral, M. Ozhavzali, Approximation by q-Szász-Mirakyan-Baskakov operators. Fasc. Math. **48**, 35–48 (2012)
40. V. Gupta, N. Malik, Th.M. Rassias, Moment generating functions and moments of linear positive operators, in *Modern Discrete Mathematics and Analysis*, ed. by N. Daras, T. Rassias. Springer Optimization and Its Applications, vol. 131 (Springer, Cham, 2018)
41. V. Gupta, Th. M. Rassias, P.N. Agrawal, A.M. Acu, *Recent Advances in Constructive Approximation Theory*. Springer Optimization and Its Applications, vol. 138 (Springer, Cham, 2018)
42. G.C. Jain, Approximation of functions by a new class of linear operators. J. Aust. Math. Soc. **13**(3), 271–276 (1972)
43. G.C. Jain, S. Pethe, On the generalizations of Bernstein and Szász Mirakjan operators. Nanta Math. **10**, 185–193 (1977)
44. V. Kac, P. Cheung, *Quantum Calculus* (Springer, New York, NY, 2002)

45. G.G. Lorentz, *Approximation of Functions* (Holt, Rinehart and Winston, New York, 1966)
46. A. Lupaş, A q-analogue of the Bernstein operator, Prepr., Babeş-Bolyai Univ., Fac. Math., Res. Semin. 9 (1987), pp. 85–92
47. N. Malik, Some approximation properties for generalized Srivastava-Gupta operators. Appl. Math. Comput. **269**, 747–758 (2015)
48. N. Malik, On approximation properties of Gupta-type operators. J. Anal. (2019). https://doi.org/10.1007/s41478-019-00195-z
49. N. Malik, V. Gupta, Approximation by (p, q)-Baskakov-Beta operators. Appl. Math. Comput. **293**, 49–56 (2017)
50. N. Malik, S. Araci, M.S. Beniwal, Approximation of Durrmeyer type operators depending on certain parameters. Abstr. Appl. Anal. (2017). Article ID 5316150, 9 pp. https://doi.org/10.1155/2017/5316150
51. C.P. May, On Phillips operator. J. Approx. Theory **20**, 315–332 (1977)
52. G.V. Milovanović, V. Gupta, N. Malik, (p, q)-Beta functions and applications in approximation. Bol. Soc. Mat. Mex., III. Ser. (2016). https://doi.org/10.1007/s40590-016-0139-1
53. M. Mursaleen, K.J. Ansari, A. Khan, On (p, q)-analogue of Bernstein operators. Appl. Math. Comput. **266**, 874–882 (2015)
54. M. Mursaleen, M. Nasiruzzaman, A. Nurgali, Some approximation results on Bernstein-Schurer operators defined by (p, q)-integers. J. Inequal. Appl. **249** (2015), 12 pp.
55. Th. M. Rassias, V. Gupta (eds.), *Mathematical Analysis, Approximation Theory and Their Applications*. Springer Optimization and Its Applications, vol. 111 (Springer, Berlin, 2016). ISBN: 978-3-319-31279-8
56. R.K.S. Rathore, Approximation of Unbounded Functions with Linear Positive Operators, D.Sc. Thesis, Technische Hogeschool Delft, Delft University Press, Delft (1974)
57. V. Sahai, S. Yadav, Representations of two parameter quantum algebras and (p, q)-special functions. J. Math. Anal. Appl. **335**(1), 268–279 (2007)
58. D. Soybaş, Approximation with modified Phillips operators. J. Nonlinear Sci. Appl. **10**(11), 5803–5812 (2017)
59. D. Soybaş, N. Malik, Convergence estimates for Gupta-Srivastava operators. Kragujevac J. Math. **45**(5), 739–749 (2021)
60. H.M. Srivastava, V. Gupta, A certain family of summation-integral type operators. Math. Comput. Modell. **37**, 1307–1315 (2003)
61. A.F. Timan, *Theory of Approximation of Functions of a Real Variable* (Macmillan, New York, 1963)

Behavior of the Solutions of Functional Equations

Ioannis P. Stavroulakis and Michail A. Xenos

Abstract In the last decades the oscillation theory of delay differential equations has been extensively developed. The oscillation theory of discrete analogues of delay differential equations has also attracted growing attention in the recent years. Consider the first-order delay differential equation,

$$x'(t) + p(t)x(\tau(t)) = 0, \quad t \geq t_0, \tag{1}$$

where $p, \tau \in C([t_0, \infty], \mathbb{R}^+)$, $\tau(t)$ is nondecreasing, $\tau(t) < t$ for $t \geq t_0$ and $\lim_{t\to\infty} \tau(t) = \infty$, and the (discrete analogue) difference equation,

$$\Delta x(n) + p(t)x(\tau(n)) = 0, \quad n = 0, 1, 2, \ldots, \tag{2}$$

where $\Delta x(n) = x(n+1) - x(n)$, $p(n)$ is a sequence of nonnegative real numbers and $\tau(n)$ is a nondecreasing sequence of integers such that $\tau(n) \leq n - 1$ for all $n \geq 0$ and $\lim_{n\to\infty} \tau(n) = \infty$.

In this review chapter, a survey of the most interesting oscillation conditions is presented, along with numerical examples of delay and difference equations. We focus our attention on these examples, to illustrate the level of improvement in the oscillation criteria and the significance of the obtained results. The numerical calculations were made with the use of MATLAB software. These examples are relevant to many physical and biological applications.

I. P. Stavroulakis · M. A. Xenos (✉)
Department of Mathematics, University of Ioannina, Ioannina, Greece
e-mail: ipstav@uoi.gr; mxenos@uoi.gr

N. J. Daras, T. M. Rassias (eds.), *Computational Mathematics and Variational Analysis*, Springer Optimization and Its Applications 159,
https://doi.org/10.1007/978-3-030-44625-3_25

1 Introduction

Ordinary differential equations appear frequently in mathematical models that attempt to describe real-life situations in which the rate of change of the system depends only on its present stage. A basic limitation behind such a model is the assumption that all interactions in the system occur instantaneously. Mathematically an ordinary differential equation, of first order, is an equation of the form,

$$x'(t) = f(t, x(t)),$$

where f is a known function and x is the unknown function. Note that both x and x' are evaluated at the same instant, t. However, in many cases the past state of the system has to be taken into consideration. Delay differential equations or differential equations with retarded argument or hysterodifferential equations provide more realistic mathematical models for such systems in which the rate of change depends not only on their present stage but also on their past history. The theory of such systems has been developed in the beginning of the last century. In his research on prey–predator population models and viscoelasticity, Volterra [70] formulated some rather general differential equations incorporating the past states of the system. A delay differential equation is an equation of the form,

$$x'(t) = F(t, x(t), x(t - \tau)),$$

where F is a known function, $\tau > 0$ is the delay or time lag or hysteresis, and x is the unknown function. Minorsky [52] was one of the first investigators to study delay differential equations and their effects on simple feedback control systems in which the communication time τ cannot be neglected. An automatic feedback control system does not respond instantaneously to input signals but requires some time τ to process the information and further react. Delay differential equations arise naturally in many other systems involving time lags such as population models, models for epidemics, economic models, nuclear reactors, collision problems in electrodynamics, and many other problems.

The oscillation theory of differential equations is not a new area. It was originated in 1836 by Sturm [65]. Since then hundreds of papers have been published studying the oscillation theory of ordinary differential equations. The reader is referred to Swanson [66] and the bibliography cited therein.

In the decade of 1970 a great number of papers were written extending known results from ordinary differential equations to delay differential equations. However, the study of oscillations which are caused by the delay and which do not appear in the corresponding ordinary differential equation has been of particular importance. In recent years there has been a great deal of interest in the study of oscillatory behavior of the solutions to delay differential equations and also the discrete analogue delay difference equations [53, 54]. The interested reader could visit the extended bibliography at the end and the references cited therein.

In this chapter, a survey of the most interesting oscillation conditions is presented, along with numerical examples of delay and difference equations. We focus our attention on these examples, to illustrate the level of improvement in the oscillation criteria and the significance of the obtained results. The numerical calculations were made with the use of MATLAB software. These examples are relevant to many physical and biological applications.

2 Delay Equation

2.1 Preliminaries and Problem Description

Consider the differential equation with a retarded argument of the form,

$$x'(t) + p(t)x(\tau(t)) = 0, \quad t \geq t_0, \tag{3}$$

where the functions $p, \tau \in C([t_0, \infty), \mathbb{R}^+)$, (here $\mathbb{R}^+ = [0, \infty)$), $\tau(t) \leq t$ for $t \geq t_0$ and $\lim_{t\to\infty} \tau(t) = \infty$.

By the solution of Equation (3) we understand a continuously differentiable function defined on $[\tau(T_0), +\infty)$ for some $T_0 \geq t$ such that Equation (3) is satisfied for $t \geq T_0$. Such a solution is called *oscillatory* if it has arbitrarily large zeros, otherwise it is called *non-oscillatory*.

It is noteworthy to observe that a first-order linear differential equation of the form (3) without delay ($\tau(t) = t$) *does not possess oscillatory solutions*. Therefore the study of oscillatory solutions is of interest for equations of the form (3). Furthermore, the mathematical modeling of several real-world problems leads to differential equations that depend on the past history rather than only the current state. For the general theory of this equation the interested reader could refer to [20, 21, 35, 53, 54].

In the following section we present a survey on the oscillation of all solutions to this equation in the case of monotone or non-monotone argument and especially in the critical case where $\liminf_{t\to\infty} p(t) = \frac{1}{e\tau}$ and also when the well-known oscillation conditions,

$$\limsup_{t\to\infty} \int_{\tau(t)}^{t} p(s)\,ds > 1 \quad \textit{and} \quad \liminf_{t\to\infty} \int_{\tau(t)}^{t} p(s)\,ds > \frac{1}{e},$$

are not satisfied.

2.2 Oscillation Criteria

The problem of establishing sufficient conditions for the oscillation of all solutions to the delay differential equation (3) has been the subject of many other studies that can be found in the recent review by Moremedi and Stavroulakis [54].

The first systematic study for the oscillation of all solutions to Equation (3) was made by Myshkis [56]. In 1950 he proved that every solution of Equation (3) oscillates if,

$$\limsup_{t\to\infty}[t-\tau(t)] < \infty \ \textit{and} \quad \liminf_{t\to\infty}[t-\tau(t)] \liminf_{t\to\infty} p(t) > \frac{1}{e}.$$

Ladas et al. in 1972 [47] proved that the same conclusion holds if

$$\tau,\ \textit{is a non-decreasing function and } A := \limsup_{t\to\infty} \int_{\tau(t)}^{t} p(s)ds > 1. \tag{4}$$

In 1979, Ladas and Lakshmikantham [45] established integral conditions for the oscillation of Equation (3) with constant delay, while Koplatadze and Canturija in 1982 established the following result [41, 53], if

$$a := \liminf_{t\to\infty} \int_{\tau(t)}^{t} p(s)ds > \frac{1}{e}, \tag{5}$$

then all solutions of Equation (3) oscillate; if

$$\limsup_{t\to\infty} \int_{\tau(t)}^{t} p(s)ds < \frac{1}{e}, \tag{6}$$

then Equation (3) has a non-oscillatory solution. Let us set,

$$P = \limsup_{t\to\infty} p(t)$$

and

$$p = \liminf_{t\to\infty} p(t).$$

Observe that in the case of the equation,

$$x'(t) + p(t)\,x(t-\tau) = 0, \quad t \geq t_0, \tag{7}$$

the results by Myshkis [56] reduce to the following conditions, if

$$p\tau > \frac{1}{e}, \tag{8}$$

then the solution of Equation (7) oscillate. while

$$P\tau < \frac{1}{e}, \tag{9}$$

implies the existence of a non-oscillatory solution of Equation (7). Thus, for the oscillation of all solutions to Equation (7), a necessary condition is the following:

$$P\tau \geq \frac{1}{e}. \tag{10}$$

It should be noted that in the special case of the delay equation with a constant positive coefficient p and a constant positive delay τ, that is in the case of the equation,

$$x'(t) + p\, x(t - \tau) = 0, \quad t \geqslant t_0, \tag{11}$$

$$p\tau > \frac{1}{e}, \tag{12}$$

is a *necessary and sufficient condition* [46] for all solutions of Equation (11) to oscillate.

Pituk in 2017 [58] studied the delay Equation (7) in the case where the function $p \in C([t_0.\infty), \mathbb{R}^+)$ is slowly varying at infinity, for every $s \in \mathbb{R}$,

$$p(t+s) - p(t) \to 0 \; as \; t \to \infty,$$

and proved the theorem.

Theorem 2.1 *Suppose that the function p is slowly varying at infinity and $p > 0$, [58]. Then,*

$$P\tau > \frac{1}{e}, \tag{13}$$

implies that all solutions of Equation (7) *oscillate.*

Remark 1 It is easy to see that [58],

$$p\tau \leqslant a \leqslant A \leqslant P\tau.$$

Thus the above oscillation results by Ladas [45] and Koplatadze and Chanturija [41] imply the results by Myshkis [56], when the function p is slowly varying, then,

$$p\tau = a \; and \; P\tau = A. \tag{14}$$

Therefore in that case both results are equivalent. Moreover, condition (4) together with (14) implies that p is slowly varying at infinity, then the condition,

$$P\tau > 1 \tag{15}$$

guarantees the oscillation of all solutions to Equation (7). Consequently, if instead of (13) the stronger condition (15) is assumed, then the uniform positively condition $p > 0$ can be omitted.

Note the analogy of the conditions (15), (4) also (13), (12), (8), (5) and (9), (6).

Remark 2 The conclusion of Theorem 2.1 does not hold if (13) is replaced by (10). Indeed, if $p(t) = \frac{1}{\tau e}$ identically for $t \geq t_0$, then function p is slowly varying at infinity with $p = P = \frac{1}{\tau e}$, so that $P\tau = \frac{1}{e}$. In this case Equation (7) admits a non-oscillatory solution given by $x(t) = e^{-t/\tau}$ for $t \geq t_0$. Furthermore in the case that $p = P = \frac{1}{\tau e}$ so that $P\tau = \frac{1}{e}$ and

$$p(t) \to \frac{1}{\tau e} \quad as \quad t \to \infty,$$

although p is slowly varying at infinity. Theorem 2.1 does not apply because in this case the oscillation of all solutions depends on the rate of convergence of $p(t)$ to the limit $\frac{1}{\tau e}$ as $t \to \infty$, as it is explained below.

Elbert and Stavroulakis in 1995 [32] established sufficient conditions under which all solutions to Equation (3) oscillate in the critical case where

$$\int_{\tau(t)}^{t} p(s)\,ds \geqslant \frac{1}{e} \text{ and } \lim_{t\to\infty} \int_{\tau(t)}^{t} p(s)\,ds = \frac{1}{e}.$$

In 1996 Domshlak [24, 25] studied Equation (7) in the *critical case* where $p = \frac{1}{\tau e}$ and sufficient conditions for the oscillation of all solutions where established in spite of the fact that the corresponding "limiting" equation,

$$x'(t) + \frac{1}{\tau e} x(t-\tau) = 0, \quad t \geqslant t_0,$$

admits a non-oscillatory solution, $x(t) = e^{-t/\tau}$. Indeed, in [24, 25] it was proved that if,

$$\liminf_{t\to\infty} p(t) = \frac{1}{\tau e} \quad and \quad \liminf_{t\to\infty} \left[\left(p(t) - \frac{1}{\tau e}\right) t^2\right] > \frac{\tau}{8e}, \tag{16}$$

then all solutions of Equation (7) oscillate.

In 1996 this result was improved by Domshlak and Stavroulakis [31] as follows.

Theorem 2.2 *Let us assume that,*

$$\liminf_{t\to\infty} p(t) = \frac{1}{\tau e}, \quad \liminf_{t\to\infty}\left[\left(p(t) - \frac{1}{\tau e}\right)t^2\right] = \frac{\tau}{8e}$$

and

$$C := \liminf_{t\to\infty}\left\{\left[\left(p(t) - \frac{1}{\tau e}\right)t^2 - \frac{\tau}{8e}\right]\ln^2 t\right\} > \frac{\tau}{8e}. \tag{17}$$

Then all solutions of Equation (7) *oscillate.*

Example 1 Consider the equation [31],

$$x'(t) + p(t)\,x(t-1) = 0, \quad t \geqslant 1,$$

where

$$p(t) = \frac{(2t-1)\ln t - 1}{2e\sqrt{t(t-1)\ln t\,\ln(t-1)}}.$$

It is easy to see that, $x(t) = e^{-t}\sqrt{t\,\ln t}$ is a non-oscillatory solution. In this case one can check that,

$$\liminf_{t\to\infty}\left\{\left[\left(p(t) - \frac{1}{\tau e}\right)t^2 - \frac{\tau}{8e}\right]\ln^2 t\right\} = \frac{1}{8e},$$

that is, condition (17) is not satisfied, as expected. Thus the inequality, $C > \frac{\tau}{8e}$, cannot be replaced by the corresponding equality.

The above results were extended by Diblik in 1998 and 2000 [20–22] using the *iterated* algorithm as follows. Call the expression, $\ln_k t$, $k \geq 1$, defined by the expression,

$$\ln_k t = \underbrace{\ln\ln\ldots\ln}_{k} t, \quad k \geqslant 1,$$

the kth *iterated* logarithm if $t > exp_{k-2}1$, where

$$\exp_k t = \underbrace{(\exp(\exp(\ldots\exp}_{k} t))), \quad k \geqslant 1,$$

$\exp_0 t \equiv t$ and $\exp_{-1} t \equiv 0$. Moreover, let us define $\ln_0 t \equiv t$ and also instead of expressions, $\ln_0 t$, $\ln_1 t$, we write only t and $\ln t$. Then the following results were established.

Theorem 2.3 *If for some integer $k \geq 0$,*

$$p(t) \leq \frac{1}{e\tau} + \frac{\tau}{8e\,t^2} + \frac{\tau}{8e(t\ \ln t)^2} + \frac{\tau}{8e(t\ \ln t \ln_2 t)^2} + \ldots + \frac{\tau}{8e(t\ \ln t \ln_2 t \ldots \ln_k t)^2},$$

as $t \to \infty$,

then there exists a positive solution, $x = x(t)$ of Equation (3) *and moreover,*

$$x(t) < e^{-t/\tau}\sqrt{t\ \ln t \ln_2 t \ldots \ln_k t}, \quad as \ \ t \to \infty,$$

while if for a constant $\theta > 1$,

$$p(t) \geq \frac{1}{e\tau} + \frac{\tau}{8e\,t^2} + \frac{\tau}{8e(t\ \ln t)^2} + \ldots + \frac{\tau}{8e(t\ \ln t \ln_2 t \ldots \ln_{k-1} t)^2} + \frac{\theta\,\tau}{8e(t\ \ln t \ln_2 t \ldots \ln_k t)^2}, \tag{18}$$

as $t \to \infty$, then all solutions of Equation (7) *oscillate.*

There is a gap between the conditions (4) and (5) when the limit $\lim_{t\to\infty} \int_{\tau(t)}^{t} p(s)\,ds$, does not exist. How to fill this gap is an interesting problem which has been studied by several authors.

Erbe and Zhang, in 1988 [33], developed new oscillation criteria by employing the upper bound of the ratio $x(\tau(t))/x(t)$ for possible non-oscillatory solution $x(t)$, of Equation (3). Their result states that all the solutions of Equation (3) are oscillatory, when $0 < a \leq \frac{1}{e}$ and,

$$A > 1 - \frac{a^2}{4}. \tag{19}$$

Since then several authors tried to obtain better results by improving the upper bound of $x(\tau(t))/x(t)$. In 1991, Jian Chao [8] derived the condition,

$$A > 1 - \frac{a^2}{2(1-a)}, \tag{20}$$

while Yu and collaborators in 1992 [71] obtained the condition,

$$A > 1 - \frac{1 - a - \sqrt{1 - 2a - a^2}}{2}. \tag{21}$$

Elbert and Stavroulakis [32] and Kwong [43], using different techniques, improved the condition (19), in the case where $0 < a \leq \frac{1}{e}$, to the conditions,

$$A > 1 - \left(1 - \frac{1}{\sqrt{\lambda_1}}\right)^2 \tag{22}$$

and

$$A > \frac{\ln \lambda_1 + 1}{\lambda_1}, \tag{23}$$

respectively, where λ_1 is the smaller real root of the equation $\lambda = e^{a\,\lambda}$.

Philos and Sficas in 1998 [57] and Zhou and Yu in 1999 [74] and Jaroš and Stavroulakis [38] improved further the above conditions in the case where, $0 < a \leq \frac{1}{e}$ as,

$$A > 1 - \frac{a^2}{2(1-a)} - \frac{a^2}{2}\lambda_1, \tag{24}$$

$$A > 1 - \frac{1 - a - \sqrt{1 - 2a - a^2}}{2} - \left(1 - \frac{1}{\sqrt{\lambda_1}}\right)^2, \tag{25}$$

and

$$A > \frac{\ln \lambda_1 + 1}{\lambda_1} - \frac{1 - a - \sqrt{1 - 2a - a^2}}{2}, \tag{26}$$

respectively.

Consider Equation (3) and assume that $\tau(t)$ is continuously differentiable and that exists $\theta > 0$, such that $p(\tau(t))\,\tau'(t) \leq \theta\, p(t)$, eventually for all t. Under this additional assumption, Kon, Sficas, and Stavroulakis in 2000 [40] and Sficas and Stavroulakis in 2003 [59] established the conditions,

$$A > 2a + \frac{2}{\lambda_1} - 1, \tag{27}$$

and

$$A > \frac{\ln \lambda_1 - 1 + \sqrt{5 - 2\lambda_1 + 2a\lambda_1}}{\lambda_1}, \tag{28}$$

respectively. In the case where $a = \frac{1}{e}$, then $\lambda_1 = e$, and (28) leads to,

$$A > \frac{\sqrt{7 - 2e}}{e} \approx 0.459987065.$$

It is to be noted that for small values of a ($a \to 0$), all the previous conditions (19)–(27) reduce to the condition (4), i.e. $A > 1$. However, the condition (28) leads to,

$$A > \sqrt{3} - 1 \approx 0.732,$$

which is an important improvement. Moreover, the condition (28) improves all the above conditions for all values of $a \in (0, \frac{1}{e}]$. Note that the value of the lower bound on A cannot be less than $\frac{1}{e} \approx 0.367879441$. Thus, the aim is to establish a condition which leads to a value *as close as possible to* $\frac{1}{e}$.

It should be pointed out that Koplatadze and Kvinikadze [42] improved (21). Let us assume,

$$\sigma(t) := \sup_{s \leq t} \tau(s), \quad t \geq 0. \tag{29}$$

Clearly $\sigma(t)$ is nondecreasing and $\tau(t) \leq \sigma(t)$, for all $t \geq 0$. Define,

$$\psi_1 = 0, \ \ \psi_i(t) = \exp\left\{ \int_{\tau(t)}^{t} p(\xi)\, \psi_{i-1}(\xi)\, d\xi \right\}, \ i = 2, 3, \ldots \ for \ t \in \mathbb{R}^+. \tag{30}$$

Then the following theorem was established in [42].

Theorem 2.4 *Let* $k \in \{1, 2, \ldots\}$ *exist such that,*

$$\limsup_{t\to\infty} \int_{\sigma(t)}^{t} p(s) \exp\left\{ \int_{\sigma(s)}^{\sigma(t)} p(\xi)\psi_k(\xi)\, d\xi \right\} ds > 1 - c(a), \tag{31}$$

where σ, ψ_k, a *are defined by Equations* (29), (30), (5), *respectively, and*

$$c(a) = \begin{cases} 0 & if \ \ a > \dfrac{1}{e}, \\ \dfrac{1}{2}\left(1 - a - \sqrt{1 - 2a - a^2}\right) & if \ \ 0 < a \leqslant \dfrac{1}{e}, \end{cases} \tag{32}$$

Then all solutions of Equation (3) *oscillate.*

Concerning the constants 1 and $\frac{1}{e}$ which appear in the conditions (4), (5), and (6), Berezansky and Braverman in 2011 [5] established the following,

Theorem 2.5 *For any* $k \in (1/e, 1)$ *there exists a non-oscillatory equation,*

$$x'(t) + p(t)\, x(t - \tau) = 0, \quad \tau > 0,$$

with $p(t) \geq 0$ *such that,*

$$\limsup_{t\to\infty} \int_{t-\tau}^{t} p(s)ds = k.$$

Braverman and Karpuz, in 2011 [6], also studied Equation (3) in the case of a general argument, τ is not assumed monotone, and proved the following,

Theorem 2.6 *There is no constant* $K > 0$ *such that,*

$$\limsup_{t\to\infty} \int_{\tau(t)}^{t} p(s)ds > K, \tag{33}$$

implies oscillation of Equation (3) *for arbitrary, not necessarily nondecreasing, argument* $\tau(t) \le t$.

Remark 3 We observe that, due to the condition (6), the constant K in the above inequality makes sense for $K > 1/e$.

Moreover, the following result was established in [6].

Theorem 2.7 *Assume that,*

$$B := \limsup_{t\to\infty} \int_{\sigma(t)}^{t} p(s) \exp\left\{\int_{\tau(s)}^{\sigma(t)} p(\xi)\,d\xi\right\} ds > 1, \tag{34}$$

where $\sigma(t)$ *is defined by* (29)*. Then all solutions of Equation* (3) *oscillate.*

Observe that condition (29) improves (4). Using the upper bound of the ratio $x(\tau(t))/x(t)$ for non-oscillatory solutions $x(t)$ of Equation (3), presented in several studies [38, 59, 61], the above result was improved in [64].

Theorem 2.8 *Assume that* $0 < a \le \frac{1}{e}$ *and,*

$$B : = \limsup_{t\to\infty} \int_{\sigma(t)}^{t} p(s) \exp\left\{! \int_{\tau(s)}^{\sigma(t)} p(\xi)\,d\xi\right\} ds > 1 - \frac{1}{2}\left(1 - a - \sqrt{1 - 2a - a^2}\right), \tag{35}$$

where $\sigma(t)$ *is defined by* (29)*. Then all solutions of Equation* (3) *oscillate.*

Remark 4 Observe that as $a \to 0$, then condition (35) reduces to (34), [64]. However, the improvement is clear as $a \to \frac{1}{e}$. Actually, when $a = \frac{1}{e}$, the value of the lower bound of B is equal to ≈ 0.863457014. That is, condition (35) essentially improves (34).

Remark 5 Note that, under the additional assumption that $\tau(t)$ is continuously differentiable and that there exists $\theta > 0$ such that $p(\tau(t))\tau'(t) \ge \theta\, p(t)$, for all t, the condition (35) of Theorem 2.8 reduces to the following (see [40, 59, 64]),

$$B > 1 - \frac{1}{2}\left(1 - a - \sqrt{(1 - a)^2 - 4M}\right), \tag{36}$$

where M is given by,

$$M = \frac{e^{\lambda_1 \theta a} - \lambda_1 \theta a - 1}{(\lambda_1 \theta)^2}$$

and λ_1 is the smaller root of the equation $\lambda = e^{\lambda a}$. When $\theta = 1$, then it follows that [59],

$$\frac{1}{2}\left(1 - a - \sqrt{(1-a)^2 - 4M}\right) = 1 - a - \frac{1}{\lambda_1}$$

and in the case that $a = \frac{1}{e}$, then $\lambda_1 = e$ and condition (36) leads to,

$$B > 1 - \left(1 - \frac{2}{e}\right) = \frac{2}{e} \approx 0.735758882.$$

So, condition (36) essentially improves condition (35) but under the additional, stronger assumptions on $\tau(t)$ and $p(t)$.

Chatzarakis [9, 10] proved that for some $j \in \mathbb{N}$,

$$\limsup_{t\to\infty} \int_{\sigma(t)}^{t} p(s) \exp\left(\int_{\tau(s)}^{\sigma(t)} p_j(u)\, du\right) ds > 1, \tag{37}$$

or

$$\limsup_{t\to\infty} \int_{\sigma(t)}^{t} p(s) \exp\left(\int_{\tau(s)}^{\sigma(t)} p_j(u)\, du\right) ds > 1 - \frac{1 - a - \sqrt{1 - 2a - a^2}}{2}, \tag{38}$$

where

$$p_j(t) = p(t)\left[1 + \int_{\tau(t)}^{t} p(s) \exp\left(\int_{\tau(s)}^{\sigma(t)} p_{j-1}(u)\, du\right) ds\right], \ \textit{with } p_0(t) = p(t), \tag{39}$$

and $0 < a \leq \frac{1}{e}$, then all solutions of Equation (3) oscillate.

Recently, Chatzarakis, Purnaras, and Stavroulakis [17] improved the above condition as follows,

Theorem 2.9 *Assume that for some* $j \in \mathbb{N}$,

$$\limsup_{t\to\infty} \int_{\sigma(t)}^{t} p(s) \exp\left\{\int_{\tau(s)}^{\sigma(t)} P_j(u)\, du\right\} ds > 1, \tag{40}$$

or

$$\limsup_{t\to\infty} \int_{\sigma(t)}^{t} p(s) \exp\left\{\int_{\tau(s)}^{\sigma(t)} P_j(u)\,du\right\} ds > 1 - \frac{1}{2}\left(1 - a - \sqrt{1 - 2a - a^2}\right), \tag{41}$$

or

$$\limsup_{t\to\infty} \int_{\sigma(t)}^{t} p(s) \exp\left\{\int_{\tau(s)}^{t} P_j(u)\,du\right\} ds > \frac{2}{1 - a - \sqrt{1 - 2a - a^2}}, \tag{42}$$

or

$$\limsup_{t\to\infty} \int_{\sigma(t)}^{t} p(s) \exp\left\{\int_{\tau(s)}^{\sigma(s)} P_j(u)\,du\right\} ds > \frac{1 + \ln \lambda_1}{\lambda_1} - \frac{1 - a - \sqrt{1 - 2a - a^2}}{2}, \tag{43}$$

where

$$P_j(t) = p(t)\left[1 + \int_{\tau(t)}^{t} p(s) \exp\left(\int_{\tau(s)}^{t} p(u) \exp\left(\int_{\tau(u)}^{u} P_{j-1}(\xi)\,d\xi\right) du\right) ds\right], \tag{44}$$

with $P_0(t) = p(t)$, $0 < a \le \frac{1}{e}$, *and* λ_1 *is the smaller root of the transcendental equation* $\lambda = e^{a\lambda}$. *Then all solutions of Equation* (3) *oscillate.*

Theorem 2.10 *Assume that for some* $j \in \mathbb{N}$,

$$\liminf_{t\to\infty} \int_{\sigma(t)}^{t} p(s) \exp\left\{\int_{\tau(s)}^{\sigma(s)} P_j(u)\,du\right\} ds > \frac{1}{e}, \tag{45}$$

where P_j *is defined by* (44). *Then all solutions of Equation* (3) *oscillate.*

We note that one can easily see that the conditions (40), (41), (43), and (45) substantially improve the previous conditions (4), (34), (37), (35), (26), and (5). An example follows for the delay equations. We focus our attention on this example, to illustrate the level of improvement in the oscillation criteria and the significance of the obtained results.

2.3 Example

The example below illustrates that the oscillation conditions presented in Theorems 2.9 and 2.10 improve known results in the literature yet indicate a type of independence among some of them. The calculations were made by the use of MATLAB software (The MathWorks, Inc).

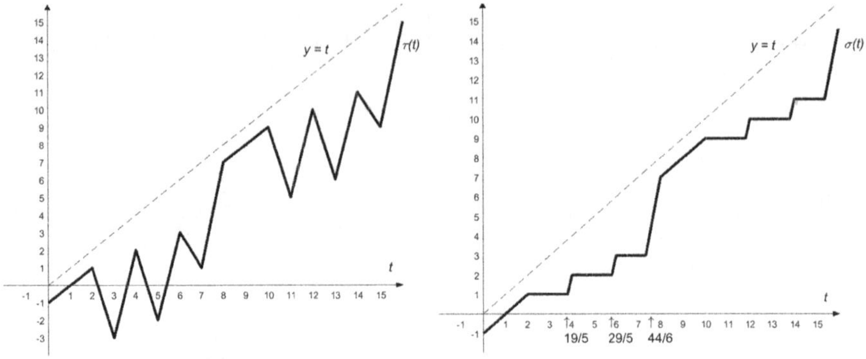

(a) (b)

Fig. 1 The graphs of $\tau(t)$ and $\sigma(t)$

Example Let us consider the retarded differential equation [17]

$$x'(t) + \frac{1}{8}x(\tau(t)) = 0, \quad t \geqslant 0, \tag{46}$$

with $\tau(t)$, $\sigma(t)$ as shown below, see also Figure 1.

$$\tau(t) = \begin{cases} t-1, & if\ t \in [8k,\ 8k+2] \\ -4t+40k+9, & if\ t \in [8k+2,\ 8k+3] \\ 5t-32k-18, & if\ t \in [8k+3,\ 8k+4] \\ -4t+40k+18, & if\ t \in [8k+4,\ 8k+5], \quad k \in \mathbb{N}_0 \\ 5t-32k-27, & if\ t \in [8k+5,\ 8k+6] \\ -2t+24k+15, & if\ t \in [8k+6,\ 8k+7] \\ 6t-40k-41, & if\ t \in [8k+7,\ 8k+8] \end{cases}$$

where $\mathbb{N}_0$ is the set of nonnegative integers.

$$\sigma(t) = \begin{cases} t-1, & if\ t \in [8k,\ 8k+2] \\ 8k+1, & if\ t \in [8k+2,\ 8k+19/5] \\ 5t-32k-18, & if\ t \in [8k+19/5,\ 8k+4] \\ 8k+2, & if\ t \in [8k+4,\ 8k+29/5], \quad k \in N_0 \\ 5t-32k-27, & if\ t \in [8k+29/5,\ 8k+6] \\ 8k+3, & if\ t \in [8k+6,\ 8k+44/6] \\ 6t-40k-41, & if\ t \in [8k+44/6,\ 8k+8]. \end{cases}$$

Let the function $F_j : \mathbb{R}_0 \to \mathbb{R}_+$ $(j \in \mathbb{N})$ be defined by

$$F_j(t) = \int_{\sigma(t)}^{t} p(s) \exp\left(\int_{\tau(s)}^{\sigma(t)} P_j(u)du\right) ds, \tag{47}$$

with P_j given by (44). Noting that F_j attains its maximum at $t = 8k + 44/6$, $k \in \mathbb{N}_0$ for every $j \in \mathbb{N}$, and using an algorithm on MATLAB software, we obtain

$$\limsup_{t\to\infty} F_1(t) = \limsup_{t\to\infty} \int_{\sigma(t)}^{t} p(s) \exp\left\{\int_{\tau(s)}^{\sigma(t)} P_1(u)\, du\right\} ds \simeq 1.0097 > 1.$$

That is, condition (40) of Theorem 2.9 is satisfied for $j = 1$, and therefore all solutions of Equation (46) oscillate. However, observe that

$$\limsup_{t\to\infty} \int_{\sigma(t)}^{t} p(s)\, ds = \limsup_{k\to\infty} \int_{8k+3}^{8k+44/6} \frac{1}{8} ds = 0.5417 < 1,$$
$$a = \liminf_{t\to\infty} \int_{\tau(t)}^{t} p(s)\, ds = \liminf_{k\to\infty} \int_{8k+1}^{8k+2} \frac{1}{8} ds = 0.125 < \frac{1}{e},$$
$$0.5417 < \frac{1 + \ln \lambda_1}{\lambda_1} - \frac{1 - a - \sqrt{1 - 2a - a^2}}{2} \simeq 0.9815,$$

where $\lambda_1 = 1.15537$ is the smaller solution of $e^{0.125\lambda} = \lambda$.
Noting that the function Φ_j defined by

$$\Phi_j(t) = \int_{\sigma(t)}^{t} p(s) \exp\left(\int_{\sigma(s)}^{\sigma(t)} p(u)\, \psi_j(u)\, du\right) ds, \; j \geq 2, \tag{48}$$

(with ψ_j defined by (30)) attains its maximum at $t = 8k + 44/6$, $k \in \mathbb{N}_0$ for every $j \geq 2$. Specifically, we find

$$\limsup_{t\to\infty} \Phi_2(t) \simeq 0.6450 < 1 - \frac{1 - a - \sqrt{1 - 2a - a^2}}{2} \simeq 0.99098.$$

Also

$$\limsup_{t\to\infty} \int_{\sigma(t)}^{t} p(s) \exp\left(\int_{\tau(s)}^{\sigma(t)} p(u)\, du\right) ds \simeq 0.74354 < 1$$

and

$$0.74354 < 1 - \frac{1 - a - \sqrt{1 - 2a - a^2}}{2} \simeq 0.99098.$$

As each one of the functions G_j, $j \in \mathbb{N}$ defined by

$$G_j(t) = \int_{\sigma(t)}^{t} p(s) \exp\left(\int_{\tau(s)}^{\sigma(t)} p(u)\, du \right) ds, \quad j \in \mathbb{N}, \tag{49}$$

attains its maximum at $t = 8k + 44/6$, $k \in \mathbb{N}_0$, for every $j \geq \mathbb{N}$ we find

$$\limsup_{t\to\infty} G_1(t) = \limsup_{t\to\infty} \int_{\sigma(t)}^{t} p(s) \exp\left(\int_{\tau(s)}^{\sigma(t)} p_1(u)\, du \right) ds \simeq 0.8626 < 1$$

and

$$0.8626 < 1 - \frac{1 - a - \sqrt{1 - 2a - a^2}}{2} \simeq 0.99098.$$

That is, none of the conditions (4), (5), (26), (31) (for $j = 2$), (34), (35), and (37) (for $j = 1$) is satisfied. In addition, observe that conditions (31) and (37) do not lead to oscillation at the first iteration. On the contrary, condition (40) is satisfied from the first iteration, which means that it is much faster than (31) and (37). Additionally,

$$\limsup_{t\to\infty} \int_{\sigma(t)}^{t} p(s) \exp\left(\int_{\tau(s)}^{t} P_1(u)\, du \right) ds \simeq 4.8243 < \frac{2}{1 - a - \sqrt{1 - 2a - a^2}} \simeq 110.85,$$

$$\limsup_{t\to\infty} \int_{\sigma(t)}^{t} p(s) \exp\left(\int_{\tau(s)}^{\sigma(s)} P_1(u)\, du \right) ds \simeq 0.7983 < \frac{1+\ln\lambda_1}{\lambda_1} - \frac{1-a-\sqrt{1-2a-a^2}}{2} \simeq 0.9815,$$

$$\liminf_{t\to\infty} \int_{\sigma(t)}^{t} p(s) \exp\left(\int_{\tau(s)}^{\sigma(s)} P_1(u)\, du \right) ds = 0.125 < \frac{1}{e},$$

that is, none of the conditions (42) (for $j = 1$), (43) (for $j = 1$) and (45) (for $j = 1$), are satisfied.

3 Difference Equation

3.1 Introduction and Problem Description

In the last decades the oscillation theory of delay differential equations has been extensively developed, as discussed in the previous sections. The oscillation theory of discrete analogues of delay differential equations has also attracted growing attention in the recent few years. The reader is referred to [54, 55] and the references cited therein. In particular, the problem of establishing sufficient conditions for the oscillation of all solutions of the delay difference equation with a constant argument,

$$\Delta x(n) + p(n)\,x(n-k) = 0, \quad n = 0, 1, 2, \ldots, \tag{50}$$

and the difference equation with a variable retarded argument [16, 55],

$$\Delta x(n) + p(n)\,x(\tau(n)) = 0, \quad n = 0, 1, 2, \ldots, \tag{51}$$

where $p(n)$ is a sequence of nonnegative real numbers, k is a positive integer, $\Delta x(n) = x(n+1) - x(n)$, and $\tau(n)$ is a sequence of integers such that $\tau(n) \leq n-1$ for all $n \geq 0$ and $\lim_{n\to\infty} \tau(n) = \infty$, has been the subject of many recent studies. Strong interest in Equations (50) and (51) is motivated by the fact that they represent the discrete analogues of the delay differential equations (see [36, 37, 49] and the references cited therein):

$$x'(t) + p(x)\,x(t-\tau) = 0, \qquad p(t) \geqslant 0, \ \tau > 0, \tag{52}$$

$$x'(t) + p(x)\,x(\tau(t)) = 0, \qquad p(t) \geqslant 0, \ \tau(t) < t. \tag{53}$$

Set $w = -\min_{n\geq 0} \tau(n)$, where w is a finite positive integer. By a solution of Equation (50) [(51)] we mean a sequence $x(n)$, which is defined for $n \geq -k$ [$n \geq -w$] and which satisfies Equation (50) [(51)], for $n \geq 0$.

A solution $x(n)$ is said to be oscillatory if the terms $x(n)$ of the solution are not eventually positive or eventually negative. Otherwise, the solution is called non-oscillatory. For convenience, we will assume that inequalities about values of sequences are satisfied eventually for all large n. For the general theory of these equations, the interested reader is referred to [1–3, 37, 50].

Besides the purely mathematical problem, the interest in the behavior of the solutions to difference equations with retarded arguments is justified by the fact that the mathematical modeling of many real-world problems leads to difference equations where the unknown function depends on the past history rather than only the present state. This interest grows stronger as difference equations naturally arise from discretization of differential equations. As a consequence, many researchers have been concerned with the study of qualitative behavior of solutions to difference equations, in particular, the study of oscillation of solutions. Pielou, in 1969 and 1974, considered the delay difference equation [37],

$$N_{t+1} = \frac{\alpha N_t}{1 + \beta N_{t-k}}, \quad \alpha \in (1, \infty), \ \ \beta \in (0, \infty), \ \ k \in \mathbb{N}, \tag{54}$$

as the discrete analogue of the delay logistic equation,

$$N'(t) = rN(t)\left[1 - \frac{N(t-\tau)}{\mathrm{K}}\right], \tag{55}$$

where r and K are the growth rate and the carrying capacity of the population, respectively. Pielou's interest in (54) was in showing that "the tendency to oscillate is a property of the populations themselves and is independent of the extrinsic factors." That is, population sizes oscillate "even though the environment remains constant." According to Pielou, "oscillations can be set up in a population if its growth rate is governed by a density dependent mechanism and if there is a delay in the response of the growth rate to density changes. When this happens the size of the population alternately overshoots and undershoots its equilibrium level."

The blowfly (*Lucilia cuprina*) studied in 1954 by Nicholson is an example of a laboratory population which behaves in the manner described above [37]. It is noteworthy that a first-order linear difference equation of the form (51) without retarded argument, $\tau(n) \equiv n$, does not possess oscillatory solutions when $p(n) < 1$. A small delay may change this situation, as one can see below, even in the case of equations with constant delays and constant coefficients, and this certainly adds interest in the investigation of the oscillatory character of Equations (50) and (51). A great part of the existing literature on the oscillation of (51) concerns the case where the argument $\tau(n)$ is nondecreasing, while only a small number of papers are dealing with the general case of arguments being not necessarily monotone. See, for example, [4, 6, 7, 11, 16, 64] and the references cited therein. The consideration of non-monotone arguments may lead to better approximation of the natural phenomena described by difference equations because quite often there appear natural disturbances, e.g., noise in communication systems, that affect all the parameters of the equation and therefore the "fair" (from a mathematical point of view) monotone arguments are, in fact, non-monotone almost always. In view of this, for the case of (51) an interesting question is whether we can state oscillation criteria considering the argument $x(n)$ to be not necessarily monotone.

In this section, we present a survey on the oscillation of solutions to Equations (50) and (51) in both cases that the argument $x(n)$ is monotone or non-monotone and examples which illustrate the significance of the obtained results.

3.2 Oscillation Criteria for Difference Equation (50)

Domshlak, in 1981, studied the problem in the case where $k = 1$ [23]. Further, Erbe and Zhang, in 1989, established the following theorems (oscillation criteria) for Equation (50) [34].

Theorem 3.11 *Assume that,*

$$\beta = \liminf_{n\to\infty} p(n) > 0, \quad \limsup_{n\to\infty} p(n) > 1 - \beta \tag{56}$$

or

$$\liminf_{n\to\infty} p(n) > \frac{k^k}{(k+1)^{k+1}} \tag{57}$$

or

$$\mathbf{A} := \limsup_{n\to\infty} \sum_{i=n-k}^{n} p(i) > 1. \tag{58}$$

Then all solutions of Equation (50) *oscillate.*

In the same year Ladas et al., proved the following Theorem [48].

Theorem 3.12 *Assume that,*

$$\liminf_{n\to\infty} \frac{1}{k} \sum_{i=n-k}^{n-1} p(i) > \frac{k^k}{(k+1)^{k+1}}. \tag{59}$$

Then all solutions of Equation (50) *oscillate.*

The above theorems improve the condition (57) by replacing the $p(n)$ of (57) by the arithmetic mean of the terms, $p(n-k), \ldots, p(n-1)$ in condition (59).

Note that this condition is sharp in the sense that the fraction on the right-hand side cannot be improved, since $p(n)$ is a constant, $p(n) = p$, then the condition reduces to,

$$p > \frac{k^k}{(k+1)^{k+1}}, \tag{60}$$

which is *necessary and sufficient condition* for the oscillation of all solutions to (50). Moreover, concerning the constant $k^k/(k+1)^{k+1}$ in (57) and (59) it should be emphasized that, as it is shown in [34], if

$$\sup p(n) < \frac{k^k}{(k+1)^{k+1}}, \tag{61}$$

then Equation (50) has a non-oscillatory solution.

Ladas in 1990 [44] conjectured that (50) has a non-oscillatory solution if

$$\frac{1}{k} \sum_{i=n-k}^{n-1} p(i) \leq \frac{k^k}{(k+1)^{k+1}}, \tag{62}$$

holds. However, this conjecture is not correct and a counterexample was given by Yu et al. in 1994 [73]. Moreover, Tang and Yu in 1999 [68], using a different technique, showed that if,

$$\sum_{i=n-k}^{n} p(i) \le \frac{1}{e} \quad \text{for all large } n, \tag{63}$$

then (50) has a non-oscillatory solution. Therefore, as a corollary, Tang and Yu presented an affirmative answer to the so-called corrected Ladas conjecture [68]. That is,

$$\sum_{i=n-k}^{n} p(i) \le \left(\frac{k}{k+1}\right)^{k+1} \quad \text{for all large } n, \tag{64}$$

then (50) has a non-oscillatory solution.

Recently, Karpuz [39] studied this problem deriving the following conditions:

$$\liminf_{n\to\infty} \inf_{\lambda\ge 1} \left[\frac{1}{\lambda} \prod_{i=n-k}^{n} [1+\lambda p(i)]\right] > 1, \tag{65}$$

then every solution of Equation (50) oscillates, while if there exists $\lambda_0 \ge 1$ such that,

$$\frac{1}{\lambda_0} \prod_{i=n-k}^{n} [1+\lambda_0 p(i)] \le 1 \quad \text{for all large } n, \tag{66}$$

then Equation (50) has a non-oscillatory solution.

Yu et al. in 1993 [72] and Lalli and Zhang [51], trying to improve (58), established the following (false) sufficient oscillating condition for Equation (50):

$$0 < \mathbf{a} := \liminf_{n\to\infty} \sum_{i=n-k}^{n-1} p(i) \le \left(\frac{k}{k+1}\right)^{k+1}, \quad \mathbf{A} > 1 - \frac{\mathbf{a}^2}{4}, \tag{67}$$

$$\begin{aligned} &\sum_{i=n-k}^{n} p(i) \ge d > 0 \quad \text{for all large } n, \\ &\mathbf{A} > 1 - \frac{d^4}{8}\left(1 - \frac{d^3}{8} + \sqrt{1 - \frac{d^3}{2}}\right)^{-1}, \end{aligned} \tag{68}$$

respectively. The above conditions are based on the following (false) discrete version of Koplatadze–Chanturia Lemma [19, 30] .

Lemma 1 (False) *Assume that $x(n)$ is an eventually positive solution of Equation* (50) *and that,*

$$\sum_{i=n-k}^{n} p(i) \geqslant M > 0 \quad \textit{for large } n. \tag{69}$$

Then

$$x(n) > \frac{M^2}{4} x(n-k) \quad \textit{for large } n. \tag{70}$$

As one can see, the erroneous proof of Lemma 1 is based of the following false statement [19, 30].

Statement 1 (False) If (69) holds, then for any large N, there exists a positive integer n such that $n - k \leq N \leq n$ and,

$$\sum_{i=n-k}^{n} p(i) \geqslant \frac{M}{2}, \quad \sum_{i=N}^{n} p(i) \geqslant \frac{M}{2}. \tag{71}$$

It is observed that all the oscillation results which have made use of Lemma 1 or *Statement 1* are incorrect. For more details on this problem see the paper by Cheng and Zhang [19]. We should point out that the following statement [48, 62] is correct and it should not be confused with *Statement 1*.

Statement 2 If,

$$\sum_{i=n-k}^{n-1} p(i) \geqslant M > 0 \quad \text{for large } n \tag{72}$$

then for any large n, there exists a positive integer, n^*, with $n - k \leq n^* \leq n$ such that,

$$\sum_{i=n-k}^{n^*} p(i) \geqslant \frac{M}{2}, \quad \sum_{i=n^*}^{n} p(i) \geqslant \frac{M}{2}. \tag{73}$$

Stavroulakis in 1995, based on *Statement 2*, proved the following theorem [62].

Theorem 3.13 *Assume that,*

$$0 < \mathbf{a} \leqslant \left(\frac{k}{k+1}\right)^{k+1}, \tag{74}$$

$$\limsup_{n\to\infty} p(n) > 1 - \frac{\mathbf{a}^2}{4}. \tag{75}$$

Then all solutions of Equation (50) *oscillate.*

Domshlak [30] and Cheng and Zhang [19] established the following lemmas, respectively, which may be looked upon as (exact) discrete versions of Koplatadze–Chanturia Lemma.

Lemma 2 *Assume that $x(n)$ is an eventually positive solution of Equation* (50) *and that condition* (72) *holds. Then,*

$$x(n) > \frac{M^2}{4} x(n-k) \quad \textit{for large } n. \tag{76}$$

Lemma 3 *Assume that $x(n)$ is an eventually positive solution of Equation* (50) *and that condition* (72) *holds. Then,*

$$x(n) > M^k x(n-k) \ \ \textit{for large } n. \tag{77}$$

Based on the above lemmas, the following theorem was established by Stavroulakis in 2004 [63].

Theorem 3.14 *Assume that,*

$$0 < \mathbf{a} \le \left(\frac{k}{k+1} \right)^{k+1}. \tag{78}$$

Then either one of the conditions,

$$\limsup_{n \to \infty} \sum_{i=n-k}^{n-1} p(i) > 1 - \frac{\mathbf{a}^2}{4} \tag{79}$$

or

$$\limsup_{n \to \infty} \sum_{i=n-k}^{n-1} p(i) > 1 - \mathbf{a}^k \tag{80}$$

implies that the solution of Equation (50) *oscillate.*

Remark 6 From the above theorem it is now clear that,

$$\begin{aligned} 0 < \mathbf{a} = \liminf_{n \to \infty} \sum_{i=n-k}^{n-1} p(i) \le \left(\frac{k}{k+1} \right)^{k+1}, \\ \limsup_{n \to \infty} \sum_{i=n-k}^{n-1} p(i) > 1 - \frac{\mathbf{a}^2}{4}, \end{aligned} \tag{81}$$

is the correct oscillation condition by which the (false) condition (67) should be replaced [63].

Remark 7 Observe the following,

(i) When $k = 1, 2,$

$$\mathbf{a}^k > \frac{\mathbf{a}^2}{4}, \tag{82}$$

(since, from the above mentioned conditions, it makes sense to study the case when $\mathbf{a} < (k/(k+1))^{k+1}$) and therefore condition (79) implies (80).

(ii) When $k = 3$,

$$\mathbf{a}^3 > \frac{\mathbf{a}^2}{4}, \quad \textit{when } \mathbf{a} > \frac{1}{4} \tag{83}$$

while,

$$\mathbf{a}^3 < \frac{\mathbf{a}^2}{4}, \quad \textit{when } \mathbf{a} < \frac{1}{4}. \tag{84}$$

Therefore, in this case conditions (79) and (80) are independent.

(iii) When $k \geq 4$,

$$\mathbf{a}^k < \frac{\mathbf{a}^2}{4}, \tag{85}$$

and therefore condition (79) implies (80).

(iv) When $k < 12$, conditions (79) or (80) implies (58).

(v) When $k \geq 12$, condition (79) may hold, but condition (58) may not hold.

We illustrate these by the following example.

Example 2 Consider the equation (see [63]),

$$x(n+1) - x(n) + p(n)\,x(n-3) = 0, \quad n = 0, 1, 2, \ldots, \tag{86}$$

where

$$\begin{aligned} p(2n) &= \frac{1}{10}, \\ p(2n+1) &= \frac{1}{10} + \frac{64}{95}\sin^2\frac{n\pi}{2}, \\ &\qquad n = 0, 1, 2, \ldots. \end{aligned} \tag{87}$$

Here $k = 3$ and it is easy to see that,

$$\begin{aligned} \mathbf{a} &= \liminf_{n\to\infty} \sum_{i=n-3}^{n-1} p(i) = \frac{3}{10} < \left(\frac{3}{4}\right)^4, \\ \limsup_{n\to\infty} \sum_{i=n-3}^{n-1} p(i) &= \frac{3}{10} + \frac{64}{95} > 1 - \mathbf{a}^3. \end{aligned} \tag{88}$$

Thus, condition (80) is satisfied and therefore all solutions oscillate. However, we can observe that condition (79) is not satisfied.

If, on the other hand, in the above equation,

$$\begin{aligned} p(2n) &= \frac{8}{100}, \\ p(2n+1) &= \frac{8}{100} + \frac{746}{1000}\sin^2\frac{n\pi}{2}, \\ & n = 0, 1, 2, \dots, \end{aligned} \tag{89}$$

then it is easy to see that,

$$\begin{aligned} \mathbf{a} &= \liminf_{n\to\infty} \sum_{i=n-3}^{n-1} p(i) = \frac{24}{100} < \left(\frac{3}{4}\right)^4, \\ \limsup_{n\to\infty} \sum_{i=n-3}^{n-1} p(i) &= \frac{24}{100} + \frac{746}{1000} > 1 - \frac{\mathbf{a}^2}{4}. \end{aligned} \tag{90}$$

In this case, condition (79) is satisfied and therefore all solutions oscillate. However, we can observe that condition (80) is not satisfied.

Chen and Yu, in 1995 [18], following the above mentioned direction, derived a condition which, formulated in terms of **a** and **A**, states that all solutions of (50) oscillate if, $0 < \mathbf{a} \le k^{k+1}/(k+1)^{k+1}$ and,

$$\mathbf{A} > 1 - \frac{1 - \mathbf{a} - \sqrt{1 - 2\mathbf{a} - \mathbf{a}^2}}{2}. \tag{91}$$

Domshlak, in 1998 [27], studied the oscillation of all solutions and the existence of non-oscillatory solutions of (50) with r-periodic positive coefficients $p(n)$, $p(n+r) = p(n)$. It is important that, in the following case where $\{r = k\}$, $\{r = k + 1\}$, $\{r = 2\}$, $\{k = 1,\ r = 3\}$, and $\{k = 1, r = 4\}$, the results obtained are stated in terms of necessary and sufficient conditions and it is easy to check them.

Tang and Yu, in 2000 [69], improved condition (91) to the condition,

$$\mathbf{A} > \lambda_2^k(1 - k\ \ln\lambda_2) - \frac{1 - \mathbf{a} - \sqrt{1 - \mathbf{a} - \mathbf{a}^2}}{2}, \tag{92}$$

where λ_2 is the greater root of the algebraic equation,

$$k\lambda^k(1-\lambda) = \mathbf{a}.$$

Shen and Stavroulakis, in 2001 [60], using new techniques, improved the previous results as follows.

Theorem 3.15 *Assume that* $0 \leq \mathbf{a} \leq k^{k+1}/(k+1)^{k+1}$ *and that there exists an integer* $l \geq 1$ *such that,*

$$\limsup_{n\to\infty} \left\{ \sum_{i=1}^{k} p(n-i) + [\bar{d}(\mathbf{a})]^{-k} \prod_{i=1}^{k} \sum_{j=1}^{k} p(n-i+j) \right. \\ \left. + \sum_{m=0}^{l-1} \left[d\left(\frac{\mathbf{a}}{k}\right)\right]^{-(n+1)k} \sum_{i=1}^{k} \prod_{j=0}^{m+1} p(n-kj+i) \right\} > 1, \tag{93}$$

where $\bar{d}(\mathbf{a})$ *and* $d(\mathbf{a}/k)$ *are the greater real roots of the equations,*

$$\begin{aligned} d^{k+1} - d^k + \mathbf{a}^k &= 0, \\ d^{k+1} - d^k + \frac{\mathbf{a}}{k} &= 0, \end{aligned} \tag{94}$$

respectively. Then all the solutions of Equation (50) *oscillate.*

Notice that when $k = 1$, $d(\mathbf{a}) = \bar{d}(\mathbf{a}) = (1+\sqrt{1-4\mathbf{a}})/2$ (see [60]), and so condition (93) reduces to,

$$\limsup_{n\to\infty} \left\{ Cp(n) + p(n-1) \right. \\ \left. + \sum_{m=0}^{l-1} C^{m+1} \prod_{j=0}^{m+1} p(n-j-1) \right\} > 1, \tag{95}$$

where $C = 2/(1+\sqrt{1-4\mathbf{a}})$, $\mathbf{a} = \liminf_{n\to\infty} p_n$. Therefore, from the above Theorem, we have the following corollary.

Corollary 1 *Assume that* $0 \leq \mathbf{a} \leq 1/4$ *and that* (95) *holds. Then all solutions of the equation,*

$$x(n+1) - x(n) + p(n)x(n-1) = 0, \tag{96}$$

oscillate.

A condition derived from (95) and which can be easily verified is given in the next corollary.

Corollary 2 *Assume that* $0 \leq \mathbf{a} \leq 1/4$ *and that,*

$$\limsup_{n\to\infty} p(n) > \left(\frac{1+\sqrt{1-4\mathbf{a}}}{2}\right)^2. \tag{97}$$

Then all solutions of (96) *oscillate.*

Corollary 3 *Assume that* $0 \le \mathbf{a} \le k^{k+1}/(k+1)^{k+1}$ *and that,*

$$\limsup_{n\to\infty} \sum_{i=n-k}^{n-1} p(i) > 1 - [\bar{d}(\mathbf{a})]^{-k}\mathbf{a}^k - \frac{k[d(\mathbf{a}/k)]^{-k}\beta^2}{1-[d(\mathbf{a}/k)]^{-k}\beta}, \tag{98}$$

where $\bar{d}(\mathbf{a})$, $d(\mathbf{a}/k)$ *are as in Theorem 3.15. Then all solution of Equation* (50) *oscillate.*

Following this chronological review we also mention that in the critical case where

$$\frac{1}{k}\sum_{i=n-k}^{n-1} p(i) \geqslant \frac{k^k}{(k+1)^{k+1}}, \qquad \lim_{n\to\infty}\frac{1}{k}\sum_{i=n-k}^{n-1} p(i) = \frac{k^k}{(k+1)^{k+1}}, \tag{99}$$

the oscillation of (50) has been studied by Domshlak, 1994 [26] and by Tang in 1998 [67]. In a case when $p(n)$ is asymptotically close to one of the periodic critical states, unimprovable results about the oscillations of the equation,

$$x(n+1) - x(n) + p(n)x(n-1) = 0, \tag{100}$$

were obtained by Domshlak in 1999 and 2000 [28, 29].

3.3 *Oscillation Criteria for Difference Equation (51)*

In this section we study the delay difference equation with variable argument (51) where $\Delta x(n) = x(n+1) - x(n)$, $p(n)$ is a sequence of nonnegative real numbers and $\tau(n)$ is a sequence of integers such that $\tau(n) \le n-1$, for all $n \ge 0$ and $\lim_{n\to\infty}\tau(n) = \infty$.

Chatzarakis et al. in 2008 [12] studied for the first time the oscillatory behavior of Equation (51) in the case of a general non-monotone delay argument $\tau(n)$ and derived the Theorem.

Theorem 3.16 *Assume that,*

$$\sigma(n) = \max_{0 \le s \le n} \tau(s), \quad n \ge 0. \tag{101}$$

If

$$\limsup_{n\to\infty} \sum_{j=\sigma(n)}^{n} p(j) > 1, \tag{102}$$

then all solutions of Equation (51) *oscillate.*

Remark 8 The sequence of integers $\sigma(n)$ is nondecreasing and $\tau(n) \le \sigma(n)$, for all $n \ge 0$.

In the same year, Chatzarakis et al. [14] derived the following theorem.

Theorem 3.17 *Assume that,*

$$\limsup_{n\to\infty} \sum_{i=\tau(n)}^{n-1} p(i) < +\infty, \tag{103}$$

$$\alpha := \liminf_{n\to\infty} \sum_{i=\tau(n)}^{n-1} p(i) > \frac{1}{e}. \tag{104}$$

Then all solutions of Equation (51) *oscillate.*

Remark 9 Note that condition (103) is not a limitation since, by condition (102), if τ is a nondecreasing function, all solutions of Equation (51) oscillate (see [14]).

Remark 10 Condition (102) is optimal for Equation (51) under the assumption that $\lim_{n\to\infty}(n - \tau(n)) = \infty$, since in this case the set of natural numbers increases infinitely in the interval $[\tau(n), n-1]$ for $n \to \infty$ (see [14]).

Chatzarakis et al. in 2008 and 2009 [12, 13, 15] derived the following conditions.

Theorem 3.18

(1) Assume that $0 < \alpha \le 1/e$. *Then either one of the conditions [12, 13, 15],*

$$\limsup_{n\to\infty} \sum_{j=\sigma(n)}^{n} p(j) > 1 - \left(1 - \sqrt{1-\alpha}\right)^2, \tag{105}$$

$$\limsup_{n\to\infty} \sum_{j=\sigma(n)}^{n} p(j) > 1 - \frac{1}{2}\left(1 - a - \sqrt{1-2\alpha}\right), \tag{106}$$

$$\limsup_{n\to\infty} \sum_{j=\sigma(n)}^{n} p(j) > 1 - \frac{1}{2}\left(1 - \alpha - \sqrt{1 - 2\alpha - \alpha^2}\right), \tag{107}$$

implies that all solutions of Equation (51) *oscillate.*

(2) *if* $0 < \alpha \le 1/e$ *and, in addition,* $p(n) \ge 1 - \sqrt{1-\alpha}$ *for all large n and*

$$\limsup_{n\to\infty} \sum_{j=\sigma(n)}^{n} p(j) > 1 - a\frac{1 - \sqrt{1-\alpha}}{\sqrt{1-\alpha}}, \tag{108}$$

or of $0 < \alpha \le 6 - 4\sqrt{2}$ *and, in addition,* $p(n) \ge \alpha/2$ *for all large n and*

$$\limsup_{n\to\infty} \sum_{j=\sigma(n)}^{n} p(j) > 1 - \frac{1}{4}\left(2 - 3\alpha - \sqrt{4 - 12\alpha - \alpha^2}\right), \tag{109}$$

then all solutions of Equation (51) *are oscillatory.*

Remark 11 Observe the following,

(i) When $0 < \alpha \le 1/e$, it is easy to verify that,

$$\begin{aligned} &\tfrac{1}{2}\left(1 - \alpha - \sqrt{1 - 2\alpha - \alpha^2}\right) > \alpha\frac{1 - \sqrt{1-\alpha}}{\sqrt{1-\alpha}} \\ &\quad > \tfrac{1}{2}\left(1 - \alpha - \sqrt{1 - 2\alpha}\right) > \left(1 - \sqrt{1-\alpha}\right)^2 \end{aligned} \tag{110}$$

and therefore condition (107) is weaker than condition (108), (106), and (105).

(ii) When $0 < \alpha \le 6 - 4\sqrt{2}$, it is easy to show that,

$$\begin{aligned} &\tfrac{1}{4}\left(2 - 3\alpha - \sqrt{4 - 12\alpha + \alpha^2}\right) \\ &\quad > \tfrac{1}{2}\left(1 - \alpha - \sqrt{1 - 2\alpha - \alpha^2}\right) \end{aligned} \tag{111}$$

and therefore, in this case, inequality (109) improves inequality (107) and. especially, when $\alpha = 6 - 4\sqrt{2} \simeq 0.3431457$, the lower bound in (107) is 0.8929094, while that in condition (109) is 0.7573593.

Braverman and Karpuz in 2011 [6] studied equation (51) also in the case of non-monotone delays. More precisely, the following were derived in [6].

Theorem 3.19 *There is no constant* $\Lambda > 0$ *such that the inequalities [6],*

$$\begin{aligned} &\limsup_{n\to\infty}\, [n - \tau(n)]\, p(n) > \Lambda, \\ &\limsup_{n\to\infty} \sum_{i=\tau(n)}^{n-1} p(i) > \Lambda, \end{aligned} \tag{112}$$

imply oscillation of Equation (51).

Remark 12 Obviously, there is no constant $\Lambda > 0$ such that,

$$\limsup_{n\to\infty} \sum_{i=\tau(n)}^{n} p(i) > \Lambda, \tag{113}$$

implies oscillation of Equation (51).

Remark 13 It should be emphasized that conditions (102) and (104) imply that all solution of Equation (51) oscillate without the assumption that $\tau(n)$ is monotone. Note that, in (102) instead of $\tau(n)$ the sequence $\sigma(n)$, defined by (101), is considered, which is nondecreasing and $\tau(n) \le \sigma(n)$, for all $n \ge 0$.

Theorem 3.20 *If,*

$$\limsup_{n\to\infty} \sum_{j=\sigma(n)}^{n} p(j) \prod_{i=\tau(j)}^{\sigma(n)-1} \frac{1}{1-p(i)} > 1, \tag{114}$$

then every solution of Equation (51) *oscillates.*

Using the upper bound of the ratio $x(\tau(x)/x(n))$ for possible non-oscillatory solutions $x(n)$ of Equation (51), presented in [12, 13, 15], the above result was improved by Stavroulakis in 2014 [64].

Theorem 3.21 *Assume that (see [64]),*

$$\alpha = \liminf_{n\to\infty} \sum_{i=\tau(n)}^{n-1} p(i), \tag{115}$$

$$\limsup_{n\to\infty} \sum_{j=\sigma(n)}^{n} p(j) \prod_{i=\tau(j)}^{\sigma(n)-1} \frac{1}{1-p(i)} > 1 - c(\alpha), \tag{116}$$

where

$$c(\alpha) = \begin{cases} \frac{1}{2}\left(1-\alpha-\sqrt{1-2\alpha-\alpha^2}\right), & if\ 0<\alpha\le 1/e, \\ \frac{1}{4}\left(2-3\alpha-\sqrt{4-12\alpha-\alpha^2}\right), & if\ 0<\alpha\le 6-4\sqrt{2},\ p(n)\ge \dfrac{\alpha}{2}. \end{cases} \tag{117}$$

Then all solutions of Equation (51) *oscillate.*

Remark 14 Observe that, as $\alpha \to 0$, condition (116) reduces to condition (114). However, the improvement is clear as $\alpha \to 1/e$. Actually, when $\alpha = 1/e$, the value of the lower bound in (116) is equal to ≈ 0.863457014, while when $p(n) \ge \alpha/2$ and $\alpha = 6-4\sqrt{2} \simeq 0.3431457$, the lower bound in (116) is 0.7573593. That is, in all cases condition (116) of Theorem 3.21 essentially improves condition (114) of Theorem 3.20.

Example 3 Consider the equation [64],

$$\Delta x(n) + p(n)\, x(\tau(n)) = 0, \quad n = 0, 1, 2, \dots, \tag{118}$$

where

$$\begin{aligned} p(n) &= \begin{cases} 10^{-4}, & n \text{ is even}, \\ \frac{1}{e}, & n \text{ is odd}, \end{cases} \\ \tau(n) &= \begin{cases} n-1, & n \text{ is even}, \\ n-2, & n \text{ is odd}. \end{cases} \end{aligned} \tag{119}$$

Observe that for this equation $\sigma(n) = \tau(n)$ and it is easy to see that,

$$\begin{aligned} \sum_{i=\tau(n)}^{n-1} p(i) &= \begin{cases} \dfrac{1}{e}, & n \text{ is even}, \\ \dfrac{1}{e} + 10^{-4}, & n \text{ is odd}, \end{cases} \\ \sum_{i=\tau(n)}^{n} p(i) &= \begin{cases} \dfrac{1}{e} + 10^{-4}, & n \text{ is even}, \\ \dfrac{2}{e} + 10^{-4}, & n \text{ is odd}, \end{cases} \\ \sum_{j=\sigma(n)}^{n} p(j) \prod_{i=\tau(j)}^{\sigma(n)-1} \frac{1}{1-p(i)} &= \begin{cases} 0.58, & n \text{ is even}, \\ 0.95, & n \text{ is odd}, \end{cases} \end{aligned} \tag{120}$$

and so,

$$\begin{aligned} &\limsup_{n\to\infty} \sum_{i=\tau(n)}^{n} p(i) = \frac{2}{e} + 10^{-4} \simeq 0.7358, \\ &\limsup_{n\to\infty} \sum_{i=\tau(n)}^{n-1} p(i) = \frac{1}{e} + 10^{-4} \simeq 0.3679, \\ &\liminf_{n\to\infty} \sum_{i=\tau(n)}^{n-1} p(i) = \frac{1}{e}, \\ &\limsup_{n\to\infty} \sum_{j=\sigma(n)}^{n} p(j) \prod_{i=\tau(j)}^{\sigma(n)-1} \frac{1}{1-p(i)} \simeq 0.95. \end{aligned} \tag{121}$$

Thus,

$$
\begin{aligned}
0.7358 &< 1 \\
\alpha &= \frac{1}{e}, \\
0.7358 &< 1 - (1 - \sqrt{1-\alpha})^2 \simeq 0.9579, \\
0.7358 &< 1 - \frac{1}{2}(1 - \alpha - \sqrt{1-2\alpha}) \simeq 0.9409, \\
0.7358 &< 1 - \frac{1}{2}(1 - \alpha - \sqrt{1-2\alpha-\alpha^2}) \simeq 0.8634, \\
0.95 &< 1
\end{aligned}
\tag{122}
$$

and therefore none of the known oscillation conditions (102), (104), (105), (106), (107), and (114) is satisfied. However,

$$
\begin{aligned}
\limsup_{n\to\infty} \sum_{j=\sigma(n)}^{n} p(j) \prod_{i=\tau(j)}^{\sigma(n)-1} \frac{1}{1-p(i)} &\simeq 0.95 \\
> 1 - \frac{1}{2}(1 - \alpha - \sqrt{1-2\alpha-\alpha^2}) &\simeq 0.8634,
\end{aligned}
\tag{123}
$$

that is, the conditions of Theorem 3.21 are satisfied and therefore all solutions of Equation (118) oscillate.

Braverman et al. in 2015 [7] established the following iterative oscillation condition. If for some $r \in \mathbb{N}$,

$$
\limsup_{n\to\infty} \sum_{j=\sigma(n)}^{n} p(j)\, a_r^{-1}(\sigma(n),\ \tau(j)) > 1, \tag{124}
$$

or

$$
\limsup_{n\to\infty} \sum_{j=\sigma(n)}^{n} p(j)\, a_r^{-1}(\sigma(n),\ \tau(j)) > 1 - \frac{1-\alpha-\sqrt{1-2\alpha-\alpha^2}}{2}, \tag{125}
$$

where

$$
a_1(n,k) = \prod_{i=k}^{n-1} [1 - p(i)],\quad a_{r+1}(n,k) = \prod_{i=k}^{n-1} [1 - p(i)\, a_r^{-1}(i,\ \tau(i))] \tag{126}
$$

and $\alpha = \liminf_{n\to\infty} \sum_{j=\tau(n)}^{n-1} p(j)$, then all solutions of Equation (51) oscillate.

Asteris and Chatzarakis [4] and Chatzarakis and Shaikhet [11] in 2017, proved that if for some $l \in \mathbb{N}$,

$$\limsup_{n\to\infty} \sum_{i=\sigma(n)}^{n} p(i) \prod_{j=\tau(i)}^{\sigma(n)-1} \frac{1}{1-p_l(j)} > 1, \tag{127}$$

or

$$\limsup_{n\to\infty} \sum_{i=\sigma(n)}^{n} p(i) \prod_{j=\tau(i)}^{\sigma(n)-1} \frac{1}{1-p_l(j)} > 1 - \frac{1}{2}\left(1-\alpha-\sqrt{1-2\alpha-\alpha^2}\right), \tag{128}$$

where

$$p_l(n) = p(n)\left[1 + \sum_{i=\tau(n)}^{n-1} p(i) \prod_{j=\tau(i)}^{\sigma(n)-1} \frac{1}{1-p_{l-1}(j)}\right], \tag{129}$$

with $p_0(n) = p(n)$ and $\alpha = \liminf_{n\to\infty} \sum_{j=\tau(n)}^{n-1} p(j)$, then all solutions of Equation (51) oscillate.

Recently, Chatzarakis et al. [16] established the following conditions which essentially improve all related conditions in the literature.

Theorem 3.22

(i) If there exists an $l \geq 1$ such that $P_l(n) \geq 1$ for sufficiently large n, then all solutions of Equation (51) *are oscillatory (see [16]).*

(ii) If for some $l \in \mathbb{N}$ we have $P_l(n) < 1$, for sufficiently large n, and

$$\limsup_{n\to\infty} \sum_{i=\sigma(n)}^{n} p(i) \prod_{j=\tau(i)}^{\sigma(n)-1} \frac{1}{1-P_l(j)} > 1, \tag{130}$$

where

$$P_l(n) = p(n)\left[1 + \sum_{i=\tau(n)}^{n-1} p(i) \exp\left(\sum_{j=\tau(i)}^{n-1} p(j) \prod_{m=\tau(j)}^{j-1} \frac{1}{1-P_{l-1}(m)}\right)\right], \tag{131}$$

then all solutions of Equation (51) *are oscillatory.*

Theorem 3.23 *Assume that,* $0 < \alpha \leq 1/e$ *and for some* $l \in \mathbb{N}$ *(see [16]),*

$$\limsup_{n\to\infty} \sum_{i=\sigma(n)}^{n} p(i) \prod_{j=\tau(i)}^{\sigma(n)-1} \frac{1}{1-P_l(j)} > 1 - \frac{1}{2}\left(1-\alpha-\sqrt{1-2\alpha-\alpha^2}\right), \tag{132}$$

or

$$\limsup_{n\to\infty} \sum_{i=\sigma(n)}^{n} p(i) \prod_{j=\tau(i)}^{n} \frac{1}{1-P_l(j)} > \frac{2}{1-\alpha-\sqrt{1-2\alpha-\alpha^2}}, \tag{133}$$

where $P_l(n)$ *is defined by* (131). *Then all solutions of Equation* (51) *are oscillatory.*

In the next section we present an example that illustrates that the conditions of Theorems 3.22 and 3.23 improve known results in the literature.

3.4 Example

The example below illustrates that the conditions improve known results in the literature and also provides independence among some of them. The calculations were made by the use of MATLAB software (The MathWorks, Inc).

Example Consider the retarded difference equation [16],

$$\Delta x(n) + p\,x(\tau(n)) = 0, \quad n \in \mathbb{N}_0, \tag{134}$$

with $0 < p \le 1/e$, see Figure 2a, and

$$\tau(n) = \begin{cases} n-1, & if\ n = 3\mu, \\ n-2, & if\ n = 3\mu+1, \quad \mu \in \mathbb{N}_0 \\ n-4, & if\ n = 3\mu+2. \end{cases} \tag{135}$$

Clearly, τ is non-monotone. For the function σ defined by (101), we have that, see also Figure 2b,

$$\sigma(n) = \max_{0\le s\le n} \tau(s) = \begin{cases} n-1, & if\ n = 3\mu, \\ n-2, & if\ n = 3\mu+1, \quad \mu \in N_0 \\ n-3, & if\ n = 3\mu+2, \end{cases} \tag{136}$$

The left-hand side in Equation (124) attains its maximum at $n = 3\mu+2$, $\mu \in \mathbb{N}_0$, for every $r \in \mathbb{N}$. Specifically,

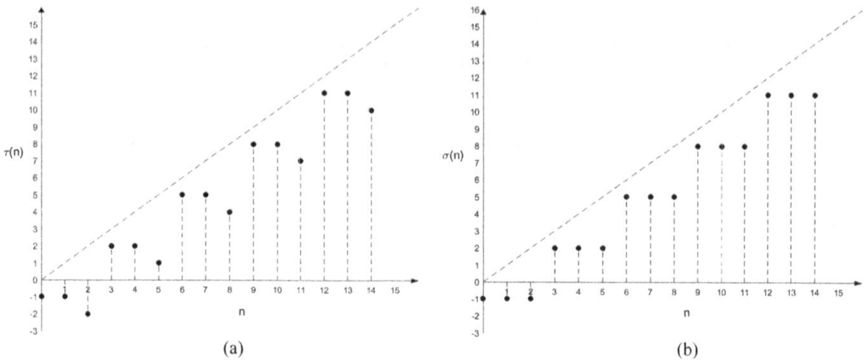

Fig. 2 The graphs of $\tau(n)$ and $\sigma(n)$

$$\begin{aligned}\sum_{j=3\mu-1}^{3\mu+2} p(j)\, a_2^{-1}(3\mu-1,\tau(j)) &= p\frac{1}{a_2(3\mu-1,\tau(3\mu-1))}+p\frac{1}{a_2(3\mu-1,\tau(3\mu))}\\ &+p\frac{1}{a_2(3\mu-1,\tau(3\mu+1))}+p\frac{1}{a_2(3\mu-1,\tau(3\mu+2))}=\\ &=p\frac{1}{a_2(3\mu-1,3\mu-5)}+p\frac{1}{a_2(3\mu-1,3\mu-1)}\\ &+p\frac{1}{a_2(3\mu-1,3\mu-1)}+p\frac{1}{a_2(3\mu-1,3\mu-2)}=\\ &=p\frac{1}{\prod_{i=3\mu-5}^{3\mu-2}[1-pa_1^{-1}(i,\tau(i))]}+p\cdot 1+p\cdot 1\\ &+p\frac{1}{\prod_{i=3\mu-2}^{3\mu-2}[1-pa_1^{-1}(i,\tau(i))]}\end{aligned} \tag{137}$$

$$\sum_{j=3\mu-1}^{3\mu+2} p(j)\, a_2^{-1}(3\mu-1,\tau(j)) = 2p+\frac{p(1-p)^9+p(1-p)^2[(1-p)^2-p][(1-p)^4-p](1-2p)}{[(1-p)^2-p]^2\,[(1-p)^4-p](1-2p)}. \tag{138}$$

The computation implies that if $p \in [0.145,\ 0.1564]$, then,

$$\limsup_{n\to\infty} \sum_{j=\sigma(n)}^{n} p(j)\, a_2^{-1}(\sigma(n),\tau(j)) < 1. \tag{139}$$

For example, for $p = 29/200 = 0.145$, we obtain

$$\alpha = \liminf_{n\to\infty} \sum_{j=\sigma(n)}^{n-1} p(j) = \liminf_{\mu\to\infty} \sum_{j=3\mu-1}^{3\mu-1} p(j) = \tfrac{29}{200} = 0.145 < \tfrac{1}{e},$$

$$\limsup_{n\to\infty} \sum_{j=\sigma(n)}^{n} p(j)\, a_2^{-1}(\sigma(n), \tau(j)) \simeq 0.8437652 < 1, \tag{140}$$

$$0.8437652 < 1 - \frac{1-\alpha-\sqrt{1-2\alpha-\alpha^2}}{2} \simeq 0.9875.$$

That is, conditions (124) and (125) are not satisfied for $r = 2$. Similarly, if $p \in [0.145,\ 0.1564]$, then,

$$\limsup_{n\to\infty} \sum_{i=\sigma(n)}^{n} p(i) \prod_{j=\tau(i)}^{\sigma(n)-1} \frac{1}{1-p_1(j)} < 1. \tag{141}$$

For example, for $p = 29/200 = 0.145$ and by using an algorithm in MATLAB program, we obtain

$$\Phi_1(n) = \sum_{i=\sigma(n)}^{n} p(i) \prod_{j=\sigma(i)}^{\sigma(n)-1} \frac{1}{1-p_1(j)} =$$

$$\sum_{i=\sigma(n)}^{n} p(i) \prod_{j=\tau(i)}^{\sigma(n)-1} \frac{1}{1-p_1(j)\left[1+\sum_{k=\tau(j)}^{j-1} p(k)\prod_{m=\tau(k)}^{\sigma(j)-1}(1/(1-p(m)))\right]}, \tag{142}$$

and therefore,

$$\Phi_1(3\mu+2) =$$

$$\sum_{i=3\mu-1}^{3\mu+2} \frac{29}{200} \prod_{j=\tau(i)}^{3\mu-2} \frac{1}{1-(29/200)[1+\sum_{k=\tau(j)}^{j-1}(29/200)\prod_{m=\tau(k)}^{\sigma(j)-1}(1/(1-29/200))]}$$

$$\simeq 0.8529. \tag{143}$$

Thus,

$$\limsup_{n\to\infty} \Phi_1(n) \simeq 0.8529 < 1,$$

$$0.8529 < 1 - \frac{1-\alpha-\sqrt{1-2\alpha-\alpha^2}}{2} \simeq 0.9875. \tag{144}$$

That is conditions (127) and (128) are not satisfied for $l = 1$. On the contrary, for every $p \in [0.145,\ 0.1564]$, we have

$$\limsup_{n\to\infty} \sum_{i=\sigma(n)}^{n} p(i) \prod_{j=\tau(i)}^{\sigma(n)-1} \frac{1}{1-P_1(j)} > 1. \tag{145}$$

For example, for $p = 29/200 = 0.145$ and by using an algorithms on MATLAB software, we obtain the following:

$$F_1(n) = \sum_{i=\sigma(n)}^{n} p(i) \prod_{j=\tau(i)}^{\sigma(n)-1} \frac{1}{1-P_1(j)} =$$
$$= \sum_{i=\sigma(n)}^{n} p(i) \prod_{j=\tau(i)}^{\sigma(n)-1} \frac{1}{1-p(j)\left[1+\sum_{k=\tau(j)}^{j-1} p(k)\exp\left(\sum_{\omega=\tau(k)}^{j-1} p(\omega)\prod_{m=\tau(\omega)}^{\omega-1}(1/(1-p(m)))\right)\right]} \tag{146}$$

and therefore,

$$F_1(3\mu+2)$$
$$= \sum_{i=3\mu-1}^{3\mu+2} \frac{29}{200} \prod_{j=\tau(i)}^{3\mu-2} \frac{1}{1-(29/200)\left[1+\sum_{k=\tau(j)}^{j-1}(29/200)\exp\left(\sum_{\omega=\tau(k)}^{j-1}(29/200)\prod_{m=\tau(\omega)}^{\omega-1}(1/(1-29/200))\right)\right]} =$$
$$= 1.0248. \tag{147}$$

Thus,

$$\limsup_{n\to\infty} F_1(n) = \limsup_{n\to\infty} \sum_{i=h(n)}^{n} p(i) \prod_{j=\tau(i)}^{\sigma(n)-1} \frac{1}{1-P_1(j)} \simeq 1.0248 > 1. \tag{148}$$

That is condition (130) of Theorem 3.22 is satisfied for $l = 1$. Therefore, all solutions of (134) are oscillatory.

Observe that, for $p = 29/200 = 0.145$, we obtain

$$\limsup_{n\to\infty} \sum_{j=\sigma(n)}^{n} p(j) = \limsup_{\mu\to\infty} \sum_{j=2\mu-1}^{3\mu+2} p(j) = 4\,\frac{29}{200} = 0.58 < 1,$$
$$\alpha = 0.145 < \frac{1}{e},$$
$$0.58 < 1 - \frac{1-\alpha-\sqrt{1-2\alpha-\alpha^2}}{2} \simeq 0.9875,$$

$$\limsup_{n\to\infty}\sum_{j=\sigma(n)}^{n} p(j)\prod_{i=\tau(j)}^{\sigma(n)-1}\frac{1}{1-p(i)} = \limsup_{\mu\to\infty}\sum_{j=3\mu-1}^{3\mu+2}\frac{29}{200}\prod_{i=\tau(j)}^{3\mu-2}\frac{1}{1-29/200} =$$

$$= \frac{29}{200}\limsup_{\mu\to\infty}\left\{\prod_{i=\tau(3\mu-1)}^{3\mu-2}\frac{1}{1-29/200}+\prod_{i=\tau(3\mu)}^{3\mu-2}\frac{1}{1-29/200}+\prod_{i=\tau(3\mu+1)}^{3\mu-2}\frac{1}{1-29/200}\right.$$
$$\left.+\prod_{i=\tau(3\mu+2)}^{3\mu-2}\frac{1}{1-29/200}\right\} =$$

$$= \frac{29}{200}\limsup_{\mu\to\infty}\left\{\prod_{i=3\mu-5}^{3\mu-2}\frac{1}{1-29/200}+\prod_{i=3\mu-1}^{3\mu-2}\frac{1}{1-29/200}+\prod_{i=3\mu-1}^{3\mu-2}\frac{1}{1-29/200}\right.$$
$$\left.+\prod_{i=3\mu-2}^{3\mu-2}\frac{1}{1-29/200}\right\} =$$

$$= \frac{29}{200}\limsup_{\mu\to\infty}\left\{\left(\frac{1}{1-29/200}\right)^4+1+1+\frac{1}{1-29/200}\right\} = 0.7309 < 1,$$

$$0.7309 < 1-\frac{1-\alpha-\sqrt{1-2\alpha-\alpha^2}}{2} \simeq 0.9875. \tag{149}$$

That is, conditions (102), (104), (107), (114)≡(124) (for $r=1$), and (117)≡(125) (for $r=1$) are not satisfied.

Also, it is clear that condition (20) is not applicable for (134). In addition, by using an algorithm in MATLAB software, we obtain

$$\limsup_{n\to\infty}\sum_{i=\sigma(n)}^{n} p(i)\prod_{j=\tau(i)}^{n}\frac{1}{1-P_1(j)} \simeq 4.6956 < \frac{2}{1-\alpha-\sqrt{1-2\alpha-\alpha^2}} \simeq 80.1448, \tag{.150}$$

which means that condition (133) is not satisfied.

The improvement of condition (130) over other conditions will be clear by comparing the values on the left side of these conditions. For example, comparison of conditions (102), (114)≡(124) (for $r=1$) (127) (for $l=1$) provides percentage improvement of 76.7%, 40.21%, and 20.15%, respectively.

While conditions (124), (125), (127), and (128) do not lead to oscillation from the first iteration, condition (130) is satisfied from the first iteration, showing that, in this example, (130) is much faster than (124), (125), (127), and (128).

References

1. R.P. Agarwal, *Difference Equations and Inequalities: Theory, Methods, and Applications* (CRC Press, Boca Raton, 2000)
2. R.P. Agarwal, P.J. Wong, *Advanced Topics in Difference Equations*, vol. 404 (Springer Science & Business Media, New York, 2013)
3. R.P. Agarwal, M. Bohner, S.R. Grace, D. O'Regan, *Discrete Oscillation Theory* (Hindawi Publ. Corp., New York, 2005)

4. P.G. Asteris, G.E. Chatzarakis, Oscillation tests for difference equations with non-monotone arguments. Dynam. Contin. Discrete Impuls. Systems Ser. A. Math. Anal. **24**(4), 287–302 (2017)
5. L. Berezansky, E. Braverman, On some constants for oscillation and stability of delay equations. Proc. Am. Math. Soc. **139**(11), 4017–4026 (2011)
6. E. Braverman, B. Karpuz, On oscillation of differential and difference equations with non-monotone delays. Appl. Math. Comput. **218**(7), 3880–3887 (2011)
7. E. Braverman, G.E. Chatzarakis, I.P. Stavroulakis, Iterative oscillation tests for difference equations with several non-monotone arguments. J. Differ. Equ. Appl. **21**(9), 854–874 (2015)
8. J. Chao, Oscillation of linear differential equations with deviating arguments. Theory Pract. Math. (Chin.) **1**(1), 99 (1991)
9. G.E. Chatzarakis, Differential equations with non-monotone arguments: iterative oscillation results. J. Math. Comput. Sci. **6**(5), 953–964 (2016)
10. G.E. Chatzarakis, On oscillation of differential equations with non-monotone deviating arguments. Mediterr. J. Math. **14**(2), 82 (2017)
11. G.E. Chatzarakis, L. Shaikhet, Oscillation criteria for difference equations with non-monotone arguments. Adv. Differ. Equ. **2017**(1), 62 (2017)
12. G.E. Chatzarakis, R. Koplatadze, I.P. Stavroulakis, Oscillation criteria of first order linear difference equations with delay argument. Nonlinear Anal. Theory Methods Appl. **68**(4), 994–1005 (2008)
13. G.E. Chatzarakis, C.G. Philos, I.P. Stavroulakis, On the oscillation of the solutions to linear difference equations with variable delay. Electron. J. Differ. Equ. [electronic only] **2008**(50), 1–15 (2008)
14. G. Chatzarakis, R. Koplatadze, I.P. Stavroulakis, Optimal oscillation criteria for first order difference equations with delay argument. Pac. J. Math. **235**(1), 15–33 (2008)
15. G. Chatzarakis, C.G. Philos, I.P. Stavroulakis, An oscillation criterion for linear difference equations with general delay argument. Port. Math. **66**(4), 513–533 (2009)
16. G.E. Chatzarakis, I.K. Purnaras, I.P. Stavroulakis, Oscillation of retarded difference equations with a non-monotone argument. J. Differ. Equ. Appl. **23**(8), 1354–1377 (2017)
17. G.E. Chatzarakis, I.K. Purnaras, I.P. Stavroulakis, Oscillation tests for differential equations with deviating argumants. Adv. Math. Sci. Appl. **27**(1), 1–28 (2018)
18. M.P. Chen, J.S. Yu, Oscillations of delay difference equations with variable coefficients, in *Proceedings of the First International Conference on Difference Equations* (Gordon and Breach, London, 1995), pp. 105–114
19. S.S. Cheng, G. Zhang, "virus" in several discrete oscillation theorems. Appl. Math. Lett. **13**(3), 9–13 (2000)
20. J. Diblík, Behaviour of solutions of linear differential equations with delay. Arch. Mathe. (BRNO) **34**, 31–47 (1998)
21. J. Diblík, Positive and oscillating solutions of differential equations with delay in critical case. J. Comput. Appl. Math. **88**(1), 185–202 (1998)
22. J. Diblík, N. Koksch, Positive solutions of the equation $\dot{x}(t) = -c(t)x(t - \tau)$ in the critical case. J. Math. Anal. Appl. **250**(2), 635–659 (2000)
23. Y. Domshlak, Discrete version of Sturmian comparison theorem for non-symmetric equations. Doklady Azerb. Acad. Sci **37**, 12–15 (1981)
24. Y. Domshlak, *Comparison Method in Investigation of the Behaviour of Solutions of Differential-Operator Equations* (ELM, Baku, 1986)
25. Y. Domshlak, On oscillation properties of delay differential equations with oscillating coefficients. Funct. Differ. Equ. Isr. Semin. **2**, 59–68 (1994)
26. Y. Domshlak, Sturmian comparison method in oscillation study for discrete difference equations. i, ii. Differ. Integr. Equ. **7**(2), 571–582 (1994)
27. Y. Domshlak, Delay-difference equations with periodic coefficients: sharp results in oscillation theory. Math. Inequal. Appl **1**(1), 998 (1998)
28. Y. Domshlak, Riccati difference equations with almost periodic coefficients in the critical state. Dynam. Syst. Appl. **8**, 389–400 (1999)

29. Y. Domshlak, The riccati difference equations near "extremal" critical states. J. Differ. Equ. Appl. **6**(4), 387–416 (2000)
30. Y. Domshlak, What should be a discrete version of the Chanturia-Koplatadze Lemma? Funct. Differ. Equ. **6**(3–4), p–299 (2004)
31. Y. Domshak, I.P. Stavroulakis, Oscillations of first-order delay differential equations in a critical state. Appl. Anal. **61**(3–4), 359–371 (1996)
32. Á. Elbert, I.P. Stavroulakis, Oscillation and nonoscillation criteria for delay differential equations. Proc. Am. Math. Soc. **123**(5), 1503–1510 (1995)
33. L.H. Erbe, B.G. Zhang, Oscillation for first order linear differential equations with deviating arguments. Differ. Integr. Equ. **1**(3), 305–314 (1988)
34. L.H. Erbe, B.G. Zhang, Oscillation of discrete analogues of delay equations. Differ. Integr. Equ. **2**(3), 300–309 (1989)
35. L.H. Erbe, Q. Kong, B.G. Zhang, *Oscillation Theory for Functional Differential Equation* (Dekker, New York, 1995)
36. K. Gopalsamy, *Stability and Oscillations in Delay Differential Equations of Population Dynamics*, vol. 74 (Springer Science & Business Media, New York, 2013)
37. I. Győri, G.E. Ladas, *Oscillation Theory of Delay Differential Equations: With Applications* (Oxford University Press, Oxford, 1991)
38. J. Jaroš, I.P. Stavroulakis, Oscillation tests for delay equations. Rocky Mt. J. Math. **29**, 197–207 (1999)
39. B. Karpuz, Sharp oscillation and nonoscillation tests for linear difference equations. J. Differ. Equ. Appl. **23**(12), 1929–1942 (2017)
40. M. Kon, Y.G. Sficas, I.P. Stavroulakis, Oscillation criteria for delay equations. Proc. Am. Math. Soc. **128**(10), 2989–2998 (2000)
41. R.G. Koplatadze, T.A. Chanturija, On the oscillatory and monotonic solutions of first order differential equations with deviating arguments. Differ. Uravn. **18**, 1463–1465 (1982)
42. R. Koplatadze, G. Kvinikadze, On the oscillation of solutions of first order delay differential inequalities and equations. J. Georgian Math. J. **1**(6), 675–685 (1994)
43. M.K. Kwong, Oscillation of first-order delay equations. J. Math. Anal. Appl. **156**(1), 274–286 (1991)
44. G. Ladas, Recent developments in the oscillation of delay difference equations, in *Proceedings of Differential Equations: Stability and Control* (1990), pp. 321–332
45. G. Ladas, V. Lakshmikantham, Sharp conditions for oscillations caused by delays. Appl. Anal. **9**(2), 93–98 (1979)
46. G. Ladas, I.P. Stavroulakis, On delay differential inequalities of first order. Funkcialaj Ekvac **25**, 105–113 (1982)
47. G. Ladas, V. Lakshmikantham, J.S. Papadakis, Oscillations of higher-order retarded differential equations generated by the retarded argument, in *Delay and Functional Differential Equations and Their Applications* (Academic Press, New York, 1972), pp. 219–231
48. G. Ladas, C.G. Philos, Y.G. Sficas, Sharp conditions for the oscillation of delay difference equations. J. Appl. Math. Stoch. Anal. **2**(2), 101–111 (1989)
49. G.S. Ladde,V. Lakshmikantham, B.G. Zhang, *Oscillation Theory of Differential Equations with Deviating Arguments*, vol. 110 (Marcel Dekker Incorporated, New York, 1987)
50. V. Lakshmikantham, V. Trigiante, *Theory of Difference Equations Numerical Methods and Applications*, vol. 251 (CRC Press, Boca Raton, 2002)
51. B.S. Lalli, B. Zhang, Oscillation of difference equations. Colloq. Math. **65**, 25–32 (1993)
52. N. Minorsky, Nonlinear oscillations, Publisher: van Nostrand (1962)
53. G.M. Moremedi, I.P. Stavroulakis, A survey on the oscillation of delay equations with a monotone or non-monotone argument, in *International Conference on Differential & Difference Equations and Applications* (Springer, New York, 2017), pp. 441–461
54. G.M. Moremedi, I.P. Stavroulakis, Oscillation conditions for difference equations with a monotone or nonmonotone argument. Discret. Dyn. Nat. Soc. (2018). https://doi.org/10.1155/2018/9416319

55. G.M. Moremedi, I.P. Stavroulakis, A survey on the oscillation of differential equations with several non-monotone arguments. Appl. Math. Inf. Sci. **12**(5), 1–7 (2018)
56. A.D. Myshkis, Linear homogeneous differential equations of first order with deviating arguments. Uspekhi Mat. Nauk **5**(36), 160–162 (1950)
57. C.G. Philos, Y.G., Sficas, An oscillation criterion for first order linear delay differential equations. Can. Math. Bull. **41**(2), 207–213 (1998)
58. M. Pituk, Oscillation of a linear delay differential equation with slowly varying coefficient. Appl. Math. Lett. **73**, 29–36 (2017)
59. Y.G. Sficas, I.P. Stavroulakis, Oscillation criteria for first-order delay equations. Bull. Lond. Math. Soc. **35**(2), 239–246 (2003)
60. J. Shen, I.P. Stavroulakis, Oscillation criteria for delay difference equations. Electron. J. Differ. Equ. **2001**(10), 1–15 (2001)
61. I.P. Stavroulakis, Oscillations of first order differential equations with deviating arguments. Recent Trends Differ. Equ. **1**, 163 (1992)
62. I.P. Stavroulakis, Oscillations of delay difference equations. Comput. Math. Appl. **29**(7), 83–88 (1995)
63. I.P. Stavroulakis, Oscillation criteria for first order delay difference equations. Mediterr. J. Math. **1**(2), 231–240 (2004)
64. I.P. Stavroulakis, Oscillation criteria for delay and difference equations with non-monotone arguments. Appl. Math. Comput. **226**, 661–672 (2014)
65. C. Sturm, Sur les équations différentielles linéaires du second ordre. J. Math. Pures Appl. **1**(1), 1836 (1836)
66. C.A. Swanson, *Comparison and Oscillation Theory of Linear Differential Equations by CA Swanson*, vol. 48 (Elsevier, Amsterdam, 2000)
67. X.H. Tang, Oscillations of delay difference equations with variable coefficients. J. Cent. South Univ. Technol. **29**, 287–288 (1998)
68. X.H. Tang, J.S. Yu, A further result on the oscillation of delay difference equations. Comput. Math. Appl. **38**(11–12), 229–237 (1999)
69. X.H. Tang, J.S. Yu, Oscillations of delay difference equations. Hokkaido Math. J. **29**(1), 213–228 (2000)
70. V. Volterra, Sur la théorie mathématique des phénomenes héréditaires. J. Math. Appl. **7**, 249–298 (1928)
71. J.S. Yu, Z.C. Wang, B.G. Zhang, X.Z. Qian, Oscillations of differential equations with deviating arguments. Panamer. J. Math. **2**(2), 59–78 (1992)
72. J.S. Yu, B.G. Zhang, X.Z. Qian, Oscillations of delay difference equations with oscillating coefficients. J. Math. Anal. Appl. **177**(2), 432–444 (1993)
73. J.S. Yu, B.G. Zhang, Z.C. Wang, Oscillation of delay difference equations. Appl. Anal. **53**(1–2), 117–124 (1994)
74. Y. Zhou, Y.H. Yu, On the oscillation of solutions of first order differential equations with deviating arguments. Acta Math. Appl. Sinica **15**(3), 297–302 (1999)

The Isometry Group of n-Dimensional Einstein Gyrogroup

Teerapong Suksumran

Abstract The space of n-dimensional relativistic velocities normalized to $c = 1$,

$$\mathbb{B} = \{\mathbf{v} \in \mathbb{R}^n \colon \|\mathbf{v}\| < 1\},$$

is naturally associated with Einstein velocity addition $\oplus_E$, which induces the rapidity metric d_E on $\mathbb{B}$ given by $d_E(\mathbf{u}, \mathbf{v}) = \tanh^{-1} \| - \mathbf{u} \oplus_E \mathbf{v}\|$. This metric is also known as the Cayley–Klein metric. We give a complete description of the isometry group of $(\mathbb{B}, d_E)$, along with its composition law.

1 Introduction

The space of n-dimensional relativistic velocities normalized to $c = 1$,

$$\mathbb{B} = \{\mathbf{v} \in \mathbb{R}^n \colon \|\mathbf{v}\| < 1\},$$

has various underlying mathematical structures, including a bounded symmetric space structure [3, 4] and a gyrovector space structure [12]. Furthermore, it is a primary object in special relativity in the case when $n = 3$ [2, 6]. Of particular importance is the composition law of Lorentz boosts:

$$L(\mathbf{u}) \circ L(\mathbf{v}) = L(\mathbf{u} \oplus_E \mathbf{v}) \circ \mathrm{Gyr}[\mathbf{u}, \mathbf{v}],$$

where $L(\mathbf{u})$ and $L(\mathbf{v})$ are Lorentz boosts parametrized by $\mathbf{u}$ and $\mathbf{v}$, respectively, $\oplus_E$ is Einstein velocity addition (defined below), and $\mathrm{Gyr}[\mathbf{u}, \mathbf{v}]$ is a rotation of spacetime coordinates induced by an Einstein addition preserving map (namely an Einstein gyroautomorphism) [12, p. 448]. Moreover, the unit ball $\mathbb{B}$ gives rise to a model

---

T. Suksumran (✉)
Department of Mathematics, Faculty of Science, Chiang Mai University, Chiang Mai, Thailand
e-mail: teerapong.suksumran@cmu.ac.th

N. J. Daras, T. M. Rassias (eds.), *Computational Mathematics and Variational Analysis*, Springer Optimization and Its Applications 159,
https://doi.org/10.1007/978-3-030-44625-3_26

for n-dimensional hyperbolic geometry when it is endowed with the Cayley–Klein metric as well as the Poincaré metric [5, 7].

The open unit ball of $\mathbb{R}^n$ admits a group-like structure when it is endowed with *Einstein addition* $\oplus_E$, defined by

$$\mathbf{u} \oplus_E \mathbf{v} = \frac{1}{1 + \langle \mathbf{u}, \mathbf{v} \rangle} \left(\mathbf{u} + \frac{1}{\gamma_{\mathbf{u}}} \mathbf{v} + \frac{\gamma_{\mathbf{u}}}{1 + \gamma_{\mathbf{u}}} \langle \mathbf{u}, \mathbf{v} \rangle \mathbf{u} \right), \tag{1}$$

where $\langle \cdot, \cdot \rangle$ denotes the usual Euclidean inner product and $\gamma_{\mathbf{u}}$ is the *Lorentz factor* normalized to $c = 1$ given by $\gamma_{\mathbf{u}} = \dfrac{1}{\sqrt{1 - \|\mathbf{u}\|^2}}$. In fact, the space $(\mathbb{B}, \oplus_E)$ satisfies the following properties [10, 12]:

I. (IDENTITY) The zero vector $\mathbf{0}$ satisfies $\mathbf{0} \oplus_E \mathbf{v} = \mathbf{v} = \mathbf{v} \oplus_E \mathbf{0}$ for all $\mathbf{v} \in \mathbb{B}$.

II. (INVERSE) For each $\mathbf{v} \in \mathbb{B}$, the negative vector $-\mathbf{v}$ belongs to $\mathbb{B}$ and satisfies

$$(-\mathbf{v}) \oplus_E \mathbf{v} = \mathbf{0} = \mathbf{v} \oplus_E (-\mathbf{v}).$$

III. (THE GYROASSOCIATIVE LAW) For all $\mathbf{u}, \mathbf{v} \in \mathbb{B}$, there are Einstein addition preserving bijective self-maps $\mathrm{gyr}[\mathbf{u}, \mathbf{v}]$ and $\mathrm{gyr}[\mathbf{v}, \mathbf{u}]$ of $\mathbb{B}$ such that

$$\mathbf{u} \oplus_E (\mathbf{v} \oplus_E \mathbf{w}) = (\mathbf{u} \oplus_E \mathbf{v}) \oplus_E \mathrm{gyr}[\mathbf{u}, \mathbf{v}]\mathbf{w}$$

and

$$(\mathbf{u} \oplus_E \mathbf{v}) \oplus_E \mathbf{w} = \mathbf{u} \oplus_E (\mathbf{v} \oplus_E \mathrm{gyr}[\mathbf{v}, \mathbf{u}]\mathbf{w})$$

for all $\mathbf{w} \in \mathbb{B}$.

IV. (THE LOOP PROPERTY) For all $\mathbf{u}, \mathbf{v} \in \mathbb{B}$,

$$\mathrm{gyr}[\mathbf{u} \oplus_E \mathbf{v}, \mathbf{v}] = \mathrm{gyr}[\mathbf{u}, \mathbf{v}] \quad \text{and} \quad \mathrm{gyr}[\mathbf{u}, \mathbf{v} \oplus_E \mathbf{u}] = \mathrm{gyr}[\mathbf{u}, \mathbf{v}].$$

V. (THE GYROCOMMUTATIVE LAW) For all $\mathbf{u}, \mathbf{v} \in \mathbb{B}$,

$$\mathbf{u} \oplus_E \mathbf{v} = \mathrm{gyr}[\mathbf{u}, \mathbf{v}](\mathbf{v} \oplus_E \mathbf{u}).$$

From properties I through V, it follows that $(\mathbb{B}, \oplus_E)$ forms a *gyrocommutative gyrogroup* (also called a *K-loop* or *Bruck loop*), which shares several properties with groups [9, 12]. However, Einstein addition is a nonassociative operation so that $(\mathbb{B}, \oplus_E)$ fails to form a group. Property III resembles the associative law in groups and property V resembles the commutative law in abelian groups. The map $\mathrm{gyr}[\mathbf{u}, \mathbf{v}]$ in property III is called an *Einstein gyroautomorphism*, which turns out to be a rotation of the unit ball. Henceforward, $(\mathbb{B}, \oplus_E)$ is referred to as the (n-dimensional) *Einstein gyrogroup*.

Recall that the *rapidity* of a vector $\mathbf{v}$ in $\mathbb{B}$ (cf. [5, p. 1229]) is defined by

$$\phi(\mathbf{v}) = \tanh^{-1} \|\mathbf{v}\|. \tag{2}$$

Theorem 1.1 *The rapidity ϕ satisfies the following properties:*

1. $\phi(\mathbf{v}) \geq 0$ *and* $\phi(\mathbf{v}) = 0$ *if and only if* $\mathbf{v} = \mathbf{0}$;
2. $\phi(-\mathbf{v}) = \phi(\mathbf{v})$;
3. $\phi(\mathbf{u} \oplus_E \mathbf{v}) \leq \phi(\mathbf{u}) + \phi(\mathbf{v})$;
4. $\phi(\mathrm{gyr}[\mathbf{u}, \mathbf{v}]\mathbf{w}) = \phi(\mathbf{w})$

for all $\mathbf{u}, \mathbf{v}, \mathbf{w} \in \mathbb{B}$.

Proof Items 1 and 2 are clear. To prove item 3, let $\mathbf{u}, \mathbf{v} \in \mathbb{B}$. By Proposition 3.3 of [5], $\|\mathbf{u} \oplus_E \mathbf{v}\| \leq \dfrac{\|\mathbf{u}\| + \|\mathbf{v}\|}{1 + \|\mathbf{u}\|\|\mathbf{v}\|}$. Set $u = \tanh^{-1} \|\mathbf{u}\|$ and $v = \tanh^{-1} \|\mathbf{v}\|$. Then

$$\|\mathbf{u} \oplus_E \mathbf{v}\| \leq \frac{\tanh u + \tanh v}{1 + (\tanh u)(\tanh v)} = \tanh(u + v),$$

which implies $\phi(\mathbf{u} \oplus_E \mathbf{v}) = \tanh^{-1} \|\mathbf{u} \oplus_E \mathbf{v}\| \leq u + v = \phi(\mathbf{u}) + \phi(\mathbf{v})$. Item 4 follows from the fact that any gyroautomorphism of the Einstein gyrogroup is indeed the restriction of an orthogonal transformation of $\mathbb{R}^n$ to $\mathbb{B}$ so that it preserves the Euclidean norm (and hence also the rapidity); see, for instance, Theorem 3 of [10] and Proposition 2.4 of [5]. □

Theorem 1.1 implies that d_E defined by

$$d_E(\mathbf{u}, \mathbf{v}) = \phi(-\mathbf{u} \oplus_E \mathbf{v}) = \tanh^{-1} \|-\mathbf{u} \oplus_E \mathbf{v}\| \tag{3}$$

for all $\mathbf{u}, \mathbf{v} \in \mathbb{B}$ is indeed a metric (or a distance function) on $\mathbb{B}$, called the *rapidity metric* of the Einstein gyrogroup. In Theorem 3.9 of [5], Kim and Lawson prove that d_E agrees with the *Cayley–Klein metric*, defined from cross-ratios, in the Beltrami–Klein model of n-dimensional hyperbolic geometry. Equation (3) includes what Ungar refers to as the (Einstein) gyrometric, which is defined by

$$\varrho_E(\mathbf{u}, \mathbf{v}) = \|-\mathbf{u} \oplus_E \mathbf{v}\| \tag{4}$$

for all $\mathbf{u}, \mathbf{v} \in \mathbb{B}$. Using Proposition 3.3 of [5], we obtain that

$$\|\mathbf{u} \oplus_E \mathbf{v}\| \leq \frac{\|\mathbf{u}\| + \|\mathbf{v}\|}{1 + \|\mathbf{u}\|\|\mathbf{v}\|} \leq \|\mathbf{u}\| + \|\mathbf{v}\|$$

for all $\mathbf{u}, \mathbf{v} \in \mathbb{B}$ and so the gyrometric ϱ_E is indeed a metric on $\mathbb{B}$. In fact, this is a consequence of Theorem 1.1 of [8]. Since $\tanh^{-1}$ is an injective function, it follows that a self-map of $\mathbb{B}$ preserves d_E if and only if it preserves ϱ_E. Hence, $(\mathbb{B}, d_E)$ and $(\mathbb{B}, \varrho_E)$ have the same isometry group.

The next theorem lists some useful algebraic properties of the Einstein gyrogroup, which will be essential in studying the geometric structure of the unit ball in Section 2.

Theorem 1.2 (See [9, 12]) *The following properties hold in* $(\mathbb{B}, \oplus_E)$*:*

1. $-\mathbf{u} \oplus_E (\mathbf{u} \oplus_E \mathbf{v}) = \mathbf{v}$; (LEFT CANCELLATION LAW)
2. $-(\mathbf{u} \oplus_E \mathbf{v}) = \mathrm{gyr}[\mathbf{u}, \mathbf{v}](-\mathbf{v} \oplus_E -\mathbf{u})$;
3. $(-\mathbf{u} \oplus_E \mathbf{v}) \oplus_E \mathrm{gyr}[-\mathbf{u}, \mathbf{v}](-\mathbf{v} \oplus_E \mathbf{w}) = -\mathbf{u} \oplus_E \mathbf{w}$;
4. $\mathrm{gyr}[-\mathbf{u}, -\mathbf{v}] = \mathrm{gyr}[\mathbf{u}, \mathbf{v}]$; (EVEN PROPERTY)
5. $\mathrm{gyr}[\mathbf{v}, \mathbf{u}] = \mathrm{gyr}^{-1}[\mathbf{u}, \mathbf{v}]$, *where* $\mathrm{gyr}^{-1}[\mathbf{u}, \mathbf{v}]$ *denotes the inverse of* $\mathrm{gyr}[\mathbf{u}, \mathbf{v}]$ *with respect to composition of functions;* (INVERSIVE SYMMETRY)
6. $L_{\mathbf{u}}\colon \mathbf{v} \mapsto \mathbf{u} \oplus_E \mathbf{v}$ *defines a bijective self-map of* $\mathbb{B}$ *and* $L_{\mathbf{u}}^{-1} = L_{-\mathbf{u}}$.

2 Main Results

Let $\mathrm{O}(\mathbb{R}^n)$ be the orthogonal group of n-dimensional Euclidean space $\mathbb{R}^n$; that is, $\mathrm{O}(\mathbb{R}^n)$ consists precisely of (bijective) Euclidean inner product preserving transformations of $\mathbb{R}^n$ (also called orthogonal transformations of $\mathbb{R}^n$). Since the unit ball $\mathbb{B}$ is invariant under orthogonal transformations of $\mathbb{R}^n$, it follows that the set

$$\mathrm{O}(\mathbb{B}) = \{\tau|_{\mathbb{B}} \colon \tau \in \mathrm{O}(\mathbb{R}^n)\}, \tag{5}$$

where $\tau|_{\mathbb{B}}$ denotes the restriction of τ to $\mathbb{B}$, forms a group under composition of functions. Note that Einstein addition is defined entirely in terms of vector addition, scalar multiplication, and the Euclidean inner product. Hence, every orthogonal transformation of $\mathbb{R}^n$ restricts to an automorphism of $\mathbb{B}$ that leaves the Euclidean norm invariant. In particular, the map $\iota\colon \mathbf{v} \mapsto -\mathbf{v}$ defines an automorphism of $\mathbb{B}$. Note that $\mathrm{O}(\mathbb{B})$ is a subgroup of the (algebraic) automorphism group of $(\mathbb{B}, d_E)$, denoted by $\mathrm{Aut}\,(\mathbb{B}, d_E)$.

Let $\mathbf{u}, \mathbf{v} \in \mathbb{B}$. It is not difficult to check that $\mathrm{gyr}[\mathbf{u}, \mathbf{v}]$ satisfies the following properties:

1. $\mathrm{gyr}[\mathbf{u}, \mathbf{v}]\mathbf{0} = \mathbf{0}$;
2. $\mathrm{gyr}[\mathbf{u}, \mathbf{v}]$ is an automorphism of $(\mathbb{B}, \oplus_E)$;
3. $\mathrm{gyr}[\mathbf{u}, \mathbf{v}]$ preserves the gyrometric ϱ_E.

Hence, by Theorem 3.1 of [1], there is an orthogonal transformation ϕ of $\mathbb{R}^n$ for which $\phi|_{\mathbb{B}} = \mathrm{gyr}[\mathbf{u}, \mathbf{v}]$. This proves the following inclusion:

$$\{\mathrm{gyr}[\mathbf{u}, \mathbf{v}] \colon \mathbf{u}, \mathbf{v} \in \mathbb{B}\} \subseteq \mathrm{O}(\mathbb{B}).$$

Theorem 2.3 *For all* $\mathbf{u} \in \mathbb{B}$*, the left gyrotranslation* $L_{\mathbf{u}}$ *defined by* $L_{\mathbf{u}}(\mathbf{v}) = \mathbf{u} \oplus_E \mathbf{v}$ *is an isometry of* $\mathbb{B}$ *with respect to* d_E.

Proof Note that $L_{\mathbf{u}}$ is a bijective self-map of $\mathbb{B}$ since $L_{-\mathbf{u}}$ acts as its inverse (see, for instance, Theorem 10 (1) of [11]). From Theorem 1.2, we have by inspection that

$$\begin{aligned}\| -(\mathbf{u}\oplus_E \mathbf{x})\oplus_E(\mathbf{u}\oplus_E \mathbf{y})\| &= \|\mathrm{gyr}[\mathbf{u},\mathbf{x}](-\mathbf{x}\oplus_E -\mathbf{u})\oplus_E(\mathbf{u}\oplus_E \mathbf{y})\|\\ &= \|(-\mathbf{x}\oplus_E -\mathbf{u})\oplus_E \mathrm{gyr}[\mathbf{x},\mathbf{u}](\mathbf{u}\oplus_E \mathbf{y})\|\\ &= \|(-\mathbf{x}\oplus_E -\mathbf{u})\oplus_E \mathrm{gyr}[-\mathbf{x},-\mathbf{u}](\mathbf{u}\oplus_E \mathbf{y})\|\\ &= \|-\mathbf{x}\oplus_E \mathbf{y}\|.\end{aligned}$$

It follows that

$$d_E(L_{\mathbf{u}}(\mathbf{x}),L_{\mathbf{u}}(\mathbf{y})) = \tanh^{-1}\|-L_{\mathbf{u}}(\mathbf{x})\oplus_E L_{\mathbf{u}}(\mathbf{y})\| = \tanh^{-1}\|-\mathbf{x}\oplus_E \mathbf{y}\| = d_E(\mathbf{x},\mathbf{y}).$$

This proves that $L_{\mathbf{u}}$ is an isometry of $(\mathbb{B}, d_E)$. □

Corollary 1 *The gyroautomorphisms of the Einstein gyrogroup are isometries of* $\mathbb{B}$ *with respect to* d_E.

Proof Let $\mathbf{u},\mathbf{v}\in\mathbb{B}$. According to Theorem 10 (3) of [11], we have

$$L_{\mathbf{u}}\circ L_{\mathbf{v}} = L_{\mathbf{u}\oplus_E\mathbf{v}}\circ \mathrm{gyr}[\mathbf{u},\mathbf{v}].$$

Hence, $\mathrm{gyr}[\mathbf{u},\mathbf{v}] = L^{-1}_{\mathbf{u}\oplus_E\mathbf{v}}\circ L_{\mathbf{u}}\circ L_{\mathbf{v}} = L_{-(\mathbf{u}\oplus_E\mathbf{v})}\circ L_{\mathbf{u}}\circ L_{\mathbf{v}}$. This implies that $\mathrm{gyr}[\mathbf{u},\mathbf{v}]$ is an isometry of $(\mathbb{B}, d_E)$, being the composite of isometries. □

In fact, Corollary 1 is a special case of the following theorem.

Theorem 2.4 *Every automorphism of* $(\mathbb{B},\oplus_E)$ *that preserves the Euclidean norm is an isometry of* $\mathbb{B}$ *with respect to* d_E. *Therefore, every transformation in* $\mathrm{O}(\mathbb{B})$ *is an isometry of* $\mathbb{B}$.

Proof Let $\tau\in\mathrm{Aut}\,(\mathbb{B},\oplus_E)$ and suppose that τ preserves the Euclidean norm. Then τ is bijective. Direct computation shows that

$$d_E(\tau(\mathbf{x}),\tau(\mathbf{y})) = \tanh^{-1}\|\tau(-\mathbf{x}\oplus_E \mathbf{y})\| = \tanh^{-1}\|-\mathbf{x}\oplus_E \mathbf{y}\| = d_E(\mathbf{x},\mathbf{y})$$

for all $\mathbf{x},\mathbf{y}\in\mathbb{B}$. Hence, τ is an isometry of $(\mathbb{B}, d_E)$. The remaining part of the theorem is immediate since $\mathrm{O}(\mathbb{B})\subseteq\mathrm{Aut}\,(\mathbb{B}, d_E)$. □

Next, we give a complete description of the isometry group of $(\mathbb{B}, d_E)$ using Abe's result [1].

Theorem 2.5 *The isometry group of* $(\mathbb{B}, d_E)$ *is given by*

$$\operatorname{Iso}(\mathbb{B}, d_E) = \{L_{\mathbf{u}} \circ \tau : \mathbf{u} \in \mathbb{B} \textit{ and } \tau \in \mathrm{O}(\mathbb{B})\}. \tag{6}$$

Proof By Theorems 2.3 and 2.4,

$$\{L_{\mathbf{u}} \circ \tau : \mathbf{u} \in \mathbb{B} \text{ and } \tau \in \mathrm{O}(\mathbb{B})\} \subseteq \operatorname{Iso}(\mathbb{B}, d_E).$$

Let $\psi \in \operatorname{Iso}(\mathbb{B}, d_E)$. By definition, ψ is a bijection from $\mathbb{B}$ to itself. By Theorem 11 of [11], $\psi = L_{\psi(\mathbf{0})} \circ \rho$, where ρ is a bijection from $\mathbb{B}$ to itself that leaves $\mathbf{0}$ fixed. As in the proof of Theorem 18 (2) of [9], $L^{-1}_{\psi(\mathbf{0})} = L_{-\psi(\mathbf{0})}$ and so $\rho = L_{-\psi(\mathbf{0})} \circ \psi$. Therefore, ρ is an isometry of $(\mathbb{B}, d_E)$. Since $d_E(\rho(\mathbf{x}), \rho(\mathbf{y})) = d_E(\mathbf{x}, \mathbf{y})$ and $\tanh^{-1}$ is injective, it follows that

$$\| -\rho(\mathbf{x}) \oplus_E \rho(\mathbf{y})\| = \| -\mathbf{x} \oplus_E \mathbf{y}\|$$

for all $\mathbf{x}, \mathbf{y} \in \mathbb{B}$. Hence, ρ preserves the Einstein gyrometric. By Theorem 3.1 of [1], $\rho = \tau|_{\mathbb{B}}$, where τ is an orthogonal transformation of $\mathbb{R}^n$. This proves the reverse inclusion. □

By Theorem 2.5, every isometry of $(\mathbb{B}, d_E)$ has a (unique) expression as the composite of a left gyrotranslation and the restriction of an orthogonal transformation of $\mathbb{R}^n$ to the unit ball. According to the commutation relation (55) of [9] for the case of the Einstein gyrogroup, one has the following composition law of isometries of $(\mathbb{B}, d_E)$:

$$(L_{\mathbf{u}} \circ \alpha) \circ (L_{\mathbf{v}} \circ \beta) = L_{\mathbf{u} \oplus_E \alpha(\mathbf{v})} \circ (\mathrm{gyr}[\mathbf{u}, \alpha(\mathbf{v})] \circ \alpha \circ \beta) \tag{7}$$

for all $\mathbf{u}, \mathbf{v} \in \mathbb{B}$, $\alpha, \beta \in \mathrm{O}(\mathbb{B})$. This reminds us of the composition law of Euclidean isometries. Note that $L_{\mathbf{u}} \circ \alpha = L_{\mathbf{v}} \circ \beta$, where $\mathbf{u}, \mathbf{v} \in \mathbb{B}$ and $\alpha, \beta \in \mathrm{O}(\mathbb{B})$, if and only if $\mathbf{u} = \mathbf{v}$ and $\alpha = \beta$. This combined with (7) implies that the map $L_{\mathbf{v}} \circ \tau \mapsto (\mathbf{v}, \tau)$ defines an isomorphism from the isometry group of $(\mathbb{B}, d_E)$ to the *gyrosemidirect product* $\mathbb{B} \rtimes_{\mathrm{gyr}} \mathrm{O}(\mathbb{B})$, which is a group consisting of the underlying set

$$\{(\mathbf{v}, \tau) : \mathbf{v} \in \mathbb{B} \text{ and } \tau \in \mathrm{O}(\mathbb{B})\}$$

and group multiplication

$$(\mathbf{u}, \alpha)(\mathbf{v}, \beta) = (\mathbf{u} \oplus_E \alpha(\mathbf{v}), \mathrm{gyr}[\mathbf{u}, \alpha(\mathbf{v})] \circ \alpha \circ \beta). \tag{8}$$

For the relevant definition of a gyrosemidirect product, see Section 2.6 of [12]. Equation (8) is an analogous result in Euclidean geometry that the isometry group of n-dimensional Euclidean space $\mathbb{R}^n$ can be realized as the semidirect product $\mathbb{R}^n \rtimes \mathrm{O}(\mathbb{R}^n)$. The result that the group of *holomorphic* automorphisms of a bounded symmetric domain can be realized as a gyrosemidirect product is proved by Friedman and Ungar in Theorem 3.2 of [3]. Furthermore, a characterization of continuous endomorphisms of the *three-dimensional* Einstein gyrogroup is obtained; see Theorem 1 of [6].

As an application of Theorem 2.5, we show that the space $(\mathbb{B}, d_E)$ is homogeneous; that is, there is an isometry of $(\mathbb{B}, d_E)$ that sends $\mathbf{u}$ to $\mathbf{v}$ for all arbitrary points $\mathbf{u}$ and $\mathbf{v}$ in $\mathbb{B}$. We also give an easy way to construct *point-reflection* symmetries of the unit ball.

Theorem 2.6 (Homogeneity) *If* $\mathbf{u}$ *and* $\mathbf{v}$ *are arbitrary points in* $\mathbb{B}$, *then there is an isometry* ψ *of* $(\mathbb{B}, d_E)$ *such that* $\psi(\mathbf{u}) = \mathbf{v}$. *In other words,* $(\mathbb{B}, d_E)$ *is homogeneous.*

Proof Let $\mathbf{u}, \mathbf{v} \in \mathbb{B}$. Define $\psi = L_{\mathbf{v}} \circ L_{-\mathbf{u}}$. Then ψ is an isometry of $(\mathbb{B}, d_E)$, being the composite of isometries. It is clear that $\psi(\mathbf{u}) = \mathbf{v} \oplus_E (-\mathbf{u} \oplus_E \mathbf{u}) = \mathbf{v}$. □

Theorem 2.7 (Symmetry) *For each point* $\mathbf{v} \in \mathbb{B}$, *there is a point-reflection* $\sigma_{\mathbf{v}}$ *of* $(\mathbb{B}, d_E)$ *corresponding to* $\mathbf{v}$*; that is,* $\sigma_{\mathbf{v}}$ *is an isometry of* $(\mathbb{B}, d_E)$ *such that* $\sigma_{\mathbf{v}}^2$ *is the identity transformation of* $\mathbb{B}$ *and* $\mathbf{v}$ *is the unique fixed point of* $\sigma_{\mathbf{v}}$.

Proof Let ι be the negative map of $\mathbb{B}$; that is, $\iota(\mathbf{w}) = -\mathbf{w}$ for all $\mathbf{w} \in \mathbb{B}$. In view of (1), it is clear that ι is an automorphism of $\mathbb{B}$ with respect to $\oplus_E$. By Theorem 2.4, ι is an isometry of $(\mathbb{B}, d_E)$. Define $\sigma_{\mathbf{v}} = L_{\mathbf{v}} \circ \iota \circ L_{-\mathbf{v}}$. Then $\sigma_{\mathbf{v}}$ is an isometry of $(\mathbb{B}, d_E)$ that is a point-reflection of $\mathbb{B}$ corresponding to $\mathbf{v}$. The uniqueness of the fixed point of $\sigma_{\mathbf{v}}$ follows from the fact that $\mathbf{0}$ is the unique fixed point of ι. □

Acknowledgments The author would like to thank Themistocles M. Rassias for his generous collaboration. He also thanks anonymous referees for useful comments.

References

1. T. Abe, Gyrometric preserving maps on Einstein gyrogroups, Möbius gyrogroups and proper velocity gyrogroups. Nonlinear Funct. Anal. Appl. **19**, 1–17 (2014)
2. Y. Friedman, T. Scarr, *Physical Applications of Homogeneous Balls, Progress in Mathematical Physics*, vol. 40 (Birkhäuser, Boston, 2005)
3. Y. Friedman, A. Ungar, Gyrosemidirect product structure of bounded symmetric domains. Results Math. **26**(1–2), 28–38 (1994)
4. S. Kim, J. Lawson, Smooth Bruck loops, symmetric spaces, and non-associative vector spaces. Demonstr. Math. **44**(4), 755–779 (2011)
5. S. Kim, J. Lawson, Unit balls, Lorentz boosts, and hyperbolic geometry. Results Math. **63**, 1225–1242 (2013)
6. L. Molnár, D. Virosztek, On algebraic endomorphisms of the Einstein gyrogroup. J. Math. Phys. **56**(8), 082302 (5 pp.) (2015)
7. Ratcliffe, J.: Foundations of hyperbolic manifolds, in *Graduate Texts in Mathematics*, vol. 149, 2nd edn. (Springer, New York, 2006)
8. T. Suksumran, On Metric Structures of Normed Gyrogroups, in *Mathematical Analysis and Applications, Springer Optimization and Its Applications*, vol. 154, ed. by Th.M. Rassias, P.M. Pardalos (Springer, Cham, 2019), pp. 529–542
9. T. Suksumran, The algebra of gyrogroups: Cayley's Theorem, Lagrange's Theorem, and Isomorphism Theorems, in *Essays in Mathematics and Its Applications: In Honor of Vladimir*

Arnold, ed. by Th.M. Rassias, P.M. Pardalos (Springer, Cham, 2016), pp. 369–437
10. T. Suksumran, K. Wiboonton, Einstein gyrogroup as a B-loop. Rep. Math. Phys. **76**, 63–74 (2015)
11. T. Suksumran, K. Wiboonton, Isomorphism theorems for gyrogroups and L-subgyrogroups. J. Geom. Symmetry Phys. **37**, 67–83 (2015)
12. A. Ungar, *Analytic Hyperbolic Geometry and Albert Einstein's Special Theory of Relativity* (World Scientific, Hackensack, 2008)

Function Variational Principles and Normed Minimizers

Mihai Turinici

Abstract The function variational principle due to El Amrouss [Rev. Col. Mat., 40 (2006), 1–14] may be obtained in a simplified manner. Further applications to existence of minimizers for Gâteaux differentiable bounded from below lsc functions over Hilbert spaces are then provided.

AMS Subject Classification 49J53 (Primary), 54H25 (Secondary)

1 Introduction

Let (M, d) be a complete metric space; and $\varphi : M \to R$ be (M, d)-*regular*; i.e.,

(a01) φ is bounded from below ($\inf \varphi(M) > -\infty$)

(a02) φ is d-lsc on M ($\liminf_n \varphi(x_n) \geq \varphi(x)$, whenever $x_n \xrightarrow{d} x$).

Define for each $\theta \geq 0$

$$\text{ulev}(\varphi; M; \theta) = \{x \in M; \varphi(x) \leq \inf \varphi(M) + \theta\};$$

this will be referred to as the θ-*upper level set* of φ with respect to M; in particular, $\text{ulev}(\varphi; M; 0)$ is nothing else than the *(global) minimizers set* of φ over M. The following 1974 statement in Ekeland [10] (referred to as Ekeland's Variational Principle; in short: EVP) is well known.

Theorem 1.1 *Let $\varepsilon > 0$ be given; as well as some $u \in \text{ulev}(\varphi; M; \varepsilon)$. Then, for each $\delta > 0$ there exists $v = v(\varepsilon, u; \delta) \in M$, with*

(11-a) $(\varepsilon/\delta)d(u, v) \leq \varphi(u) - \varphi(v)$; *hence* $[\varphi(u) \geq \varphi(v), d(u, v) \leq \delta]$

(11-b) $(\varepsilon/\delta)d(v, x) > \varphi(v) - \varphi(x)$, *for all* $x \in M \setminus \{v\}$.

M. Turinici (✉)
A. Myller Mathematical Seminar, A. I. Cuza University, Iaşi, Romania
e-mail: mturi@uaic.ro

N. J. Daras, T. M. Rassias (eds.), *Computational Mathematics and Variational Analysis*, Springer Optimization and Its Applications 159,
https://doi.org/10.1007/978-3-030-44625-3_27

This principle found some basic applications to control and optimization, generalized differential calculus, critical point theory and global analysis; we refer to the 1979 paper by Ekeland [11] for a survey of these. So, it cannot be surprising that, soon after its formulation, many extensions of (EVP) were proposed. For example, the *abstract (order)* one starts from the fact that, with respect to the Brøndsted order [6] (on M)

$(x, y \in M)$: $x \leq y$ iff $(\varepsilon/\delta)d(x, y) \leq \varphi(x) - \varphi(y)$

the point $v \in M$ appearing in the second conclusion above is *maximal*; so that, (EVP) is nothing but a variant of the Zorn-Bourbaki maximal principle [4, 35]; see Hyers et al. [16, Ch 5] for a number of technical aspects. The *dimensional* way of extension refers to the ambient space (R) of $\varphi(M)$ being substituted by a (topological or not) vector space; an account of the results in this area is to be found in the 2003 monograph by Goepfert et al. [14, Ch 3]. Further, the *(pseudo) metrical* one consists in conditions imposed to the ambient metric over M being relaxed. The basic result in this direction was obtained in 1992 by Tataru [27], via Ekeland type techniques; subsequent extensions of it may be found in the 1996 paper by Kada et al. [17]. Finally, we must add to this list the *functional* extension of (EVP) obtained in 1997 by Zhong [34] (and referred to as Zhong's Variational Principle; in short: ZVP). Let again $\varphi : M \to R$ be a (M, d)-regular function. Further, take a *normal* function $t \mapsto b(t)$ from $R_+ := [0, \infty[$ to $R_+^0 :=]0, \infty[$; i.e.,

(a03) $b(.)$ is increasing and continuous
(a04) $B(\infty) = \infty$, where $B(t) = \int_0^t (1/b(\tau))d\tau$, $t \geq 0$.

Theorem 1.2 *Let* $\varepsilon > 0$ *be given, as well as some* $u \in \text{ulev}(\varphi; M; \varepsilon)$. *Further, let* $x_0 \in M$ *and* $\delta, \rho > 0$ *be taken according to*

$\delta \leq B(r + \rho) - B(r)$, *where* $r := d(x_0, u)$.

There exists then $v = v(\varepsilon, u; x_0, \delta, \rho)$ *in* M, *with*

(12-a) $\varphi(u) \geq \varphi(v)$, $d(x_0, v) \leq r + \rho$
(12-b) $(\varepsilon/\delta)d(v, x)/b(d(x_0, v)) > \varphi(v) - \varphi(x)$, *for each* $x \in M \setminus \{v\}$.

Now, evidently, (ZVP) includes (for $b = 1$, $x_0 = u$, and $\delta = \rho$) the local version of (EVP) based upon (11-a) (the second half). The relative form of the same, based upon (11-a) (the first half) also holds (but indirectly); see Bao and Khanh [2] for details. Summing up, (ZVP) includes (EVP), but the provided argument is rather involved. A simplification of the proposed reasoning was carried out in Turinici [30], by a technique developed in Park and Bae [24]; note that, as a consequence of this, (ZVP) is nothing but a logical equivalent of (EVP).

Recently, a local functional version of (EVP) was established by El Amrouss [12]. Let $\varphi : M \to R$ be a (M, d)-regular function. Further, let the function $a : R_+ \to R_+^0$ be *admissible*, in the sense

(a05) $a(.)$ is increasing and continuous
(a06) $a(.)$ is a comparison function, of order $k > 0$
$(\forall q \geq k, \exists \lambda, \mu \geq 0 : a((t + 1)s) \leq a(t)[\lambda s^q + \mu], \forall t, s \geq 0)$.

Denote also, for $u \in M$, $\rho > 0$,

$$M[u, \rho] = \{x \in M; d(u, x) \leq \rho\}, M(u, \rho) = \{x \in M; d(u, x) < \rho\}$$

(referred to as: the *closed/open* sphere with center x and radius ρ). Finally, take some function $\gamma : M \to R^0_+$, with

(a07) $u \mapsto \gamma(u)/(1 + d(x_0, u))$ is bounded on M, for some $x_0 \in M$.

The following result (referred to as El Amrouss Ordering Principle; in short: EAOP) is now available.

Theorem 1.3 *Let $\varepsilon, \delta > 0$ be given, as well as some $u \in \text{ulev}(\varphi; M; \varepsilon)$. There exists then a sequence $(z_n; n \geq 0)$ in $M[u, \gamma(u)]$ and a point $v \in M[u, \gamma(u)]$, with*

(13-a) $z_0 = u$, $\lim_n z_n = v$ *and* $(d(x_0, z_n); n \geq 0)$ *is ascending*
(13-b) $\sum_{n=0}^{j} d(z_n, z_{n+1})/a(d(x_0, z_{n+1})) < 2\delta$, *for all* $j \geq 0$
(13-c) $\varphi(u) \geq \varphi(v)$ *and* $d(u, v) \leq \min\{\gamma(u), \delta a(d(x_0, v))\}$
(13-d) $(\varepsilon/\delta)d(v, w)/a(d(x_0, w)) \geq \varphi(v) - \varphi(w)$,
for all $w \in M[u, \gamma(u)] \setminus M(x_0, d(x_0, v))$.

In particular, the constant function $a = 1$ is admissible. Note that the variational conclusion (13-d) above (retainable for certain points of a closed sphere in M) is weaker than the variational conclusion (12-b) (valid for all elements of M); nevertheless, it allows us to get genuine differential translations of the involved facts. Consequently, it may be not surprising that (EAOP) found some nice applications to Variational Analysis; see the quoted paper for details. So, a technical analysis of its basic lines may be not without profit. It is our aim in this exposition to show that a simplification of author's reasoning is possible, by reducing (EAOP) to the metrical version of (EVP) in Turinici [28]. This, among others, yields a reconsideration of admissibility and boundedness concepts; in fact, we simply show that the conditions (a06) and (a07) may be dropped. Then, a differential version of our main result is given, in the realm of Hilbert spaces. Finally, an application of the obtained differential result to existence of minimizers for Gâteaux differentiable regular functions is considered. Some other aspects of this problem will be discussed elsewhere.

2 Metrical Ordering Principles

Let (M, d) be a metric space; and $(\preceq)$ be an *order* (that is: reflexive, transitive, antisymmetric relation) over it; the resulting triple $(M, d; \preceq)$ will be termed an *ordered metric space*. A point $z \in M$ is called $(\preceq)$-*maximal*, when

$$w \in M, z \preceq w \Longrightarrow z = w;$$

and the order $(\preceq)$ is termed a *Zorn* one, when:

for each $x \in M$ there exists a $(\preceq)$-maximal $z \in M$ with $x \preceq z$.

For a number of both practical and theoretical reasons, it would be useful to establish under which conditions is such a property retainable. The standard way of solving this question is based upon the *chains* (i.e.: *totally ordered* subsets) of the structure $(M, \preceq)$; cf. Bourbaki [4]. However, under the precise metrical setting, a denumerable version of such principles is more appropriate for our purposes. This will necessitate a few conventions and auxiliary facts. Let (x_n) be a sequence in M; we shall term it *ascending* (resp.: *descending*), if

$x_i \preceq x_j$ (resp.: $x_i \succeq x_j$) when $i \leq j$.

(Here, $(\succeq)$ is the *dual* of the order $(\preceq)$.) Further, a point $u \in M$ is called an *upper bound* of our sequence (x_n), provided

$x_n \preceq u$, for all n (written as: $(x_n) \preceq u$);

when such elements u exist, we say that (x_n) is *bounded above* (modulo $(\preceq)$).

The following 1984 answer to the above problem (referred to as: *metrical Zorn-Bourbaki principle*; in short: (ZB-m)) is provided in Turinici [29]:

Theorem 2.4 *Suppose that*

(b01) $(\preceq)$ is inductive:
each ascending sequence is bounded above (modulo $(\preceq)$)
(b02) $(\preceq)$ is regular: each ascending sequence is d-Cauchy.

Then, $(\preceq)$ is a Zorn order.

Technically speaking, (ZB-m) is deductible from the Principle of Dependent Choices (in short: DC) due to Bernays [3] and Tarski [26] (discussed a bit further). Precisely, the following inclusion holds:

Proposition 2.1 *Under these conventions, we have*

(DC) implies (ZB-m), in (ZF-AC);
or, equivalently: (ZB-m) is deductible in (ZF-AC+DC).

Proof (See also Cârjă et al. [9, Ch 2, Sect 2.1])
Let the premises above be in use. We claim that

(21-1) $\forall x \in M, \forall \varepsilon > 0$, there exists $y \succeq x$ such that $y \preceq u \preceq v \Longrightarrow d(u, v) < \varepsilon$.

Indeed, assume this would be false; that is, for some $x \in M, \varepsilon > 0$,

(21-2) for each $y \succeq x$ there exist $u, v \in M$ with $y \preceq u \preceq v, d(u, v) \geq \varepsilon$.

Denote, for simplicity

$G(\varepsilon) = \{(a, b) \in M \times M; x \preceq a \preceq b\}$ (the *graph* of $(\preceq)$ beyond x);

and then, let the relation $\mathscr{S}$ over $G(\varepsilon)$ be introduced as

$(x_1, x_2)\mathscr{S}(y_1, y_2)$ iff $x_2 \preceq y_1$ and $d(y_1, y_2) \geq \varepsilon$.

By the working assumption (21-1), we must have

$G(\varepsilon)((a, b), \mathscr{S})$ is nonempty, for each $(a, b) \in G(\varepsilon)$.

This, along with Dependent Choice principle (DC), assures us that there must be a sequence $(z_n := (x_{2n}, x_{2n+1}); n \geq 0)$ in $G(\varepsilon)$, with

$(\forall n)$: $z_n \mathscr{S} z_{n+1}$; that is, $x_{2n+1} \preceq x_{2n+2}, d(x_{2n+2}, x_{2n+3}) \geq \varepsilon$.

But then, $(x_n; n \geq 0)$ is an ascending sequence in M, that is not endowed with the d-Cauchy property; in contradiction with the regularity of $(\preceq)$. Consequently, our claim follows; and then

(21-3) for each $x \in M, \varepsilon > 0$, there exists $y \in M$ such that $x \preceq y, \varphi(y) \leq \varepsilon$;

where, by definition,

$\varphi(y) = \sup\{d(u, v); y \preceq u \preceq v\}, y \in M$.

For each $\varepsilon > 0$, let $\mathscr{R}(\varepsilon)$ stand for the relation

$x\mathscr{R}(\varepsilon)y$ iff $x \preceq y, \varphi(y) \leq \varepsilon$.

Further, let $(\varepsilon_n; n \geq 0)$ be a strictly descending sequence in R^0_+ with $\varepsilon_n \to 0$. (For example, we may take $(\varepsilon_n = 2^{-n}; n \geq 0)$; so, no choice techniques are needed here.) Put $(\mathscr{R}_n := \mathscr{R}(\varepsilon_n); n \geq 0)$; and note that, by (21-3), we have

$M(x, \mathscr{R}_n)$ is nonempty, for each $x \in M$ and each $n \geq 0$.

Combining with Diagonal Dependent Choice principle (DDC) it follows that, given $x \in X$, there exists a sequence (x_n) in M with $x_0 = x$ and

(21-4) $(\forall n)$: $x_n \mathscr{R}_n x_{n+1}$; hence, $x_n \preceq x_{n+1}$ and $\varphi(x_{n+1}) \leq \varepsilon_{n+1}$
(wherefrom, $x_n \preceq x_{n+1}$ and $(x_{n+1} \preceq u \preceq v$ implies $d(u, v) \leq \varepsilon_{n+1}$)).

Let z be an upper bound of (x_n) (existing by the inductive assumption upon $(\preceq)$). We claim that this is our desired element. Firstly, it is clear that $x \preceq z$. Secondly, take $u, v \in M$ in accordance with $z \preceq u \preceq v$. By (21-4), one gets $d(u, v) = 0$ (hence, $u = v$); wherefrom, z is $(\preceq)$-maximal. The proof is thereby complete.

Concerning our precise relationship, it would be natural to ask whether the reciprocal inclusion (ZB-m) $\Longrightarrow$ (DC) is available. A positive answer to this will be developed at the end of this exposition.

A basic particular case of these developments is stated as follows. Let again $(M, d; \preceq)$ be an ordered metric space. Call the subset $Z \subseteq M$, $(\preceq)$-*closed* if:

the limit of each ascending (modulo $(\preceq)$) sequence in Z belongs to Z.

For example, this holds whenever Z is closed; but the reciprocal is not in general true. Further, let us say that $(\preceq)$ is *self-closed*, when;

$M(x, \preceq) := \{y \in M; x \preceq y\}$ is $(\preceq)$-closed, for each $x \in M$;
or, equivalently:
the limit of each ascending sequence is an upper bound of it (modulo $(\preceq)$).

In particular, this holds whenever (cf. Nachbin [22, Appendix]):

$(\preceq)$ is *closed from the right*: $M(x, \preceq)$ is closed, for each $x \in M$;

but, the converse is not in general valid. Finally, call d, *$(\preceq)$-complete* if

each ascending (modulo $(\preceq)$) d-Cauchy sequence in M is d-convergent.

For example, this happens when d is complete. The reciprocal is not in general true; just take $M =]0, 1]$, endowed with the standard metric and order.

The following maximal type statement (referred to as: *strong metrical Zorn-Bourbaki principle*; in short: (ZB-m-s)) established in 1981 by Turinici [28] will be useful for us.

Theorem 2.5 *Suppose that*

(b03) $(\preceq)$ is regular and self-closed
(b04) d is $(\preceq)$-complete (over M).

Then, for each $u \in M$ there exists $v \in M$ with

(22-a) $u \preceq v$; as well as
(22-b) $w \in M$ and $v \preceq w$ imply $v = w$.

The proof consists in verifying that, under (b03) and (b04), one gets the couple (b01)+(b02); we do not give details.

Finally, let us clarify the relationships between this result and (EVP) above.

Proposition 2.2 *Under the accepted setting, one has*

(DC) $\Longrightarrow$ (ZB-m) $\Longrightarrow$ (ZB-m-s) $\Longrightarrow$ (EVP), in (ZF-AC);
so that, (EVP) is deductible in (ZF-AC+DC).

Proof Let the premises of (EVP) be accepted; so, let (M, d) be a complete metric space and $\varphi : M \to R$ be a (M, d)-regular function. Define a relation $(\preceq)$ over M as

$$x \preceq y \text{ iff } (\varepsilon/\delta)d(x, y) \leq \varphi(x) - \varphi(y).$$

Clearly, $(\preceq)$ acts as an *order* (antisymmetric quasi-order) on M. We claim that conditions of (ZB-m-s) are fulfilled on $(M, d; \preceq)$ and φ. In fact, by the imposed setting, d is $(\preceq)$-complete over M. Further, let (x_n) be a $(\preceq)$-ascending sequence in M:

(b05) $(\varepsilon/\delta)d(x_n, x_m) \leq \varphi(x_n) - \varphi(x_m)$, if $n \leq m$.

The sequence $(\varphi(x_n))$ is descending and bounded from below; hence a Cauchy one. This, along with (b05), tells us that (x_n) is a d-Cauchy sequence in M; and therefore, $(\preceq)$ is regular. By completeness, there must be some $y \in M$ with $x_n \xrightarrow{d} y$. Passing to limit as $m \to \infty$ in (b05) one derives (by the d-lsc property)

$$(\varepsilon/\delta)d(x_n, y) \leq \varphi(x_n) - \varphi(y) \text{ (i.e.: } x_n \preceq y\text{), for all } n.$$

In other words, $y \in M$ is an upper bound (modulo $(\preceq)$) of (x_n), and this shows that $(\preceq)$ is self-closed. From (ZB-m-s) it then follows that, for the starting point $u \in M$ there exists some other point $v \in M$, with

$$u \preceq v \text{ and } [v \preceq x \in M \text{ implies } v = x].$$

The former of these is just our first conclusion. Moreover, the latter one gives our second conclusion. In fact, let $y \in M$ be such that

$$(\varepsilon/\delta)d(v, y) \le \varphi(v) - \varphi(y); \text{ that is (by definition), } v \preceq y.$$

By the maximal property, we then have $v = y$. The proof is complete.

3 Extension of (EAOP)

With this information at hand, we may now return to the posed questions of introductory part.

Let (X, d) be a complete metric space; and $\varphi : X \to R$ be a function with the (X, d)-regular properties; i.e.:

(c01) φ is bounded below ($\inf \varphi(X) > -\infty$)

(c02) φ is d-lsc on X ($\liminf_n \varphi(x_n) \ge \varphi(x)$, whenever $x_n \xrightarrow{d} x$).

Further, take some function $\Gamma : X \to R_+$, with

(c03) Γ is d-Lipschitz: $|\Gamma(x) - \Gamma(y)| \le Ld(x, y), \forall x, y \in X$, for some $L > 0$.

For example, a natural choice of such functions is

$$\Gamma(x) = \text{dist}(x, G), x \in X, \text{ for some } G \in \exp(X);$$

where dist is the usual point to set distance. Finally, let the function $a : R_+ \to R_+$ be taken as

(c04) $a(.)$ is increasing ($t_1 \le t_2$ implies $a(t_1) \le a(t_2)$).

(A) Let $(\le)$ stand for the relation (in X)

$$(z, w \in X)\text{: } z \le w \text{ iff } \Gamma(z) \le \Gamma(w).$$

Clearly, $(\le)$ is reflexive and transitive; hence, a *quasi-order* on X. Further, let us fix $\varepsilon, \delta > 0$ and consider the relation

$$(z, w \in X)\text{: } z \perp w \text{ iff } (\varepsilon/\delta)d(z, w) \le [\varphi(z) - \varphi(w)]a(\Gamma(w)).$$

For the moment, $(\perp)$ is reflexive and antisymmetric. Further properties of this (and the preceding) object are contained in

Proposition 3.3 *The couple* $(\le, \perp)$ *is product transitive, in the sense*

$$u \le v \le w, u \perp v, v \perp w \Longrightarrow u \perp w.$$

Proof Let $u, v, w \in X$ be as in the premise of this implication; that is

(31-1) $\Gamma(u) \leq \Gamma(v) \leq \Gamma(w)$
(31-2) $(\varepsilon/\delta)d(u, v) \leq [\varphi(u) - \varphi(v)]a(\Gamma(v))$,
$(\varepsilon/\delta)d(v, w) \leq [\varphi(v) - \varphi(w)]a(\Gamma(w))$.

By the former of these conditions (and $a(.)$=increasing), the latter one gives

$(\varepsilon/\delta)d(u, v) \leq [\varphi(u) - \varphi(v)]a(\Gamma(w))$,
$(\varepsilon/\delta)d(v, w) \leq [\varphi(v) - \varphi(w)]a(\Gamma(w))$;

wherefrom (by the triangular property of d)

$$(\varepsilon/\delta)d(u, w) \leq (\varepsilon/\delta)(d(u, v) + d(v, w)) \leq [\varphi(u) - \varphi(w)]a(\Gamma(w));$$

that is: $u \perp w$; hence the conclusion.

As a consequence of this, the "product" relation $(\preceq)$ over X

$(z, w \in X)$: $z \preceq w$ iff $z \leq w$ and $z \perp w$

is reflexive, transitive, and antisymmetric; hence, an ordering on X. Our next objective is to establish that the strong metrical Zorn-Bourbaki principle (ZB-m-s) is applicable on each structure $(M, d; \preceq)$, where M is a (nonempty) closed bounded part of X. This is clearly the case with the $(\preceq)$-completeness property, because (as M is closed) we necessarily have (by the completeness assumption)

d is complete on M; hence, d is $(\preceq)$-complete on M.

So, it remains to verify that

$(\preceq)$ is regular and self-closed (over this class of subsets).

(B) We start with the verification of regularity condition.

Proposition 3.4 *Let the couple (Γ, a) be taken as before. Then, the introduced order $(\preceq)$ is boundedly regular; i.e.,*

(32-1) each ascending (modulo $(\preceq)$) bounded sequence in X is d-Cauchy.

And, as such,

(32-2) $(\preceq)$ is regular over each (nonempty) closed bounded part M of X.

Proof Let (y_n) be some bounded sequence in X with

$y_n \preceq y_{n+1}, \forall n$; or, equivalently: $y_n \preceq y_m$, whenever $n \leq m$.

By the very definition of our order $(\preceq)$, this amounts to the couple of conditions below being fulfilled

(asc-1) $\Gamma(y_n) \leq \Gamma(y_m)$, whenever $n \leq m$
(asc-2) $(\varepsilon/\delta)d(y_n, y_{n+1}) \leq [\varphi(y_n) - \varphi(y_{n+1})]a(\Gamma(y_{n+1}))$, $\forall n$.

The d-Lipschitz condition upon $\Gamma(.)$ and ($a(.)$=increasing) yields

$\mu := \sup\{\Gamma(y_i); i \geq 0\} < \infty$;
whence, $\nu := \sup\{a(\Gamma(y_i)); i \geq 0\} \leq a(\mu) < \infty$.

Combining with (asc-2), we then get

$(\varepsilon/\delta)d(y_n, y_{n+1}) \leq \nu[\varphi(y_n) - \varphi(y_{n+1})], \forall n.$

The (real) sequence $(\nu\varphi(y_n))$ is descending and bounded from below; hence

the series $\sum_n \nu[\varphi(y_n) - \varphi(y_{n+1})]$ converges (in R_+).

This, added to the previous relation, assures us that

the series $\sum_n d(y_n, y_{n+1})$ converges (in R_+);

whence, (y_n) is d-Cauchy; as claimed.

(C) Finally, we are passing to the verification of self-closeness condition. An appropriate answer to this question is contained in

Proposition 3.5 *Let the couple (Γ, a) be taken as before. Then,*

(33-1) $(\preceq)$ is self-closed on X; hence, a fortiori,
(33-2) $(\preceq)$ is self-closed over each (nonempty) closed bounded part M of X.

Proof Let (y_n) be some ascending (modulo $(\preceq)$) sequence in X; i.e.,

conditions (asc-1) and (asc-2) (see above) are fulfilled.

In addition, assume that (for some $y \in X$)

$y_n \xrightarrow{d} y$ (i.e.: $d(y_n, y) \to 0$) as $n \to \infty$.

By the d-Lipschitz property of Γ, we necessarily derive ($\Gamma(y_n) \to \Gamma(y)$ as $n \to \infty$). Taking (asc-1) into account, one gets

$\Gamma(y_n) \leq \Gamma(y)$ (i.e.: $y_n \leq y$), for all n;

so, combining with (asc-2), one gets (along with $a(.)$=increasing)

$(\varepsilon/\delta)d(y_n, y_{n+1}) \leq [\varphi(y_n) - \varphi(y_{n+1})]a(\Gamma(y))$, for all n
(whence, $(\varphi(y_n))$ is descending and $\lim_n(\varphi(y_n))$ exists in R).

This, by the triangular property of d, gives

$(\varepsilon/\delta)d(y_n, y_m) \leq [\varphi(y_n) - \varphi(y_m)]a(\Gamma(y))$, for $n \leq m$.

Passing to limit as $m \to \infty$, one derives (by the d-lsc property of φ)

$(\varepsilon/\delta)d(y_n, y) \leq [\varphi(y_n) - \varphi(y)]a(\Gamma((y))$ (i.e.: $y_n \perp y$), for all n.

Hence, $(y_n) \preceq y$, and this ends the argument.

We are now in position to formulate an appropriate variational result involving these data. Let the general conditions above be accepted. Precisely, let (X, d) be a complete metric space; and $\varphi : X \to R$ be a function with (regu) φ is (X, d)-regular (see above).

Remember that, for each $\theta \geq 0$ we denoted

$$\text{ulev}(\varphi; X; \theta) = \{x \in X; \varphi(x) \leq \inf \varphi(X) + \theta\};$$

this will be referred to as the θ-upper level set of φ with respect to X; in particular, ulev$(\varphi; X; 0)$ is nothing else than the (global) minimizers set of φ over X. Further, let the mapping $\Gamma : X \to R_+$ and the function $a : R_+ \to R_+$ be chosen according to

(incr) $\Gamma(.)$ is d-Lipschitz and $a(.)$ is increasing.

Finally, pick some map $\gamma : X \to R^0_+$. The following extended El Amrouss Ordering Principle (in short: (EAOP-ext)) is now available:

Theorem 3.6 *Let $\varepsilon, \delta > 0$ be given, as well as some point $u \in$ ulev$(\varphi; X; \varepsilon)$. There exists then another point $v = v(\varepsilon, \delta; u) \in X[u, \gamma(u)]$, with*

(31-a) $\Gamma(u) \leq \Gamma(v)$, $(\varepsilon/\delta)d(u, v) \leq [\varphi(u) - \varphi(v)]a(\Gamma(v))$;
hence $\varphi(u) \geq \varphi(v)$, $d(u, v) \leq \min\{\gamma(u), \delta a(\Gamma(v))\}$
(31-b) $w \in X[u, \gamma(u)] \setminus \{v\}$ and $\Gamma(v) \leq \Gamma(w)$
imply $(\varepsilon/\delta)d(v, w) > [\varphi(v) - \varphi(w)]a(\Gamma(w))$;
hence, $(\varepsilon/\delta)d(v, w) > [\varphi(v) - \varphi(w)]a(\Gamma(v))$.

Proof Denote for simplicity $M = X[u, \gamma(u)]$, where u is as before. Clearly, M is closed bounded; as well as nonempty (since $u \in M$). Let also $(\preceq)$ stands for the (product) order

$(z, w \in X)$: $z \preceq w$ iff $z \leq w$ (i.e.: $\Gamma(z) \leq \Gamma(w)$), and
$z \perp w$ [i.e.: $(\varepsilon/\delta)d(z, w) \leq [\varphi(z) - \varphi(w)]a(\Gamma(w))$].

By the preliminary facts we just exposed, the strong metrical Zorn-Bourbaki principle (ZB-m-s) applies to $(M, d; \preceq)$; so that (the restriction to M of) $(\preceq)$ is a Zorn order (over M). This means that, for the starting point $u \in M$, there exists another point $v = v(u) \in M$, with the properties

(31-c) $u \preceq v$ (i.e.: $u \leq v$ and $u \perp v$)
(31-d) $v \preceq w \in M \Longrightarrow v = w$ (i.e.: $w \in M, v \leq w, v \perp w \Longrightarrow v = w$).

The former of these means, by definition

(31-e) $\Gamma(u) \leq \Gamma(v)$, and $(\varepsilon/\delta)d(u, v) \leq [\varphi(u) - \varphi(v)]a(\Gamma(v))$.

Note that, as a consequence of its second half,

$\varphi(u) \geq \varphi(v)$; hence, $\varphi(u) \geq \varphi(v) \geq \varphi_*$, where $\varphi_* := \inf \varphi(X)$;
so that (by the choice of u): $\varphi(u) - \varphi(v) \leq \varphi(u) - \varphi_* \leq \varepsilon$.

(In fact, the assertion is clear if $a(\Gamma(v)) > 0$. And, if $a(\Gamma(v)) = 0$, we must have $u = v$; whence, $\varphi(u) = \varphi(v)$.) This, again combined with (31-e), gives

$d(u, v) \leq \delta a(\Gamma(v))$; hence $d(u, v) \leq \min\{\gamma(u), \delta a(\Gamma(v))\}$;

which, along with (31-e), is just our first conclusion in the statement. On the other hand, (31-d) may be written as (by definition)

(31-f) whenever $w \in M$ fulfills $\Gamma(v) \le \Gamma(w)$ and
$(\varepsilon/\delta)d(v, w) \le [\varphi(v) - \varphi(w)]a(\Gamma(w))$, then $v = w$;

or, equivalently,

for each $w \in M \setminus \{v\}$ with $\Gamma(v) \le \Gamma(w)$,
we must have $(\varepsilon/\delta)d(v, w) > [\varphi(v) - \varphi(w)]a(\Gamma(w))$;

which is just the first half of second conclusion in the statement. The second half of the same is immediate, in view of

$\Gamma(v) \le \Gamma(w)$ implies $a(\Gamma(v)) \le a(\Gamma(w))$.

The proof is thereby complete.

Technically speaking, this result may be viewed as a refinement of El Amrouss Ordering Principle (EAOP). Remember that, in the quoted statement, the extra regularity conditions below were considered (in our notations)

(c05) $a(.)$ is continuous on R_+
(c06) $a(.)$ is a comparison function of order $k > 0$
$(\forall q \ge k, \exists \lambda, \mu \ge 0 : a((t+1)s) \le a(t)[\lambda s^q + \mu], \forall t, s \ge 0)$
(c07) $u \mapsto \gamma(u)/(1 + d(x_0, u))$ is bounded on X, for some $x_0 \in X$.

The proposed argument tells us that, in the refined principle (EAOP-ext), all regularity conditions (c05)–(c07) may be dropped. This may have some theoretical impact upon the quoted result; but, in general, not a practical one.

Finally, the obtained "relaxed" principle (EAOP-ext) allows us (in a limited sense) certain comparison type operations with Zhong's Variational Principle (ZVP). For, as precise, a direct consequence of conclusion (31-b) is

(31-g) $w \in X[u, \gamma(u)] \setminus \{v\}$, $\Gamma(v) \le \Gamma(w)$
imply $(\varepsilon/\delta)d(v, w) > [\varphi(v) - \varphi(w)]a(\Gamma(v))$.

This, added to the fact that (with x_0 taken as before)

(12-b) $\Longrightarrow$ (31-g), under $(\Gamma(x) = d(x_0, x); x \in X)$ and $b = a$,

tells us that the variant of (EAOP-ext) with (31-g) in place of (31-b) is obtainable from (ZVP), whenever

(c08) the function $a(.)$ is normal (see above).

However, a general inclusion of this type is not accessible for the moment. Further aspects will be discussed elsewhere.

4 Differential Versions

The usefulness of our main result above is best manifested in a differential context. For, as we shall see further, this setting has an essential role in solving the minimizers problem for such (nonlinear) functions. Precisely, it is our aim in the following to establish a differential version of extended El Amrouss Ordering Principle (EAOP-ext), within the class of Gâteaux differentiable regular maps acting on a Hilbert space. Note that, such a context is not the most general one; but, for the applications to be considered, this will suffice.

Let H be a (real) Hilbert space with respect to the scalar product $(x, y) \mapsto \langle x, y\rangle$. As usual, we denote by $||.||$ the norm induced by this scalar product; and by $d(.,.)$, its associated metric:

$$(||x|| = \langle x, x\rangle^{1/2}; x \in H), (d(x, y) = ||x - y||; x, y \in H);$$

remember that, by the very definition of our structure,

$(H, ||.||)$ is a Banach space; hence, (H, d) is a complete metric space.

Put also, for simplicity,

$H_1 = \{h \in H; ||h|| \leq 1\}$ (the *unitary closed sphere* in H)
$\partial H_1 := \{h \in H_1; ||h|| = 1\}$ (the *boundary* of unitary closed sphere).

Let the function $\varphi : H \to R$ be (H, d)-regular; i.e. (see above)

(d01) φ is bounded below ($\inf \varphi(H) > -\infty$)
(d02) φ is d-lsc on H ($\liminf_n \varphi(x_n) \geq \varphi(x)$, whenever $x_n \xrightarrow{d} x$).

Define for each $\theta \geq 0$

$$\text{ulev}(\varphi, H; \theta) = \{x \in H; \varphi(x) \leq \inf \varphi(H) + \theta\};$$

this will be referred to as the θ-upper level set of φ with respect to H; in particular, ulev$(\varphi; H; 0)$ is nothing else than the (global) minimizers set of φ over H. The differential setting we just evoked is to be introduced as follows. Let $z \in H$ be arbitrary fixed. We say that φ is *Gâteaux differentiable* at z, when there exists an element $\varphi'(z) \in H$ (the *Gâteaux differential* of φ at z), with

$$\lim_{t \to 0+}(1/t)[\varphi(z + th) - \varphi(z)] = \langle \varphi'(z), h\rangle, \text{ for all } h \in H.$$

Suppose that (in addition to (d01)+(d02))

(d03) φ is Gâteaux differentiable over H:
$\varphi'(z)$ exists (cf. this definition), for all $z \in H$.

Denote, for each $z \in H$,

$$C(z) = \{h \in H; \langle z, h\rangle \leq 0\}; C_1(z) = C(z) \cap \partial H_1.$$

Clearly, $C(z)$ is a *convex cone* in H; i.e.:

$$C(z) + C(z) \subseteq C(z), R_+ C(z) \subseteq C(z).$$

On the other hand, $C_1(z)$ is always non-degenerate; precisely,

$$C_1(0) = \partial H_1;\ C_1(z) \ni -z/||z|| \neq 0, \forall z \in H \setminus \{0\}.$$

Denote also (for each z in H)

$$\Delta(\varphi'(z)) = \sup\{\langle \varphi'(z), h\rangle; h \in C_1(z)\},\ |\Delta|(\varphi'(z)) = \max\{\Delta(\varphi'(z)), 0\}.$$

Note that, in view of

$$|\langle \varphi'(z), h\rangle| \leq ||\varphi'(z)|| \cdot ||h|| = ||\varphi'(z)||, \forall h \in C_1(z), \forall z \in H,$$

we must have (for all such elements z)

$$\Delta(\varphi'(z)) \in R;\ \text{hence, } |\Delta|(\varphi'(z)) \in R_+.$$

Finally, take the map $\Gamma : X \to R_+$ and the function $a : R_+ \to R_+$, according to

(d04) $(\Gamma(x) = ||x||; x \in X)$ and $a(.)$ is increasing;

and let the couple of mappings (from H to R^0_+) be defined under

$$\beta(u) = ||u|| + 1,\ \ \gamma(u) = 2\beta(u) = 2||u|| + 2,\ \ u \in H.$$

As a direct application of extended El Amrouss Ordering Principle (EAOP-ext), we have the following statement, referred to as: differential El Amrouss Ordering Principle (in short: (EAOP-dif)):

Theorem 4.7 *Let* $\varepsilon > 0$ *be given, as well as some* $u \in \text{ulev}(\varphi; H; \varepsilon)$. *Further, let* $\delta > 0$ *be taken according to*

(d05) $\delta a(3\beta(u)) < \beta(u)$.

There exists then $v = v(\varepsilon, u; \delta) \in H$, *with the properties*

(41-a) $||u|| \leq ||v||, \varphi(u) \geq \varphi(v)$, *and* $||u - v|| \leq \min\{\gamma(u), \delta a(||v||)\}$
(41-b) $||u - v|| < \beta(u) < \gamma(u)$; *hence,* $v \in H(u, \beta(u)) \subseteq H(u, \gamma(u))$
(41-c) $(\varepsilon/\delta)t > [\varphi(v) - \varphi(v - th)]a(||v||)$, *if* $t \in]0, \beta(u)[$ *and* $h \in C_1(v)$
(41-d) $\varepsilon/\delta \geq |\Delta|(\varphi'(v))a(||v||) \geq \Delta(\varphi'(v))a(||v||)$.

Proof By the imposed conditions, (EAOP-ext) is applicable to the initial data $(H, d; \varphi; \Gamma, a; \gamma)$. So, for (ε, u) and δ as before, there exists $v = v(\varepsilon, u; \delta) \in H[u, \gamma(u)]$, with the properties (determined by its conclusion)

(41-e) $||u|| \leq ||v||, (\varepsilon/\delta)||u - v|| \leq [\varphi(u) - \varphi(v)]a(||v||)$;
hence (see above) $\varphi(u) \geq \varphi(v)$ and $||u - v|| \leq \min\{\gamma(u), \delta a(||v||)\}$
(41-f) $w \in H[u, \gamma(u)] \setminus \{v\}$ and $||v|| \leq ||w||$
imply $(\varepsilon/\delta)||v - w|| > [\varphi(v) - \varphi(w)]a(||w||)$;
hence, $(\varepsilon/\delta)||v - w|| > [\varphi(v) - \varphi(w)]a(||v||)$.

By the former of these one gets (41-a). As a direct consequence,

$$||v|| \leq ||v - u|| + ||u|| \leq \gamma(u) + ||u|| = 3||u|| + 2 < 3\beta(u);$$

and this, in combination with (d05), gives (via (41-a) above)

$$||u - v|| \leq \delta a(3\beta(u)) < \beta(u) < \gamma(u);$$

i.e., (41-b) is verified. This, in particular, tells us that

$$v \in H(u, \gamma(u)) \text{ (open sphere in } H);$$

as well as (from our previous facts),

$$v + tk \in H(u, \gamma(u)),\ \forall t \in [-\beta(u), \beta(u)], \forall k \in \partial H_1.$$

Indeed, let the couple (t, k) be as before; then

$$||v + tk - u|| \leq ||v - u|| + |t| < \beta(u) + \beta(u) = \gamma(u);$$

and the claim follows. On the other hand,

$$||v - th|| > ||v|| \text{ for } t \in]0, \beta(u)[, \text{ whenever } h \in C_1(v);$$

because (from the properties of scalar product)

$$||v - th||^2 = ||v||^2 + t^2 - 2t\langle v, h\rangle > ||v||^2, \text{ for all such } (t, h).$$

Putting these together gives (41-c), if one takes (41-f) into account. Finally, let $h \in C_1(v)$ be arbitrary fixed; hence, $h \in \partial H_1$, $\langle v, h\rangle \leq 0$. From (41-c), one gets

$$\varepsilon/\delta > (1/t)[\varphi(v) - \varphi(v - th)]a(||v||), \text{ when } t \in]0, \beta(u)[.$$

So, passing to limit as $t \to 0+$ and taking the Gâteaux differentiable property of φ into account, one derives

$$\varepsilon/\delta \geq \langle\varphi'(v), h\rangle a(||v||).$$

This, by the arbitrariness of h in $C_1(v)$, yields

$$\varepsilon/\delta \geq \Delta(\varphi'(v))a(||v||); \text{ hence, } \varepsilon/\delta \geq |\Delta|(\varphi'(v))a(||v||);$$

and establishes the final conclusion (41-d) in the statement.

The differential principle (EAOP-dif) is comparable with a related one in El Amrouss and Tsouli [13]. However, some basic differences between these occur.

I) As precise, the extra regularity conditions imposed by the quoted authors

(d06) $a(.)$ is continuous (over the whole of R_+)
(d07) $a(.)$ is a comparison function of order $k > 0$
$(\forall q \geq k, \exists\lambda, \mu \geq 0: a((t + 1)s) \leq a(t)[\lambda s^q + \mu], \forall t, s \geq 0)$

are not needed here.

II) The final differential relation above is written by the quoted authors as

(41-d-var) $\varepsilon/\delta \geq ||\varphi'(v)||a(||v||)$.

Formally, this is better that (41-d) above, in view of

$$||\varphi'(z)|| \geq |\Delta|(\varphi'(z)), \forall z \in H.$$

Unfortunately, (41-d-var) is not true under authors' directional context; and this is retainable as well for our statement we just exposed.

Note finally that some extensions of these results are possible, within the class of quasi-ordered normed spaces; see Turinici [31] for details.

5 Existence of Minimizers

In the following, an application of the result above to existence of minimizers for Gâteaux differentiable regular functionals is considered. The basic instrument for our investigations is the well known Palais–Smale condition [23].

Let H be a (real) Hilbert space with respect to the scalar product $(x, y) \mapsto \langle x, y\rangle$. Further, let $\varphi : H \to R$ be a (H, d)-regular function; i.e.,

(e01) φ is bounded below ($\inf \varphi(X) > -\infty$)

(e02) φ is d-lsc on X ($\liminf_n \varphi(x_n) \geq \varphi(x)$, whenever $x_n \xrightarrow{d} x$).

Remember that, for each $\theta \geq 0$ we denoted

$$\text{ulev}(\varphi; H; \theta) = \{x \in H; \varphi(x) \leq \inf \varphi(H) + \theta\};$$

this will be referred to as the θ-upper level set of φ with respect to H; note that ulev(φ; M; 0) is nothing else than the (global) minimizers set of φ over H.

In addition, suppose that

(e03) φ is Gâteaux differentiable over H.

Further, take the map $\Gamma : X \to R_+$ and the function $a : R_+ \to R_+$, according to

(e04) ($\Gamma(x) = ||x||; x \in X$) and $a(.)$ is increasing.

Roughly speaking, the differential El Amrouss Ordering Principle (EAOP-dif) we just established is a local one, because, given the starting point $u \in H$, the associated variational point $v = v(u)$ is to be found in a (closed) sphere $H[u, \gamma(u)]$ around u. But, for an appropriate solving of our problem, a global version of this ordering principle, relative to the class of (nonempty) bounded parts of H is needed. Call the subset K of H, *admissible* (modulo φ) provided

$$\text{ulev}(\varphi; H; \varepsilon) \cap K \text{ is nonempty, for each } \varepsilon > 0;$$

hence, necessarily, K is nonempty. Assume in the following that

(e05) (H, φ) is admissible:
there exist admissible (modulo φ) bounded parts of H.

Fix in the following such an object, K; as well as some number $\delta > 0$ with

(e06) $\delta a(3(\mu + 1)) < 1$; where $\mu = \sup\{||u||; u \in K\}$.

A direct consequence of this is the following. Let us introduce the mappings

$$\beta(u) = ||u|| + 1, \ \gamma(u) = 2\beta(u) = 2||u|| + 2, \ u \in H.$$

By the imposed condition, we have

$$\beta(u) \le \mu + 1, \ \gamma(u) \le 2\mu + 2, \forall u \in K;$$

and this yields (via $a(.)$=increasing)

$$\delta a(3\beta(u)) < 1, \text{ for each } u \in K.$$

The following global differential type variational statement (referred to as: global differential El Amrouss Ordering Principle; in short: (EAOP-dif-g)) is our main step towards the desired answer.

Theorem 5.8 *Let $\varepsilon > 0$ be given, as well as some $u \in \text{ulev}(\varphi; H; \varepsilon) \cap K$. Further, let $\delta > 0$ be taken as before. There exists then some $v \in H$, such that*

(51-a) $\varphi_* \le \varphi(v) \le \varphi(u) \le \varphi_* + \varepsilon$, *where* $\varphi_* := \inf \varphi(H)$
(51-b) $||u - v|| \le \beta(u) \le \mu + 1$; *hence* $||v|| \le ||u|| + \beta(u) \le 2\mu + 1$
(51-c) $|\Delta|(\varphi'(v))a(||v||) \le \varepsilon/\delta$.

Proof Let $\varepsilon > 0$, $u \in \text{ulev}(\varphi; H; \varepsilon) \cap K$ and $\delta > 0$ be taken as before. From the above relations involving the constant δ and the couple of functions $(\beta(.), \gamma(.))$, it follows that the (local) differential El Amrouss Ordering Principle (EAOP-dif) is applicable to the data $(H; \varepsilon, u; \delta; \beta(.), \gamma(.))$, and gives us all conclusions in the statement. The proof is thereby complete.

In particular, when $a : R_+ \to R_+$ fulfills the extra conditions

(ec-1) $a(.)$ is continuous (over the whole of R_+)
(ec-2) $a(.)$ is a comparison function, of order $k > 0$
$(\forall q \ge k, \exists \lambda, \mu \ge 0 : \ a((t + 1)s) \le a(t)[\lambda s^q + \mu], \forall t, s \ge 0)$

the global differential El Amrouss Ordering Principle (EAOP-dif-g) is nothing but the statement in El Amrouss [12] proved under different methods. However, its differential conclusion

(51-c-var) $||\varphi'(v)||a(||v||) \le \varepsilon/\delta$

is not true under author's directional context; we do not give further details.

An application of (EAOP-dif-g) to existence of (global) minimizers for the functional φ may now be given along the lines below. Let us say that (H, φ) satisfies the *Palais–Smale condition* (modulo a), when

(PS-a) each bounded sequence (x_n) in H such that $\varphi(x_n) \to \varphi_*$
and $|\Delta|(\varphi'(x_n))a(||x_n||) \to 0$ has a convergent subsequence.

In particular, when $a = 1$, this is referred to as (H, φ) fulfilling the *standard Palais–Smale condition*:

(PS-st) each bounded sequence (x_n) in H such that $\varphi(x_n) \to \varphi_*$
and $|\Delta|(\varphi'(x_n)) \to 0$ has a convergent subsequence.

Concerning the relationship between these, note that (under the choice of our function $a(.)$), the following inclusion holds

(P-S-incl) whenever (PS-a) holds then (PS-st) holds too.

Indeed, let the bounded sequence (x_n) in H be such that the premises of (PS-st) hold:

$$\varphi(x_n) \to \varphi_* \text{ and } |\Delta|(\varphi'(x_n)) \to 0.$$

Then, under the notation $\sigma := \sup\{||x_n||; n \geq 0\}$,

$$|\Delta|(\varphi'(x_n))a(||x_n||) \leq |\Delta|(\varphi'(x_n))a(\sigma), \forall n,$$

which tells us that the premises of (PS-a) hold. By the accepted hypothesis, it then follows that (x_n) has a convergent subsequence; and we are done. However, the reciprocal of (P-S-incl) is not in general true.

We are now in position to formulate the announced answer. Let the general conditions above be accepted.

Theorem 5.9 *Suppose, in addition, that (H, φ) is admissible and fulfills the Palais–Smale condition (PS-a). Then, φ admits at least one minimizer on H.*

Proof As (H, φ) is admissible, it admits at least a bounded admissible (hence, nonempty) subset K, in the sense:

(adm-1) K is bounded; hence, $\mu := \sup\{||u||; u \in K\} < \infty$
(adm-2) K is admissible: $\text{ulev}(\varphi; H; \varepsilon) \cap K \neq \emptyset, \forall \varepsilon > 0$.

Further, pick the number $\delta > 0$ in accordance with

$\delta a(3(\mu + 1)) < 1$ (where $\mu \geq 0$ is the above precise number).

For the arbitrary fixed $\varepsilon > 0$, take some (starting) point $u_\varepsilon \in \text{ulev}(\varphi; H; \varepsilon) \cap K$. From the global differential El Amrouss Ordering Principle (EAOP-dif-g), there must be some associated point $v_\varepsilon \in H$ fulfilling its conclusions (51-a)–(51-c); that is (for any such $\varepsilon > 0$)

(52-a) $\varphi_* \leq \varphi(v_\varepsilon) \leq \varphi(u_\varepsilon) \leq \varphi_* + \varepsilon$, where $\varphi_* := \inf \varphi(H)$
(52-b) $||u_\varepsilon - v_\varepsilon|| \leq \beta(u_\varepsilon)$; hence $||v_\varepsilon|| \leq ||u_\varepsilon|| + \beta(u_\varepsilon) \leq 2\mu + 1$
(52-c) $|\Delta|(\varphi'(v_\varepsilon))a(||v_\varepsilon||) \leq \varepsilon/\delta$.

In particular, taking the sequence $(\varepsilon_n = 2^{-n}; n \geq 0)$, it results that, for each $n \geq 0$ and each starting $u_n := u_{\varepsilon_n} \in \text{ulev}(\varphi; H; \varepsilon_n) \cap K$, there exists $v_n := v_{\varepsilon_n}$ in H with

(52-d) $\varphi_* \leq \varphi(v_n) \leq \varphi(u_n) \leq \varphi_* + 2^{-n}$
(52-e) $||u_n - v_n|| \leq \beta(u_n)$; hence $||v_n|| \leq ||u_n|| + \beta(u_n) \leq 2\mu + 1$
(52-f) $|\Delta|(\varphi'(v_n))a(||v_n||) \leq 2^{-n}/\delta$.

Now, (52-d)+(52-e) give us that (v_n) is a bounded sequence in H with $\varphi(v_n) \to \varphi_*$; and, from (52-f), one derives that $|\Delta|(\varphi'(v_n))a(||v_n||) \to 0$. This, along with the Palais–Smale condition (PS-a), yields a subsequence $(y_n := v_{i(n)})$ of (v_n) and an element $y \in H$, with:

$$(\varphi(y_n) \to \varphi_* \text{ and}) \ y_n \xrightarrow{d} y.$$

Combining with the d-lsc condition imposed upon φ, we thus get

$$\varphi_* \leq \varphi(y) \leq \lim_n \varphi(y_n) = \varphi_*; \text{ hence } \varphi(y) = \varphi_*;$$

or, in other words: $y \in H$ is a minimizer for φ. The proof is thereby complete.

In particular, under the extra regularity conditions

(er-1) $a(.)$ is continuous (over the whole of R_+)
(er-2) $a(.)$ is a comparison function of order $k > 0$ (see above)

the obtained existence result yields a related statement in El Amrouss [12]. Further aspects may be found in Motreanu et al. [21, Ch 5].

6 Dependent Choice Principles

In what follows, a lot of technical facts involving the (already used) Dependent Choice principle and its equivalents will be presented. The axiomatic system of this exposition is Zermelo-Fraenkel's (abbreviated: ZF), as described by Cohen [8, Ch 2]. All notations and basic facts to be considered are standard; some important ones are discussed below.

(A) Let X be a nonempty set. By a *relation* over X, we mean any (nonempty) part $\mathscr{R} \subseteq X \times X$; then, $(X, \mathscr{R})$ will be referred to as a *relational structure*. Note that $\mathscr{R}$ may be regarded as a mapping between X and $\exp[X]$ (=the class of all subsets in X). In fact, write $(x, y) \in \mathscr{R}$ as $x\mathscr{R}y$; and put, for $x \in X$,

$$X(x, \mathscr{R}) = \{y \in X; x\mathscr{R}y\} \text{ (the } \textit{section} \text{ of } \mathscr{R} \text{ through } x);$$

then, the desired mapping representation is $(\mathscr{R}(x) = X(x, \mathscr{R}); x \in X)$. A basic example of such object is

$$\mathscr{I} = \{(x, x); x \in X\} \text{ [the } \textit{identical relation} \text{ over } X].$$

Given the relations $\mathscr{R}$, $\mathscr{S}$ over X, define their *product* $\mathscr{R} \circ \mathscr{S}$ as

$$(x, z) \in \mathscr{R} \circ \mathscr{S}, \text{ if there exists } y \in X \text{ with } (x, y) \in \mathscr{R}, (y, z) \in \mathscr{S}.$$

Also, for each relation $\mathscr{R}$ over X, denote

$$\mathscr{R}^{-1} = \{(x, y) \in X \times X; (y, x) \in \mathscr{R}\} \text{ (the } \textit{inverse} \text{ of } \mathscr{R}).$$

Finally, given the relations $\mathscr{R}$ and $\mathscr{S}$ on X, let us say that $\mathscr{R}$ is *coarser* than $\mathscr{S}$ (or, equivalently: $\mathscr{S}$ is *finer* than $\mathscr{R}$), provided

$$\mathscr{R} \subseteq \mathscr{S}; \text{ i.e.: } x\mathscr{R}y \text{ implies } x\mathscr{S}y.$$

Given a relation $\mathscr{R}$ on X, the following properties are to be discussed here:

(P1) $\mathscr{R}$ is *reflexive*: $\mathscr{I} \subseteq \mathscr{R}$
(P2) $\mathscr{R}$ is *irreflexive*: $\mathscr{I} \cap \mathscr{R} = \emptyset$
(P3) $\mathscr{R}$ is *transitive*: $\mathscr{R} \circ \mathscr{R} \subseteq \mathscr{R}$

(P4) $\mathscr{R}$ is *symmetric*: $\mathscr{R}^{-1} = \mathscr{R}$
(P5) $\mathscr{R}$ is *antisymmetric*: $\mathscr{R}^{-1} \cap \mathscr{R} \subseteq \mathscr{I}$.

This yields the classes of relations to be used; the following ones are important for our developments:

(C0) $\mathscr{R}$ is *amorphous* (i.e.: it has no specific properties)
(C1) $\mathscr{R}$ is a *quasi-order* (reflexive and transitive)
(C2) $\mathscr{R}$ is a *strict order* (irreflexive and transitive)
(C3) $\mathscr{R}$ is an *equivalence* (reflexive, transitive, symmetric)
(C4) $\mathscr{R}$ is a *(partial) order* (reflexive, transitive, antisymmetric)
(C5) $\mathscr{R}$ is the *trivial* relation (i.e.: $\mathscr{R} = X \times X$).

(B) A basic example of relational structure is to be constructed as below. Let

$N = \{0, 1, 2, \ldots\}$, where $(0 = \emptyset, 1 = \{0\}, 2 = \{0, 1\}, \ldots)$

denote the set of *natural* numbers. Technically speaking, the basic (algebraic and order) structures over N may be obtained by means of the *(immediate) successor* function $\mathrm{suc} : N \to N$, and the following Peano properties (deductible in our axiomatic system (ZF)):

(pea-1) $(0 \in N$ and$)$ $0 \notin \mathrm{suc}(N)$
(pea-2) $\mathrm{suc}(.)$ is injective ($\mathrm{suc}(n) = \mathrm{suc}(m)$ implies $n = m$)
(pea-3) if $M \subseteq N$ fulfills $[0 \in M]$ and $[\mathrm{suc}(M) \subseteq M]$, then $M = N$.

(Note that, in the absence of our axiomatic setting, these properties become the well known Peano axioms, as described in Halmos [15, Ch 12]; we do not give details.) In fact, starting from these properties, one may construct, in a recurrent way, an *addition* $(a, b) \mapsto a + b$ over N, according to

$(\forall m \in N)$: $m + 0 = m$; $m + \mathrm{suc}(n) = \mathrm{suc}(m + n)$.

This, in turn, makes possible the introduction of a relation $(\leq)$ over N, as

$(m, n \in N)$: $m \leq n$ iff $m + p = n$, for some $p \in N$.

Concerning the properties of this structure, the most important one writes

$(N, \leq)$ is well ordered:
any (nonempty) subset of N has a first element;

hence (in particular), $(N, \leq)$ is (partially) ordered.

Having these precise, let the notion of *sequence* (in X) be used to designate any mapping $x : N \to X$. For simplicity reasons, it will be useful to denote it as $(x(n); n \geq 0)$, or $(x_n; n \geq 0)$; moreover, when no confusion can arise, we further simplify this notation as $(x(n))$ or (x_n), respectively. Also, any sequence $(y_n := x_{i(n)}; n \geq 0)$ with

$(i(n); n \geq 0)$ is *strictly ascending* (hence: $i(n) \to \infty$ as $n \to \infty$)

will be referred to as a *subsequence* of $(x_n; n \geq 0)$. Note that, under such a convention, the relation "subsequence of" is transitive; i.e.:

(z_n)=subsequence of (y_n) and (y_n)=subsequence of (x_n)
imply (z_n)=subsequence of (x_n).

(B) Remember that, an outstanding part of (ZF) is the *Axiom of Choice* (abbreviated: AC); which, in a convenient manner, may be written as

(AC) For each couple (J, X) of nonempty sets and each function
$F : J \to \exp(X)$, there exists a (selective) function
$f : J \to X$, with $f(\nu) \in F(\nu)$, for each $\nu \in J$.

(Here, $\exp(X)$ stands for the class of all nonempty elements in $\exp[X]$.) Sometimes, when the ambient set X is endowed with denumerable type structures, the existence of such a selective function (over $J = N$) may be determined by using a weaker form of (AC), referred to as: *Dependent Choice* principle (in short: DC). Call the relation $\mathscr{R}$ over X, *proper* when

$(X(x, \mathscr{R}) =)\mathscr{R}(x)$ is nonempty, for each $x \in X$.

Then, $\mathscr{R}$ is to be viewed as a mapping between X and $\exp(X)$, and the couple $(X, \mathscr{R})$ will be referred to as a *proper relational structure*. Further, given $a \in X$, let us say that the sequence $(x_n; n \geq 0)$ in X is *$(a; \mathscr{R})$-iterative*, provided

$x_0 = a$, and $x_n \mathscr{R} x_{n+1}$ (i.e.: $x_{n+1} \in \mathscr{R}(x_n)$), for all n.

Proposition 6.6 *Let the relational structure $(X, \mathscr{R})$ be proper. Then, for each $a \in X$ there is at least an $(a; \mathscr{R})$-iterative sequence in X.*

This principle—proposed, independently, by Bernays [3] and Tarski [26]—is deductible from (AC), but not conversely; cf. Wolk [33]. Moreover, by the developments in Moskhovakis [20, Ch 8], and Schechter [25, Ch 6], the *reduced system* (ZF-AC+DC) it comprehensive enough so as to cover the "usual" mathematics; see also Moore [19, Appendix 2].

Let $(\mathscr{R}_n; n \geq 0)$ be a sequence of relations on X. Given $a \in X$, let us say that the sequence $(x_n; n \geq 0)$ in X is *$(a; (\mathscr{R}_n; n \geq 0))$-iterative*, provided

$x_0 = a$, and $x_n \mathscr{R}_n x_{n+1}$ (i.e.: $x_{n+1} \in \mathscr{R}_n(x_n)$), for all n.

The following *Diagonal Dependent Choice* principle (in short: DDC) is available.

Proposition 6.7 *Let $(\mathscr{R}_n; n \geq 0)$ be a sequence of proper relations on X. Then, for each $a \in X$ there exists at least one $(a; (\mathscr{R}_n; n \geq 0))$-iterative sequence in X.*

Clearly, (DDC) includes (DC), to which it reduces when $(\mathscr{R}_n; n \geq 0)$ is constant. The reciprocal of this is also true. In fact, letting the premises of (DDC) hold, put $P = N \times X$; and let $\mathscr{S}$ be the relation over P introduced as

$$\mathscr{S}(i, x) = \{i + 1\} \times \mathscr{R}_i(x), \ \ (i, x) \in P.$$

It will suffice applying (DC) to $(P, \mathscr{S})$ and $b := (0, a) \in P$ to get the conclusion in our statement; we do not give details.

Summing up, (DDC) is provable in (ZF-AC+DC). This is valid as well for its variant, referred to as: the *Selected Dependent Choice* principle (in short: SDC).

Proposition 6.8 *Let the map* $F : N \to \exp(X)$ *and the relation* $\mathscr{R}$ *over* X *fulfill*

$$(\forall n \in N)\text{: } \mathscr{R}(x) \cap F(n+1) \neq \emptyset, \text{ for all } x \in F(n).$$

Then, for each $a \in F(0)$ *there exists a sequence* $(x(n); n \geq 0)$ *in* X*, with*

$$x(0) = a,\ x(n) \in F(n),\ x(n)\mathscr{R}x(n+1),\ \forall n.$$

As before, (SDC) $\Longrightarrow$ (DC) ($\Longleftrightarrow$ (DDC)); just take $(F(n) = X; n \in N)$. But, the reciprocal is also true, in the sense: (DDC) $\Longrightarrow$ (SDC). This follows from

Proof (Proposition 6.8) Let the premises of (SDC) be true. Define a sequence of relations $(\mathscr{R}_n; n \geq 0)$ over X as: for each $n \geq 0$,

$$\mathscr{R}_n(x) = \mathscr{R}(x) \cap F(n+1), \text{ if } x \in F(n),$$
$$\mathscr{R}_n(x) = \{x\}, \text{ otherwise } (x \in X \setminus F(n)).$$

Clearly, $\mathscr{R}_n$ is proper, for all $n \geq 0$. So, by (DDC), it follows that for the starting $a \in F(0)$, there exists an $(a, (R_n; n \geq 0))$-iterative sequence $(x(n); n \geq 0)$ in X. Combining with the very definition above, one derives that conclusion in our statement is holding.

In particular, when $\mathscr{R} = X \times X$, the regularity condition imposed in (SDC) holds. The corresponding variant of our underlying statement is just (AC(N)) (=the *Denumerable Axiom of Choice*). Precisely, we have

Proposition 6.9 *Let* $F : N \to \exp(X)$ *be a function. Then, for each* $a \in F(0)$ *there exists a function* $f : N \to X$ *with* $f(0) = a$ *and* $f(n) \in F(n)$*,* $\forall n \in N$.

As a consequence of the above facts, (DC) $\Longrightarrow$ (AC(N)) in (ZF-AC). A direct verification of this is obtainable by taking $Q = N \times X$ and introducing the relation $\mathscr{R}$ over it, according to:

$$\mathscr{R}(n, x) = \{n+1\} \times F(n+1),\ \ n \in N, x \in X;$$

then, an application of (DC) to $(Q, \mathscr{R})$ gives all desired facts. The reciprocal of the written inclusion is not true; see Moskhovakis [20, Ch 8, Sect 8.25] for details.

7 Equivalence Results

In the following, the relationships between our maximal principles used above and the Dependent Choice Principle (DC) are to be clarified. Further aspects involving these facts will be also discussed.

(I) Let M be a nonempty set, and $d : M \times M \to R_+$ be a *metric* over it; the couple (M, d) will be then referred to as a *metric space*. The following 1974 Ekeland variational principle [10] (in short: EVP) is now entering into our discussion.

Theorem 7.10 *Let the metric space* (M, d) *be complete and the function* $\varphi : M \to R$ *be* (M, d)*-regular. Then, for each* $u \in M$ *there exists* $v \in M$*, with the properties*

(71-a) $d(u, v) \leq \varphi(u) - \varphi(v)$ *(hence* $\varphi(u) \geq \varphi(v)$*)*
(71-b) $d(v, x) > \varphi(v) - \varphi(x)$*, for all* $x \in M \setminus \{v\}$.

Clearly, this principle is just the one of introductory part, under $\varepsilon = \delta = 1$. However, for technical reasons, it will be convenient taking it as our starting point.

Concerning the relationships between this and the choice/maximal principles we already stated, one has for the moment

(DC) $\Longrightarrow$ (ZB-m) $\Longrightarrow$ (ZB-m-s) $\Longrightarrow$ (EVP), in (ZF-AC).

So, we may ask whether these may be reversed. As we shall see, a positive answer to this is essentially possible; and may be done along the lines below.

Let $(X, \leq)$ be a partially ordered structure. We say that $(\leq)$ has the *inf-lattice* property, provided:

$x \wedge y := \inf(x, y)$ exists, for all $x, y \in X$.

Further, call $z \in X$, $(\leq)$-*maximal* if $X(z, \leq) = \{z\}$; the class of all these points will be denoted as $\max(X, \leq)$. In this case, $(\leq)$ is termed a *Zorn order* when

$\max(X, \leq)$ is nonempty and *cofinal* in X
(for each $u \in X$ there exists a $(\leq)$-maximal $v \in X$ with $u \leq v$).

Further aspects are to be described in a metrical setting. Let $d : X \times X \to R_+$ be a metric over X; and $\varphi : X \to R_+$ be some function. Then, the natural choice for $(\leq)$ above is

$x \leq_{(d,\varphi)} y$ iff $d(x, y) \leq \varphi(x) - \varphi(y)$;

referred to as the Brøndsted order [6] attached to (d, φ). Denote

$X(x, \rho) = \{u \in X; d(x, u) < \rho\}, x \in X, \rho > 0$
(the *open sphere* with center x and radius ρ).

Call the ambient metric space (X, d), *discrete* when

for each $x \in X$ there exists $\rho = \rho(x) > 0$ such that $X(x, \rho) = \{x\}$.

Note that, under such a hypothesis, any function $\psi : X \to R$ is continuous over X. However, the d-Lipschitz property

$|\psi(x) - \psi(y)| \leq Ld(x, y), x, y \in X$, for some $L > 0$

as well as the d-nonexpansive one ($L = 1$) cannot be retained, in general.

The following maximal/variational statement (referred to as: *discrete Lipschitz countable* version of (EVP) (in short: (EVP-dLc)) is now coming into discussion.

Theorem 7.11 *Let the metric space* (X, d) *and the function* $\varphi : X \to R_+$ *satisfy*

(72-i) (X, d) *is discrete bounded and complete*
(72-ii) $(\leq_{(d,\varphi)})$ *has the inf-lattice property*
(72-iii) φ *is* d*-nonexpansive and* $\varphi(X)$ *is countable.*

Then, $(\leq_{(d,\varphi)})$ *is a Zorn order.*

Clearly, (EVP) $\Longrightarrow$ (EVP-dLc). The remarkable fact to be added is that this last principle yields (DC); so, it completes the circle between all these.

Proposition 7.10 *Under the precise conventions,*

(71-1) (EVP-dLc) $\Longrightarrow$ (DC), in (ZF-AC)
(71-2) The maximal/variational principles (ZB-m), (ZB-m-s) and (EVP) are all equivalent with (DC); hence, mutually equivalent
(71-3) Each maximal principle (MP) with (DC) $\Longrightarrow$ (MP) $\Longrightarrow$ (EVP) is equivalent with both (DC) and (EVP).

For a complete proof, we refer to the 2014 survey paper by Turinici [32]. In particular, when the boundedness and Lipschitz properties are ignored, this result is just the one in Brunner [7].

Summing up, all variational principles in this exposition (derived from (DC))—as well as the ones described in Altman [1], Brezis and Browder [5], Kang and Park [18] or Turinici [29]—are nothing but logical equivalents of (EVP). Note that the list of these is rather comprehensive; see the 1997 monograph by Hyers et al. [16, Ch 5] for details. Further aspects will be delineated elsewhere.

References

1. M. Altman, A generalization of the Brezis-Browder principle on ordered sets. Nonlinear Anal. **6**, 157–165 (1982)
2. T.Q. Bao, P.Q. Khanh, Are several recent generalizations of Ekeland's variational principle more general than the original principle?. Acta Math. Vietnam. **28**, 345–350 (2003)
3. P. Bernays, A system of axiomatic set theory: Part III. Infinity and enumerability analysis. J. Symb. Log. **7**, 65–89 (1942)
4. N. Bourbaki, Sur le théorème de Zorn. Arch. Math. **2**, 434–437 (1949/1950)
5. H. Brezis, F.E. Browder, A general principle on ordered sets in nonlinear functional analysis. Adv. Math. **21**, 355–364 (1976)
6. A. Brøndsted, Fixed points and partial orders. Proc. Am. Math. Soc. **60**, 365–366 (1976)
7. N. Brunner, Topologische Maximalprinzipien. Z. Math. Logik Grundl. Math. **33**, 135–139 (1987)
8. P.J. Cohen, *Set Theory and the Continuum Hypothesis* (Benjamin, New York, 1966)
9. O. Cârjă, M. Necula, I.I. Vrabie, *Viability, Invariance and Applications*. North Holland Math. Studies, vol. 207 (Elsevier, Amsterdam, 2007)
10. I. Ekeland, On the variational principle. J. Math. Anal. Appl. **47**, 324–353 (1974)
11. I. Ekeland, Nonconvex minimization problems. Bull. Amer. Math. Soc. (N. S.) **1**, 443–474 (1979)
12. A.R. El Amrouss, Variantes du principle variationnel d'Ekeland et applications. Rev. Colomb. Mat. **40**, 1–14 (2006)
13. A.R. El Amrouss, N. Tsouli, A generalization of Ekeland's variational principle with applications. Electron. J. Diff. Eqs. Conference **14**, 173–180 (2006)
14. A. Goepfert, H. Riahi, C. Tammer, C. Zălinescu, *Variational Methods in Partially Ordered Spaces*. Canad. Math. Soc. Books Math., vol. 17 (Springer, New York, 2003)
15. P.R. Halmos, *Naive Set Theory* (Van Nostrand Reinhold Co., New York, 1960)
16. D.H. Hyers, G. Isac, T.M. Rassias, *Topics in Nonlinear Analysis and Applications* (World Sci. Publ., Singapore, 1997)

17. O. Kada, T. Suzuki, W. Takahashi, Nonconvex minimization theorems and fixed point theorems in complete metric spaces. Math. Jpn. **44**, 381–391 (1996)
18. B.G. Kang, S. Park, On generalized ordering principles in nonlinear analysis. Nonlinear Anal. **14**, 159–165 (1990)
19. G.H. Moore, *Zermelo's Axiom of Choice: Its Origin, Development and Influence* (Springer, New York, 1982)
20. Y. Moskhovakis, *Notes on Set Theory* (Springer, New York, 2006)
21. D. Motreanu, V.V. Motreanu, N. Papageorgiou, *Topological and Variational Methods with Applications to Nonlinear Boundary Value Problems* (Springer, New York, 2014)
22. L. Nachbin, *Topology and Order* (van Nostrand, Princeton, 1965)
23. R.S. Palais, S. Smale, A generalized Morse theory. Bull. Amer. Math. Soc. **70**, 165–171 (1964)
24. S. Park, J.S. Bae, On the Ray-Walker extension of the Caristi-Kirk fixed point theorem. Nonlinear Anal. **9**, 1135–1136 (1985)
25. E. Schechter, *Handbook of Analysis and Its Foundation* (Academic Press, New York, 1997)
26. A. Tarski, Axiomatic and algebraic aspects of two theorems on sums of cardinals. Fundam. Math. **35**, 79–104 (1948)
27. D. Tataru, Viscosity solutions of Hamilton-Jacobi equations with unbounded nonlinear terms. J. Math. Anal. Appl. **163**, 345–392 (1992)
28. M. Turinici, Maximality principles and mean value theorems. An. Acad. Bras. Cienc. **53**, 653–655 (1981)
29. M. Turinici, A generalization of Altman's ordering principle. Proc. Am. Math. Soc. **90**, 128–132 (1984)
30. M. Turinici, Function variational principles and coercivity. J. Math. Anal. Appl. **304**, 236–248 (2005)
31. M. Turinici, Normed coercivity for monotone functionals. Romai **7**(2), 169–179 (2011)
32. M. Turinici, Sequential maximality principles, in *Mathematics Without Boundaries*, ed. by T.M. Rassias, P.M. Pardalos (Springer, New York, 2014), pp. 515–548
33. E.S. Wolk, On the principle of dependent choices and some forms of Zorn's lemma. Can. Math. Bull. **26**, 365–367 (1983)
34. C.K. Zhong, A generalization of Ekeland's variational principle and application to the study of the relation between the weak P.S. condition and coercivity. Nonlinear Anal. **29**, 1421–1431 (1997)
35. M. Zorn, A remark on method in transfinite algebra. Bull. Amer. Math. Soc. **41**, 667–670 (1935)

Nadler-Liu Functional Contractions in Metric Spaces

Mihai Turinici

Abstract A technical extension is given for the fixed point result in Liu et al. [J. Appl. Math., Volume 2012, Article ID: 786061].

AMS Subject Classification 47H10 (Primary), 54H25 (Secondary)

1 Introduction

Let X be a nonempty set. Call the subset Y of X, *almost-singleton* (in short: *asingleton*) provided $y_1, y_2 \in Y$ implies $y_1 = y_2$; and *singleton* if, in addition, Y is nonempty; note that in this case $Y = \{y\}$, for some $y \in X$. Take a metric $d : X \times X \to R_+ := [0, \infty[$ over X; as well as a selfmap $T \in \mathscr{F}(X)$. [Here, for each couple A, B of nonempty sets, $\mathscr{F}(A, B)$ denotes the class of all functions from A to B; when $A = B$, we write $\mathscr{F}(A)$ in place of $\mathscr{F}(A, A)$.] Denote $\text{Fix}(T) = \{x \in X; x = Tx\}$; each point of this set is called *fixed* under T. Concerning the existence and uniqueness of such points, a basic result (referred to as: Banach fixed point theorem; in short: (B-fpt)) may be stated as follows. Call the selfmap T, $(d; \alpha)$-*contractive* (where $\alpha \geq 0$), if

(con) $d(Tx, Ty) \leq \alpha d(x, y)$, for all $x, y \in X$.

Theorem 1.1 *Assume that T is Banach $(d; \alpha)$-contractive, for some $\alpha \in [0, 1[$. In addition, let X be d-complete. Then,*

(11-a) $\text{Fix}(T)$ *is a singleton,* $\{z\}$

(11-b) $T^n x \xrightarrow{d} z$ *as $n \to \infty$, for each $x \in X$.*

This result, established in 1922 by Banach [2], found some important applications to the operator equations theory. Consequently, a multitude of extensions for

M. Turinici (✉)
A. Myller Mathematical Seminar, A. I. Cuza University, Iaşi, Romania
e-mail: mturi@uaic.ro

N. J. Daras, T. M. Rassias (eds.), *Computational Mathematics and Variational Analysis*, Springer Optimization and Its Applications 159,
https://doi.org/10.1007/978-3-030-44625-3_28

it were proposed. Here, we shall be interested in the *relational* way of enlarging (B-fpt), based on implicit contractive conditions like

(i-con-r) $F(d(Tx, Ty), d(x, y), d(x, Tx), d(y, Ty), d(x, Ty), d(Tx, y)) \leq 0$, for all $x, y \in X$ with $x\mathscr{R}y$,

where $F : R_+^6 \to R$ is a function and $\mathscr{R}$ is a relation over X. Note that, when $\mathscr{R} = X \times X$ (the *trivial* relation over X), some basic contributions in the area were obtained by Boyd and Wong [5], Meir and Keeler [30], Leader [25], Matkowski [29], and Rhoades [38]. Further, when $\mathscr{R}$ is an *order* on X, a first couple of 1986 results was formulated—in the realm of Matkowski type contractions—by Turinici [47, 48]. Two decades later, these fixed point statements have been rediscovered—over the Banach contractive setting—by Ran and Reurings [36]; see also Nieto and Rodriguez-Lopez [35]; and since then, the number of papers devoted to the precise topic increased rapidly. Finally, when $\mathscr{R}$ is an *amorphous* relation over X, an appropriate statement of this type was obtained in 2012 by Samet and Turinici [40]; see also Jachymski [16]. It is our aim in the following to give further extensions of these last results, via functional type contractive concepts involving multivalued maps over relational metric spaces. As a by-product of these, some related statements in this area obtained by Du et al. [11]—based on some techniques appearing in Khan et al. [21], Nadler [34], and Suzuki [42]—are being derived. Further aspects involving extensions of related fixed point statements due to Feng and Liu [12], Klim and Wardowski [23], or Liu et al. [26] will be delineated elsewhere.

2 Dependent Choice Principles

Throughout this exposition, the axiomatic system in use is Zermelo-Fraenkel's (abbreviated: ZF), as described by Cohen [9, Ch 2]. The notations and basic facts to be considered are standard; some important ones are discussed below.

(A) Let X be a nonempty set. By a *relation* over X, we mean any (nonempty) part $\mathscr{R} \subseteq X \times X$; then, $(X, \mathscr{R})$ will be referred to as a *relational structure*. Note that $\mathscr{R}$ may be regarded as a mapping between X and $\exp[X]$ (=the class of all subsets in X). In fact, let us simplify the string $(x, y) \in \mathscr{R}$ as $x\mathscr{R}y$; and put

$X(x, \mathscr{R}) = \{y \in X; x\mathscr{R}y\}$ (the *section* of $\mathscr{R}$ through x), $x \in X$;

then, the desired mapping representation is $(\mathscr{R}(x) = X(x, \mathscr{R}); x \in X)$. A basic example of such object is

$\mathscr{I} = \{(x, x); x \in X\}$ [the *identical relation* over X].

Given the relations $\mathscr{R}, \mathscr{S}$ over X, define their *product* $\mathscr{R} \circ \mathscr{S}$ as

$(x, z) \in \mathscr{R} \circ \mathscr{S}$, if there exists $y \in X$ with $(x, y) \in \mathscr{R}$, $(y, z) \in \mathscr{S}$.

Also, for each relation $\mathscr{R}$ on X, denote

$\mathscr{R}^{-1} = \{(x, y) \in X \times X; (y, x) \in \mathscr{R}\}$ (the *inverse* of $\mathscr{R}$).

Finally, given the relations $\mathscr{R}$ and $\mathscr{S}$ on X, let us say that $\mathscr{R}$ is *coarser* than $\mathscr{S}$ (or, equivalently: $\mathscr{S}$ is *finer* than $\mathscr{R}$), provided

$\mathscr{R} \subseteq \mathscr{S}$; i.e.: $x\mathscr{R}y$ implies $x\mathscr{S}y$.

Given a relation $\mathscr{R}$ on X, the following properties are to be discussed here:

(P1) $\mathscr{R}$ is *reflexive*: $\mathscr{I} \subseteq \mathscr{R}$
(P2) $\mathscr{R}$ is *irreflexive*: $\mathscr{I} \cap \mathscr{R} = \emptyset$
(P3) $\mathscr{R}$ is *transitive*: $\mathscr{R} \circ \mathscr{R} \subseteq \mathscr{R}$
(P4) $\mathscr{R}$ is *symmetric*: $\mathscr{R}^{-1} = \mathscr{R}$
(P5) $\mathscr{R}$ is *antisymmetric*: $\mathscr{R}^{-1} \cap \mathscr{R} \subseteq \mathscr{I}$.

This yields the classes of relations to be used; the following ones are important for our developments:

(C0) $\mathscr{R}$ is *amorphous* (i.e.: it has no properties at all)
(C1) $\mathscr{R}$ is a *quasi-order* (reflexive and transitive)
(C2) $\mathscr{R}$ is a *strict order* (irreflexive and transitive)
(C3) $\mathscr{R}$ is an *equivalence* (reflexive, transitive, symmetric)
(C4) $\mathscr{R}$ is a *(partial) order* (reflexive, transitive, antisymmetric)
(C5) $\mathscr{R}$ is the *trivial* relation (i.e.: $\mathscr{R} = X \times X$).

(B) A basic example of relational structure is to be constructed as below. Let

$N = \{0, 1, 2, \dots\}$, where $(0 = \emptyset, 1 = \{0\}, 2 = \{0, 1\}, \dots)$

denote the set of *natural* numbers. Technically speaking, the basic (algebraic and order) structures over N may be obtained by means of the *(immediate) successor* function suc : $N \to N$, and the following Peano properties (deductible in our axiomatic system (ZF)):

(pea-1) $(0 \in N$ and) $0 \notin \text{suc}(N)$
(pea-2) suc(.) is injective ($\text{suc}(n) = \text{suc}(m)$ implies $n = m$)
(pea-3) if $M \subseteq N$ fulfills $[0 \in M]$ and $[\text{suc}(M) \subseteq M]$, then $M = N$.

(Note that, in the absence of our axiomatic setting, these properties become the well known Peano axioms, as described in Halmos [13, Ch 12]; we do not give details.) In fact, starting from these properties, one may construct, in a recurrent way, an *addition* $(a, b) \mapsto a + b$ over N, according to

$(\forall m \in N)$: $m + 0 = m$; $m + \text{suc}(n) = \text{suc}(m + n)$.

This, in turn, makes possible the introduction of a relation $(\leq)$ over N, as

$(m, n \in N)$: $m \leq n$ iff $m + p = n$, for some $p \in N$.

Concerning the properties of this structure, the most important one writes

$(N, \leq)$ is well ordered:
any (nonempty) subset of N has a first element;

hence (in particular), $(N, \leq)$ is (partially) ordered. Denote, for simplicity

$$N(r, \leq) = \{n \in N; r \leq n\} = \{r, r+1, \ldots, \}, r \geq 0,$$
$$N(r, >) = \{n \in N; r > n\} = \{0, \ldots, r-1\}, r \geq 1;$$

the latter one is referred to as the *initial interval* (in N) induced by r. Any set P with $N \sim P$ (in the sense: there exists a bijection from N to P) will be referred to as *effectively denumerable*. In addition, given the natural number $n \geq 1$, any (nonempty) set Q with $N(n, >) \sim Q$ will be said to be *n-finite*; when n is generic here, we say that Q is *finite*. As a combination of these, we say that the (nonempty) set Y is (at most) *denumerable* iff it is either effectively denumerable or finite.

Having these precise, let the notion of *sequence* (in X) be used to designate any mapping $x : N \to X$. For simplicity reasons, it will be useful to denote it as $(x(n); n \geq 0)$, or $(x_n; n \geq 0)$; moreover, when no confusion can arise, we further simplify this notation as $(x(n))$ or (x_n), respectively. Also, any sequence $(y_n := x_{i(n)}; n \geq 0)$ with

$(i(n); n \geq 0)$ is *strictly ascending* (hence: $i(n) \to \infty$ as $n \to \infty$)

will be referred to as a *subsequence* of $(x_n; n \geq 0)$. Note that, under such a convention, the relation "subsequence of" is transitive; i.e.:

(z_n)=subsequence of (y_n) and (y_n)=subsequence of (x_n)
imply (z_n)=subsequence of (x_n).

(C) Remember that, an outstanding part of (ZF) is the *Axiom of Choice* (abbreviated: AC), which, in a convenient manner, may be written as

(AC) For each couple (J, X) of nonempty sets and each function
$F : J \to \exp(X)$, there exists a (selective) function
$f : J \to X$, with $f(\nu) \in F(\nu)$, for each $\nu \in J$.

(Here, $\exp(X)$ stands for the class of all nonempty elements in $\exp[X]$.) Sometimes, when the ambient set X is endowed with denumerable type structures, the existence of such a selective function (over $J = N$) may be determined by using a weaker form of (AC), referred to as: *Dependent Choice* principle (in short: DC). Call the relation $\mathscr{R}$ over X, *proper* when

$(X(x, \mathscr{R}) =)\mathscr{R}(x)$ is nonempty, for each $x \in X$.

Then, $\mathscr{R}$ is to be viewed as a mapping between X and $\exp(X)$; and the couple $(X, \mathscr{R})$ will be referred to as a *proper relational structure*. Further, given $a \in X$, let us say that the sequence $(x_n; n \geq 0)$ in X is *$(a; \mathscr{R})$-iterative*, provided

$x_0 = a$, and $x_n \mathscr{R} x_{n+1}$ (i.e.: $x_{n+1} \in \mathscr{R}(x_n)$), for all n.

Proposition 2.1 *Let the relational structure $(X, \mathscr{R})$ be proper. Then, for each $a \in X$ there is at least an $(a; \mathscr{R})$-iterative sequence in X.*

This principle—proposed, independently, by Bernays [4] and Tarski [43]—is deductible from (AC), but not conversely; cf. Wolk [53]. Moreover, by the developments in Moskhovakis [33, Ch 8], and Schechter [41, Ch 6], the *reduced system*

(ZF-AC+DC) is comprehensive enough so as to cover the "usual" mathematics; see also Moore [32, Appendix 2].

Let $(\mathscr{R}_n; n \geq 0)$ be a sequence of relations on X. Given $a \in X$, let us say that the sequence $(x_n; n \geq 0)$ in X is $(a; (\mathscr{R}_n; n \geq 0))$-*iterative*, provided

$$x_0 = a, \text{ and } x_n \mathscr{R}_n x_{n+1} \text{ (i.e.: } x_{n+1} \in \mathscr{R}_n(x_n)\text{), for all } n.$$

The following *Diagonal Dependent Choice* principle (in short: DDC) is available.

Proposition 2.2 *Let $(\mathscr{R}_n; n \geq 0)$ be a sequence of proper relations on X. Then, for each $a \in X$ there exists at least one $(a; (\mathscr{R}_n; n \geq 0))$-iterative sequence in X.*

Clearly, (DDC) includes (DC), to which it reduces when $(\mathscr{R}_n; n \geq 0)$ is constant. The reciprocal of this is also true. In fact, letting the premises of (DDC) hold, put $P = N \times X$; and let $\mathscr{S}$ be the relation over P introduced as

$$\mathscr{S}(i, x) = \{i + 1\} \times \mathscr{R}_i(x), \ (i, x) \in P.$$

It will suffice applying (DC) to $(P, \mathscr{S})$ and $b := (0, a) \in P$ to get the conclusion in our statement; we do not give details.

Summing up, (DDC) is provable in (ZF-AC+DC). This is valid as well for its variant, referred to as: the *Selected Dependent Choice* principle (in short: SDC).

Proposition 2.3 *Let the map $F : N \to \exp(X)$ and the relation $\mathscr{R}$ over X fulfill*

$$(\forall n \in N)\text{: } \mathscr{R}(x) \cap F(n + 1) \neq \emptyset, \text{ for all } x \in F(n).$$

Then, for each $a \in F(0)$ there exists a sequence $(x(n); n \geq 0)$ in X, with

$$x(0) = a, \ x(n) \in F(n), \ x(n + 1) \in \mathscr{R}(x(n)), \ \forall n.$$

As before, (SDC) $\Longrightarrow$ (DC) ($\Longleftrightarrow$ (DDC)); just take $(F(n) = X; n \geq 0)$. But, the reciprocal is also true, in the sense: (DDC) $\Longrightarrow$ (SDC). This follows from

Proof (Proposition 2.3) Let the premises of (SDC) be true. Define a sequence of relations $(\mathscr{R}_n; n \geq 0)$ over X as: for each $n \geq 0$,

$$\mathscr{R}_n(x) = \mathscr{R}(x) \cap F(n + 1), \ \text{if } x \in F(n),$$
$$\mathscr{R}_n(x) = \{x\}, \text{ otherwise } (x \in X \setminus F(n)).$$

Clearly, $\mathscr{R}_n$ is proper, for all $n \geq 0$. So, by (DDC), it follows that for the starting $a \in F(0)$, there exists an $(a, (R_n; n \geq 0))$-iterative sequence $(x(n); n \geq 0)$ in X. Combining with the very definition above, one derives that conclusion in the statement is holding.

In particular, when $\mathscr{R} = X \times X$, the regularity condition imposed in (SDC) holds. The corresponding variant of the underlying statement is just (AC(N)) (=the *Denumerable Axiom of Choice*). Precisely, we have

Proposition 2.4 *Let $F : N \to \exp(X)$ be a function. Then, for each $a \in F(0)$ there exists a function $f : N \to X$ with $f(0) = a$ and $f(n) \in F(n)$, $\forall n \in N$.*

As a consequence of the above facts, (DC) $\Longrightarrow$ (AC(N)) in (ZF-AC). A direct verification of this is obtainable by taking $Q = N \times X$ and introducing the relation $\mathscr{S}$ over it, according to:

$$\mathscr{S}(n, x) = \{n + 1\} \times F(n + 1), \ (n, x) \in Q;$$

we do not give details. The reciprocal of the written inclusion is not true; see, for instance, Moskhovakis [33, Ch 8, Sect 8.25].

3 Conv-Cauchy Structures

Let X be a nonempty set, and $\mathscr{S}(X)$ stand for the class of all sequences (x_n) in X. By a (sequential) *convergence structure* on X we mean any part $\mathscr{C}$ of $\mathscr{S}(X) \times X$, with the properties (cf. Kasahara [20]):

(conv-1) $\mathscr{C}$ is *hereditary*:
$((x_n); x) \in \mathscr{C} \Longrightarrow ((y_n); x) \in \mathscr{C}$, for each subsequence (y_n) of (x_n)
(conv-2) $\mathscr{C}$ is *reflexive*: for each $u \in X$,
the constant sequence $(x_n = u; n \geq 0)$ fulfills $((x_n); u) \in \mathscr{C}$.

For each sequence (x_n) in $\mathscr{S}(X)$ and each $x \in X$, we write $((x_n); x) \in \mathscr{C}$ as $x_n \xrightarrow{\mathscr{C}} x$; this reads:

(x_n), $\mathscr{C}$-converges to x (also referred to as: x is the $\mathscr{C}$*-limit* of (x_n)).

The set of all such x is denoted $\lim_n(x_n)$; when it is nonempty, we say that (x_n) is $\mathscr{C}$*-convergent*. The following condition is to be optionally considered here:

(conv-3) $\mathscr{C}$ is separated:
$\lim_n(x_n)$ is an asingleton, for each sequence (x_n);

when it holds, $x_n \xrightarrow{\mathscr{C}} z$ will be also written as $\lim_n(x_n) = z$.

Further, by a (sequential) *Cauchy structure* on X we shall mean any part $\mathscr{H}$ of $\mathscr{S}(X)$ with (cf. Turinici [49])

(Cauchy-1) $\mathscr{H}$ is *hereditary*:
$(x_n) \in \mathscr{H} \Longrightarrow (y_n) \in \mathscr{H}$, for each subsequence (y_n) of (x_n)
(Cauchy-2) $\mathscr{H}$ is *reflexive*: for each $u \in X$,
the constant sequence $(x_n = u; n \geq 0)$ fulfills $(x_n) \in \mathscr{H}$.

Each element of $\mathscr{H}$ will be referred to as a $\mathscr{H}$*-Cauchy* sequence in X.

Finally, given the couple $(\mathscr{C}, \mathscr{H})$ as before, we shall say that it is a *conv-Cauchy structure* on X. The optional conditions about the conv-Cauchy structure $(\mathscr{C}, \mathscr{H})$ to be considered here are

(CC-1) $(\mathscr{C}, \mathscr{H})$ is *regular*: each $\mathscr{C}$-convergent sequence is $\mathscr{H}$-Cauchy
(CC-2) $(\mathscr{C}, \mathscr{H})$ is *complete*: each $\mathscr{H}$-Cauchy sequence is $\mathscr{C}$-convergent.

A standard way of introducing such structures is the *(pseudo) metrical* one. By a *pseudometric* over X we shall mean any map $d : X \times X \to R_+$. Fix such an object; with, in addition,

(r-s) d is *reflexive sufficient*: $x = y \Longleftrightarrow d(x, y) = 0$;

in this case, (X, d) is called a *rs-pseudometric space*. Given the sequence (x_n) in X and the point $x \in X$, we say that (x_n), *d-converges* to x (written as: $x_n \overset{d}{\longrightarrow} x$) provided $d(x_n, x) \to 0$ as $n \to \infty$; i.e.,

$\forall \varepsilon > 0, \exists i = i(\varepsilon)\text{: } i \leq n \Longrightarrow d(x_n, x) < \varepsilon.$

By this very definition, we have the hereditary and reflexive properties:

(d-conv-1) $(\overset{d}{\longrightarrow})$ is hereditary:
$x_n \overset{d}{\longrightarrow} x$ implies $y_n \overset{d}{\longrightarrow} x$, for each subsequence (y_n) of (x_n)
(d-conv-2) $(\overset{d}{\longrightarrow})$ is reflexive: for each $u \in X$,
the constant sequence $(x_n = u; n \geq 0)$ fulfills $x_n \overset{d}{\longrightarrow} u$;

hence, $(\overset{d}{\longrightarrow})$ is a sequential convergence on X. The set of all such limit points of (x_n) will be denoted $\lim_n(x_n)$; if it is nonempty, then (x_n) is called *d-convergent*. Finally, note that $(\overset{d}{\longrightarrow})$ is not separated, in general. However, this property holds, provided (in addition)

(sym) d is *symmetric*: $d(x, y) = d(y, x)$, for all $x, y \in X$
(tri) d is triangular: $d(x, y) \leq d(x, z) + d(z, y), \forall x, y, z \in X$;

i.e.: when d is a *metric* on X.

Further, call the sequence (x_n), *d-Cauchy* when $d(x_m, x_n) \to 0$ as $m, n \to \infty$, $m < n$; i.e.,

$\forall \varepsilon > 0, \exists j = j(\varepsilon)\text{: } j \leq m < n \Longrightarrow d(x_m, x_n) < \varepsilon$;

the class of all these will be denoted as $Cauchy(d)$. As before, we have the hereditary and reflexive properties

(d-Cauchy-1) $Cauchy(d)$ is hereditary: (x_n) is d-Cauchy
implies (y_n) is d-Cauchy, for each subsequence (y_n) of (x_n)
(d-Cauchy-2) $Cauchy(d)$ is reflexive: for each $u \in X$,
the constant sequence $(x_n = u; n \geq 0)$ is d-Cauchy;

hence, $Cauchy(d)$ is a Cauchy structure on X.

Now, the couple $((\overset{d}{\longrightarrow}), Cauchy(d))$ will be referred to as a *conv-Cauchy structure* on X generated by d. Note that, by the imposed (upon d) conditions, this conv-Cauchy structure is not (regular or complete), in general. But, when d is symmetric triangular (hence, a metric) the regularity condition holds, as it can be directly seen.

We close this section with a few remarks involving convergent real sequences. For each sequence (r_n) in R, and each element $r \in R$, denote

$r_n \to r+$ (resp., $r_n \to r-$), when $r_n \to r$ and [$r_n > r$ (resp., $r_n < r$), $\forall n$].

Proposition 3.5 *Let the sequence $(r_n; n \geq 0)$ in R and the number $\varepsilon \in R$ be such that $r_n \to \varepsilon+$. Then, there exists a subsequence $(r_n^* := r_{i(n)}; n \geq 0)$ of $(r_n; n \geq 0)$ with the properties*

$(r_n^*; n \geq 0)$ is strictly descending and $r_n^* \to \varepsilon+$.

Proof Put $i(0) = 0$. As $\varepsilon < r_{i(0)}$ and $r_n \to \varepsilon+$, we have that

$A(i(0)) := \{n > i(0); r_n < r_{i(0)}\}$ is not empty;
hence, $i(1) := \min(A(i(0)))$ is an element of it, and $r_{i(1)} < r_{i(0)}$.

Likewise, as $\varepsilon < r_{i(1)}$ and $r_n \to \varepsilon+$, we have that

$A(i(1)) := \{n > i(1); r_n < r_{i(1)}\}$ is not empty;
hence, $i(2) := \min(A(i(1)))$ is an element of it, and $r_{i(2)} < r_{i(1)}$.

This procedure may continue indefinitely, and yields (without any choice technique) a strictly ascending rank sequence $(i(n); n \geq 0)$ (hence, $i(n) \to \infty$ as $n \to \infty$) for which the attached subsequence $(r_n^* := r_{i(n)}; n \geq 0)$ of $(r_n; n \geq 0)$ fulfills

$r_{n+1}^* < r_n^*$, for all n; hence, (r_n^*) is (strictly) descending.

On the other hand, by this very subsequence property,

$(r_n^* > \varepsilon, \forall n)$, and $\lim_n r_n^* = \lim_n r_n = \varepsilon$.

Putting these together, we get the desired fact.

A bi-dimensional counterpart of these facts may be given along the lines below. Let $\pi(t, s)$ (where $t, s \in R$) be a logical property involving pairs or real numbers. Given the couple of real sequences $(t_n; n \geq 0)$ and $(s_n; n \geq 0)$, call the subsequences $(t_n^*; n \geq 0)$ of (t_n) and $(s_n^*; n \geq 0)$ of (s_n), *compatible* when

$(t_n^* = t_{i(n)} n \geq 0)$, and $(s_n^* = s_{i(n)}; n \geq 0)$, for the same strictly ascending rank sequence $(i(n); n \geq 0)$.

Proposition 3.6 *Let the couple of real sequences $(t_n; n \geq 0)$, $(s_n; n \geq 0)$ and the pair of real numbers (a, b) be such that*

$t_n \to a+$, $s_n \to b+$ as $n \to \infty$ and ($\pi(t_n, s_n)$ is true, $\forall n$).

There exists then a compatible couple of subsequences $(t_n^*; n \geq 0)$ of $(t_n; n \geq 0)$ and $(s_n^*; n \geq 0)$ of $(s_n; n \geq 0)$, respectively, with

(32-1) $(t_n^*; n \geq 0)$ and $(s_n^*; n \geq 0)$ are strictly descending, compatible
(32-2) ($t_n^* \to a+$, $s_n^* \to b+$, as $n \to \infty$), and ($\pi(t_n^*, s_n^*)$ holds, for all n).

Proof By the preceding statement, $(t_n; n \geq 0)$ admits a subsequence $(T_n := t_{i(n)}; n \geq 0)$, with

$(T_n; n \geq 0)$ is strictly descending, and ($T_n \to a+$, as $n \to \infty$).

Denote $(S_n := s_{i(n)}; n \geq 0)$; clearly,

$(S_n; n \geq 0)$ is a subsequence of $(s_n; n \geq 0)$ with $S_n \to b+$ as $n \to \infty$.

Moreover, by this very construction

$\pi(T_n, S_n)$ holds, for all n.

Again by the statement above, there exists a subsequence $(s_n^* := S_{j(n)} = s_{i(j(n))}; n \geq 0)$ of $(S_n; n \geq 0)$ (hence, of $(s_n; n \geq 0)$ as well), with

$(s_n^*; n \geq 0)$ is strictly descending, and $(s_n^* \to b+$, as $n \to \infty)$.

Denote further $(t_n^* := T_{j(n)} = t_{i(j(n))}; n \geq 0)$; this is a subsequence of $(T_n; n \geq 0)$ (hence, of $(t_n; n \geq 0)$ as well), with

$(t_n^*; n \geq 0)$ is strictly descending, and $(t_n^* \to a+$, as $n \to \infty)$;

Finally, by this very construction (and a previous relation)

$\pi(t_n^*, s_n^*)$ holds, for all n.

Summing up, the couple of subsequences $(t_n^*; n \geq 0)$ and $(s_n^*; n \geq 0)$ has all needed properties; and the conclusion follows.

Note that further extensions of this result are possible, in the framework of quasi-metric spaces, taken as in Hitzler [14, Ch 1, Sect 1.2]; we shall discuss them in a separate paper.

4 Admissible Functions

Let $R_+^0 :=]0, \infty[$ stand for the class of strictly positive real numbers. Denote by $\mathscr{F}(re)(R_+^0, R)$ the family of all functions $\varphi \in \mathscr{F}(R_+^0, R)$ with

φ is *regressive*: $\varphi(t) < t$, for all $t \in R_+^0$.

For each $\varphi \in \mathscr{F}(re)(R_+^0, R)$, let us introduce the sequential properties

(n-d-a) φ is *non-diagonally admissible*:
there are no strictly descending sequences $(t_n; n \geq 0)$ in R_+^0
and no elements $\varepsilon \in R_+^0$ with $t_n \to \varepsilon+$, $\varphi(t_n) \to \varepsilon+$
(M-a) φ is *Matkowski admissible*:
for each $(t_n; n \geq 0)$ in R_+^0 with $(t_{n+1} \leq \varphi(t_n), \forall n)$ we have $\lim_n t_n = 0$
(str-M-a) φ is *strongly Matkowski admissible*:
for each $(t_n; n \geq 0)$ in R_+^0 with $(t_{n+1} \leq \varphi(t_n), \forall n)$ we have $\sum_n t_n < \infty$.

(The conventions (M-a) and (str-M-a) are taken from Matkowski [29] and Turinici [50], respectively.) Clearly, for each $\varphi \in \mathscr{F}(re)(R_+^0, R)$,

strongly Matkowski admissible implies Matkowski admissible;

but the converse is not in general true.

To get concrete circumstances under which such properties hold, we need some conventions. Given $\varphi \in \mathscr{F}(re)(R_+^0, R)$, let us introduce the global properties

(R-a) φ is *Rhoades admissible*:
for each $\varepsilon > 0$, there exists $\delta > 0$, such that
$t, s > 0$, $t \leq \varphi(s)$ and $\varepsilon < s < \varepsilon + \delta$ imply $t \leq \varepsilon$
(MK-a) φ is *Meir–Keeler admissible*:
for each $\varepsilon > 0$, there exists $\delta > 0$, such that
$\varepsilon < s < \varepsilon + \delta$ implies $\varphi(s) \leq \varepsilon$.

Theorem 4.2 *For each $\varphi \in \mathscr{F}(re)(R_+^0, R)$, we have in (ZF-AC+DC)*

(41-a) (MK-a) $\Longrightarrow$ (R-a) $\Longrightarrow$ (M-a)
(41-b) (M-a) $\Longrightarrow$ (n-d-a) $\Longrightarrow$ (MK-a).

Hence, for each $\varphi \in \mathscr{F}(re)(R_+^0, R)$, the properties (MK-a), (R-a), (M-a), (n-d-a) are equivalent to each other.

Proof

i) Suppose that φ is Meir–Keeler admissible; we claim that φ is Rhoades admissible, with the same couple (ε, δ). In fact, let $\varepsilon > 0$ be given; and $\delta > 0$ be assured by the Meir–Keeler admissible property. Take the numbers $t, s > 0$ with $t \leq \varphi(s)$ and $\varepsilon < s < \varepsilon + \delta$. As φ is Meir–Keeler admissible, this yields $\varphi(s) \leq \varepsilon$; wherefrom (as $t \leq \varphi(s)$), we must have $t \leq \varepsilon$; hence the claim.
ii) Suppose that φ is Rhoades admissible; we have to establish that φ is Matkowski admissible. Let $(s_n; n \geq 0)$ be a sequence in R_+^0 with the property $(s_{n+1} \leq \varphi(s_n); n \geq 0)$. Clearly, (s_n) is strictly descending in R_+^0; hence, $\sigma := \lim_n s_n$ exists in R_+. Suppose by contradiction that $\sigma > 0$; and let $\rho > 0$ be given by the Rhoades admissible property of φ. By the above convergence relations, there exists some rank $n(\rho)$, such that

   $$n \geq n(\rho) \text{ implies } \sigma < s_n < \sigma + \rho.$$

   But then, under the notation $(t_n := \varphi(s_n); n \geq 0)$, we get (for the same ranks)

   $$\sigma < s_{n+1} \leq t_n < s_n < \sigma + \rho;$$

   in contradiction with the Rhoades admissible property. Hence, necessarily, $\sigma = 0$; and conclusion follows.
iii) Suppose that φ is Matkowski admissible; we assert that φ is non-diagonally admissible. For, if φ is not endowed with such a property, there must be a strictly descending sequence $(t_n; n \geq 0)$ in R_+^0 and an $\varepsilon > 0$, such that

   $$t_n \to \varepsilon+ \text{ and } \varphi(t_n) \to \varepsilon+, \text{ as } n \to \infty.$$

   Put $i(0) = 0$. As $\varepsilon < \varphi(t_{i(0)})$ and $t_n \to \varepsilon+$, we have that

   $A(i(0)) := \{n > i(0); t_n < \varphi(t_{i(0)})\}$ is not empty;
   hence, $i(1) := \min(A(i(0)))$ is an element of it, and $t_{i(1)} < \varphi(t_{i(0)})$.

Likewise, as $\varepsilon < \varphi(t_{i(1)})$ and $t_n \to \varepsilon+$, we have that

$A(i(1)) := \{n > i(1); t_n < \varphi(t_{i(1)})\}$ is not empty;
hence, $i(2) := \min(A(i(1)))$ is an element of it, and $t_{i(2)} < \varphi(t_{i(1)})$.

This procedure may continue indefinitely; and yields (without any choice technique) a strictly ascending rank sequence $(i(n); n \geq 0)$ (hence, $i(n) \to \infty$ as $n \to \infty$) for which the attached subsequence $(s_n := t_{i(n)}; n \geq 0)$ of (t_n) fulfills

$s_{n+1} < \varphi(s_n)(< s_n)$, for all n.

On the other hand, by this very subsequence property,

$(s_n > \varepsilon, \forall n)$ and $\lim_n s_n = \lim_n t_n = \varepsilon$.

The obtained relations are in contradiction with the Matkowski property of φ; hence, the working condition cannot be true; and we are done.

iv) Suppose that φ is non-diagonally admissible; we show that, necessarily, φ is Meir–Keeler admissible. For, if φ is not endowed with such a property, we must have (for some $\varepsilon > 0$)

$H(\delta) := \{t \in R_+^0; \varepsilon < t < \varepsilon + \delta, \varphi(t) > \varepsilon\}$ is not empty, for each $\delta > 0$.

Taking a strictly descending sequence $(\delta_n; n \geq 0)$ in R_+^0 with $\delta_n \to 0$, we get by the Denumerable Axiom of Choice (AC(N)) [deductible, as precise, in (ZF-AC+DC)], a sequence $(t_n; n \geq 0)$ in R_+^0, so as

$(\forall n)$: t_n is an element of $H(\delta_n)$;

or, equivalently (by the very definition above and φ=regressive)

$(\forall n)$: $\varepsilon < \varphi(t_n) < t_n < \varepsilon + \delta_n$;
hence, in particular: $\varphi(t_n) \to \varepsilon+$ and $t_n \to \varepsilon+$.

By a previous result, there exists a subsequence $(r_n := t_{i(n)}; n \geq 0)$ of $(t_n; n \geq 0)$, such that

(r_n) is strictly descending and $r_n \to \varepsilon+$; hence, necessarily, $\varphi(r_n) \to \varepsilon+$.

But, this relation is in contradiction with the non-diagonal admissible property of our function. Hence, the assertion follows; and we are done.

In the following, some sequential counterparts of the Rhoades admissible and Meir–Keeler admissible properties are provided.

For any $\varphi \in \mathscr{F}(re)(R_+^0, R)$ and any $s \in R_+^0$, put

$\Lambda^+\varphi(s) = \inf_{\varepsilon > 0} \Phi(s+)(\varepsilon)$, where $\Phi(s+)(\varepsilon) = \sup \varphi(]s, s+\varepsilon[)$, $\varepsilon > 0$.

From the regressiveness of φ,

$-\infty \leq \Lambda^+\varphi(s) \leq s, \quad \forall s \in R_+^0$;

but the case of these extremal values being attained cannot be avoided.

The following consequence of this convention will be useful.

Proposition 4.7 *Let $\varphi \in \mathscr{F}(re)(R^0_+, R)$ and $s \in R^0_+$ be arbitrary fixed. Then,*

(41-1) $\limsup_n(\varphi(t_n)) \le \Lambda^+\varphi(s)$,
for each sequence (t_n) in R^0_+ with $t_n \to s+$
(41-2) there exists a strictly descending sequence (r_n) in R^0_+ with $r_n \to s+$ and $\varphi(r_n) \to \Lambda^+\varphi(s)$.

Proof Denote, for simplicity,

$\alpha = \Lambda^+\varphi(s)$; hence, $\alpha = \inf_{\varepsilon>0} \Phi(s+)(\varepsilon)$, and $-\infty \le \alpha \le s$.

i) Given $\varepsilon > 0$, there exists a rank $p(\varepsilon) \ge 0$ such that $s < t_n < s + \varepsilon$, for all $n \ge p(\varepsilon)$; hence

$$\limsup_n(\varphi(t_n)) \le \sup\{\varphi(t_n); n \ge p(\varepsilon)\} \le \Phi(s+)(\varepsilon).$$

Passing to infimum over $\varepsilon > 0$ yields (see above)

$$\limsup_n(\varphi(t_n)) \le \inf_{\varepsilon>0} \Phi(s+)(\varepsilon) = \alpha;$$

and the claim follows.

ii) Define $(\beta_n := \Phi(s+)(2^{-n}); n \ge 0)$; clearly,

(p-1) $(\beta_n; n \ge 0)$ is a sequence in $]-\infty, s+1] \subseteq R$; because $(\forall n)$: $-\infty < \varphi(t) < t < s + 2^{-n} \le s + 1$, whenever $s < t < s + 2^{-n}$
(p-2) (β_n) is descending, $(\beta_n \ge \alpha, \ \forall n)$, $\inf_n \beta_n = \alpha$; hence $\lim_n \beta_n = \alpha$.

By these properties, there may be constructed a sequence $(\gamma_n; n \ge 0)$ in R, with

$$\gamma_n < \beta_n, \ \forall n; \ \lim_n \gamma_n = \lim_n \beta_n = \alpha.$$

(For example, we may take $(\gamma_n = \beta_n - 3^{-n}; n \ge 0)$; but this is not the only possible choice.) Let $n \ge 0$ be arbitrary fixed. By the supremum definition, there exists $t_n \in]s, s + 2^{-n}[$ such that $\varphi(t_n) > \gamma_n$; moreover (again by definition), $\varphi(t_n) \le \beta_n$. Putting these together yields

$s < t_n < s + 2^{-n}$, $\gamma_n < \varphi(t_n) \le \beta_n$, for all n;
whence, $t_n \to s+$ and $\varphi(t_n) \to \alpha$, as $n \to \infty$.

By a previous result, there exists a subsequence $(r_n := t_{i(n)}; n \ge 0)$ of $(t_n; n \ge 0)$, such that

(r_n) is strictly descending and $r_n \to \varepsilon+$; hence, $\varphi(r_n) \to \alpha$.

In other words: the obtained sequence $(r_n; n \ge 0)$ has all needed properties; wherefrom, the conclusion follows.

Call $\varphi \in \mathscr{F}(re)(R^0_+, R)$, *Boyd–Wong admissible* [5], if

(BW-a) $\Lambda^+\varphi(s) < s$, for all $s > 0$.

In particular, $\varphi \in \mathscr{F}(re)(R^0_+, R)$ is Boyd–Wong admissible provided

φ is *upper semicontinuous at the right* on R_+^0:
$\Lambda^+\varphi(s) \leq \varphi(s)$, for each $s \in R_+^0$.

This, e.g., is fulfilled when φ is *continuous at the right* on R_+^0; for, in such a case,

$\Lambda^+\varphi(s) = \varphi(s)$, for each $s \in R_+^0$.

Some sequential counterparts of this convention may be described along the lines below. Call $\varphi \in \mathscr{F}(re)(R_+^0, R)$, *sequentially Boyd–Wong admissible* provided

(s-BW-a) for each strictly descending sequence $(t_n; n \geq 0)$ in R_+^0 and each $\varepsilon > 0$ with $t_n \to \varepsilon+$, we have $\limsup_n \varphi(t_n) < \varepsilon$.

Further, call $\varphi \in \mathscr{F}(re)(R_+^0, R)$, *sequentially Rhoades admissible* provided

(s-R-a) for each strictly descending sequence $(t_n; n \geq 0)$ in R_+^0 and each $\varepsilon > 0$ with $t_n \to \varepsilon+$, we have $\liminf_n \varphi(t_n) < \varepsilon$.

Theorem 4.3 *The following inclusions are valid, in (ZF-AC+DC), for a generic function $\varphi \in \mathscr{F}(re)(R_+^0, R)$*

(42-a) (BW-a) $\Longrightarrow$ (s-BW-a) $\Longrightarrow$ (BW-a); hence, (BW-a) $\Longleftrightarrow$ (s-BW-a)
(42-b) (s-BW-a) $\Longrightarrow$ (s-R-a) $\Longrightarrow$ (s-BW-a); so, (s-BW-a) $\Longleftrightarrow$ (s-R-a)
(42-c) (s-R-a) $\Longrightarrow$ (n-d-a), (s-R-a) $\Longrightarrow$ (R-a), (s-R-a) $\Longrightarrow$ (MK-a)
(42-d) (s-BW-a) $\Longrightarrow$ (n-d-a), (s-BW-a) $\Longrightarrow$ (R-a), (s-BW-a) $\Longrightarrow$ (MK-a).

Proof

i) The first half is immediate, by the preceding statement. This is also true for the second half of the same; however, for completeness reasons, we shall supply an argument. Suppose that (BW-a) is not true. As φ=regressive, we have

$\Lambda^+\varphi(\varepsilon) = \varepsilon$, for some $\varepsilon > 0$.

By the quoted auxiliary fact, there exists a strictly descending sequence (r_n) in R_+^0, with the properties

$r_n \to \varepsilon+$ and $(\limsup_n \varphi(r_n) =) \lim_n \varphi(r_n) = \Lambda^+\varphi(\varepsilon) = \varepsilon$.

But then, φ does not satisfy (s-BW-a); contradiction; hence the assertion.

ii) The first half of this chain is immediate, by definition. For the second part of the same, one may proceed as below. Suppose by contradiction that $\varphi \in \mathscr{F}(re)(R_+^0, R)$ is not sequentially Boyd–Wong admissible; hence,

$\limsup_n \varphi(t_n) = \varepsilon$, for some strictly descending sequence $(t_n; n \geq 0)$ in R_+^0 and some $\varepsilon > 0$ with $t_n \to \varepsilon+$.

Combining with

$\varepsilon = \limsup_n \varphi(t_n) \leq \Lambda^+\varphi(\varepsilon) \leq \varepsilon$ (see above),

one derives $\Lambda^+\varphi(\varepsilon) = \varepsilon$. This, by a previous auxiliary fact, yields

$(\liminf_n \varphi(r_n) =) \lim_n \varphi(r_n) = \varepsilon$, for some strictly descending sequence $(r_n; n \geq 0)$ in R^0_+ with $r \to \varepsilon+$.

But then, the sequential Rhoades admissible property of φ will be contradicted. Hence, our working assumption is not true; and the assertion follows.

iii) Let us establish the first relation in this series. Suppose by contradiction that (n-d-a) is not holding. By definition, there must be a strictly descending sequence $(t_n; n \geq 0)$ in R^0_+ and a number $\varepsilon > 0$, such that

$$t_n \to \varepsilon+, \varphi(t_n) \to \varepsilon+, \text{ as } n \to \infty.$$

This, along with

$$\liminf_n \varphi(t_n) = \limsup_n \varphi(t_n) = \varepsilon,$$

contradicts the property (s-R-a). Hence, our working assumption is not true; and the conclusion is clear. The remaining relations follow at once in view of

(n-d-a) $\Longleftrightarrow$ (R-a) $\Longleftrightarrow$ (MK-a) (see above);

however, for completeness we provide a direct proof of the second one. Suppose that $\varphi \in \mathscr{F}(re)(R^0_+, R)$ is sequentially Rhoades admissible; we have to establish that φ is Rhoades admissible. Suppose not; that is, for some $\varepsilon > 0$,

$H(\delta) := \{(t, s) \in R^0_+ \times R^0_+; t \leq \varphi(s), \varepsilon < s < \varepsilon + \delta, t > \varepsilon\}$
is not empty, for each $\delta > 0$.

Taking a strictly descending sequence $(\delta_n; n \geq 0)$ in R^0_+ with $\delta_n \to 0$, we get by the Denumerable Axiom of Choice (AC(N)) [deductible, as precise, in (ZF-AC+DC)], a sequence $((t_n, s_n); n \geq 0)$ in R^0_+, so as

$(\forall n)$: (t_n, s_n) is an element of $H(\delta_n)$;

or, equivalently (by the very definition above)

$$(\forall n): \varepsilon < t_n \leq \varphi(s_n) < s_n < \varepsilon + \delta_n.$$

As a direct consequence of this, we have

$$t_n \to \varepsilon+, s_n \to \varepsilon+, \text{ as } n \to \infty.$$

On the other hand,

$\varepsilon < \varphi(s_n) < \varepsilon + \delta_n$, for all n implies $\varepsilon = \lim_n \varphi(s_n) = \liminf_n \varphi(s_n)$.

By a previous result, there exists a subsequence $(s^*_n := s_{i(n)}; n \geq 0)$ of $(s_n; n \geq 0)$, such that

(s^*_n) is strictly descending and $s^*_n \to \varepsilon+$;
hence, necessarily, $(\liminf_n \varphi(s^*_n) =) \lim_n \varphi(s^*_n) = \varepsilon$;

in contradiction with the sequential Rhoades admissible property of φ. The remaining inclusions are obtainable in a similar way.

In the following, a special version of these facts is formulated, in terms of increasing functions from the class $\mathscr{F}(re)(R^0_+, R)$.

Let $\mathscr{F}(re, in)(R^0_+, R)$ stand for the class of all $\varphi \in \mathscr{F}(re)(R^0_+, R)$, with

φ is increasing on R^0_+ $(0 < t_1 \le t_2$ implies $\varphi(t_1) \le \varphi(t_2))$.

The following characterization of our previous Matkowski properties imposed to φ is now available. Given $t > 0$, let

$$\varphi^0(t) = t, \varphi^1(t) = \varphi(t), \ldots, \varphi^{n+1}(t) = \varphi(\varphi^n(t)) \ (n \ge 0)$$

stand for the iterates sequence of φ at this point. Note that such a construction may be non-effective; for, e.g.,

$\varphi^2(t) = \varphi(\varphi(t))$ is undefined whenever $\varphi(t) \le 0$.

Proposition 4.8 *For each $\varphi \in \mathscr{F}(re, in)(R^0_+, R)$, we have*

(42-1) φ is Matkowski admissible, iff
$(\forall t > 0)$: $\lim_n \varphi^n(t) = 0$, whenever $(\varphi^n(t); n \ge 0)$ exists
(42-2) φ is strongly Matkowski admissible, iff
$(\forall t > 0)$: $\sum_n \varphi^n(t) < \infty$, whenever $(\varphi^n(t); n \ge 0)$ exists.

The proof is immediate, by the increasing property of φ and

$(\forall t > 0)$: $(\varphi^n(t); n \ge 0)$ exists iff $\{\varphi^n(t); n \ge 0\} \subseteq R^0_+$;

so, further details are not necessary.

As before, we need sufficient conditions (involving the class $\mathscr{F}(re, in)(R^0_+, R)$) under which this property holds. For each $\varphi \in \mathscr{F}(re, in)(R^0_+, R)$, denote

$\varphi(s + 0) := \lim_{t \to s+} \varphi(t)$, $s \in R^0_+$ (the *right limit* of φ at s);

clearly, the following evaluation holds

$\varphi(s) \le \varphi(s + 0) \le s$, for all $s > 0$.

Proposition 4.9 *Suppose that the function $\varphi \in \mathscr{F}(re, in)(R^0_+, R)$ fulfills*

φ is strongly regressive: $\varphi(s + 0) < s$, for each $s > 0$.

Then, φ is Matkowski admissible.

Proof Clearly,

$\forall \varphi \in \mathscr{F}(re, in)(R^0_+, R)$: $\Lambda^+\varphi(t) = \varphi(t + 0)$, $\forall t > 0$;
whence: strongly regressive implies Boyd–Wong admissible.

But then, the desired fact follows at once from a previous one, developed over the class $\mathscr{F}(re)(R^0_+, R)$; however, for completeness reasons, we shall provide an argument for this. Given $s_0 > 0$, assume that the iterative sequence $(s_n = \varphi^n(s_0); n \ge 0)$ exists (in R^0_+). By the regressive property of φ, (s_n) is strictly descending; hence, $s := \lim_n s_n$ exists, with (in addition) $s_n > s$, for all n. Suppose by contradiction that $s > 0$. Combining with

$$\varphi(s+0) = \lim_n \varphi(s_n) = \lim_n s_{n+1},$$

yields $\varphi(s+0) = s$; contradiction. Hence, $s = 0$; and we are done.

Remark 1 The reverse inclusion is not in general true. Indeed, let us consider the function $\varphi \in \mathscr{F}(re, in)(R^0_+, R)$, according to (for some $r > 0$):

$$(\varphi(t) = t/2, \text{ if } t \leq r), (\varphi(t) = r, \text{ if } t > r).$$

Clearly, φ is Matkowski admissible, as it can be directly seen. On the other hand,

$$\varphi(r+0) = r; \text{ whence, } \varphi \text{ is not strongly regressive;}$$

and this proves our claim. For an extended example of this type, see Turinici [45] and the references therein.

Now, it is natural to establish the connection between the introduced class and the Meir–Keeler one. An appropriate answer to this is contained in

Theorem 4.4 *Under these conventions, we have, for each $\varphi \in \mathscr{F}(re, in)(R^0_+, R)$: Matkowski admissible is equivalent with Meir–Keeler admissible.*

Proof The verification of this fact is already performed over the class $\mathscr{F}(re)(R^0_+, R)$; however, for completeness reasons, we shall give a (different) reasoning for it.

i) (cf. Jachymski [15]). Assume that $\varphi \in \mathscr{F}(re, in)(R^0_+, R)$ is Matkowski admissible; we have to establish that it is Meir–Keeler admissible. If the underlying property fails, then (for some $\gamma > 0$):

$$\forall \beta > 0, \exists t \in]\gamma, \gamma + \beta[, \text{ such that } \varphi(t) > \gamma.$$

As φ is increasing, this yields (by the arbitrariness of β)

$$(\varphi(t) > \gamma, \forall t > \gamma); \text{ whence, by induction: } (\varphi^n(t) > \gamma, \forall n, \forall t > \gamma).$$

Taking some $t > \gamma$ and passing to limit as $n \to \infty$, one gets $0 \geq \gamma$; contradiction.

ii) Assume that $\varphi \in \mathscr{F}(re, in)(R^0_+, R)$ is Meir–Keeler admissible; we have to establish that it is Matkowski admissible. Given $s_0 > 0$, suppose that the iterative sequence $(s_n = \varphi^n(s_0); n \geq 0)$ exists (in R^0_+). By the regressive property of φ, (s_n) is strictly descending; hence, $s := \lim_n s_n$ exists, with (in addition) $s_n > s$, for all n. Suppose by contradiction that $s > 0$; and let $r > 0$ be the number assured by the Meir–Keeler admissible property of φ. By definition, there exists a rank $n(r) \geq 0$, such that

$$n \geq n(r) \text{ implies } s < s_n < s + r.$$

This, by the underlying property, gives (for the same ranks)

$$s < s_{n+1} = \varphi(s_n) \leq s; \text{ contradiction.}$$

Hence, $s = 0$; wherefrom φ is Matkowski admissible.

We close this section with a lot of criteria for the strong Matkowski admissible property. Given $\varphi \in \mathcal{F}(re)(R_+^0, R)$, let us associate it the function

$(g \in \mathcal{F}(R_+^0))$: $g(t) = t/(t - \varphi(t)), t > 0$; in short: $g = I/(I - \varphi)$,

where $(I(t) = t; t > 0)$ is the *identical function* of $\mathcal{F}(R_+^0)$. Now, call the function $h \in \mathcal{F}(R_+^0)$ *globally normal* provided

(gn-1) $h(.)$ is decreasing on R_+^0
(gn-2) $H(t) := \int_0^t h(\xi)d\xi < \infty$, for each $t > 0$.

Note that, by the former condition (gn-1),

$\int_0^t h(\xi)d\xi := \lim_{s\to 0+} \int_s^t g(\xi)d\xi$ exists in $R_+ \cup \{\infty\}$, for each $t > 0$;

so, the latter condition (gn-2) is meaningful; moreover,

$H(.)$ is strictly increasing on R_+^0 $(t_1 < t_2 \Longrightarrow H(t_1) < H(t_2))$.

Further, let us say that $g \in \mathcal{F}(R_+^0)$ is *globally subnormal*, provided

$g(t) \leq h(t), t \in R_+^0$, where $h \in \mathcal{F}(R_+^0)$ is globally normal.

Proposition 4.10 *Let the function $\varphi \in \mathcal{F}(re)(R_+^0, R)$ be such that*

the associated function $g = I/(I - \varphi)$ is globally subnormal.

Then, φ is strongly Matkowski admissible (see above).

Proof By the imposed condition,

$g(t) \leq h(t), t \in R_+$, for some globally normal function $h \in \mathcal{F}(R_+^0)$.

Moreover, by the very definitions above,

$H(.)$ is continuous on R_+^0 and $H(0+) := \lim_{t\to 0+} H(t) = 0$.

Let the sequence $(t_n; n \geq 0)$ in R_+^0 be such that

$t_{n+1} \leq \varphi(t_n)$, for all $n \geq 0$; clearly, $(t_n; n \geq 0)$ is strictly descending.

Further, let $i \geq 0$ be arbitrary fixed. By the above choice,

$t_i - \varphi(t_i) \leq t_i - t_{i+1}$; whence, $1 \leq (t_i - t_{i+1})/(t_i - \varphi(t_i))$.

Combining with $g \leq h$ and the decreasing property of $h(.)$ yields

$t_i \leq (t_i - t_{i+1})g(t_i) \leq (t_i - t_{i+1})h(t_i) \leq H(t_i) - H(t_{i+1})$.

Passing to limit in the relation involving extremal members gives

$(\lim_n t_n = 0$, and$)$ $\sum_n t_n \leq H(t_0) - H(0+) < \infty$;

i.e.: the series $\sum_n t_n$ converges. The proof is complete.

In particular, when g is globally normal, this result is just the related one in Altman [1].

A local version of this result may be given along the lines below. Given $h \in \mathcal{F}(R_+^0)$, $\gamma \in]0, \infty]$, let us say that h is γ*-locally normal* provided

(locn-1) $h(.)$ is decreasing on $]0, \gamma[$
(locn-2) $H(t) := \int_0^t h(\xi)d\xi < \infty$, for each $t \in]0, \gamma[$.

Note that, by the former condition (locn-1),

$\int_0^t h(\xi)d\xi := \lim_{s\to 0+} \int_s^t g(\xi)d\xi$ exists in $R_+ \cup \{\infty\}$, for each $t \in]0, \gamma[$;

so, the latter condition (locn-2) is meaningful; moreover,

$H(.)$ is strictly increasing on $]0, \gamma[$ $(t_1 < t_2 < \gamma \Longrightarrow H(t_1) < H(t_2))$.

Given $g \in \mathscr{F}(R_+^0)$, $\gamma \in]0, \infty]$, let us say that g is *γ-locally subnormal*, provided

$g(t) \le h(t), t \in]0, \gamma[$, where $h \in \mathscr{F}(R_+^0)$ is γ-locally normal.

Proposition 4.11 *Let the function $\varphi \in \mathscr{F}(re)(R_+^0, R)$ be such that*

(45-i) φ is Boyd–Wong admissible ($\Lambda^+\varphi(s) < s$, for all $s > 0$)
(45-ii) the associated function $g = I/(I - \varphi)$ in $\mathscr{F}(R_+^0)$
is γ-locally subnormal, for some $\gamma \in]0, \infty]$.

Then, φ is strongly Matkowski admissible (see above).

Proof By the imposed condition,

$g(t) \le h(t), t \in]0, \gamma[$, for some γ-locally normal function $h \in \mathscr{F}(R_+^0)$.

Moreover, by the very definitions above,

$H(.)$ is continuous on $]0, \gamma[$ and $H(0+) := \lim_{t\to 0+} H(t) = 0$.

Let the sequence $(t_n; n \ge 0)$ in R_+^0 be such that

(iter) $t_{n+1} \le \varphi(t_n)$, for all $n \ge 0$;

clearly, $(t_n; n \ge 0)$ is strictly descending; whence, $\tau := \lim_n t_n$ exists in R_+. Suppose by contradiction that $\tau > 0$. Passing to superior limit in (iter) gives (by a previous auxiliary fact)

$\tau \le \limsup_n \varphi(t_n) \le \Lambda^+\varphi(\tau) < \tau$; contradiction.

Hence, the working hypothesis above is not acceptable; so that

$\tau = 0$; i.e.: $t_n \to 0$ as $n \to \infty$.

Note that, as a consequence of this,

$t_n < \gamma, \forall n \ge n(\gamma)$, for some $n(\gamma) \in N$.

Having these precise, let $i \ge n(\gamma)$ be arbitrary fixed. By the above choice,

$t_i - \varphi(t_i) \le t_i - t_{i+1}$; whence, $1 \le (t_i - t_{i+1})/(t_i - \varphi(t_i))$.

Combining with $g \le h$ and the decreasing property of $h(.)$ yields

$t_i \le (t_i - t_{i+1})g(t_i) \le (t_i - t_{i+1})h(t_i) \le H(t_i) - H(t_{i+1})$.

Taking the relation between extremal members gives (by the preceding fact)

$(\lim_n t_n = 0$, and$)\ \sum\{t_n; n \geq n(\gamma)\} \leq H(t_{n(\gamma)}) - H(0+) < \infty$;

whence: the series $\sum_n t_n$ converges. The proof is complete.

A useful particular case of this local result is the following. Let us say that $\varphi \in \mathscr{F}(re)(R^0_+, R)$ is *Reich admissible*, provided

(re-1) φ is Boyd–Wong admissible $(\Lambda^+\varphi(s) < s, \forall s > 0)$
(re-2) φ is *zero-contractive* $(\limsup_{t\to 0+}[\varphi(t)/t] < 1)$.

Proposition 4.12 *Let the function $\varphi \in \mathscr{F}(re)(R^0_+, R)$ be Reich admissible. Then, φ is strongly Matkowski admissible (see above).*

Proof Let $g = I/(I - \varphi)$ be the associated function; clearly, $g \in \mathscr{F}(R^0_+)$. By the zero-contractive property, there exists $\gamma > 0$ and $\lambda \in]0, 1[$, such that

(for each $t \in]0, \gamma[$): $\varphi(t) \leq \lambda t$; whence, $g(t) \leq \mu := 1/(1 - \lambda)$.

The function $h \in \mathscr{F}(R^0_+)$ defined as

$h(t) = \mu, 0 < t < \gamma; h(t) = g(t), t \geq \gamma$

is clearly γ-locally normal. This along with $(g(t) \leq h(t), t \in R^0_+)$ tells us that $g(.)$ is γ-locally subnormal; so that, the preceding statement is applicable to the underlying context. The proof is thereby complete.

Note, finally, that these results are not the most general in the area; but, they may be useful in practice. Further aspects are to be found in Timofte [44].

5 Main Result

Let X be a nonempty set. Take a metric $d : X \times X \to R_+$ on X; the couple (X, d) will be referred to as a *metric space*. Denote

$C(X)$=the class of (nonempty) d-closed parts of X.

Further, let the point to set distance in X be introduced as:

$d(x, Y) = \inf\{d(x, y); y \in Y\}, x \in X, Y \in \exp(X)$.

Note that, for each $Y \in \exp(X)$, we have

$x \mapsto d(x, Y)$ is *nonexpansive*:
$|d(x_1, Y) - d(x_2, Y)| \leq d(x_1, x_2), x_1, x_2 \in X$.

By a *d-closed multivalued map* over X, we mean any map $T \in \mathscr{F}(X, C(X))$. As usual, we identify T with its *graph* in $X \times X$; that is

$(x, y \in X)$: xTy iff $y \in Tx$.

Remember that $z \in X$ is a fixed point of T iff

$z \in Tz$ (i.e.: zTz); or, equivalently: $d(z, Tz) = 0$ (as Tz is d-closed);

the set of all such points will be denoted as $\mathrm{Fix}(T)$.

It is our aim in the sequel to determine elements of $\mathrm{Fix}(T)$ by means of orbital iterative sequences and contractive type techniques. Some preliminaries are in order. Let $x_0 \in X$ be arbitrary fixed. As the relation T is proper, it follows—by the Dependent Choice principle—that there may be constructed an orbital iterative sequence $(x_n; n \geq 0)$, according to the iterative type construction

$(x_n, x_{n+1}) \in T$ (i.e.: $x_{n+1} \in Tx_n$), for all $n \geq 0$;

this will be referred to as a (x_0, T)-*iterative sequence*.

The following directions under which our problem is to be solved (comparable with the ones in Rus [39, Ch 2, Sect 2.2]) will be considered:

np-0) We say that T is *fix-asingleton* when $\mathrm{Fix}(T)$ is an asingleton; and *fix-singleton* when $\mathrm{Fix}(T)$ is a singleton

np-2) We say that $x_0 \in X$ is a *Picard point* (modulo (d, T)), if each (x_0, T)-iterative sequence $(x_n; n \geq 0)$ is endowed with the d-Cauchy property; when this holds for all $x_0 \in X$, then T is called a *Picard map* (modulo d)

np-3) We say that $x_0 \in X$ is a *strongly Picard point* (modulo (d, T)), if each (x_0, T)-iterative sequence $(x_n; n \geq 0)$ is d-convergent and $\lim_n(x_n) \in \mathrm{Fix}(T)$; when this holds for all $x_0 \in X$, then T is called a *strongly Picard map* (modulo d).

The sufficient (regularity) conditions for such properties involve *orbitally* concepts. Namely, given the sequence $(z_n; n \geq 0)$ in X, define the property

$(z_n; n \geq 0)$ is *orbital*: $(z_n, z_{n+1}) \in T$, for all n.

reg-1) Call X, *orbitally d-complete*, provided

(z_n)=orbital d-Cauchy sequence implies (z_n) is d-convergent.

reg-2) Call T, *orbitally d-closed*, provided

(z_n)=orbital sequence, $z \in X$, and $z_n \xrightarrow{d} z$ imply $(z, z) \in T$.

Finally, when the orbital properties are ignored, these conventions may be written in the usual way; we do not give details.

As a completion of these, we describe the metrical type contractive conditions to be considered. Let $E : C(X) \times C(X) \to R_+(\infty) := R_+ \cup \{\infty\}$ be the mapping

$$E(Y, Z) = \sup\{d(y, Z); y \in Y\}, \; Y, Z \in C(X).$$

It is not hard to see (cf. Kuratowski [24, Ch II, Sect 21/VII]) that

(hp-1) (E is *reflexive*): $E(Y, Y) = 0$, $Y \in C(X)$
(hp-2) (E is *triangular*): $E(Y, W) \leq E(Y, Z) + E(Z, W)$, $Y, Z, W \in C(X)$
(hp-3) (E is *almost sufficient*): $E(Y, Z) = 0$ iff $Y \subseteq Z$.

The mapping $(Y, Z) \mapsto E(Y, Z)$ will be referred to as the *Hausdorff–Pompeiu almost generalized metric* of $C(X)$. This is a consequence of its associated map

$$D(Y, Z) = \max\{E(Y, Z), E(Z, Y)\}, Y, Z \in C(X)$$

having all properties of a generalized metric over $C(X)$ in the Luxemburg–Jung sense [18, 28], and usually referred to as the *Hausdorff–Pompeiu generalized metric* of the space $C(X)$.

Now, letting $M : X \times X \to R_+(\infty)$ be a mapping, define the concept

M is T*-compatible*: $E(Tx, Ty) \leq M(x, y)$, for all $x, y \in X$.

In particular, we have that

$(M_0(x, y) = E(Tx, Ty); x, y \in X)$ is T-compatible.

This is not the only mapping with such a property; because the mapping

$$M_1(x, y) = \max\{E(Tx, Ty), d(x, Tx), d(y, Ty)\}, x, y \in X$$

is also T-compatible, as it can be directly seen.

Fix in the following a T-compatible map $M : X \times X \to R_+(\infty)$. Then, let $G : X \times X \to R_+$ stand for the map

$$G(x, y) = \min\{d(x, Tx), d(y, Ty), d(x, y)\}, x, y \in X;$$

and $K : R_+^0 \times R_+^0 \times R_+^0 \to R$ be a real function; for simplicity, we write $K(t, s; \lambda)$ as $K_\lambda(t, s)$. For each $\lambda > 0$, we shall denote by $[G; K; \lambda]$ the relation over X

$(x, y) \in [G; K; \lambda]$ iff $G(x, y) > 0$ and $K_\lambda(d(x, Tx), d(x, y)) \leq 0$.

Further, take a function $\psi : R_+^0 \to R_+^0$ according to the *normal* properties below

(nor-1) ψ is increasing, right continuous (on R_+^0)
(nor-2) ψ is *asymptotic expansive* (on R_+^0):
for each strictly descending sequence $(t_n; n \geq 0)$ in R_+^0
with $\sum_n \psi(t_n) < \infty$, we have $\sum_n t_n < \infty$;

and $\Delta : R_+^0 \times R_+^0 \to R$ be a function with

(u-d-pos) Δ is upper diagonal positive:
$\Delta(t, s) \geq 0$, for all $t, s \in R_+^0$ with $t < s$.

Denote by $[\psi; \Delta]$ the relation over R_+^0

$(t, s \in R_+^0)$: $(t, s) \in [\psi; \Delta]$ iff $t < s$, $\Delta(t, s) < 1$ and $\psi(t) \leq \Delta(t, s)\psi(s)$.

We say that (ψ, Δ) is *asymptotic subunitary*, when

(a-sub) $\limsup_n \Delta(t_n, s_n) < 1$, for each sequence $((t_n, s_n); n \geq 0)$
in $[\psi; \Delta]$, with (t_n)=bounded, and (s_n)=strictly descending.

Finally, given $\lambda > 0$, we say that T is $(M; K; \lambda; \psi, \Delta)$*-contractive*, provided

(contr-K) $(x, y) \in [G; K; \lambda]$ (hence, $d(x, y) > 0$) and $M(x, y) > 0$ imply

$(M(x, y), d(x, y)) \in [\psi; \Delta]$; that is: $M(x, y) < d(x, y)$,
$\Delta(M(x, y), d(x, y)) < 1$, and $\psi(M(x, y)) \leq \Delta(M(x, y), d(x, y))\psi(d(x, y))$.

As we shall see below, the function $K : R_+^0 \times R_+^0 \times R_+^0 \to R$ has an essential role in our arguments. Two particular cases occur.

I) The basic choice of this function is $K = A$, where $A : R_+^0 \times R_+^0 \times R_+^0 \to R$ denotes the real function

$$A(t, s; \lambda) = t - \lambda s, t, s > 0, \lambda > 0;$$

for simplicity, we write $A(t, s; \lambda)$ as $A_\lambda(t, s)$.

II) Another basic choice of this function is $K = B$, where $B : R_+^0 \times R_+^0 \times R_+^0 \to R$ is a real function with

B is *subordinated* to A: $B(t, s; \lambda) \leq A(t, s; \lambda), t, s > 0, \lambda > 0$;

as before, we write $B(t, s; \lambda)$ as $B_\lambda(t, s)$.

The following relative statement is useful in the sequel.

Proposition 5.13 *Suppose that B is subordinated to A. Then, for each $\lambda > 0$,*

T is $(M; B; \lambda; \psi, \Delta)$-contractive implies T is $(M; A; \lambda; \psi, \Delta)$-contractive.

Proof Suppose that T is $(M; B; \lambda; \psi, \Delta)$-contractive; and let $x, y \in X$ be taken as in the premise of (contr-A); that is,

(contr-A-pre) $(x, y) \in [G, A; \lambda]$
(i.e.: $G(x, y) > 0$, $A_\lambda(d(x, Tx), d(x, y)) \leq 0$), and $M(x, y) > 0$.

As B is subordinated to A, we get

(contr-B-pre) $(x, y) \in [G, B; \lambda]$
(i.e., $G(x, y) > 0$, $B_\lambda(d(x, Tx), d(x, y)) \leq 0$), and $M(x, y) > 0$.

This, along with T being $(M; B; \lambda; \psi, \Delta)$-contractive, gives

(contr-BA) $(M(x, y), d(x, y)) \in [\psi; \Delta]$; that is: $M(x, y) < d(x, y)$,
$\Delta(M(x, y), d(x, y)) < 1$, and $\psi(M(x, y)) \leq \Delta(M(x, y), d(x, y))\psi(d(x, y))$;

which tells us that T is $(M; A; \lambda; \psi, \Delta)$-contractive.

Let the above conventions be in use. The main result in this exposition is

Theorem 5.5 *Suppose that the map $T : X \to C(X)$ is $(M; A; \lambda; \psi, \Delta)$-contractive, for some $\lambda > 0$, some normal function $\psi : R_+^0 \to R_+^0$, and some upper diagonal positive function $\Delta : R_+^0 \times R_+^0 \to R$ with (ψ, Δ)=asymptotic subunitary. Then,*

(51-a) If, in addition, $\lambda \geq 1$, then either Fix$(T) \neq \emptyset$, *or (under* Fix$(T) = \emptyset$*), T is a Picard map (modulo d)*
(51-b) If, in addition, $\lambda \geq 2$ and X is orbitally d-complete, then, necessarily, Fix(T) *is nonempty.*

Proof There are two steps to be passed.

Step 1.) Suppose that $\lambda \geq 1$ and

(non-fix-1) $\text{Fix}(T) = \emptyset$; whence, $d(x, Tx) > 0$, for all $x \in X$.

Note that, as a direct consequence of this, we have (by definition)

(non-fix-2) $(\forall (x, y) \in T)$: T is regular at (x, y); that is:
$d(x, Tx) > 0, d(y, Ty) > 0, d(x, y) \geq d(x, Tx) > 0$,
$M(x, y) \geq E(Tx, Ty) \geq d(y, Ty) > 0$; so that:
$(x, y) \in [G, A; \lambda]$, and $M(x, y) \geq d(y, Ty) > 0$.

Let $(y_0, y_1) \in T$ be arbitrary fixed. By the imposed hypothesis,

(51-c-1) T is regular at (y_0, y_1); whence:
$(y_0, y_1) \in [G, A; \lambda]$, and $M(y_0, y_1) \geq d(y_1, Ty_1) > 0$.

Taking the contractive condition into account yields

(51-c-2) T is admissible at (y_0, y_1), in the sense:
$(M(y_0, y_1), d(y_0, y_1)) \in [\psi, \Delta]$
(that is: $M(y_0, y_1) < d(y_0, y_1)$, $\Delta(M(y_0, y_1), d(y_0, y_1)) < 1$,
and $\psi(M(y_0, y_1)) \leq \Delta(M(y_0, y_1), d(y_0, y_1))\psi(d(y_0, y_1))$).

Note that, as a direct consequence, we have

(51-d-1) $0 < \psi(d(y_1, Ty_1)) \leq \psi(M(y_0, y_1)) \leq$
$\Delta(M(y_0, y_1), d(y_0, y_1))\psi(d(y_0, y_1))$.

Now, in view of $\psi(d(y_1, Ty_1)) > 0$,

(51-d-2) $\psi(d(y_1, Ty_1)) < \Delta(M(y_0, y_1), d(y_0, y_1))^{-1/2}\psi(d(y_1, Ty_1))$;
because $0 < \Delta(M(y_0, y_1), d(y_0, y_1)) < 1$.

This (by definition) tells us that

(51-d-3) there exists $y_2 \in Ty_1$ (hence, $y_2 \neq y_1$), with
$\psi(d(y_1, y_2)) < \Delta(M(y_0, y_1), d(y_0, y_1))^{-1/2}\psi(d(y_1, Ty_1))$.

For, if the opposite relation holds

$\psi(d(y_1, y_2)) \geq \Delta(M(y_0, y_1), d(y_0, y_1))^{-1/2}\psi(d(y_1, Ty_1))$,
for all $y_2 \in Ty_1$,

one gets (passing to infimum and noting that ψ is right continuous)

$$\psi(d(y_1, Ty_1)) \geq \Delta(M(y_0, y_1), d(y_0, y_1))^{-1/2}\psi(d(y_1, Ty_1));$$

in contradiction with our previous evaluation (51-d-2). Hence, (51-d-3) holds; and then (combining with (51-d-1) above)

(51-d-4) T is transitive, in the sense:
for each $(y_0, y_1) \in T$ (hence, $y_0 \neq y_1$), there exists
$(y_1, y_2) \in T$ (hence, $y_1 \neq y_2$), with $(M(y_0, y_1), d(y_0, y_1)) \in [\psi; \Delta]$
(that is: $M(y_0, y_1) < d(y_0, y_1)$, $\Delta(M(y_0, y_1), d(y_0, y_1)) < 1$,

$\psi(M(y_0, y_1)) \leq \Delta(M(y_0, y_1), d(y_0, y_1))\psi(d(y_0, y_1)))$,
and $\psi(d(y_1, y_2)) \leq \Delta(M(y_0, y_1), d(y_0, y_1))^{1/2}\psi(d(y_0, y_1))$; so, in particular:
$\psi(d(y_1, y_2)) < \psi(d(y_0, y_1))$ (wherefrom, $d(y_1, y_2) < d(y_0, y_1)$).

This tells us that the relation $\mathscr{R}$ over T introduced as

$((u_1, v_1), (u_2, v_2) \in T)$: $(u_1, v_1)\mathscr{R}(u_2, v_2)$ iff $v_1 = u_2$,
$(M(u_1, v_1), d(u_1, v_1)) \in [\psi; \Delta]$ (that is: $M(u_1, v_1) < d(u_1, v_1)$,
$\Delta(M(u_1, v_1), d(u_1, v_1)) < 1$,
$\psi(M(u_1, v_1)) \leq \Delta(M(u_1, v_1), d(u_1, v_1))\psi(d(u_1, v_1)))$,
and $\psi(d(u_2, v_2)) \leq \Delta(M(u_1, v_1), d(u_1, v_1))^{1/2}\psi(d(u_1, v_1))$

is proper, in the sense

$\mathscr{R}(u, v)$ is nonempty, for each $(u, v) \in T$.

By the Dependent Choice principle, it then follows that, given $(x_0, x_1) \in T$, there must be a sequence $(x_n; n \geq 0)$ in X, such that

$(x_n, x_{n+1}) \in T$ and $(x_n, x_{n+1})\mathscr{R}(x_{n+1}, x_{n+2})$, for all n.

This, under the notations

$(t_n := M(x_n, x_{n+1}); n \geq 0)$, $(s_n := d(x_n, x_{n+1}); n \geq 0)$

and regular admissible properties, tells us that we must have, $\forall n \geq 1$

(51-e-1) $x_n \in Tx_{n-1}$, and $0 < d(x_n, Tx_n) \leq t_{n-1}$
(51-e-2) $(t_{n-1}, s_{n-1}) \in [\psi; \Delta]$; that is:
$t_{n-1} < s_{n-1}$, $\Delta(t_{n-1}, s_{n-1}) < 1$, and $\psi(t_{n-1}) \leq \Delta(t_{n-1}, s_{n-1})\psi(s_{n-1})$
(51-e-3) $\psi(s_n) \leq \Delta(t_{n-1}, s_{n-1})^{1/2}\psi(s_{n-1})$; whence, $s_n < s_{n-1}$.

By these relations, it is clear that

$(t_n; n \geq 0)$ is bounded and $(s_n; n \geq 0)$ is strictly descending in R^0_+.

As (ψ, Δ) is asymptotic subunitary,

$\alpha := \limsup_n \rho_n < 1$, where $(\rho_n := \Delta(t_n, s_n)^{1/2}; n \geq 0)$.

This yields directly

$\beta := (1/2)(\alpha + 1) \in]\alpha, 1[$; so that $[\exists n(\beta): \rho_n \leq \beta, \forall n \geq n(\beta)]$.

Finally, by the same relations, one gets (under these notations)

$\psi(s_n) \leq \rho_{n-1}\psi(s_{n-1}), \forall n \geq 1$;

so that, combining with the previous relation,

$\psi(s_n) \leq \beta\psi(s_{n-1}), \forall n \geq n(\beta) + 1$.

This yields at once

$\sum_n \psi(d(x_n, x_{n+1})) < \infty$; whence, $\sum_n d(x_n, x_{n+1}) < \infty$

if we remember that ψ is asymptotic expansive. Consequently, $(x_n; n \geq 0)$ is d-Cauchy; whence, T is a Picard map (modulo d).

Step 2.) Suppose that $\lambda \geq 2$ and X is orbitally d-complete. We claim that

(fix-non) $\text{Fix}(T) = \emptyset$; that is: $d(x, Tx) > 0, x \in X$

yields a contradiction. In fact, assume that this alternative holds; and let $x_0 \in X$ be arbitrary fixed. By the preceding step, we get an (x_0, T)-iterative sequence $(x_n; n \geq 0)$ in X with

(x_n) is d-Cauchy; hence, $z := \lim_n(x_n)$ exists
(as X is orbitally d-complete).

The argument will be divided into two parts.

Step-2-1) We claim that

(51-f) $d(z, Ty) \leq d(z, y)$, for each $y \in X \setminus \{z\}$.

In fact, let $y \in X \setminus \{z\}$ be arbitrary fixed. The case of $z \in Ty$ is clear; so, we may assume that $z \notin Ty$. As

$$\lim_n d(x_n, y) = d(z, y) > 0, \lim_n d(x_n, Ty) = d(z, Ty) > 0,$$

there must be some index $n(y)$ such that

$$(\forall n \geq n(y)): d(x_n, x_{n+1}) \leq d(x_n, y), \text{ and } d(x_{n+1}, Ty) > 0.$$

The former of these yields, for the same ranks

(p-1) $(0 <)d(x_n, Tx_n) \leq d(x_n, x_{n+1}) \leq d(x_n, y) \leq \lambda d(x_n, y)$;
wherefrom: $A_\lambda(d(x_n, Tx_n), d(x_n, y)) \leq 0$;
(p-2) $G(x_n, y) > 0$ (i.e.: $d(x_n, Tx_n) > 0, d(y, Ty) > 0, d(x_n, y) > 0$).

And, the latter of these gives (via M=T-compatible), again for the same ranks

(p-3) $M(x_n, y) \geq E(Tx_n, Ty) \geq d(x_{n+1}, Ty) > 0$.

Putting these together gives (by the imposed contractive condition)

$$(d(x_{n+1}, Ty) \leq)M(x_n, y) < d(x_n, y), \forall n \geq n(y).$$

Passing to limit as $n \to \infty$, one derives the stated assertion.

Step-2-2) We start by noting that the alternative

(alter-1) $(\exists k)$: $x_n = z$, for all $n > k$

yields $(z, z) \in T$; that is: $z \in \text{Fix}(T)$; in contradiction with our working hypothesis (fix-non). So, it remains to discuss the alternative

(alter-2) for each k, there exists $j > k$ with $x_j \neq z$.

As a consequence of this, there must be a sequence $(i(n); n \geq 0)$ in N, with

$(i(n))$ is strictly ascending (hence, $\lim_n i(n) = \infty$), and $x_{i(n)} \neq z, \forall n$.

Moreover, as $\lim_n d(x_n, Tz) = d(z, Tz) > 0$, there must be some index $m(z)$, with

$(\forall n \geq m(z))$: $d(x_{n+1}, Tz) > 0$.

Note that, by the divergence property of $(i(n); n \geq 0)$, one may assume that

(for all n): $i(n) \geq m(z)$; whence, $d(x_{i(n)+1}, Tz) > 0$.

Having these precise, we claim that the contractive property applies to all pairs $((x_{i(n)}, z); n \geq 0)$. For the moment, one has

(51-g-1) (for each n): $d(x_{i(n)}, Tx_{i(n)}) \leq \lambda d(x_{i(n)}, z)$;
whence, $A_\lambda(d(x_{i(n)}, Tx_{i(n)}), d(x_{i(n)}, z)) \leq 0$.

In fact, let $n \geq 0$ be arbitrary fixed. We have, by the preceding step (and the Lipschitz property of point to set distance)

(for each n): $d(x_{i(n)}, Tx_{i(n)}) \leq d(z, Tx_{i(n)}) + d(z, x_{i(n)}) \leq$
$2d(z, x_{i(n)}) \leq \lambda d(z, x_{i(n)})$; and the assertion follows.

On the other hand, again by the choice of our subsequence $(x_{i(n)}; n \geq 0)$,

(51-g-2) (for each n): $G(x_{i(n)}, z) > 0$
(i.e.: $d(x_{i(n)}, Tx_{i(n)}) > 0, d(z, Tz) > 0, d(x_{i(n)}, z) > 0$).

Finally, by a preceding observation,

(51-g-3) $0 < d(x_{i(n)+1}, Tz) \leq E(Tx_{i(n)}, Tz) \leq M(x_{i(n)}, z)$.

Summing up, the contractive property upon T is indeed applicable to the pairs $((x_{i(n)}, z); n \geq 0)$. As a consequence of this, we derive

$d(x_{i(n)+1}, Tz) \leq E(Tx_{i(n)}, Tz) \leq M(x_{i(n)}, z) < d(x_{i(n)}, z)$, for all n.

Passing to limit as $n \to \infty$ yields $d(z, Tz) = 0$; that is, $z \in Tz$; in contradiction with our working condition (fix-non). The proof is thereby complete.

Note that an extended setting of these developments is possible, under the lines developed by Hitzler [14, Ch 1, Sect 1.2]. On the other hand, this result may be stated in the context of relational metric spaces; and then, our main result includes the statement in Javahernia et al. [17]. We shall discuss all these facts elsewhere.

6 Particular Versions

Let X be a nonempty set, and $d : X \times X \to R_+$ be a metric over X; the couple (X, d) will be called a *metric space*. Denote

$C(X)$=the class of (nonempty) d-closed parts of X.

Remember that the point to set distance in X is introduced as:

$d(x, Y) = \inf\{d(x, y); y \in Y\}, x \in X, Y \in \exp(X)$.

Let $E : C(X) \times C(X) \to R_+(\infty) := R_+ \cup \{\infty\}$ stand for the Hausdorff–Pompeiu almost generalized metric

$$E(Y, Z) = \sup\{d(y, Z);\ y \in Y\},\ Y, Z \in C(X).$$

We say that the map $M : X \times X \to R_+(\infty)$ is *T-compatible*, if

$$E(Tx, Ty) \le M(x, y), \text{ for all } x, y \in X.$$

Fix in the following such an object. Then, let $G : X \times X \to R_+$ stand for the map

$$G(x, y) = \min\{d(x, Tx), d(y, Ty), d(x, y)\},\ x, y \in X;$$

and let $K : R^0_+ \times R^0_+ \times R^0_+ \to R$ be a real function; for simplicity, we write $K(t, s; \lambda)$ as $K_\lambda(t, s)$. For each $\lambda > 0$, we shall denote by $[G; K; \lambda]$ the relation over X

$$(x, y) \in [G; K; \lambda] \text{ iff } G(x, y) > 0 \text{ and and } K_\lambda(d(x, Tx), d(x, y)) \le 0.$$

Further, let $\psi : R^0_+ \to R^0_+$ be a *normal* function, in the sense

(nor-1) ψ is increasing, right continuous (on R^0_+)
(nor-2) ψ is *asymptotic expansive* (on R^0_+):
for each strictly descending sequence $(t_n; n \ge 0)$ in R^0_+
with $\sum_n \psi(t_n) < \infty$, we have $\sum_n t_n < \infty$;

and $\Delta : R^0_+ \times R^0_+ \to R$ be a function with

(u-d-pos) Δ is upper diagonal positive:
$\Delta(t, s) \ge 0$, for all $t, s \in R^0_+$ with $t < s$.

Denote by $[\psi; \Delta]$ the relation over R^0_+

$$(t, s \in R^0_+)\text{: } (t, s) \in [\psi; \Delta] \text{ iff } t < s,\ \Delta(t, s) < 1 \text{ and } \psi(t) \le \Delta(t, s)\psi(s).$$

We say that (ψ, Δ) is *asymptotic subunitary*, when

(a-sub) $\limsup_n \Delta(t_n, s_n) < 1$, for each sequence $((t_n, s_n); n \ge 0)$
in $[\psi; \Delta]$, with (t_n)=bounded, (s_n)=strictly descending.

Then, given $\lambda > 0$, we say that T is *$(M; K; \lambda; \psi, \Delta)$-contractive*, provided

(contr-K) $(x, y) \in [G; K; \lambda]$ (hence, $d(x, y) > 0$) and $M(x, y) > 0$ imply
$(M(x, y), d(x, y)) \in [\psi; \Delta]$; that is: $M(x, y) < d(x, y)$,
$\Delta(M(x, y), d(x, y)) < 1$, and $\psi(M(x, y)) \le \Delta(M(x, y), d(x, y))\psi(d(x, y))$.

As already precise, the function $K : R_+ \times R_+ \times R_+ \to R_+$ we just introduced has an essential role in our arguments. Two particular choices of it are of interest.

I) The basic choice of this function is $K = A$, where $A : R^0_+ \times R^0_+ \times R^0_+ \to R$ denotes the real function

$$A(t, s; \lambda) = t - \lambda s,\ t, s > 0,\ \lambda > 0;$$

for simplicity, we write $A(t, s; \lambda)$ as $A_\lambda(t, s)$. In fact, the main result of this exposition is essentially related to the choice $K = A$.

II) Another basic choice of this function is $K = B$, where $B : R^0_+ \times R^0_+ \times R^0_+ \to R$ is a real function with

B is *subordinated* to A: $B(t, s; \lambda) \le A(t, s; \lambda)$, $t, s > 0$, $\lambda > 0$.

as before, we write $B(t, s; \lambda)$ as $B_\lambda(t, s)$.

Concerning this aspect, the following particular version of our main result is to be noted.

Theorem 6.6 *Suppose that the map* $T : X \to C(X)$ *is* $(M; B; \lambda; \psi, \Delta)$*-contractive, for some subordinated to A function* $B : R^0_+ \times R^0_+ \times R^0_+ \to R$*, some* $\lambda > 0$*, some normal function* $\psi : R^0_+ \to R^0_+$*, and some upper diagonal positive function* $\Delta : R^0_+ \times R^0_+ \to R$ *with* (ψ, Δ)*=asymptotic subunitary. Then,*

(61-a) If, in addition, $\lambda \ge 1$*, then either* $\mathrm{Fix}(T) \neq \emptyset$ *or*
(under $\mathrm{Fix}(T) = \emptyset$*)* T *is a Picard map (modulo d)*
(61-b) If, in addition, $\lambda \ge 2$ *and X is orbitally d-complete,*
then, necessarily, $\mathrm{Fix}(T)$ *is nonempty.*

Proof By a previous auxiliary fact, it follows that

T is $(M; A; \lambda; \psi, \Delta)$-contractive (see above).

This tells us that the main result applies to our data; wherefrom, all is clear.

Technically speaking, the normal property is essential for all these results. So, it would be useful for us to give a lot of normality criteria.

Proposition 6.14 *Let the increasing and right continuous function* $\psi : R^0_+ \to R^0_+$ *be such that one of the additional conditions below is holding*

(61-1) ψ *is strongly asymptotic expansive:*
($\psi(t) \ge \alpha t$*, if* $\varphi(t) \le \beta$*), where* $\alpha, \beta > 0$ *are constants*
(61-2) ψ *is subadditive and coercive:* $(\psi(\infty) :=) \lim_{t\to\infty} \psi(t) = \infty$.

Then, necessarily,

ψ *is asymptotic expansive; hence, normal.*

Proof

i) Let the strictly descending sequence $(t_n; n \ge 0)$ in R^0_+ be such that

$\sum_n \psi(t_n) < \infty$; hence, in particular, $\lim_n \psi(t_n) = 0$.

By this last observation, there must be some rank $n(\beta)$ such that

$\psi(t_n) \le \beta$, for all $n \ge n(\beta)$.

Combining with the strong asymptotic expansive property, gives

$\psi(t_n) \ge \alpha t_n$, for all $n \ge n(\beta)$; wherefrom, $\sum_n t_n < \infty$.

ii) Let the strictly descending sequence $(t_n; n \geq 0)$ in R^0_+ be such that

$C := \sum_n \psi(t_n) < \infty.$

By the subadditive property,

$\psi(\gamma_n) \leq C, \forall n$, where $(\gamma_n := \sum_{i \leq n} t_i; n \geq 0)$.

If $\lim_n \gamma_n = \sum_n t_n = \infty$, we must have (by coerciveness) $\psi(\infty) = \infty \leq C$; contradiction. Hence,

$\lim_n \gamma_n = \sum_n t_n < \infty;$

and conclusion follows.

In the following, two basic particular cases of Theorem 6.6 are discussed. Let the T-compatible map $M : X \times X \to R_+(\infty)$, and the functions $G : X \times X \to R_+$, $A : R_+ \times R_+ \times R_+ \to R$ be introduced as above.

Part-Case 1 Let $\psi : R^0_+ \to R^0_+$ be a normal function; and $\Lambda : R^0_+ \times R^0_+ \to R$ be a function with

Λ is upper diagonal positive: $\Lambda(t, s) \geq 0$, for all $t, s \in R^0_+, t < s$.

Then, let $[[\psi; \Lambda]]$ stand for the relation over R^0_+

$(t, s \in R^0_+)$: $(t, s) \in [[\psi; \Lambda]]$ iff $t < s$ and $\psi(t) - \psi(s) + \Lambda(t, s) < 0$.

As a consequence of this, the mapping $\Delta : R^0_+ \times R^0_+ \to R$ introduced as

$\Delta(t, s) = (\psi(t) + \Lambda(t, s))/\psi(s), t, s > 0$

is well defined and upper diagonal positive. In addition, we have the relative fact

$(t, s \in R^0_+)$: $(t, s) \in [[\psi; \Lambda]]$ implies $(t, s) \in [\psi; \Delta]$
(that is: $t < s$, $\Delta(t, s) < 1$, $\psi(t) \leq \Delta(t, s)\psi(s)$).

It will suffice then assuming that $(\psi; \Delta)$=asymptotic subunitary to apply Theorem 6.6 upon these data. Combining with the first normality criterion being applicable (under $\alpha = \beta = 1$) when

$\psi(t) = t^\nu, t > 0$, for some $\nu \in]0, 1]$,

we derive that the corresponding form of Theorem 6.6 includes the related fixed point result in Du et al. [11]. Further aspects may be found in Mizoguchi and Takahashi [31]; see also Nadler [34], Reich [37], and Turinici [46]. In fact, a *complete* additive form of this result is available, under the model in Wardowski [52]. However, the multivalued versions of the statements in Turinici [51] are not available via these methods; further aspects may be found in Choudhury and Metiya [6].

Part-Case 2 Let $\psi : R^0_+ \to R^0_+$ be a normal function; and $\lambda > 0$ be fixed. Further, let $\alpha, \beta \in \mathscr{F}(R^0_+, R_+)$ be a couple of functions. We say that T is $(M; \psi; \lambda; \alpha, \beta)$-*contractive*, provided

(contr-fct) $G(x, y) > 0$ (hence, $d(x, y) > 0$), $d(x, Tx) \leq \beta(d(x, y))d(x, y)$ and $M(x, y) > 0$
$(0 <)M(x, y) < d(x, y)$ and $\psi(M(x, y)) \leq \alpha(d(x, y))\psi(d(x, y))$.

Theorem 6.7 *Suppose that the map* $T : X \to C(X)$ *is* $(M; \psi; \lambda; \alpha, \beta)$-*contractive, for some normal function* $\psi : R_+^0 \to R_+^0$, *some* $\lambda > 0$, *and some functional couple* $\alpha, \beta \in \mathscr{F}(R_+^0, R_+)$, *with*

(62-i) (α, β) *is unitary separated:* $\alpha(t) < 1 \leq \beta(t)$, *for all* $t > 0$
(62-ii) $\alpha(.)$ *is asymptotic subunitary:* $\limsup_n \alpha(s_n) < 1$,
for each strictly descending sequence $(s_n; n \geq 0)$ *in* R_+^0.

Then,

(62-a) If, in addition, $\lambda \geq 1$, *then either* $\mathrm{Fix}(T) \neq \emptyset$ *or*
(under $\mathrm{Fix}(T) = \emptyset$*)* T *is a Picard map (modulo* d*)*
(62-b) If, in addition, $\lambda \geq 2$ *and* X *is orbitally* d-*complete,*
then, necessarily, $\mathrm{Fix}(T)$ *is nonempty.*

Proof Let the functions $B : R_+^0 \times R_+^0 \times R_+^0 \to R$, $\Delta : R_+^0 \times R_+^0 \to R$ be defined as

$$B(t, s; \lambda) = t - \lambda\beta(s)s,\ t, s > 0, \lambda > 0;\ \Delta(t, s) = \alpha(s),\ t, s > 0;$$

as usual, we write $B(t, s; \lambda)$ as $B_\lambda(t, s)$. By the unitary separated property of (α, β),

$$B(t, s; \lambda) \leq t - \lambda s = A(t, s; \lambda),\ t, s > 0, \lambda > 0;$$

so, B is subordinated to A. On the other hand, Δ is upper diagonal positive; and

(ψ, Δ)=asymptotic subunitary (since $\alpha(.)$=asymptotic subunitary).

Finally, letting $x, y \in X$ be taken so as

$$G(x, y) > 0,\ B_\lambda(d(x, Tx), d(x, y)) \leq 0,\ M(x, y) > 0,$$

we get by the imposed contractive property

$(M(x, y), d(x, y)) \in [\psi; \Delta]$; that is: $M(x, y) < d(x, y)$,
$\Delta(M(x, y), d(x, y)) < 1$, and $\psi(M(x, y)) \leq \Delta(M(x, y), d(x, y))\psi(d(x, y))$;

or, in other words: T is $(M; B; \lambda; \psi, \Delta)$-contractive. This tells us that Theorem 6.6 applies here; wherefrom, all is clear.

In particular, when the increasing right continuous function $\psi : R_+^0 \to R_+^0$ is subadditive and coercive the second normality criterion we just proved assures us that ψ is endowed with such a property. Note that, in this case, the corresponding version of Theorem 6.7 is just the main result in Liu et al. [27].

Finally, by similar procedures, it is possible to extend the results in Berinde and Păcurar [3], Cirić [7, 8], Daffer and Kaneko [10], Kamran [19], Klim and Wardowski [23], or Khojasteh and Rakočević [22]; we shall discuss them elsewhere.

References

1. M. Altman, An integral test for series and generalized contractions. Am. Math. Mon. **82**, 827–829 (1975)
2. S. Banach, Sur les opérations dans les ensembles abstraits et leur application aux équations intégrales. Fundam. Math. **3**, 133–181 (1922)
3. V. Berinde, M. Păcurar, The role of the Pompeiu-Hausdorff metric in fixed point theory. Creat. Math. Inform. **22**, 35–42 (2013)
4. P. Bernays, A system of axiomatic set theory: Part III. Infinity and enumerability analysis. J. Symb. Log. **7**, 65–89 (1942)
5. D.W. Boyd, J.S.W. Wong, On nonlinear contractions. Proc. Am. Math. Soc. **20**, 458–464 (1969)
6. B.S. Choudhury, N. Metiya, Fixed point theorems for almost contractions in partially ordered metric spaces. Ann. Univ. Ferrara **58**, 21–36 (2012)
7. L.B. Cirić, Fixed point theorems for multi-valued contractions in complete metric spaces. J. Math. Anal. Appl. **348**, 499–507 (2008)
8. L.B. Cirić, Multi-valued nonlinear contraction mappings. Nonlinear Anal. **71**, 2716–2723 (2009)
9. P.J. Cohen, *Set Theory and the Continuum Hypothesis* (Benjamin, New York, 1966)
10. P.Z. Daffer, H. Kaneko, Fixed points of generalized contractive multi-valued mappings. J. Math. Anal. Appl. **192**, 655–666 (1995)
11. W.-S. Du, F. Khojasteh, Y.-N. Chiu, Some generalizations of Mizoguchi-Takahashi's fixed point theorem with new local constraints. Fixed Point Theory Appl. **2014**, 31 (2014)
12. Y. Feng, S. Liu, Fixed point theorems for multi-valued contractive mappings and multi-valued Caristi type mappings. J. Math. Anal. Appl. **317**, 103–112 (2006)
13. P.R. Halmos, *Naive Set Theory* (Van Nostrand Reinhold Co., New York, 1960)
14. P. Hitzler, Generalized metrics and topology in logic programming semantics. PhD Thesis, Natl. Univ. Ireland, Univ. College Cork, 2001
15. J. Jachymski, Common fixed point theorems for some families of mappings. Indian J. Pure Appl. Math. **25**, 925–937 (1994)
16. J. Jachymski, The contraction principle for mappings on a metric space with a graph. Proc. Am. Math. Soc. **136**, 1359–1373 (2008)
17. M. Javahernia, A. Razani, F. Khojasteh, Fixed point of multi-valued contractions via manageable functions and Liu's generalization. Cogent Math. **3**, 1276818 (2016)
18. C.F.K. Jung, On generalized complete metric spaces. Bull. Amer. Math. Soc. **75**, 113–116 (1969)
19. T. Kamran, Mizoguchi-Takahashi's type fixed point theorem. Comput. Math. Appl. **57**, 507–511 (2009)
20. S. Kasahara, On some generalizations of the Banach contraction theorem. Publ. Res. Inst. Math. Sci. Kyoto Univ. **12**, 427–437 (1976)
21. M.S. Khan, M. Swaleh, S. Sessa, Fixed point theorems by altering distances between the points. Bull. Aust. Math. Soc. **30**, 1–9 (1984)
22. F. Khojasteh, V. Rakočević, Some new common fixed point results for generalized contractive multi-valued nonself-mappings. Appl. Math. Lett. **25**, 287–293 (2012)
23. D. Klim, D. Wardowski, Fixed point theorems for set-valued contractions in complete metric spaces. J. Math. Anal. Appl. **334**, 132–139 (2007)
24. K. Kuratowski, *Topology*, vol. I (Academic Press, New York, 1966)
25. S. Leader, Fixed points for general contractions in metric spaces. Math. Jpn. **24**, 17–24 (1979)
26. Z. Liu, W. Sun, S.M. Kang, J.S. Ume, On fixed point theorems for multivalued contractions. Fixed Point Theory Appl. **2010**, Article ID: 870980 (2010)
27. Z. Liu, Z. Wu, S.M. Kang, S. Lee, Some fixed point theorems for nonlinear set-valued contractive mappings. J. Appl. Math. **2012**, Article ID: 786061 (2012)
28. W.A.J. Luxemburg, On the convergence of successive approximations in the theory of ordinary differential equations (II). Indag. Math. **20**, 540–546 (1958)

29. J. Matkowski, *Integrable Solutions of Functional Equations*. Dissertationes Math., vol. 127 (Polish Sci. Publ., Warsaw, 1975)
30. A. Meir, E. Keeler, A theorem on contraction mappings. J. Math. Anal. Appl. **28**, 326–329 (1969)
31. N. Mizoguchi, W. Takahashi, Fixed point theorems for multivalued mappings on complete metric spaces. J. Math. Anal. Appl. **141**, 177–188 (1989)
32. G.H. Moore, *Zermelo's Axiom of Choice: Its Origin, Development and Influence* (Springer, New York, 1982)
33. Y. Moskhovakis, *Notes on Set Theory* (Springer, New York, 2006)
34. S.B. Nadler Jr., Multi-valued contraction mappings. Pac. J. Math. **30**, 475–488 (1969)
35. J.J. Nieto, R. Rodriguez-Lopez, Contractive mapping theorems in partially ordered sets and applications to ordinary differential equations. Order **22**, 223–239 (2005)
36. A.C.M. Ran, M.C. Reurings, A fixed point theorem in partially ordered sets and some applications to matrix equations. Proc. Am. Math. Soc. **132**, 1435–1443 (2004)
37. S. Reich, Fixed points of contractive functions. Boll. Un. Mat. Ital. **5**, 26–42 (1972)
38. B.E. Rhoades, A comparison of various definitions of contractive mappings. Trans. Am. Math. Soc. **226**, 257–290 (1977)
39. I.A. Rus, *Generalized Contractions and Applications* (Cluj University Press, Cluj-Napoca, 2001)
40. B. Samet, M. Turinici, Fixed point theorems on a metric space endowed with an arbitrary binary relation and applications. Commun. Math. Anal. **13**, 82–97 (2012)
41. E. Schechter, *Handbook of Analysis and Its Foundation* (Academic Press, New York, 1997)
42. T. Suzuki, A generalized Banach contraction principle that characterizes metric completeness. Proc. Am. Math. Soc. **136**, 1861–1869 (2008)
43. A. Tarski, Axiomatic and algebraic aspects of two theorems on sums of cardinals. Fundam. Math. **35**, 79–104 (1948)
44. V. Timofte, New tests for positive iteration series. Real Anal. Exchange **30**, 799–812 (2004/2005)
45. M. Turinici, Nonlinear contractions and applications to Volterra functional equations. An. Şt. Univ. "Al. I. Cuza" Iaşi (S I-a, Mat) **23**, 43–50 (1977)
46. M. Turinici, Multivalued contractions and applications to functional differential equations. Acta Math. Acad. Sci. Hung. **37**, 147–151 (1981)
47. M. Turinici, Fixed points for monotone iteratively local contractions. Demonstratio Math. **19**, 171–180 (1986)
48. M. Turinici, Abstract comparison principles and multivariable Gronwall-Bellman inequalities. J. Math. Anal. Appl. **117**, 100–127 (1986)
49. M. Turinici, Function pseudometric VP and applications. Bul. Inst. Polit. Iaşi (S. Mat. Mec. Teor. Fiz.) **53**(57), 393–411 (2007)
50. M. Turinici, Wardowski implicit contractions in metric spaces. arXiv 1211-3164-v2, 15 Sept 2013
51. M. Turinici, Contraction maps in pseudometric structures, in *Essays in Mathematics and Its Applications*, ed. by T.M. Rassias, P.M. Pardalos (Springer Intl. Publ., Switzerland, 2016), pp. 513–562
52. D. Wardowski, Fixed points of a new type of contractive mappings in complete metric spaces. Fixed Point Theory Appl. **2012**, 94 (2012)
53. E.S. Wolk, On the principle of dependent choices and some forms of Zorn's lemma. Can. Math. Bull. **26**, 365–367 (1983)

The manufacturer's authorised representative in the EU is Springer Nature Customer Service Centre GmbH, Europaplatz 3, 69115 Heidelberg, Germany. If you have any concerns regarding our products, please contact ProductSafety@springernature.com

Printed and bound by CPI Group (UK) Ltd, Croydon, CR0 4YY
28/09/2026
02224067-0009